Prealgebra

Prealgebra

5th EDITION

Marvin L. Bittinger

Indiana University Purdue University Indianapolis

David J. Ellenbogen

Community College of Vermont

Barbara L. Johnson

Indiana University Purdue University Indianapolis

PEARSON

Addison Wesley

Boston   San Francisco   New York
London   Toronto   Sydney   Tokyo   Singapore   Madrid
Mexico City   Munich   Paris   Cape Town   Hong Kong   Montreal

Publisher	Greg Tobin
Editor in Chief	Maureen O'Connor
Executive Editor	Jennifer Crum
Acquisitions Editor	Randy Welch
Executive Project Manager	Kari Heen
Assistant Editor	Joanna Doxey
Editorial Assistant	Antonio Arvelo
Production Manager	Ron Hampton
Senior Designer	Dennis Schaefer
Cover Designer	Dennis Schaefer
Digital Assets Manager	Marianne Groth
Media Producer	Ceci Fleming
Software Development	Mary Durnwald, TestGen; Rebecca Williams, MathXL
Marketing Manager	Jay Jenkins
Marketing Coordinator	Alexandra Waibel
Senior Author Support/ Technology Specialist	Joseph K. Vetere
Senior Prepress Supervisor	Caroline Fell
Media Buyer	Ginny Michaud
Rights and Permissions	Dana Weightman
Manufacturing Manager	Evelyn M. Beaton
Art and Design Services	Geri Davis/The Davis Group, Inc.
Production Coordination	Kathy Diamond
Composition	BeaconPMG
Illustrations	William Melvin, Network Graphics
Cover Photo	© Alec Pytlowany/MasterFile

About the cover: The aurora borealis, or northern lights, as photographed in the skies over British Columbia, Canada.

Photo credits appear on page P-1.

Library of Congress Cataloging-in-Publication Data

Bittinger, Marvin L.
    Prealgebra.—5th ed./Marvin L. Bittinger, David J. Ellenbogen, Barbara L. Johnson.
      p. cm.
    ISBN-10: 0-321-33190-7 (Student's Edition)
    ISBN-13: 978-0-321-33190-8
    ISBN-10: 0-321-45602-5 (Hardback Student's Edition)
    ISBN-13: 978-0-321-45602-1
    1. Algebra. I. Ellenbogen, David, II. Johnson, Barbara L. (Barbara Loreen), 1962– III. Title.

QA39.3.B58 2007
513'.14—dc22   2006045792

6 7 8 9 10–CKV–10 09

Contents

1   WHOLE NUMBERS

2   INTRODUCTION TO INTEGERS AND ALGEBRAIC EXPRESSIONS

3 FRACTION NOTATION: MULTIPLICATION AND DIVISION

4 FRACTION NOTATION: ADDITION, SUBTRACTION, AND MIXED NUMERALS

DEVELOPMENTAL UNITS

Index of Applications

Index of Study Tips

CHAPTER 9

CHAPTER 10

Preface

It is with great pride and excitement that we present to you the fifth edition of *Prealgebra*. The text has evolved dramatically over the years in response to your comments, responses, and opinions. This feedback, combined with our overall objective of presenting the material in a clear and accurate manner, drives each revision. It is our hope that *Prealgebra*, Fifth Edition, and the supporting supplements will help provide an improved teaching and learning experience by meeting the needs of instructors and successfully preparing students for their futures.

This text is part of a complete series that includes the following:

Bittinger: *Fundamental Mathematics,* Fourth Edition

Bittinger/Penna: *Basic Mathematics with Early Integers*

Bittinger: *Basic Mathematics,* Tenth Edition

Bittinger/Ellenbogen/Johnson: *Prealgebra,* Fifth Edition

Bittinger: *Introductory Algebra,* Tenth Edition

Bittinger: *Intermediate Algebra,* Tenth Edition

Bittinger/Beecher: *Developmental Mathematics: College Mathematics and Introductory Algebra,* Seventh Edition

Bittinger/Ellenbogen: *Prealgebra and Introductory Algebra,* Second Edition

Bittinger/Beecher: *Introductory and Intermediate Algebra,* Third Edition

Building Understanding through an Interactive Approach

The pedagogy of this text is designed to provide an interactive learning experience between the student and the exposition, annotated examples, art, margin exercises, and exercise sets. This unique approach, which has been developed and refined over many editions and is illustrated at right, provides students with a clear set of learning

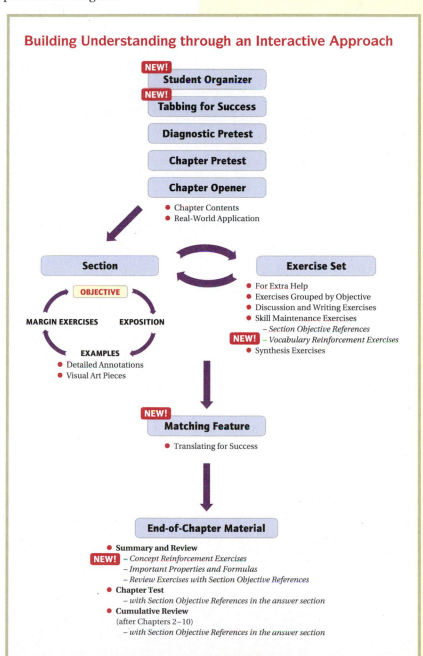

Building Understanding through an Interactive Approach

NEW! **Student Organizer**

NEW! **Tabbing for Success**

Diagnostic Pretest

Chapter Pretest

Chapter Opener
- Chapter Contents
- Real-World Application

Section

OBJECTIVE

MARGIN EXERCISES   EXPOSITION

EXAMPLES
- Detailed Annotations
- Visual Art Pieces

Exercise Set
- For Extra Help
- Exercises Grouped by Objective
- Discussion and Writing Exercises
- Skill Maintenance Exercises
  – *Section Objective References*
  NEW! – *Vocabulary Reinforcement Exercises*
- Synthesis Exercises

NEW! **Matching Feature**
- Translating for Success

End-of-Chapter Material
- **Summary and Review**
  NEW! – *Concept Reinforcement Exercises*
  – *Important Properties and Formulas*
  – *Review Exercises with Section Objective References*
- **Chapter Test**
  – *with Section Objective References in the answer section*
- **Cumulative Review**
  (after Chapters 2–10)
  – *with Section Objective References in the answer section*

objectives, involves them with the development of the material, and provides immediate and continual reinforcement and assessment through the margin exercises.

Let's Visit the Fifth Edition

The style, format, and approach of the fifth edition have been strengthened in a number of ways. However, the accuracy that the Bittinger books are known for has not changed. This edition, as with all editions, has gone through an exhaustive checking process to ensure accuracy in the problem sets, mathematical art, and accompanying supplements. We know what a critical role the accuracy of a book plays in student learning, and we value the reputation for accuracy that we have earned.

NEW! IN THE FIFTH EDITION

Each revision gives us the opportunity to incorporate new elements and refine existing elements to provide a better experience for students and teachers alike. Below are five new features designed to help students succeed.

- Student Organizer
- Tabbing for Success
- Translating for Success matching exercises
- Vocabulary Reinforcement exercises
- Concept Reinforcement exercises

These features, along with the hallmark features of this book, are discussed in the pages that follow.

In addition, the fifth edition has been designed to be open and flexible, helping students focus their attention on details that are critical at this level through prominent headings, boxed definitions and rules, and clearly labeled objectives. Chapter Pretests, now located along with the Diagnostic Pretest in the *Printed Test Bank* and in MyMathLab, diagnose at the section and objective level and can be used to place students in a specific section, or objective, of the chapter, allowing them to concentrate on topics with which they have particular difficulty. Answers to these pretests are available in the *Printed Test Bank* and in MyMathLab.

NEW! STUDENT ORGANIZER AND TABBING FOR SUCCESS

Along with study tips found throughout the text, a pull-out schedule card helps students stay organized. Students can schedule time on the card for study, commuting, work, family, and relaxation, and can also record important dates, useful contact information, and technology references.

In addition to the new Student Organizer, the new Tabbing for Success page provides students with forty color-coded reusable tabs to quickly locate key examples, review important summaries, flag text topics, and highlight areas where they need help. Together, these features will help students better use their time and their textbooks to succeed in the course.

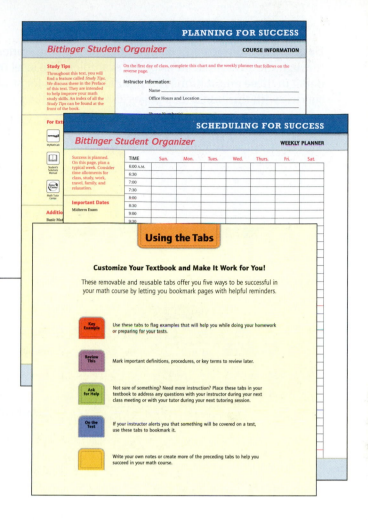

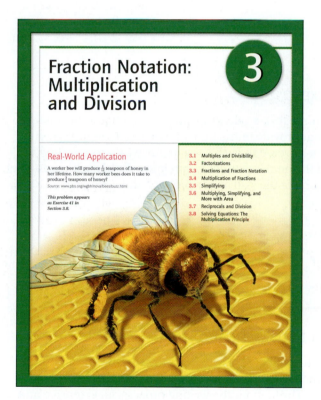

CHAPTER OPENERS

To engage students and prepare them for the upcoming chapter material, gateway chapter openers are designed with exceptional artwork that is tied to a motivating real-world application. (See pages 1, 95, and 309.)

OBJECTIVE BOXES

At the beginning of each section, a boxed list of objectives is keyed by letter not only to section subheadings, but also to the exercises in the Pretest (located in the *Printed Test Bank* and MyMathLab), the section exercise sets, and the Summary and Review exercises, as well as to the answers to the questions in the Chapter Tests and Cumulative Reviews. This correlation enables students to easily find appropriate review material if they need help with a particular exercise or skill at the objective level. (See pages 2, 129, and 310.)

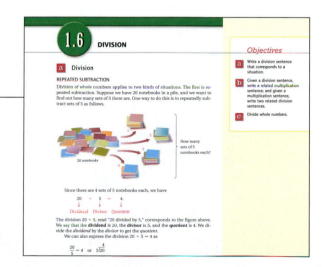

ANNOTATED EXAMPLES

Detailed annotations and color highlights lead the student through the structured steps of the examples. The level of detail in these annotations is a significant reason for students' success with this book. (See pages 253, 320, and 424.)

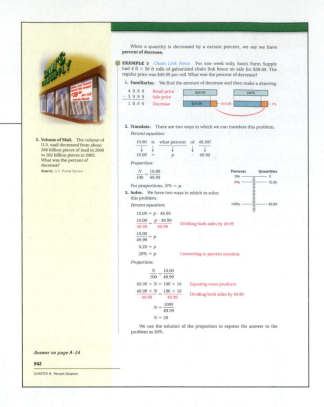

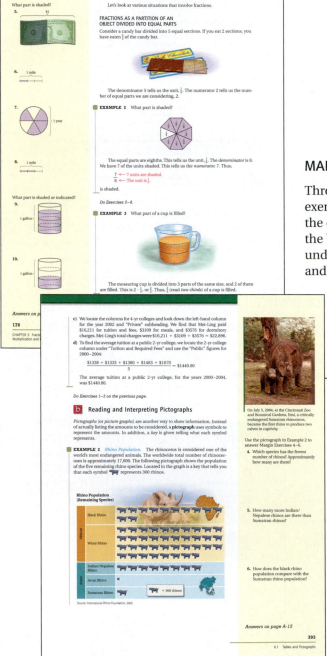

MARGIN EXERCISES

Throughout the text, students are directed to numerous margin exercises that provide immediate practice and reinforcement of the concepts covered in each section. Answers are provided at the back of the book so students can immediately self-assess their understanding of the skill or concept at hand. (See pages 110, 316, and 437.)

REAL-DATA APPLICATIONS

This text encourages students to see and interpret the mathematics that appears every day in the world around them. Throughout the writing process, an extensive and energetic search for real-data applications was conducted, and the result is a variety of examples and exercises that connect the mathematical content with the real world. A large number of the applications are new to this edition, and many are drawn from the fields of business and economics, life and physical sciences, social sciences, and areas of general interest such as sports and daily life. To further encourage students to understand the relevance of mathematics, many applications are enhanced by graphs and drawings similar to those found in today's newspapers and magazines, and feature source lines as well. (See pages 296, 404, and 413.)

TRANSLATING FOR SUCCESS

Translating for Success The goal of the matching exercises in this new feature is to practice step (2), *Translate*, of the five-step problem-solving process. Students translate each of ten problems to an equation and select the correct translation from fifteen given equations. This feature appears once in each chapter and reviews skills and concepts with problems from all preceding chapters. (See pages 140, 219, and 376.)

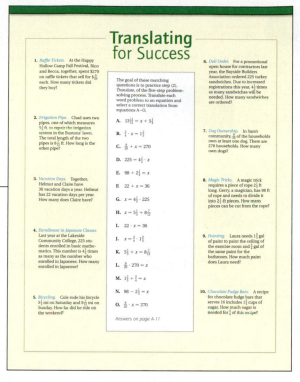

ART PROGRAM

Today's students are often visually oriented and their approach to a printed page is no exception. The art program is designed to improve the visualization of the mathematical concepts and to enhance the real-data applications. (See pages 115, 177, and 412.)

The use of color is carried out in a methodical and precise manner so that it conveys a consistent meaning, which enhances the readability of the text. For example, when perimeter is considered, figures have a red border to emphasize the perimeter. When area is considered, figures are outlined in black and screened with amber to emphasize the area. Similarly, when volume is considered, figures are three-dimensional and airbrushed blue. When fractional parts are illustrated, those parts are shown in purple.

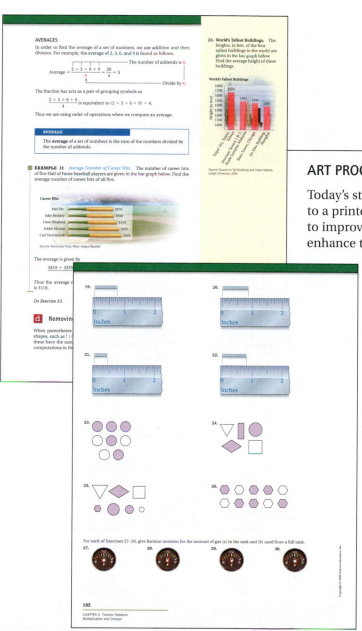

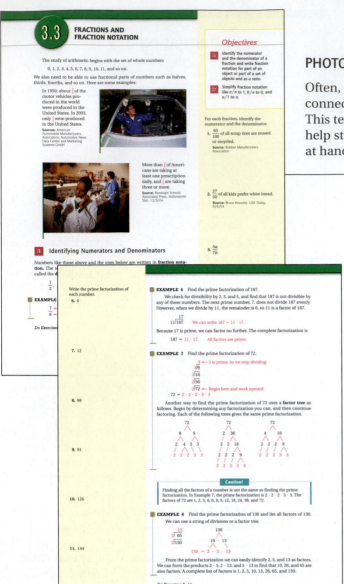

PHOTOGRAPHS

Often, an application becomes relevant to students when the connection to the real world is illustrated with a photograph. This text has numerous photographs throughout in order to help students see the relevance and visualize the application at hand. (See pages 114, 239, and 311.)

CAUTION BOXES

Found at relevant points throughout the text, boxes with the "Caution!" heading warn students of common misconceptions or errors made in performing a particular mathematics operation or skill. (See pages 201, 262, and 316.)

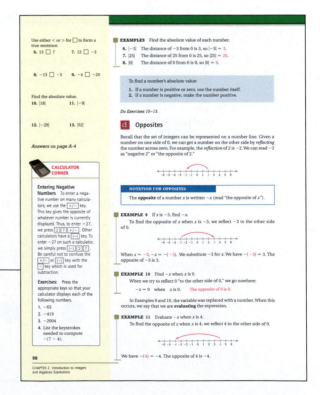

CALCULATOR CORNERS

Where appropriate throughout the text, students will find optional Calculator Corners. These Calculator Corners have been written to be accessible to students and to represent current calculators. (See pages 253, 364, and 529.)

STUDY TIPS

A variety of Study Tips throughout the text give students pointers on how to develop good study habits as they progress through the course. Presented at times as brief suggestions and at other times as more lengthy discussions, these Study Tips encourage students to get involved in the learning process. (See pages 162, 233, and 395.)

EXERCISE SETS

The exercise sets are a critical part of any math book. To give students ample opportunity to practice what they have learned, each section is followed by an extensive exercise set designed to reinforce the section concepts. In addition, students also have the opportunity to synthesize the objectives from the current section with those from preceding sections.

For Extra Help   Many valuable study aids accompany this text. Located before each exercise set, For Extra Help references list appropriate Video Lectures on CD, tutorial, and Web resources so that students can easily find related support materials.

Exercises Grouped by Objective   Exercises in the section exercise sets are keyed by letter to the section objectives for easy review and remediation. This reinforces the objective-based structure of the book. (See pages 173, 294, and 317.)

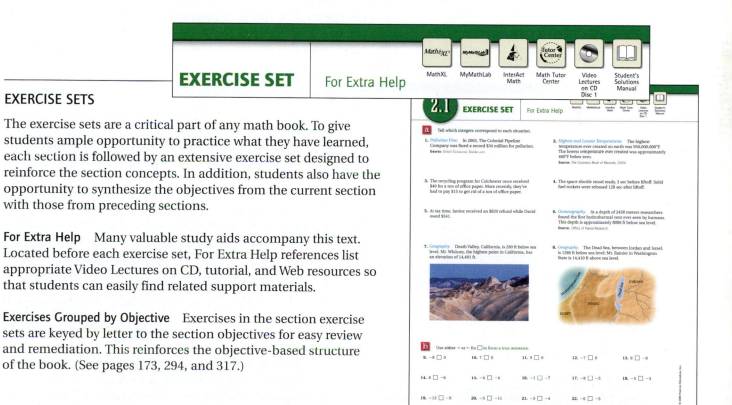

Discussion and Writing Exercises   Designed to help students develop a deeper comprehension of critical concepts, Discussion and Writing exercises (indicated by $\mathbf{D_W}$) are suitable for individual or group work. These exercises encourage students to both think and write about key mathematical ideas in the chapter. (See pages 89, 284, and 382.)

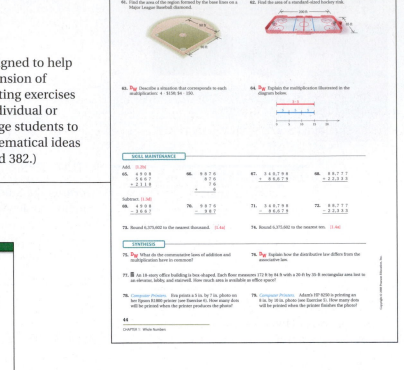

Skill Maintenance Exercises   Found in each exercise set, these exercises review concepts from other sections in the text to prepare students for their final examination. Section and objective codes appear next to each Skill Maintenance exercise for easy reference. (See pages 128, 185, and 209.)

NEW! **Vocabulary Reinforcement Exercises**
This new feature checks and reviews students' understanding of the vocabulary introduced throughout the text. It appears once in every chapter, in the Skill Maintenance portion of an exercise set, and is intended to provide a continuing review of the terms that students must know in order to be able to communicate effectively in the language of mathematics. (See pages 57, 214, and 362.)

Synthesis Exercises   In most exercise sets, Synthesis exercises help build critical-thinking skills by requiring students to synthesize or combine learning objectives from the current section as well as from preceding text sections. (See pages 92, 153, and 414.)

END-OF-CHAPTER MATERIAL

At the end of each chapter, students can practice all they have learned as well as tie the current chapter content to material covered in earlier chapters.

SUMMARY AND REVIEW

A three-part *Summary and Review* appears at the end of each chapter. The first part includes the Concept Reinforcement Exercises described below. The second part is a list of important properties and formulas, when applicable, and the third part provides an extensive set of review exercises.

NEW! **Concept Reinforcement Exercises** Found in the Summary and Review of every chapter, these true/false exercises are designed to increase understanding of the concepts rather than merely assess students' skill at memorizing procedures. (See pages 300, 452, and 504.)

Important Properties and Formulas A list of the important properties and formulas discussed in the chapter is provided for students in an organized manner to help them prioritize topics learned and prepare for chapter tests. This list is provided only in those chapters in which new properties or formulas are presented. (See pages 578, 671, and 719.)

Review Exercises At the end of each chapter, students are provided with an extensive set of review exercises. Reference codes beside each exercise or direction line allow students to easily refer back to specific, objective-level content for remediation. (See pages 90, 224, and 383.)

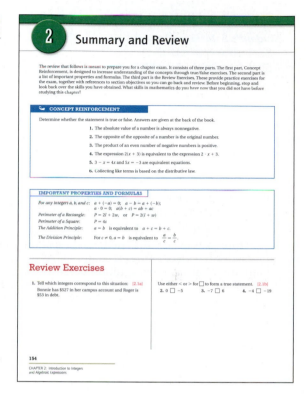

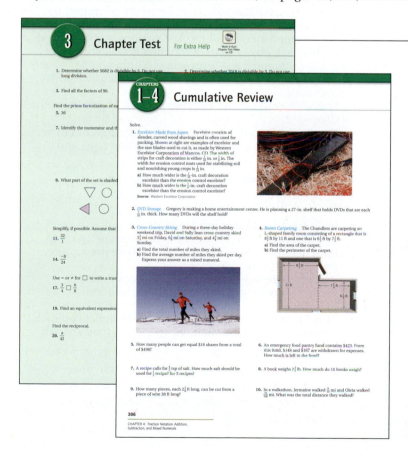

CHAPTER TEST

Following the Review Exercises, a sample Chapter Test allows students to review and test comprehension of chapter skills prior to taking an instructor's exam. Answers to all questions in the Chapter Test are given at the back of the book. Section and objective references for each question are included with the answers. (See pages 157, 304, and 386.)

CUMULATIVE REVIEW

Following Chapters 2 through 10, students encounter a Cumulative Review. This exercise set reviews skills and concepts from all preceding chapters to help students recall previously learned material and prepare for a final exam. At the back of the book are answers to all Cumulative Review exercises, together with section and objective references, so that students know exactly what material to study if they miss a review exercise. Additional Cumulative Review Tests for every chapter are available in the *Printed Test Bank*. (See pages 159, 306, and 461.)

Ancillaries

The following ancillaries are available to help both instructors and students use this text more effectively.

Student Supplements

Student's Solutions Manual
(ISBN-10: 0-321-33707-7)
(ISBN-13: 978-0-321-33707-8)

- By Judith A. Penna, *Indiana University Purdue University Indianapolis*
- Contains completely worked-out solutions with step-by-step annotations for all the odd-numbered exercises in the text, with the exception of the Discussion and Writing exercises, as well as completely worked-out solutions to all the exercises in the Chapter Reviews, Chapter Tests, and Cumulative Reviews.

Collaborative Learning Activities Manual
(ISBN-10: 0-321-33710-7)
(ISBN-13: 978-0-321-33710-8)

- Features group activities tied to text sections and includes the focus, time estimate, suggested group size and materials, and background notes for each activity.
- Available as a stand-alone supplement sold in the bookstore, as a textbook bundle component for students, or as a classroom activity resource for instructors.

Video Lectures on CD
(ISBN-10: 0-321-33706-9)
(ISBN-13: 978-0-321-33706-1)

- Complete set of digitized videos on CD-ROMs for student use at home or on campus.
- To tie student learning to the pedagogy of the text, lectures are organized by objectives, which are indicated on the screen at the start of each new objective.
- Presents a series of lectures correlated directly to the content of each section of the text.
- Features an engaging team of instructors including authors Barbara Johnson and David Ellenbogen who present material in a format that stresses student interaction, often using examples and exercises from the text.
- Ideal for distance learning or supplemental instruction.
- Videos on CD include an expandable window that shows text captioning. Captions can be turned on or off.

NEW! Work It Out! Chapter Test Video Solutions on CD
(ISBN-10: 0-321-45091-4)
(ISBN-13: 978-0-321-45091-3)

- Presented by author Barbara Johnson and by Judith A. Penna
- Provides step-by-step solutions to every exercise in each Chapter Test from the text.
- Helps students prepare for chapter tests and synthesize content.

Instructor Supplements

Annotated Instructor's Edition
(ISBN-10: 0-321-33712-3)
(ISBN-13: 978-0-321-33712-2)

- Includes answers to all exercises printed in blue on the same page as those exercises.

Instructor's Solutions Manual
(ISBN-10: 0-321-33714-X)
(ISBN-13: 978-0-321-33714-6)

- By Judith A. Penna, *Indiana University Purdue University Indianapolis*
- Contains brief solutions to the even-numbered exercises in the exercise sets, answers to all of the Discussion and Writing exercises, and the completely worked-out solutions to all the exercises in the Chapter Reviews, Chapter Tests, and Cumulative Reviews.

Printed Test Bank
(ISBN-10: 0-321-33711-5)
(ISBN-13: 978-0-321-33711-5)

- By Laurie Hurley
- Contains one diagnostic test.
- Contains one pretest for each chapter.
- Provides 13 new test forms for every chapter and 8 new test forms for the final exam.
- For the chapter tests, 5 test forms are modeled after the chapter tests in the text, 3 test forms are organized by topic order following the text objectives, 3 test forms are designed for 50-minute class periods and organized so that each objective in the chapter is covered on one of the tests, and 2 test forms are multiple-choice. Chapter tests also include more challenging synthesis questions.
- Contains 2 cumulative tests per chapter beginning with Chapter 2.
- For the final exam, 3 test forms are organized by chapter, 3 test forms are organized by question type, and 2 test forms are multiple-choice.

NEW! Instructor and Adjunct Support Manual
(ISBN-10: 0-321-33718-2)
(ISBN-13: 978-0-321-33718-4)

- Includes *Adjunct Support Manual* material.
- Features resources and teaching tips designed to help both new and adjunct faculty with course preparation and classroom management.
- Resources include chapter reviews, extra practice sheets, conversion guide, video index, and transparency masters.
- Also available electronically so course/adjunct coordinators can customize material specific to their schools.

Student Supplements

Math Study Skills for Students Video on CD
(ISBN-10: 0-321-29745-8)
(ISBN-13: 978-0-321-29745-7)

- Presented by author Marvin Bittinger
- Designed to help students make better use of their math study time and improve their retention of concepts and procedures taught in classes from basic mathematics through intermediate algebra.
- Through carefully crafted graphics and comprehensive on-camera explanation, focuses on study skills that are commonly overlooked.

Addison-Wesley Math Tutor Center
www.aw-bc.com/tutorcenter

- The Addison-Wesley Math Tutor Center is staffed by qualified mathematics instructors who provide students with tutoring on examples and odd-numbered exercises from the textbook. Tutoring is available via toll-free telephone, toll-free fax, e-mail, or the Internet. White Board technology allows tutors and students to actually see problems worked while they "talk" in real time over the Internet during tutoring sessions.

MathXL® Tutorials on CD
(ISBN-10: 0-321-33713-1)
(ISBN-13: 978-0-321-33713-9)

- Provides algorithmically generated practice exercises that correlate at the objective level to the content of the text.
- Includes an example and a guided solution to accompany every exercise and video clips for selected exercises.
- Recognizes student errors and provides feedback; generates printed summaries of students' progress.

Instructor Supplements

TestGen with Quizmaster
(ISBN-10: 0-321-33708-5)
(ISBN-13: 978-0-321-33708-5)

- Enables instructors to build, edit, print, and administer tests.
- Features a computerized bank of questions developed to cover all text objectives.
- Algorithmically based content allows instructors to create multiple but equivalent versions of the same question or test with a click of a button.
- Instructors can also modify test-bank questions or add new questions by using the built-in question editor, which allows users to create graphs, input graphics, and insert math notation, variable numbers, or text.
- Tests can be printed or administered online via the Internet or another network. Quizmaster allows students to take tests on a local area network.
- Available on a dual-platform Windows/Macintosh CD-ROM.

NEW! *PowerPoint Lecture Presentation*

- Classroom presentation software geared specifically to this textbook.
- Available within MyMathLab or on the Addison-Wesley catalog www.aw-bc.com/math.

MathXL®   www.mathxl.com

MathXL is a powerful online homework, tutorial, and assessment system that accompanies Addison-Wesley textbooks in mathematics or statistics. With MathXL, instructors can create, edit, and assign online homework and tests using algorithmically generated exercises correlated at the objective level to the textbook. They can also create and assign their own online exercises and import TestGen tests for added flexibility. All student work is tracked in MathXL's online gradebook. Students can take chapter tests in MathXL and receive personalized study plans based on their test results. The study plan diagnoses weaknesses and links students directly to tutorial exercises for the objectives they need to study and retest. Students can also access supplemental animations and video clips directly from selected exercises. MathXL is available to qualified adopters. For more information, visit our Web site at www.mathxl.com or contact your Addison-Wesley representative.

MyMathLab   www.mymathlab.com

MyMathLab is a series of text-specific, easily customizable online courses for Addison-Wesley textbooks in mathematics and statistics. Powered by CourseCompass™ (Pearson Education's online teaching and learning environment) and MathXL® (our online homework, tutorial, and assessment system), MyMathLab gives instructors the tools they need to deliver all or a portion of their course online, whether students are in a lab setting or working from home. MyMathLab provides a rich and flexible set of course materials, featuring free-response exercises that are algorithmically generated for unlimited practice and mastery. Students can also use online tools, such as video lectures, animations, and a multimedia textbook, to independently improve their understanding and performance. Instructors can use MyMathLab's homework and test managers to select and assign online exercises correlated directly to the textbook, and they can also create and assign their own online exercises and import TestGen tests for added flexibility. MyMathLab's online gradebook—designed specifically for mathematics and statistics—automatically tracks students' homework and test results and gives the instructor control over how to calculate final grades. Instructors can also add offline (paper-and-pencil) grades to the gradebook. MyMathLab is available to qualified adopters. For more information, visit our Web site at www.mymathlab.com or contact your Addison-Wesley representative.

InterAct Math® Tutorial Web site   www.interactmath.com

Get practice and tutorial help online! This interactive tutorial Web site provides algorithmically generated practice exercises that correlate directly to the exercises in the textbook. Students can retry an exercise as many times as they like with new values each time for unlimited practice and mastery. Every exercise is accompanied by an interactive guided solution that provides helpful feedback for incorrect answers, and students can also view a worked-out sample problem that steps them through an exercise similar to the one they're working on.

ADDISON-WESLEY MATH ADJUNCT SUPPORT CENTER

The Addison-Wesley Math Adjunct Support Center is staffed by qualified mathematics instructors with over 50 years of combined experience at both the community college and university level. Assistance is provided for faculty in the following areas:

- Suggested syllabus consultation
- Tips on using materials packaged with your book
- Book-specific content assistance
- Teaching suggestions including advice on classroom strategies

For more information, visit www.aw-bc.com/tutorcenter/math-adjunct.html

Acknowledgments and Reviewers

Many of you helped to shape *Prealgebra,* Fifth Edition, by reviewing and spending time with us on your campuses. Our deepest appreciation to all of you and in particular to the following:

Lucio Della Vecchia, *Daytona Beach Community College*

David Whittlesey, *Valencia Community College*

Babette Dickelman, *Jackson Community College*

Wayne Browne, *Oklahoma State University–Oklahoma City*

James Vogel, *Sanford-Brown College*

John Close, *Salt Lake Community College*

Gail Burkett, *Palm Beach Community College*

Linda Spears, *Rock Valley College*

We wish to express our heartfelt appreciation to a number of people who have contributed in special ways to the development of this textbook. Our editors, Jennifer Crum and Randy Welch, and marketing manager, Jay Jenkins, encouraged our vision and provided marketing insight. Kari Heen, the project manager, deserves special recognition for overseeing every phase of the project and keeping it moving. The unwavering support of the Developmental Math group, including Antonio Arvelo, editorial assistant, Joanna Doxey, assistant editor, Ron Hampton, production manager, Dennis Schaefer, cover designer, and Sharon Smith and Ceci Fleming, media producers, plus the endless hours of hard work by Kathy Diamond and Geri Davis have led to products of which we are immensely proud.

Other strong support has come from Dawn Mulheron, Elina Niemelä, Holly Martinez, and Jennifer Rosenberg for their accuracy checking of the manuscript and page proofs. We also wish to recognize those who wrote scripts, presented lessons on camera, and checked the accuracy of the videos.

To the Student

Whatever your past experiences, we encourage you to look at this mathematics course as a fresh start and to approach it with a positive attitude. Understanding mathematics will enrich and improve your life. Mathematics is the basis for making many important decisions and for controlling your personal finances and is essential for many careers.

You are the most important factor in the success of your learning experience. In earlier situations, you might have just allowed yourself to sit back and let the instructor "pour in" the learning, with little or no effort on your part. One of the biggest adjustments you might have to make in college is to realize that now you must take a more assertive and proactive role. For example, as soon as possible after class, study the textbook and do the homework assigned, making use of the supplementary materials that accompany the text. Take responsibility for your own learning. This will put you in the best possible position to be successful in this course.

One of the most important suggestions we can make is that you allow yourself enough *time* to learn. You can have an outstanding instructor, an excellent textbook, and the best supplementary materials, but if you do not give yourself time to learn, how can they be of benefit? Suggestions like this are found throughout the book under the heading *Study Tips*. You might want to read all the Study Tips and devise a comprehensive study plan before you begin your course. An Index of Study Tips appears immediately before this Preface.

We wish you success.

M.L.B.
D.J.E.
B.L.J.

Bittinger Student Organizer

Study Tips

Throughout this text, you will find a feature called *Study Tips*. We discuss these in the Preface of this text. They are intended to help improve your math study skills. An index of all the *Study Tips* can be found at the front of the book.

For Extra Help

MyMathLab

MathXL

Student's Solutions Manual

Video Lectures on CD

Math Tutor Center

InterAct Math

Additional Resources

Basic Math Review Card
(ISBN-10: 0-321-39476-3)
(ISBN-13: 978-0-321-39476-7)

Algebra Review Card
(ISBN-10: 0-321-39473-9)
(ISBN-13: 978-0-321-39473-6)

Math for Allied Health Reference Card
(ISBN-10: 0-321-39474-7)
(ISBN-13: 978-0-321-39474-3)

Graphing Calculator Reference Card
(ISBN-10: 0-321-39475-5)
(ISBN-13: 978-0-321-39475-0)

Spanish Basic Math Study Card
(ISBN-10: 0-321-43858-2)
(ISBN-13: 978-0-321-43854-4)

Math Study Skills for Students Video on CD
(ISBN-10: 0-321-29745-8)
(ISBN-13: 978-0-321-29745-7)

Go to www.aw-bc.com/math for more information.

On the first day of class, complete this chart and the weekly planner that follows on the reverse page.

Instructor Information:

Name _____

Office Hours and Location _____

_____

Phone Number(s) _____

Fax Number _____

E-mail Address _____

Find the names of two students whom you could contact for class information or study questions:

1. Name _____

   Phone Number(s) _____

   E-mail Address _____

   IM Name/Host _____

2. Name _____

   Phone Number(s) _____

   E-mail Address _____

   IM Name/Host _____

Math Lab on Campus:

Location _____

Hours _____

Phone Number(s) _____

Tutoring:

Campus Location _____

Hours _____

To order the Addison-Wesley Math Tutor Center, call 1-888-777-0463.

(*See the Preface for important information concerning this tutoring.*)

Important Supplements: (*See the Preface for a complete list of available supplements.*)

Supplements recommended by the instructor _____

_____

Online Log-in Information (*include access code, password, Web address, etc.*)

_____

_____

Bittinger Student Organizer

WEEKLY PLANNER

Success is planned. On this page, plan a typical week. Consider time allotments for class, study, work, travel, family, and relaxation.

Important Dates

Midterm Exam

_____

Final Exam

_____

Holidays

_____

_____

_____

_____

_____

Other

_____

_____

_____

_____

TIME	Sun.	Mon.	Tues.	Wed.	Thurs.	Fri.	Sat.
6:00 A.M.							
6:30							
7:00							
7:30							
8:00							
8:30							
9:00							
9:30							
10:00							
10:30							
11:00							
11:30							
12:00 P.M.							
12:30							
1:00							
1:30							
2:00							
2:30							
3:00							
3:30							
4:00							
4:30							
5:00							
5:30							
6:00							
6:30							
7:00							
7:30							
8:00							
8:30							
9:00							
9:30							
10:00							
10:30							
11:00							
11:30							
12:00 A.M.							

Tabbing for Success

Use these tabs to mark important pages of
your textbook for quick reference and review.

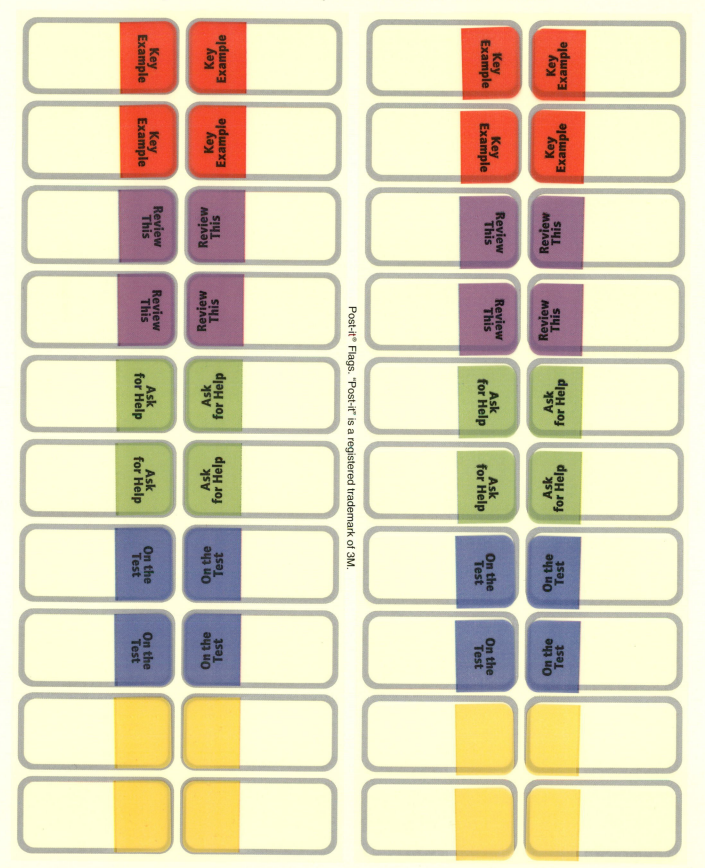

Post-it® Flags. "Post-it" is a registered trademark of 3M.

Using the Tabs

Customize Your Textbook and Make It Work for You!

These removable and reusable tabs offer you five ways to be successful in your math course by letting you bookmark pages with helpful reminders.

 Key Example Use these tabs to flag examples that will help you while doing your homework or preparing for your tests.

 Review This Mark important definitions, procedures, or key terms to review later.

 Ask for Help Not sure of something? Need more instruction? Place these tabs in your textbook to address any questions with your instructor during your next class meeting or with your tutor during your next tutoring session.

 On the Test If your instructor alerts you that something will be covered on a test, use these tabs to bookmark it.

 Write your own notes or create more of the preceding tabs to help you succeed in your math course.

ISBN-13: 978-0-321-50072-4
ISBN-10:   0-321-50072-5

EAN

9 780321 500724

90000

Whole Numbers

Real-World Application

Races in which runners climb the steps inside a building are called "run-up" races. There are 2058 steps in the International Towerthon, Kuala Lumpur, Malaysia. Write expanded notation for the number of steps.

This problem appears as Exercise 13 in Section 1.1.

1.1 STANDARD NOTATION

Objectives

a Give the meaning of digits in standard notation.

b Convert from standard notation to expanded notation.

c Convert between standard notation and word names.

What does the digit 2 mean in each number?

1. 526,555

2. 265,789

3. 42,789,654

4. 24,789,654

5. 8924

6. 5,643,201

Answers on page A-1

To the student:

In the Preface, at the front of the text, you will find a Student Organizer card. This pullout card will help you keep track of important dates and useful contact information. You can also use it to plan time for class, study, work, and relaxation. By managing your time wisely, you will provide yourself the best possible opportunity to be successful in this course.

We study mathematics in order to be able to solve problems. In this section, we study how numbers are named. We begin with the concept of place value.

a Place Value

Consider the numbers in the following table.

Three Most Populous Countries in the World

COUNTRY	POPULATION 2005
China	1,306,313,812
India	1,080,264,388
United States	295,734,134

Source: The World Factbook, July 2005 estimates

A **digit** is a number 0, 1, 2, 3, 4, 5, 6, 7, 8, or 9 that names a place-value location. For large numbers, digits are separated by commas into groups of three, called **periods.** Each period has a name: *ones, thousands, millions, billions, trillions,* and so on. To understand the population of China in the table above, we can use a **place-value chart,** as shown below.

PLACE-VALUE CHART														
Periods → Trillions			Billions			Millions			Thousands			Ones		
					1	3	0	6	3	1	3	8	1	2
Hundreds	Tens	Ones	Hundreds	Tens	Ones	Hundreds	Tens	Ones	Hundreds	Tens	Ones	Hundreds	Tens	Ones

1 billion   306 millions 313 thousands   812 ones

EXAMPLES What does the digit 8 mean in each number?

1. 27**8**,342   8 thousands
2. **8**72,342   8 hundred thousands
3. 2**8**,343,399,223   8 billions
4. 1,023,**8**50   8 hundreds
5. 9**8**,413,099   8 millions
6. 632**8**   8 ones

Do Margin Exercises 1–6.

EXAMPLE 7 *American Red Cross.* In 2003, private donations to the American Red Cross totaled about $587,492,000. What does each digit name?

Source: *The Chronicle of Philanthropy*

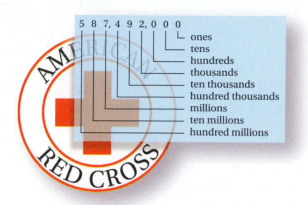

5 8 7, 4 9 2, 0 0 0
- ones
- tens
- hundreds
- thousands
- ten thousands
- hundred thousands
- millions
- ten millions
- hundred millions

Do Exercise 7.

b Converting from Standard Notation to Expanded Notation

To answer questions such as "How many?", "How much?", and "How tall?", we use whole numbers. The set, or collection, of **whole numbers** is

$$0, 1, 2, 3, 4, 5, 6, 7, 8, 9, 10, 11, 12, \ldots.$$

The set goes on indefinitely. There is no largest whole number, and the smallest whole number is 0. Each whole number can be named using various notations. The set $1, 2, 3, 4, 5, \ldots$, without 0, is called the set of **natural numbers.**

Let's look at the data from the bar graph shown here.

Fewer Computer Majors

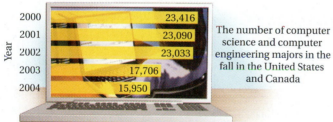

Year	
2000	23,416
2001	23,090
2002	23,033
2003	17,706
2004	15,950

The number of computer science and computer engineering majors in the fall in the United States and Canada

Source: Computing Research Association Taulbee Survey

The number of computer majors in 2003 was 17,706. **Standard notation** for the number of computer majors is 17,706. We write **expanded notation** for 17,706 as follows:

$$17,706 = 1 \text{ ten thousand} + 7 \text{ thousands}$$
$$+ 7 \text{ hundreds} + 0 \text{ tens} + 6 \text{ ones}.$$

7. Presidential Library. In the first year of operation, 280,219 people visited the Ronald Reagan Presidential Library in Simi Valley, California. What does each digit name?

Sources: *USA Today* research by Bruce Rosenstein; National Archives & Records Administration; Associated Press

Write expanded notation.

8. 1895

9. 23,416, the number of computer majors in 2000

10. 4218 mi (miles), the diameter of Mars

Answers on page A-1

11. 4180 mi, the length of the Nile River, the longest river in the world

12. 146,692, the number of Labrador retrievers registered in 2004

Source: The American Kennel Club

Write a word name. (Refer to the figure below right.)

13. 49, the total number of medals won by Australia

14. 16, the number of silver medals won by Germany

15. 38, the number of bronze medals won by Russia

EXAMPLE 8 Write expanded notation for 3031 mi (miles), the diameter of Mercury.

$$3031 = 3 \text{ thousands} + 0 \text{ hundreds} + 3 \text{ tens} + 1 \text{ one, or}$$
$$3 \text{ thousands} + 3 \text{ tens} + 1 \text{ one}$$

EXAMPLE 9 Write expanded notation for 563,384, the population of Washington, D.C.

$$563,384 = 5 \text{ hundred thousands} + 6 \text{ ten thousands}$$
$$+ 3 \text{ thousands} + 3 \text{ hundreds} + 8 \text{ tens} + 4 \text{ ones}$$

Do Exercises 8–12 (8–10 are on the preceding page).

C Converting Between Standard Notation and Word Names

We often use **word names** for numbers. When we pronounce a number, we are speaking its word name. Russia won 92 medals in the 2004 Summer Olympics in Athens, Greece. A word name for 92 is "ninety-two." Word names for some two-digit numbers like 27, 39, and 92 use hyphens. Others like 17 use only one word, "seventeen."

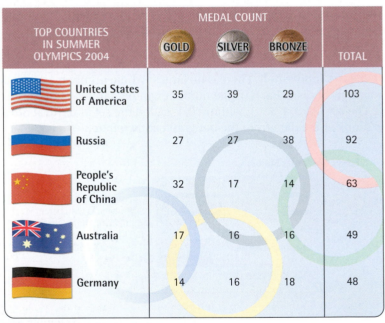

TOP COUNTRIES IN SUMMER OLYMPICS 2004	MEDAL COUNT			TOTAL
	GOLD	SILVER	BRONZE	
United States of America	35	39	29	103
Russia	27	27	38	92
People's Republic of China	32	17	14	63
Australia	17	16	16	49
Germany	14	16	18	48

Source: 2004 Olympics, Athens, Greece

EXAMPLES Write a word name.

10. 35, the total number of gold medals won by the United States

Thirty-five

11. 17, the number of silver medals won by the People's Republic of China

Seventeen

Do Exercises 13–15.

For word names for larger numbers, we begin at the left with the largest period. The number named in the period is followed by the name of the period; then a comma is written and the next period is named.

EXAMPLE 12   Write a word name for 46,605,314,732.

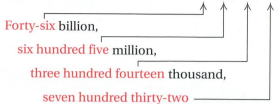

Forty-six billion,

six hundred five million,

three hundred fourteen thousand,

seven hundred thirty-two

The word "and" *should not* appear in word names for whole numbers. Although we commonly hear such expressions as "two hundred *and* one," the use of "and" is not, strictly speaking, correct in word names for whole numbers. For decimal notation, it is appropriate to use "and" for the decimal point. For example, 317.4 is read as "three hundred seventeen *and* four tenths."

Do Exercises 16–19.

EXAMPLE 13   Write standard notation.

Five hundred six million,

three hundred forty-five thousand,

two hundred twelve

Standard notation is 506,345,212.

Do Exercise 20.

Write a word name.

16. 204

17. $51,206, the average salary in 2004 for those who have a bachelor's degree or more
Source: U.S. Bureau of the Census

18. 1,879,204

19. 6,449,000,000, the world population in 2005
Source: U.S. Bureau of the Census

20. Write standard notation.

Two hundred thirteen million, one hundred five thousand, three hundred twenty-nine

Answers on page A-1

Study Tips

USING THIS TEXTBOOK

Throughout this textbook, you will find a feature called "Study Tips." One of the most important ways in which to improve your math study skills is to learn the proper use of the textbook. Here we highlight a few points that we consider most helpful.

- ■ **Be sure to note the symbols** [a], [b], [c], **and so on, that correspond to the objectives you are to master in each section.**   The first time you see them is in the margin at the beginning of the section; the second time is in the subheadings of each section; and the third time is in the exercise set for the section. You will also find symbols like [1.1a] or [1.2c] next to the skill maintenance exercises in each exercise set and the review exercises at the end of the chapter, as well as in the answers to the chapter tests and the cumulative reviews. These objective symbols allow you to refer to the appropriate place in the text when you need to review a topic.

- ■ **Read and study each step of each example.**   The examples include important side comments that explain each step. These examples and annotations have been carefully chosen so that you will be fully prepared to do the exercises.

- ■ **Stop and do the margin exercises as you study a section.**   This gives you immediate reinforcement of each concept as it is introduced and is one of the most effective ways to master the mathematical skills in this text. Don't deprive yourself of this benefit!

- ■ **Note the icons listed at the top of each exercise set.**   These refer to the many distinctive multimedia study aids that accompany the book.

a What does the digit 5 mean in each case?

1. 235,888

2. 253,777

3. 1,488,526

4. 500,736

Used Cars. 1,582,370 certified used cars were sold in 2004 in the United States. **Source:** *Motor Trend,* April 2005, p. 26
In the number 1,582,370, what digit names the number of:

5. thousands?

6. ones?

7. millions?

8. hundred thousands?

b Write expanded notation.

9. 5702

10. 3097

11. 93,986

12. 38,453

Step-Climbing Races. Races in which runners climb the steps inside a building are called "run-up" races. The graph below shows the number of steps in four buildings. In Exercises 13–16, write expanded notation for the number of steps in each race.

Step-Climbing Races

2058 — International Towerthon, Kuala Lumpur, Malaysia
1776 — CN Tower Run-Up, Toronto
1268 — World Financial Center, New York
1081 — Skytower Run-Up, Auckland, New Zealand

Source: New York Road Runners Club

13. 2058 steps in the International Towerthon, Kuala Lumpur, Malaysia

14. 1776 steps in the CN Tower Run-Up, Toronto, Ontario, Canada

15. 1268 steps in the World Financial Center, New York City, New York

16. 1081 steps in the Skytower Run-Up, Auckland, New Zealand

Overseas Travelers. The chart below shows the residence and number of overseas travelers to the United States in 2004. In Exercises 17–22, write expanded notation for the number of travelers from each country.

OVERSEAS TRAVELERS TO U.S., 2004

Australia	519,955
Brazil	384,734
United Kingdom	4,302,737
India	308,845
Japan	3,747,620
Spain	333,432

Source: U.S. Department of Commerce, ITA, Office of Travel and Tourism Industries

17. 519,955 from Australia

18. 3,747,620 from Japan

19. 308,845 from India

20. 333,432 from Spain

21. 4,302,737 from the United Kingdom

22. 384,734 from Brazil

 C Write a word name.

23. 85

24. 48

25. 88,000

26. 45,987

27. 123,765

28. 111,013

29. 7,754,211,577

30. 43,550,651,808

Write standard notation.

31. Two million, two hundred thirty-three thousand, eight hundred twelve

32. Three hundred fifty-four thousand, seven hundred two

33. Eight billion

34. Seven hundred million

Write a word name for the number in each sentence.

35. *Great Pyramid.* The area of the base of the Great Pyramid in Egypt is 566,280 square feet.

36. *Population of the United States.* The population of the United States in April 2006 was estimated to be 298,509,533.
Source: U.S. Bureau of the Census

37. *Busiest Airport.* In 2004, the world's busiest airport, Hartsfield in Atlanta, had 83,578,906 passengers.

Source: Airports Council International World Headquarters, Geneva, Switzerland

Write standard notation for the number in each sentence.

39. Light travels nine trillion, four hundred sixty billion kilometers in one year.

41. *Pacific Ocean.* The area of the Pacific Ocean is sixty-four million, one hundred eighty-six thousand square miles.

38. *Prisoners.* There were 2,131,180 total prisoners, federal, state, and local, in the United States in 2004.

Source: *Prison and Jail Inmates at Midyear 2004,* U.S. Bureau of Justice Statistics

40. The distance from the sun to Pluto is three billion, six hundred sixty-four million miles.

42. *Internet Users.* In a recent year, there were fifty-four million, five hundred thousand Internet users in China.

Source: Computer Industry Almanac, Inc.

To the student and the instructor: The Discussion and Writing exercises, denoted by the symbol **Dw**, are meant to be answered with one or more sentences. They can be discussed and answered collaboratively by the entire class or by small groups. Because of their open-ended nature, the answers to these exercises do not appear at the back of the book.

43. **Dw** Explain why we use commas when writing large numbers.

44. **Dw** Write an English sentence in which the number 370,000,000 is used.

SYNTHESIS

To the student and the instructor: The Synthesis exercises found at the end of every exercise set challenge students to combine concepts or skills studied in that section or in preceding parts of the text. Exercises marked with a ▦ symbol are meant to be solved using a calculator.

45. How many whole numbers between 100 and 400 contain the digit 2 in their standard notation?

46. ▦ What is the largest number that you can name on your calculator? How many digits does that number have? How many periods?

1.2 ADDITION

Objectives

a Write an addition sentence that corresponds to a situation.

b Add whole numbers.

c Use addition in finding perimeter.

a Addition and Related Sentences

Addition of whole numbers corresponds to combining or putting things together.

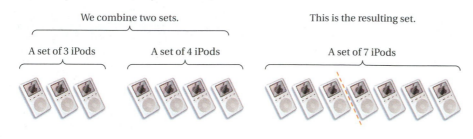

We combine two sets.   This is the resulting set.

A set of 3 iPods   A set of 4 iPods   A set of 7 iPods

The addition that corresponds to the figure above is

$$3 + 4 = 7.$$ This is read "3 plus 4 equals 7."

Addend Addend Sum

We say that the **sum** of 3 and 4 is 7. The numbers added are called **addends.**

EXAMPLE 1 Write an addition sentence that corresponds to this situation.

Kelly has $3 and earns $10 more. How much money does she have?

An addition that corresponds is $3 + $10 = $13. This sentence is also called an *equation*.

Do Exercises 1 and 2.

Addition also corresponds to combining distances or lengths.

EXAMPLE 2 Write an addition sentence that corresponds to this situation.

A car is driven 44 mi from San Francisco to San Jose. It is then driven 42 mi from San Jose to Oakland. How far is it from San Francisco to Oakland along the same route?

$$44 \text{ mi} + 42 \text{ mi} = 86 \text{ mi}$$

It is 42 miles from San Jose to Oakland.

It is 44 miles from San Francisco to San Jose.

Do Exercises 3 and 4.

Write an addition sentence that corresponds to each situation.

1. John has 8 music CD-ROMs in his backpack. Then he buys 2 educational CD-ROMs at the bookstore. How many CD-ROMs does John have in all?

2. Sue earns $20 in overtime pay on Thursday and $13 on Friday. How much overtime pay does she earn altogether on the two days?

Write an addition sentence that corresponds to each situation.

3. A car is driven 100 mi from Austin to Waco. It is then driven 93 mi from Waco to Dallas. How far is it from Austin to Dallas along the same route?

4. A coaxial cable 5 ft (feet) long is connected to a cable 7 ft long. How long is the resulting cable?

Answers on page A-1

Add.

5.
```
    7 9 6 8
  + 5 4 9 7
```

6. 6203 + 3542

7.
```
    9 8 0 4
  + 6 3 7 8
```

8.
```
    1 9 3 2
    6 7 2 3
    9 8 7 8
  + 8 9 4 1
```

Answers on page A-1

b   Addition of Whole Numbers

To add whole numbers, we add the ones digits first, then the tens, then the hundreds, then the thousands, and so on.

EXAMPLE 3   Add: 6878 + 4995.

Place values are lined up in columns.

```
        1
    6 8 7 8
  + 4 9 9 5
          3
```
Add ones. We get 13 ones, or 1 ten + 3 ones. Write 3 in the ones column and 1 above the tens. This is called *carrying,* or *regrouping.*

```
      1 1
    6 8 7 8
  + 4 9 9 5
        7 3
```
Add tens. We get 17 tens, or 1 hundred + 7 tens. Write 7 in the tens column and 1 above the hundreds.

```
    1 1 1
    6 8 7 8
  + 4 9 9 5
      8 7 3
```
Add hundreds. We get 18 hundreds, or 1 thousand + 8 hundreds. Write 8 in the hundreds column and 1 above the thousands.

```
    1 1 1
    6 8 7 8
  + 4 9 9 5
  1 1 8 7 3
```
Add thousands. We get 11 thousands.

We show you these steps for explanation. You need write only this.

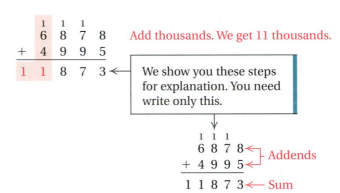

EXAMPLE 4   Add: 391 + 276 + 789 + 498.

```
        2
    3 9 1
    2 7 6
    7 8 9
  + 4 9 8
        4
```
Add ones. We get 24, so we have 2 tens + 4 ones. Write 4 in the ones column and 2 above the tens.

```
    3 2
    3 9 1
    2 7 6
    7 8 9
  + 4 9 8
      5 4
```
Add tens. We get 35 tens, so we have 30 tens + 5 tens. This is also 3 hundreds + 5 tens. Write 5 in the tens column and 3 above the hundreds.

```
    3 2
    3 9 1
    2 7 6
    7 8 9
  + 4 9 8
  1 9 5 4
```
Add hundreds. We get 19 hundreds.

Do Exercises 5–8 on the preceding page.

How do we do an addition of three numbers, like $2 + 3 + 6$? We do so by adding 3 and 6, and then 2. We can show this with parentheses:

$2 + (3 + 6) = 2 + 9 = 11$.   Parentheses tell what to do first.

We could also add 2 and 3, and then 6:

$(2 + 3) + 6 = 5 + 6 = 11$.

Either way we get 11. It does not matter how we group the numbers. This illustrates the **associative law of addition,** $a + (b + c) = (a + b) + c$.

EXAMPLE 5 Insert parentheses to illustrate the associative law of addition:

$5 + (1 + 7) = \ \ 5 \ \ + \ \ 1 \ \ + \ \ 7$.

We group as follows:

$5 + (1 + 7) = (5 + 1) + 7$.

Do Exercises 9 and 10.

We can also add whole numbers in any order. That is, $2 + 3 = 3 + 2$. This illustrates the **commutative law of addition,** $a + b = b + a$.

EXAMPLE 6 Complete the following to illustrate the commutative law of addition:

$5 + 4 = \square + \square$

We reverse the appearance of the two addends:

$5 + 4 = 4 + 5$.

Do Exercises 11 and 12.

THE ASSOCIATIVE LAW OF ADDITION
For any numbers a, b, and c, $(a + b) + c = a + (b + c)$.

THE COMMUTATIVE LAW OF ADDITION
For any numbers a and b, $a + b = b + a$.

Together the commutative and associative laws tell us that to add more than two numbers, we can use any order and grouping we wish.

Insert parentheses to illustrate the associative law of addition.

9. $2 + (6 + 3) = \ \ 2 \ \ + \ \ 6 \ \ + \ \ 3$

10. $(5 + 1) + 4 = \ \ 5 \ \ + \ \ 1 \ \ + \ \ 4$

Complete the following to illustrate the commutative law of addition.

11. $2 + 6 = \square + \square$

12. $7 + 1 = \square + \square$

Answers on page A-1

Add from the top.

13.
```
    9
    9
    4
 +  5
```

14.
```
    8
    6
    9
    7
 +  4
```

15. Add from the bottom.
```
    9
    9
    4
 +  5
```

Answers on page A-1

■ **EXAMPLE 7** Add from the top.
```
    8
    9
    7
 +  6
```

We first add 8 and 9, getting 17; then 17 and 7, getting 24; then 24 and 6, getting 30.

```
 8
 9  ──→  17
 7        7  ──→  24
+6        6        6  ──→  30
─────
30  ←
```
| Try to write only this. |

■ **EXAMPLE 8** Add from the bottom.

```
 8          8          8  ──→  30
 9          9  ──→  22
 7  ──→  13
+6
─────
30  ←
```
| You still write the answer here. |

Do Exercises 13–15.

C Finding Perimeter

Addition can be used when finding perimeter.

> **PERIMETER**
>
> The distance around an object is its **perimeter.**

■ **EXAMPLE 9** Find the perimeter of the octagonal (eight-sided) resort swimming pool.

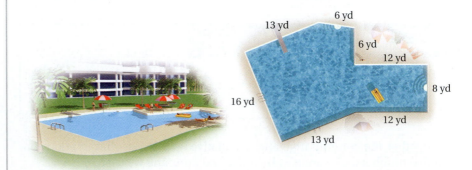

Perimeter = 13 yd + 6 yd + 6 yd + 12 yd + 8 yd
          + 12 yd + 13 yd + 16 yd

The perimeter of the pool is 86 yd.

EXAMPLE 10 Find the perimeter of the soccer field shown.

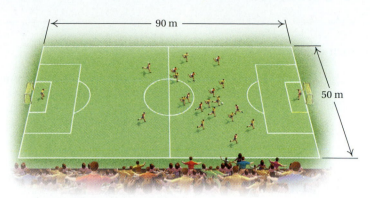

The letter *m* denotes *meters* (a meter is slightly more than 3 ft). Note that for any rectangle, the opposite sides are the same length. Thus

50 m + 90 m + 50 m + 90 m = Perimeter

$\underbrace{50 \text{ m} + 50 \text{ m}}$ + $\underbrace{90 \text{ m} + 90 \text{ m}}$ = Perimeter   Re-ordering the addition

   100 m   +   180 m   = 280 m

The perimeter is 280 m.

Do Exercises 16–18.

CALCULATOR CORNER

Adding Whole Numbers   *To the student and the instructor:* This is the first of a series of *optional* discussions on using a calculator. A calculator is *not* a requirement for this textbook. There are many kinds of calculators and different instructions for their usage. We have included instructions here for a minimum-cost calculator. Be sure to consult your user's manual as well. Also, check with your instructor about whether you are allowed to use a calculator in the course.

To add whole numbers on a calculator, we use the $+$ and $=$ keys. For example, to add 57 and 34, we press 5 7 $+$ 3 4 $=$. The calculator displays 91 , so 57 + 34 = 91. To find 314 + 259 + 478, we press 3 1 4 $+$ 2 5 9 $+$ 4 7 8 $=$. The display reads 1051 , so 314 + 259 + 478 = 1051.

Exercises: Use a calculator to find each sum.

1. 19 + 36

2. 73 + 48

3. 925 + 677

4. 276 + 458

5.　　8 2 6
　　　4 1 5
　　+ 6 9 1

6.　　2 5 3
　　　4 9 0
　　+ 1 2 1

Find the perimeter of each figure.

16.

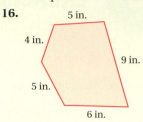

5 in.
4 in.
9 in.
5 in.
6 in.

17.

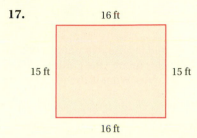

16 ft
15 ft   15 ft
16 ft

Solve.

18. Index Cards.   Two standard sizes for index cards are 3 in. (inches) by 5 in. and 5 in. by 8 in. Find the perimeter of each card.

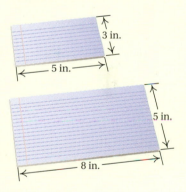

3 in.
5 in.

5 in.
8 in.

Answers on page A-1

To the instructor and the student: This section presented a review of addition of whole numbers. Students who are successful should go on to Section 1.3. Those who have trouble should study developmental unit A near the back of this text and then repeat Section 1.2.

a Write an addition sentence that corresponds to each situation.

1. Two trucks haul sand to a construction site to use in a driveway. One carries 6 cu yd (cubic yards) and the other 8 cu yd. Altogether, how many cubic yards of sand are they hauling to the site?

2. At a construction site, there are two gasoline containers to be used by earth-moving vehicles. One contains 400 gal (gallons) and the other 200 gal. How many gallons do both contain together?

3. A builder buys two parcels of land to build a housing development. One contains 500 acres and the other 300 acres. What is the total number of acres purchased?

4. During March and April, Deron earns extra money doing income taxes part time. In March he earned $220, and in April he earned $340. How much extra did he earn altogether in March and April?

b Add.

5.
$$\begin{array}{r} 3\ 6\ 4 \\ +\quad 2\ 3 \\ \hline \end{array}$$

6.
$$\begin{array}{r} 1\ 5\ 2\ 1 \\ +\quad 3\ 4\ 8 \\ \hline \end{array}$$

7.
$$\begin{array}{r} 1\ 7\ 1\ 6 \\ +\ 3\ 4\ 8\ 2 \\ \hline \end{array}$$

8.
$$\begin{array}{r} 7\ 5\ 0\ 3 \\ +\ 2\ 6\ 8\ 3 \\ \hline \end{array}$$

9.
$$\begin{array}{r} 8\ 6 \\ +\ 7\ 8 \\ \hline \end{array}$$

10.
$$\begin{array}{r} 7\ 3 \\ +\ 6\ 9 \\ \hline \end{array}$$

11.
$$\begin{array}{r} 9\ 9 \\ +\quad 1 \\ \hline \end{array}$$

12.
$$\begin{array}{r} 9\ 9\ 9 \\ +\quad 1\ 1 \\ \hline \end{array}$$

13. 8113 + 390

14. 271 + 3338

15. 356 + 4910

16. 280 + 34,702

17. 3870 + 92 + 7 + 497

18. 10,120 + 12,989 + 5738

19.
$$\begin{array}{r} 4\ 8\ 2\ 5 \\ +\ 1\ 7\ 8\ 3 \\ \hline \end{array}$$

20.
$$\begin{array}{r} 3\ 6\ 5\ 4 \\ +\ 2\ 7\ 0\ 0 \\ \hline \end{array}$$

21.
$$\begin{array}{r} 2\ 3{,}4\ 4\ 3 \\ +\ 1\ 0{,}9\ 8\ 9 \\ \hline \end{array}$$

22.
$$\begin{array}{r} 4\ 5{,}8\ 7\ 9 \\ +\ 2\ 1{,}7\ 8\ 6 \\ \hline \end{array}$$

23.
$$
\begin{array}{r}
7\,7,5\,4\,3 \\
+\;2\,3,7\,6\,7 \\
\hline
\end{array}
$$

24.
$$
\begin{array}{r}
9\,9,9\,9\,9 \\
+\;\quad 1\,1\,2 \\
\hline
\end{array}
$$

25.
$$
\begin{array}{r}
4\,5 \\
2\,5 \\
3\,6 \\
4\,4 \\
+\;8\,0 \\
\hline
\end{array}
$$

26.
$$
\begin{array}{r}
3\,8 \\
2\,7 \\
3\,2 \\
1\,4 \\
+\;7\,6 \\
\hline
\end{array}
$$

27.
$$
\begin{array}{r}
1\,2,0\,7\,0 \\
2,9\,5\,4 \\
+\;\quad 3,4\,0\,0 \\
\hline
\end{array}
$$

28.
$$
\begin{array}{r}
4\,2,4\,8\,7 \\
8\,3,1\,4\,1 \\
+\;3\,6,7\,1\,2 \\
\hline
\end{array}
$$

29.
$$
\begin{array}{r}
4\,8\,3\,5 \\
7\,2\,9 \\
9\,2\,0\,4 \\
8\,9\,8\,6 \\
+\;7\,9\,3\,1 \\
\hline
\end{array}
$$

30.
$$
\begin{array}{r}
9\,8\,9 \\
5\,6\,6 \\
8\,3\,4 \\
9\,2\,0 \\
+\;7\,0\,3 \\
\hline
\end{array}
$$

Insert parentheses to illustrate the associative law of addition.

31. $(2 + 5) + 4 = 2 + 5 + 4$

32. $(7 + 1) + 5 = 7 + 1 + 5$

33. $6 + (3 + 2) = 6 + 3 + 2$

34. $5 + (1 + 4) = 5 + 1 + 4$

Complete each equation to illustrate the commutative law of addition.

35. $2 + 7 = \square + \square$

36. $5 + 2 = \square + \square$

37. $6 + 1 = \square + \square$

38. $1 + 9 = \square + \square$

39. $2 + 9 = \square + \square$

40. $7 + 5 = \square + \square$

Add from the top. Then check by adding from the bottom.

41.
$$
\begin{array}{r}
7 \\
9 \\
4 \\
+\;8 \\
\hline
\end{array}
$$

42.
$$
\begin{array}{r}
4 \\
3 \\
9 \\
1 \\
+\;8 \\
\hline
\end{array}
$$

43.
$$
\begin{array}{r}
8 \\
6 \\
2 \\
3 \\
+\;7 \\
\hline
\end{array}
$$

44.
$$
\begin{array}{r}
9 \\
4 \\
7 \\
8 \\
+\;7 \\
\hline
\end{array}
$$

C Find the perimeter of each figure.

45.

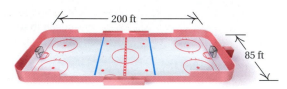

46.

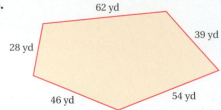

47. Find the perimeter of a standard hockey rink.

200 ft

85 ft

48. In Major League Baseball, how far does a batter travel in circling the bases when a home run has been hit?

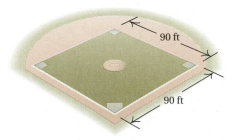

90 ft

90 ft

49. **D**_W_ Explain in your own words what the associative law of addition means.

50. **D**_W_ Describe a situation that corresponds to this mathematical expression:

80 mi + 245 mi + 336 mi.

SKILL MAINTENANCE

The exercises that follow begin an important feature called *Skill Maintenance exercises.* These exercises provide an ongoing review of topics previously covered in the book. You will see them in virtually every exercise set. It has been found that this kind of continuing review can significantly improve your performance on a final examination.

51. What does the digit 8 mean in 486,205?   [1.1a]

52. Write a word name for the number in the following sentence: [1.1c]

In fiscal year 2004, Starbucks Corporation had total net revenues of $5,294,247,000.

Source: Starbucks Corporation

SYNTHESIS

53. **D**_W_ Is it possible for a narrower soccer field to have a greater perimeter than the one in Example 10? Why or why not?

54. **D**_W_ Is it possible for a rectangle to have a perimeter of 12 cu yd? Why or why not?

Add.

55. 🖩 5,987,943 + 328,959 + 49,738,765

56. 🖩 39,487,981 + 8,709,486 + 989,765

57. A fast way to add all the numbers from 1 to 10 inclusive is to pair 1 with 9, 2 with 8, and so on. Use a similar approach to add all numbers from 1 to 100 inclusive.

16

1.3 SUBTRACTION

a Subtraction and Related Sentences

TAKE AWAY

Subtraction of whole numbers applies to two kinds of situations. The first is called "take away." Consider the following example.

> A bowler starts with 10 pins and knocks down 8 of them.

From 10 pins, the bowler "takes away" 8 pins. There are 2 pins left. The subtraction is $10 - 8 = 2$.

 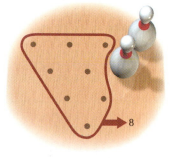

10                    $10 - 8 = 2$

We use the following terminology with subtraction:

$$10 \quad - \quad 8 \quad = \quad 2.$$

Minuend   Subtrahend   Difference

The **minuend** is the number from which another number is being subtracted. The **subtrahend** is the number being subtracted. The **difference** is the result of subtracting the subtrahend from the minuend.

EXAMPLES Write a subtraction sentence that corresponds to each situation.

1. Juan goes to a music store and chooses 10 CDs to take to the listening station. He rejects 7 of them, but buys the rest. How many CDs does Juan buy?

There are 10 CDs to begin with.   He rejects 7 of them.   He buys the remaining 3.

10   −   7   =   3

Objectives

a Write a subtraction sentence that corresponds to a situation involving "take away."

b Given a subtraction sentence, write a related addition sentence; and given an addition sentence, write two related subtraction sentences.

c Write a subtraction sentence that corresponds to a situation of "How much do I need?"

d Subtract whole numbers.

Study Tips

HIGHLIGHTING

- **Highlight important points.** You are probably used to highlighting key points as you study. If that works for you, continue to do so. But you will notice many design features throughout this book that already highlight important points. Thus you may not need to highlight as much as you generally do.

- **Highlight points that you do not understand.** Use a unique mark to indicate trouble spots that can lead to questions to be asked during class, in a tutoring session, or when calling or contacting the AW Math Tutor Center.

Write a subtraction sentence that corresponds to each situation.

1. A contractor removes 5 cu yd of sand from a pile containing 67 cu yd. How many cubic yards of sand are left in the pile?

2. Sparks Electronics owns a field next door that has an area of 20,000 sq ft (square feet). Deciding they need more room for parking, the owners have 12,000 sq ft paved. How many square feet of field are left unpaved?

2. Kaitlin has $300 and spends $85 for office supplies. How much money is left?

Amount to begin with   Amount spent for office supplies   Amount left

$300   − $85   = $215

Do Exercises 1 and 2.

b   Related Sentences

Subtraction is defined in terms of addition. For example, $7 - 4$ is that number which when added to 4 gives 7. Thus for the subtraction sentence

$7 - 4 = 3$ Taking away 4 from 7 gives 3.

there is a *related addition* sentence

$7 = 3 + 4.$ Putting back the 4 gives 7 again.

This can be illustrated using a number line. Both addition and subtraction correspond to moving distances on a number line. The number lines below are marked with tick marks at equal distances of 1 *unit*. On the left, the sum $3 + 4$ is shown. We start at 3 and move 4 units to the right, to end up at 7. The addition that corresponds to the situation is $3 + 4 = 7$.

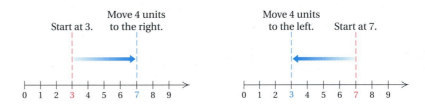

Start at 3.   Move 4 units to the right.   Move 4 units to the left.   Start at 7.

On the right above, the difference $7 - 4$ is shown. We start at 7 and move 4 units to the left, to end up at 3. The subtraction that corresponds to the situation is $7 - 4 = 3$.

This leads us to the following definition of subtraction.

SUBTRACTION

The difference $a - b$ is that unique whole number c for which $a = c + b$.

For example, $13 - 4$ is the number 9 since $13 = 9 + 4$.

We know that answers we find to subtractions are correct only because of the related addition, which provides a handy way to *check* a subtraction.

Answers on page A-1

EXAMPLE 3 Write a related addition sentence: $8 - 5 = 3$.

$$8 - \underset{\uparrow}{5} = 3$$

This number
gets added (to 3).

$$8 = 3 + \underset{\downarrow}{5}$$

> By the commutative law of addition, there is also another addition sentence:
> $$8 = 5 + 3.$$

The related addition sentence is $8 = 3 + 5$.

Do Exercises 3 and 4.

EXAMPLE 4 Write two related subtraction sentences: $4 + 3 = 7$.

$$4 + \underset{\uparrow}{3} = 7 \qquad\qquad \underset{\uparrow}{4} + 3 = 7$$

This addend gets
subtracted from
the sum.

This addend gets
subtracted from
the sum.

$$4 = 7 - 3 \qquad\qquad 3 = 7 - 4$$

(7 take away 3 is 4.)   (7 take away 4 is 3.)

The related subtraction sentences are $4 = 7 - 3$ and $3 = 7 - 4$.

Do Exercises 5 and 6.

C How Much Do I Need?

The second kind of situation to which subtraction can apply is called "how many more." You have 2 notebooks, but you need 7. You can think of this as "how many do I need to add to 2 to get 7?" Finding the answer can be thought of as finding a missing addend, and can be found by subtracting 2 from 7.

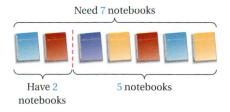

Need 7 notebooks

Have 2
notebooks

5 notebooks

What must be added to 2 to get 7? The answer is 5.

Missing addend   Difference

$$2 + \square = 7 \qquad\qquad 7 - 2 = \square$$

EXAMPLES Write a subtraction sentence that corresponds to each situation.

5. Jillian wants to buy the roll-on luggage shown in this ad. She has $30. She needs $79. How much more does she need in order to buy the luggage?

Write a related addition sentence.

3. $7 - 5 = 2$

4. $17 - 8 = 9$

Write two related subtraction sentences.

5. $5 + 8 = 13$

6. $11 + 3 = 14$

Answers on page A-1

Write an addition sentence and a related subtraction sentence corresponding to each situation.

7. It is 348 mi from Miami to Jacksonville. Alice has completed 200 miles of the drive. How much farther does she need to travel?

8. Cedric estimates that it will take 1200 bricks to complete the side of a building, but he has only 800 bricks. How many more bricks will be needed?

9. Subtract.

$$\begin{array}{r} 7\ 8\ 9\ 3 \\ -\ 4\ 0\ 9\ 2 \\ \hline \end{array}$$

Answers on pages A-1–A-2

Thinking of this situation in terms of a missing addend, we have:

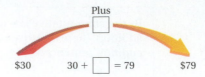

Plus

$30 \quad 30 + \boxed{} = 79 \quad \$79$

To find the answer, we think of the related subtraction sentence:

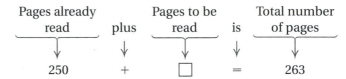

$$30 + \boxed{} = 79$$

$$\boxed{} = 79 - 30, \quad \text{or} \quad \$79 - \$30 = \$49.$$

6. Cathy is reading *Ishmael* by Daniel Quinn as part of her philosophy class. It contains 263 pages, of which she has read 250. How many more pages does she have left?

Pages already read	plus	Pages to be read	is	Total number of pages
↓	↓	↓	↓	↓
250	+	$\boxed{}$	=	263

Now we write a related subtraction sentence:

$263 - 250 = 13.$ 250 gets subtracted.

Do Exercises 7 and 8.

d Subtraction of Whole Numbers

To subtract numbers, we subtract the ones digits first, then the tens digits, then the hundreds, then the thousands, and so on.

EXAMPLE 7 Subtract: $9768 - 4320$.

$$\begin{array}{r} 9\ 7\ 6\ 8 \\ -\ 4\ 3\ 2\ 0 \\ \hline 8 \end{array}$$ Subtract ones.

$$\begin{array}{r} 9\ 7\ 6\ 8 \\ -\ 4\ 3\ 2\ 0 \\ \hline 4\ 8 \end{array}$$ Subtract tens.

$$\begin{array}{r} 9\ 7\ 6\ 8 \\ -\ 4\ 3\ 2\ 0 \\ \hline 4\ 4\ 8 \end{array}$$ Subtract hundreds.

This is for explanation.

$$\begin{array}{r} 9\ 7\ 6\ 8 \\ -\ 4\ 3\ 2\ 0 \\ \hline 5\ 4\ 4\ 8 \end{array}$$ Subtract thousands.

$$\begin{array}{r} 9\ 7\ 6\ 8 \\ -\ 4\ 3\ 2\ 0 \\ \hline 5\ 4\ 4\ 8 \end{array}$$ You should write only this.

Do Exercise 9.

Sometimes we need to borrow.

EXAMPLE 8 Subtract: 6246 − 1879.

$$\begin{array}{r}{\scriptstyle 3\ \ 16}\\ 6\ 2\ \cancel{4}\ \cancel{6}\\ -\ 1\ 8\ 7\ 9\\ \hline 7\end{array}$$ We cannot subtract 9 ones from 6 ones, but we can subtract 9 ones from 16 ones. We borrow 1 ten to get 16 ones.

$$\begin{array}{r}{\scriptstyle 13}\\ {\scriptstyle 1\ \ \cancel{3}\ \ 16}\\ 6\ \cancel{2}\ \cancel{4}\ \cancel{6}\\ -\ 1\ 8\ 7\ 9\\ \hline 6\ 7\end{array}$$ We cannot subtract 7 tens from 3 tens, but we can subtract 7 tens from 13 tens. We borrow 1 hundred to get 13 tens.

$$\begin{array}{r}{\scriptstyle 11\ \ 13}\\ {\scriptstyle 5\ \ \cancel{1}\ \ \cancel{3}\ \ 16}\\ \cancel{6}\ \cancel{2}\ \cancel{4}\ \cancel{6}\\ -\ 1\ 8\ 7\ 9\\ \hline 4\ 3\ 6\ 7\end{array}$$ We cannot subtract 8 hundreds from 1 hundred, but we can subtract 8 hundreds from 11 hundreds. We borrow 1 thousand to get 11 hundreds.

We can always check the answer by adding it to the number being subtracted.

| This is what you should write. | $\begin{array}{r}{\scriptstyle 11\ 13}\\ {\scriptstyle 5\ \cancel{1}\ \cancel{3}\ 16}\\ \cancel{6}\ \cancel{2}\ \cancel{4}\ \cancel{6}\\ -\ 1\ 8\ 7\ 9\\ \hline 4\ 3\ 6\ 7\end{array}$ | *Check:* | $\begin{array}{r}{\scriptstyle 1\ 1\ 1}\\ 4\ 3\ 6\ 7\\ +\ 1\ 8\ 7\ 9\\ \hline 6\ 2\ 4\ 6\end{array}$ | The answer checks because this is the top number in the subtraction. |

Do Exercises 10 and 11.

EXAMPLE 9 Subtract: 902 − 477.

$$\begin{array}{r}{\scriptstyle 8\ \ 9\ \ 12}\\ \cancel{9}\ \cancel{0}\ \cancel{2}\\ -\ 4\ 7\ 7\\ \hline 4\ 2\ 5\end{array}$$ We cannot subtract 7 ones from 2 ones. We have 9 hundreds, or 90 tens. We borrow 1 ten to get 12 ones. We then have 89 tens.

Do Exercises 12 and 13.

EXAMPLE 10 Subtract: 8003 − 3667.

$$\begin{array}{r}{\scriptstyle 7\ \ 9\ \ 9\ \ 13}\\ \cancel{8}\ \cancel{0}\ \cancel{0}\ \cancel{3}\\ -\ 3\ 6\ 6\ 7\\ \hline 4\ 3\ 3\ 6\end{array}$$ We have 8 thousands, or 800 tens. We borrow 1 ten to get 13 ones. We then have 799 tens.

EXAMPLES

11. Subtract: 6000 − 3762.

$$\begin{array}{r}{\scriptstyle 5\ \ 9\ \ 9\ \ 10}\\ \cancel{6}\ \cancel{0}\ \cancel{0}\ \cancel{0}\\ -\ 3\ 7\ 6\ 2\\ \hline 2\ 2\ 3\ 8\end{array}$$

12. Subtract: 6024 − 2968.

$$\begin{array}{r}{\scriptstyle 11}\\ {\scriptstyle 5\ \ 9\ \ \cancel{1}\ \ 14}\\ \cancel{6}\ \cancel{0}\ \cancel{2}\ \cancel{4}\\ -\ 2\ 9\ 6\ 8\\ \hline 3\ 0\ 5\ 6\end{array}$$

Do Exercises 14–16.

Subtract. Check by adding.

10.
$$\begin{array}{r}8\ 6\ 8\ 6\\ -\ 2\ 3\ 5\ 8\end{array}$$

11.
$$\begin{array}{r}7\ 1\ 4\ 5\\ -\ 2\ 3\ 9\ 8\end{array}$$

Subtract.

12.
$$\begin{array}{r}7\ 0\\ -\ 1\ 4\end{array}$$

13.
$$\begin{array}{r}5\ 0\ 3\\ -\ 2\ 9\ 8\end{array}$$

Subtract.

14.
$$\begin{array}{r}7\ 0\ 0\ 7\\ -\ 6\ 3\ 4\ 9\end{array}$$

15.
$$\begin{array}{r}6\ 0\ 0\ 0\\ -\ 3\ 1\ 4\ 9\end{array}$$

16.
$$\begin{array}{r}9\ 0\ 3\ 5\\ -\ 7\ 4\ 8\ 9\end{array}$$

Answers on page A-2

To the instructor and the student: This section presented a review of subtraction of whole numbers. Students who are successful should go on to Section 1.4. Those who have trouble should study developmental unit S near the back of this text and then repeat Section 1.3.

1.3 EXERCISE SET

For Extra Help

a   Write a subtraction sentence that corresponds to each situation.

1. Lauren arrives at the Evergreen State Fair with 20 ride tickets and uses 4 to ride the Upside Down Ride. How many tickets remain?

2. A host pours 5 oz (ounces) of salsa from a jar containing 16 oz. How many ounces are left?

3. *Frozen Yogurt.* A dispenser at a frozen yogurt store contains 126 oz of strawberry yogurt. A 13-oz cup is sold to a customer. How much is left in the dispenser?

4. *Chocolate Cake.* One slice of chocolate cake with fudge frosting contains 564 cal (calories). One cup of hot cocoa made with skim milk contains 188 calories. How many more calories are in the cake than in the cocoa?

b   Write a related addition sentence.

5. $7 - 4 = 3$

6. $12 - 5 = 7$

7. $13 - 8 = 5$

8. $9 - 9 = 0$

9. $23 - 9 = 14$

10. $20 - 8 = 12$

11. $43 - 16 = 27$

12. $51 - 18 = 33$

Write two related subtraction sentences.

13. $6 + 9 = 15$

14. $7 + 9 = 16$

15. $8 + 7 = 15$

16. $8 + 0 = 8$

17. $17 + 6 = 23$

18. $11 + 8 = 19$

19. $23 + 9 = 32$

20. $42 + 10 = 52$

c   Write an addition sentence and a related subtraction sentence corresponding to each situation.

21. *Kangaroos.* There are 32 million kangaroos in Australia and 17 million people. How many more kangaroos are there than people?

22. *Interstate Speeds.* Speed limits on interstate highways in many western states were raised from 65 mph (miles per hour) to 75 mph. By how many miles per hour were they raised?

23. A set of drapes requires 23 yd of material. The decorator has 10 yd of material in stock. How much more must be ordered?

24. Marv needs to bowl a score of 223 in order to win his tournament. His score with one frame to go is 195. How many pins does Marv need in the last frame to win the tournament?

d Subtract.

25.
$$
\begin{array}{r}
6\ 5 \\
-\ 2\ 1 \\
\hline
\end{array}
$$

26.
$$
\begin{array}{r}
8\ 7 \\
-\ 3\ 4 \\
\hline
\end{array}
$$

27.
$$
\begin{array}{r}
8\ 6\ 6 \\
-\ 3\ 3\ 3 \\
\hline
\end{array}
$$

28.
$$
\begin{array}{r}
5\ 2\ 6 \\
-\ 3\ 2\ 3 \\
\hline
\end{array}
$$

29. $86 - 47$

30. $73 - 28$

31. $981 - 747$

32. $887 - 698$

33.
$$
\begin{array}{r}
7\ 7\ 6\ 9 \\
-\ 2\ 3\ 8\ 7 \\
\hline
\end{array}
$$

34.
$$
\begin{array}{r}
6\ 4\ 3\ 1 \\
-\ 2\ 8\ 9\ 6 \\
\hline
\end{array}
$$

35.
$$
\begin{array}{r}
7\ 6\ 4\ 0 \\
-\ 3\ 8\ 0\ 9 \\
\hline
\end{array}
$$

36.
$$
\begin{array}{r}
8\ 0\ 0\ 3 \\
-\ \ \ \ 5\ 9\ 9 \\
\hline
\end{array}
$$

37.
$$
\begin{array}{r}
1\ 2,6\ 4\ 7 \\
-\ \ \ 4,8\ 9\ 9 \\
\hline
\end{array}
$$

38.
$$
\begin{array}{r}
1\ 6,2\ 2\ 2 \\
-\ \ \ 5,8\ 8\ 8 \\
\hline
\end{array}
$$

39. $90,237 - 47,209$

40. $84,703 - 298$

41.
$$
\begin{array}{r}
8\ 0 \\
-\ 2\ 4 \\
\hline
\end{array}
$$

42.
$$
\begin{array}{r}
9\ 0 \\
-\ 7\ 8 \\
\hline
\end{array}
$$

43.
$$
\begin{array}{r}
6\ 9\ 0 \\
-\ 2\ 3\ 6 \\
\hline
\end{array}
$$

44.
$$
\begin{array}{r}
8\ 0\ 3 \\
-\ 4\ 1\ 8 \\
\hline
\end{array}
$$

45.
$$
\begin{array}{r}
6\ 8\ 0\ 8 \\
-\ 3\ 0\ 5\ 9 \\
\hline
\end{array}
$$

46.
$$
\begin{array}{r}
6\ 4\ 0\ 8 \\
-\ \ \ 2\ 5\ 8 \\
\hline
\end{array}
$$

47.
$$
\begin{array}{r}
2\ 3\ 0\ 0 \\
-\ \ \ 1\ 0\ 9 \\
\hline
\end{array}
$$

48.
$$
\begin{array}{r}
6\ 0\ 0\ 7 \\
-\ 1\ 5\ 8\ 9 \\
\hline
\end{array}
$$

49.
$$
\begin{array}{r}
1\ 6\ 0 \\
-\ \ \ 7\ 4 \\
\hline
\end{array}
$$

50.
$$
\begin{array}{r}
4\ 7\ 0 \\
-\ 1\ 8\ 8 \\
\hline
\end{array}
$$

51.
$$
\begin{array}{r}
7\ 8\ 4\ 0 \\
-\ 3\ 0\ 2\ 7 \\
\hline
\end{array}
$$

52.
$$
\begin{array}{r}
8\ 0\ 9\ 2 \\
-\ 1\ 0\ 7\ 3 \\
\hline
\end{array}
$$

53. 5843 − 98

54. 10,002 − 398

55. 101,734 − 5760

56. 15,017 − 7809

57. 10,004 − 29

58. 21,043 − 8909

59. 83,907 − 89

60. 311,568 − 19,394

61.
```
   7 0 0 0
 − 2 7 9 4
```

62.
```
   8 0 0 1
 − 6 5 4 3
```

63.
```
   4 8,0 0 0
 − 3 7,6 9 5
```

64.
```
   1 7,0 4 3
 − 1 1,5 9 8
```

65. **D**_{**W**} Describe two situations that correspond to the subtraction $20 − $17, one "take away" and one "missing addend."

66. **D**_{**W**} Is subtraction commutative (is there a commutative law of subtraction)? Why or why not?

SKILL MAINTENANCE

Add. [1.2b]

67.
```
   9 4 6
 +   7 8
```

68.
```
   9 0 7 8
 + 3 6 5 4
```

69.
```
   5 7,8 7 7
 + 3 2,4 0 6
```

70.
```
   8 0 0 4
   6 7 8 9
   7 7 2 0
 + 6 8 5 1
```

71. 567 + 778

72. 901 + 23

73. 12,885 + 9807

74. 9909 + 1011

75. Write a word name for 6,375,602. [1.1c]

76. What does the digit 7 mean in 6,375,602? [1.1a]

SYNTHESIS

77. **D**_{**W**} Describe the one situation in which subtraction *is* commutative (see Exercise 66).

78. **D**_{**W**} Explain what it means to "borrow" in subtraction.

Subtract.

79. ▦ 3,928,124 − 1,098,947

80. ▦ 21,431,206 − 9,724,837

81. Fill in the missing digits to make the subtraction true:

9,☐48,621 − 2,097,☐81 = 7,251,140.

1.4 ROUNDING AND ESTIMATING; ORDER

Objectives

a Round to the nearest ten, hundred, or thousand.

b Estimate sums and differences by rounding.

c Use < or > for ☐ to write a true sentence in a situation like 6 ☐ 10.

a Rounding

We round numbers in various situations when we do not need an exact answer. For example, we might round to see if we are being charged the correct amount in a store. We might also round to check if an answer to a problem is reasonable or to check a calculation done by hand or on a calculator.

To understand how to round, we first look at some examples using number lines, even though this is not the way we generally do rounding.

EXAMPLE 1 Round 47 to the nearest ten.

Here is a part of a number line; 47 is between 40 and 50. Since 47 is closer to 50, we round up to 50.

EXAMPLE 2 Round 42 to the nearest ten.

42 is between 40 and 50. Since 42 is closer to 40, we round down to 40.

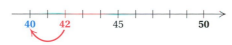

Do Exercises 1–4.

EXAMPLE 3 Round 45 to the nearest ten.

45 is halfway between 40 and 50. We could round 45 down to 40 or up to 50. We agree to round up to 50.

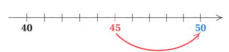

> When a number is halfway between rounding numbers, round up.

Do Exercises 5–7.

Here is a rule for rounding.

ROUNDING WHOLE NUMBERS

To round to a certain place:

a) Locate the digit in that place.
b) Consider the next digit to the right.
c) If the digit to the right is 5 or higher, round up. If the digit to the right is 4 or lower, round down.
d) Change all digits to the right of the rounding location to zeros.

Round to the nearest ten.

1. 37

2. 52

3. 73

4. 98

Round to the nearest ten.

5. 35

6. 75

7. 85

Answers on page A-2

25

Round to the nearest ten.

8. 137

9. 473

10. 235

11. 285

Round to the nearest hundred.

12. 641

13. 759

14. 750

15. 9325

Round to the nearest thousand.

16. 7896

17. 8459

18. 19,343

19. 68,500

Answers on page A-2

EXAMPLE 4   Round 6485 to the nearest ten.

a) Locate the digit in the tens place, 8.

$$6\ 4\ \underset{\uparrow}{8}\ 5$$

b) Consider the next digit to the right, 5.

$$6\ 4\ 8\ \underset{\uparrow}{5}$$

c) Since that digit, 5, is 5 or higher, round 8 tens up to 9 tens.

d) Change all digits to the right of the tens digit to zeros.

$$6\ 4\ 9\ 0\ \leftarrow \text{This is the answer.}$$

EXAMPLE 5   Round 6485 to the nearest hundred.

a) Locate the digit in the hundreds place, 4.

$$6\ \underset{\uparrow}{4}\ 8\ 5$$

b) Consider the next digit to the right, 8.

$$6\ 4\ \underset{\uparrow}{8}\ 5$$

c) Since that digit, 8, is 5 or higher, round 4 hundreds up to 5 hundreds.

d) Change all digits to the right of hundreds to zeros.

$$6\ 5\ 0\ 0\ \leftarrow \text{This is the answer.}$$

EXAMPLE 6   Round 6485 to the nearest thousand.

a) Locate the digit in the thousands place, 6.

$$\underset{\uparrow}{6}\ 4\ 8\ 5$$

b) Consider the next digit to the right, 4.

$$6\ \underset{\uparrow}{4}\ 8\ 5$$

c) Since that digit, 4, is 4 or lower, round down, meaning that 6 thousands stays as 6 thousands.

d) Change all digits to the right of thousands to zeros.

$$6\ 0\ 0\ 0\ \leftarrow \text{This is the answer.}$$

Do Exercises 8–19.

Caution!

7000 is not a correct answer to Example 6. It is incorrect to round from the ones digit over, as follows:

$$6485, \rightarrow 6490, \rightarrow 6500, \rightarrow 7000.$$

Note that 6485 is closer to 6000 than it is to 7000.

We can use the symbol ≈, read **"is approximately equal to,"** to indicate that we have rounded 6485 to 6490. Thus, in Example 4, we can write

$$6485 \approx 6490.$$

Sometimes rounding involves changing more than one digit in a number.

EXAMPLE 7 Round 78,595 to the nearest ten.

a) Locate the digit in the tens place, 9.

     7 8,5 9 5
            ↑

b) Consider the next digit to the right, 5.

     7 8,5 9 5
           ↑

c) Since that digit, 5, is 5 or higher, round 9 tens to 10 tens. To carry this out, we think of 10 tens as 1 hundred + 0 tens, and increase the hundreds digit by 1, to get 6 hundreds + 0 tens. We then write 6 in the hundreds place and 0 in the tens place.

d) Change the digit to the right of the tens digit to zero.

     7 8,6 0 0 ← This is the answer.

Note that if we round this number to the nearest hundred, we get the same answer.

Do Exercises 20 and 21.

b Estimating

Estimating can be done in many ways. In general, an estimate made by rounding to the nearest ten is more accurate than one rounded to the nearest hundred, and an estimate rounded to the nearest hundred is more accurate than one rounded to the nearest thousand, and so on.

In the following example, we see how estimation can be used in making a purchase.

EXAMPLE 8 *Estimating the Cost of an Automobile Purchase.* Ethan and Olivia Benson are shopping for a new car. They are considering a Saturn ION. There are three basic models of this car, and each has options beyond the basic price, as shown in the chart on the following page. Ethan and Olivia have allowed themselves a budget of $16,500. They look at the list of options and want to make a quick estimate of the cost of model ION·2 with all the options.

Estimate by rounding to the nearest hundred the cost of the ION·2 with all the options and decide whether it will fit into their budget.

20. Round 48,968 to the nearest ten, hundred, and thousand.

21. Round 269,582 to the nearest ten, hundred, and thousand.

Refer to the chart on the next page to answer Margin Exercises 22 and 23.

22. By eliminating options, find a way that Ethan and Olivia can buy the ION·2 and stay within their $16,500 budget.

23. Tara and Alex are shopping for a new car. They are considering a Saturn ION·3 and have allowed a budget of $19,000.

a) Estimate by first rounding to the nearest hundred the cost of an ION·3 with all the options.

b) Can they afford this car with a budget of $19,000?

Answers on page A-2

MODEL ION·1 SEDAN (4 DOOR) 2.2-LITER ENGINE, 4-SPEED AUTOMATIC TRANSMISSION	MODEL ION·2 SEDAN (4 DOOR) 2.2-LITER ENGINE, 5-SPEED MANUAL TRANSMISSION	MODEL ION·3 SEDAN (4 DOOR) 2.2-LITER ENGINE, 5-SPEED MANUAL TRANSMISSION
Base Price: $12,975	Base Price: $14,945	Base Price: $16,470

Each of these vehicles comes with several options. Note that some of the options are standard on certain models. Others are not available for all models.

Antilock Braking System with Traction Control:	$400
Head Curtain Side Air Bags:	$395
Power Sunroof (Not available for ION·1):	$725
Rear Spoiler (Not available for ION·1):	$250
Air Conditioning with Dust and Pollen Filtration (Standard on ION·2 and ION·3):	$960
CD/MP3 Player with AM/FM Stereo and 4 Coaxial Speakers (Standard on ION·3):	ION·1—$510 ION·2—$220
Power Package: Power Windows, Power Exterior Mirrors, Remote Keyless Entry, and Cruise Control (Not available for ION·1 and Standard for ION·3):	$825

Source: Saturn

24. Estimate the sum by first rounding to the nearest ten. Show your work.

$$\begin{array}{r} 7\ 4 \\ 2\ 3 \\ 3\ 5 \\ +\ 6\ 6 \\ \hline \end{array}$$

25. Estimate the sum by first rounding to the nearest hundred. Show your work.

$$\begin{array}{r} 6\ 5\ 0 \\ 6\ 8\ 5 \\ 2\ 3\ 8 \\ +\ 1\ 6\ 8 \\ \hline \end{array}$$

Answers on page A-2

First, we list the base price of the ION·2 and then the cost of each of the options. We then round each number to the nearest hundred and add.

$$\begin{array}{r} 1\ 4{,}9\ 4\ 5 \\ 4\ 0\ 0 \\ 3\ 9\ 5 \\ 7\ 2\ 5 \\ 2\ 5\ 0 \\ 2\ 2\ 0 \\ +\ \ \ \ 8\ 2\ 5 \\ \hline \end{array} \qquad \begin{array}{r} 1\ 4{,}9\ 0\ 0 \\ 4\ 0\ 0 \\ 4\ 0\ 0 \\ 7\ 0\ 0 \\ 3\ 0\ 0 \\ 2\ 0\ 0 \\ +\ \ \ \ 8\ 0\ 0 \\ \hline 1\ 7{,}7\ 0\ 0 \end{array}$$

← Estimated answer

Air conditioning is standard on the ION·2, so we do not include that cost. The estimated cost is $17,700. Since Ethan and Olivia have allowed themselves a budget of $16,500 for the car, they will need to forgo some options.

Do Exercises 22 and 23 on the preceding page.

EXAMPLE 9 Estimate this sum by first rounding to the nearest ten:

$$78 + 49 + 31 + 85.$$

We round each number to the nearest ten. Then we add.

$$\begin{array}{r} 7\ 8 \\ 4\ 9 \\ 3\ 1 \\ +\ 8\ 5 \\ \hline \end{array} \qquad \begin{array}{r} 8\ 0 \\ 5\ 0 \\ 3\ 0 \\ +\ 9\ 0 \\ \hline 2\ 5\ 0 \end{array}$$

← Estimated answer

Do Exercises 24 and 25.

EXAMPLE 10 Estimate the difference by first rounding to the nearest thousand: $9324 - 2849$.

We have

$$
\begin{array}{r}
9\ 3\ 2\ 4 \\
-\ 2\ 8\ 4\ 9 \\
\end{array}
\qquad
\begin{array}{r}
9\ 0\ 0\ 0 \\
-\ 3\ 0\ 0\ 0 \\
\hline
6\ 0\ 0\ 0 \leftarrow \text{Estimated answer}
\end{array}
$$

Do Exercises 26 and 27.

C   Order

We know that 2 is not the same as 5, that is, 2 is not equal to 5. We express this by the sentence $2 \neq 5$. We also know that 2 is less than 5. We symbolize this by the expression $2 < 5$. We can see this order on the number line: 2 is to the left of 5. The number 0 is the smallest whole number, so $0 < a$ for any whole number a.

ORDER OF WHOLE NUMBERS

For any whole numbers a and b:

1. $a < b$ (read "a is less than b") is true when a is to the left of b on the number line.
2. $a > b$ (read "a is greater than b") is true when a is to the right of b on the number line.

We call $<$ and $>$ **inequality symbols.**

EXAMPLE 11 Use $<$ or $>$ for ☐ to write a true sentence: $7 \ \square\ 11$.

Since 7 is to the left of 11 on the number line, $7 < 11$.

EXAMPLE 12 Use $<$ or $>$ for ☐ to write a true sentence: $92 \ \square\ 87$.

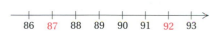

Since 92 is to the right of 87 on the number line, $92 > 87$.

A sentence like $8 + 5 = 13$ is called an **equation.** It is a *true* equation. The equation $4 + 8 = 11$ is a *false* equation. A sentence like $7 < 11$ is called an **inequality.** The sentence $7 < 11$ is a *true* inequality. The sentence $23 > 69$ is a *false* inequality.

Do Exercises 28–33.

---

26. Estimate the difference by first rounding to the nearest hundred. Show your work.

$$
\begin{array}{r}
9\ 2\ 8\ 5 \\
-\ 6\ 7\ 3\ 9 \\
\end{array}
$$

27. Estimate the difference by first rounding to the nearest thousand. Show your work.

$$
\begin{array}{r}
2\ 3,2\ 7\ 8 \\
-\ 1\ 1,6\ 9\ 8 \\
\end{array}
$$

Use $<$ or $>$ for ☐ to write a true sentence. Draw a number line if necessary.

28. $8 \ \square\ 12$

29. $12 \ \square\ 8$

30. $76 \ \square\ 64$

31. $64 \ \square\ 76$

32. $217 \ \square\ 345$

33. $345 \ \square\ 217$

Answers on page A-2

a   Round to the nearest ten.

1. 48   **2.** 532   **3.** 467   **4.** 8945

5. 731   **6.** 17   **7.** 895   **8.** 798

Round to the nearest hundred.

9. 146   **10.** 874   **11.** 957   **12.** 650

13. 9079   **14.** 4645   **15.** 32,850   **16.** 198,402

Round to the nearest thousand.

17. 5876   **18.** 4500   **19.** 7500   **20.** 2001

21. 45,340   **22.** 735,562   **23.** 373,405   **24.** 6,713,855

b   Estimate the sum or difference by first rounding to the nearest ten. Show your work.

25.
```
   7 8
 + 9 7
```

26.
```
   6 2
   9 7
   4 6
 + 8 8
```

27.
```
   8 0 7 4
 − 2 3 4 7
```

28.
```
   6 7 3
 −   2 8
```

Estimate the sum by first rounding to the nearest ten. Do any of the given sums seem to be incorrect when compared to the estimate? Which ones?

29.
```
   4 5
   7 7
   2 5
 + 5 6
   3 4 3
```

30.
```
   4 1
   2 1
   5 5
 + 6 0
   1 7 7
```

31.
```
   6 2 2
     7 8
     8 1
 + 1 1 1
   9 3 2
```

32.
```
   8 3 6
   3 7 4
   7 9 4
 + 9 3 8
   3 9 4 7
```

Estimate the sum or difference by first rounding to the nearest hundred. Show your work.

33.
```
   7 3 4 8
 + 9 2 4 7
```

34.
```
   5 6 8
   4 7 2
   9 3 8
 + 4 0 2
```

35.
```
   6 8 5 2
 − 1 7 4 8
```

36.
```
   9 4 3 8
 − 2 7 8 7
```

Planning a Kitchen. Perfect Kitchens offers custom kitchen packages with three choices for each of four items: cabinets, countertops, appliances, and flooring. The chart below lists the price for each choice.

Perfect Kitchens lets you customize your kitchen with one choice from each of the following four features:

CABINETS		TYPE	PRICE
	(a)	Oak	$7450
	(b)	Cherry	8820
	(c)	Painted	9630

COUNTERTOPS		TYPE	PRICE
	(d)	Laminate	$1595
	(e)	Solid surface	2870
	(f)	Granite	3528

APPLIANCES		PRICE RANGE	PRICE
	(g)	Low	$1540
	(h)	Medium	3575
	(i)	High	6245

FLOORING		TYPE	PRICE
	(j)	Vinyl	$ 625
	(k)	Ceramic tile	985
	(l)	Hardwood	1160

37. Estimate the cost of remodeling a kitchen with choices (a), (d), (g), and (j) by rounding to the nearest hundred dollars.

38. Estimate the cost of a kitchen with choices (c), (f), (i), and (l) by rounding to the nearest hundred dollars.

39. Sara and Ben are planning to remodel their kitchen and have a budget of $17,700. Estimate by rounding to the nearest hundred dollars the cost of their kitchen remodeling project if they choose options (b), (e), (i), and (k). Can they afford their choices?

40. The Davidsons must make a final decision on the kitchen choices for their new home. The allotted kitchen budget is $16,000. Estimate by rounding to the nearest hundred dollars the kitchen cost if they choose options (a), (f), (h), and (l). Does their budget allotment cover the cost?

41. Suppose you are planning a new kitchen and must stay within a budget of $14,500. Decide on the options you would like and estimate the cost by rounding to the nearest hundred dollars. Does your budget support your choices?

42. Suppose you are planning a new kitchen and must stay within a budget of $18,500. Decide on the options you would like and estimate the cost by rounding to the nearest hundred dollars. Does your budget support your choices?

Estimate the sum by first rounding to the nearest hundred. Do any of the given sums seem to be incorrect when compared to the estimate? Which ones?

43.
```
    2 1 6
      8 4
    7 4 5
  + 5 9 5
  ─────────
  1 6 4 0
```

44.
```
      4 8 1
      7 0 2
      6 2 3
  + 1 0 4 3
  ─────────
    1 8 4 9
```

45.
```
    7 5 0
    4 2 8
      6 3
  + 2 0 5
  ─────────
  1 4 4 6
```

46.
```
    3 2 6
    2 7 5
    7 5 8
  + 9 4 3
  ─────────
  2 3 0 2
```

Estimate the sum or difference by first rounding to the nearest thousand. Show your work.

47.
```
    9 6 4 3
    4 8 2 1
    8 9 4 3
  + 7 0 0 4
```

48.
```
    7 6 4 8
    9 3 4 8
    7 8 4 2
  + 2 2 2 2
```

49.
```
    9 2,1 4 9
  − 2 2,5 5 5
```

50.
```
    8 4,8 9 0
  − 1 1,1 1 0
```

C   Use < or > for ☐ to write a true sentence. Draw a number line if necessary.

51. 0 ☐ 17

52. 32 ☐ 0

53. 34 ☐ 12

54. 28 ☐ 18

55. 1000 ☐ 1001

56. 77 ☐ 117

57. 133 ☐ 132

58. 999 ☐ 997

59. 460 ☐ 17

60. 345 ☐ 456

61. 37 ☐ 11

62. 12 ☐ 32

Daily Newspapers.   The top three daily newspapers in the United States are *USA Today,* the *Wall Street Journal,* and the *New York Times.* The daily circulation of each as of September 30, 2005, is listed in the table below. Use this table when answering Exercises 63 and 64.

NEWSPAPER	DAILY CIRCULATION
USA Today	2,296,335
Wall Street Journal	2,083,660
New York Times	1,126,190

Source: Audit Bureau of Circulations

63. Use an inequality to compare the daily circulation of the *Wall Street Journal* and *USA Today.*

64. Use an inequality to compare the daily circulation of *USA Today* and the *New York Times.*

65. Pedestrian Fatalities. The annual number of pedestrians killed when hit by a motor vehicle has declined from 6482 in 1990 to 4641 in 2004. Use an inequality to compare these annual totals of pedestrian fatalities.

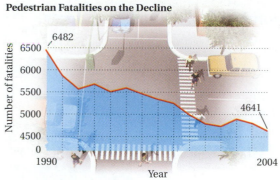

Pedestrian Fatalities on the Decline

Source: National Center for Statistics and Analysis

66. Life Expectancy. The life expectancy of a female in 2010 is predicted to be about 82 yr (years) and of a male about 76 yr. Use an inequality to compare these life expectancies.

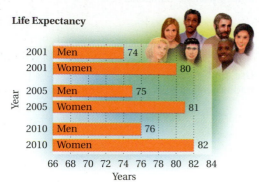

Life Expectancy

Source: U.S. Bureau of the Census

67. D_W Explain how estimating and rounding can be useful when shopping for groceries.

68. D_W When rounding 748 to the nearest hundred, a student rounds to 750 and then to 800. What mistake is he making?

SKILL MAINTENANCE

Write expanded notation.

69. 7992   [1.1b]

70. 23,000,000   [1.1b]

71. Write a word name for 246,605,004,032.   [1.1c]

72. Write a word name for 1,005,100.   [1.1c]

Add.   [1.2b]

73.
```
  6 7,7 8 9
+ 1 8,9 6 5
```

74.
```
  9 0 0 2
+ 4 5 8 7
```

Subtract.   [1.3d]

75.
```
  6 7,7 8 9
− 1 8,9 6 5
```

76.
```
  9 0 0 2
− 4 5 8 7
```

SYNTHESIS

77. D_W Consider two numbers *a* and *b*, with $a < b$. Is it possible that when *a* and *b* are each rounded down, the result of rounding *a* is greater than the result of rounding *b*? Why or why not?

78. D_W Why do you think we round *up* when a 5 is the next digit to the right of the place in which we are rounding?

79.–82. ▦ Use a calculator to find the sums and differences in each of Exercises 47–50. Then compare your answers with those found using estimation. Even when using a calculator it is possible to make an error if you press the wrong buttons, so it is a good idea to check by estimating.

Objectives

a Multiplication of Whole Numbers

REPEATED ADDITION

The multiplication 3×5 corresponds to this repeated addition:

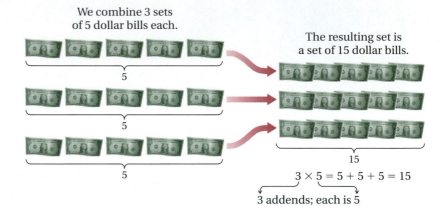

We combine 3 sets of 5 dollar bills each.

The resulting set is a set of 15 dollar bills.

$$3 \times 5 = 5 + 5 + 5 = 15$$
3 addends; each is 5

The numbers that we multiply are called **factors.** The result of the multiplication is called a **product.**

$$\begin{array}{ccccc} 3 & \times & 5 & = & 15 \\ \downarrow & & \downarrow & & \downarrow \\ \text{Factor} & & \text{Factor} & & \text{Product} \end{array}$$

RECTANGULAR ARRAYS

Multiplications can also be thought of as rectangular arrays. Each of the following corresponds to the multiplication 3×5.

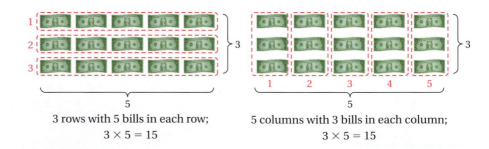

3 rows with 5 bills in each row;
$3 \times 5 = 15$

5 columns with 3 bills in each column;
$3 \times 5 = 15$

When you write a multiplication sentence corresponding to a real-world situation, you should think of either a rectangular array or repeated addition. In some cases, it may help to think both ways.

We have used an "$\times$" to denote multiplication. A dot "$\cdot$" is also commonly used. (Use of the dot is attributed to the German mathematician Gottfried Wilhelm von Leibniz in 1698.) Parentheses are also used to denote multiplication. For example,

$$3 \times 5 = 3 \cdot 5 = (3)(5) = 3(5) = 15.$$

EXAMPLES Write a multiplication sentence that corresponds to each situation.

1. It is known that Americans drink 24 million gal of soft drinks per day (*per day* means *each day*). What quantity of soft drinks is consumed every 5 days?

   We draw a picture in which  = 1 million gallons or we can simply visualize the situation. Repeated addition fits best in this case.

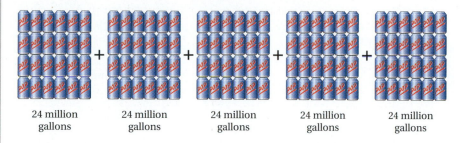

| | 24 million gallons | 24 million gallons | 24 million gallons | 24 million gallons | 24 million gallons |

$5 \cdot 24$ million gallons $= 120$ million gallons

2. One side of a building has 6 floors with 7 windows on each floor. How many windows are on that side of the building?

   We have a rectangular array and can easily draw a sketch.

   $6 \cdot 7 = 42$

6 floors

7 windows

Do Exercises 1–3.

Repeated addition of 0 results in 0, so $0 \cdot a = a \cdot 0 = 0$. When 1 is added to itself a times, the result is a, so $1 \cdot a = a \cdot 1 = a$. We say that the number 1 is the **multiplicative identity.**

EXAMPLE 3 Multiply: 37×2.

   We have

$$
\begin{array}{r}
3\ 7 \\
\times\quad 2 \\
\hline
1\ 4 \\
6\ 0 \\
\hline
7\ 4
\end{array}
$$

1 4 ← Multiply the 7 ones by 2: $2 \times 7 = 14$.
6 0 ← Multiply the 3 tens by 2: $2 \times 30 = 60$.
7 4 ← Add.

1. Marv practices for the U.S. Open bowling tournament. He bowls 8 games each day for 7 days. How many games does he bowl altogether for practice?

2. A lab technician pours 75 mL (milliliters) of acid into each of 10 beakers. How much acid is poured in all?

3. **Checkerboard.** A checkerboard consists of 8 rows with 8 squares in each row. How many squares in all are there on a checkerboard?

Answers on page A-2

Multiply.

4. 5 8
    ×   2

5. 3 7
    ×   4

6. 8 2 3
    ×   6

7. 1 3 4 8
    ×   5

Multiply.

8. 4 5
    × 2 3

9. 48 × 63

Answers on page A-2

We can simplify what we write by writing just one line for the products.

$$\begin{array}{r} {\scriptstyle 1} \\ 3\ 7 \\ \times\ \ \ 2 \\ \hline 4 \end{array}$$

Multiply the ones by 2: $2 \cdot (7 \text{ ones}) = 14 \text{ ones} = 1 \text{ ten} + 4 \text{ ones}$. Write 4 in the ones column and 1 above the tens.

$$\begin{array}{r} {\scriptstyle 1} \\ 3\ 7 \\ \times\ \ \ 2 \\ \hline 7\ 4 \end{array}$$

Multiply the 3 tens by 2 and add 1 ten: $2 \cdot (3 \text{ tens}) = 6 \text{ tens}$, 6 tens + 1 ten = 7 tens. Write 7 in the tens column.

$$\left.\begin{array}{r} {\scriptstyle 1} \\ 3\ 7 \\ \times\ \ \ 2 \\ \hline 7\ 4 \end{array}\right\}$$ Try to write only this.

Do Exercises 4–7.

The fact that we can do this is based on a property called the **distributive law.**

THE DISTRIBUTIVE LAW

For any numbers *a*, *b*, and *c*,

$$a \cdot (b + c) = (a \cdot b) + (a \cdot c).$$

Applied to the example above, the distributive law gives us

$$2 \cdot 37 = 2 \cdot (30 + 7) = (2 \cdot 30) + (2 \cdot 7).$$

EXAMPLE 4 Multiply: 43×57.

$$\begin{array}{r} {\scriptstyle 2} \\ 5\ 7 \\ \times\ 4\ 3 \\ \hline 1\ 7\ 1 \end{array}$$ Multiplying by 3

$$\begin{array}{r} {\scriptstyle 2} \\ {\scriptstyle 2} \\ 5\ 7 \\ \times\ 4\ 3 \\ \hline 1\ 7\ 1 \\ 2\ 2\ 8\ 0 \end{array}$$ Multiplying by 40. (We write a 0 and then multiply 57 by 4.)

$$\begin{array}{r} {\scriptstyle 2} \\ {\scriptstyle 2} \\ 5\ 7 \\ \times\ 4\ 3 \\ \hline 1\ 7\ 1 \\ 2\ 2\ 8\ 0 \\ \hline 2\ 4\ 5\ 1 \end{array}$$

You may have learned that such a 0 does not have to be written. You may omit it if you wish. If you do omit it, remember, when multiplying by tens, to put the answer in the tens place.

Adding to obtain the product

Do Exercises 8 and 9.

EXAMPLE 5 Multiply: 457×683.

```
        5  2
     6  8  3
  ×  4  5  7
  ─────────────
  4  7  8  1     Multiplying 683 by 7
```

```
        4  1
        5  2
     6  8  3
  ×  4  5  7
  ─────────────
     4  7  8  1
  3  4  1  5  0     Multiplying 683 by 50
```

```
     3  1
     4  1
     5  2
     6  8  3
  ×  4  5  7
  ─────────────
        4  7  8  1
     3  4  1  5  0
  2  7  3  2  0  0     Multiplying 683 by 400
  ─────────────
  3  1  2 , 1  3  1     Adding
```

Do Exercises 10 and 11.

EXAMPLE 6 Multiply: 306×274.

Note that 306 = 3 hundreds + 6 ones.

```
        2  7  4
     ×  3  0  6
  ─────────────
        1  6  4  4  ← Multiplying by 6 ones
     8  2  2  0  0  ← Multiplying by 3 hundreds. (We write 00
  ─────────────        and then multiply 274 by 3.)
     8  3 , 8  4  4  ← Adding
```

Do Exercises 12–14.

EXAMPLE 7 Multiply: 360×274.

Note that 360 = 3 hundreds + 6 tens.

```
        2  7  4   ┌ Multiplying by 6 tens. (We write 0 and
     ×     3  6  0 │   then multiply 274 by 6.)
  ─────────────
     1  6  4  4  0 ← Multiplying by 3 hundreds. (We write 00
     8  2  2  0  0 ←   and then multiply 274 by 3.)
  ─────────────
     9  8 , 6  4  0   Adding
```

Do Exercises 15–18.

Multiply.

10.
```
     7  4  6
  ×     6  2
```

11. 245×837

Multiply.

12.
```
     4  7  2
  ×  3  0  6
```

13. 408×704

14.
```
     2  3  4  4
  ×  6  0  0  5
```

Multiply.

15.
```
     4  7  2
  ×  8  3  0
```

16.
```
     2  3  4  4
  ×  7  4  0  0
```

17. 100×562

18. 1000×562

Answers on page A-2

Complete the following to illustrate the commutative law of multiplication.

19. $8 \cdot 7 = \square \cdot \square$

20. $2 \cdot 6 = \square \cdot \square$

Insert parentheses to illustrate the associative law of multiplication.

21. $3 \cdot (7 \cdot 9) = 3 \cdot 7 \cdot 9$

22. $(5 \cdot 4) \cdot 8 = 5 \cdot 4 \cdot 8$

Multiply.

23. $5 \cdot 2 \cdot 4$

24. $5 \cdot 1 \cdot 3$

Check on your own that $17 \cdot 37 = 629$ and that $37 \cdot 17 = 629$. This illustrates the **commutative law of multiplication.** It says that we can multiply two numbers in any order, $a \cdot b = b \cdot a$, and get the same answer.

EXAMPLE 8 Complete the following to illustrate the commutative law of multiplication:

$$6 \cdot 9 = \square \cdot \square.$$

We reverse the order of the multiplication:

$$6 \cdot 9 = 9 \cdot 6.$$

Do Exercises 19 and 20.

To multiply three or more numbers, we group them so that we multiply two at a time. Consider $2 \cdot (3 \cdot 4)$ and $(2 \cdot 3) \cdot 4$. The parentheses tell what to do first:

$$2 \cdot (3 \cdot 4) = 2 \cdot (12) = 24. \qquad \text{We multiply 3 and 4, then 2.}$$

We can also multiply 2 and 3, then 4:

$$(2 \cdot 3) \cdot 4 = (6) \cdot 4 = 24.$$

Either way we get 24. It does not matter how we group the numbers. This illustrates that **multiplication is associative:** $a \cdot (b \cdot c) = (a \cdot b) \cdot c$.

EXAMPLE 9 Insert parentheses to illustrate the associative law of multiplication:

$$6 \cdot (2 \cdot 5) = 6 \cdot 2 \cdot 5.$$

We regroup as follows:

$$6 \cdot (2 \cdot 5) = (6 \cdot 2) \cdot 5.$$

Do Exercises 21–24.

> **THE COMMUTATIVE LAW OF MULTIPLICATION**
>
> For any numbers a and b,
> $$a \cdot b = b \cdot a.$$

> **THE ASSOCIATIVE LAW OF MULTIPLICATION**
>
> For any numbers a, b, and c,
> $$a \cdot (b \cdot c) = (a \cdot b) \cdot c.$$

Caution!

Do not confuse the associative law with the distributive law. When you multiply $2 \cdot (3 \cdot 4)$, do *not* use the 2 twice. When you multiply $2 \cdot (3 + 4)$ using the distributive law, use the 2 twice: $2 \cdot 3 + 2 \cdot 4$.

Answers on page A-2

b Estimating Products by Rounding

EXAMPLE 10 *Lawn Tractors.* Leisure Lawn Care is buying new lawn tractors that cost $1534 each. By rounding to the nearest ten, estimate the cost of purchasing 18 tractors.

Exact	*Nearest ten*
1 5 3 4	1 5 3 0
× 1 8	× 2 0
2 7,6 1 2	3 0,6 0 0

The lawn tractors will cost about $30,600.

Do Exercise 25.

EXAMPLE 11 Estimate the following product by first rounding to the nearest ten and to the nearest hundred: 683 × 457.

Nearest ten	*Nearest hundred*	*Exact*
6 8 0	7 0 0	6 8 3
× 4 6 0	× 5 0 0	× 4 5 7
4 0 8 0 0	3 5 0,0 0 0	4 7 8 1
2 7 2 0 0 0		3 4 1 5 0
3 1 2,8 0 0		2 7 3 2 0 0
		3 1 2,1 3 1

Do Exercise 26.

c Finding Area

The area of a rectangular region can be considered to be the number of square units needed to fill it. Here is a rectangle 4 cm (centimeters) long and 3 cm wide. It takes 12 sq cm (square centimeters) to fill it.

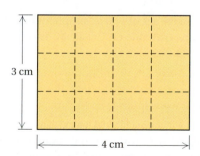

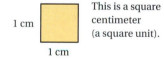
1 cm

This is a square centimeter (a square unit).

  In this case, we have a rectangular array of 3 rows, each of which contains 4 squares. The number of square units is given by 4 · 3, or 12. That is,
$$A = l \cdot w = 4 \text{ cm} \cdot 3 \text{ cm} = 12 \text{ sq cm.}$$

AREA

The area of a shape is a measure of its surface using **square units**.

CALCULATOR CORNER

Multiplying Whole Numbers To multiply whole numbers on a calculator, we use the ☒ and ＝ keys. For example, to find 13 × 47, we press ①③☒④⑦＝. The calculator displays 611, so 13 × 47 = 611.

Exercises: Use a calculator to find each product.

1. 56 × 8
2. 845 × 26
3. 5 · 1276
4. 126(314)
5.   3 7 6 0
      ×   4 8
6.   5 2 1 8
      ×   4 5 3

25. Lawn Tractors. By rounding to the nearest ten, estimate the cost to Leisure Lawn Care of 12 lawn tractors.

26. Estimate the product by first rounding to the nearest ten and to the nearest hundred. Show your work.

     8 3 7
     × 2 4 5

Answers on page A-2

27. Table Tennis. Find the area of a standard table tennis table that has dimensions of 9 ft by 5 ft.

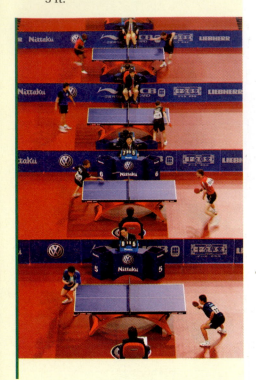

Answer on page A-2

To the instructor and the student: This section presented a review of multiplication of whole numbers. Students who are successful should go on to Section 1.6. Those who have trouble should study developmental unit M near the back of this text and then repeat Section 1.5.

EXAMPLE 12 *Professional Pool Table.* The playing area of a standard pool table has dimensions of 50 in. by 100 in. (There are rails 6 in. wide on the outside that are not included in the playing area.) Find the playing area.

If we think of filling the rectangle with square inches, we have a rectangular array. The length $l = 100$ in. and the width $w = 50$ in. Thus the area A is given by the formula

$$A = l \cdot w = 100 \cdot 50 = 5000 \text{ sq in.}$$

Do Exercise 27.

Study Tips

Are you aware of all the learning resources that exist for this textbook? Many details are given in the Preface.

LEARNING RESOURCES

- The *Student's Solutions Manual* contains worked-out solutions to the odd-numbered exercises in the exercise sets.

- An extensive set of *videos* supplements this text. These are available on CD-ROM by calling 1-800-282-0693.

- *InterAct Math* offers unlimited practice exercises that are correlated to the text. It is available online at www.interactmath.com.

- *Tutorial software* called MathXL Tutorials on CD is available with this text. If it is not available in the campus learning center, you can order it by calling 1-800-282-0693.

- The Addison-Wesley *Math Tutor Center* has experienced instructors available to help with the odd-numbered exercises. You can order this service by calling 1-800-824-7799.

- Extensive help is available online via MyMathLab and/or MathXL. Ask your instructor for information about these or visit MyMathLab.com and MathXL.com.

a   Write a multiplication sentence that corresponds to each situation.

1. The *Los Angeles Sunday Times* crossword puzzle is arranged rectangularly with squares in 21 rows and 21 columns. How many squares does the puzzle have altogether?

2. *Pixels.* A digital phototgraph is stored using small rectangular dots called *pixels*. How many pixels are there in a digital photo that has 600 rows with 400 pixels in each row?

3. A new soft drink beverage carton contains 8 cans, each of which holds 12 oz. How many ounces are there in the carton?

4. There are 7 days in a week. How many days are there in 18 weeks?

5. *Computer Printers.* The HP Photosmart 8250 can print 4800 × 1200 dots per square inch (dpi), optimized. How many dots can it print in one square inch?
Source: www.hp.com

6. *Computer Printers.* The Epson Stylus Photo R1800 can print 5760 × 1440 dots per square inch (dpi), optimized. How many dots can it print in one square inch?
Source: www.epson.com

Multiply.

7.
$$\begin{array}{r} 87 \\ \times\ 10 \\ \hline \end{array}$$

8.
$$\begin{array}{r} 100 \\ \times\ 96 \\ \hline \end{array}$$

9.
$$\begin{array}{r} 2340 \\ \times\ 1000 \\ \hline \end{array}$$

10.
$$\begin{array}{r} 800 \\ \times\ 70 \\ \hline \end{array}$$

11.
$$\begin{array}{r} 65 \\ \times\ 8 \\ \hline \end{array}$$

12.
$$\begin{array}{r} 87 \\ \times\ 4 \\ \hline \end{array}$$

13.
$$\begin{array}{r} 94 \\ \times\ 6 \\ \hline \end{array}$$

14.
$$\begin{array}{r} 76 \\ \times\ 9 \\ \hline \end{array}$$

15. $3 \cdot 509$

16. $7 \cdot 806$

17. $7(9229)$

18. $4(7867)$

19. $90(53)$

20. $60(78)$

21. $(47)(85)$

22. $(34)(87)$

23.
```
    6 4 0
  ×   7 2
```

24.
```
    7 7 7
  ×   7 7
```

25.
```
    4 4 4
  ×   3 3
```

26.
```
    5 0 9
  ×   8 8
```

27.
```
    5 0 9
  × 4 0 8
```

28.
```
    4 3 2
  × 3 7 5
```

29.
```
    8 5 3
  × 9 3 6
```

30.
```
    3 4 6
  × 6 5 0
```

31.
```
    6 4 2 8
  × 3 2 2 4
```

32.
```
    8 9 2 8
  × 3 1 7 2
```

33.
```
    3 4 8 2
  ×   1 0 4
```

34.
```
    6 4 0 8
  × 6 0 6 4
```

35.
```
    5 0 0 6
  × 4 0 0 8
```

36.
```
    6 7 8 9
  × 2 3 3 0
```

37.
```
    5 6 0 8
  × 4 5 0 0
```

38.
```
    4 5 6 0
  × 7 8 9 0
```

39.
```
    8 7 6
  × 3 4 5
```

40.
```
    3 5 5
  × 2 9 9
```

41.
```
    7 8 8 9
  × 6 2 2 4
```

42.
```
    6 5 0 1
  × 3 4 4 9
```

b Estimate the product by first rounding to the nearest ten. Show your work.

43.
```
    4 5
  × 6 7
```

44.
```
    5 1
  × 7 8
```

45.
```
    3 4
  × 2 9
```

46.
```
    6 3
  × 5 4
```

Estimate the product by first rounding to the nearest hundred. Show your work.

47.
```
    8 7 6
  × 3 4 5
```

48.
```
    3 5 5
  × 2 9 9
```

49.
```
    4 3 2
  × 1 9 9
```

50.
```
    7 8 9
  × 4 3 4
```

51. *Toyota Sienna.* A pharmaceutical company buys a Toyota Sienna for each of its 112 sales representatives. Each car costs $27,896 plus an additional $540 per car in destination charges.

a) Estimate the total cost of the purchase by rounding the cost of each car, the destination charge, and the number of sales representatives to the nearest hundred.

b) Estimate the total cost of the purchase by rounding the cost of each car to the nearest thousand and the destination charge and the number of reps to the nearest hundred.

Source: Toyota

52. A travel club of 176 people decides to fly from Los Angeles to Tokyo. The cost of a round-trip ticket is $643.

a) Estimate the total cost of the trip by rounding the cost of the airfare and the number of travelers to the nearest ten.

b) Estimate the total cost of the trip by rounding the cost of the airfare and the number of travelers to the nearest hundred.

C Find the area of each region.

53.

728 mi

728 mi

54.

129 yd

65 yd

55.

3 ft

6 ft

56.

7 mi

7 mi

57.

11 yd

11 yd

58.

16 cm

9 cm

59.

3 mm

48 mm

60.

247 mi

19 mi

61. Find the area of the region formed by the base lines on a Major League Baseball diamond.

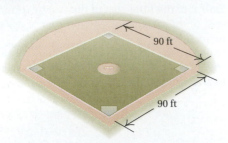

90 ft

90 ft

62. Find the area of a standard-sized hockey rink.

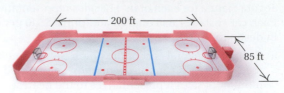

200 ft

85 ft

63. **D**_W Describe a situation that corresponds to each multiplication: 4 · $150; $4 · 150.

64. **D**_W Explain the multiplication illustrated in the diagram below.

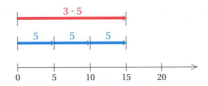

3 · 5

5   5   5

0   5   10   15   20

SKILL MAINTENANCE

Add. [1.2b]

65.
```
  4 9 0 8
  5 6 6 7
+ 2 1 1 0
```

66.
```
  9 8 7 6
    8 7 6
      7 6
+      6
```

67.
```
  3 4 0,7 9 8
+   8 6,6 7 9
```

68.
```
  8 8,7 7 7
+ 2 2,3 3 3
```

Subtract. [1.3d]

69.
```
  4 9 0 8
− 3 6 6 7
```

70.
```
  9 8 7 6
−   9 8 7
```

71.
```
  3 4 0,7 9 8
−   8 6,6 7 9
```

72.
```
  8 8,7 7 7
− 2 2,3 3 3
```

73. Round 6,375,602 to the nearest thousand. [1.4a]

74. Round 6,375,602 to the nearest ten. [1.4a]

SYNTHESIS

75. **D**_W What do the commutative laws of addition and multiplication have in common?

76. **D**_W Explain how the distributive law differs from the associative law.

77. An 18-story office building is box-shaped. Each floor measures 172 ft by 84 ft with a 20-ft by 35-ft rectangular area lost to an elevator, lobby, and stairwell. How much area is available as office space?

78. *Computer Printers.* Eva prints a 5 in. by 7 in. photo on her Epson R1800 printer (see Exercise 6). How many dots will be printed when the printer produces the photo?

79. *Computer Printers.* Adam's HP 8250 is printing an 8 in. by 10 in. photo (see Exercise 5). How many dots will be printed when the printer finishes the photo?

44

1.6 DIVISION

a Division

REPEATED SUBTRACTION

Division of whole numbers applies to two kinds of situations. The first is repeated subtraction. Suppose we have 20 notebooks in a pile, and we want to find out how many sets of 5 there are. One way to do this is to repeatedly subtract sets of 5 as follows.

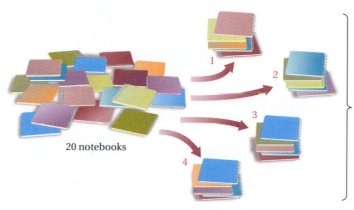

20 notebooks

How many sets of 5 notebooks each?

Since there are 4 sets of 5 notebooks each, we have

$$20 \div 5 = 4.$$

Dividend Divisor Quotient

The division $20 \div 5$, read "20 divided by 5," corresponds to the figure above. We say that the **dividend** is 20, the **divisor** is 5, and the **quotient** is 4. We divide the *dividend* by the *divisor* to get the *quotient*.

We can also express the division $20 \div 5 = 4$ as

$$\frac{20}{5} = 4 \quad \text{or} \quad 5\overline{)20}^{\,4}$$

EXAMPLE 1 Write a division sentence that corresponds to this situation.

A parent directs 3 children to share $24, with each child getting the same amount. How much does each child get?

We think of an array with 3 rows. Each row will go to a child. How many dollars will be in each row?

3 rows with 8 in each row

$$24 \div 3 = 8$$

Objectives

a Write a division sentence that corresponds to a situation.

b Given a division sentence, write a related multiplication sentence; and given a multiplication sentence, write two related division sentences.

c Divide whole numbers.

Write a division sentence that corresponds to each situation. You need not carry out the division.

1. There are 112 students in a college band, and they are marching with 14 in each row. How many rows are there?

2. A college band is in a rectangular array. There are 112 students in the band, and they are marching in 8 rows. How many students are there in each row?

We can also think of division in terms of rectangular arrays. Consider again the pile of 20 notebooks and division by 5. We can arrange the notebooks in a rectangular array with 5 rows and ask, "How many are in each row?"

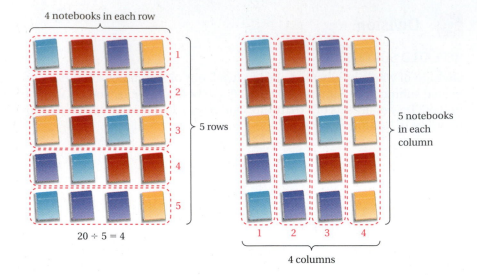

4 notebooks in each row

5 rows

$20 \div 5 = 4$

5 notebooks in each column

4 columns

We can also consider a rectangular array with 5 notebooks in each column and ask, "How many columns are there?" The answer is still 4.

In each case, we are asking, "What do we multiply 5 by in order to get 20?"

Missing factor

$5 \cdot \square = 20$

Quotient

$20 \div 5 = \square$

This leads us to the following definition of division.

DIVISION

The quotient $a \div b$, where $b \neq 0$, is that unique whole number c for which $a = b \cdot c$.

EXAMPLE 2 Write a division sentence that corresponds to this situation. You need not carry out the division.

How many uniforms that cost $45 each can be purchased for $495?

We think of an array with 45 dollar bills in each row. The money in each row will buy one uniform. How many rows will there be?

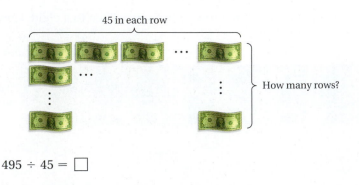

45 in each row

How many rows?

$495 \div 45 = \square$

Do Exercises 1 and 2.

Answers on page A-2

b Related Sentences

By looking at rectangular arrays, we can see how multiplication and division are related. The array of notebooks on the preceding page shows that $5 \cdot 4 = 20$. The array also shows the following:

$$20 \div 5 = 4 \quad \text{and} \quad 20 \div 4 = 5.$$

The division $20 \div 5$ is defined to be the number that when multiplied by 5 gives 20. Thus, for every division sentence, there is a related multiplication sentence.

$20 \div 5 = 4$ Division sentence

$20 = 4 \cdot 5$ Related multiplication sentence

> To get the related multiplication sentence, we use
> Dividend = Quotient · Divisor.

EXAMPLE 3 Write a related multiplication sentence: $12 \div 6 = 2$.

We have

$12 \div 6 = 2$ Division sentence

$12 = 2 \cdot 6.$ Related multiplication sentence

The related multiplication sentence is $12 = 2 \cdot 6$.

> By the commutative law of multiplication, there is also another multiplication sentence: $12 = 6 \cdot 2$.

Do Exercises 3 and 4.

For every multiplication sentence, we can write related division sentences.

EXAMPLE 4 Write two related division sentences: $7 \cdot 8 = 56$.

We have

$7 \cdot 8 = 56$ $7 \cdot 8 = 56$

This factor   This factor
becomes   becomes
a divisor.   a divisor.

$7 = 56 \div 8.$ $8 = 56 \div 7.$

The related division sentences are $7 = 56 \div 8$ and $8 = 56 \div 7$.

Do Exercises 5 and 6.

Write a related multiplication sentence.

3. $15 \div 3 = 5$

4. $72 \div 8 = 9$

Write two related division sentences.

5. $6 \cdot 2 = 12$

6. $7 \cdot 6 = 42$

Answers on page A-2

C Division of Whole Numbers

Before we consider division with remainders, let's recall four basic facts about division.

DIVIDING BY 1

Any number divided by 1 is that same number:

$$a \div 1 = \frac{a}{1} = a.$$

DIVIDING A NUMBER BY ITSELF

Any nonzero number divided by itself is 1:

$$\frac{a}{a} = 1, \quad a \neq 0.$$

DIVIDENDS OF 0

Zero divided by any nonzero number is 0:

$$\frac{0}{a} = 0, \quad a \neq 0.$$

EXCLUDING DIVISION BY 0

Division by 0 is not defined. (We agree not to divide by 0.)

$$\frac{a}{0} \text{ is } \textbf{not defined.}$$

Why can't we divide by 0? Suppose the number 4 could be divided by 0. Then if $\square$ were the answer,

$$4 \div 0 = \square$$

and since 0 times any number is 0, we would have

$$4 = \square \cdot 0 = 0. \qquad \text{False!}$$

Thus, $a \div 0$ would be some number $\square$ such that $a = \square \cdot 0 = 0$. So the only possible number that could be divided by 0 would be 0 itself.

But such a division would give us any number we wish, for we could say

$$0 \div 0 = 8 \quad \text{because} \quad 0 = 8 \cdot 0;$$
$$0 \div 0 = 3 \quad \text{because} \quad 0 = 3 \cdot 0; \qquad \text{All true!}$$
$$0 \div 0 = 7 \quad \text{because} \quad 0 = 7 \cdot 0.$$

We avoid the preceding difficulties by agreeing to exclude division by 0.

Suppose we have 18 cans of soda and want to pack them in cartons of 6 cans each. How many cartons will we fill? We can determine this by repeated subtraction. We keep track of the number of times we subtract. We stop when the number of objects remaining, the **remainder,** is smaller than the divisor.

EXAMPLE 5   Divide by repeated subtraction: 18 ÷ 6.

$$
\begin{array}{r}
1\ 8 \\
-\ \ 6 \\
\hline
1\ 2
\end{array}
\qquad
\left.
\begin{array}{r}
1\ 2 \\
-\ \ 6 \\
\hline
6 \\
\end{array}
\right\}
$$

Subtracting 3 times

$$
\begin{array}{r}
6 \\
-\ 6 \\
\hline
0
\end{array}
$$
← The remainder, 0, is smaller than the divisor, 6.

Thus, 18 ÷ 6 = 3.

Suppose we have 22 cans of soda and want to pack them in cartons of 6 cans each. We end up with 3 cartons with 4 cans left over.

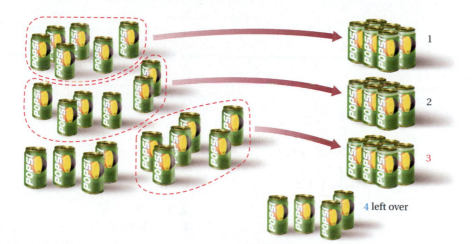

1

2

3

4 left over

EXAMPLE 6   Divide by repeated subtraction: 22 ÷ 6.

$$
\begin{array}{r}
2\ 2 \\
-\ \ 6 \\
\hline
1\ 6
\end{array}
\qquad
\left.
\begin{array}{r}
1\ 6 \\
-\ \ 6 \\
\hline
1\ 0 \\
\end{array}
\right\}
$$

Subtracting 3 times

$$
\begin{array}{r}
1\ 0 \\
-\ \ 6 \\
\hline
4
\end{array}
$$
← Remainder

Check: 3 · 6 = 18, 18 + 4 = 22.

We can write this division as follows.

$$
\begin{array}{r}
3 \\
6\overline{)2\ 2} \\
1\ 8 \\
\hline
4
\end{array}
$$

Divide by repeated subtraction. Then check.

7. 54 ÷ 9

8. 61 ÷ 9

9. 53 ÷ 12

10. 157 ÷ 24

Answers on page A-2

Divide and check.

11. 4) 2 3 9

Note that

$$\text{Quotient} \cdot \text{Divisor} + \text{Remainder} = \text{Dividend}.$$

We write answers to a division sentence as follows.

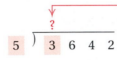

$$22 \div 6 = 3 \, \text{R} \, 4$$

Dividend   Divisor   Quotient   Remainder

Do Exercises 7–10 on the preceding page.

EXAMPLE 7   Divide and check: $3642 \div 5$.

```
        ?
   5 ) 3 6 4 2
```

1. Find the number of thousands in the quotient. Consider 3 thousands ÷ 5 and think 3 ÷ 5. Since 3 ÷ 5 is not a whole number, move to hundreds.

12. 6) 8 8 5 5

```
        7
   5 ) 3 6 4 2
       3 5 0 0
         1 4 2  ← The remainder is larger than the divisor.
```

2. Find the number of hundreds in the quotient. Consider 36 hundreds ÷ 5 and think 36 ÷ 5. The estimate is about 7 hundreds. Multiply 700 by 5 and subtract.

```
        7 2
   5 ) 3 6 4 2
       3 5 0 0
         1 4 2
         1 0 0
           4 2  ← The remainder is larger than the divisor.
```

3. Find the number of tens in the quotient using 142, the first remainder. Consider 14 tens ÷ 5 and think 14 ÷ 5. The estimate is about 2 tens. Multiply 20 by 5 and subtract. (If our estimate had been 3 tens, we could not have subtracted 150 from 142.)

13. 5) 5 0 7 5

```
        7 2 8
   5 ) 3 6 4 2
       3 5 0 0
         1 4 2
         1 0 0
           4 2
           4 0
             2  ← The remainder is less than the divisor.
```

4. Find the number of ones in the quotient using 42, the second remainder. Consider 42 ones ÷ 5 and think 42 ÷ 5. The estimate is about 8 ones. Multiply 8 by 5 and subtract. The remainder, 2, is less than the divisor, 5, so we are finished.

You may have learned to divide like this, not writing the extra zeros. You may omit them if desired.

```
        7 2 8
   5 ) 3 6 4 2
       3 5
         1 4
         1 0
           4 2
           4 0
             2
```

Check: $728 \cdot 5 = 3640,$
$3640 + 2 = 3642.$

The answer is 728 R 2.

Do Exercises 11–13.

Answers on page A-2

We can summarize our division procedure as follows.

> **To divide whole numbers:**
>
> **a)** Estimate.
> **b)** Multiply.
> **c)** Subtract.

Sometimes rounding the divisor helps us find estimates.

EXAMPLE 8 Divide: $8904 \div 42$.

We mentally round 42 to 40.

```
         2
  4 2 ) 8 9 0 4   ← Think: 89 hundreds ÷ 40.
        8 4 0 0      Estimate 2 hundreds. Multiply 200 · 42
        -------      and subtract.
          5 0 4
```

```
         2 1
  4 2 ) 8 9 0 4
        8 4 0 0
        -------
          5 0 4   ← Think: 50 tens ÷ 40.
          4 2 0      Estimate 1 ten. Multiply 10 · 42
          -----      and subtract.
            8 4
```

```
         2 1 2
  4 2 ) 8 9 0 4
        8 4 0 0
        -------
          5 0 4
          4 2 0
          -----
            8 4   ← Think: 84 ones ÷ 40.
            8 4      Estimate 2 ones. Multiply 2 · 42
            ---      and subtract.
              0
```

Caution!

Be careful to keep the digits lined up correctly.

The answer is 212. *Remember*: If after estimating and multiplying you get a number that is larger than the number from which it is being subtracted, lower your estimate.

Do Exercises 14 and 15.

Divide.

14. $4\,5\,)\overline{\,6\,0\,3\,0\,}$

15. $5\,2\,)\overline{\,3\,2\,8\,8\,}$

Divide.

16. $6\,)\overline{\,4\,8\,4\,6\,}$

17. $7\,)\overline{\,7\,6\,1\,6\,}$

Answers on page A-2

Divide.

18. $27\overline{\smash{\big)}9724}$

19. $56\overline{\smash{\big)}44{,}847}$

Answers on page A-2

To the instructor and the student: This section presented a review of division of whole numbers. Students who are successful should go on to Section 1.7. Those who have trouble should study developmental unit D near the back of this text and then repeat Section 1.6.

ZEROS IN QUOTIENTS

EXAMPLE 9 Divide: $6341 \div 7$.

$$
\begin{array}{r}
9 \\
7\,\overline{)\,6\;3\;4\;1} \\
6\;3\;0\;0 \\
\hline
4\;1
\end{array}
$$
← *Think*: 63 hundreds ÷ 7. Estimate 9 hundreds. Multiply $900 \cdot 7$ and subtract.

$$
\begin{array}{r}
9\;\;0 \\
7\,\overline{)\,6\;3\;4\;1} \\
6\;3\;0\;0 \\
\hline
4\;1
\end{array}
$$
← *Think*: 4 tens ÷ 7. The tens digit of the quotient is 0.

$$
\begin{array}{r}
9\;\;0\;\;5 \\
7\,\overline{)\,6\;3\;4\;1} \\
6\;3\;0\;0 \\
\hline
4\;1 \\
3\;5 \\
\hline
6
\end{array}
$$
← *Think*: 41 ones ÷ 7. Estimate 5 ones. Multiply $5 \cdot 7$ and subtract.

← The remainder, 6, is less than the divisor, 7.

The answer is 905 R 6.

Do Exercises 16 and 17 on the preceding page.

EXAMPLE 10 Divide: $8889 \div 37$.

We round 37 to 40.

$$
\begin{array}{r}
2 \\
37\,\overline{)\,8\;8\;8\;9} \\
7\;4\;0\;0 \\
\hline
1\;4\;8\;9
\end{array}
$$
← *Think*: $37 \approx 40$; 88 hundreds ÷ 40. Estimate 2 hundreds. Multiply $200 \cdot 37$ and subtract.

$$
\begin{array}{r}
2\;\;4 \\
37\,\overline{)\,8\;8\;8\;9} \\
7\;4\;0\;0 \\
\hline
1\;4\;8\;9 \\
1\;4\;8\;0 \\
\hline
9
\end{array}
$$
← *Think*: 148 tens ÷ 40. Estimate 4 tens. Multiply $40 \cdot 37$ and subtract.

$$
\begin{array}{r}
2\;\;4\;\;0 \\
37\,\overline{)\,8\;8\;8\;9} \\
7\;4\;0\;0 \\
\hline
1\;4\;8\;9 \\
1\;4\;8\;0 \\
\hline
9
\end{array}
$$
← The remainder, 9, is less than the divisor, 37.

The answer is 240 R 9.

Do Exercises 18 and 19.

Dividing Whole Numbers: Finding Remainders   To divide whole numbers on a calculator, we use the $\div$ and $=$ keys. For example, to divide 711 by 9, we press $\boxed{7}\,\boxed{1}\,\boxed{1}\,\boxed{\div}\,\boxed{9}\,\boxed{=}$. The display reads $\boxed{\quad 79\quad}$, so 711 $\div$ 9 = 79.

When we enter 453 $\div$ 15, the display reads $\boxed{\quad 30.2\quad}$. Note that the result is not a whole number. This tells us that there is a remainder. The number 30.2 is expressed in decimal notation. The symbol "." is called a decimal point. (Decimal notation will be studied in Chapter 5.) The number to the left of the decimal point, 30, is the quotient. We can use the remaining part of the result to find the remainder. To do this, first subtract 30 from 30.2 to find the remaining part of the result. Then multiply the difference by the divisor, 15. We get 3. This is the remainder. Thus, 453 $\div$ 15 = 30 R 3. The steps that we performed to find this result can be summarized as follows:

$$453 \div 15 = 30.2,$$
$$30.2 - 30 = .2,$$
$$0.2 \times 15 = 3.$$

To follow these steps on a calculator, we press $\boxed{4}\,\boxed{5}\,\boxed{3}\,\boxed{\div}$ $\boxed{1}\,\boxed{5}\,\boxed{=}$ and write the number that appears to the left of the decimal point. This is the quotient. Then we continue by pressing $\boxed{-}\,\boxed{3}\,\boxed{0}\,\boxed{=}$ $\boxed{\times}\,\boxed{1}\,\boxed{5}\,\boxed{=}$. The last number that appears is the remainder. In some cases, it will be necessary to round the remainder to the nearest one.

To check this result, we multiply the quotient by the divisor and then add the remainder.

$$30 \times 15 = 450,$$
$$450 + 3 = 453$$

Exercises: Use a calculator to perform each division. Check the results with a calculator also.

1. $92 \div 27$
2. $9\overline{)532}$
3. $6\overline{)746}$
4. $3817 \div 29$
5. $126\overline{)35,715}$
6. $308\overline{)259,831}$

1.6

EXERCISE SET

For Extra Help

MathXL   MyMathLab   InterAct   Math Tutor   Video   Student's
Math   Center   Lectures   Solutions
on CD   Manual
Disc 1

a   Write a division sentence that corresponds to each situation.

1. *Canyonlands.*   The trail boss for a trip into Canyonlands National Park divides 760 lb (pounds) of equipment among 4 mules. How many pounds does each mule carry?

2. *Surf Expo.*   In a swimwear showing at Surf Expo, a trade show for retailers of beach supplies, each swimsuit test takes 8 min (minutes). If the show runs for 240 min, how many tests can be scheduled?

3. A lab technician pours 455 mL of sulfuric acid into 5 beakers, putting the same amount in each. How much acid is in each beaker?

4. A computer screen is made up of a rectangular array of pixels. There are 480,000 pixels in all, with 800 pixels in each row. How many rows are there on the screen?

b   Write a related multiplication sentence.

5. $18 \div 3 = 6$

6. $72 \div 9 = 8$

7. $22 \div 22 = 1$

8. $32 \div 1 = 32$

9. $54 \div 6 = 9$

10. $40 \div 8 = 5$

11. $37 \div 1 = 37$

12. $28 \div 28 = 1$

Write two related division sentences.

13. $9 \times 5 = 45$

14. $2 \cdot 7 = 14$

15. $37 \cdot 1 = 37$

16. $4 \cdot 12 = 48$

17. $8 \times 8 = 64$

18. $9 \cdot 7 = 63$

19. $11 \cdot 6 = 66$

20. $1 \cdot 43 = 43$

C   Divide, if possible. If not possible, write "not defined."

21. $72 \div 6$

22. $54 \div 9$

23. $\dfrac{23}{23}$

24. $\dfrac{37}{37}$

25. $22 \div 1$

26. $\dfrac{56}{1}$

27. $\dfrac{16}{0}$

28. $74 \div 0$

Divide.

29. $277 \div 5$

30. $699 \div 3$

31. $864 \div 8$

32. $869 \div 8$

33. $4 \overline{)1228}$

34. $3 \overline{)2124}$

35. $6 \overline{)4521}$

36. $9 \overline{)9110}$

37. $5 \overline{)8515}$

38. $3 \overline{)6027}$

39. $9 \overline{)8888}$

40. $8 \overline{)4139}$

41. $127,000 \div 10$

42. $127,000 \div 100$

43. $127,000 \div 1000$

44. $4260 \div 10$

45. $70 \overline{)3692}$

46. $20 \overline{)5798}$

47. $30 \overline{)875}$

48. $40 \overline{)987}$

49. $111 \overline{)3219}$ **50.** $102 \overline{)5612}$ **51.** $8 \overline{)843}$ **52.** $7 \overline{)749}$

53. $5 \overline{)8047}$ **54.** $9 \overline{)7273}$ **55.** $5 \overline{)5036}$ **56.** $7 \overline{)7074}$

57. $1058 \div 46$ **58.** $7242 \div 24$ **59.** $3425 \div 32$ **60.** $48 \overline{)4899}$

61. $24 \overline{)8880}$ **62.** $36 \overline{)7563}$ **63.** $28 \overline{)17,067}$ **64.** $36 \overline{)28,929}$

65. $80 \overline{)24,320}$ **66.** $90 \overline{)88,560}$ **67.** $285 \overline{)999,999}$

68. $306 \overline{)888,888}$ **69.** $456 \overline{)3,679,920}$ **70.** $803 \overline{)5,622,606}$

71. **D_W** Is division associative? Why or why not? Give an example.

72. **D_W** Suppose a student asserts that "$0 \div 0 = 0$ because nothing divided by nothing is nothing." Devise an explanation to persuade the student that the assertion is false.

SKILL MAINTENANCE

 VOCABULARY REINFORCEMENT

In each of Exercises 73–80, fill in the blank with the correct term from the given list. Some of the choices may not be used and some may be used more than once.

73. The distance around an object is its _____ .
[1.2c]

74. A sentence like $10 - 3 = 7$ is called a(n) _____ ;
a sentence like $31 < 33$ is called a(n) _____ .
[1.4c]

75. For large numbers, _____ are separated by commas into groups of three, called _____ .
[1.1a]

76. Because changing the order of addends does not change the sum, addition is _____ . [1.2b]

77. In the sentence $28 \div 7 = 4$, the _____ is 28.
[1.6a]

78. In the sentence $10 \times 1000 = 10,000$, 10 and 1000 are called _____ and 10,000 is called the _____ .
[1.5a]

79. The _____ is the number from which another number is being subtracted. [1.3a]

80. The sentence $3 \times (6 \times 2) = (3 \times 6) \times 2$ illustrates the _____ law of multiplication. [1.5a]

associative	product
commutative	minuend
addends	subtrahend
factors	digits
perimeter	periods
dividend	equation
quotient	inequality

SYNTHESIS

81. **D_W** Describe a situation that corresponds to the division $1180 \div 295$. (See Examples 1 and 2.)

82. **D_W** What is it about division that makes it more difficult than addition, subtraction, or multiplication?

83. Complete the following table.

a	b	$a \cdot b$	$a + b$
	68	3672	
84			117
		32	12

84. Find a pair of factors whose product is 36 and:
   a) whose sum is 13.
   b) whose difference is 0.
   c) whose sum is 20.
   d) whose difference is 9.

85. A group of 1231 college students is going to take buses for a field trip. Each bus can hold only 42 students. How many buses are needed?

86. ▦ Fill in the missing digits to make the equation true:
$34,584,132 \div 76\square = 4\square,386.$

Objectives

a Solve simple equations by trial.

b Solve equations like $x + 28 = 54$, $28 \cdot x = 168$, and $98 \cdot 2 = y$.

Find a number that makes the sentence true.

1. $8 = 1 + \square$

2. $\square + 2 = 7$

3. Determine whether 7 is a solution of $\square + 5 = 9$.

4. Determine whether 4 is a solution of $\square + 5 = 9$.

Solve by trial.

5. $n + 3 = 8$

6. $x - 2 = 8$

7. $45 \div 9 = y$

8. $10 + t = 32$

Answers on page A-3

a Solutions by Trial

Let's find a number that we can put in the blank to make this sentence true:

$$9 = 3 + \square.$$

We are asking "9 is 3 plus what number?" The answer is 6.

$$9 = 3 + \;6$$

Do Exercises 1 and 2.

A sentence with = is called an **equation.** A **solution** of an equation is a number that makes the sentence true. Thus, 6 is a solution of

$$9 = 3 + \square \quad \text{because} \quad 9 = 3 + \;6 \quad \text{is true.}$$

However, 7 is not a solution of

$$9 = 3 + \square \quad \text{because} \quad 9 = 3 + \;7 \quad \text{is false.}$$

Do Exercises 3 and 4.

We can use a letter instead of a blank. For example,

$$9 = 3 + n.$$

We call n a **variable** because it can represent any number. If a replacement for a variable makes an equation true, it is a **solution** of the equation.

> **SOLUTIONS OF AN EQUATION**
>
> A **solution** is a replacement for the variable that makes the equation true. When we find all the solutions, we say that we have **solved** the equation.

EXAMPLE 1 Solve $y + 12 = 27$ by trial.

We replace y with several numbers.

If we replace y with 13, we get a false equation: $13 + 12 = 27$.

If we replace y with 14, we get a false equation: $14 + 12 = 27$.

If we replace y with 15, we get a true equation: $15 + 12 = 27$.

No other replacement makes the equation true, so the solution is 15.

EXAMPLES Solve.

2. $7 + n = 22$
(7 plus what number is 22?)
The solution is 15.

3. $63 = 3 \cdot x$
(63 is 3 times what number?)
The solution is 21.

Do Exercises 5–8.

b   Solving Equations

We now begin to develop more efficient ways to solve certain equations. When an equation has a variable alone on one side, it is easy to see the solution or to compute it. When a calculation is on one side and the variable is alone on the other, we can find the solution by carrying out the calculation.

EXAMPLE 4   Solve: $x = 245 \times 34$.

To solve the equation, we carry out the calculation.

$$
\begin{array}{r}
2\ 4\ 5 \\
\times\ \ \ 3\ 4 \\
\hline
9\ 8\ 0 \\
7\ 3\ 5\ 0 \\
\hline
8\ 3\ 3\ 0
\end{array}
$$

$x = 245 \times 34$
$x = 8330$

The solution is 8330.

Do Exercises 9–12.

Consider the equation

$$x + 12 = 27.$$

We can get x alone by writing a related subtraction sentence:

$x = 27 - 12$ 12 gets subtracted to find the related subtraction sentence.

$x = 15.$ Doing the subtraction

It is useful in our later study of algebra to think of this as "subtracting 12 *on both sides.*" Thus

$x + 12 - 12 = 27 - 12$ Subtracting 12 on both sides

$x + 0 = 15$ Carrying out the subtraction

$x = 15.$

SOLVING $x + a = b$

To solve $x + a = b$, subtract a on both sides.

If we can get an equation in a form with the variable alone on one side, we can "see" the solution.

EXAMPLE 5   Solve: $t + 28 = 54$.

We have

$t + 28 = 54$

$t + 28 - 28 = 54 - 28$ Subtracting 28 on both sides

$t + 0 = 26$

$t = 26.$

Solve.

9. $346 \times 65 = y$

10. $x = 2347 + 6675$

11. $4560 \div 8 = t$

12. $x = 6007 - 2346$

Answers on page A-3

Study Tips

STAYING AHEAD

Try to keep one section ahead of your syllabus. If you study ahead of your lectures, you can concentrate on what is being explained in them, rather than trying to write everything down. You can then write notes on only special points or questions related to what is happening in class.

Solve. Be sure to check.

13. $x + 9 = 17$

14. $77 = m + 32$

15. Solve: $155 = t + 78$. Be sure to check.

Solve. Be sure to check.

16. $4566 + x = 7877$

17. $8172 = h + 2058$

Answers on page A-3

To check the answer, we substitute 26 for t in the original equation.

Check:
$$t + 28 = 54$$
$$26 + 28 \ ? \ 54$$
$$54 \ | \quad \text{TRUE}$$

The solution is 26.

Do Exercises 13 and 14.

EXAMPLE 6 Solve: $182 = 65 + n$.

We have

$$182 = 65 + n$$
$$182 - 65 = 65 + n - 65 \qquad \text{Subtracting 65 on both sides}$$
$$117 = 0 + n \qquad \text{65 plus } n \text{ minus 65 is } 0 + n.$$
$$117 = n.$$

Check:
$$182 = 65 + n$$
$$182 \ ? \ 65 + 117$$
$$| \ 182 \qquad \text{TRUE}$$

The solution is 117.

Do Exercise 15.

EXAMPLE 7 Solve: $7381 + x = 8067$.

We have

$$7381 + x = 8067$$
$$7381 + x - 7381 = 8067 - 7381 \qquad \text{Subtracting 7381 on both sides}$$
$$x = 686.$$

The check is left to the student. The solution is 686.

Do Exercises 16 and 17.

We now learn to solve equations like $8 \cdot n = 96$. Look at

$$8 \cdot n = 96.$$

We can get n alone by writing a related division sentence:

$$n = 96 \div 8 = \frac{96}{8} \qquad \text{96 is divided by 8.}$$
$$n = 12. \qquad \text{Doing the division}$$

It is useful in our later study of algebra to think of the preceding as "dividing by 8 *on both sides*." Thus,

$$\frac{8 \cdot n}{8} = \frac{96}{8} \qquad \text{Dividing both sides by 8}$$
$$n = 12. \qquad \text{8 times } n \text{ divided by 8 is } n.$$

SOLVING $a \cdot x = b$

To solve $a \cdot x = b$, divide both sides by a.

EXAMPLE 8 Solve: $10 \cdot x = 240$.

We have

$$10 \cdot x = 240$$

$$\frac{10 \cdot x}{10} = \frac{240}{10} \qquad \text{Dividing both sides by 10}$$

$$x = 24.$$

Check:

$$\frac{10 \cdot x = 240}{10 \cdot 24 \; ? \; 240}$$
$$240 \; | \qquad \text{TRUE}$$

The solution is 24.

EXAMPLE 9 Solve: $5202 = 9 \cdot t$.

We have

$$5202 = 9 \cdot t$$

$$\frac{5202}{9} = \frac{9 \cdot t}{9} \qquad \text{Dividing both sides by 9}$$

$$578 = t.$$

The check is left to the student. The solution is 578.

Do Exercises 18–20.

EXAMPLE 10 Solve: $14 \cdot y = 1092$.

We have

$$14 \cdot y = 1092$$

$$\frac{14 \cdot y}{14} = \frac{1092}{14} \qquad \text{Dividing both sides by 14}$$

$$y = 78.$$

The check is left to the student. The solution is 78.

EXAMPLE 11 Solve: $n \cdot 56 = 4648$.

We have

$$n \cdot 56 = 4648$$

$$\frac{n \cdot 56}{56} = \frac{4648}{56} \qquad \text{Dividing both sides by 56}$$

$$n = 83.$$

The check is left to the student. The solution is 83.

Do Exercises 21 and 22.

Solve. Be sure to check.

18. $8 \cdot x = 64$

19. $144 = 9 \cdot n$

20. $5152 = 8 \cdot t$

Solve. Be sure to check.

21. $18 \cdot y = 1728$

22. $n \cdot 48 = 4512$

Answers on page A-3

a Solve by trial.

1. $x + 0 = 14$

2. $x - 7 = 18$

3. $y \cdot 17 = 0$

4. $56 \div m = 7$

b Solve. Be sure to check.

5. $13 + x = 42$

6. $15 + t = 22$

7. $12 = 12 + m$

8. $16 = t + 16$

9. $3 \cdot x = 24$

10. $6 \cdot x = 42$

11. $112 = n \cdot 8$

12. $162 = 9 \cdot m$

13. $45 \cdot 23 = x$

14. $23 \cdot 78 = y$

15. $t = 125 \div 5$

16. $w = 256 \div 16$

17. $p = 908 - 458$

18. $9007 - 5667 = m$

19. $x = 12{,}345 + 78{,}555$

20. $5678 + 9034 = t$

21. $3 \cdot m = 96$

22. $4 \cdot y = 96$

23. $715 = 5 \cdot z$

24. $741 = 3 \cdot t$

25. $10 + x = 89$

26. $20 + x = 57$

27. $61 = 16 + y$

28. $53 = 17 + w$

29. $6 \cdot p = 1944$

30. $4 \cdot w = 3404$

31. $5 \cdot x = 3715$

32. $9 \cdot x = 1269$

33. $47 + n = 84$

34. $56 + p = 92$

35. $x + 78 = 144$

36. $z + 67 = 133$

37. $165 = 11 \cdot n$

38. $660 = 12 \cdot n$

39. $624 = t \cdot 13$

40. $784 = y \cdot 16$

41. $x + 214 = 389$

42. $x + 221 = 333$

43. $567 + x = 902$

44. $438 + x = 807$

45. $18 \cdot x = 1872$

46. $19 \cdot x = 6080$

47. $40 \cdot x = 1800$

48. $20 \cdot x = 1500$

49. $2344 + y = 6400$

50. $9281 = 8322 + t$

51. $8322 + 9281 = x$

52. $9281 - 8322 = y$

53. $234 \cdot 78 = y$

54. $10{,}534 \div 458 = q$

55. $58 \cdot m = 11{,}890$

56. $233 \cdot x = 22{,}135$

57. D_W Describe a procedure that can be used to convert any equation of the form $a \cdot b = c$ to a related division equation.

58. D_W Describe a procedure that can be used to convert any equation of the form $a + b = c$ to a related subtraction equation.

SKILL MAINTENANCE

59. Write two related subtraction sentences: $7 + 8 = 15$. [1.3b]

60. Write two related division sentences: $6 \cdot 8 = 48$. [1.6b]

Use $>$ or $<$ for ☐ to write a true sentence. [1.4c]

61. 123 ☐ 789

62. 342 ☐ 339

63. 688 ☐ 0

64. 0 ☐ 11

Divide. [1.6c]

65. $1283 \div 9$

66. $1278 \div 9$

67. $1\,7\,\overline{)\,5\,6\,7\,8}$

68. $1\,7\,\overline{)\,5\,6\,8\,9}$

SYNTHESIS

69. D_W Give an example of an equation in which a variable appears but for which there is no solution. Then explain why no solution exists.

70. D_W Is it possible for an equation to have many solutions? If not, explain why, and if so, explain how to write such equations.

Solve.

71. ▦ $23{,}465 \cdot x = 8{,}142{,}355$

72. ▦ $48{,}916 \cdot x = 14{,}332{,}388$

Objective

1.8 APPLICATIONS AND PROBLEM SOLVING

a A Problem-Solving Strategy

Applications and problem solving are the most important uses of mathematics. To solve a problem, we first look at the situation and try to translate the problem to an equation. Then we solve the equation. We check to see if the solution of the equation is a solution of the original problem. We are using the following five-step strategy.

FIVE STEPS FOR PROBLEM SOLVING

1. *Familiarize* yourself with the situation.

   a) Carefully read and reread until you understand *what* you are being asked to find.
   b) Draw a diagram or see if there is a formula that applies to the situation.
   c) Assign a letter, or *variable,* to the unknown.

2. *Translate* the problem to an equation using the letter or variable.
3. *Solve* the equation.
4. *Check* the answer in the original wording of the problem.
5. *State* the answer to the problem clearly with appropriate units.

EXAMPLE 1 *International Adoptions.* The top ten countries of origin for United States international adoptions in 2005 and 2004 are listed in the table below. Find the total number of adoptions from China, South Korea, Ethiopia, and India in 2005.

INTERNATIONAL ADOPTIONS

	2005		2004	
RANK	COUNTRY OF ORIGIN	NUMBER	COUNTRY OF ORIGIN	NUMBER
1	China (mainland)	7906	China (mainland)	7044
2	Russia	4639	Russia	5865
3	Guatemala	3783	Guatemala	3264
4	South Korea	821	South Korea	1716
5	Ukraine	755	Kazakhstan	826
6	Kazakhstan	441	Ukraine	723
7	Ethiopia	323	India	406
8	India	291	Haiti	356
9	Colombia	271	Ethiopia	289
10	Philippines	231	Colombia	287

Source: U.S. Department of State

1. **Familiarize.** We can make a drawing or at least visualize the situation.

7906	+	821	+	323	+	291
from		from		from		from
China		South Korea		Ethiopia		India

Since we are combining numbers of adoptions, addition can be used. First, we define the unknown. We let n = the total number of adoptions from China, South Korea, Ethiopia, and India.

2. **Translate.** We translate to an equation:

$$7906 + 821 + 323 + 291 = n.$$

3. **Solve.** We solve the equation by carrying out the addition.

$$
\begin{array}{r}
\overset{2}{} \overset{1}{} \overset{1}{} \\
7\ 9\ 0\ 6 \\
8\ 2\ 1 \\
3\ 2\ 3 \\
+\ \ \ 2\ 9\ 1 \\
\hline
9\ 3\ 4\ 1
\end{array}
$$

$$7906 + 821 + 323 + 291 = n$$
$$9341 = n$$

4. **Check.** We check 9341 in the original problem. There are many ways in which this can be done. For example, we can repeat the calculation. (We leave this to the student.) Another way is to check whether the answer is reasonable. In this case, we would expect the total to be greater than the number of adoptions from any of the individual countries, which it is. We can also estimate by rounding. Here we round to the nearest hundred.

$$7906 + 821 + 323 + 291 \approx 7900 + 800 + 300 + 300$$
$$= 9300$$

Since $9300 \approx 9341$, we have a partial check. If we had an estimate like 4800 or 7500, we might be suspicious that our calculated answer is incorrect. Since our estimated answer is close to our calculation, we are further convinced that our answer checks.

5. **State.** The total number of adoptions from China, South Korea, Ethiopia, and India in 2005 is 9341.

Do Exercises 1–3.

Refer to the table on the preceding page to answer Margin Exercises 1–3.

1. Find the total number of adoptions from China, Russia, Kazakhstan, and India in 2004.

2. Find the total number of adoptions from Russia, Guatemala, South Korea, Ukraine, and Colombia in 2005.

3. Find the total number of adoptions from the top ten countries in 2005.

Answers on page A-3

Study Tips

TIME MANAGEMENT

Time is the most critical factor in your success in learning mathematics. Have reasonable expectations about the time you need to study math.

- **Juggling time.** Working 40 hours per week and taking 12 credit hours is equivalent to working two full-time jobs. Can you handle such a load? Your ratio of number of work hours to number of credit hours should be about 40/3, 30/6, 20/9, 10/12, or 5/14.

- **A rule of thumb on study time.** Budget about 2–3 hours for homework and study for every hour you spend in class each week.

- **Scheduling your time.** Use the Student Organizer card to make an hour-by-hour schedule of your typical week. Include work, school, home, sleep, study, and leisure times. Try to schedule time for study when you are most alert. Choose a setting that will enable you to focus and concentrate. Plan for success and it will happen!

4. Checking Account Balance.
The balance in Heidi's checking account is $2003. She uses her debit card to buy the same Roto Zip Spiral Saw Combo, featured in Example 2, that Tyler did. Find the new balance in her checking account.

EXAMPLE 2 *Checking Account Balance.* The balance in Tyler's checking account is $528. He uses his debit card to buy the Roto Zip Spiral Saw Combo shown in this ad. Find the new balance in his checking account.

NOW
$**129**⁰⁰

Source: Reproduced with permission from the copyright owner (©2004 by Robert Bosch Tool Corporation). Further reproductions strictly prohibited.

1. **Familiarize.** We first make a drawing or at least visualize the situation. We let M = the new balance in his account. This gives us the following:

Take away
$129

$528   New balance

2. **Translate.** We can think of this as a "take-away" situation. We translate to an equation.

Money in the account	minus	Money spent	is	New balance
↓	↓	↓	↓	↓
528	−	129	=	M

3. **Solve.** This sentence tells us what to do. We subtract.

$$\begin{array}{r} \overset{11}{} \\ \overset{4}{5}\,\overset{\cancel{2}}{}\,\overset{18}{8} \\ 5\;2\;8 \\ -\;1\;2\;9 \\ \hline 3\;9\;9 \end{array}$$

$$528 - 129 = M$$
$$399 = M$$

4. **Check.** To check our answer of $399, we can repeat the calculation. We note that the answer should be less than the original amount, $528, which it is. We can add the difference, 399, to the subtrahend, 129: $129 + 399 = 528$. We can also estimate:

$$528 - 129 \approx 530 - 130 = 400 \approx 399.$$

5. **State.** Tyler has a new balance of $399 in his checking account.

Answer on page A-3

Do Exercise 4.

In the real world, problems may not be stated in written words. You must still become familiar with the situation before you can solve the problem.

EXAMPLE 3 *Travel Distance.* Abigail is driving from Indianapolis to Salt Lake City to interview for a news anchor position. The distance from Indianapolis to Salt Lake City is 1634 mi. She travels 1154 mi to Denver. How much farther must she travel?

1. **Familiarize.** We first make a drawing or at least visualize the situation. We let x = the remaining distance to Salt Lake City.

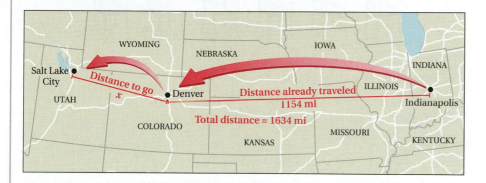

2. **Translate.** We see that this is a "how much more" situation. We translate to an equation.

Distance already traveled	plus	Distance to go	is	Total distance of trip
↓	↓	↓	↓	↓
1154	+	x	=	1634

3. **Solve.** To solve the equation, we subtract 1154 on both sides:

$$1154 + x = 1634$$
$$1154 + x - 1154 = 1634 - 1154$$
$$x = 480.$$

$$\begin{array}{r} \overset{5\ 13}{1\ 6\ \cancel{3}\ 4} \\ -\ 1\ 1\ 5\ 4 \\ \hline 4\ 8\ 0 \end{array}$$

4. **Check.** We check our answer of 480 mi in the original problem. This number should be less than the total distance, 1634 mi, which it is. We can add the difference, 480, to the subtrahend, 1154: 1154 + 480 = 1634. We can also estimate:

$$1634 - 1154 \approx 1600 - 1200$$
$$= 400 \approx 480.$$

The answer, 480 mi, checks.

5. **State.** Abigail must travel 480 mi farther to Salt Lake City.

Do Exercise 5.

Answer on page A-3

6. Total Cost of Gas Grills. What is the total cost of 14 gas grills, each with 520 sq in. (square inches) of total cooking surface and porcelain cast-iron cooking grates, if each one costs $398?

EXAMPLE 4 *Total Cost of Chairs.* What is the total cost of 6 Logan side chairs from Restoration Hardware if each one costs $210?

1. **Familiarize.** We first make a drawing or at least visualize the situation. We let $T =$ the cost of 6 chairs. Repeated addition works well in this case.

$210 $210 $210 $210 $210 $210

2. **Translate.** We translate to an equation.

Number of chairs	times	Cost of each chair	is	Total cost
6	$\times$	$210	$=$	T

3. **Solve.** This sentence tells us what to do. We multiply.

$$\begin{array}{r} 2\ 1\ 0 \\ \times\qquad 6 \\ \hline 1\ 2\ 6\ 0 \end{array}$$

$6 \times 210 = T$
$1260 = T$

4. **Check.** We have an answer, 1260, that is much greater than the cost of any individual chair, which is reasonable. We can repeat our calculation. We can also check by estimating:

$6 \times 210 \approx 6 \times 200 = 1200 \approx 1260.$

The answer checks.

5. **State.** The total cost of 6 chairs is $1260.

Do Exercise 6.

EXAMPLE 5 *Truck Bed Cover.* The dimensions of a fiberglass truck bed cover for a pickup truck are 79 in. by 68 in. What is the area of the cover?

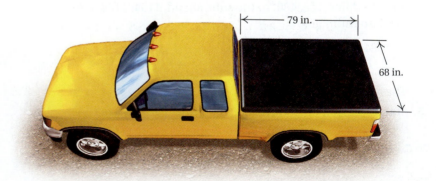

79 in.

68 in.

Answer on page A-3

1. **Familiarize.** The truck bed cover is a rectangle that measures 79 in. by 68 in. We let A = the area of the cover, in sq in. (square inches), and use the area formula $A = l \cdot w$.

2. **Translate.** Using this formula, we have

$$A = \text{length} \cdot \text{width} = l \cdot w = 79 \cdot 68.$$

3. **Solve.** We carry out the multiplication.

$$
\begin{array}{r}
7\ 9 \\
\times\ 6\ 8 \\
\hline
6\ 3\ 2 \\
4\ 7\ 4\ 0 \\
\hline
5\ 3\ 7\ 2
\end{array}
$$

$A = 79 \cdot 68$
$A = 5372$

4. **Check.** We repeat our calculation. We also note that the answer is greater than either the length or the width, which it should be. (This might not be the case if we were using fractions or decimals.) The answer checks.

5. **State.** The area of the truck bed cover is 5372 sq in.

Do Exercise 7.

EXAMPLE 6 *Cartons of Soda.* A bottling company produces 3304 cans of soda. How many 12-can cartons can be filled? How many cans will be left over?

1. **Familiarize.** We first make a drawing. We let n = the number of 12-can cartons that can be filled. The problem can be considered as repeated subtraction, taking successive sets of 12 cans and putting them into n cartons.

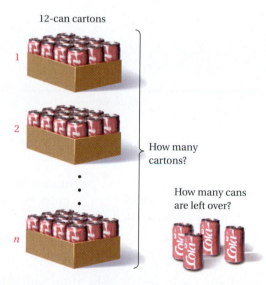

12-can cartons

1

2

How many cartons?

How many cans are left over?

n

2. **Translate.** We translate to an equation.

Number of cans	divided by	Number in each carton	is	Number of cartons
↓	↓	↓	↓	↓
3304	÷	12	=	n

7. **Bed Sheets.** The dimensions of a flat sheet for a queen-size bed are 90 in. by 102 in. What is the area of the sheet?

Answer on page A-3

8. Cartons of Soda. The bottling company in Example 6 also uses 6-can cartons. How many 6-can cartons can be filled with 2269 cans of cola? How many will be left over?

3. Solve. We solve the equation by carrying out the division.

```
        2 7 5
  1 2 ) 3 3 0 4
        2 4 0 0
          9 0 4
          8 4 0
            6 4
            6 0
               4
```

$$3304 \div 12 = n$$
$$275 \text{ R } 4 = n$$

4. Check. We can check by multiplying the number of cartons by 12 and adding the remainder, 4:

$$12 \cdot 275 = 3300,$$
$$3300 + 4 = 3304.$$

5. State. Thus, 275 twelve-can cartons can be filled. There will be 4 cans left over.

Do Exercise 8.

EXAMPLE 7 *Automobile Mileage.* The Pontiac G6 GT gets 21 miles to the gallon (mpg) in city driving. How many gallons will it use in 3843 mi of city driving?

Source: General Motors

1. Familiarize. We first make a drawing. It is often helpful to be descriptive about how we define a variable. In this case, we let $g =$ the number of gallons ("g" comes from "gallons").

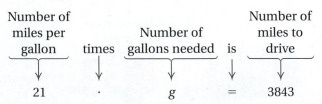

3843 mi to drive

2. Translate. Repeated addition applies here. Thus the following multiplication applies to the situation.

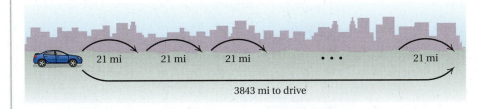

Number of miles per gallon	times	Number of gallons needed	is	Number of miles to drive
21	$\cdot$	g	$=$	3843

3. Solve. To solve the equation, we divide by 21 on both sides.

$$21 \cdot g = 3843$$
$$\frac{21 \cdot g}{21} = \frac{3843}{21}$$
$$g = 183$$

```
          1 8 3
  2 1 ) 3 8 4 3
        2 1 0 0
        1 7 4 3
        1 6 8 0
            6 3
            6 3
               0
```

Answer on page A-3

70

4. Check. To check, we multiply 183 by 21: $21 \times 183 = 3843$.

5. State. The Pontiac G6 GT will use 183 gal.

Do Exercise 9.

Multistep Problems

Sometimes we must use more than one operation to solve a problem, as in the following example.

EXAMPLE 8 *Aircraft Seating.* Boeing Corporation builds commercial aircraft. A Boeing 767 has a seating configuration with 4 rows of 6 seats across in first class and 35 rows of 7 seats across in economy class. Find the total seating capacity of the plane.

Sources: The Boeing Company; Delta Airlines

1. Familiarize. We first make a drawing.

Economy class:
35 rows of 7 seats

First class:
4 rows of
6 seats

2. Translate. There are three parts to the problem. We first find the number of seats in each class. Then we add.

First class: Repeated addition applies here. Thus the following multiplication corresponds to the situation. We let F = the number of seats in first class.

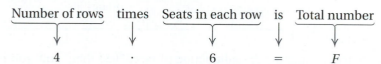

Number of rows	times	Seats in each row	is	Total number
4	·	6	=	F

Economy class: Repeated addition applies here. Thus the following multiplication corresponds to the situation. We let E = the number of seats in economy class.

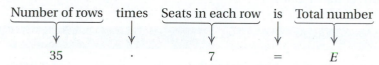

Number of rows	times	Seats in each row	is	Total number
35	·	7	=	E

We let T = the total number of seats in both classes.

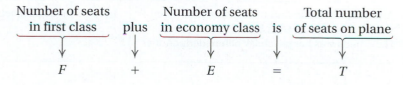

Number of seats in first class	plus	Number of seats in economy class	is	Total number of seats on plane
F	+	E	=	T

9. Automobile Mileage. The Pontiac G6 GT gets 29 miles to the gallon (mpg) in highway driving. How many gallons will it take to drive 2291 mi of highway driving?

Source: General Motors

Answer on page A-3

10. Aircraft Seating. A Boeing 767 used for foreign travel has three classes of seats. First class has 3 rows of 5 seats across; business class has 6 rows with 6 seats across and 1 row with 2 seats on each of the outside aisles. Economy class has 18 rows with 7 seats across. Find the total seating capacity of the plane.

Sources: The Boeing Company; Delta Airlines

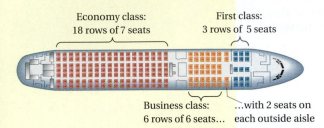

Economy class: 18 rows of 7 seats

First class: 3 rows of 5 seats

Business class: 6 rows of 6 seats…

…with 2 seats on each outside aisle

3. Solve. We solve each equation and add the solutions.

$$4 \cdot 6 = F \qquad 35 \cdot 7 = E \qquad F + E = T$$
$$24 = F \qquad 245 = E \qquad 24 + 245 = T$$
$$269 = T$$

4. Check. To check, we repeat our calculations. (We leave this to the student.) We could also check by rounding, multiplying, and adding.

5. State. There are 269 seats in a Boeing 767.

Do Exercise 10.

As you consider the following exercises, here are some words and phrases that may be helpful to look for when you are translating problems to equations.

KEY WORDS, PHRASES, AND CONCEPTS

ADDITION (+)	SUBTRACTION (−)
add	subtract
added to	subtracted from
sum	difference
total	minus
plus	less than
more than	decreased by
increased by	take away
	how much more
	missing addend

MULTIPLICATION (·)	DIVISION (÷)
multiply	divide
multiplied by	divided by
product	quotient
times	repeated subtraction
of	missing factor
repeated addition	split into equal
rectangular arrays	quantities

Sentences containing the subtraction phrases "less than" and "subtracted from" can be difficult to translate correctly. Since subtraction is not commutative (for example, $5 - 2$ is not the same as $2 - 5$), the order in which the subtraction is written is important. Listed below are some sample problems and correct translations.

PROBLEM	TRANSLATION
Bryan weighs 98 lb. This is 7 lb less than Cory's weight. How much does Cory weigh?	Let c = Cory's weight. $98 = c - 7$
After payroll deductions of $89 were subtracted from her paycheck, Gina brought home $255. How much was her paycheck before deductions?	Let p = Gina's paycheck. $p - 89 = 255$

Answer on page A-3

Translating
for Success

1. *Brick-mason Expense.* A commercial contractor is building 30 two-unit condominiums in a retirement community. The brick-mason expense for each building is $10,860. What is the total cost of bricking the buildings?

2. *Heights.* Dean's sons are on the high school basketball team. Their heights are 73 in., 69 in., and 76 in. How much taller is the tallest son than the shortest son?

3. *Account Balance.* You have $423 in your checking account. Then you deposit $73 and use your debit card for purchases of $76 and $69. How much is left in your account?

4. *Purchasing Camcorder.* A camcorder is on sale for $423. Jenny has only $69. How much more does she need to buy the camcorder?

5. *Purchasing Coffee Makers.* Sara purchases 8 coffee makers for the newly remodeled bed-and-breakfast inn that she manages. If she pays $52 for each coffee maker, what is the total cost of her purchase?

The goal of these matching questions is to practice step (2), *Translate,* of the five-step problem-solving process. Translate each word problem to an equation and select a correct translation from equations A–O.

A. $8 \cdot 52 = n$

B. $69 \cdot n = 76$

C. $73 - 76 - 69 = n$

D. $423 + 73 - 76 - 69 = n$

E. $30 \cdot 10{,}860 = n$

F. $15 \cdot n = 195$

G. $69 + n = 423$

H. $n = 10{,}860 - 300$

I. $n = 423 \div 69$

J. $30 \cdot n = 10{,}860$

K. $15 \cdot 195 = n$

L. $n = 52 - 8$

M. $69 + n = 76$

N. $15 \div 195 = n$

O. $52 + n = 60$

Answers on page A-3

---

6. *Hourly Rate.* Miller Auto Repair charges $52 an hour for labor. Jackson Auto Care charges $60 per hour. How much more does Jackson charge than Miller?

7. *College Band.* A college band with 195 members marches in a 15-row formation in the homecoming halftime performance. How many members are in each row?

8. *Cleats Purchase.* A professional football team purchases 15 pairs of cleats at $195 a pair. What is the total cost of this purchase?

9. *Loan Payment.* Kendra borrows $10,860 for a new boat. The loan is to be paid off in 30 payments. How much is each payment (excluding interest)?

10. *College Enrollment.* At the beginning of the fall term, the total enrollment in Lakeview Community College was 10,860. By the end of the first two weeks, 300 students withdrew. How many students were then enrolled?

a Solve.

Longest Broadway Run.   In January 2006, *The Phantom of the Opera* became the longest-running broadway show. The bar graph below lists the five broadway shows with the greatest number of performances as of January 9, 2006. Use the graph for Exercises 1–4.

Source: League of American Theatres and Producers, Inc.

1. What was the total number of performances of all five shows?

2. What was the total number of performances of the top three shows?

3. How many more performances of *The Phantom of the Opera* were there than performances of *A Chorus Line*?

4. How many more performances of *Les Misérables* were there than performances of *Oh! Calcutta!*?

5. *Boundaries Between Countries.*   The boundary between mainland United States and Canada including the Great Lakes is 3987 miles long. The length of the boundary between the United States and Mexico is 1933 miles. How much longer is the Canadian border?
   Source: U.S. Geological Survey

6. *Caffeine.*   Hershey's 6-oz milk chocolate almond bar contains 25 milligrams of caffeine. A 20-oz bottle of Coca-Cola has 32 more milligrams of caffeine than the Hershey bar. How many milligrams of caffeine does the 20-oz bottle of Coca-Cola have?
   Source: *National Geographic*, "Caffeine," by T. R. Reid, January 2005

7. *Carpentry.*   A carpenter drills 216 holes in a rectangular array in a pegboard. There are 12 holes in each row. How many rows are there?

8. *Spreadsheets.*   Lou works as a CPA. He arranges 504 entries on a spreadsheet in a rectangular array that has 36 rows. How many entries are in each row?

Bachelor's Degree.   The line graph below illustrates data about bachelor's degrees awarded to men and women from 1970 to 2003. Use this graph when answering Exercises 9–12.

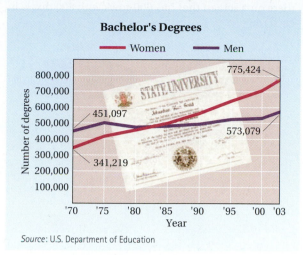

Bachelor's Degrees

—— Women   —— Men

Number of degrees

800,000
700,000
600,000
500,000
400,000
300,000
200,000
100,000

451,097
341,219
775,424
573,079

'70 '75 '80 '85 '90 '95 '00 '03
Year

Source: U.S. Department of Education

9. Find the total number of bachelor's degrees awarded in 1970 and the total number awarded in 2003.

10. Determine how many more bachelor's degrees were awarded in 2003 than in 1970.

11. How many more bachelor's degrees were awarded to women than to men in 2003?

12. How many more bachelor's degrees were awarded to men than to women in 1970?

13. *Median Mortgage Debt.*   The median mortgage debt in 2004 was $48,388 more than the median mortgage debt in 1989. The debt in 1989 was $39,802. What was the median mortgage debt in 2004?

Source: Federal Reserve Board Survey of Consumer Finances

14. *Olympics in Athens.*   In the first modern Olympics in Athens, Greece, in 1896, there were 43 events. In the 2004 Summer Olympics, also in Athens, there were 258 more events than in 1896. How many events were there in 2004?

Source: *USA Today* research; The Olympic Games: Athens 1896–Athens 2004

15. *Longest Rivers.*   The longest river in the world is the Nile in Egypt at 4180 mi. The longest river in the United States is the Missouri–Mississippi–Red Rock at 3710 mi. How much longer is the Nile?

Source: *Time Almanac 2006*

16. *Speeds on Interstates.*   Recently, speed limits on interstate highways in many Western states were raised from 65 mph to 75 mph. By how many miles per hour were they raised?

17. There are 24 hr (hours) in a day and 7 days in a week. How many hours are there in a week?

18. There are 60 min in an hour and 24 hr in a day. How many minutes are there in a day?

19. *Crossword.* The *USA Today* crossword puzzle is a rectangle containing 15 rows with 15 squares in each row. How many squares does the puzzle have altogether?

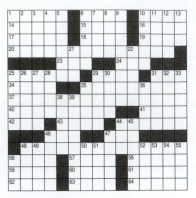

20. *Pixels.* A computer screen consists of small rectangular dots called *pixels.* How many pixels are there on a screen that has 600 rows with 800 pixels in each row?

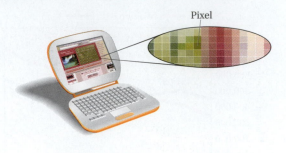

Pixel

21. *Refrigerator Purchase.* Gourmet Deli has a chain of 24 restaurants. It buys a commercial refrigerator for each store at a cost of $1019 each. Find the total cost of the purchase.

22. *Microwave Purchase.* Bridgeway College is constructing new dorms, in which each room has a small kitchen. It buys 96 microwave ovens at $88 each. Find the total cost of the purchase.

23. *"Seinfeld" Episodes.* "Seinfeld" was a long-running television comedy with 177 episodes. A local station picks up the syndicated reruns. If the station runs 5 episodes per week, how many full weeks will pass before it must start over with past episodes? How many episodes will be left for the last week?

24. *Laboratory Test Tubes.* A lab technician separates a vial containing 70 cc (cubic centimeters) of blood into test tubes, each of which contains 3 cc of blood. How many test tubes can be filled? How much blood is left over?

25. *Automobile Mileage.* The 2005 Hyundai Tucson GLS gets 26 miles to the gallon (mpg) in highway driving. How many gallons will it use in 6136 mi of highway driving?

Source: Hyundai

26. *Automobile Mileage.* The 2005 Volkswagen Jetta (5 cylinder) gets 24 miles to the gallon (mpg) in city driving. How many gallons will it use in 3960 mi of city driving?

Source: Volkswagen of America, Inc.

Boeing Jets. Use the chart below in Exercises 27–32 to compare the Boeing 747 jet with its main competitor, the Boeing 777 jet.

232 ft

209 ft

BOEING 747–400	
Passenger capacity	416
Nonstop flight distance	8826 miles
Cruising speed	567 mph
Gallons of fuel used per hour	3201
Costs (to fly one hour):	
Crew	$1948
Fuel	$2867

BOEING 777–200	
Passenger capacity	368
Nonstop flight distance	5210 miles
Cruising speed	615 mph
Gallons of fuel used per hour	2021
Costs (to fly one hour):	
Crew	$1131
Fuel	$1816

Source: Éclat Consulting; Boeing

27. The nonstop flight distance of the Boeing 747 jet is how much greater than the nonstop flight distance of the Boeing 777 jet?

28. How much larger is the passenger capacity of the Boeing 747 jet than the passenger capacity of the Boeing 777 jet?

29. How many gallons of fuel are needed for a 4-hour flight of the Boeing 747?

30. How much longer is the Boeing 747 than the Boeing 777?

31. What is the total cost for the crew and fuel for a 3-hour flight of the Boeing 747 jet?

32. What is the total cost for the crew and fuel for a 2-hour flight of the Boeing 777 jet?

33. *Car Payments.* Dana borrows $5928 for a used car. The loan is to be paid off in 24 equal monthly payments. How much is each payment (excluding interest)?

34. *Loan Payments.* A family borrows $7824 to build a sunroom on the back of their home. The loan is to be paid off in equal monthly payments of $163 (excluding interest). How many months will it take to pay off the loan?

35. *High School Court.* The standard basketball court used by high school players has dimensions of 50 ft by 84 ft.
a) What is its area?
b) What is its perimeter?

50 ft

84 ft

36. *NBA Court.* The standard basketball court used by college and NBA players has dimensions of 50 ft by 94 ft.
a) What is its area?
b) What is its perimeter?
c) How much greater is the area of an NBA court than a high school court? (See Exercise 35.)

94 ft

50 ft

37. *Clothing Imports and Exports.*   In the United States, the exports of clothing in 2003 totaled $2,596,000,000 while the imports totaled $31,701,000,000. How much more were the imports than the exports?

Source: U.S. Bureau of the Census, Foreign Trade Division

38. *Corn Imports and Exports.*   In the United States, the exports of corn in 2003 totaled $2,264,000,000 while the imports totaled $130,000,000. How much more were the exports than the imports?

Source: U.S. Bureau of the Census, Foreign Trade Division

39. *Colonial Population.*   Before the establishment of the U.S. Census in 1790, it was estimated that the Colonial population in 1780 was 2,780,400. This was an increase of 2,628,900 from the population in 1680. What was the Colonial population in 1680?

Source: *Time Almanac,* 2005

40. *Deaths by Firearms.*   Deaths by firearms totaled 30,242 in 2002. This was a decrease of 9353 from the deaths by firearms in 1993. How many deaths by firearms were there in 1993?

Source: Centers for Disease Control and Prevention, National Center for Health Statistics Mortality Report online, 2005

41. *Hershey Bars.*   Hershey Chocolate USA makes small, fun-size chocolate bars. How many 20-bar packages can be filled with 11,267 bars? How many bars will be left over?

42. *Reese's Peanut Butter Cups.*   H. B. Reese Candy Co. makes small, fun-size peanut butter cups. The company manufactures 23,579 cups and fills 1025 packages. How many cups are in a package? How many cups will be left over?

43. *Map Drawing.*   A map has a scale of 64 mi to the inch. How far apart *in reality* are two cities that are 6 in. apart on the map? How far apart *on the map* are two cities that, in reality, are 1728 mi apart?

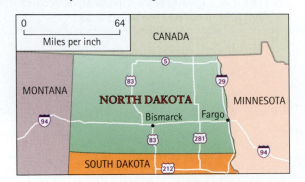

44. *Map Drawing.*   A map has a scale of 150 mi to the inch. How far apart *on the map* are two cities that, in reality, are 2400 mi apart? How far apart *in reality* are two cities that are 13 in. apart on the map?

45. *Crossword.* The *Los Angeles Times* crossword puzzle is a rectangle containing 441 squares arranged in 21 rows. How many columns does the puzzle have?

46. *Sheet of Stamps.* A sheet of 100 stamps typically has 10 rows of stamps. How many stamps are in each row?

47. Copies of this book are generally shipped from the warehouse in cartons containing 24 books each. How many cartons are needed to ship 1355 books?

48. According to the H. J. Heinz Company, 16-oz bottles of catsup are generally shipped in cartons containing 12 bottles each. How many cartons are needed to ship 528 bottles of catsup?

49. Elena buys 5 video games at $64 each and pays for them with $10 bills. How many $10 bills does it take?

50. Pedro buys 5 video games at $64 each and pays for them with $20 bills. How many $20 bills does it take?

51. You have $568 in your bank account. You use your debit card for $46, $87, and $129. Then you deposit $94 back in the account after the return of some books. How much is left in your account?

52. The balance in your bank account is $749. You use your debit card for $34 and $65. Then you make a deposit of $123 from your paycheck. What is your new balance?

Weight Loss. Many Americans exercise for weight control. It is known that one must burn off about 3500 calories in order to lose one pound. The chart shown here details how much of certain types of exercise is required to burn 100 calories. Use this chart for Exercises 53–56.

To burn off 100 calories, you must:

- Run for 8 minutes at a brisk pace, or
- Swim for 2 minutes at a brisk pace, or
- Bicycle for 15 minutes at 9 mph, or
- Do aerobic exercises for 15 minutes.

53. How long must you run at a brisk pace in order to lose one pound?

54. How long must you swim in order to lose one pound?

55. How long must you do aerobic exercises in order to lose one pound?

56. How long must you bicycle at 9 mph in order to lose one pound?

57. *Bones in the Hands and Feet.* There are 27 bones in each human hand and 26 bones in each human foot. How many bones are there in all in the hands and feet?

  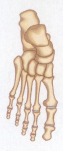 

58. *Subway Travel.* The distance to Mars is about 303,000,000 miles. The number of miles that people in the United States traveled on subways in 2003 approximately equaled 22 round trips to Mars. What was the total distance traveled on subways?

Source: American Public Transportation Association, NASA

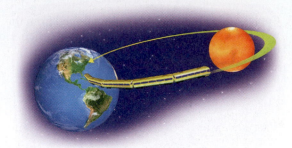

59. *Index Cards.* Index cards of dimension 3 in. by 5 in. are normally shipped in packages containing 100 cards each. How much writing area is available if one uses the front and back sides of a package of these cards?

60. An office for adjunct instructors at a community college has 6 bookshelves, each of which is 3 ft wide. The office is moved to a new location that has dimensions of 16 ft by 21 ft. Is it possible for the bookshelves to be put side by side on the 16-ft wall?

61. **D_W** In the newspaper article "When Girls Play, Knees Fail," the author discusses the fact that female athletes have six times the number of knee injuries that male athletes have. What information would be needed if you were to write a math problem based on the article? What might the problem be?

Source: *The Arizona Republic*, 2/9/00, p. C1

62. **D_W** Write a problem for a classmate to solve. Design the problem so that the solution is "The driver still has 329 mi to travel."

SKILL MAINTENANCE

Round 234,562 to the nearest: [1.4a]

63. Hundred.

64. Ten.

65. Thousand.

Estimate the computation by rounding to the nearest thousand. [1.4b]

66. 2783 + 4602 + 5797 + 8111

67. 28,430 − 11,977

68. 2100 + 5800

69. 5800 − 2100

Estimate the product by rounding to the nearest hundred. [1.5b]

70. 787 · 363

71. 887 · 799

72. 10,362 · 4531

SYNTHESIS

73. **D_W** Karen translates a problem into a multiplication equation, whereas Don translates the same problem into a division equation. Can they both be correct? Explain.

74. **D_W** Of the five problem-solving steps listed at the beginning of this section, which is the most difficult for you? Why?

75. 🖩 *Speed of Light.* Light travels about 186,000 mi/sec (miles per second) in a vacuum as in outer space. In ice it travels about 142,000 mi/sec, and in glass it travels about 109,000 mi/sec. In 18 sec, how many more miles will light travel in a vacuum than in ice? than in glass?

76. Carney Community College has 1200 students. Each professor teaches 4 classes and each student takes 5 classes. There are 30 students and 1 teacher in each classroom. How many professors are there at Carney Community College?

1.9 EXPONENTIAL NOTATION AND ORDER OF OPERATIONS

Objectives

a Write exponential notation for products such as $4 \cdot 4 \cdot 4$.

b Evaluate exponential notation.

c Simplify expressions using the rules for order of operations.

d Remove parentheses within parentheses.

a Writing Exponential Notation

Consider the product $3 \cdot 3 \cdot 3 \cdot 3$. Such products occur often enough that mathematicians have found it convenient to create a shorter notation, called **exponential notation,** for them. For example,

$\underbrace{3 \cdot 3 \cdot 3 \cdot 3}_{\text{4 factors}}$ is shortened to $3^{4} \leftarrow$ exponent

base

We read exponential notation as follows.

NOTATION	WORD DESCRIPTION
3^4	"three to the fourth power," or "the fourth power of three"
5^3	"five cubed," or "the cube of five," or "five to the third power," or "the third power of five"
7^2	"seven squared," or "the square of seven," or "seven to the second power," or "the second power of seven"

The wording "seven squared" for 7^2 comes from the fact that a square with side s has area A given by $A = s^2$.

$A = s^2$ s

An expression like $3 \cdot 5^2$ is read "three times five squared," or "three times the square of five."

EXAMPLE 1 Write exponential notation for $10 \cdot 10 \cdot 10 \cdot 10 \cdot 10$.

Exponential notation is 10^5.   5 is the *exponent.*
10 is the *base.*

EXAMPLE 2 Write exponential notation for $2 \cdot 2 \cdot 2$.

Exponential notation is 2^3.

Do Exercises 1–4.

Write exponential notation.

1. $5 \cdot 5 \cdot 5 \cdot 5$

2. $5 \cdot 5 \cdot 5 \cdot 5 \cdot 5$

3. $10 \cdot 10$

4. $10 \cdot 10 \cdot 10 \cdot 10$

Evaluate.

5. 10^4 **6.** 10^2

7. 8^3 **8.** 2^5

Answers on page A-3

Simplify.

9. $93 - 14 \cdot 3$

10. $104 \div 4 + 4$

11. $25 \cdot 26 - (56 + 10)$

12. $75 \div 5 + (83 - 14)$

Exponents are often used for units of area. For example, 10 square feet can be written 10 ft^2. Lengths are measured in units like ft, cm, and in. and areas are measured in units like ft^2, cm^2, and in^2. We can think of multiplying units like we multiply numbers. For example, the area of a square with sides of length 5 cm can be thought of as

$$\text{Area} = s^2 = (5 \text{ cm})^2$$
$$= (5 \text{ cm}) \cdot (5 \text{ cm})$$
$$= (5 \cdot 5) \cdot (\text{cm} \cdot \text{cm})$$
$$= 25 \text{ cm}^2.$$

Knowing that we must multiply two units of length to get a unit of area can serve as a check when calculating the area of a figure.

b Evaluating Exponential Notation

We evaluate exponential notation by rewriting it as a product and computing the product.

EXAMPLE 3 Evaluate: 10^3.

$$10^3 = 10 \cdot 10 \cdot 10 = 1000$$

> **Caution!**
>
> 10^3 does not mean $10 \cdot 3$.

EXAMPLE 4 Evaluate: 5^4.

$$5^4 = 5 \cdot 5 \cdot 5 \cdot 5 = 625$$

Do Exercises 5–8 on the preceding page.

c Simplifying Expressions

Suppose we have a calculation like the following:

$$3 + 4 \cdot 8.$$

How do we find the answer? Do we add 3 to 4 and then multiply by 8, or do we multiply 4 by 8 and then add 3? In the first case, the answer is 56. In the second, the answer is 35. We agree to compute as in the second case.

Consider the calculation

$$7 \cdot 14 - (12 + 18).$$

What do the parentheses mean? To deal with these questions, we must make some agreement regarding the order in which we perform operations. The rules are as follows.

RULES FOR ORDER OF OPERATIONS

1. Do all calculations within parentheses ( ), brackets [ ], or braces { } before operations outside.
2. Evaluate all exponential expressions.
3. Do all multiplications and divisions in order from left to right.
4. Do all additions and subtractions in order from left to right.

It is worth noting that these are the rules that computers and most scientific calculators use to do computations.

EXAMPLE 5 Simplify: $16 \div 8 \cdot 2$.

There are no parentheses or exponents, so we start with the third step.

$$16 \div 8 \cdot 2 = 2 \cdot 2 \quad \left.\begin{array}{l} \\ \\ \end{array}\right\} \quad \text{Doing all multiplications and}$$
$$= 4 \qquad \qquad \text{divisions in order from left to right}$$

EXAMPLE 6 Simplify: $7 \cdot 14 - (12 + 18)$.

$$7 \cdot 14 - (12 + 18) = 7 \cdot 14 - 30 \qquad \text{Carrying out operations inside parentheses}$$
$$= 98 - 30 \qquad \text{Doing all multiplications and divisions}$$
$$= 68 \qquad \text{Doing all additions and subtractions}$$

Do Exercises 9–12 on the preceding page.

EXAMPLE 7 Simplify and compare: $23 - (10 - 9)$ and $(23 - 10) - 9$.

We have

$$23 - (10 - 9) = 23 - 1 = 22;$$
$$(23 - 10) - 9 = 13 - 9 = 4.$$

We can see that $23 - (10 - 9)$ and $(23 - 10) - 9$ represent different numbers. Thus subtraction is not associative.

Do Exercises 13 and 14.

EXAMPLE 8 Simplify: $7 \cdot 2 - (12 + 0) \div 3 - (5 - 2)$.

$$7 \cdot 2 - (12 + 0) \div 3 - (5 - 2) = 7 \cdot 2 - 12 \div 3 - 3 \qquad \text{Carrying out operations inside parentheses}$$
$$= 14 - 4 - 3 \qquad \text{Doing all multiplications and divisions in order from left to right}$$
$$= 7 \qquad \text{Doing all additions and subtractions in order from left to right}$$

Do Exercise 15.

EXAMPLE 9 Simplify: $15 \div 3 \cdot 2 \div (10 - 8)$.

$$15 \div 3 \cdot 2 \div (10 - 8) = 15 \div 3 \cdot 2 \div 2 \qquad \text{Carrying out operations inside parentheses}$$
$$= 5 \cdot 2 \div 2 \quad \left.\begin{array}{l} \\ \\ \\ \end{array}\right\} \quad \text{Doing all multiplications}$$
$$= 10 \div 2 \qquad \text{and divisions in order}$$
$$= 5 \qquad \qquad \text{from left to right}$$

Do Exercises 16–18.

Simplify and compare.

13. $64 \div (32 \div 2)$ and $(64 \div 32) \div 2$

14. $(28 + 13) + 11$ and $28 + (13 + 11)$

15. Simplify:

$9 \times 4 - (20 + 4) \div 8 - (6 - 2)$.

Simplify.

16. $5 \cdot 5 \cdot 5 + 26 \cdot 71 - (16 + 25 \cdot 3)$

17. $30 \div 5 \cdot 2 + 10 \cdot 20 + 8 \cdot 8 - 23$

18. $95 - 2 \cdot 2 \cdot 2 \cdot 5 \div (24 - 4)$

Answers on page A-3

Simplify.

19. $5^3 + 26 \cdot 71 - (16 + 25 \cdot 3)$

20. $(1 + 3)^3 + 10 \cdot 20 + 8^2 - 23$

21. $81 - 3^2 \cdot 2 \div (12 - 9)$

22. Simplify: $2^3 \cdot 2^5 \div 2^6$.

EXAMPLE 10 Simplify: $4^2 \div (10 - 9 + 1)^3 \cdot 3 - 5$.

$$4^2 \div (10 - 9 + 1)^3 \cdot 3 - 5$$

$= 4^2 \div (1 + 1)^3 \cdot 3 - 5$	Subtracting inside parentheses
$= 4^2 \div 2^3 \cdot 3 - 5$	Adding inside parentheses
$= 16 \div 8 \cdot 3 - 5$	Evaluating exponential expressions
$= 2 \cdot 3 - 5$ $\Big\}$ $= 6 - 5$	Doing all multiplications and divisions in order from left to right
$= 1$	Subtracting

Do Exercises 19–21.

EXAMPLE 11 Simplify: $2^6 \div 2^4 \cdot 2^3$.

$2^6 \div 2^4 \cdot 2^3 = 64 \div 16 \cdot 8$	There are no parentheses. Evaluating exponential expressions
$= 4 \cdot 8$ $\Big\}$ $= 32$	Doing all multiplications and divisions in order from left to right

Do Exercise 22.

CALCULATOR CORNER

Order of Operations To determine whether a calculator is programmed to follow the rules for order of operations, we can enter a simple calculation that requires using those rules. For example, we enter ⎡3⎤ ⎡+⎤ ⎡4⎤ ⎡×⎤ ⎡2⎤ ⎡=⎤. If the result is 11, we know that the rules for order of operations have been followed. That is, the multiplication $4 \times 2 = 8$ was performed first and then 3 was added to produce a result of 11. If the result is 14, we know that the calculator performs operations as they are entered rather than following the rules for order of operations. That means, in this case, that 3 and 4 were added first to get 7 and then that sum was multiplied by 2 to produce the result of 14. For such calculators, we would have to enter the operations in the order in which we want them performed. In this case, we would press ⎡4⎤ ⎡×⎤ ⎡2⎤ ⎡+⎤ ⎡3⎤ ⎡=⎤.

Many calculators have parenthesis keys that can be used to enter an expression containing parentheses. To enter $5(4 + 3)$, for example, we press ⎡5⎤ ⎡(⎤ ⎡4⎤ ⎡+⎤ ⎡3⎤ ⎡)⎤ ⎡=⎤. The result is 35.

Exercises: Simplify.

1. $84 - 5 \cdot 7$

2. $80 + 50 \div 10$

3. $3^2 + 9^2 \div 3$

4. $4^4 \div 64 - 4$

5. $15 \cdot 7 - (23 + 9)$

6. $(4 + 3)^2$

Answers on page A-3

AVERAGES

In order to find the average of a set of numbers, we use addition and then division. For example, the average of 2, 3, 6, and 9 is found as follows.

$$\text{Average} = \frac{2 + 3 + 6 + 9}{4} = \frac{20}{4} = 5$$

The number of addends is 4.

Divide by 4.

The fraction bar acts as a pair of grouping symbols so

$$\frac{2 + 3 + 6 + 9}{4} \text{ is equivalent to } (2 + 3 + 6 + 9) \div 4.$$

Thus we are using order of operations when we compute an average.

> **AVERAGE**
>
> The **average** of a set of numbers is the sum of the numbers divided by the number of addends.

EXAMPLE 12 *Average Number of Career Hits.* The number of career hits of five Hall of Fame baseball players are given in the bar graph below. Find the average number of career hits of all five.

Career Hits

Player	Hits
Mel Ott	2876
Jake Beckley	2930
Dave Winfield	3110
Eddie Murray	3255
Carl Yastrzemski	3419

Sources: Associated Press; Major League Baseball

The average is given by

$$\frac{3419 + 3255 + 3110 + 2930 + 2876}{5} = \frac{15{,}590}{5} = 3118.$$

Thus the average number of career hits of these five Hall of Fame players is 3118.

Do Exercise 23.

d Removing Parentheses within Parentheses

When parentheses occur within parentheses, we can make them different shapes, such as [ ] (also called "brackets") and { } (also called "braces"). All of these have the same meaning. When parentheses occur within parentheses, computations in the innermost ones are to be done first.

---

23. World's Tallest Buildings. The heights, in feet, of the four tallest buildings in the world are given in the bar graph below. Find the average height of these buildings.

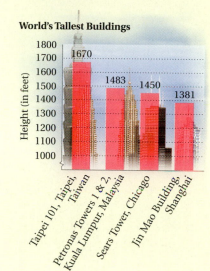

World's Tallest Buildings

Source: Council on Tall Buildings and Urban Habitat, Lehigh University, 2004

Answer on page A-3

1.9 Exponential Notation and Order of Operations

Simplify.

24. $9 \times 5 + \{6 \div [14 - (5 + 3)]\}$

25. $[18 - (2 + 7) \div 3]$
$- (31 - 10 \times 2)$

Answers on page A-3

EXAMPLE 13 Simplify: $[25 - (4 + 3) \cdot 3] \div (11 - 7)$.

$$[25 - (4 + 3) \cdot 3] \div (11 - 7)$$
$$= [25 - 7 \cdot 3] \div (11 - 7) \qquad \text{Doing the calculations in the innermost parentheses first}$$
$$= [25 - 21] \div (11 - 7) \qquad \text{Doing the multiplication in the brackets}$$
$$= 4 \div 4 \qquad \text{Subtracting}$$
$$= 1 \qquad \text{Dividing}$$

EXAMPLE 14 Simplify: $16 \div 2 + \{40 - [13 - (4 + 2)]\}$.

$$16 \div 2 + \{40 - [13 - (4 + 2)]\}$$
$$= 16 \div 2 + \{40 - [13 - 6]\} \qquad \text{Doing the calculations in the innermost parentheses first}$$
$$= 16 \div 2 + \{40 - 7\} \qquad \text{Again, doing the calculations in the innermost parentheses}$$
$$= 16 \div 2 + 33 \qquad \text{Subtracting inside the braces}$$
$$= 8 + 33 \qquad \text{Doing all multiplications and divisions in order from left to right}$$
$$= 41 \qquad \text{Adding}$$

Do Exercises 24 and 25.

Study Tips

TEST PREPARATION

You are probably ready to begin preparing for your first test. Here are some test-taking study tips.

- **Make up your own test questions as you study.**   After you have done your homework over a particular objective, write one or two questions on your own that you think might be on a test. You will be amazed at the insight this will provide.

- **Do an overall review of the chapter, focusing on the objectives and the examples.**   This should be accompanied by a study of any class notes you may have taken.

- **Do the review exercises at the end of the chapter.**   Check your answers at the back of the book. If you have trouble with an exercise, use the objective symbol as a guide to go back and do further study of that objective.

- **Call the AW Math Tutor Center at 1-888-777-0463 if you need extra help.**

- **Take the chapter test at the end of the chapter.**   Check the answers and use the objective symbols at the back of the book as a reference for where to review.

- **Ask former students for old exams.**   Working such exams can be very helpful and allows you to see what various professors think is important.

- **When taking a test, read each question carefully and try to do all the questions the first time through, but pace yourself.**   Answer all the questions, and mark those to recheck if you have time at the end. Very often, your first hunch will be correct.

- **Try to write your test in a neat and orderly manner.**   Very often, your instructor tries to give you partial credit when grading an exam. If your test paper is sloppy and disorderly, it is difficult to verify the partial credit. Doing your work neatly can ease such a task for the instructor.

EXERCISE SET

For Extra Help

a   Write exponential notation.

1. $3 \cdot 3 \cdot 3 \cdot 3$ **2.** $2 \cdot 2 \cdot 2 \cdot 2 \cdot 2$ **3.** $5 \cdot 5$ **4.** $13 \cdot 13 \cdot 13$

5. $7 \cdot 7 \cdot 7 \cdot 7 \cdot 7$ **6.** $10 \cdot 10$ **7.** $10 \cdot 10 \cdot 10$ **8.** $1 \cdot 1 \cdot 1 \cdot 1$

b   Evaluate.

9. 7^2 **10.** 5^3 **11.** 9^3 **12.** 10^2

13. 12^4 **14.** 10^5 **15.** 11^2 **16.** 6^3

c   Simplify.

17. $12 + (6 + 4)$ **18.** $(12 + 6) + 18$ **19.** $52 - (40 - 8)$

20. $(52 - 40) - 8$ **21.** $1000 \div (100 \div 10)$ **22.** $(1000 \div 100) \div 10$

23. $(256 \div 64) \div 4$ **24.** $256 \div (64 \div 4)$ **25.** $(2 + 5)^2$

26. $2^2 + 5^2$ **27.** $(11 - 8)^2 - (18 - 16)^2$ **28.** $(32 - 27)^3 + (19 + 1)^3$

29. $16 \cdot 24 + 50$ **30.** $23 + 18 \cdot 20$ **31.** $83 - 7 \cdot 6$

32. $10 \cdot 7 - 4$ **33.** $10 \cdot 10 - 3 \cdot 4$ **34.** $90 - 5 \cdot 5 \cdot 2$

35. $4^3 \div 8 - 4$

36. $8^2 - 8 \cdot 2$

37. $17 \cdot 20 - (17 + 20)$

38. $1000 \div 25 - (15 + 5)$

39. $6 \cdot 10 - 4 \cdot 10$

40. $3 \cdot 8 + 5 \cdot 8$

41. $300 \div 5 + 10$

42. $144 \div 4 - 2$

43. $3 \cdot (2 + 8)^2 - 5 \cdot (4 - 3)^2$

44. $7 \cdot (10 - 3)^2 - 2 \cdot (3 + 1)^2$

45. $4^2 + 8^2 \div 2^2$

46. $6^2 - 3^4 \div 3^3$

47. $10^3 - 10 \cdot 6 - (4 + 5 \cdot 6)$

48. $7^2 + 20 \cdot 4 - (28 + 9 \cdot 2)$

49. $6 \cdot 11 - (7 + 3) \div 5 - (6 - 4)$

50. $8 \times 9 - (12 - 8) \div 4 - (10 - 7)$

51. $120 - 3^3 \cdot 4 \div (5 \cdot 6 - 6 \cdot 4)$

52. $80 - 2^4 \cdot 15 \div (7 \cdot 5 - 45 \div 3)$

53. $2^3 \cdot 2^8 \div 2^6$

54. $2^7 \div 2^5 \cdot 2^4 \div 2^2$

55. Find the average of $64, $97, and $121.

56. Find the average of four test grades of 86, 92, 80, and 78.

57. Find the average of 320, 128, 276, and 880.

58. Find the average of $1025, $775, $2062, $942, and $3721.

d   Simplify.

59. $8 \times 13 + \{42 \div [18 - (6 + 5)]\}$

60. $72 \div 6 - \{2 \times [9 - (4 \times 2)]\}$

61. $[14 - (3 + 5) \div 2] - [18 \div (8 - 2)]$

62. $[92 \times (6 - 4) \div 8] + [7 \times (8 - 3)]$

63. $(82 - 14) \times [(10 + 45 \div 5) - (6 \cdot 6 - 5 \cdot 5)]$

64. $(18 \div 2) \cdot \{[(9 \cdot 9 - 1) \div 2] - [5 \cdot 20 - (7 \cdot 9 - 2)]\}$

65. $4 \times \{(200 - 50 \div 5) - [(35 \div 7) \cdot (35 \div 7) - 4 \times 3]\}$

66. $15(23 - 4 \cdot 2)^3 \div (3 \cdot 25)$

67. $\{[18 - 2 \cdot 6] - [40 \div (17 - 9)]\} + \{48 - 13 \times 3 + [(50 - 7 \cdot 5) + 2]\}$

68. $(19 - 2^4)^5 - (141 \div 47)^2$

69. $\mathbf{D_W}$ Consider the problem in Example 8 of Section 1.8. How can you translate the problem to a single equation involving what you have learned about order of operations? How does the single equation relate to how we solved the problem?

70. $\mathbf{D_W}$ Consider the expressions $9 - (4 \cdot 2)$ and $(3 \cdot 4)^2$. Are the parentheses necessary in each case? Explain.

SKILL MAINTENANCE

Solve. [1.7b]

71. $x + 341 = 793$

72. $4197 + x = 5032$

73. $7 \cdot x = 91$

74. $1554 = 42 \cdot y$

75. $3240 = y + 898$

76. $6000 = 1102 + t$

77. $25 \cdot t = 625$

78. $10,000 = 100 \cdot t$

Solve. [1.8a]

79. *Colorado.* The state of Colorado is roughly the shape of a rectangle that is 273 mi by 382 mi. What is its area?

80. On a four-day trip, a family bought the following amounts of gasoline for their motor home:

23 gallons,   24 gallons,
26 gallons,   25 gallons.

How much gasoline did they buy in all?

SYNTHESIS

81. $\mathbf{D_W}$ Is it possible to compute an average of several numbers on a calculator without using parentheses or the $\boxed{=}$ key?

82. $\mathbf{D_W}$ Is the average of two sets of numbers the same as the average of the two averages? Why or why not?

Each of the answers in Exercises 83–85 is incorrect. First find the correct answer. Then place as many parentheses as needed in the expression in order to make the incorrect answer correct.

83. $1 + 5 \cdot 4 + 3 = 36$

84. $12 \div 4 + 2 \cdot 3 - 2 = 2$

85. $12 \div 4 + 2 \cdot 3 - 2 = 4$

86. Use one occurrence each of 1, 2, 3, 4, 5, 6, 7, 8, and 9 and any of the symbols $+$, $-$, $\times$, $\div$, and ( ) to represent 100.

The review that follows is meant to prepare you for a chapter exam. It consists of two parts. The first part, Concept Reinforcement, is designed to increase understanding of the concepts through true/false exercises. The second part is the Review Exercises. These provide practice exercises for the exam, together with references to section objectives so you can go back and review. Before beginning, stop and look back over the skills you have obtained. What skills in mathematics do you have now that you did not have before studying this chapter?

↪ CONCEPT REINFORCEMENT

Determine whether the statement is true or false. Answers are given at the back of the book.

_____ **1.** The product of two natural numbers is always greater than either of the factors.

_____ **2.** Zero divided by any nonzero number is 0.

_____ **3.** Each member of the set of natural numbers is a member of the set of whole numbers.

_____ **4.** The sum of two natural numbers is always greater than either of the addends.

_____ **5.** Any number divided by 1 is the number 1.

_____ **6.** The number 0 is the smallest natural number.

Review Exercises

The review exercises that follow are for practice. Answers are given at the back of the book. If you miss an exercise, restudy the objective indicated in red next to the exercise or direction line that precedes it.

1. What does the digit 8 mean in 4,678,952? [1.1a]

2. In 13,768,940, what digit tells the number of millions? [1.1a]

Write expanded notation. [1.1b]

3. 2793

4. 56,078

5. 4,007,101

Write a word name. [1.1c]

6. 67,819

7. 2,781,427

8. 1,065,070,607, the population of India in 2004.
Source: U.S. Bureau of the Census, International Database

Write standard notation. [1.1c]

9. Four hundred seventy-six thousand, five hundred eighty-eight

10. *e-books.* The publishing industry predicted that sales of digital books would reach two billion, four hundred thousand by 2005.
Source: Andersen Consulting

Add. [1.2b]

11. 7304 + 6968

12. 27,609 + 38,415

13. 2703 + 4125 + 6004 + 8956

14.
$$\begin{array}{r} 9\ 1,4\ 2\ 6 \\ +\quad 7,4\ 9\ 5 \\ \hline \end{array}$$

15. Write a related addition sentence: [1.3b]
$$10 - 6 = 4.$$

16. Write two related subtraction sentences: [1.3b]
$$8 + 3 = 11.$$

Subtract. [1.3d]

17. 8045 − 2897

18. 9001 − 7312

19. 6003 − 3729

20.
$$\begin{array}{r} 3\ 7,4\ 0\ 5 \\ -\ 1\ 9,6\ 4\ 8 \\ \hline \end{array}$$

Round 345,759 to the nearest: [1.4a]

21. Hundred.   **22.** Ten.

23. Thousand.   **24.** Hundred thousand.

Estimate the sum, difference, or product by first rounding to the nearest hundred. Show your work. [1.4b], [1.5b]

25. 41,348 + 19,749   **26.** 38,652 − 24,549

27. 396 · 748

Use < or > for ☐ to write a true sentence. [1.4c]

28. 67 ☐ 56   **29.** 1 ☐ 23

Multiply. [1.5a]

30. 17,000 · 300   **31.** 7846 · 800

32. 726 · 698   **33.** 587 · 47

34.
$$\begin{array}{r} 8\ 3\ 0\ 5 \\ \times\quad 6\ 4\ 2 \\ \hline \end{array}$$

35. Write a related multiplication sentence: [1.6b]
$$56 \div 8 = 7.$$

36. Write two related division sentences: [1.6b]
$$13 \cdot 4 = 52.$$

Divide. [1.6c]

37. 63 ÷ 5   **38.** 80 ÷ 16

39. $7\overline{)6\ 3\ 9\ 4}$ **40.** 3073 ÷ 8

41. $6\ 0\overline{)2\ 8\ 6}$ **42.** 4266 ÷ 79

43. $3\ 8\overline{)1\ 7,1\ 7\ 6}$ **44.** $1\ 4\overline{)7\ 0,1\ 1\ 2}$

45. 52,668 ÷ 12

Solve. [1.7b]

46. $46 \cdot n = 368$ **47.** $47 + x = 92$

48. $1 \cdot y = 58$ **49.** $24 = x + 24$

50. Write exponential notation: $4 \cdot 4 \cdot 4$. [1.9a]

Evaluate. [1.9b]

51. 10^4 **52.** 6^2

Simplify. [1.9c, d]

53. $8 \cdot 6 + 17$

54. $10 \cdot 24 - (18 + 2) \div 4 - (9 - 7)$

55. $7 + (4 + 3)^2$

56. $7 + 4^2 + 3^2$

57. $(80 \div 16) \times [(20 - 56 \div 8) + (8 \cdot 8 - 5 \cdot 5)]$

58. Find the average of 157, 170, and 168.

Solve. [1.8a]

59. *Workstation.* Natasha has $196 and wants to buy a computer workstation for $698. How much more does she need?

Computer Workstation
Raised monitor platform, sliding keyboard shelf, and mobile CPU shelf. Locking 3-drawer file cabinet. Maple and honey finish with durable melamine work surface.

Workstation just... $698

60. Taneesha has $406 in her checking account. She is paid $78 for a part-time job and deposits that in her checking account. How much is then in her account?

61. *Lincoln-Head Pennies.* In 1909, the first Lincoln-head pennies were minted. Seventy-three years later, these pennies began to be minted with a decreased copper content. In what year was the copper content reduced?

62. A beverage company packed 228 cans of soda into 12-can cartons. How many cartons did they fill?

63. An apple farmer keeps bees in her orchard to help pollinate the apple blossoms so more apples will be produced. The bees from an average beehive can pollinate 30 surrounding trees during one growing season. A farmer has 420 trees. How many beehives does she need to pollinate them all?
Source: Jordan Orchards, Westminster, PA

64. An apartment builder bought 13 gas stoves at $425 each and 13 refrigerators at $620 each. What was the total cost?

65. A family budgets $7825 for food and clothing and $2860 for entertainment. The yearly income of the family was $38,283. How much of this income remained after these two allotments?

66. A chemist has 2753 mL of alcohol. How many 20-mL beakers can be filled? How much will be left over?

67. *Olympic Trampoline.* Shown below is an Olympic trampoline. Find the area and the perimeter of the trampoline. [1.2c], [1.5c]
Source: International Trampoline Industry Association, Inc.

├─── 14 ft ───┤

7 ft

68. **D**_W Write a problem for a classmate to solve. Design the problem so that the solution is "Each of the 144 bottles will contain 8 oz of hot sauce." [1.8a]

69. **D**_W Is subtraction associative? Why or why not? [1.3d]

SYNTHESIS

70. ▦ Determine the missing digit d. [1.5a]

$$
\begin{array}{r}
9\ d \\
\times\ \ d\ 2 \\
\hline
8\ 0\ 3\ 6
\end{array}
$$

71. ▦ Determine the missing digits a and b. [1.6c]

$$
2\ b\ 1\ \overline{)\ 2\ 3\ 6{,}4\ 2\ 1}\quad 9\ a\ 1
$$

72. A mining company estimates that a crew must tunnel 2000 ft into a mountain to reach a deposit of copper ore. Each day the crew tunnels about 500 ft. Each night about 200 ft of loose rocks roll back into the tunnel. How many days will it take the mining company to reach the copper deposit? [1.8a]

1. In the number 546,789, which digit tells the number of hundred thousands?

2. Write expanded notation: 8843.

3. Write a word name: 38,403,277.

Add.

4.
```
  6 8 1 1
+ 3 1 7 8
```

5.
```
  4 5,8 8 9
+ 1 7,9 0 2
```

6.
```
  1 2
    8
    3
    7
+   4
```

7.
```
  6 2 0 3
+ 4 3 1 2
```

Subtract.

8.
```
  7 9 8 3
- 4 3 5 3
```

9.
```
  2 9 7 4
- 1 9 3 5
```

10.
```
  8 9 0 7
- 2 0 5 9
```

11.
```
  2 3,0 6 7
- 1 7,8 9 2
```

Multiply.

12.
```
  4 5 6 8
×       9
```

13.
```
  8 8 7 6
×     6 0 0
```

14.
```
  6 5
× 3 7
```

15.
```
  6 7 8
× 7 8 8
```

Divide.

16. $15 \div 4$

17. $420 \div 6$

18. $89\overline{)8633}$

19. $44\overline{)35,428}$

Solve.

20. *Hostess Ding Dongs®.* Hostess packages its Ding Dong® snack products in 12-packs. It manufactures 22,231 cakes. How many 12-packs can it fill? How many will be left over?

21. *Largest States.* The following table lists the five largest states in terms of their land area. Find the total land area of these states.

STATE	AREA (In Square Miles)
Alaska	571,951
Texas	261,797
California	155,959
Montana	145,552
New Mexico	121,356

Source: U.S. Department of Commerce, U.S. Bureau of the Census

22. *Pool Tables.* The Hartford™ pool table made by Brunswick Billiards comes in three sizes of playing area, 50 in. by 100 in., 44 in. by 88 in., and 38 in. by 76 in.

a) Find the perimeter and the area of the playing area of each table.

b) By how much playing area does the large table exceed the small table?

Source: Brunswick Billiards

23. *Voting Early.* In the 2004 Presidential Election, 345,689 Nevada voters voted early. This was 139,359 more than in the 2000 election. How many voted early in Nevada in 2000?

Source: National Association of Secretaries of State

24. A sack of oranges weighs 27 lb. A sack of apples weighs 32 lb. Find the total weight of 16 bags of oranges and 43 bags of apples.

25. A box contains 5000 staples. How many staplers can be filled from the box if each stapler holds 250 staples?

Solve.

26. $28 + x = 74$

27. $169 \div 13 = n$

28. $38 \cdot y = 532$

29. $381 = 0 + a$

Round 34,578 to the nearest:

30. Thousand.

31. Ten.

32. Hundred.

Estimate the sum, difference, or product by first rounding to the nearest hundred. Show your work.

33.
$$\begin{array}{r} 2\ 3,6\ 4\ 9 \\ +\ 5\ 4,7\ 4\ 6 \\ \hline \end{array}$$

34.
$$\begin{array}{r} 5\ 4,7\ 5\ 1 \\ -\ 2\ 3,6\ 4\ 9 \\ \hline \end{array}$$

35.
$$\begin{array}{r} 8\ 2\ 4 \\ \times\ 4\ 8\ 9 \\ \hline \end{array}$$

Use < or > for ☐ to write a true sentence.

36. 34 ☐ 17

37. 117 ☐ 157

38. Write exponential notation: $12 \cdot 12 \cdot 12 \cdot 12$.

Evaluate.

39. 7^3

40. 10^5

41. 25^2

Simplify.

42. $35 - 1 \cdot 28 \div 4 + 3$

43. $10^2 - 2^2 \div 2$

44. $(25 - 15) \div 5$

45. $8 \times \{(20 - 11) \cdot [(12 + 48) \div 6 - (9 - 2)]\}$

46. $2^4 + 24 \div 12$

47. Find the average of 97, 98, 87, and 86.

---

48. An open cardboard container is 8 in. wide, 12 in. long, and 6 in. high. How many square inches of cardboard are used?

49. Use trials to find the single-digit number a for which
$$359 - 46 + a \div 3 \times 25 - 7^2 = 339.$$

50. Cara spends $229 a month to repay her student loan. If she has already paid $9160 on the 10-yr loan, how many payments remain?

51. Jennie scores three 90's, four 80's, and a 74 on her eight quizzes. Find her average.

Introduction to Integers and Algebraic Expressions

Real-World Application

Surface temperatures on Mars vary from −128°C during polar night to 27°C at the equator during midday at the closest point in orbit to the sun. Find the difference between the highest value and the lowest value in this temperature range.

Source: Mars Institute

This problem appears as Exercise 73 in Section 2.3.

2.1 INTEGERS AND THE NUMBER LINE

To the student:

In the Preface, at the front of the text, you will find a Student Organizer card. This pullout card will help you keep track of important dates and useful contact information. You can also use it to plan time for class, study, work, and relaxation. By managing your time wisely, you will provide yourself the best possible opportunity to be successful in this course.

Study Tips

LEARN DEFINITIONS

Take time to learn the definitions in each section. Try to go beyond memorizing the words of a definition to understanding the meaning. Asking yourself questions about a definition can aid in understanding. For example, for the word *integers*, first memorize the list of integers, and then see if you can think of some numbers that are not integers.

In this section, we extend the set of whole numbers to form the set of *integers*. You have probably already used negative numbers. For example, the outside temperature could drop to *negative five* degrees and a credit card statement could indicate activity of *negative forty-eight* dollars.

To create the set of integers, we begin with the set of whole numbers, 0, 1, 2, 3, and so on. For each number 1, 2, 3, and so on, we obtain a new number the same number of units to the left of zero on a number line.

For the number 1, there is the *opposite* number -1 (negative 1).

For the number 2, there is the *opposite* number -2 (negative 2).

For the number 3, there is the *opposite* number -3 (negative 3), and so on.

The **integers** consist of the whole numbers and these new numbers. We illustrate them on a number line as follows.

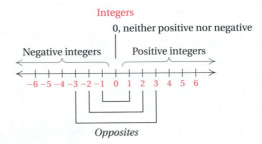

Integers

0, neither positive nor negative

Negative integers | Positive integers

$-6\ -5\ -4\ -3\ -2\ -1\ \ 0\ \ 1\ \ 2\ \ 3\ \ 4\ \ 5\ \ 6$

Opposites

The integers to the left of zero on the number line are called **negative integers** and those to the right of zero are called **positive integers.** Zero is neither positive nor negative and serves as its own opposite.

> **INTEGERS**
>
> The **integers:** $\ldots, -5, -4, -3, -2, -1, 0, 1, 2, 3, 4, 5, \ldots$

a Integers and the Real World

Integers correspond to many real-world problems and situations. The following examples will help you get ready to translate problem situations to mathematical language.

EXAMPLE 1 Tell which integer corresponds to this situation: Researcher Robert Ballard discovered the wreck of the *Titanic* 12,500 ft below sea level.

Source: Office of Naval Research

12,500 ft below sea level corresponds to the integer $-12,500$.

EXAMPLE 2 Tell which integers correspond to this situation: Elaine reversed the disc in her DVD player 17 min and then advanced it 25 min.

The integers −17 and 25 correspond to the situation. The integer −17 corresponds to the reversing and 25 corresponds to the advancing.

Do Exercises 1–5.

b Order on the Number Line

Numbers are written in order on the number line, increasing as we move to the right. For any two numbers on the line, the one to the left is *less than* the one to the right.

Since the symbol < means "is less than," the sentence −5 < 9 is read "−5 is less than 9." The symbol > means "is greater than," so the sentence −4 > −8 is read "−4 is greater than −8."

EXAMPLES Use either < or > for ☐ to form a true sentence.

3. −9 ☐ 2   Since −9 is to the left of 2, we have −9 < 2.
4. 7 ☐ −13   Since 7 is to the right of −13, we have 7 > −13.
5. −19 ☐ −6   Since −19 is to the left of −6, we have −19 < −6.

Do Exercises 6–9 on the next page.

c Absolute Value

From the number line, we see that some integers, like 5 and −5, are the same distance from zero.

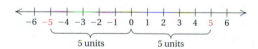

How far is 5 from 0? How far is −5 from 0? Since distance is never negative (it is "nonnegative," that is, either positive or zero), it follows that both 5 and −5 are 5 units from 0.

> ### ABSOLUTE VALUE
>
> The **absolute value** of a number is its distance from zero on a number line. We use the symbol $|x|$ to represent the absolute value of a number x.

Like distance, the absolute value of a number is never negative; it is always either positive or zero.

Tell which integers correspond to each situation.

1. The halfback gained 8 yd on first down. The quarterback was sacked for a 5-yd loss on second down.

2. **Temperature High and Low.** The highest recorded temperature in Nevada is 125°F (degrees Fahrenheit) on June 29, 1994, in Laughlin. The lowest recorded temperature in Nevada is 50°F below zero on January 8, 1937, in San Jacinto.

   Sources: National Climatic Data Center, Asheville, NC, and Storm Phillips, STORMFAX, INC.

3. **Stock Decrease.** The stock of Wendy's decreased from $41 per share to $38 per share over a recent period.

   Source: The New York Stock Exchange

4. At 10 sec (seconds) before liftoff, ignition occurs. At 148 sec after liftoff, the first stage is detached from the rocket.

5. Jacob owes $137 to the bookstore. Fortunately, he has $289 in a savings account.

Answers on page A-4

Use either < or > for ☐ to form a true sentence.

6. 13 ☐ 7   **7.** 12 ☐ −3

8. −13 ☐ −3   **9.** −4 ☐ −20

Find the absolute value.

10. |18| **11.** |−9|

12. |−29| **13.** |52|

Answers on page A-4

EXAMPLES   Find the absolute value of each number.

6. |−3| The distance of −3 from 0 is 3, so |−3| = 3.
7. |25| The distance of 25 from 0 is 25, so |25| = 25.
8. |0| The distance of 0 from 0 is 0, so |0| = 0.

> To find a number's absolute value:
> 1. If a number is positive or zero, use the number itself.
> 2. If a number is negative, make the number positive.

Do Exercises 10–13.

d   Opposites

Recall that the set of integers can be represented on a number line. Given a number on one side of 0, we can get a number on the other side by *reflecting* the number across zero. For example, the *reflection* of 2 is −2. We can read −2 as "negative 2" or "the opposite of 2."

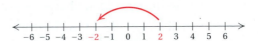

> **NOTATION FOR OPPOSITES**
>
> The **opposite** of a number x is written $-x$ (read "the opposite of x").

EXAMPLE 9   If x is −3, find $-x$.

To find the opposite of x when x is −3, we reflect −3 to the other side of 0.

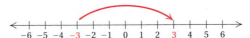

When $x = -3$, $-x = -(-3)$. We substitute −3 for x. We have $-(-3) = 3$. The opposite of −3 is 3.

EXAMPLE 10   Find $-x$ when x is 0.

When we try to reflect 0 "to the other side of 0," we go nowhere:

$-x = 0$ when x is 0.   The opposite of 0 is 0.

In Examples 9 and 10, the variable was replaced with a number. When this occurs, we say that we are **evaluating** the expression.

EXAMPLE 11   Evaluate $-x$ when x is 4.

To find the opposite of x when x is 4, we reflect 4 to the other side of 0.

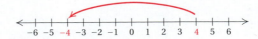

We have $-(4) = -4$. The opposite of 4 is −4.

Do Exercises 14–16.

A negative number is sometimes said to have a negative *sign*. A positive number is said to have a positive sign, even though it rarely is written in.

EXAMPLES   Determine the sign of each number.

12. -7 Negative   **13.** 23   Positive

Replacing a number with its opposite, or *additive inverse*, is sometimes called *changing the sign*.

EXAMPLES   Change the sign (find the opposite, or additive inverse) of each number.

14. -6 $-(-6) = 6$ **15.** -10 $-(-10) = 10$
16. 0 $-(0) = 0$ **17.** 14 $-(14) = -14$

Do Exercises 17–20.

EXAMPLE 18   If x is 2, find $-(-x)$.

We replace x with 2:

$\qquad -(-x)$ Read "the opposite of the opposite of x"

$\qquad = -(-2).$ We copy the expression, replacing x with 2

The opposite of the opposite of 2 is 2, or $-(-2) = 2$.

EXAMPLE 19   Evaluate $-(-x)$ for $x = -4$.

We replace x with -4:

$\qquad -(-x)$

$\qquad = -(-(-4))$ Using an extra set of parentheses to avoid notation like $--4$

$\qquad = -(\;\;4\;\;)$ Changing the sign of -4

$\qquad = -4.$ Changing the sign of 4

Thus, $-(-(-4)) = -4$.

When we change a number's sign twice, we return to the original number.

Do Exercises 21–24.

It is important not to confuse parentheses with absolute-value symbols.

EXAMPLE 20   Evaluate $-|-x|$ for $x = 2$.

We replace x with 2:

$\qquad -|-x|$

$\qquad = -|-2|$ Replacing x with 2

$\qquad = -2.$ The absolute value of -2 is 2.

Thus, $-|-2| = -2$.

Note that $-(-2) = $ 2, whereas $-|-2| = -2$.

Do Exercises 25 and 26.

In each case draw a number line, if necessary.

14. Find $-x$ when x is 1.

15. Find $-x$ when x is -2.

16. Evaluate $-x$ when x is 0.

Change the sign. (Find the opposite, or additive inverse.)

17. -4 **18.** -13

19. 39   **20.** 0

21. If x is 7, find $-(-x)$.

22. If x is 1, find $-(-x)$.

23. Evaluate $-(-x)$ for $x = -6$.

24. Evaluate $-(-x)$ for $x = -2$.

25. Find $-|-7|$.

26. Find $-|-39|$.

Answers on page A-4

a Tell which integers correspond to each situation.

1. *Pollution Fine.* In 2003, The Colonial Pipeline Company was fined a record $34 million for pollution.
Source: Green Consumer Guide.com

2. *Highest and Lowest Temperatures.* The highest temperature ever created on earth was 950,000,000°F. The lowest temperature ever created was approximately 460°F below zero.
Source: *The Guinness Book of Records,* 2004

3. The recycling program for Colchester once received $40 for a ton of office paper. More recently, they've had to pay $15 to get rid of a ton of office paper.

4. The space shuttle stood ready, 3 sec before liftoff. Solid fuel rockets were released 128 sec after liftoff.

5. At tax time, Janine received an $820 refund while David owed $541.

6. *Oceanography.* At a depth of 2438 meters researchers found the first hydrothermal vent ever seen by humans. This depth is approximately 8000 ft below sea level.
Source: Office of Naval Research

7. *Geography.* Death Valley, California, is 280 ft below sea level. Mt. Whitney, the highest point in California, has an elevation of 14,491 ft.

8. *Geography.* The Dead Sea, between Jordan and Israel, is 1286 ft below sea level; Mt. Rainier in Washington State is 14,410 ft above sea level.

b Use either < or > for ☐ to form a true sentence.

9. −8 ☐ 0

10. 7 ☐ 0

11. 9 ☐ 0

12. −7 ☐ 0

13. 8 ☐ −8

14. 6 ☐ −6

15. −6 ☐ −4

16. −1 ☐ −7

17. −8 ☐ −5

18. −5 ☐ −3

19. −13 ☐ −9

20. −5 ☐ −11

21. −3 ☐ −4

22. −6 ☐ −5

CHAPTER 2: Introduction to Integers and Algebraic Expressions

c Find the absolute value.

23. $|57|$ **24.** $|11|$ **25.** $|0|$ **26.** $|-4|$ **27.** $|-24|$

28. $|-36|$ **29.** $|53|$ **30.** $|54|$ **31.** $|-8|$ **32.** $|-79|$

d Find $-x$ when x is each of the following.

33. -7 **34.** -6 **35.** 7 **36.** 6 **37.** 0

38. -1 **39.** -19 **40.** 50 **41.** 42 **42.** -73

Change the sign. (Find the opposite, or additive inverse.)

43. -8 **44.** -7 **45.** 7 **46.** 10 **47.** -29

48. -14 **49.** -22 **50.** 0 **51.** 1 **52.** -53

Evaluate $-(-x)$ when x is each of the following.

53. 7 **54.** -8 **55.** -9 **56.** 3 **57.** -17 **58.** -19

59. 23 **60.** 0 **61.** -1 **62.** 73 **63.** 85 **64.** -37

Evaluate $-|-x|$ when x is each of the following.

65. 47 **66.** 92 **67.** 345 **68.** 729

69. 0 **70.** 1 **71.** -8 **72.** -3

73. D_W Does $-x$ always represent a negative number? Why or why not?

74. D_W Explain in your own words why $-(-x) = x$.

SKILL MAINTENANCE

75. Add: $327 + 498$. [1.2b]

76. Evaluate: 5^3. [1.9b]

77. Multiply: $209 \cdot 34$. [1.5a]

78. Solve: $300 \cdot x = 1200$. [1.7b]

79. Evaluate: 9^2. [1.9b]

80. Multiply: $31 \cdot 50$. [1.5a]

81. Simplify: $5(8 - 6)$. [1.9c]

82. Simplify: $7(9 - 3)$. [1.9c]

SYNTHESIS

83. D_W If $a > b$ is true, does it follow that $-b > -a$ is also true? Why or why not?

84. D_W Does $|x|$ always represent a positive number? Why or why not?

85. ▨ On your calculator list the sequence of keystrokes needed to find the opposite of the sum of 549 and 387.

86. ▨ On your calculator list the sequence of keystrokes needed to find the opposite of the product of 438 and 97.

Use either $<$, $>$, or $=$ for $\square$ to write a true sentence.

87. $|-5| \; \square \; |-2|$

88. $|4| \; \square \; |-7|$

89. $|-8| \; \square \; |8|$

Simplify.

90. $-|3|$

91. $-|-8|$

92. $-|-2|$

93. $-|7|$

Solve. Consider only integer replacements.

94. $|x| = 7$

95. $|x| < 2$

96. Simplify $-(-x)$, $-(-(-x))$, and $-(-(-(-x)))$.

97. List these integers in order from least to greatest.

$$2^{10}, \; -5, \; |-6|, \; 4, \; |3|, \; -100, \; 0, \; 2^7, \; 7^2, \; 10^2$$

CHAPTER 2: Introduction to Integers
and Algebraic Expressions

2.2 ADDITION OF INTEGERS

Objective

a Add integers without using a number line.

a Addition

To explain addition of integers, we can use the number line. Once our understanding is developed, we will streamline our approach.

> **ADDING INTEGERS**
>
> To perform the addition $a + b$, we start at a, and then move according to b.
>
> **a)** If b is positive, we move to the right.
> **b)** If b is negative, we move to the left.
> **c)** If b is 0, we stay at a.

EXAMPLE 1 Add: $2 + (-5)$.

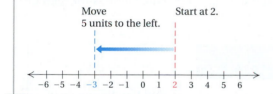

$$2 + (-5) = -3$$

EXAMPLE 2 Add: $-1 + (-3)$.

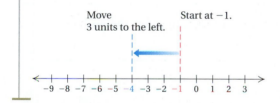

$$-1 + (-3) = -4$$

EXAMPLE 3 Add: $-4 + 9$.

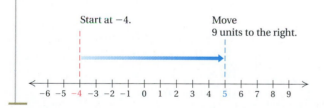

$$-4 + 9 = 5$$

Do Exercises 1–7.

You may have noticed a pattern in Example 2 and Margin Exercises 2 and 6. When two negative integers are added, the result is negative.

> **ADDING NEGATIVE INTEGERS**
>
> To add two negative integers, add their absolute values and change the sign (making the answer negative).

Add, using a number line.

1. $3 + (-4)$

2. $-3 + (-5)$

3. $-3 + 7$

4. $-5 + 5$

For each illustration, write a corresponding addition sentence.

5.

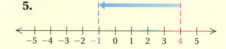

6.

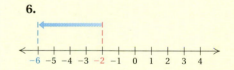

7.

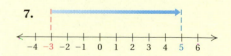

Answers on page A-4

Add. Do not use a number line except as a check.

8. $-5 + (-6)$

9. $-9 + (-3)$

10. $-20 + (-14)$

11. $-11 + (-11)$

Add.

12. $0 + (-17)$

13. $49 + 0$

14. $-56 + 0$

Add, using a number line only as a check.

15. $-4 + 6$

16. $-7 + 3$

17. $5 + (-7)$

18. $10 + (-7)$

EXAMPLES Add.

4. $-5 + (-7) = -12$ *Think*: Add the absolute values: $5 + 7 = 12$. Make the answer negative, -12.

5. $-8 + (-2) = -10$ We can visualize the number line without actually drawing it.

Do Exercises 8–11.

Note that the sum of two positive integers is positive, and the sum of two negative integers is negative.

When the number 0 is added to any number, that number remains unchanged. For this reason, the number 0 is referred to as the **additive identity.**

EXAMPLES Add.

6. $-4 + 0 = -4$ **7.** $0 + (-9) = -9$ **8.** $17 + 0 = 17$

Do Exercises 12–14.

When we add a positive integer and a negative integer, as in Examples 1 and 3, the sign of the number with the greater absolute value is the sign of the answer.

> ### ADDING POSITIVE AND NEGATIVE INTEGERS
>
> To add a positive integer and a negative integer, find the difference of their absolute values.
>
> **a)** If the negative integer has the greater absolute value, the answer is negative.
> **b)** If the positive integer has the greater absolute value, the answer is positive.

EXAMPLES Add.

9. $3 + (-5) = -2$ *Think*: The absolute values are 3 and 5. The difference is 2. Since the negative number has the larger absolute value, the answer is *negative*, -2.

10. $11 + (-8) = 3$ *Think*: The absolute values are 11 and 8. The difference is 3. The positive number has the larger absolute value, so the answer is *positive*, 3.

11. $1 + (-6) = -5$ **12.** $-7 + 4 = -3$

13. $7 + (-3) = 4$ **14.** $-6 + 10 = 4$

Do Exercises 15–18.

Sometimes $-a$ is referred to as the **additive inverse** of a. This terminology is used because adding any number to its additive inverse always results in the additive identity, 0.

$$-8 + 8 = 0, \quad 14 + (-14) = 0, \quad \text{and} \quad 0 + 0 = 0.$$

Answers on page A-4

ADDING OPPOSITES

For any integer a,

$$a + (-a) = -a + a = 0.$$

(The sum of any number and its additive inverse, or opposite, is 0.)

Do Exercises 19–22.

Suppose we wish to add several numbers, positive and negative:

$$15 + (-2) + 7 + 14 + (-5) + (-12).$$

Because of the commutative and associative laws for addition, we can group the positive numbers together and the negative numbers together and add them separately. Then we add the two results.

EXAMPLE 15 Add: $15 + (-2) + 7 + 14 + (-5) + (-12)$.

First add the positive numbers: $15 + 7 + 14 = 36$.

Then add the negative numbers: $-2 + (-5) + (-12) = -19$.

Finally, add the results: $36 + (-19) = 17$.

We can also add in any other order we wish, say, from left to right:

$$
\begin{aligned}
15 + (-2) + 7 + 14 + (-5) + (-12) &= 13 + 7 + 14 + (-5) + (-12) \\
&= 20 + 14 + (-5) + (-12) \\
&= 34 + (-5) + (-12) \\
&= 29 + (-12) \\
&= 17.
\end{aligned}
$$

Do Exercises 23–25.

Study Tips   HELP SESSIONS

Make the most of tutoring sessions by doing what you can ahead of time and knowing the topics with which you need help.

Often students find that a tutoring session would be helpful. The following comments may help you to make the most of such sessions.

■ **Work on the topics *before* you go to the help or tutoring session.**   Do not regard yourself as an empty cup that the tutor will fill with knowledge. The primary source of your ability to learn is within you. When students go to help or tutoring sessions unprepared, they waste time and, in many cases, money. Go to class, study the textbook, work exercises, and mark trouble spots. *Then* use the help and tutoring sessions to work on the trouble spots.

■ **Do not be afraid to ask questions in these sessions!**   The more you talk to your tutor, the more the tutor can help you.

■ **Try being a "tutor" yourself.**   Explaining a topic to someone else—a classmate, your instructor— is often the best way to master it.

Add, using a number line only as a check.

19. $5 + (-5)$

20. $-6 + 6$

21. $-10 + 10$

22. $89 + (-89)$

Add.

23. $(-15) + (-37) + 25 + 42 + (-59) + (-14)$

24. $42 + (-81) + (-28) + 24 + 18 + (-31)$

25. $-35 + 17 + 14 + (-27) + 31 + (-12)$

Answers on page A-4

a   Add, using a number line.

1. $-7 + 2$

2. $1 + (-5)$

3. $-9 + 5$

4. $8 + (-3)$

5. $-3 + 9$

6. $9 + (-9)$

7. $-7 + 7$

8. $-8 + (-5)$

9. $-3 + (-1)$

10. $-2 + (-9)$

11. $4 + (-9)$

12. $-4 + 13$

13. $-7 + 12$

14. $-3 + 2$

Add. Use a number line only as a check.

15. $-3 + (-9)$

16. $-3 + (-7)$

17. $-6 + (-5)$

18. $-10 + (-14)$

19. $5 + (-5)$

20. $10 + (-10)$

21. $-2 + 2$

22. $-3 + 3$

23. $0 + 6$

24. $7 + 0$

25. $13 + (-13)$

26. $-17 + 17$

27. $-25 + 0$

28. $-43 + 0$

29. $0 + (-27)$

30. $0 + (-19)$

31. $-31 + 31$

32. $12 + (-12)$

33. $-8 + 0$

34. $-11 + 0$

35. $9 + (-4)$

36. $-7 + 8$

37. $-4 + (-5)$

38. $0 + (-3)$

39. $0 + (-5)$

40. $10 + (-12)$

41. $14 + (-5)$

42. $-3 + 14$

43. $-11 + 8$

44. $0 + (-34)$

45. $-19 + 19$

46. $-10 + 3$

47. $-16 + 6$

48. $-15 + 5$

49. $-17 + (-7)$

50. $-15 + (-5)$

51. $11 + (-16)$

52. $-8 + 14$

53. $-15 + (-6)$

54. $-8 + 8$

55. $11 + (-9)$

56. $-14 + (-19)$

57. $-11 + 17$

58. $19 + (-19)$

59. $-15 + (-7) + 1$

60. $23 + (-5) + 4$

61. $30 + (-10) + 5$

62. $40 + (-8) + 5$

63. $-23 + (-9) + 15$

64. $-25 + 25 + (-9)$

65. $40 + (-40) + 6$

66. $63 + (-18) + 12$

67. $12 + (-65) + (-12)$

68. $-35 + (-63) + 35$

69. $-24 + (-37) + (-19) + (-45) + (-35)$

70. $75 + (-14) + (-17) + (-5)$

71. $28 + (-44) + 17 + 31 + (-94)$

72. $27 + (-54) + (-32) + 65 + 46$

73. $-19 + 73 + (-23) + 19 + (-73)$

74. $35 + (-51) + 29 + 51 + (-35)$

75. D_W Explain in your own words why the sum of two negative numbers is always negative.

76. D_W A student states "-45 is bigger than -21." What mistake do you think the student is making?

SKILL MAINTENANCE

Subtract. [1.3d]

77.
$$\begin{array}{r} 5\ 4\ 3 \\ -\ 2\ 1\ 9 \\ \hline \end{array}$$

78.
$$\begin{array}{r} 6\ 3\ 1\ 4 \\ -\ 2\ 6\ 8\ 9 \\ \hline \end{array}$$

79.
$$\begin{array}{r} 2\ 8\ 9\ 1 \\ -\ 1\ 4\ 0\ 7 \\ \hline \end{array}$$

80.
$$\begin{array}{r} 4\ 3,2\ 1\ 3 \\ -\ 1\ 9,8\ 7\ 6 \\ \hline \end{array}$$

81. Write in expanded notation: 39,417. [1.1b]

82. Round to the nearest hundred: 746. [1.4a]

83. Round to the nearest thousand: 32,831. [1.4a]

84. Multiply: $42 \cdot 56$. [1.5a]

85. Divide: $288 \div 9$. [1.6c]

86. Round to the nearest ten: 3496. [1.4a]

SYNTHESIS

87. D_W Without using the words "absolute value," explain how to find the sum of a positive number and a negative number.

88. D_W Why is it important to understand the associative and commutative laws when adding more than two integers at a time?

Add.

89. $-|27| + (-|-13|)$

90. $|-32| + (-|15|)$

91. ▦ $-3496 + (-2987)$

92. ▦ $497 + (-3028)$

93. ▦ $-7846 + 5978$

94. ▦ $-7623 + 4839$

95. For what numbers x is $-x$ positive?

96. For what numbers x is $-x$ negative?

Tell whether each sum is positive, negative, or zero.

97. If n is positive and m is negative, then $-n + m$ is _____ .

98. If $n = m$ and n is negative, then $-n + (-m)$ is _____ .

99. If n is negative and m is less than n, then $n + m$ is _____ .

100. If n is positive and m is greater than n, then $n + m$ is _____ .

2.3 SUBTRACTION OF INTEGERS

Objectives

a Subtract integers and simplify combinations of additions and subtractions.

b Solve applied problems involving addition and subtraction of integers.

a Subtraction

We now consider subtraction of integers. To find the difference $a - b$, we look for a number to add to b that gives us a.

> **THE DIFFERENCE $a - b$**
>
> The difference $a - b$ is the number that when added to b gives a.

For example, $45 - 17 = 28$ because $28 + 17 = 45$. Let's consider an example in which the answer is a negative number.

EXAMPLE 1 Subtract: $5 - 8$.

Think: $5 - 8$ is the number that when added to 8 gives 5. What number can we add to 8 to get 5? The number must be negative. The number is -3:

$$5 - 8 = -3.$$

That is, $5 - 8 = -3$ because $8 + (-3) = 5$.

Do Exercises 1–3.

The definition of $a - b$ above does not always provide the most efficient way to subtract. To understand a faster way to subtract, consider finding $5 - 8$ using a number line. We start at 5. Then we move 8 units to the *left* to do the subtracting. Note that this is the same as adding the opposite of 8, or -8, to 5.

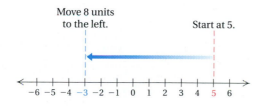

Move 8 units to the left.   Start at 5.

$$5 - 8 = -3$$

Look for a pattern in the following table.

SUBTRACTIONS	ADDING AN OPPOSITE
$5 - 8 = -3$	$5 + (-8) = -3$
$-6 - 4 = -10$	$-6 + (-4) = -10$
$-7 - (-10) = 3$	$-7 + 10 = 3$
$-7 - (-2) = -5$	$-7 + 2 = -5$

Do Exercises 4–7.

Perhaps you have noticed that we can subtract by adding the opposite of the number being subtracted. This can always be done.

Subtract.

1. $-6 - 4$

Think: What number can be added to 4 to get -6?

2. $-7 - (-10)$

Think: What number can be added to -10 to get -7?

3. $-7 - (-2)$

Think: What number can be added to -2 to get -7?

Complete the addition and compare with the subtraction.

4. $4 - 6 = -2$;
$4 + (-6) = \underline{\hspace{1.5cm}}$

5. $-3 - 8 = -11$;
$-3 + (-8) = \underline{\hspace{1.5cm}}$

6. $-5 - (-9) = 4$;
$-5 + 9 = \underline{\hspace{1.5cm}}$

7. $-5 - (-3) = -2$;
$-5 + 3 = \underline{\hspace{1.5cm}}$

Answers on page A-4

Equate each subtraction with a corresponding addition. Then write the equation in words.

8. $3 - 10$

9. $13 - 5$

10. $-12 - (-9)$

11. $-12 - 10$

12. $-14 - (-14)$

Subtract.

13. $7 - 11$

14. $-6 - 10$

15. $13 - 8$

16. $-7 - (-9)$

17. $-8 - (-2)$

18. $5 - (-8)$

Answers on page A-4

SUBTRACTING BY ADDING THE OPPOSITE

To subtract, add the opposite, or additive inverse, of the number being subtracted:

$$a - b = a + (-b).$$

This is the method generally used for quick subtraction of integers.

EXAMPLES Equate each subtraction with a corresponding addition. Then write the equation in words.

2. $-12 - 30$;

$-12 - 30 = -12 + (-30)$ Adding the opposite of 30

Negative twelve minus thirty is negative twelve plus negative thirty.

3. $-20 - (-17)$;

$-20 - (-17) = -20 + 17$ Adding the opposite of -17

Negative twenty minus negative seventeen is negative twenty plus seventeen.

Do Exercises 8–12.

Once the subtraction has been rewritten as addition, we add as in Section 2.2.

EXAMPLES Subtract.

4. $2 - 6 = 2 + (-6)$ The opposite of 6 is -6. We change the subtraction to addition and add the opposite. Instead of subtracting 6, we add -6.

$\quad\quad\quad = -4$

5. $4 - (-9) = 4 + 9$ The opposite of -9 is 9. We change the subtraction to addition and add the opposite. Instead of subtracting -9, we add 9.

$\quad\quad\quad = 13$

6. $-4 - 8 = -4 + (-8)$ We change the subtraction to addition and add the opposite. Instead of subtracting 8, we add -8.

$\quad\quad\quad = -12$

7. $10 - 7 = 10 + (-7)$ We change the subtraction to addition and add the opposite. Instead of subtracting 7, we add -7.

$\quad\quad\quad = 3$

8. $-4 - (-9) = -4 + 9$ Instead of subtracting -9, we add 9.

$\quad\quad\quad = 5$ To check, note that $5 + (-9) = -4$.

9. $-7 - (-3) = -7 + 3$ Instead of subtracting -3, we add 3.

$\quad\quad\quad = -4$ *Check*: $-4 + (-3) = -7$.

Do Exercises 13–18.

When several additions and subtractions occur together, we can make them all additions. The commutative law for addition can then be used.

EXAMPLE 10 Simplify: $-3 - (-5) - 9 + 4 - (-6)$.

$$-3 - (-5) - 9 + 4 - (-6) = -3 + 5 + (-9) + 4 + 6 \qquad \text{Adding opposites}$$
$$= -3 + (-9) + 5 + 4 + 6 \qquad \text{Using a}$$
$$\text{commutative law}$$
$$= -12 + 15$$
$$= 3$$

Do Exercises 19 and 20.

b Applications and Problem Solving

We need addition and subtraction of integers to solve a variety of applied problems.

EXAMPLE 11 *Toll Roads.* The E-Z Pass program allows drivers in the Northeast to travel certain toll roads without having to stop to pay. Instead, a transponder attached to the vehicle is scanned as the vehicle rolls through a toll booth. Recently the Ramones began a trip to New York City with a balance of $12 in their E-Z Pass account. Their trip accumulated $15 in tolls, and because they overspent their balance, the Ramones had to pay $80 in fines and administrative fees. By how much were the Ramones in debt as a result of their travel on the toll roads?

Source: State of New Jersey

We solve by first subtracting the cost of the tolls from the original balance in the account. Then we subtract the cost of the fees and fines from the new balance in the account:

$$12 - 15 = 12 + (-15) \qquad \text{Adding the opposite of 15}$$
$$= -3,$$

and

$$-3 - 80 = -3 + (-80) \qquad \text{Adding the opposite of 80}$$
$$= -83.$$

The Ramones were $83 in debt as a result of their travel on toll roads.

Do Exercises 21 and 22.

Simplify.

19. $-6 - (-2) - (-4) - 12 + 3$

20. $9 - (-6) + 7 - 9 - 8 - (-20)$

21. E-Z Pass. (See Example 11.) Suppose the Ramones had a balance of $11 in their account, accumulated $25 in tolls, and had to pay $85 in fines and administrative fees. By how much would the Ramones be in debt?

22. Temperature Extremes. In Churchill, Manitoba, Canada, the average daily low temperature in January is $-31°C$ (degrees Celsius). The average daily low temperature in Key West, Florida, is 50° warmer. What is the average daily low temperature in Key West, Florida?

Answers on page A-4

a   Subtract.

1. $2 - 7$

2. $3 - 8$

3. $0 - 8$

4. $0 - 9$

5. $-7 - (-4)$

6. $-6 - (-8)$

7. $-11 - (-11)$

8. $-6 - (-6)$

9. $13 - 17$

10. $14 - 19$

11. $20 - 27$

12. $30 - 4$

13. $-9 - (-4)$

14. $-7 - (-9)$

15. $-40 - (-40)$

16. $-9 - (-9)$

17. $7 - 7$

18. $9 - 9$

19. $7 - (-7)$

20. $4 - (-4)$

21. $8 - (-3)$

22. $-7 - 4$

23. $-6 - 8$

24. $6 - (-10)$

25. $-3 - (-9)$

26. $-14 - 2$

27. $1 - 9$

28. $2 - 8$

29. $-6 - (-5)$

30. $-4 - (-3)$

31. $8 - (-10)$

32. $5 - (-6)$

33. $0 - 10$

34. $0 - 23$

35. $-5 - (-2)$

36. $-3 - (-1)$

37. $-7 - 14$

38. $-9 - 16$

39. $0 - (-5)$

40. $0 - (-1)$

41. $-8 - 0$

42. $-9 - 0$

43. $7 - (-5)$

44. $7 - (-4)$

CHAPTER 2: Introduction to Integers and Algebraic Expressions

45. $6 - 25$

46. $18 - 63$

47. $-42 - 26$

48. $-18 - 63$

49. $-72 - 9$

50. $-49 - 3$

51. $24 - (-92)$

52. $48 - (-73)$

53. $-50 - (-50)$

54. $-70 - (-70)$

55. $-30 - (-85)$

56. $-25 - (-15)$

Simplify.

57. $7 - (-5) + 4 - (-3)$

58. $-5 - (-8) + 3 - (-7)$

59. $-31 + (-28) - (-14) - 17$

60. $-43 - (-19) - (-21) + 25$

61. $-34 - 28 + (-33) - 44$

62. $39 + (-88) - 29 - (-83)$

63. $-93 - (-84) - 41 - (-56)$

64. $84 + (-99) + 44 - (-18) - 43$

65. $-5 - (-30) + 30 + 40 - (-12)$

66. $14 - (-50) + 20 - (-32)$

67. $132 - (-21) + 45 - (-21)$

68. $81 - (-20) - 14 - (-50) + 53$

b   Solve.

69. *Reading.*   Before falling asleep, Alicia read from the top of page 37 to the top of page 62 of her book. How many pages did she read?

70. *Writing.*   During a weekend retreat, James wrote from the bottom of page 29 to the bottom of page 37 of his memoirs. How many pages did he write?

71. Through exercise, Rod went from 8 lb above his "ideal" body weight to 9 lb below it. How many pounds did Rod lose?

72. Laura has a charge of $476.89 on her credit card, but she then returns a sweater that cost $128.95. How much does she now owe on her credit card?

73. *Surface Temperatures on Mars.* Surface temperatures on Mars vary from −128°C during polar night to 27°C at the equator during midday at the closest point in orbit to the sun. Find the difference between the highest value and the lowest value in this temperature range.

Source: Mars Institute

74. Carla is completing the production work on a track that is to appear on her band's upcoming CD. In doing so, she resets the digital recorder to 0, advances the recording 16 sec, and then reverses the recording 25 sec. What reading will the recorder then display?

75. While recording a 60-minute television show, the reading on Kate's VCR changes from −21 min to 29 min. How many minutes have been recorded? Has she recorded the entire show?

76. As a result of coaching, Cedric's average golf score improved from 3 over par to 2 under. By how many strokes did his score change?

77. *Temperature Changes.* One day the temperature in Lawrence, Kansas, is 32° at 6:00 A.M. It rises 15° by noon, but falls 50° by midnight when a cold front moves in. What is the final temperature?

78. Midway through a movie, Lisa resets the counter on her DVD player to 0. She then reverses the disc 8 min, and then advances the movie 11 min. What does the counter now read?

79. *Profit.* Teapots and Treasures lost $5000 in 2004. In 2005, the store made a profit of $8000. How much more did the store make in 2005 than in 2004?

80. *Tallest Mountain.* The tallest mountain in the world, when measured from base to peak, is Mauna Kea (White Mountain) in Hawaii. From its base 19,684 ft below sea level in the Hawaiian Trough, it rises 33,480 ft. What is the elevation of the peak?

Source: *The Guinness Book of Records*

81. *Offshore Oil.* In 1998, the elevation of the world's deepwater drilling record was −7718 ft. In 2005, the deepwater drilling record was 2293 ft deeper. What was the elevation of the deepwater drilling record in 2005?

Source: www.deepwater.com/FactsandFirsts.cfm

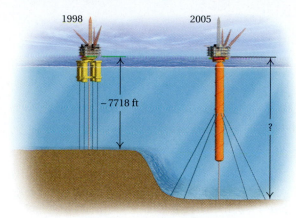

82. *Oceanography.* The deepest point in the Pacific Ocean is the Marianas Trench, with a depth of 11,033 m. The deepest point in the Atlantic Ocean is the Puerto Rico Trench, with a depth of 8648 m. What is the difference in the elevation of the two trenches?

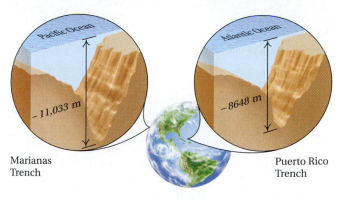

83. *Toll Roads.* The Murrays began a trip with $13 in their E-Z Pass account (see Example 11). They accumulated $20 in tolls and had to pay $80 in fines and administrative fees. By how much were the Murrays in debt as a result of their travel on toll roads?

84. *Toll Roads.* Suppose the Murrays (see Exercise 83) incurred $25 in tolls and $85 in fines and administrative fees. By how much would the Murrays be in debt?

85. **Dw** Write a subtraction problem for a classmate to solve. Design the problem so that the solution is "Clara ends up $15 in debt."

86. **Dw** If a negative number is subtracted from a positive number, will the result always be positive? Why or why not?

SKILL MAINTENANCE

Evaluate.

87. 4^3 [1.9b]

88. $68 \cdot 72$ [1.5a]

89. 1^7 [1.9b]

90. $143 \cdot 29$ [1.5a]

91. How many 12-oz cans of soda can be filled with 96 oz of soda? [1.8a]

92. A case of soda contains 24 bottles. If each bottle contains 12 oz, how many ounces of soda are in the case? [1.8a]

Simplify.

93. $5 + 4^2 + 2 \cdot 7$ [1.9c]

94. $45 \div (2^2 + 11)$ [1.9c]

95. $(9 + 7)(9 - 7)$ [1.9c]

96. $(13 - 2)(13 + 2)$ [1.9c]

97. D_W Explain why the commutative law was used in Example 10.

98. D_W Is subtraction of integers associative? Why or why not?

Subtract.

99. 🔢 $123{,}907 - 433{,}789$

100. 🔢 $23{,}011 - (-60{,}432)$

For Exercises 101–106, tell whether each statement is true or false for all integers a and b. If false, show why.

101. $a - 0 = 0 - a$

102. $0 - a = a$

103. If $a \neq b$, then $a - b \neq 0$.

104. If $a = -b$, then $a + b = 0$.

105. If $a + b = 0$, then a and b are opposites.

106. If $a - b = 0$, then $a = -b$.

107. If $a - 54$ is -37, find the value of a.

108. If $x - 48$ is -15, find the value of x.

109. Doreen is a stockbroker. She kept track of the weekly changes in the stock market over a period of 5 weeks. By how many points (pts) had the market risen or fallen over this time?

WEEK 1	WEEK 2	WEEK 3	WEEK 4	WEEK 5
Down 13 pts	Down 16 pts	Up 36 pts	Down 11 pts	Up 19 pts

110. *Blackjack Counting System.* The casino game of blackjack makes use of many card-counting systems to give players an advantage if the count becomes negative. One such system is called *High–Low*, first developed by Harvey Dubner in 1963. Each card counts as -1, 0, or 1 as follows:

2, 3, 4, 5, 6 count as $+1$;

7, 8, 9 count as 0;

10, J, Q, K, A count as -1.

Source: Patterson, Jerry L. *Casino Gambling.* New York: Perigee, 1982

a) Find the total count on the sequence of cards

K, A, 2, 4, 5, 10, J, 8, Q, K, 5.

b) Does the player have a winning edge?

2.4 MULTIPLICATION OF INTEGERS

Objectives

a Multiply integers.

b Find products of three or more integers and simplify powers of integers.

a Multiplication

Multiplication of integers is like multiplication of whole numbers. The difference is that we must determine whether the answer is positive or negative.

MULTIPLICATION OF A POSITIVE INTEGER AND A NEGATIVE INTEGER

To see how to multiply a positive integer and a negative integer, consider the following pattern.

This number decreases by 1 each time.

$$
\begin{aligned}
4 \cdot 5 &= 20 \\
3 \cdot 5 &= 15 \\
2 \cdot 5 &= 10 \\
1 \cdot 5 &= 5 \\
0 \cdot 5 &= 0 \\
-1 \cdot 5 &= -5 \\
-2 \cdot 5 &= -10 \\
-3 \cdot 5 &= -15
\end{aligned}
$$

This number decreases by 5 each time.

Do Exercise 1.

According to this pattern, it looks as though the product of a negative integer and a positive integer is negative. To confirm this, use repeated addition:

$$-1 \cdot 5 = 5 \cdot (-1) = -1 + (-1) + (-1) + (-1) + (-1) = -5$$
$$-2 \cdot 5 = 5 \cdot (-2) = -2 + (-2) + (-2) + (-2) + (-2) = -10$$
$$-3 \cdot 5 = 5 \cdot (-3) = -3 + (-3) + (-3) + (-3) + (-3) = -15$$

> **MULTIPLYING A POSITIVE AND A NEGATIVE INTEGER**
>
> To multiply a positive integer and a negative integer, multiply their absolute values and make the answer negative.

EXAMPLES Multiply.

1. $8(-5) = -40$ **2.** $50(-1) = -50$ **3.** $-7 \cdot 6 = -42$

Do Exercises 2–4.

MULTIPLICATION OF TWO NEGATIVE INTEGERS

How do we multiply two negative integers? Again we look for a pattern.

This number decreases by 1 each time.

$$
\begin{aligned}
4 \cdot (-5) &= -20 \\
3 \cdot (-5) &= -15 \\
2 \cdot (-5) &= -10 \\
1 \cdot (-5) &= -5 \\
0 \cdot (-5) &= 0 \\
-1 \cdot (-5) &= 5 \\
-2 \cdot (-5) &= 10 \\
-3 \cdot (-5) &= 15
\end{aligned}
$$

This number increases by 5 each time.

Do Exercise 5.

---

1. Complete, as in the example.

$$
\begin{aligned}
4 \cdot 10 &= 40 \\
3 \cdot 10 &= 30 \\
2 \cdot 10 &= \\
1 \cdot 10 &= \\
0 \cdot 10 &= \\
-1 \cdot 10 &= \\
-2 \cdot 10 &= \\
-3 \cdot 10 &=
\end{aligned}
$$

Multiply.

2. $-3 \cdot 6$

3. $20 \cdot (-5)$

4. $9(-1)$

5. Complete, as in the example.

$$
\begin{aligned}
3 \cdot (-10) &= -30 \\
2 \cdot (-10) &= -20 \\
1 \cdot (-10) &= \\
0 \cdot (-10) &= \\
-1 \cdot (-10) &= \\
-2 \cdot (-10) &= \\
-3 \cdot (-10) &=
\end{aligned}
$$

Answers on page A-5

Multiply.

6. $(-3)(-4)$

7. $-9(-5)$

8. $(-1)(-6)$

Multiply.

9. $0 \cdot (-5)$

10. $-23 \cdot 0$

Answers on page A-5

According to the pattern, the product of two negative integers is positive. This leads to the second part of the rule for multiplying integers.

> **MULTIPLYING TWO NEGATIVE INTEGERS**
>
> To multiply two negative integers, multiply their absolute values. The answer is positive.

EXAMPLES Multiply.

4. $(-2)(-4) = 8$

5. $(-10)(-7) = 70$

6. $(-9)(-1) = 9$

Do Exercises 6–8.

The following is another way to state the rules for multiplication.

> To multiply two integers:
>
> **a)** Multiply the absolute values.
> **b)** If the signs are the same, the answer is positive.
> **c)** If the signs are different, the answer is negative.

MULTIPLICATION BY ZERO

No matter how many times 0 is added to itself, the answer is 0. This leads to the following result.

> For any integer a,
>
> $$a \cdot 0 = 0.$$
>
> (The product of 0 and any integer is 0.)

EXAMPLES Multiply.

7. $-19 \cdot 0 = 0$

8. $0(-7) = 0$

Do Exercises 9 and 10.

b Multiplication of More Than Two Integers

Because of the commutative and the associative laws, to multiply three or more integers, we can group as we please.

EXAMPLES Multiply.

9. a) $-8 \cdot 2(-3) = -16(-3)$ Multiplying the first two numbers

$\qquad\qquad\qquad = 48$ Multiplying the results

b) $-8 \cdot 2(-3) = 24 \cdot 2$ Multiplying the negatives

$\qquad\qquad\qquad = 48$ The result is the same as above.

10. $7(-1)(-4)(-2) = (-7)8$ Multiplying the first two numbers and the last two numbers

$$= -56$$

11. a) $-5 \cdot (-2) \cdot (-3) \cdot (-6) = 10 \cdot 18$ Each pair of negatives gives a positive product.

$$= 180$$

   b) $-5 \cdot (-2) \cdot (-3) \cdot (-6) \cdot (-1) = 10 \cdot 18 \cdot (-1)$ Making use of Example 11(a)

$$= -180$$

We can see the following pattern in the results of Examples 9–11.

> The product of an even number of negative integers is positive.
> The product of an odd number of negative integers is negative.

Do Exercises 11–13.

POWERS OF INTEGERS

A positive number raised to any power is positive. When a negative number is raised to a power, the sign of the result depends upon whether the exponent is even or odd.

EXAMPLES Simplify.

12. $(-7)^2 = (-7)(-7) = 49$ The result is positive.

13. $(-4)^3 = (-4)(-4)(-4)$

$$= 16(-4)$$

$$= -64$$ The result is negative.

14. $(-3)^4 = (-3)(-3)(-3)(-3)$

$$= 9 \cdot 9$$

$$= 81$$ The result is positive.

15. $(-2)^5 = (-2)(-2)(-2)(-2)(-2)$

$$= 4 \cdot 4 \cdot (-2)$$

$$= 16(-2)$$

$$= -32$$ The result is negative.

Perhaps you noted the following.

> When a negative number is raised to an even exponent, the result is positive.
> When a negative number is raised to an odd exponent, the result is negative.

Do Exercises 14–17.

When an integer is multiplied by -1, the result is the opposite of that integer.

> For any integer a,
>
> $$-1 \cdot a = -a$$

Multiply.

11. $-2 \cdot (-5) \cdot (-4) \cdot (-3)$

12. $(-4)(-5)(-2)(-3)(-1)$

13. $(-1)(-1)(-2)(-3)(-1)(-1)$

Simplify.

14. $(-2)^3$

15. $(-9)^2$

16. $(-1)^9$

17. 2^5

Answers on page A-5

18. Simplify: -5^2.

EXAMPLE 16 Simplify: -7^2.

Since -7^2 lacks parentheses, the base is 7, not -7. Thus we regard -7^2 as $-1 \cdot 7^2$:

$$-7^2 = -1 \cdot 7^2$$
$$= -1 \cdot 7 \cdot 7 \qquad \text{The rules for order of operations tell us to square first.}$$
$$= -1 \cdot 49$$
$$= -49.$$

Compare Examples 12 and 16 and note that $(-7)^2 \neq -7^2$. In fact, the expressions $(-7)^2$ and -7^2 are not even read the same way: $(-7)^2$ is read "negative seven squared," whereas -7^2 is read "the opposite of seven squared."

Do Exercises 18–20.

19. Simplify: $(-5)^2$.

CALCULATOR CORNER

Exponential Notation When using a calculator to calculate expressions like $(-39)^4$ or -39^4, it is important to use the correct sequence of keystrokes.

Calculators with $\boxed{+/-}$ **key:** On some calculators, a $\boxed{+/-}$ key must be pressed after a number is entered to make the number negative. For these calculators, appropriate keystrokes for $(-39)^4$ are

$$\boxed{3}\ \boxed{9}\ \boxed{+/-}\ \boxed{x^y}\ \boxed{4}\ \boxed{=}.$$

To calculate -39^4, we must first raise 39 to the power 4. Then the sign of the result must be changed. This can be done with the keystrokes

$$\boxed{3}\ \boxed{9}\ \boxed{x^y}\ \boxed{4}\ \boxed{=}\ \boxed{+/-}$$

or by multiplying 39^4 by -1:

$$\boxed{3}\ \boxed{9}\ \boxed{x^y}\ \boxed{4}\ \boxed{=}\ \boxed{\times}\ \boxed{1}\ \boxed{+/-}\ \boxed{=}.$$

Calculators with $\boxed{(-)}$ **key:** On some calculators, the $\boxed{(-)}$ key is pressed before a number to indicate that the number is negative. This is similar to the way the expression is written on paper. For these calculators, $(-39)^4$ is found by pressing

$$\boxed{(}\ \boxed{(-)}\ \boxed{3}\ \boxed{9}\ \boxed{)}\ \boxed{\wedge}\ \boxed{4}\ \boxed{\text{ENTER} =}$$

20. Write $(-8)^2$ and -8^2 in words.

and -39^4 is found by pressing

$$\boxed{(-)}\ \boxed{3}\ \boxed{9}\ \boxed{\wedge}\ \boxed{4}\ \boxed{\text{ENTER} =}.$$

You can either experiment or consult a user's manual if you are unsure of the proper keystrokes for your calculator.

Exercises: Use a calculator to determine each of the following.

1. $(-23)^6$ 5. -9^6

2. $(-17)^5$ 6. -7^6

3. $(-104)^3$ 7. -6^5

4. $(-4)^{10}$ 8. -3^9

Answers on page A-5

2.4 EXERCISE SET

a Multiply.

1. $-2 \cdot 8$

2. $-7 \cdot 3$

3. $-9 \cdot 2$

4. $-7 \cdot 7$

5. $8 \cdot (-6)$

6. $8 \cdot (-3)$

7. $-10 \cdot 3$

8. $-9 \cdot 8$

9. $-3(-5)$

10. $-8 \cdot (-2)$

11. $-9 \cdot (-2)$

12. $(-8)(-9)$

13. $(-6)(-7)$

14. $-8 \cdot (-3)$

15. $-10(-2)$

16. $-9(-8)$

17. $12(-10)$

18. $15(-8)$

19. $-6(-50)$

20. $-25(-8)$

21. $(-72)(-1)$

22. $41(-3)$

23. $(-20)17$

24. $(-1)43$

25. $-47 \cdot 0$

26. $-17 \cdot 0$

27. $0(-14)$

28. $0(-38)$

b Multiply.

29. $3 \cdot (-8) \cdot (-1)$

30. $(-7) \cdot (-4) \cdot (-1)$

31. $7(-4)(-3)5$

32. $9(-2)(-6)7$

33. $-2(-5)(-7)$

34. $(-2)(-5)(-3)(-5)$

35. $(-5)(-2)(-3)(-1)$

36. $-6(-5)(-9)$

37. $(-15)(-29)0 \cdot 8$

38. $19(-7)(-8)0 \cdot 6$

39. $(-7)(-1)(7)(-6)$

40. $(-5)6(-4)5$

Simplify.

41. $(-6)^2$ **42.** $(-8)^2$ **43.** $(-5)^3$ **44.** $(-2)^4$

45. $(-10)^4$ **46.** $(-1)^5$ **47.** -2^4 **48.** $(-2)^6$

49. $(-3)^5$ **50.** -10^4 **51.** $(-1)^{12}$ **52.** $(-1)^{13}$

53. -3^6 **54.** -2^6 **55.** -4^3 **56.** -2^5

Write each of the following expressions in words.

57. -8^4 **58.** $(-6)^8$ **59.** $(-9)^{10}$ **60.** -5^4

61. **D$_W$** Explain in your own words why $(-9)^{10}$ is positive.

62. **D$_W$** Explain in your own words why -9^{10} is negative.

63. Round 532,451 to the nearest hundred.   [1.4a]

64. Write standard notation for sixty million.   [1.1c]

65. Divide: $2880 \div 36$.   [1.6c]

66. Multiply: 75×34.   [1.5a]

67. Simplify: $10 - 2^3 + 6 \div 2$.   [1.9c]

68. Simplify: $2 \cdot 5^2 - 3 \cdot 2^3 \div (3 + 3^2)$.   [1.9c]

69. A rectangular rug measures 5 ft by 8 ft. What is the area of the rug?   [1.5c], [1.8a]

70. How many 12-egg cartons can be filled with 2880 eggs?   [1.8a]

71. A ferry can accommodate 12 cars and 53 cars are waiting. How many trips will be required to ferry them all?   [1.8a]

72. An elevator can hold 16 people and 50 people are waiting to go up. How many trips will be required to transport all of them?   [1.8a]

73. **D$_W$** Which number is larger, $(-3)^{79}$ or $(-5)^{79}$? Why?

74. **D$_W$** Describe all conditions for which a^x is negative.

Simplify.

75. $(-3)^5(-1)^{379}$ **76.** $(-2)^3 \cdot [(-1)^{29}]^{46}$

77. $-9^4 + (-9)^4$ **78.** $-5^2(-1)^{29}$

79. $|(-2)^5 + 3^2| - (3 - 7)^2$

80. $|-12(-3)^2 - 5^3 - 6^2 - (-5)^2|$

81. 🖩 -47^2 **82.** 🖩 -53^2 **83.** 🖩 $(-19)^4$ **84.** 🖩 $(-23)^4$

85. 🖩 $(73 - 86)^3$ **86.** 🖩 $(-49 + 34)^3$ **87.** 🖩 $-935(238 - 243)^3$ **88.** 🖩 $(-17)^4(129 - 133)^5$

89. Jo wrote seven checks for $13 each. If she had a balance of $68 in her account, what was her balance after writing the checks?

90. After diving 95 m below the surface, a diver rises at a rate of 7 meters per minute for 9 min. What is the diver's new elevation?

91. What must be true of m and n if $[(-5)^m]^n$ is to be **(a)** negative? **(b)** positive?

92. What must be true of m and n if $-mn$ is to be **(a)** positive? **(b)** zero? **(c)** negative?

2.5 DIVISION OF INTEGERS AND ORDER OF OPERATIONS

Objectives

a Divide integers.

b Use the rules for order of operations with integers.

We now consider division of integers. Because of the way in which division is defined, its rules are similar to those for multiplication.

a Division of Integers

> ### THE QUOTIENT a/b
>
> The quotient $\dfrac{a}{b}$ (or $a \div b$, or a/b) is the number, if there is one, that when multiplied by b gives a.

Let's use the definition to divide integers.

EXAMPLES Divide, if possible. Check each answer.

1. $14 \div (-7) = -2$ *Think*: What number multiplied by -7 gives 14? The number is -2. *Check*: $(-2)(-7) = 14$.

2. $\dfrac{-32}{-4} = 8$ *Think*: What number multiplied by -4 gives -32? The number is 8. *Check*: $8(-4) = -32$.

3. $-21 \div 7 = -3$ *Think*: What number multiplied by 7 gives -21? The number is -3. *Check*: $(-3) \cdot 7 = -21$.

4. $\dfrac{0}{-5} = 0$ *Think*: What number multiplied by -5 gives 0? The number is 0. *Check*: $0(-5) = 0$.

5. $\dfrac{-5}{0}$ is **not defined.** *Think*: What number multiplied by 0 gives -5? There is no such number because the product of 0 and *any* number is 0.

The rules for determining the sign of a quotient are the same as for determining the sign of a product. We state them together.

> To multiply or divide two integers:
>
> **a)** Multiply or divide the absolute values.
> **b)** If the signs are the same, the answer is positive.
> **c)** If the signs are different, the answer is negative.

Do Exercises 1–6.

Recall that, in general, $a \div b$ and $b \div a$ are different numbers. In Example 4, we divided *into* 0. In Example 5, we attempted to divide *by* 0. Since any number times 0 gives 0, not -5, we say that $-5 \div 0$ is **not defined** or is **undefined.** Also, since *any* number times 0 gives 0, $0 \div 0$ is also not defined.

Divide.

1. $6 \div (-2)$

Think: What number multiplied by -2 gives 6?

2. $\dfrac{-15}{-3}$

Think: What number multiplied by -3 gives -15?

3. $-24 \div 8$

Think: What number multiplied by 8 gives -24?

4. $\dfrac{0}{-4}$

5. $\dfrac{30}{-5}$

6. $\dfrac{-45}{9}$

Answers on page A-5

Divide, if possible.

7. $34 \div 0$

8. $0 \div (-4)$

9. $-52 \div 0$

Answers on page A-5

124

> **EXCLUDING DIVISION BY 0**
>
> Division by 0 is not defined:
>
> $$a \div 0, \text{ or } \frac{a}{0}, \text{ is undefined for all real numbers } a.$$

DIVIDING 0 BY OTHER NUMBERS

Note that $0 \div 8 = 0$ because $0 = 0 \cdot 8$.

> **DIVIDENDS OF 0**
>
> Zero divided by any nonzero real number is 0:
>
> $$\frac{0}{a} = 0, \qquad a \neq 0.$$

EXAMPLES Divide.

6. $0 \div (-6) = 0$ **7.** $\dfrac{0}{12} = 0$ **8.** $\dfrac{-3}{0}$ is undefined.

Do Exercises 7–9.

b   Rules for Order of Operations

When more than one operation appears in a calculation or problem, we apply the same rules that were used in Section 1.9. We repeat them here for review, now including absolute-value symbols.

> **RULES FOR ORDER OF OPERATIONS**
>
> 1. Do all calculations within parentheses, brackets, braces, absolute-value symbols, numerators, or denominators.
> 2. Evaluate all exponential expressions.
> 3. Do all multiplications and divisions in order from left to right.
> 4. Do all additions and subtractions in order from left to right.

EXAMPLES Simplify.

9. $17 - 10 \div 2 \cdot 4$

With no grouping symbols or exponents, we begin with the third rule.

$$\begin{aligned}
17 - 10 \div 2 \cdot 4 &= 17 - 5 \cdot 4 \\
&= 17 - 20 \\
&= -3
\end{aligned}$$

Carrying out all multiplications and divisions in order from left to right

10. $|(-2)^3 \div 4| - 5(-2)$

We first simplify within the absolute-value symbols.

$$\begin{aligned}
|(-2)^3 \div 4| - 5(-2) &= |-8 \div 4| - 5(-2) &\quad (-2)(-2)(-2) = -8 \\
&= |-2| - 5(-2) &\quad \text{Dividing} \\
&= 2 - 5(-2) &\quad |-2| = 2 \\
&= 2 - (-10) &\quad \text{Multiplying} \\
&= 12 &\quad \text{Subtracting}
\end{aligned}$$

EXAMPLE 11 Simplify: $2^4 + 51 \cdot 4 - (37 + 23 \cdot 2)$.

$2^4 + 51 \cdot 4 - (37 + 23 \cdot 2)$

$= 2^4 + 51 \cdot 4 - (37 + 46)$ Carrying out all operations inside parentheses first, multiplying 23 by 2, following the rules for order of operations within the parentheses

$= 2^4 + 51 \cdot 4 - 83$ Completing the addition inside parentheses

$= 16 + 51 \cdot 4 - 83$ Evaluating exponential expressions

$= 16 + 204 - 83$ Doing all multiplications

$= 220 - 83$ Doing all additions and subtractions in order from left to right

$= 137$

Always regard a fraction bar as a grouping symbol. It separates any calculations in the numerator from those in the denominator.

EXAMPLE 12 Simplify: $\dfrac{5 - (-3)^2}{-2}$.

$\dfrac{5 - (-3)^2}{-2} = \dfrac{5 - 9}{-2}$

$= \dfrac{-4}{-2}$ Calculating within the numerator: $(-3)^2 = (-3)(-3) = 9$ and $5 - 9 = -4$

$= 2$ Dividing

Do Exercises 10–13.

CALCULATOR CORNER

Grouping Symbols Most calculators now provide grouping symbols. Such keys may appear as $($ and $)$ or $[\![\,...\,$ and $\,...\,]\!]$. Grouping symbols can be useful when we are simplifying expressions written in fraction form. For example, to simplify

$$\frac{38 + 142}{2 - 47},$$

we press $($ 3 8 $+$ 1 4 2 $)$ $\div$ $($ 2 $-$ 4 7 $)$ $=$. Failure to include grouping symbols in the above keystrokes would mean that we are simplifying a different expression:

$$38 + \frac{142}{2} - 47.$$

Exercises: Use a calculator with grouping symbols to simplify each of the following.

1. $\dfrac{38 - 178}{5 + 30}$ 2. $\dfrac{311 - 17^2}{2 - 13}$ 3. $785 - \dfrac{285 - 5^4}{17 + 3 \cdot 51}$

Simplify.

10. $5 - (-7)(-3)^2$

11. $(-2) \cdot |3 - 2^2| + 5$

12. $52 \cdot 5 + 5^3 - (4^2 - 48 \div 4)$

13. $\dfrac{(-5)(-9)}{1 - 2 \cdot 2}$

Answers on page A-5

a   Divide, if possible, and check each answer by multiplying. If an answer is undefined, state so.

1. $28 \div (-4)$

2. $\dfrac{35}{-7}$

3. $\dfrac{28}{-2}$

4. $26 \div (-13)$

5. $\dfrac{18}{-2}$

6. $-22 \div (-2)$

7. $\dfrac{-48}{-12}$

8. $-63 \div (-9)$

9. $\dfrac{-72}{8}$

10. $\dfrac{-50}{25}$

11. $-100 \div (-50)$

12. $\dfrac{-400}{8}$

13. $-344 \div 8$

14. $\dfrac{-128}{8}$

15. $\dfrac{200}{-25}$

16. $-651 \div (-31)$

17. $\dfrac{-56}{0}$

18. $\dfrac{0}{-5}$

19. $\dfrac{88}{-11}$

20. $\dfrac{-145}{-5}$

21. $-\dfrac{276}{12}$

22. $-\dfrac{217}{7}$

23. $\dfrac{0}{-2}$

24. $\dfrac{-13}{0}$

25. $\dfrac{19}{-1}$

26. $\dfrac{-17}{1}$

27. $-41 \div 1$

28. $23 \div (-1)$

b   Simplify, if possible. If an answer is undefined, state so.

29. $5 - 2 \cdot 3 - 6$

30. $5 - (2 \cdot 3 - 7)$

31. $9 - 2(3 - 8)$

32. $(8 - 2)(3 - 9)$

33. $16 \cdot (-24) + 50$

34. $10 \cdot 20 - 15 \cdot 24$

35. $40 - 3^2 - 2^3$

36. $2^4 + 2^2 - 10$

37. $4 \cdot (6 + 8)/(4 + 3)$

38. $4^3 + 10 \cdot 20 + 8^2 - 23$

39. $4 \cdot 5 - 2 \cdot 6 + 4$

40. $5^3 + 4 \cdot 9 - (8 + 9 \cdot 3)$

41. $\dfrac{9^2 - 1}{1 - 3^2}$

42. $\dfrac{100 - 6^2}{(-5)^2 - 3^2}$

43. $8(-7) + 6(-5)$

44. $10(-5) \div 1(-1)$

45. $20 \div 5(-3) + 3$

46. $14 \div 2(-6) + 7$

47. $18 - 0(3^2 - 5^2 \cdot 7 - 4)$

48. $9 \cdot 0 \div 5 \cdot 4$

49. $4 \cdot 5^2 \div 10$

50. $(2 - 5)^2 \div (-9)$

51. $(3 - 8)^2 \div (-1)$

52. $3 - 3^2$

53. $17 - 10^3$

54. $30 + (-5)^3$

55. $2 + 10^2 \div 5 \cdot 2^2$

56. $5 + 6^2 \div 3 \cdot 2^2$

57. $12 - 20^3$

58. $20 + 4^3 \div (-8)$

59. $2 \times 10^3 - 5000$

60. $-7(3^4) + 18$

61. $6[9 - (3 - 4)]$

62. $8[(6 - 13) - 11]$

63. $-1000 \div (-100) \div 10$

64. $256 + (-32) \div (-4)$

65. $8 - |7 - 9| \cdot 3$

66. $|8 - 7 - 9| \cdot 2 + 1$

67. $9 - |7 - 3^2|$

68. $9 - |5 - 7|^3$

69. $\dfrac{6^3 - 7 \cdot 3^4 - 2^5 \cdot 9}{(1 - 2^3)^3 + 7^3}$

70. $\dfrac{4 \div 2 \cdot 4^2 - 3 \cdot 2}{(7 - 4)^3 - 2 \cdot 5 - 4}$

71. $\dfrac{2 \cdot 3^2 \div (3^2 - (2 + 1))}{5^2 - 6^2 - 2^2(-3)}$

72. $\dfrac{5 \cdot 6^2 \div (2^2 \cdot 5) - 7^2}{3^2 - 4^2 - (-2)^3 - 2}$

73. $\dfrac{(-5)^3 + 17}{10(2 - 6) - 2(5 + 2)}$

74. $\dfrac{(3 - 5)^2 - (7 - 13)}{(2 - 5)3 + 2 \cdot 4}$

75. $\dfrac{2 \cdot 4^3 - 4 \cdot 32}{19^3 - 17^4}$

76. $\dfrac{-16 \cdot 28 \div 2^2}{5 \cdot 25 - 5^3}$

77. **D_W** Explain in your own words why $17 \div 0$ is undefined.

78. **D_W** Without performing any calculations, Stefan reports that $(19^2 - 17^2)/(16^2 - 18^2)$ is negative. How do you think he reached this conclusion?

79. Fabrikant Fine Diamonds ran a 4-in. by 7-in. advertisement in *The New York Times*. Find the area of the ad.   [1.5c], [1.8a]

80. A classroom contains 7 rows of chairs with 6 chairs in each row. How many chairs are there in the classroom?   [1.8a]

81. Cindi's Ford Focus gets 32 mpg (miles per gallon). How many gallons will it take to travel 384 mi?   [1.8a]

82. Craig's Chevy Blazer gets 14 mpg. How many gallons will it take to travel 378 mi?   [1.8a]

83. A 7-oz bag of tortilla chips contains 1050 calories. How many calories are in a 1-oz serving?   [1.8a]

84. A 7-oz bag of tortilla chips contains 8 g (grams) of fat per ounce. How many grams of fat are in a carton containing 12 bags of chips?   [1.8a]

85. There are 18 sticks in a large pack of Trident gum. If 4 people share the pack equally, how many whole pieces will each person receive? How many extra pieces will remain?   [1.8a]

86. A bag of Ricola throat lozenges contains 24 cough drops. If 5 people share the bag equally, how many lozenges will each person receive? How many extra lozenges will remain?   [1.8a]

87. **D$_W$** Ty claims that $8 - 3^2 + 1$ is -2. What mistake do you think he is making?

88. **D$_W$** Bryn contends that $13 - 10/2 - 5$ is -1. What mistake do you think she is making?

Simplify, if possible.

89. $\dfrac{9 - 3^2}{2 \cdot 4^2 - 5^2 \cdot 9 + 8^2 \cdot 7}$

90. $\dfrac{7^3 \cdot 9 - 6^2 \cdot 8 + 4^3 \cdot 6}{5^2 - 25}$

91. $\dfrac{(25 - 4^2)^3}{17^2 - 16^2} \cdot ((-6)^2 - 6^2)$

92. $\dfrac{(7 - 8)^{37}}{7^2 - 8^2} \cdot (98 - 7^2 \cdot 2)$

93. $\dfrac{19 - 17^2}{13^2 - 34}$

94. $\dfrac{195 + (-15)^3}{195 - 7 \cdot 5^2}$

95. $28^2 - 36^2/4^2 + 17^2$

96. $9^3 - 36^3/12^2 + 9^2$

97. Write down the keystrokes needed to calculate $\dfrac{15^2 - 5^3}{3^2 + 4^2}$.

98. Write down the keystrokes needed to calculate $\dfrac{16^2 - 24 \cdot 23}{3 \cdot 4 + 5^2}$.

99. Evaluate the expression for which the keystrokes are as follows: $\boxed{4}\ \boxed{-}\ \boxed{1}\ \boxed{0}\ \boxed{\div}\ \boxed{2}\ \boxed{+}\ \boxed{6}$.

100. Evaluate the expression for which the keystrokes are $\boxed{4}\ \boxed{-}\ \boxed{1}\ \boxed{6}\ \boxed{\div}\ \boxed{(}\ \boxed{2}\ \boxed{+}\ \boxed{6}\ \boxed{)}$.

Determine the sign of each expression if m is negative and n is positive.

101. $\dfrac{-n}{m}$

102. $\dfrac{-n}{-m}$

103. $-\left(\dfrac{-n}{m}\right)$

104. $-\left(\dfrac{n}{-m}\right)$

105. $-\left(\dfrac{-n}{-m}\right)$

In this section, we will learn to write *equivalent expressions* by making use of the *distributive law*, both of which are very important concepts.

a Algebraic Expressions

In arithmetic, we work with expressions such as

$$37 + 86, \quad 7 \cdot 8, \quad 19 - 7, \quad \text{and} \quad \frac{3}{8}.$$

In algebra, we use both numbers and letters and work with *algebraic expressions* such as

$$x + 86, \quad 7 \cdot t, \quad 19 - y, \quad \text{and} \quad \frac{a}{b}.$$

We have already worked with expressions like these.

When a letter can stand for various numbers, we call the letter a **variable.** A number or a letter that stands for just one number is called a **constant.** Let $c =$ the speed of light. Then c is a constant. Let $a =$ the speed of an Amtrak metroliner. Then a is a variable since the value of a can vary.

An **algebraic expression** consists of variables, constants, numerals, and operation signs. When we replace a variable with a number, we say that we are **substituting** for the variable. This process is called **evaluating the expression.**

EXAMPLE 1 Evaluate $x + y$ for $x = 37$ and $y = 29$.

We substitute 37 for x and 29 for y and carry out the addition:

$$x + y = 37 + 29 = 66.$$

The number 66 is called the **value** of the expression.

Algebraic expressions involving multiplication, like "8 times a," can be written as $8 \times a$, $8 \cdot a$, $8(a)$, or simply $8a$. Two letters written together without an operation symbol, such as ab, also indicate multiplication.

EXAMPLE 2 Evaluate $3y$ for $y = -14$.

$$3y = 3(-14) = -42 \qquad \text{Parentheses are required here.}$$

Do Exercises 1–3.

Algebraic expressions involving division can also be written several ways. For example, "8 divided by t" can be written as $8 \div t$, $8/t$, or $\dfrac{8}{t}$.

EXAMPLE 3 Evaluate $\dfrac{a}{b}$ and $\dfrac{-a}{-b}$ for $a = 35$ and $b = 7$.

We substitute 35 for a and 7 for b:

$$\frac{a}{b} = \frac{35}{7} = 5; \qquad \frac{-a}{-b} = \frac{-35}{-7} = 5.$$

1. Evaluate $a + b$ for $a = 38$ and $b = 26$.

2. Evaluate $x - y$ for $x = 57$ and $y = 29$.

3. Evaluate $5t$ for $t = -14$.

Answers on page A-5

Study Tips

READING A MATH TEXT

Do not expect a math text to read like a magazine or novel. On one hand, most assigned readings in a math text consist of only a few pages. On the other hand, every sentence and word is important and should make sense. If they don't, seek help as soon as possible.

For each number, find two equivalent expressions with negative signs in different places.

4. $\dfrac{-6}{x}$

5. $-\dfrac{m}{n}$

6. $\dfrac{r}{-4}$

7. Evaluate $\dfrac{a}{-b}, \dfrac{-a}{b}$, and $-\dfrac{a}{b}$ for $a = 28$ and $b = 4$.

8. Find the Fahrenheit temperature that corresponds to 10 degrees Celsius (see Example 5).

9. Evaluate $3x^2$ for $x = 4$ and $x = -4$.

10. Evaluate a^4 for $a = 3$ and $a = -3$.

11. Evaluate $(-x)^2$ and $-x^2$ for $x = 3$.

12. Evaluate $(-x)^2$ and $-x^2$ for $x = 2$.

13. Evaluate x^5 for $x = 2$ and $x = -2$.

Answers on page A-5

130

CHAPTER 2: Introduction to Integers and Algebraic Expressions

■ **EXAMPLE 4** Evaluate $-\dfrac{a}{b}, \dfrac{-a}{b}$, and $\dfrac{a}{-b}$ for $a = 15$ and $b = 3$.

We substitute 15 for a and 3 for b:

$$-\frac{a}{b} = -\frac{15}{3} = -5; \qquad \frac{-a}{b} = \frac{-15}{3} = -5; \qquad \frac{a}{-b} = \frac{15}{-3} = -5.$$

Examples 3 and 4 illustrate the following.

$\dfrac{-a}{-b}$ and $\dfrac{a}{b}$ represent the same number.

$-\dfrac{a}{b}, \dfrac{-a}{b}$, and $\dfrac{a}{-b}$ all represent the same number.

Do Exercises 4–7.

■ **EXAMPLE 5** Evaluate $\dfrac{9C}{5} + 32$ for $C = 20$.

This expression can be used to find the Fahrenheit temperature that corresponds to 20 degrees Celsius:

$$\frac{9C}{5} + 32 = \frac{9 \cdot 20}{5} + 32 = \frac{180}{5} + 32 = 36 + 32 = 68.$$

Do Exercise 8.

■ **EXAMPLE 6** Evaluate $5x^2$ for $x = 3$ and $x = -3$.

The rules for order of operations specify that the replacement for x be squared. That result is then multiplied by 5:

$$5x^2 = 5(3)^2 = 5(9) = 45;$$
$$5x^2 = 5(-3)^2 = 5(9) = 45.$$

Example 6 illustrates that when opposites are raised to an even power, the results are the same.

Do Exercises 9 and 10.

■ **EXAMPLE 7** Evaluate $(-x)^2$ and $-x^2$ for $x = 7$.

When we evaluate $(-x)^2$ for $x = 7$, we have

$$(-x)^2 = (-7)^2 = (-7)(-7) = 49. \qquad \text{Substitute 7 for } x. \text{ Then evaluate the power.}$$

To evaluate $-x^2$, we again substitute 7 for x. We must recall that taking the opposite of a number is the same as multiplying that number by -1.

$$-x^2 = -1 \cdot x^2 \qquad \text{The opposite of a number is the same as multiplying by } -1.$$
$$-7^2 = -1 \cdot 7^2 \qquad \text{Substituting 7 for } x$$
$$= -1 \cdot 49 = -49. \qquad \text{Using the rules for order of operations; calculating the power before multiplying}$$

Example 7 shows that $(-x)^2 \neq -x^2$.

Do Exercises 11–13.

b Equivalent Expressions and the Distributive Law

It is useful to know when two algebraic expressions will represent the same number. In many situations, this will help with problem solving.

EXAMPLE 8 Evaluate $x + x$ and $2x$ for $x = 3$ and $x = -5$ and compare the results.

We substitute 3 for x in $x + x$ and again in $2x$:

$$x + x = 3 + 3 = 6; \qquad 2x = 2 \cdot 3 = 6.$$

Next we repeat the procedure, substituting -5 for x:

$$x + x = -5 + (-5) = -10; \qquad 2x = 2(-5) = -10.$$

The results can be shown in a table. It appears that $x + x$ and $2x$ represent the same number.

	$x + x$	$2x$
$x = 3$	6	6
$x = -5$	-10	-10

Do Exercises 14 and 15.

Example 8 suggests that $x + x$ and $2x$ represent the same number for any replacement of x. When this is known to be the case, we can say that $x + x$ and $2x$ are **equivalent expressions.**

EQUIVALENT EXPRESSIONS

Two expressions that have the same value for all allowable replacements are called **equivalent.**

In Examples 3 and 7 we saw that the expressions $\dfrac{-a}{-b}$ and $\dfrac{a}{b}$ are equivalent but that the expressions $(-x)^2$ and $-x^2$ are *not* equivalent.

An important concept, known as the **distributive law,** is useful for finding equivalent algebraic expressions. The distributive law involves two operations: multiplication and either addition or subtraction.

To review how the distributive law works, consider the following:

$$
\begin{array}{r}
4\ 5 \\
\times\ \ 7 \\
\hline
3\ 5 \\
2\ 8\ 0 \\
3\ 1\ 5 \\
\end{array}
$$

3 5 ← This is $7 \cdot 5$.
2 8 0 ← This is $7 \cdot 40$.
3 1 5 ← This is the sum $7 \cdot 40 + 7 \cdot 5$.

To carry out the multiplication, we actually added two products. That is,

$$7 \cdot 45 = 7(40 + 5) = 7 \cdot 40 + 7 \cdot 5.$$

The distributive law says that if we want to multiply a sum of several numbers by a number, we can either add within the grouping symbols and then multiply, or multiply each of the terms separately and then add.

Complete each table by evaluating each expression for the given values.

14.

	$3x + 2x$	$5x$
$x = 4$		
$x = -2$		
$x = 0$		

15.

	$4x - x$	$3x$
$x = 2$		
$x = -2$		
$x = 0$		

Answers on page A-5

CALCULATOR CORNER

Evaluating Powers To evaluate an expression like $-x^3$ for $x = -14$ with a calculator, it is imperative that we keep in mind the rules for order of operations. On some calculators, this expression is evaluated by pressing

[1] [4] [+/−] [x^y] [3] [=] [+/−] . Other calculators use the keystrokes [(−)] [(] [(−)] [1] [4] [)] [∧] [3] [ENTER =] . Consult your owner's manual, an instructor, or simply experiment if your calculator behaves differently.

Exercises: Evaluate.

1. $-a^5$ for $a = -3$
2. $-x^5$ for $x = -4$
3. $-x^5$ for $x = 2$
4. $-x^5$ for $x = 5$

Use the distributive law to write an equivalent expression.

16. $5(a + b)$

17. $6(x + y + z)$

Use the distributive law to write an equivalent expression.

18. $4(x - y)$

19. $3(a - b + c)$

20. $(m - 4)6$

21. $-8(2a - b + 3c)$

Answers on page A-5

132

CHAPTER 2: Introduction to Integers
and Algebraic Expressions

THE DISTRIBUTIVE LAW

For any numbers a, b, and c,
$$a(b + c) = ab + ac.$$

EXAMPLE 9 Evaluate $a(b + c)$ and $ab + ac$ for $a = 3$, $b = 4$, and $c = 2$.

We have

$$a(b + c) = 3(4 + 2) = 3 \cdot 6 = 18 \quad \text{and}$$
$$ab + ac = 3 \cdot 4 + 3 \cdot 2 = 12 + 6 = 18.$$

It is impossible to overemphasize the importance of the parentheses in the statement of the distributive law. Were we to omit the parentheses, we would have $ab + c$. To see that $a(b + c) \neq ab + c$, note that $3(4 + 2) = 18$, but $3 \cdot 4 + 2 = 14$.

EXAMPLE 10 Use the distributive law to write an expression equivalent to $2(l + w)$.

$2(l + w) = 2 \cdot l + 2 \cdot w$ Note that the $+$ sign between l and w now appears between $2 \cdot l$ and $2 \cdot w$.

$\quad\quad\quad = 2l + 2w.$ Try to go directly to this step.

Do Exercises 16 and 17.

Since subtraction can be regarded as addition of the opposite, it follows that the distributive law holds in cases involving subtraction.

EXAMPLE 11 Use the distributive law to write an expression equivalent to each of the following:

a) $7(a - b)$;   **b)** $9(x - 5)$;   **c)** $(a - 7)b$;   **d)** $-4(x - 2y + 3z)$

a) $7(a - b) = 7 \cdot a - 7 \cdot b$

$\quad\quad\quad\quad = 7a - 7b$ Try to go directly to this step.

b) $9(x - 5) = 9x - 9(5)$

$\quad\quad\quad\quad = 9x - 45$ Again, try to go directly to this step.

c) $(a - 7)b = b(a - 7)$ Using a commutative law

$\quad\quad\quad\quad = b \cdot a - b \cdot 7$ Using the distributive law

$\quad\quad\quad\quad = ab - 7b$ Using a commutative law to write ba alphabetically and $b \cdot 7$ with the constant first

d) $-4(x - 2y + 3z) = -4 \cdot x - (-4)(2y) + (-4)(3z)$ Using the distributive law

$\quad\quad\quad\quad = -4x - (-4 \cdot 2)y + (-4 \cdot 3)z$ Using an associative law (twice)

$\quad\quad\quad\quad = -4x - (-8y) + (-12z)$

$\quad\quad\quad\quad = -4x + 8y - 12z$ Try to go directly to this step.

Do Exercises 18–21.

2.6

EXERCISE SET

For Extra Help

Math XL   MyMathLab   InterAct   Math Tutor   Video   Student's
                        Math   Center   Lectures Solutions
MathXL   MyMathLab   on CD   Manual
                                                  Disc 1

a   Evaluate.

1. $12n$, for $n = 2$
   (The cost, in cents, of sending 2 text messages)

2. $39n$, for $n = 2$
   (The cost, in cents, of sending 2 letters)

3. $\dfrac{x}{y}$, for $x = 6$ and $y = -3$

4. $\dfrac{m}{n}$, for $m = 18$ and $n = 2$

5. $\dfrac{2q}{p}$, for $p = 6$ and $q = 3$

6. $\dfrac{5y}{z}$, for $y = 15$ and $z = -25$

7. $\dfrac{72}{r}$, for $r = 4$
   (The approximate doubling time, in years, for an investment earning 4% interest per year)

8. $\dfrac{72}{i}$, for $i = 2$
   (The approximate doubling time, in years, for an investment earning 2% interest per year)

9. $3 + 5 \cdot x$, for $x = 2$

10. $9 - 2 \cdot x$, for $x = 3$

11. $2l + 2w$, for $l = 3$ and $w = 4$
   (The perimeter, in feet, of a 3-ft by 4-ft rectangle)

12. $3(a + b)$, for $a = 2$ and $b = 4$

13. $2(l + w)$, for $l = 3$ and $w = 4$
   (The perimeter, in feet, of a 3-ft by 4-ft rectangle)

14. $3a + 3b$, for $a = 2$ and $b = 4$

15. $7a - 7b$, for $a = 5$ and $b = 2$

16. $4x - 4y$, for $x = 6$ and $y = 1$

17. $7(a - b)$, for $a = 5$ and $b = 2$

18. $4(x - y)$, for $x = 6$ and $y = 1$

19. $16t^2$, for $t = 5$
   (The distance, in feet, that an object falls in 5 sec)

20. $\dfrac{49t^2}{10}$, for $t = 10$
   (The distance, in meters, that an object falls in 10 sec)

21. $a + (b - a)^2$, for $a = 6$ and $b = 4$

22. $(x + y)^2$, for $x = 2$ and $y = 3$

23. $9a + 9b$, for $a = 13$ and $b = -13$

24. $8x + 8y$, for $x = 17$ and $y = -17$

25. $\dfrac{n^2 - n}{2}$, for $n = 9$

(For determining the number of handshakes possible among 9 people)

26. $\dfrac{5(F - 32)}{9}$, for $F = 50$

(For converting 50 degrees Fahrenheit to degrees Celsius)

27. $m^3 - m^2$, for $m = 5$

28. $a^6 - a$, for $a = -2$

For each expression, write two equivalent expressions with negative signs in different places.

29. $-\dfrac{5}{t}$

30. $\dfrac{7}{-x}$

31. $\dfrac{-n}{b}$

32. $-\dfrac{3}{r}$

33. $\dfrac{9}{-p}$

34. $\dfrac{-u}{5}$

35. $\dfrac{-14}{w}$

36. $\dfrac{-23}{m}$

Evaluate $\dfrac{-a}{b}, \dfrac{a}{-b}$, and $-\dfrac{a}{b}$ for the given values.

37. $a = 45, b = 9$

38. $a = 40, b = 5$

39. $a = 81, b = 3$

40. $a = 56, b = 7$

Evaluate.

41. $(-3x)^2$ and $-3x^2$, for $x = 2$

42. $(-2x)^2$ and $-2x^2$, for $x = 3$

43. $5x^2$, for $x = 3$ and $x = -3$

44. $2x^2$, for $x = 5$ and $x = -5$

45. x^3, for $x = 6$ and $x = -6$

46. x^6, for $x = 2$ and $x = -2$

47. x^8, for $x = 1$ and $x = -1$

48. x^5, for $x = 3$ and $x = -3$

49. a^5, for $a = 2$ and $a = -2$

50. a^7, for $a = 1$ and $a = -1$

b   Use the distributive law to write an equivalent expression.

51. $5(a + b)$

52. $7(x + y)$

53. $4(x + 1)$

54. $6(a + 1)$

55. $2(b + 5)$

56. $3(x - 6)$

57. $7(1 - t)$

58. $4(1 - y)$

59. $6(5x - 2)$

60. $9(6m + 7)$

61. $8(x + 7 + 6y)$

62. $4(5x + 8 + 3p)$

63. $-7(y - 2)$

64. $-9(y - 7)$

65. $(x + 2)3$

66. $(x + 4)2$

67. $-4(x - 3y - 2z)$

68. $8(2x - 5y - 8z)$

69. $8(a - 3b + c)$

70. $-6(a + 2b - c)$

71. $4(x - 3y - 7z)$

72. $5(9x - y + 8z)$

73. $(4a - 5b + c - 2d)5$

74. $(9a - 4b + 3c - d)7$

75. **Dw** Does $-\dfrac{x}{y}$ always represent a negative number? Why or why not?

76. **Dw** Is $-x^2$ always negative? Why or why not?

77. Write a word name for 23,043,921.   [1.1c]

78. Multiply: $17 \cdot 53$.   [1.5a]

79. Estimate by rounding to the nearest ten. Show your work.   [1.4b]

$$\begin{array}{r} 5\ 2\ 8\ 3 \\ -\ 2\ 4\ 7\ 5 \\ \hline \end{array}$$

80. Divide: $2982 \div 3$.   [1.6c]

81. On January 6, it snowed 9 in., and on January 7, it snowed 8 in. How much did it snow altogether?   [1.8a]

82. On March 9, it snowed 12 in., but on March 10, the sun melted 7 in. How much snow remained?   [1.8a]

83. For Brett's party, his wife ordered two cheese pizzas at $11 apiece and two pepperoni pizzas for $13 apiece. How much did she pay for the pizza?   [1.8a]

84. For Tania's graduation party, her husband ordered three buckets of chicken wings at $12 apiece and 3 trays of nachos at $9 a tray. How much did he pay for the wings and nachos?   [1.8a]

85. **D$_W$** Under what condition(s) will the expression ax^2 be nonnegative? Explain.

86. **D$_W$** Ted evaluates $a + a^2$ for $a = 5$ and gets 100 as the result. What mistake did he probably make?

87. A car's catalytic converter works most efficiently after it is heated to about 370°C. To what Fahrenheit temperature does this correspond? (*Hint*: see Example 5.)

Evaluate.

88. $x^2 - xy^2 \div 2 \cdot y$, for $x = 24$ and $y = 6$

89. $a - b^3 + 17a$, for $a = 19$ and $b = -16$

90. $x^2 - 23y + y^3$, for $x = 18$ and $y = -21$

91. $r^3 + r^2t - rt^2$, for $r = -9$ and $t = 7$

92. $a^3b - a^2b^2 + ab^3$, for $a = -8$ and $b = -6$

93. $a^{1996} - a^{1997}$, for $a = -1$

94. $x^{1492} - x^{1493}$, for $x = -1$

95. $(m^3 - mn)^m$, for $m = 4$ and $n = 6$

96. $5a^{3a-4}$, for $a = 2$

Replace the blanks with $\boxed{+}$, $\boxed{-}$, $\boxed{\times}$, or $\boxed{\div}$ to make each statement true.

97. $-32\ \square\ (88\ \square\ 29) = -1888$

98. $59\ \square\ 17\ \square\ 59\ \square\ 8 = 1475$

Classify each statement as true or false. If false, write an example showing why.

99. For any choice of x , $x^2 = (-x)^2$.

100. For any choice of x , $x^3 = -x^3$.

101. For any choice of x , $x^6 + x^4 = (-x)^6 + (-x)^4$.

102. For any choice of x , $(-3x)^2 = 9x^2$.

103. **D$_W$** If the Fahrenheit temperature is doubled, does it follow that the corresponding Celsius temperature is also doubled? (*Hint*: see Example 5.)

2.7

LIKE TERMS AND PERIMETER

Objectives

a   Combine like terms.

b   Determine the perimeter of a polygon.

One common way in which equivalent expressions are formed is by *combining like terms*. In this section we learn how this is accomplished and apply the concept to geometry.

a   Combining Like Terms

A **term** is a number, a variable, a product of numbers and/or variables, or a quotient of numbers and/or variables. Terms are separated by addition signs. If there are subtraction signs, we can find an equivalent expression that uses addition signs.

EXAMPLE 1   What are the terms of $3xy - 4y + \dfrac{2}{z}$?

$$3xy - 4y + \frac{2}{z} = 3xy + (-4y) + \frac{2}{z} \qquad \text{Separating parts with + signs}$$

The terms are $3xy$, $-4y$, and $\dfrac{2}{z}$.

Do Exercises 1 and 2.

What are the terms of each expression?

1. $5x - 4y + 3$

2. $-4y - 2x + \dfrac{x}{y}$

Terms in which the variable factors are exactly the same, such as $9x$ and $-4x$, are called **like,** or **similar, terms.** For example, $3y^2$ and $7y^2$ are like terms, whereas $5x$ and $6x^2$ are not. Constants, like 7 and 3, are also like terms.

EXAMPLES   Identify the like terms.

2. $7x + 5x^2 + 2x + 8 + 5x^3 + 1$

$7x$ and $2x$ are like terms;   8 and 1 are like terms.

3. $5ab + a^3 - a^2b - 2ab + 7a^3$

$5ab$ and $-2ab$ are like terms; a^3 and $7a^3$ are like terms.

Do Exercises 3 and 4.

Identify the like terms.

3. $9a^3 + 4ab + a^3 + 3ab + 7$

When an algebraic expression contains like terms, an equivalent expression can be formed by **combining,** or **collecting, like terms.** To combine like terms, we rely on the distributive law.

EXAMPLE 4   Combine like terms to form an equivalent expression.

a) $4x + 3x$ **b)** $6mn - 7mn$

c) $7y - 2 - 3y + 5$ **d)** $2a^5 + 9ab + 3 + a^5 - 7 - 4ab$

a) $4x + 3x = (4 + 3)x$ Using the distributive law (in "reverse")

$\qquad\quad = 7x$ We usually go directly to this step.

b) $6mn - 7mn = (6 - 7)mn$ Try to do this mentally.

$\qquad\qquad\quad = -1mn,$ or simply $-mn$

c) $7y - 2 - 3y + 5 = 7y + (-2) + (-3y) + 5$ Rewriting as addition

$\qquad\qquad\qquad\quad = 7y + (-3y) + (-2) + 5$ Using a commutative law

$\qquad\qquad\qquad\quad = 4y + 3$ Try to go directly to this step.

4. $3xy - 5x^2 + y^2 - 4xy + y$

Answers on page A-6

Combine like terms to form an equivalent expression.

5. $2a + 7a$

6. $5x^2 - 9 + 2x^2 + 3$

7. $4m - 2n^2 + 5 + n^2 + m - 9$

Find the perimeter of each polygon.

8.

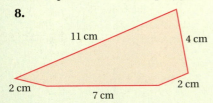

9.

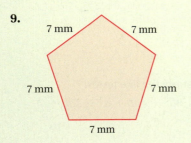

10. Find the perimeter of a rectangle that is 2 cm by 4 cm.

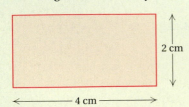

Answers on page A-6

d) $2a^5 + 9ab + 3 + a^5 - 7 - 4ab$

$= 2a^5 + 9ab + 3 + a^5 + (-7) + (-4ab)$

$= 2a^5 + a^5 + 9ab + (-4ab) + 3 + (-7)$ Rearranging terms

$= 3a^5 + 5ab + (-4)$ Think of a^5 as $1a^5$; $2a^5 + a^5 = 3a^5$

$= 3a^5 + 5ab - 4$

Do Exercises 5–7.

b   Perimeter

> **PERIMETER OF A POLYGON**
>
> A **polygon** is a closed geometric figure with three or more sides. The **perimeter** of a polygon is the distance around it, or the sum of the lengths of its sides.

EXAMPLE 5   Find the perimeter of this polygon.

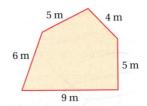

> A polygon with five sides is called a pentagon.

We add the lengths of all sides. Since all the units are the same, we are effectively combining like terms.

Perimeter $= 6\,\text{m} + 5\,\text{m} + 4\,\text{m} + 5\,\text{m} + 9\,\text{m}$

$= (6 + 5 + 4 + 5 + 9)\,\text{m}$ Using the distributive law

$= 29\,\text{m}$ Try to go directly to this step.

> **Caution!**
>
> When units of measurement are given in the statement of a problem, as in Example 5, the solution should also contain units of measurement.

Do Exercises 8 and 9.

A **rectangle** is a polygon with four sides and four 90° angles. Opposite sides of a rectangle have the same measure. The symbol ⌐ or ⌐ indicates a 90° angle. A 90° angle is often referred to as a **right angle.**

EXAMPLE 6   Find the perimeter of a rectangle that is 3 cm by 4 cm.

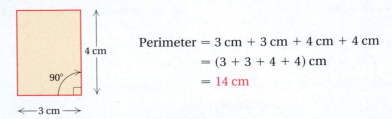

Perimeter $= 3\,\text{cm} + 3\,\text{cm} + 4\,\text{cm} + 4\,\text{cm}$

$= (3 + 3 + 4 + 4)\,\text{cm}$

$= 14\,\text{cm}$

Do Exercise 10.

The perimeter of the rectangle in Example 6 is 2 · 3 cm + 2 · 4 cm, or equivalently 2(3 cm + 4 cm). This can be generalized, as follows.

> **PERIMETER OF A RECTANGLE**
>
> The **perimeter P of a rectangle** of length l and width w is given by
>
> $$P = 2l + 2w, \quad \text{or} \quad P = 2 \cdot (l + w).$$
>
>

EXAMPLE 7 A common door size is 3 ft by 7 ft. Find the perimeter of such a door.

$$P = 2l + 2w \qquad \text{We could also use } P = 2(l + w).$$
$$= 2 \cdot 7\,\text{ft} + 2 \cdot 3\,\text{ft}$$
$$= (2 \cdot 7)\,\text{ft} + (2 \cdot 3)\,\text{ft} \qquad \text{Try to do this mentally.}$$
$$= 14\,\text{ft} + 6\,\text{ft}$$
$$= 20\,\text{ft} \qquad \text{Combining like terms}$$

The perimeter of the door is 20 ft.

Do Exercise 11.

A **square** is a rectangle in which all sides have the same length.

EXAMPLE 8 Find the perimeter of a square with sides of length 9 mm.

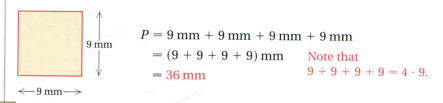

$$P = 9\,\text{mm} + 9\,\text{mm} + 9\,\text{mm} + 9\,\text{mm}$$
$$= (9 + 9 + 9 + 9)\,\text{mm} \qquad \text{Note that}$$
$$= 36\,\text{mm} \qquad \qquad \quad 9 + 9 + 9 + 9 = 4 \cdot 9.$$

Do Exercise 12.

> **PERIMETER OF A SQUARE**
>
> The **perimeter P of a square** is four times s, the length of a side:
>
> $$P = s + s + s + s$$
> $$= 4s.$$
>
>

EXAMPLE 9 Find the perimeter of a square garden with sides of length 12 ft.

$$P = 4s$$
$$= 4 \cdot 12\,\text{ft}$$
$$= 48\,\text{ft}$$

The perimeter of the garden is 48 ft.

Do Exercise 13.

---

11. Find the perimeter of a 4-ft by 8-ft sheet of plywood.

12. Find the perimeter of a square with sides of length 10 km.

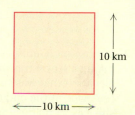

13. Find the perimeter of a square sandbox with sides of length 6 ft.

Answers on page A-6

Study Tips

UNDERSTAND YOUR MISTAKES

When you receive a graded quiz, test, or assignment back from your instructor, it is important to review and understand what your mistakes were. Too often students simply file away old papers without first making an effort to learn from their mistakes.

Translating for Success

1. **Yarn Cost.** It costs $8 for the yarn for each scarf that Annette knits. If she used $120 worth of yarn, how many scarves did she knit?

2. **Elevation.** Genine started hiking a 10-mi trail at an elevation that was 150 ft below sea level. At the end of the trail, she was 75 ft above sea level. How many feet higher was she at the end of the trail than at the beginning?

3. **Community Service.** In order to fulfill the requirements for a sociology class, Glen must log 120 hr of community service. So far, he has spent 75 hr volunteering at a youth center. How many more hours must he serve?

4. **Disaster Relief.** Each package that is prepared for a disaster relief effort contains 15 meal bars. How many packages can be filled from a donation of 750 bars?

5. **Perimeter.** A rectangular building lot is 75 ft wide and 150 ft long. What is the perimeter of the lot?

The goal of these matching questions is to practice step (2), *Translate*, of the five-step problem-solving process. Translate each word problem to an equation and select a correct translation from equations A–O.

A. $75 + x = 120$

B. $15 \div 750 = x$

C. $-10 - 15 = x$

D. $8 \cdot 120 = x$

E. $150 - 75 = x$

F. $15 \cdot 750 = x$

G. $75 - (-150) = x$

H. $2 \cdot 150 + 2 \cdot 75 = x$

I. $-10 - (-15) = x$

J. $8 \cdot x = 120$

K. $750 \div 15 = x$

L. $15 - (-10) = x$

M. $75 + 120 = x$

N. $75 - 150 = x$

O. $75 = 120 + x$

Answers on page A-6

6. **Account Balance.** Lorenzo had $75 in his checking account. He then wrote a check for $150. What was the balance in his account?

7. **Laptop Computers.** Great Graphics purchased a laptop computer for each of its 15 employees. If each laptop costs $750, how much did the computers cost?

8. **Basketball.** A basketball team scored 75 points in one game. In the next game, the team scored a record 120 points. How many points did the team score in the two games?

9. **Pizza Sales.** A youth club sold 120 pizzas for a fundraiser. If each pizza sold for $8, how much money was taken in?

10. **Temperature.** The temperature in Fairbanks was $-10°$ at 6:00 P.M. and fell another 15° by midnight. What was the temperature at midnight?

a   List the terms of each expression.

1. $2a + 5b - 7c$

2. $4x - 6y + 7z$

3. $9mn - 6n + 8$

4. $7rs + 4s - 5$

5. $3x^2y - 4y^2 - 2z^3$

6. $4a^3b + ab^2 - 9b^3$

Combine like terms to form an equivalent expression.

7. $5x + 9x$

8. $9a + 7a$

9. $10a - 13a$

10. $-16x + x$

11. $2x + 6z + 9x$

12. $3a - 5b + 7a$

13. $27a + 70 - 40a - 8$

14. $42x - 6 - 4x + 2$

15. $9 + 5t + 7y - t - y - 13$

16. $8 - 4a + 9b + 7a - 3b - 15$

17. $a + 3b + 5a - 2 + b$

18. $x + 7y + 5 - 2y + 3x$

19. $-8 + 11a - 5b + 6a - 7b + 7$

20. $8x - 5x + 6 + 3y - y - 4$

21. $8x^2 + 3y - x^2$

22. $8y^3 - 3z + 4y^3$

23. $11x^4 + 2y^3 - 4x^4 - y^3$

24. $13a^5 + 9b^4 - 2a^5 - 4b^4$

25. $9a^2 - 4a + a - 3a^2$

26. $3a^2 + 7a^3 - a^2 + 5 + a^3$

27. $x^3 - 5x^2 + 2x^3 - 3x^2 + 4$

28. $9xy + 4y^2 - 2xy + 2y^2 - 1$

29. $7a^3 + 4ab - 5 - 7ab + 8$

30. $8a^2b - 3ab^2 - 4a^2b + 2ab$

31. $9x^3y + 4xy^3 - 6xy^3 + 3xy$

32. $3x^4 - 2y^4 + 8x^4y^4 - 7x^4 + 8y^4$

33. $3a^6 - 9b^4 + 2a^6b^4 - 7a^6 - 2b^4$

34. $9x^6 - 5y^5 + 3x^6y - 8x^6 + 4y^5$

b Find the perimeter of each polygon.

35.
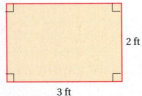
2 ft
3 ft

36.
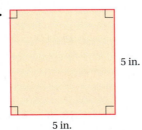
5 in.
5 in.

37.
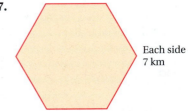
Each side
7 km

38.
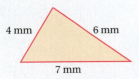
4 mm   6 mm
7 mm

39.
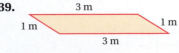
3 m
1 m   1 m
3 m

40.
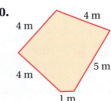
4 m
4 m
5 m
4 m
1 m

Tennis Court. A tennis court contains many rectangles. Use the diagram of a regulation tennis court to calculate the perimeters in Exercises 41–44.

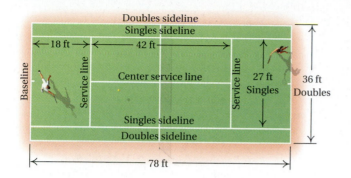

41. The perimeter of a singles court

42. The perimeter of a doubles court

43. The perimeter of the rectangle formed by the service lines and the singles sidelines

44. The perimeter of the rectangle formed by a service line, a baseline, and the singles sidelines

45. Find the perimeter of a rectangular 8-ft by 10-ft bedroom.

46. Find the perimeter of a rectangular 3-ft by 4-ft doghouse.

47. Find the perimeter of a checkerboard that is 14 in. on each side.

48. Find the perimeter of a square skylight that is 2 m on each side.

49. Find the perimeter of a square frame that is 65 cm on each side.

50. Find the perimeter of a square garden that is 12 yd on each side.

51. Find the perimeter of a 12-ft by 20-ft rectangular deck.

52. Find the perimeter of a 40-ft by 35-ft rectangular backyard.

53. **Dw** Explain in your own words what it means for two algebraic expressions to be equivalent.

54. **Dw** Can the formula for the perimeter of a rectangle be used to find the perimeter of a square? Why or why not?

55. A box of Shaw's Corn Flakes contains 510 grams (g) of corn flakes. A serving of corn flakes weighs 30 g. How many servings are in one box?   [1.8a]

56. Estimate the difference by rounding to the nearest ten. [1.4b]

$$\begin{array}{r} 7\ 0\ 4 \\ -\ 4\ 8\ 6 \\ \hline \end{array}$$

Simplify.   [1.9c]

57. $5 + 3 \cdot 2^3$

58. $(9 - 7)^4 - 3^2$

59. $12 \div 3 \cdot 2$

60. $27 \div 3(2 + 1)$

61. $15 - 3 \cdot 2 + 7$

62. $30 - 4^2 \div 8 \cdot 2$

Solve.   [1.7b]

63. $25 = t + 9$

64. $19 = x + 6$

65. $45 = 3x$

66. $50 = 2t$

67. **D**_W Does doubling the length of a square's side double the perimeter of the original square? Why or why not?

68. **D**_W Why was it necessary to introduce the distributive law before discussing how to combine like terms?

Simplify. (Multiply and then combine like terms.)

69. $5(x + 3) + 2(x - 7)$

70. $3(a - 7) + 7(a + 4)$

71. $2(3 - 4a) + 5(a - 7)$

72. $7(2 - 5x) + 3(x - 8)$

73. $-5(2 + 3x + 4y) + 7(2x - y)$

74. $3(4 - 2x) + 5(9x - 3y + 1)$

75. In order to save energy, Andrea plans to run a bead of caulk sealant around 3 exterior doors and 13 windows. Each window measures 3 ft by 4 ft, each door measures 3 ft by 7 ft, and there is no need to caulk the bottom of each door. If each cartridge of caulk seals 56 ft and costs $5.95, how much will it cost Andrea to seal the windows and doors?

76. Eric is attaching lace trim to small tablecloths that are 5 ft by 5 ft, and to large tablecloths that are 7 ft by 7 ft. If the lace costs $1.95 per yard, how much will the trim cost for 6 small tablecloths and 6 large tablecloths?

77. 🖩 A square wooden rack is used to store the 15 numbered balls as well as the cue ball in pool. If a pool ball has a diameter of 57 mm, find the inside perimeter of the storage rack.

78. A rectangular box is used to store six Christmas ornaments. Find the perimeter of such a box if each ornament has a diameter of 72 mm.

2.8 SOLVING EQUATIONS

Objectives

a Use the addition principle to solve equations.

b Use the division principle to solve equations.

c Decide which principle should be used to solve an equation.

d Solve equations that require use of both the addition principle and the division principle.

In Section 1.7, we learned to solve certain equations by writing a "related equation." We now extend this approach to include negative integers, as well as equations that involve both addition and multiplication.

a   The Addition Principle

In Section 1.7, we learned to solve an equation like $x + 12 = 27$ by writing the related subtraction, $x = 27 - 12$, or $x = 15$. Note that $x = 15$ is an equation, not a solution. Of course, the solution of the equation $x = 15$ is obviously 15. The solution of $x + 12 = 27$ is also 15. Because their solutions are identical, $x = 15$ and $x + 12 = 27$ are said to be **equivalent equations.**

> **EQUIVALENT EQUATIONS**
>
> Equations with the same solutions are called **equivalent equations.**

It is important to be able to distinguish between equivalent *expressions* and equivalent *equations*.

- $6a$ and $4a + 2a$ are equivalent *expressions* because, for any replacement of a, both expressions represent the same number.
- $3x = 15$ and $4x = 20$ are equivalent *equations* because any solution of one equation is also a solution of the other equation.

EXAMPLE 1   Classify each pair as either equivalent equations or equivalent expressions:

a) $5x + 1$; $2x - 4 + 3x + 5$ **b)** $x = -7$; $x + 2 = -5$.

a) First note that these are expressions, not equations. To see if they are equivalent, we combine like terms in the second expression:

$$2x - 4 + 3x + 5 = (2 + 3)x + (-4 + 5) \quad \text{Regrouping and using the distributive law}$$

$$= 5x + 1.$$

We see that $2x - 4 + 3x + 5$ and $5x + 1$ are *equivalent expressions*.

b) First note that both $x = -7$ and $x + 2 = -5$ are equations. The solution of $x = -7$ is -7. We substitute to see if -7 is also the solution of $x + 2 = -5$:

$$x + 2 = -5$$
$$-7 + 2 = -5 \quad \text{TRUE}$$

Since $x = -7$ and $x + 2 = -5$ have the same solution, they are *equivalent equations*.

Do Exercises 1 and 2.

There are principles that enable us to begin with one equation and create an equivalent equation similar to $x = 15$, for which the solution is obvious. One such principle, the *addition principle*, is stated on the next page.

Suppose that a and b stand for the same number and some number c is added to a. We get the same result if we add c to b, because a and b are equal.

Classify each pair as equivalent expressions or equivalent equations.

1. $a - 5 = -3$; $a = 2$

2. $a - 9 + 6a$; $7a - 9$

Answers on page A-6

Solve.

3. $x - 5 = 19$

4. $x - 9 = -12$

Solve.

5. $42 = x + 17$

6. $a + 8 = -6$

Answers on page A-6

Study Tips

USE A PENCIL

It is no coincidence that the students who experience the greatest success in this course work in pencil. We all make mistakes and by using pencil and eraser we are more willing to admit to ourselves that something needs to be rewritten. Please work with a pencil and eraser if you aren't doing so already.

THE ADDITION PRINCIPLE

For any numbers a, b, and c,

$$a = b \quad \text{is equivalent to} \quad a + c = b + c.$$

■ **EXAMPLE 2** Solve: $x - 7 = -2$.

We have

$$x - 7 = -2$$
$$x - 7 + 7 = -2 + 7 \qquad \text{Using the addition principle: adding 7 to both sides}$$
$$x + 0 = 5 \qquad \text{Adding 7 "undoes" the subtraction of 7.}$$
$$x = 5.$$

The solution appears to be 5. To check, we use the original equation.

Check:
$$\frac{x - 7 = -2}{5 - 7 \ ? \ -2}$$
$$-2 \ | \qquad \text{TRUE}$$

The solution is 5.

Do Exercises 3 and 4.

Recall from Section 2.3 that we can subtract by adding the opposite of the number being subtracted. Because of this, the addition principle allows us to subtract the same number from both sides of an equation.

■ **EXAMPLE 3** Solve: $23 = t + 7$.

We have

$$23 = t + 7$$
$$23 - 7 = t + 7 - 7 \qquad \text{Using the addition principle to add } -7 \text{ or to subtract 7 on both sides}$$
$$16 = t + 0 \qquad \text{Subtracting 7 "undoes" the addition of 7.}$$
$$16 = t.$$

The solution is 16. The check is left to the student.

To visualize the addition principle, think of a jeweler's balance. When both sides of the balance hold equal amounts of weight, the balance is level. If weight is added or removed, equally, on both sides, the balance remains level.

Do Exercises 5 and 6.

b The Division Principle

In Section 1.7, we solved $8n = 96$ by dividing both sides by 8:

$$8 \cdot n = 96$$

$$\frac{8 \cdot n}{8} = \frac{96}{8} \qquad \text{Dividing both sides by 8}$$

$$n = 12. \qquad \text{8 times } n \text{, divided by 8, is } n. \ 96 \div 8 \text{ is 12.}$$

You can check that $8 \cdot n = 96$ and $n = 12$ are equivalent. We can divide both sides of an equation by any nonzero number in order to find an equivalent equation.

THE DIVISION PRINCIPLE

For any numbers a, b, and c ($c \neq 0$),

$$a = b \quad \text{is equivalent to} \quad \frac{a}{c} = \frac{b}{c}.$$

In Chapter 3, after we have discussed multiplication of fractions, we will use an equivalent form of this principle: the multiplication principle.

EXAMPLE 4 Solve: $9x = 63$.

We have

$$9x = 63$$

$$\frac{9x}{9} = \frac{63}{9} \qquad \text{Using the division principle to divide both sides by 9}$$

$$x = 7.$$

Check:
$$\begin{array}{c|c} 9x = 63 \\ \hline 9 \cdot 7 \ ? \ 63 \\ 63 \ | & \text{TRUE} \end{array}$$

The solution is 7.

Do Exercises 7 and 8.

EXAMPLE 5 Solve: $48 = -8n$.

It is important to distinguish between an opposite, as we have in $-8n$, and subtraction, as we had in $x - 5 = 19$ (margin exercise 3). To undo multiplication by -8, we use the division principle:

$$48 = -8n$$

$$\frac{48}{-8} = \frac{-8n}{-8} \qquad \text{Dividing both sides by } -8$$

$$-6 = n.$$

Check:
$$\begin{array}{c|c} 48 = -8n \\ \hline 48 \ ? \ -8(-6) \\ | \ 48 & \text{TRUE} \end{array}$$

The solution is -6.

Solve.

7. $7x = 42$

8. $-24 = 3t$

Solve.

9. $63 = -7n$

10. $-6x = 72$

Answers on page A-6

Solve.

11. $-x = 23$

12. $-t = -3$

Be sure that you understand why the addition principle is used in Example 2 and the division principle is used in Example 5.

Do Exercises 9 and 10 on the previous page.

Equations like $-x = 7$ or $-t = -3$ often give students difficulty. One way to handle problems of this sort is to multiply both sides of the equation by -1.

EXAMPLE 6 Solve: $-x = 7$.

To solve an equation like $-x = 7$, remember that when an expression is multiplied or divided by -1, its sign is changed. Here we multiply on both sides by -1 to change the sign of $-x$:

$$-x = 7$$
$$(-1)(-x) = (-1) \cdot 7 \qquad \text{Multiplying both sides by } -1$$
$$x = -7. \qquad \text{Note that } (-1)(-x) \text{ is the same as } (-1)(-1)x.$$

Check: $$\frac{-x = 7}{-(-7) \; ? \; 7}$$
$$7 \mid \quad \text{TRUE}$$

The solution is -7.

Another way to solve Example 6 is to note that $-x = -1 \cdot x$. Then we can divide both sides by -1:

$$-x = 7$$
$$-1 \cdot x = 7$$
$$\frac{-1 \cdot x}{-1} = \frac{7}{-1}$$
$$x = -7.$$

Do Exercises 11 and 12.

C Selecting the Correct Approach

It is important for you to be able to determine which principle should be used to solve a particular equation.

EXAMPLES Solve.

7. $39 = -3 + t$

Note that -3 is added to t. To undo addition of -3, we subtract -3 or simply add 3 on both sides:

$$3 + 39 = 3 + (-3) + t \qquad \text{Using the addition principle}$$
$$42 = 0 + t$$
$$42 = t.$$

Check: $$\frac{39 = -3 + t}{39 \; ? \; -3 + 42}$$
$$\mid \; 39 \qquad \text{TRUE}$$

The solution is 42.

8. $39 = -3t$

To undo multiplication by -3, we divide by -3 on both sides:

$$39 = -3t$$

$$\frac{39}{-3} = \frac{-3t}{-3} \qquad \text{Using the division principle}$$

$$-13 = t.$$

Check:
$$\begin{array}{c|c} \multicolumn{2}{c}{39 = -3t} \\ \hline 39 \;\; ? \;\; -3(-13) \\ 39 & \text{TRUE} \end{array}$$

The solution is -13.

Do Exercises 13–15.

d Using the Principles Together

Suppose we want to determine whether 7 is the solution of $5x - 8 = 27$. To check, we replace x with 7 and simplify.

Check:
$$\begin{array}{c|c} \multicolumn{2}{c}{5x - 8 = 27} \\ \hline 5 \cdot 7 - 8 \;\; ? \;\; 27 \\ 35 - 8 \\ 27 & \text{TRUE} \end{array}$$

This shows that 7 *is* the solution.

Do Exercises 16 and 17.

In the check above, note that the rules for order of operations require that we multiply before we subtract (or add).

The rules for order of operations dictate that unless grouping symbols indicate otherwise, multiplication and division are performed before any addition or subtraction. Thus, to evaluate $5x - 8$,

we *select* a value: x

then *multiply* by 5: $5x$

and then *subtract* 8: $5x - 8$.

In Example 9, which follows, these steps are reversed to solve for x:

We will *add* 8: $5x - 8 + 8$,

then *divide* by 5: $\dfrac{5x}{5}$

and *isolate* x: x.

In general, the *last* step performed when calculating is the *first* step to be reversed when finding a solution.

Solve.

13. $-2x = -52$

14. $-2 + x = -52$

15. $x \cdot 7 = -28$

16. Determine whether -9 is the solution of $7x + 8 = -55$.

17. Determine whether -6 is the solution of $4x + 3 = -25$.

Answers on page A-6

18. Solve: $2x - 9 = 43$

EXAMPLE 9 Solve: $5x - 8 = 27$.

We first note that the term containing x is $5x$. To isolate $5x$, we add 8 on both sides:

$$5x - 8 = 27$$
$$5x - 8 + 8 = 27 + 8 \qquad \text{Using the addition principle}$$
$$5x + 0 = 35 \qquad \text{Try to do this step mentally.}$$
$$5x = 35.$$

Next, we isolate x by dividing by 5 on both sides:

$$5x = 35$$
$$\frac{5x}{5} = \frac{35}{5} \qquad \text{Using the division principle}$$
$$1x = 7 \qquad \text{Try to do this step mentally.}$$
$$x = 7.$$

The check was performed on the previous page. The solution is 7.

Do Exercise 18.

EXAMPLE 10 Solve: $38 = -9x + 2$.

We first isolate $-9x$ by subtracting 2 on both sides:

$$38 = -9x + 2$$
$$38 - 2 = -9x + 2 - 2 \qquad \text{Subtracting 2 (or adding } -2 \text{) from both sides}$$
$$36 = -9x + 0 \qquad \text{Try to do this step mentally.}$$
$$36 = -9x.$$

19. Solve: $-3x + 2 = 47$

Now that we have isolated $-9x$ on one side of the equation, we can divide by -9 to isolate x:

$$36 = -9x$$
$$\frac{36}{-9} = \frac{-9x}{-9} \qquad \text{Dividing both sides by } -9$$
$$-4 = x. \qquad \text{Simplifying}$$

Check:

$$\begin{array}{c|c} 38 = -9x + 2 \\ \hline 38 \ ? \ -9 \cdot (-4) + 2 \\ & 36 + 2 \\ & 38 \qquad \text{TRUE} \end{array}$$

The solution is -4.

Do Exercise 19.

a   Classify each pair as either equivalent expressions or equivalent equations.

1. $2x = 10$; $5x = 25$

2. $4x + 1$; $6 + 4x - 5$

3. $7a - 3$; $4a - 3 + 3a$

4. $7t = 14$; $4t = 8$

5. $4r + 3$; $8 + 4r - 5$

6. $2r - 7$; $r - 7 + r$

7. $x - 9 = 8$; $x + 3 = 20$

8. $t + 4 = 19$; $t - 6 = 9$

9. $3(t + 2)$; $5 + 3t + 1$

10. $2x = -14$; $x - 2 = -9$

11. $x + 4 = -8$; $2x = -24$

12. $4(x - 7)$; $3x - 28 + x$

Solve.

13. $x - 6 = -9$

14. $x - 5 = -7$

15. $x - 4 = -12$

16. $x - 7 = 5$

17. $a + 7 = 25$

18. $x + 9 = -3$

19. $x + 8 = -6$

20. $t + 5 = 13$

21. $24 = t - 8$

22. $-9 = x + 3$

23. $-12 = x + 5$

24. $17 = n - 6$

25. $-5 + a = 12$

26. $3 = 17 + x$

27. $-8 = -8 + t$

28. $-7 + t = -7$

b   Solve.

29. $6x = -24$

30. $-8t = 40$

31. $-3t = 42$

32. $3x = 24$

33. $-7n = -35$

34. $64 = -2t$

35. $0 = 6x$

36. $-5n = -65$

37. $55 = -5t$

38. $-x = 83$

39. $-x = 56$

40. $-2x = 0$

41. $n(-4) = -48$

42. $-x = -475$

43. $-x = -390$

44. $n(-6) = -42$

C Solve.

45. $t - 6 = -2$

46. $3t = -45$

47. $6x = -54$

48. $x + 9 = -15$

49. $15 = -x$

50. $-13 = x - 4$

51. $-21 = x + 5$

52. $-42 = -x$

53. $35 = -7t$

54. $7 + t = -18$

55. $-17x = 68$

56. $-34 = x + 10$

57. $18 + t = -160$

58. $-48 = t(-12)$

59. $-27 = x + 23$

60. $-135 = -9t$

d Solve.

61. $5x - 1 = 34$

62. $7x - 3 = 25$

63. $4t + 2 = 14$

64. $3t + 5 = 26$

65. $6a + 1 = -17$

66. $8a + 3 = -37$

67. $2x - 9 = -23$

68. $3x - 5 = -35$

69. $-2x + 1 = 17$

70. $-4t + 3 = -17$

71. $-8t - 3 = -67$

72. $-7x - 4 = -46$

73. $-x + 9 = -15$

74. $-x - 6 = 8$

75. $7 = 2x - 5$

76. $9 = 4x - 7$

77. $13 = 3 + 2x$

78. $33 = 5 - 4x$

79. $13 = 5 - x$

80. $12 = 7 - x$

81. **D**_{**W**} To solve $-5x = 13$, Eva decides to add 5 to both sides of the equation. Is there anything wrong with her doing this? Why or why not?

82. **D**_{**W**} Gary decides to solve $x - 9 = -5$ by adding 5 to both sides of the equation. Is there anything wrong with his doing this? Why or why not?

🐦 **VOCABULARY REINFORCEMENT**

In each of Exercises 83–90, fill in the blank with the correct term from the given list. Some of the choices may not be used and some may be used more than once.

83. A(n) _____ is a closed geometric figure. [2.7b]

84. Terms are _____ if they have the same variable factor(s). [2.7a]

85. Numbers we multiply together are called _____ . [1.5a]

86. Equations are _____ if they have the same solutions. [2.8a]

87. The result of an addition is a(n) _____ . [1.2a]

88. A(n) _____ is a letter that can stand for various numbers. [2.6a]

89. The _____ of a number is its distance from zero on a number line. [2.1c]

90. We _____ for a variable when we replace it with a number. [2.6a]

absolute value	substitute
opposite	polygon
constant	perimeter
variable	equivalent
factors	similar
addends	sum
evaluate	product

SYNTHESIS

91. $\mathbf{D_W}$ Explain how equivalent expressions can be used to write equivalent equations.

92. $\mathbf{D_W}$ To solve $2x + 8 = 24$, Wilma divides both sides by 2. Can this first step lead to a solution? Why or why not?

Solve.

93. $2x - 7x = -40$

94. $9 + x - 5 = 23$

95. $17 - 3^2 = 4 + t - 5^2$

96. $(-9)^2 = 2^3 t + (3 \cdot 6 + 1)t$

97. $(-7)^2 - 5 = t + 4^3$

98. ▦ $(-42)^3 = 14^2 t$

99. ▦ $x - (19)^3 = -18^3$

100. ▦ $23^2 = x + 22^2$

101. ▦ $35^3 = -125t$

102. ▦ $248 = 24 - 32x$

103. ▦ $529 - 143x = -1902$

The review that follows is meant to prepare you for a chapter exam. It consists of three parts. The first part, Concept Reinforcement, is designed to increase understanding of the concepts through true/false exercises. The second part is a list of important properties and formulas. The third part is the Review Exercises. These provide practice exercises for the exam, together with references to section objectives so you can go back and review. Before beginning, stop and look back over the skills you have obtained. What skills in mathematics do you have now that you did not have before studying this chapter?

⤳ CONCEPT REINFORCEMENT

Determine whether the statement is true or false. Answers are given at the back of the book.

_____ **1.** The absolute value of a number is always nonnegative.

_____ **2.** The opposite of the opposite of a number is the original number.

_____ **3.** The product of an even number of negative numbers is positive.

_____ **4.** The expression $2(x + 3)$ is equivalent to the expression $2 \cdot x + 3$.

_____ **5.** $3 - x = 4x$ and $5x = -3$ are equivalent equations.

_____ **6.** Collecting like terms is based on the distributive law.

IMPORTANT PROPERTIES AND FORMULAS

For any integers a, b, and c:	$a + (-a) = 0;\quad a - b = a + (-b);$ $a \cdot 0 = 0;\quad a(b + c) = ab + ac$
Perimeter of a Rectangle:	$P = 2l + 2w,\quad$ or $\quad P = 2(l + w)$
Perimeter of a Square:	$P = 4s$
The Addition Principle:	$a = b\quad$ is equivalent to $\quad a + c = b + c.$
The Division Principle:	For $c \neq 0$, $a = b\quad$ is equivalent to $\quad \dfrac{a}{c} = \dfrac{b}{c}.$

Review Exercises

1. Tell which integers correspond to this situation: [2.1a]

Bonnie has $527 in her campus account and Roger is $53 in debt.

Use either $<$ or $>$ for ☐ to form a true statement. [2.1b]

2. 0 ☐ -5 **3.** -7 ☐ 6 **4.** -4 ☐ -19

Find the absolute value.   [2.1c]

5. $|-39|$ **6.** $|23|$ **7.** $|0|$

8. Find $-x$ when $x = -72$.   [2.1d]

9. Find $-(-x)$ when $x = 59$.   [2.1d]

Compute and simplify.   [2.2a], [2.3a], [2.4a, b], [2.5a, b]

10. $-14 + 5$ **11.** $-5 + (-6)$

12. $14 + (-8)$ **13.** $0 + (-24)$

14. $17 - 29$ **15.** $9 - (-14)$

16. $-8 - (-7)$ **17.** $-3 - (-10)$

18. $-3 + 7 + (-8)$ **19.** $8 - (-9) - 7 + 2$

20. $-23 \cdot (-4)$ **21.** $7(-12)$

22. $2(-4)(-5)(-1)$ **23.** $15 \div (-5)$

24. $\dfrac{-55}{11}$ **25.** $\dfrac{0}{7}$

26. $7 \div 1^2 \cdot (-3) - 4$ **27.** $(-3)|4 - 3^2| - 5$

28. Evaluate $3a + b$ for $a = 4$ and $b = -5$.   [2.6a]

29. Evaluate $\dfrac{-x}{y}$, $\dfrac{x}{-y}$, and $-\dfrac{x}{y}$ for $x = 30$ and $y = 5$.
   [2.6a]

Use the distributive law to write an equivalent expression.
[2.6b]

30. $4(5x + 9)$ **31.** $3(2a - 4b + 5)$

Combine like terms.   [2.7a]

32. $5a + 12a$ **33.** $-7x + 13x$

34. $9m + 14 - 12m - 8$

35. Find the perimeter of a rectangular frame that is 8 in. by 10 in.   [2.7b]

36. Find the perimeter of a square pane of glass that is 25 cm on each side.   [2.7b]

Solve.   [2.8a, b, c, d]

37. $x - 9 = -17$

38. $-4t = 36$

39. $13 = -x$

40. $56 = 6x - 10$

41. $-x + 3 = -12$

42. $18 = 4 - 2x$

43. **D_W** Explain the difference between equivalent expressions and equivalent equations.   [2.8a]

44. **D_W** Is a number's absolute value ever less than the number itself? Why or why not?   [2.1c]

SYNTHESIS

45. **D_W** A classmate insists on reading $-x$ as "negative x." When asked why, the response is "because $-x$ is negative." What mistake is this student making?   [2.1d]

46. **D_W** Are $(a - b)^2$ and $(b - a)^2$ equivalent for all choices of a and b? Why or why not? Experiment with different replacements for a and b.   [2.6a]

Simplify.   [2.5b]

47. 🔢 $87 \div 3 \cdot 29^3 - (-6)^6 + 1957$

48. 🔢 $1969 + (-8)^5 - 17 \cdot 15^3$

49. 🔢 $\dfrac{113 - 17^3}{15 + 8^3 - 507}$

50. For what values of x will $8 + x^3$ be negative?   [2.6a]

51. For what values of x is $|x| > x$?   [2.1b, c]

1. Tell which integers correspond to this situation: The Tee Shop sold 542 fewer muscle shirts than expected in January and 307 more than expected in February.

2. Use either $<$ or $>$ for $\square$ to form a true statement.

$$-14 \;\square\; -21$$

3. Find the absolute value: $|-739|$.

4. Find $-(-x)$ when $x = -19$.

Compute and simplify.

5. $6 + (-17)$

6. $-9 + (-12)$

7. $-8 + 17$

8. $0 - 12$

9. $7 - 22$

10. $-5 - 19$

11. $-8 - (-27)$

12. $31 - (-3) - 5 + 9$

13. $(-4)^3$

14. $27(-10)$

15. $-9 \cdot 0$

16. $-72 \div (-9)$

17. $\dfrac{-56}{7}$

18. $8 \div 2 \cdot 2 - 3^2$

19. $29 - (3 - 5)^2$

20. *Antarctica Highs and Lows.* The continent of Antarctica, which lies in the southern hemisphere, experiences winter in July. The average high temperature is $-67°$F and the average low temperature is $-81°$F. How much higher is the average high than the average low?

Source: National Climatic Data Center

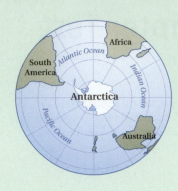

21. Jeannie rewound a tape in her video camera from the 8 minute mark to the -15 minute mark. How many minutes of tape were rewound?

22. Evaluate $\dfrac{a - b}{6}$ for $a = -8$ and $b = 10$.

23. Use the distributive law to write an equivalent expression.
$7(2x + 3y - 1)$

24. Combine like terms.
$9x - 14 - 5x - 3$

Solve.

25. $-7x = -35$

26. $a + 9 = -3$

SYNTHESIS

27. Monty plans to attach trim around the doorway and along the base of the walls in a 12-ft by 14-ft room. If the doorway is 3 ft by 7 ft, how many feet of trim are needed? (Only three sides of a doorway get trim.)

Simplify.

28. $9 - 5[x + 2(3 - 4x)] + 14$

29. $15x + 3(2x - 7) - 9(4 + 5x)$

30. ▦ $49 \cdot 14^3 \div 7^4 + 1926^2 \div 6^2$

31. ▦ $3487 - 16 \div 4 \cdot 4 \div 2^8 \cdot 14^4$

CHAPTER 2: Introduction to Integers
and Algebraic Expressions

Cumulative Review

1. Write standard notation for the number written in words in the following sentence: In 2003 there were about one hundred eighty-one million, five hundred ninety-nine thousand, nine hundred telephone lines in use in the United States.

2. Write a word name for 5,380,001,437.

Add.

3. $\begin{array}{r} 15{,}892 \\ + 2{,}935 \\ \hline \end{array}$

4. $\begin{array}{r} 7989 \\ 789 \\ + 79 \\ \hline \end{array}$

Subtract.

5. $\begin{array}{r} 8276 \\ - 430 \\ \hline \end{array}$

6. $\begin{array}{r} 3006 \\ - 578 \\ \hline \end{array}$

Multiply.

7. $\begin{array}{r} 621 \\ \times 27 \\ \hline \end{array}$

8. $\begin{array}{r} 2505 \\ \times 3300 \\ \hline \end{array}$

9. $43 \cdot (-8)$

10. $-12(-6)$

Divide.

11. $63\overline{)6552}$

12. $62\overline{)3844}$

13. $0 \div (-67)$

14. $60 \div (-12)$

15. Round 427,931 to the nearest thousand.

16. Round 5309 to the nearest hundred.

Estimate each sum or product by rounding to the nearest hundred. Show your work.

17. $\begin{array}{r} 749{,}559 \\ + 301{,}362 \\ \hline \end{array}$

18. $\begin{array}{r} 749 \\ \times 531 \\ \hline \end{array}$

19. Use $<$ or $>$ for $\square$ to form a true sentence:
$-26 \ \square \ 19.$

20. Find the absolute value: $|-279|$.

Simplify.

21. $35 - 25 \div 5 + 2 \times 3$

22. $\{17 - [8 - (5 - 2 \times 2)]\} \div (3 + 12 \div 6)$

23. $10 \div 1(-5) - 6^2$

24. 5^3

25. Evaluate $\dfrac{x + y}{5}$ for $x = 11$ and $y = 4$.

26. Evaluate $7x^2$ for $x = -2$.

Use the distributive law to write an equivalent expression.

27. $-2(x + 5)$

28. $6(3x - 2y + 4)$

Simplify.

29. $-12 + (-14)$

30. $(-3)(-10)$

31. $23 - 38$

32. $64 \div (-2)$

33. $-12 - (-25)$

34. $(-2)(-3)(-5)$

35. $3 - (-8) + 2 - (-3)$

36. $16 \div 2(-8) + 7$

Solve.

37. $x + 8 = 35$

38. $-12t = 36$

39. $6 - x = -9$

40. $-39 = 4x - 7$

Solve.

41. In the movie *Little Big Man,* Dustin Hoffman plays a character who ages from 17 to 121. This represents the greatest age range depicted by one actor in one film. How many years did Hoffman's character age?
Source: Guinness Book of World Records

42. The ten largest hotels in the United States are in Las Vegas. Of these, the four largest are the MGM Grand, the Luxor, the Excalibur, and the Circus Circus. These have 5034 rooms, 4408 rooms, 4008 rooms, and 3770 rooms, respectively. What is the total number of rooms in these four hotels?
Source: http://govegas.about.com/cs/hotels/tp/largesthotels.htm

43. Amanda is offered a part-time job paying $4940 a year. How much is each weekly paycheck?

44. Eastside Appliance sells a refrigerator for $600 and $30 tax with no delivery charge. Westside Appliance sells the same model for $560 and $28 tax plus a $25 delivery charge. Which is the better buy?

45. Write an equivalent expression by combining like terms: $7x - 9 + 3x - 5$.

SYNTHESIS

46. A soft-drink distributor has 166 loose cans of cola. The distributor wishes to form as many 24-can cases as possible and then, with any remaining cans, as many six-packs as possible. How many cases will be filled? How many six-packs? How many loose cans will remain?

47. Simplify: $a - \{3a - [4a - (2a - 4a)]\}$.

48. ▦ Simplify: $37 \cdot 64 \div 4^2 \cdot 2 - (7^3 - (-4)^5)$.

49. Find two solutions of $5|x| - 2 = 13$.

Fraction Notation: Multiplication and Division

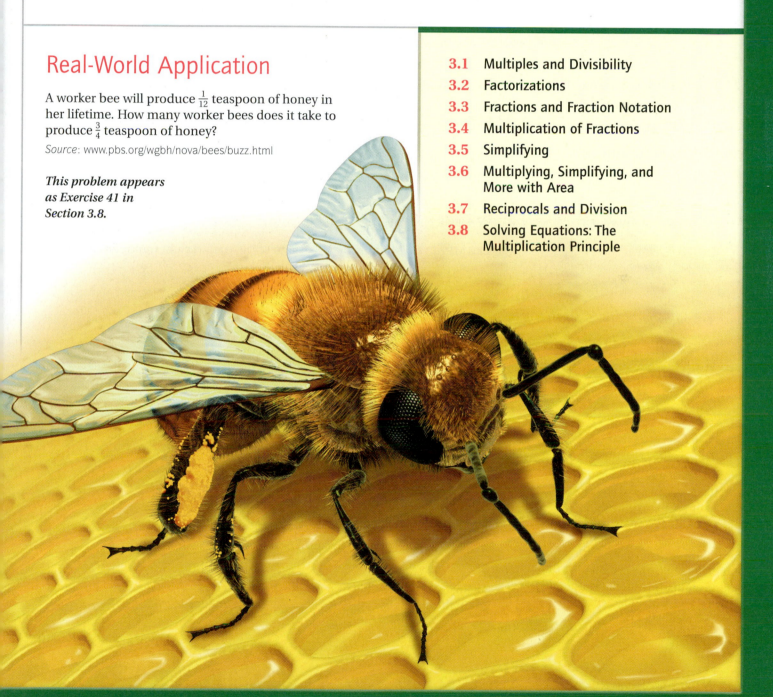

Real-World Application

A worker bee will produce $\frac{1}{12}$ teaspoon of honey in her lifetime. How many worker bees does it take to produce $\frac{3}{4}$ teaspoon of honey?

Source: www.pbs.org/wgbh/nova/bees/buzz.html

This problem appears as Exercise 41 in Section 3.8.

Objectives

a Find some multiples of a number, and determine whether a number is divisible by another number.

b Test to see if a number is divisible by 2, 3, 5, 6, 9, or 10.

1. Show that each of the numbers 5, 45, and 100 is a multiple of 5.

2. Show that each of the numbers 10, 60, and 110 is a multiple of 10.

3. Multiply by 1, 2, 3, and so on, to find ten multiples of 5.

Answers on page A-7

Study Tips

CREATE YOUR OWN GLOSSARY

Understanding the meaning of mathematical terminology is essential for success in any math course. To assist with this, try writing your own glossary of important words toward the back of your note-book. Often, just the act of writing out a word's definition can help you remember what the word means.

Before we can begin our work with fractions, we must discuss products and factorizations. In this section, we learn about *multiples* and *divisibility* in order to be able to simplify fractions like $\frac{117}{225}$.

a Multiples

A **multiple** of a number is a product of that number and an integer. For example, some multiples of 2 are:

$$2 \quad (\text{because } 2 = 1 \cdot 2);$$
$$4 \quad (\text{because } 4 = 2 \cdot 2);$$
$$6 \quad (\text{because } 6 = 3 \cdot 2);$$
$$8 \quad (\text{because } 8 = 4 \cdot 2);$$
$$10 \quad (\text{because } 10 = 5 \cdot 2).$$

We can also find multiples of 2 by counting by twos: 2, 4, 6, 8, and so on.

EXAMPLE 1 Show that each of the numbers 3, 6, 9, and 15 is a multiple of 3.

We show that each of 3, 6, 9, and 15 can be expressed as a product of 3 and some integer:

$$3 = 1 \cdot 3; \qquad 6 = 2 \cdot 3; \qquad 9 = 3 \cdot 3; \qquad 15 = 5 \cdot 3.$$

Do Exercises 1 and 2.

EXAMPLE 2 Multiply by 1, 2, 3, and so on, to find eight multiples of six.

$$1 \cdot 6 = 6 \qquad 5 \cdot 6 = 30$$
$$2 \cdot 6 = 12 \qquad 6 \cdot 6 = 36$$
$$3 \cdot 6 = 18 \qquad 7 \cdot 6 = 42$$
$$4 \cdot 6 = 24 \qquad 8 \cdot 6 = 48$$

Do Exercise 3.

> **DIVISIBILITY**
>
> A number b is said to be **divisible** by another number a if b is a multiple of a.

Thus,

6 is divisible by 2 because 6 is a multiple of 2 ($6 = 3 \cdot 2$);

27 is divisible by 3 because 27 is a multiple of 3 ($27 = 9 \cdot 3$);

100 is divisible by 25 because 100 is a multiple of 25 ($100 = 4 \cdot 25$).

Saying that b is divisible by a means that $b \div a$ results in a remainder of zero. When this happens, we sometimes say that a divides b "evenly."

EXAMPLE 3 Determine whether 24 is divisible by 3.

We divide 24 by 3:

$$
\begin{array}{r}
8 \\
3\overline{)24} \\
\underline{24} \\
0
\end{array}
$$

The remainder of 0 indicates that 24 is divisible by 3.

EXAMPLE 4 Determine whether 98 is divisible by 4.

We divide 98 by 4:

$$
\begin{array}{r}
24 \\
4\overline{)98} \\
\underline{8} \\
18 \\
\underline{16} \\
2 \leftarrow \text{Not } 0
\end{array}
$$

Since the remainder is not 0 we know that 98 is *not* divisible by 4.

Do Exercises 4–6.

b Tests for Divisibility

We now learn quick ways of checking for divisibility by 2, 3, 5, 6, 9, and 10 without actually performing long division. Tests do exist for divisibility by 4, 7, and 8, but they can be as difficult to perform as the actual long division.

To test for divisibility by 2, 5, and 10, we examine the ones digit.

DIVISIBILITY BY 2

All even numbers are divisible by 2.

BY 2
A number is **divisible by 2** (is *even*) if it has a ones digit of 0, 2, 4, 6, or 8 (that is, it has an even ones digit).

To see why this test works, start counting by 2's: 2, 4, 6, 8, 10, 12, 14, 16, 18, 20, 22, ... Note that the ones digit will always be 0, 2, 4, 6, or 8, no matter how high we count.

EXAMPLES Determine whether the number is divisible by 2.

5. 355 *is not* divisible by 2; 5 is not even.

6. 4786 *is* divisible by 2; 6 is even.

7. 8990 *is* divisible by 2; 0 is even.

8. 4261 *is not* divisible by 2; 1 is not even.

Do Exercises 7–10.

4. Determine whether 16 is divisible by 2.

5. Determine whether 125 is divisible by 5.

6. Determine whether 125 is divisible by 6.

Determine whether the number is divisible by 2.

7. 84 **8.** 59

9. 998 **10.** 2225

Answers on page A-7

CALCULATOR CORNER

Divisibility Rather than list remainders, most calculators display quotients using decimal notation. Although decimal notation is not studied until Chapter 5, it is possible for us now to check for divisibility using a calculator.

To see if a number, like 551, is divisible by another number, like 19, we simply press 5 5 1 ÷ 1 9 = . If the resulting quotient contains no digits to the right of the decimal point, the first number is divisible by the second. Thus, since 551 ÷ 19 = 29, we know that 551 *is* divisible by 19. On the other hand, since 551 ÷ 20 = 27.55, we know that 551 is *not* divisible by 20.

Exercises: For each pair of numbers, determine whether the first number is divisible by the second number.

1. 731; 17
2. 1502; 79
3. 1053; 36
4. 4183; 47
5. 3875; 15
6. 32,768; 256

Determine whether the number is divisible by 5.

11. 5780

12. 3427

13. 34,678

14. 7775

Determine whether the number is divisible by 10.

15. 305

16. 300

17. 847

18. 8760

Determine whether the number is divisible by 3.

19. 111

20. 1111

21. 309

22. 17,216

Answers on page A-7

164

DIVISIBILITY BY 5

To determine the test for divisibility by 5, we start counting by 5's: 5, 10, 15, 20, 25, 30, 35, ... Note that the ones digit will always be 5 or 0, no matter how high we count.

BY 5
A number is **divisible by 5** if its ones digit is 0 or 5.

EXAMPLES Determine whether the number is divisible by 5.

9. 220 *is* divisible by 5 because the ones digit is 0.

10. 475 *is* divisible by 5 because the ones digit is 5.

11. 6514 *is not* divisible by 5 because the ones digit is neither 0 nor 5.

Do Exercises 11–14.

DIVISIBILITY BY 10

BY 10
A number is **divisible by 10** if its ones digit is 0.

We know that this test works because the product of 10 and *any* number has a ones digit of 0.

EXAMPLES Determine whether the number is divisible by 10.

12. 3440 *is* divisible by 10 because the ones digit is 0.

13. 3447 *is not* divisible by 10 because the ones digit is not 0.

Do Exercises 15–18.

To test for divisibility by 3 and 9, we examine the sum of a number's digits.

DIVISIBILITY BY 3

BY 3
A number is **divisible by 3** if the sum of its digits is divisible by 3.

An explanation of why this test works is outlined in Exercise 77 at the end of this section.

EXAMPLES Determine whether the number is divisible by 3.

14. 18 $1 + 8 = 9$
15. 93 $9 + 3 = 12$ Each *is* divisible by 3 because the sum of its digits is divisible by 3.
16. 201 $2 + 0 + 1 = 3$

17. 256 $2 + 5 + 6 = 13$ The sum of the digits, 13, *is not* divisible by 3, so 256 *is not* divisible by 3.

Do Exercises 19–22.

DIVISIBILITY BY 9

The test for divisibility by 9 is similar to the test for divisibility by 3.

> **BY 9**
>
> A number is **divisible by 9** if the sum of its digits is divisible by 9.

EXAMPLES   Determine whether the number is divisible by 9.

18. 6984

Because $6 + 9 + 8 + 4 = 27$ and 27 is divisible by 9, 6984 *is* divisible by 9.

19. 322

Because $3 + 2 + 2 = 7$ and 7 is not divisible by 9, 322 *is not* divisible by 9.

Do Exercises 23–26.

DIVISIBILITY BY 6

A number divisible by 6 is a multiple of 6. But $6 = 2 \cdot 3$, so the number is also a multiple of 2 and 3. Since 2 and 3 have no factors in common, a number is divisible by 6 if it is divisible by 2 *and* by 3.

> **BY 6**
>
> A number is **divisible by 6** if its ones digit is 0, 2, 4, 6, or 8 (is even) and the sum of its digits is divisible by 3.

EXAMPLES   Determine whether the number is divisible by 6.

20. 720

Because 720 is even, it is divisible by 2. Also, $7 + 2 + 0 = 9$, so 720 is divisible by 3. Thus, 720 *is* divisible by 6.

720    $7 + 2 + 0 = 9$

↑   ↑

Even   Divisible by 3

21. 73

73 *is not* divisible by 6 because it is not even.

22. 256

Although 256 is even, it *is not* divisible by 6 because the sum of its digits, $2 + 5 + 6$, or 13, is not divisible by 3.

Do Exercises 27–30.

Determine whether the number is divisible by 9.

23. 16

24. 117

25. 930

26. 29,223

Determine whether the number is divisible by 6.

27. 420

28. 106

29. 321

30. 444

Answers on page A-7

a Multiply by 1, 2, 3, and so on, to find ten multiples of each number.

1. 7 **2.** 4 **3.** 20 **4.** 50 **5.** 3 **6.** 8

7. 12 **8.** 15 **9.** 10 **10.** 11 **11.** 25 **12.** 100

13. Determine whether 61 is divisible by 3.

14. Determine whether 29 is divisible by 2.

15. Determine whether 527 is divisible by 7.

16. Determine whether 336 is divisible by 8.

17. Determine whether 8127 is divisible by 9.

18. Determine whether 4144 is divisible by 4.

b For Exercises 19–30, answer Yes or No and give a reason based on the tests for divisibility.

19. Determine whether 84 is divisible by 3.

20. Determine whether 467 is divisible by 9.

21. Determine whether 5553 is divisible by 5.

22. Determine whether 2004 is divisible by 6.

23. Determine whether 671,500 is divisible by 10.

24. Determine whether 6120 is divisible by 5.

25. Determine whether 1773 is divisible by 9.

26. Determine whether 3286 is divisible by 3.

27. Determine whether 21,687 is divisible by 2.

28. Determine whether 64,091 is divisible by 10.

29. Determine whether 32,109 is divisible by 6.

30. Determine whether 9840 is divisible by 2.

For Exercises 31–38, test each number for divisibility by 2, 3, 5, 6, 9, and 10.

31. 6825 **32.** 12,600 **33.** 119,117 **34.** 2916

35. 127,575 **36.** 25,088 **37.** 9360 **38.** 143,507

To answer Exercises 39–44, consider the following numbers. Use the tests for divisibility.

46	300	85	256
224	36	711	8064
19	45,270	13,251	1867
555	4444	254,765	21,568

39. Which of the above are divisible by 3?

40. Which of the above are divisible by 2?

41. Which of the above are divisible by 10?

42. Which of the above are divisible by 5?

43. Which of the above are divisible by 6?

44. Which of the above are divisible by 9?

To answer Exercises 45–50, consider the following numbers.

56	200	75	35
324	42	812	402
784	501	2345	111,111
55,555	3009	2001	1005

45. Which of the above are divisible by 2?

46. Which of the above are divisible by 3?

47. Which of the above are divisible by 5?

48. Which of the above are divisible by 10?

49. Which of the above are divisible by 9?

50. Which of the above are divisible by 6?

51. **D**_W Describe a test that can be used to determine whether a number is divisible by 25.

52. **D**_W Is every counting number a multiple of 1? Why or why not?

Solve.

53. $16 \cdot t = 848$ [1.7b], [2.8b]

54. $m + 9 = 14$ [1.7b], [2.8a]

55. $23 + x = 15$ [1.7b], [2.8a]

56. $24 \cdot m = -576$ [1.7b], [2.8b]

57. Find the total cost of 12 sweaters at $37 each and 4 jackets at $59 each. [1.8a]

58. Add: $-34 + 76$. [2.2a]

Evaluate. [1.9b]

59. 5^3

60. 7^3

61. 4^5

62. 3^6

Write in exponential notation. [1.9a]

63. $9 \cdot 9 \cdot 9 \cdot 9 \cdot 9$

64. $7 \cdot 7 \cdot 7 \cdot 7 \cdot 7 \cdot 7$

SYNTHESIS

65. **D$_W$** Describe a test that can be used to determine whether a number is divisible by 18.

66. **D$_W$** Describe a test for determining whether a number is divisible by 30.

67. ▦ Find the largest five-digit number that is divisible by 47.

68. ▦ Find the largest six-digit number that is divisible by 53.

Find the smallest number that is simultaneously a multiple of the given numbers.

69. 2, 3, and 5

70. 3, 5, and 7

71. 4, 6, and 10

72. 6, 10, and 14

73. ▦ 17, 43, and 85

74. ▦ 26, 57, and 130

75. 30, 70, and 120

76. 25, 100, and 175

77. **D$_W$** To help see why the tests for division by 3 and 9 work, note that any four-digit number $abcd$ can be rewritten as $1000 \cdot a + 100 \cdot b + 10 \cdot c + d$, or $999a + 99b + 9c + a + b + c + d$.

  a) Explain why $999a + 99b + 9c$ is divisible by both 9 and 3 for all choices of a, b, c, and d.

  b) Explain why the four-digit number $abcd$ is divisible by 9 if $a + b + c + d$ is divisible by 9 and is divisible by 3 if $a + b + c + d$ is divisible by 3.

78. A passenger in a taxicab asks for the cab number. The driver says abruptly, "Sure—it's the smallest multiple of 11 that, when divided by 2, 3, 4, 5, or 6, has a remainder of 1." What is the number?

79. ▦ Fill in the missing digits of the number

  95, ☐ ☐ 8

so that it is divisible by 99.

3.2 FACTORIZATIONS

Objectives

a   Find the factors of a number.

b   Given a number from 1 to 100, tell whether it is prime, composite, or neither.

c   Find the prime factorization of a composite number.

In Section 3.1, we saw that both 28 and 35 are multiples of 7. This is the same as saying that 7 is a *factor* of both 28 and 35. When a number is expressed as a product of factors, we say that we have *factored* the original number. Thus, "factor" can be used as either a noun or a verb. Being able to factor is an important skill needed for a solid understanding of fractions.

a   Factoring Numbers

From the equation $3 \cdot 4 = 12$, we can say that 3 and 4 are *factors* of 12. Since $12 = 12 \cdot 1$, we know that 12 and 1 are also factors of 12.

> **FACTORS AND FACTORIZATIONS**
>
> A number c is a **factor** of a if a is divisible by c.
>
> A **factorization** of a expresses a as a product of two or more numbers.

For example, each of the following gives a factorization of 12.

$12 = 4 \cdot 3$ ← This factorization shows that 4 and 3 are factors of 12.

$12 = 12 \cdot 1$ ← This factorization shows that 12 and 1 are factors of 12.

$12 = 6 \cdot 2$ ← This factorization shows that 6 and 2 are factors of 12.

$12 = 2 \cdot 3 \cdot 2$ ← This factorization shows that 2 and 3 are factors of 12.

Thus, 1, 2, 3, 4, 6, and 12 are all factors of 12. Note that since $n = n \cdot 1$, every number has a factorization, and every number has itself and 1 as factors.

EXAMPLE 1   Find all the factors of 24.

To get started, we can use some of the tests for divisibility. For example, since 24 is even, we know that 2 is a factor. Since the sum of the digits in 24 is 6 and 6 is divisible by 3, we know that 3 is a factor. We can use trial and error to determine that 4 is also a factor, but that 5 is not. A list of factorizations can then be used to make a complete list of factors.

Factorizations: $1 \cdot 24$; $2 \cdot 12$; $3 \cdot 8$; $4 \cdot 6$

Factors: 1, 2, 3, 4, 6, 8, 12, 24

It is useful to note that when two different numbers are factors of a number, the product of those numbers is also a factor. For instance, in Example 1, since 2 and 3 are both factors of 24, their product, 6, is also a factor.

EXAMPLE 2   Find all the factors of 105.

Since 105 is odd, 2 is not a factor. Because the sum of the digits is 6 and 6 is divisible by 3, we know that 3 is a factor. Since the ones digit is 5, we also know that 5 is a factor.

Factorizations: $1 \cdot 105$; $3 \cdot 35$; $5 \cdot 21$; $7 \cdot 15$

Factors: 1, 3, 5, 7, 15, 21, 35, 105

Do Exercises 1–4.

List all the factors of each number. (*Hint*: Find some factorizations of the number.)

1. 14

2. 10

3. 8

4. 32

Answers on page A-7

5. Classify each number as prime, composite, or neither.

1, 2, 6, 12, 13, 19, 41, 65, 73, 99

b   Prime and Composite Numbers

PRIME AND COMPOSITE NUMBERS

A natural number that has exactly two different factors, itself and 1, is called a **prime number.**

- The number 1 is *not* prime.
- A natural number, other than 1, that is not prime is **composite.**

To see if a number is prime or composite, we can check to see if it is divisible by a number other than itself or 1. If such a number exists, then the original number is composite.

EXAMPLE 3   Determine which of the numbers 2, 3, 4, 5, 7, 9, 10, 11, 27, and 63 are prime, which are composite, and which are neither.

Because 2, 3, 5, 7, and 11 are each divisible only by 1 and the number itself, they are *prime.*

Because 4 and 10 are each divisible by 2 (as well as 1 and the number itself), they are *composite.*

Because 9, 27, and 63 are each divisible by 3 (as well as 1 and the number itself), they are *composite.*

Thus we have:

Prime: 2, 3, 5, 7, 11;

Composite: 4, 9, 10, 27, 63.

Because 0 is not a natural number, it is neither prime nor composite. Note that 1 is not prime because it does not have two different factors. The number 2 is the smallest prime, and the only even prime, since 2 is a factor of all even numbers. If you do not recognize immediately whether an odd number is prime or composite, begin by checking divisibility by 3 and 5. You need to check divisibility by prime numbers only, so continue with 7, 11, 13, and so on. If you reach a point where the quotient is less than the divisor, and none of the primes up to that point are factors, the number you are checking is prime.

Do Exercise 5.

Lists of numbers known to be prime are available in math handbooks. The following is a table of the prime numbers from 2 to 157. Although you need not memorize the entire list, remembering at least the first nine or ten is important.

A TABLE OF PRIMES FROM 2 TO 157

2, 3, 5, 7, 11, 13, 17, 19, 23, 29, 31, 37, 41, 43, 47, 53, 59, 61, 67, 71, 73, 79, 83, 89, 97, 101, 103, 107, 109, 113, 127, 131, 137, 139, 149, 151, 157

There are infinitely many prime numbers. It takes extensive computer operations to determine whether very large numbers are prime. At present, the largest known prime is $2^{25964951} - 1$. This number has over 7 million digits!

Source: http://primes.utm.edu//largest.html

Answer on page A-7

Study Tips

PACE YOURSELF

Most instructors agree that it is better for a student to study for one hour four days in a week, than to study once a week for four hours. Of course, the total weekly study time will vary from student to student. It is common to expect an average of two hours of homework for each hour of class time.

C Prime Factorizations

When we factor a composite number into a product of primes, we find the **prime factorization** of the number. We can do this by a series of successive divisions, or by using a *factor tree*.

To use division, we consider the prime numbers 2, 3, 5, 7, 11, 13, and so on. We check for divisibility by the first prime number, 2. If the number is even, we perform the division. We then check the quotient of the division for divisibility by 2. If the quotient is even, we divide it by 2. We continue dividing by 2 until the quotient is no longer even. Now we repeat the process for the second prime number, 3, then for 5, 7, and so on, until the quotient is itself a prime number.

EXAMPLE 4 Find the prime factorization of 39.

We check for divisibility by the first prime, 2. Since 39 is not even, 2 is not a factor of 39. Since the sum of the digits in 39 is 12 and 12 is divisible by 3, we know that 39 is divisible by 3. We then perform the division.

$$\frac{13}{3)\overline{39}} \quad R = 0 \qquad \text{A remainder of 0 confirms that 3 is a factor of 39.}$$

Because 13 is a prime, we can now write the prime factorization:

$$39 = 3 \cdot 13.$$

EXAMPLE 5 Find the prime factorization of 76.

Since 76 is even, it must have 2 as a factor.

$$\frac{38}{2)\overline{76}} \qquad \text{We can write } 76 = 2 \cdot 38.$$

Because 38 is also even, we see that 76 contains a second factor of 2.

$$\frac{19}{2)\overline{38}} \qquad \text{Note that } 38 = 2 \cdot 19, \text{ so } 76 = 2 \cdot 2 \cdot 19.$$

Because 19 is prime, we can factor no further. The complete factorization is

$$76 = 2 \cdot 2 \cdot 19. \qquad \text{All factors are prime.}$$

We can abbreviate our procedure as follows.

$$\frac{19}{2)\overline{38}} \longleftarrow \text{We stop dividing when the quotient is prime.}$$
$$2)\overline{76} \longleftarrow \text{We begin here and work upward.}$$
$$76 = 2 \cdot 2 \cdot 19$$

A factorization like $2 \cdot 2 \cdot 19$ can be written as $2^2 \cdot 19$ or $2 \cdot 19 \cdot 2$ or $19 \cdot 2 \cdot 2$ or $19 \cdot 2^2$. In any case, the prime factors are the same. For this reason, we agree that any of these may be considered "the" prime factorization of 76.

> Each composite number is uniquely determined by its prime factorization.

This last result is sometimes called the Fundamental Theorem of Arithmetic.

Write the prime factorization of each number.

6. 6

7. 12

8. 98

9. 91

10. 126

11. 144

Answers on page A-7

EXAMPLE 6 Find the prime factorization of 187.

We check for divisibility by 2, 3, and 5, and find that 187 is not divisible by any of these numbers. The next prime number, 7, does not divide 187 evenly. However, when we divide by 11, the remainder is 0, so 11 is a factor of 187.

$$11\overline{)187}^{17}$$ We can write $187 = 11 \cdot 17$.

Because 17 is prime, we can factor no further. The complete factorization is

$$187 = 11 \cdot 17.$$ All factors are prime.

EXAMPLE 7 Find the prime factorization of 72.

$$
\begin{aligned}
&3 \leftarrow \text{3 is prime, so we stop dividing.}\\
&3\overline{)9}\\
&2\overline{)18}\\
&2\overline{)36}\\
&2\overline{)72} \leftarrow \text{Begin here and work upward.}
\end{aligned}
$$
$$72 = 2 \cdot 2 \cdot 2 \cdot 3 \cdot 3$$

Another way to find the prime factorization of 72 uses a **factor tree** as follows. Begin by determining any factorization you can, and then continue factoring. Each of the following trees gives the same prime factorization.

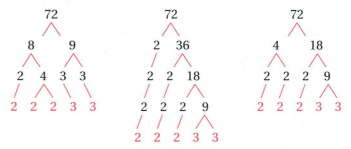

Caution!

Finding all the factors of a number is not the same as finding the prime factorization. In Example 7, the prime factorization is $2 \cdot 2 \cdot 2 \cdot 3 \cdot 3$. The factors of 72 are 1, 2, 3, 4, 6, 8, 9, 12, 18, 24, 36, and 72.

EXAMPLE 8 Find the prime factorization of 130 and list all factors of 130.

We can use a string of divisions or a factor tree.

$$
\begin{aligned}
&13\\
&5\overline{)\,65}\\
&2\overline{)130}
\end{aligned}
\qquad
\begin{aligned}
&130\\
&\diagup\!\diagdown\\
&10 \cdot 13\\
\end{aligned}
$$
$$130 = 2 \cdot 5 \cdot 13$$

From the prime factorization we can easily identify 2, 5, and 13 as factors. We can form the products $2 \cdot 5$, $2 \cdot 13$, and $5 \cdot 13$ to find that 10, 26, and 65 are also factors. A complete list of factors is 1, 2, 5, 10, 13, 26, 65, and 130.

Do Exercises 6–11.

Finding a number's prime factorization can be quite challenging, especially when the prime factors themselves are large. This difficulty is used worldwide as a way of securing transactions over the Internet.

a List all the factors of each number.

1. 18   **2.** 16   **3.** 54   **4.** 48

5. 9   **6.** 4   **7.** 13   **8.** 11

b Classify each number as prime, composite, or neither.

9. 17   **10.** 24   **11.** 22   **12.** 31

13. 48   **14.** 43   **15.** 53   **16.** 54

17. 1   **18.** 2   **19.** 81   **20.** 37

21. 47   **22.** 51   **23.** 29   **24.** 49

c Find the prime factorization of each number.

25. 27   **26.** 16   **27.** 14   **28.** 15   **29.** 80

30. 32   **31.** 25   **32.** 40   **33.** 62   **34.** 169

35. 100   **36.** 110   **37.** 143   **38.** 50   **39.** 121

40. 170   **41.** 273   **42.** 675   **43.** 175   **44.** 196

45. 209   **46.** 133   **47.** 86   **48.** 78   **49.** 217

50. 497   **51.** 7000   **52.** 5000   **53.** 1122   **54.** 6435

a , **c** List all the factors of each number.

55. 100   **56.** 135   **57.** 385   **58.** 110

59. 81   **60.** 196   **61.** 225   **62.** 441

63. **Dw** Is it possible for two consecutive natural numbers other than 2 and 3 to both be prime? Why or why not?

64. **Dw** Is it necessary to try dividing 41 by all primes through 37 in order to demonstrate that 41 is prime?

Multiply.

65. $-2 \cdot 13$ [2.4a]

66. $(-8)(-32)$ [2.4a]

Add.

67. $-17 + 25$ [2.2a]

68. $-9 + (-14)$ [2.2a]

Divide.

69. $53 \div 53$ [1.6c]

70. $73 \div 1$ [1.6c]

71. $0 \div 22$ [2.5a]

72. $22 \div 22$ [1.6c]

73. $-42 \div 1$ [2.5a]

74. $0 \div (-42)$ [2.5a]

75. **Dw** Is it possible for the sum or product of two prime numbers to be prime? Why or why not?

76. **Dw** It was mentioned that the largest known prime is $2^{25964951} - 1$. How can you tell that it is not even?

77. **Dw** If a and b are both factors of c, does it follow that $a \cdot b$ is also a factor of c? Why or why not?

Find the prime factorization of each number.

78. 🖩 136,097

79. 🖩 102,971

80. 🖩 473,073,361

81. 🖩 168,840

82. Describe an arrangement of 54 objects that corresponds to the factorization $54 = 6 \times 9$.

83. Describe an arrangement of 24 objects that corresponds to the factorization $24 = 2 \cdot 3 \cdot 4$.

84. Two numbers are **relatively prime** if there is no prime number that is a factor of both numbers. For example, 10 and 21 are relatively prime but 15 and 18 are not. List five pairs of composite numbers that are relatively prime.

85. *Factors and Sums.* In the table below, the top number in each column has been factored in such a way that the sum of the factors is the bottom number in the column. For example, in the first column, 56 has been factored as $7 \cdot 8$, and $7 + 8 = 15$, the bottom number. Such thinking will be important in understanding the meaning of a factor and in algebra.

Product	56	63	36	72	140	96	48	168	110	90	432	63
Factor	7											
Factor	8											
Sum	15	16	20	38	24	20	14	29	21	19	42	24

Find the missing numbers in the table.

3.3 FRACTIONS AND FRACTION NOTATION

Objectives

a Identify the numerator and the denominator of a fraction and write fraction notation for part of an object or part of a set of objects and as a ratio.

b Simplify fraction notation like n/n to 1, $0/n$ to 0, and $n/1$ to n.

The study of arithmetic begins with the set of whole numbers

0, 1, 2, 3, 4, 5, 6, 7, 8, 9, 10, 11, and so on.

We also need to be able to use fractional parts of numbers such as halves, thirds, fourths, and so on. Here are some examples:

In 1950, about $\frac{3}{4}$ of the motor vehicles produced in the world were produced in the United States. In 2003, only $\frac{1}{5}$ were produced in the United States.

Sources: American Automobile Manufacturers Association; Automotive News Data Center and Marketing Systems GmbH

More than $\frac{2}{5}$ of Americans are taking at least one prescription daily, and $\frac{1}{6}$ are taking three or more.

Source: Randolph Schmid, Associated Press, *Indianapolis Star*, 12/3/04

For each fraction, identify the numerator and the denominator.

1. $\frac{83}{100}$ of all scrap tires are reused or recycled.

   Source: Rubber Manufacturers Association

2. $\frac{27}{50}$ of all kids prefer white bread.

   Source: Bruce Horovitz, *USA Today*, 8/9/04

3. $\frac{5a}{7b}$

4. $\frac{-22}{3}$

a Identifying Numerators and Denominators

Numbers like those above and the ones below are written in **fraction notation.** The top number is called the **numerator** and the bottom number is called the **denominator.**

$$\frac{1}{2}, \quad \frac{3}{4}, \quad \frac{-8}{5}, \quad \frac{x}{y}, \quad -\frac{4}{25}, \quad \frac{2a}{7b}$$

EXAMPLE 1 Identify the numerator and the denominator.

$\dfrac{7 \leftarrow \text{Numerator}}{8 \leftarrow \text{Denominator}}$

Do Exercises 1–4.

Answers on page A-7

What part is shaded?

5.

$1

6.   1 mile

7.

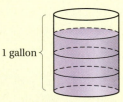

1 year

8.   1 mile

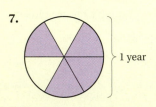

What part is shaded or indicated?

9.

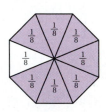

1 gallon

10.

1 gallon

Answers on page A-7

Let's look at various situations that involve fractions.

FRACTIONS AS A PARTITION OF AN OBJECT DIVIDED INTO EQUAL PARTS

Consider a candy bar divided into 5 equal sections. If you eat 2 sections, you have eaten $\frac{2}{5}$ of the candy bar.

The denominator 5 tells us the unit, $\frac{1}{5}$. The numerator 2 tells us the number of equal parts we are considering, 2.

EXAMPLE 2 What part is shaded?

The equal parts are eighths. This tells us the unit, $\frac{1}{8}$. The *denominator* is 8. We have 7 of the units shaded. This tells us the *numerator*, 7. Thus,

$$\frac{7}{8} \quad \begin{array}{l}\leftarrow \text{7 units are shaded.} \\ \leftarrow \text{The unit is } \frac{1}{8}.\end{array}$$

is shaded.

Do Exercises 5–8.

EXAMPLE 3 What part of a cup is filled?

The measuring cup is divided into 3 parts of the same size, and 2 of them are filled. This is $2 \cdot \frac{1}{3}$, or $\frac{2}{3}$. Thus, $\frac{2}{3}$ (read *two-thirds*) of a cup is filled.

Do Exercises 9 and 10.

The markings on a ruler use fractions.

EXAMPLE 4 What part of an inch is indicated?

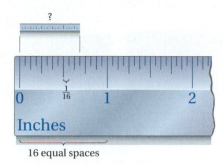

Inches

16 equal spaces

Each inch on the ruler shown above is divided into 16 equal parts. The measured length extends to the 11th mark. Thus, $\frac{11}{16}$ of an inch is indicated.

Do Exercise 11.

EXAMPLE 5 What part of this set, or collection, of people are actresses? U.S. Senators?

| Nicole Kidman | Halle Berry | Elizabeth Dole | Hilary Swank | Gwyneth Paltrow | Julia Roberts | Diane Feinstein |

There are 7 people in the set. We know that 5 of them, Nicole Kidman, Halle Berry, Julia Roberts, Hilary Swank, and Gwyneth Paltrow, are actresses. Thus, 5 of 7, or $\frac{5}{7}$, are actresses. The 2 remaining are U.S. Senators. Thus, $\frac{2}{7}$ are U.S. Senators.

Do Exercise 12.

Fractions greater than 1 correspond to situations like the following.

EXAMPLE 6 What part is shaded?

4 loaves

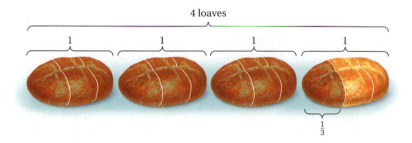

Each loaf of bread is divided into 3 equal parts. The unit is $\frac{1}{3}$. The *denominator* is 3. We have 10 of the units shaded. This tells us the *numerator* is 10. Thus, $\frac{10}{3}$ are shaded.

11. What part of an inch is indicated?

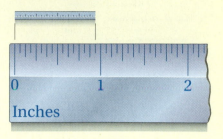

Inches

12. What part of this set, or collection, are clocks? thermometers?

Answers on page A-7

177

What part is shaded?

13.

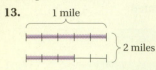

14.

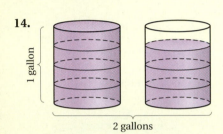

15. Baseball Standings. Refer to the table in Example 8. The Minnesota Twins finished third in the American League Central Division in 2005. Find the ratio of Twins wins to losses, wins to total number of games, and losses to total number of games.

Source: Major League Baseball

EXAMPLE 7 What part is shaded?

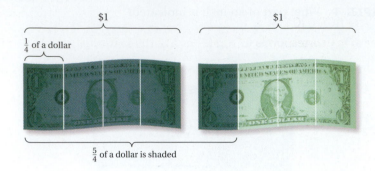

$\frac{5}{4}$ of a dollar is shaded

We can regard this as two objects of 4 parts each and take 5 of those parts. We have more than one whole object. Thus, $5 \cdot \frac{1}{4}$, or $\frac{5}{4}$ (also, 5 quarters) is shaded.

Do Exercises 13 and 14.

FRACTIONS AS RATIOS

A **ratio** is a quotient of two quantities. We can express a ratio with fraction notation. (We will consider ratios in more detail in Chapter 7.)

EXAMPLE 8 *Baseball Standings.* The following are the final standings in the American League Central Division for 2005, when the division was won by the Chicago White Sox. Find the ratio of White Sox wins to losses, wins to total number of games, and losses to total number of games.

CENTRAL	W	L	Pct.	GB	HOME	ROAD	LAST 10	STREAK
White Sox	99	63	.611	—	47-34	52-29	8-2	Won 5
Indians	93	69	.574	6.0	43-38	50-31	4-6	Lost 3
Twins	83	79	.512	16.0	45-36	38-43	5-5	Won 3
Tigers	71	91	.438	28.0	39-42	32-49	4-6	Lost 5
Royals	56	106	.346	43.0	34-47	22-59	4-6	Lost 1

Source: Major League Baseball

The White Sox won 99 games and lost 63 games. They played a total of $99 + 63$, or 162 games. Thus we have the following.

The ratio of wins to losses is $\frac{99}{63}$.

The ratio of wins to total number of games is $\frac{99}{162}$.

The ratio of losses to total number of games is $\frac{63}{162}$.

Do Exercise 15.

b Some Fraction Notation for Integers

FRACTION NOTATION FOR 1

The number 1 corresponds to situations like the following.

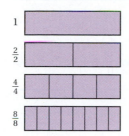

If we divide an object into *n* parts and take *n* of them, we get all of the object (1 whole object). Since a negative divided by a negative is a positive, the following is stated for *all* nonzero integers.

> **WRITING 1 AS A FRACTION**
>
> $\dfrac{n}{n} = 1,$ for any integer *n* that is not 0.

EXAMPLE 9 Simplify: **a)** $\dfrac{5}{5}$; **b)** $\dfrac{-9}{-9}$; **c)** $\dfrac{17x}{17x}$ (assume $x \neq 0$).

a) $\dfrac{5}{5} = 1$ **b)** $\dfrac{-9}{-9} = 1$ **c)** $\dfrac{17x}{17x} = 1$

Do Exercises 16–21.

FRACTION NOTATION FOR 0

Consider $\frac{0}{4}$. This corresponds to dividing an object into 4 parts and taking none of them. We get 0. This result also extends to all nonzero integers.

> **THE NUMBER 0 AS A FRACTION**
>
> $\dfrac{0}{n} = 0,$ for any integer *n* that is not 0.

EXAMPLE 10 Simplify: **a)** $\dfrac{0}{9}$; **b)** $\dfrac{0}{1}$; **c)** $\dfrac{0}{5a}$ (assume that $a \neq 0$); **d)** $\dfrac{0}{-23}$.

a) $\dfrac{0}{9} = 0$ **b)** $\dfrac{0}{1} = 0$

c) $\dfrac{0}{5a} = 0$ **d)** $\dfrac{0}{-23} = 0$

Simplify. Assume that $a \neq 0$.

16. $\dfrac{8}{8}$

17. $\dfrac{a}{a}$

18. $\dfrac{-52}{-52}$

19. $\dfrac{1}{1}$

20. $\dfrac{-2347}{-2347}$

21. $\dfrac{54a}{54a}$

Answers on page A-7

Simplify, if possible. Assume that $x \neq 0$.

22. $\dfrac{0}{8}$

23. $\dfrac{0}{-6}$

24. $\dfrac{0}{4x}$

25. $\dfrac{4-4}{236}$

26. $\dfrac{7}{0}$

27. $\dfrac{-4}{0}$

Simplify.

28. $\dfrac{8}{1}$

29. $\dfrac{-10}{1}$

30. $\dfrac{-346}{1}$

31. $\dfrac{17-2}{1}$

Fraction notation with a denominator of 0, such as $n/0$, does not represent a number because we cannot divide an object into *zero* parts. (If it is not divided at all, then we say that it is undivided and remains in one part.)

> **DIVISION BY 0**
>
> $\dfrac{n}{0}$ is not defined.
>
> $\left(\text{When asked to simplify } \dfrac{n}{0}, \text{ we write } undefined.\right)$

Do Exercises 22–27.

OTHER INTEGERS

Consider $\frac{4}{1}$. This corresponds to dividing an object into one part (leaving it whole) and taking four of them. We have 4 objects.

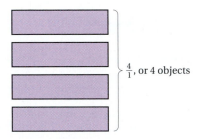

$\frac{4}{1}$, or 4 objects

> **DIVISION BY 1**
>
> Any integer divided by 1 is the original integer. That is,
>
> $\dfrac{n}{1} = n, \quad \text{for any integer } n.$

EXAMPLE 11 Simplify: **a)** $\dfrac{2}{1}$; **b)** $\dfrac{-9}{1}$; **c)** $\dfrac{3x}{1}$.

a) $\dfrac{2}{1} = 2$

b) $\dfrac{-9}{1} = -9$

c) $\dfrac{3x}{1} = 3x$

Do Exercises 28–31.

3.3 EXERCISE SET   For Extra Help

a   Identify the numerator and the denominator of each fraction.

1. $\dfrac{3}{4}$

2. $\dfrac{-9}{10}$

3. $\dfrac{7}{-9}$

4. $\dfrac{15}{8}$

5. $\dfrac{2x}{3z}$

6. $\dfrac{9a}{2b}$

What part of the object or set of objects is shaded? In Exercises 19–22, what part of an inch is indicated?

7.

$1

8.
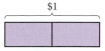
$1

9.
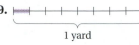
1 yard

10.
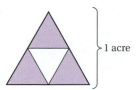
1 mile

11.
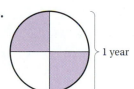
1 window

12.
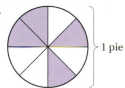
1 square yard

13.
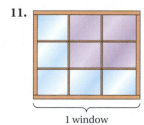
1 acre

14.

1 year

15.
1 pie

16.
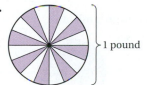
1 pound

17.
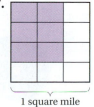
1 square mile

18.
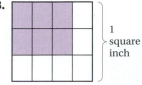
1 square inch

19.

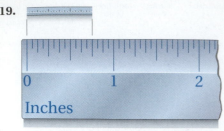

20.

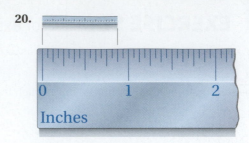

21.

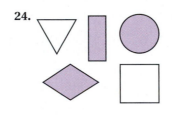

22.

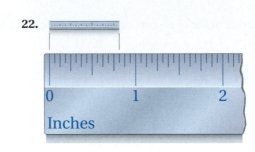

23.

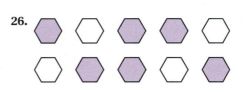

24.

25.

26.

For each of Exercises 27–30, give fraction notation for the amount of gas (a) in the tank and (b) used from a full tank.

27.

28.

29.

30.

CHAPTER 3: Fraction Notation:
Multiplication and Division

What part is shaded?

31.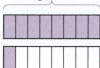

32.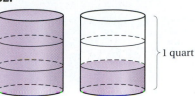

33.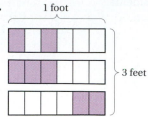

34.

35.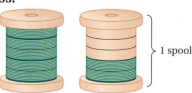

36.

37. *Gas Mileage.* A 2005 Mini Cooper S will travel 390 mi on 13 gal of gasoline. What is the ratio of:

  a) miles driven to gasoline burned?
  b) gasoline burned to miles driven?

38. Randy delivers furniture to customers. On Thursday he had 15 deliveries. By noon he had delivered only 4 orders. What is the ratio of:

  a) orders delivered to total number of orders?
  b) orders delivered to orders not delivered?
  c) orders not delivered to total number of orders?

39. *Veterinary Care.* Of every 1000 households that own dogs, 850 obtain veterinary care. What is the ratio of households who own dogs that seek veterinary care to all households that own dogs?

Source: *U.S. Pet Ownership and Demographics Sourcebook,* 2002

40. *Moviegoers.* Of every 1000 people who attend movies, 340 are in the 18–24 age group. What is the ratio of moviegoers in the 18–24 age group to all moviegoers?

Source: American Demographics

41. For the following set of people, what is the ratio of:

  a) women to the total number of people?
  b) women to men?
  c) men to the total number of people?
  d) men to women?

42. For the following set of nuts and bolts, what is the ratio of:

  a) nuts to bolts?
  b) bolts to nuts?
  c) nuts to the total number of elements?
  d) total number of elements to nuts?

For Exercises 43 and 44, use the bar graph below that lists the police agency and the number of full-time police officers per 10,000 residents.

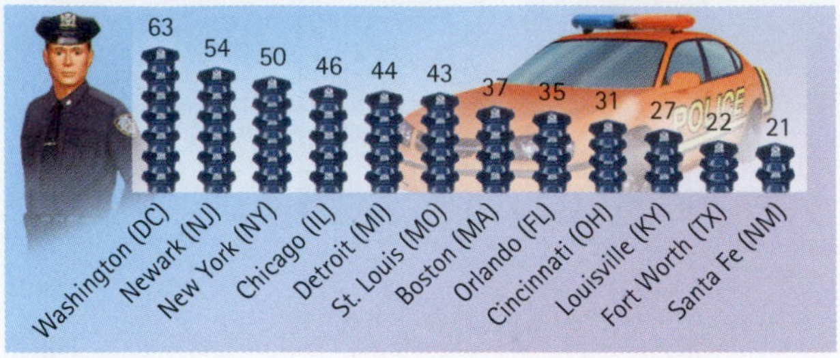

Source: Law Enforcement and Administrative Statistics, 2000

43. What is the ratio of police officers per 10,000 residents in the given city?

a) Orlando   b) New York
c) Detroit   d) Washington
e) St. Louis   f) Santa Fe

44. What is the ratio of police officers per 10,000 residents in the given city?

a) Chicago   b) Boston
c) Newark   d) Louisville
e) Cincinnati   f) Fort Worth

b   Simplify, if possible. Assume that all variables are nonzero.

45. $\dfrac{0}{17}$

46. $\dfrac{19}{19}$

47. $\dfrac{15}{1}$

48. $\dfrac{10}{1}$

49. $\dfrac{20}{20}$

50. $\dfrac{-20}{1}$

51. $\dfrac{-14}{-14}$

52. $\dfrac{4a}{1}$

53. $\dfrac{0}{-234}$

54. $\dfrac{37a}{37a}$

55. $\dfrac{3n}{3n}$

56. $\dfrac{0}{-1}$

57. $\dfrac{9x}{9x}$

58. $\dfrac{-12a}{1}$

59. $\dfrac{-63}{1}$

60. $\dfrac{-3x}{-3x}$

61. $\dfrac{0}{2a}$

62. $\dfrac{0}{8}$

63. $\dfrac{52}{0}$

64. $\dfrac{8-8}{1247}$

65. $\dfrac{7n}{1}$

66. $\dfrac{247}{0}$

67. $\dfrac{6}{7-7}$

68. $\dfrac{15}{9-9}$

69. **D**_W Explain in your own words why $\dfrac{n}{1} = n$, for any integer n.

70. **D**_W Explain in your own words why $\dfrac{0}{n} = 0$, for any nonzero integer n.

70. **D**$_W$ Explain in your own words why $\dfrac{0}{n} = 0$, for any nonzero integer n.

CHAPTER 3: Fraction Notation: Multiplication and Division

Multiply.

71. $-7(30)$ [2.4a]

72. $23 \cdot (-14)$ [2.4a]

73. $(-71)(-12)0$ [2.4b]

74. $32(-29)0$ [2.4b]

75. A Burger King Double Whopper Value Meal with cheese has 2050 calories. A Wendy's Classic Triple Combo Meal with cheese has 1750 calories. How many more calories does the Burger King meal have than the Wendy's meal? [1.8a]

Source: *Nutrition Action Healthletter, March 2005*

76. Sandy can type 62 words per minute. How long will it take Sandy to type 12,462 words? [1.8a]

77. $\mathbf{D_W}$ Explain in your own words why $\dfrac{n}{0}$ is undefined for any integer n.

78. $\mathbf{D_W}$ What is the ratio of negative integers to positive integers? Why?

79. The year 2006 began on a Sunday. What fractional part of 2006 were Mondays?

80. The year 2007 began on a Monday. What fractional part of 2007 were Mondays?

81. The surface of Earth is 3 parts water and 1 part land. What fractional part of Earth is water? land?

82. A couple had 3 sons, each of whom had 3 daughters. If each daughter gave birth to 3 sons, what fractional part of the couple's descendants is female?

83. On the average, a U.S. household has 2.8 working television sets. Out of all households, $\frac{2}{100}$ have no working TV, $\frac{15}{100}$ have one TV, $\frac{29}{100}$ have two TVs, and $\frac{54}{100}$ have three or more TVs. Label each sector of the graph with the most appropriate number of television sets.

Source: Frank N. Magid Associates, in *USA TODAY*

Televisions per Household

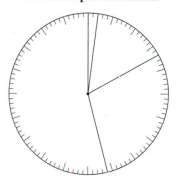

What part of each object is shaded?

84.

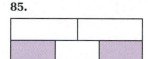

85.

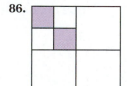

86.

87.

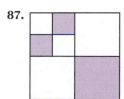

Objectives

a Multiply an integer and a fraction.

b Multiply using fraction notation.

c Solve problems involving multiplication of fractions.

Before discussing how to simplify fraction notation (Section 3.5), it is essential that we study multiplication of fractions.

a Multiplication by an Integer

We can find $3 \cdot \frac{1}{4}$ by thinking of repeated addition. We add three $\frac{1}{4}$'s.

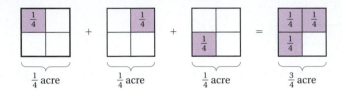

$\frac{1}{4}$ acre $\quad$ $\frac{1}{4}$ acre $\quad$ $\frac{1}{4}$ acre $\quad$ $\frac{3}{4}$ acre

We see that $3 \cdot \frac{1}{4} = \frac{1}{4} + \frac{1}{4} + \frac{1}{4} = \frac{3}{4}$.

Do Exercises 1 and 2.

1. Find $2 \cdot \frac{1}{3}$.

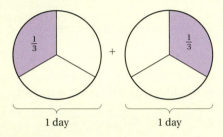

1 day $\qquad$ 1 day

To multiply a fraction by an integer,

a) multiply the top number (the numerator) by the integer and

b) keep the same denominator.

$$6 \cdot \frac{4}{5} = \frac{6 \cdot 4}{5} = \frac{24}{5}$$

2. Find $5 \cdot \frac{1}{8}$.

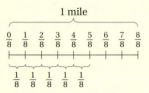

EXAMPLES Multiply.

1. $5 \times \frac{3}{8} = \frac{5 \times 3}{8} = \frac{15}{8}$ $\qquad$ We generally replace the $\times$ symbol with $\cdot$

Skip this step when you feel comfortable doing so.

2. $\frac{2}{5} \cdot 13 = \frac{2 \cdot 13}{5} = \frac{26}{5}$

3. $-10 \cdot \frac{1}{3} = \frac{-10}{3}$, or $-\frac{10}{3}$ $\qquad$ Recall that $\frac{-a}{b} = -\frac{a}{b}$.

4. $a \cdot \frac{4}{7} = \frac{4a}{7}$ $\qquad$ Recall that $a \cdot 4 = 4 \cdot a$.

Do Exercises 3–6.

Multiply.

3. $7 \times \frac{2}{3}$

4. $(-11) \cdot \frac{3}{10}$

5. $34 \cdot \frac{2}{5}$

6. $x \cdot \frac{4}{9}$

Answers on page A-8

b Multiplication Using Fraction Notation

To illustrate the meaning of an expression like $\frac{1}{2} \cdot \frac{1}{3}$, we first represent $\frac{1}{3}$, and then shade half of that region. Note that $\frac{1}{2} \cdot \frac{1}{3}$ is the same as $\frac{1}{2}$ of $\frac{1}{3}$.

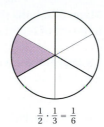

$\frac{1}{3}$

$\frac{1}{2} \cdot \frac{1}{3} = \frac{1}{6}$

Do Exercise 7.

To visualize $\frac{2}{5} \cdot \frac{3}{4}$, we first represent $\frac{3}{4}$, and then shade $\frac{2}{5}$ of that region.

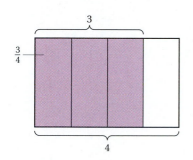

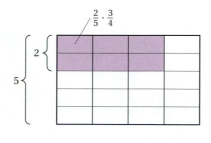

Since 6 of the 20 sections are now shaded, we have

$$\frac{2}{5} \cdot \frac{3}{4} = \frac{6}{20}$$
← This is the product of the numerators.
← This is the product of the denominators.

Do Exercise 8.

Notice that the product of two fractions is the product of the numerators over the product of the denominators.

> To multiply a fraction by a fraction,
>
> **a)** multiply the numerators and
>
> $$\frac{9}{7} \cdot \frac{3}{4} = \frac{9 \cdot 3}{7 \cdot 4} = \frac{27}{28}$$
>
> **b)** multiply the denominators.

EXAMPLES Multiply.

5. $\dfrac{5}{6} \cdot \dfrac{7}{4} = \dfrac{5 \cdot 7}{6 \cdot 4} = \dfrac{35}{24}$

> Skip this step when you feel comfortable doing so.

6. $\dfrac{3}{5} \cdot \dfrac{7}{8} = \dfrac{3 \cdot 7}{5 \cdot 8} = \dfrac{21}{40}$

7. $\dfrac{4}{x} \cdot \dfrac{y}{9} = \dfrac{4y}{9x}$

8. $(-6)\left(-\dfrac{4}{5}\right) = \dfrac{-6}{1} \cdot \dfrac{-4}{5} = \dfrac{24}{5}$ Recall that $\dfrac{n}{1} = n$.

Do Exercises 9–12.

7. Draw diagrams to illustrate $\frac{1}{4}$ and $\frac{1}{2} \cdot \frac{1}{4}$.

8. Draw diagrams to illustrate $\frac{1}{3}$ and $\frac{4}{5} \cdot \frac{1}{3}$.

Multiply.

9. $\dfrac{3}{8} \cdot \dfrac{5}{7}$

10. $\dfrac{4}{3} \cdot \dfrac{8}{5}$

11. $\left(-\dfrac{3}{10}\right)\left(-\dfrac{1}{10}\right)$

12. $(-7)\dfrac{a}{b}$

Answers on page A-8

13. Camp Mohawk uses $\frac{3}{4}$ of its extra land for recreational purposes. Of that, $\frac{1}{2}$ is used for swimming pools. What part of the extra land is used for swimming pools?

C Applications and Problem Solving

Many problems that can be solved by multiplying fractions can be thought of in terms of rectangular arrays.

EXAMPLE 9 A dude ranch owns a square mile of land. The owner gives $\frac{4}{5}$ of it to her daughter who, in turn, gives $\frac{2}{3}$ of her share to her son. How much land goes to the daughter's son?

1. **Familiarize.** We first make a drawing to help solve the problem. The land may not be square. It could be in a shape like A or B below, or it could even be in more than one piece. But to think about the problem, we can visualize a square, as shown by shape C.

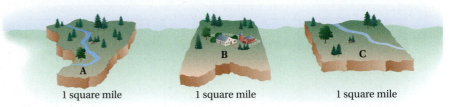

| 1 square mile | 1 square mile | 1 square mile |

The daughter gets $\frac{4}{5}$ of the land. We shade $\frac{4}{5}$.

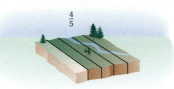

Her son gets $\frac{2}{3}$ of her part. We "raise" that.

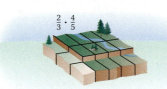

2. **Translate.** We let $n =$ the part of the land that goes to the daughter's son. We are taking "two-thirds of four-fifths." The word "of" corresponds to multiplication. Thus, the following multiplication sentence corresponds to the situation:

$$\frac{2}{3} \cdot \frac{4}{5} = n.$$

3. **Solve.** The number sentence tells us what to do. We have

$$\frac{2}{3} \cdot \frac{4}{5} = n, \quad \text{or} \quad \frac{8}{15} = n.$$

4. **Check.** We can check this in the figure above, where we see that 8 of 15 equally sized parts will go to the daughter's son.

5. **State.** The daughter's son gets $\frac{8}{15}$ of a square mile of land.

Do Exercise 13.

Answer on page A-8

Study Tips

USE YOUR TEXT FOR NOTES

If you own your text, consider using it as a notebook. Since many instructors closely parallel the book, it is often useful to make notes on the appropriate page as he or she is lecturing.

We have seen that the area of a rectangular region is found by multiplying length by width. That is true whether length and width are whole numbers or not. Remember, the area of a rectangular region is given by the formula

$$A = l \cdot w \quad (Area = length \cdot width).$$

EXAMPLE 10 The length of a computer's Delete key is $\frac{9}{16}$ in. The width is $\frac{3}{8}$ in. What is the area?

1. **Familiarize.** Recall that area is length times width. We make a drawing, letting $A =$ the area of the computer key.

2. **Translate.** Next, we translate.

Area	is	Length	times	Width
↓	↓	↓	↓	↓
A	$=$	$\frac{9}{16}$	$\cdot$	$\frac{3}{8}$

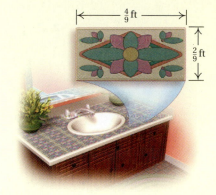

3. **Solve.** The sentence tells us what to do. We multiply:

$$\frac{9}{16} \cdot \frac{3}{8} = \frac{9 \cdot 3}{16 \cdot 8} = \frac{27}{128}.$$

4. **Check.** To check, we can repeat the calculation. This is left to the student.

5. **State.** The area of the key is $\frac{27}{128}$ square inches, or $\frac{27}{128}$ in^2.

Do Exercise 14.

EXAMPLE 11 A recipe for Banana Oat Pancakes calls for $\frac{3}{4}$ cup of old-fashioned oats. A chef is making $\frac{1}{2}$ of the recipe. How much oats should the chef use?

Source: Reprinted with permission from *Taste of Home* magazine, www.tasteofhome.com

1. **Familiarize.** We first make a drawing or at least visualize the situation. We let $n =$ the amount of oats the chef should use.

$\frac{3}{4}$ cup in recipe

$\frac{1}{2}$ of $\frac{3}{4}$ cup

2. **Translate.** We are finding $\frac{1}{2}$ of $\frac{3}{4}$, so the multiplication sentence $\frac{1}{2} \cdot \frac{3}{4} = n$ corresponds to the situation.

3. **Solve.** We carry out the multiplication:

$$\frac{1}{2} \cdot \frac{3}{4} = \frac{1 \cdot 3}{2 \cdot 4} = \frac{3}{8}.$$

4. **Check.** We check by repeating the calculation. This is left to the student.

5. **State.** The chef should use $\frac{3}{8}$ cup of oats.

Do Exercise 15.

14. Area of a Ceramic Tile. The length of a rectangular ceramic tile on an inlaid ceramic counter is $\frac{4}{9}$ ft. The width is $\frac{2}{9}$ ft. What is the area of one tile?

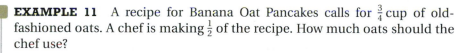

15. Of the students at Overton Junior College, $\frac{1}{8}$ participate in sports and $\frac{3}{5}$ of these play soccer. What fractional part of the students play soccer?

Answers on page A-8

3.4 EXERCISE SET

For Extra Help

MathXL MyMathLab InterAct Math Math Tutor Center Video Lectures on CD Disc 2 Student's Solutions Manual

a Multiply.

1. $3 \cdot \dfrac{1}{8}$

2. $2 \cdot \dfrac{1}{5}$

3. $(-5) \times \dfrac{1}{6}$

4. $(-4) \times \dfrac{1}{7}$

5. $\dfrac{2}{3} \cdot 7$

6. $\dfrac{2}{5} \cdot 7$

7. $(-1)\dfrac{7}{9}$

8. $(-1)\dfrac{4}{11}$

9. $\dfrac{5}{6} \cdot x$

10. $\dfrac{5}{6} \cdot y$

11. $\dfrac{2}{5}(-3)$

12. $\dfrac{3}{5}(-4)$

13. $a \cdot \dfrac{2}{7}$

14. $b \cdot \dfrac{3}{8}$

15. $17 \times \dfrac{m}{6}$

16. $\dfrac{n}{7} \cdot 30$

17. $-3 \cdot \dfrac{-2}{5}$

18. $-4 \cdot \dfrac{-5}{7}$

19. $-\dfrac{2}{7}(-x)$

20. $-\dfrac{3}{4}(-a)$

b Multiply.

21. $\dfrac{1}{3} \cdot \dfrac{1}{5}$

22. $\dfrac{1}{4} \cdot \dfrac{1}{2}$

23. $\left(-\dfrac{1}{4}\right) \times \dfrac{1}{10}$

24. $\left(-\dfrac{1}{3}\right) \times \dfrac{1}{10}$

25. $\dfrac{2}{3} \times \dfrac{1}{5}$

26. $\dfrac{3}{5} \times \dfrac{1}{5}$

27. $\dfrac{2}{y} \cdot \dfrac{x}{9}$

28. $\left(-\dfrac{3}{4}\right)\left(-\dfrac{3}{5}\right)$

29. $\left(-\dfrac{3}{4}\right)\left(-\dfrac{3}{4}\right)$

30. $\dfrac{3}{b} \cdot \dfrac{a}{7}$

31. $\dfrac{2}{3} \cdot \dfrac{7}{13}$

32. $\dfrac{3}{11} \cdot \dfrac{4}{5}$

33. $\dfrac{1}{10}\left(\dfrac{-3}{5}\right)$

34. $\dfrac{3}{10}\left(\dfrac{-7}{5}\right)$

35. $\dfrac{7}{8} \cdot \dfrac{a}{8}$

36. $\dfrac{4}{5} \cdot \dfrac{7}{x}$

37. $\dfrac{1}{y} \cdot \dfrac{1}{100}$

38. $\dfrac{b}{10} \cdot \dfrac{13}{100}$

39. $\dfrac{-14}{15} \cdot \dfrac{13}{19}$

40. $\dfrac{-12}{13} \cdot \dfrac{12}{13}$

41. *Tossed Salad.*   The recipe for Cherry Brie Tossed Salad calls for $\frac{3}{4}$ cup of sliced almonds. How much is needed to make $\frac{1}{2}$ of the recipe?

Source: Reprinted with permission from *Taste of Home* magazine, www.tasteofhome.com

42. It takes $\frac{2}{3}$ yd of ribbon to make a bow. How much ribbon is needed to make 5 bows?

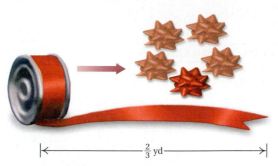

43. A rectangular table top measures $\frac{4}{5}$ m long by $\frac{3}{5}$ m wide. What is its area?

44. If each piece of pie is $\frac{1}{6}$ of a pie, how much of the pie is $\frac{1}{2}$ of a piece?

45. *Municipal Waste.*   Four-fifths of all municipal waste is landfill. One-tenth of this is landscape trimmings. What fractional part of municipal waste is landscape trimmings that are landfill?

Source: Based on information from the *Statistical Abstract of the United States* and the Chittenden, VT, Solid Waste District.

46. *Municipal Waste.*   About $\frac{19}{50}$ of all municipal waste is paper and paperboard. Of this, approximately $\frac{2}{5}$ is recycled. What fractional part of municipal waste is paper and paperboard that will be recycled?

Source: *Statistical Abstract of the United States*

47. *Floor Tiling.*   A bathroom floor is being tiled. An area $\frac{3}{5}$ of the length and $\frac{3}{4}$ of the width is covered. What fraction of the floor has been tiled?

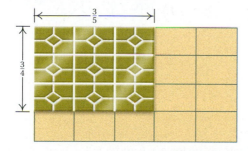

48. A gasoline can holds $\frac{7}{8}$ L (liter). How much will the can hold when it is $\frac{1}{2}$ full?

49. *Basketball: High School to Pro.*   One of 35 high school basketball players plays college basketball. One of 75 college players plays professional basketball. What fractional part of high school players play professional basketball?

Source: National Basketball Association

50. *Basement Carpet.*   A basement floor is being carpeted. An area $\frac{7}{8}$ of the length and $\frac{3}{4}$ of the width is covered by lunch time. What fraction of the floor has been completed?

51. **D_W** Following Example 4, we explained, using words and pictures, why $\frac{2}{5} \cdot \frac{3}{4}$ equals $\frac{6}{20}$. Present a similar explanation of why $\frac{2}{3} \cdot \frac{4}{7}$ equals $\frac{8}{21}$.

52. **D_W** Write a problem for a classmate to solve. Design the problem so that the solution is "About $\frac{1}{30}$ of the students are left-handed women."

SKILL MAINTENANCE

Divide.

53. $180 \div 20$ [1.6c]

54. $280 \div 40$ [1.6c]

55. $450 \div (-9)$ [2.5a]

56. $540 \div (-6)$ [2.5a]

57. $\dfrac{-35}{5}$ [2.5a]

58. $\dfrac{-60}{12}$ [2.5a]

59. $\dfrac{-65}{-5}$ [2.5a]

60. $\dfrac{-42}{-7}$ [2.5a]

What does the digit 8 mean in each number?   [1.1a]

61. 4,678,952

62. 8,473,901

63. 7148

64. 23,803

SYNTHESIS

65. **D_W** Is multiplication of fractions commutative? Why or why not?

66. **D_W** Is multiplication of fractions associative? Why or why not?

67. *Forestry.* A chain saw holds $\frac{1}{5}$ gal of fuel. Chain saw fuel is $\frac{1}{16}$ two-cycle oil and $\frac{15}{16}$ unleaded gasoline. How much two-cycle oil is in a freshly filled chain saw?

Multiply. Write each answer using fraction notation.

68. ▦ $\dfrac{341}{517} \cdot \dfrac{209}{349}$

69. ▦ $\left(-\dfrac{57}{61}\right)^3$

70. $\left(\dfrac{2}{5}\right)^3 \left(-\dfrac{7}{9}\right)$

71. $\left(-\dfrac{1}{2}\right)^5 \left(\dfrac{3}{5}\right)$

72. $\left(-\dfrac{3}{4}\right)^2 \left(-\dfrac{5}{7}\right)^2$

73. Evaluate $-\dfrac{2}{3}xy$ for $x = \dfrac{2}{5}$ and $y = -\dfrac{1}{7}$.

74. Evaluate $-\dfrac{3}{4}ab$ for $a = \dfrac{2}{5}$ and $b = \dfrac{7}{5}$.

75. ▦ Evaluate $-\dfrac{4}{7}ab$ for $a = \dfrac{93}{107}$ and $b = \dfrac{13}{41}$.

76. ▦ Evaluate $-\dfrac{19}{73}xy$ for $x = \dfrac{103}{105}$ and $y = \dfrac{47}{61}$.

3.5 SIMPLIFYING

Objectives

a Multiply by 1 to find an equivalent expression using a different denominator.

b Simplify fraction notation.

c Test to determine whether two fractions are equivalent.

a   Multiplying by 1

Recall the following:

$$1 = \frac{1}{1} = \frac{2}{2} = \frac{3}{3} = \frac{4}{4} = \frac{-13}{-13} = \frac{45}{45} = \frac{100}{100} = \frac{n}{n}.$$

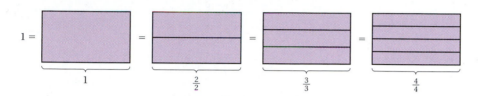

$$1 = \qquad = \qquad = \qquad =$$

$$1 \qquad \frac{2}{2} \qquad \frac{3}{3} \qquad \frac{4}{4}$$

Any nonzero number divided by itself is 1. (See Section 1.6.)

Now recall that for any whole number a, we have $1 \cdot a = a \cdot 1 = a$. This holds for fractions as well.

> ### MULTIPLICATIVE IDENTITY FOR FRACTIONS
>
> When we multiply a number by 1, we get the same number:
>
> $$\frac{3}{5} = \frac{3}{5} \cdot 1 = \frac{3}{5} \cdot \frac{4}{4} = \frac{12}{20}.$$

Since $\frac{3}{5} = \frac{12}{20}$, we know that $\frac{3}{5}$ and $\frac{12}{20}$ are two names for the same number. We also say that $\frac{3}{5}$ and $\frac{12}{20}$ are **equivalent.** (See Section 2.6).

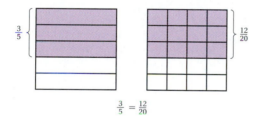

$$\frac{3}{5} = \frac{12}{20}$$

Do Exercises 1–4.

Suppose we want to rename $\frac{2}{3}$, using a denominator of 15. We can multiply by 1 to find a number equivalent to $\frac{2}{3}$:

$$\frac{2}{3} = \frac{2}{3} \cdot \frac{5}{5} = \frac{2 \cdot 5}{3 \cdot 5} = \frac{10}{15}.$$

We chose $\frac{5}{5}$ for 1 because $15 \div 3$ is 5.

EXAMPLE 1   Find a number equivalent to $\frac{1}{4}$ with a denominator of 24.

Since $24 \div 4 = 6$, we multiply by 1, using $\frac{6}{6}$:

$$\frac{1}{4} = \frac{1}{4} \cdot \frac{6}{6} = \frac{1 \cdot 6}{4 \cdot 6} = \frac{6}{24}.$$

Multiply.

1. $\dfrac{1}{2} \cdot \dfrac{8}{8}$

2. $\dfrac{3}{7} \cdot \dfrac{a}{a}$

3. $-\dfrac{8}{25} \cdot \dfrac{4}{4}$

4. $\dfrac{8}{3}\left(\dfrac{-2}{-2}\right)$

Answers on page A-8

Find an equivalent expression for each number, using the denominator indicated. Use multiplication by 1.

5. $\dfrac{4}{3} = \dfrac{?}{9}$

6. $\dfrac{3}{4} = \dfrac{?}{-24}$

7. $\dfrac{9}{10} = \dfrac{?}{10x}$

8. $\dfrac{7}{15} = \dfrac{?}{45}$

9. $\dfrac{-8}{7} = \dfrac{?}{49}$

Answers on page A-8

EXAMPLE 2 Find a number equivalent to $\frac{2}{5}$ with a denominator of -35.

Since $-35 \div 5 = -7$, we multiply by 1, using $\frac{-7}{-7}$:

$$\frac{2}{5} = \frac{2}{5}\left(\frac{-7}{-7}\right) = \frac{2(-7)}{5(-7)} = \frac{-14}{-35}.$$

EXAMPLE 3 Find an expression equivalent to $\frac{9}{8}$ with a denominator of $8a$.

Since $8a \div 8 = a$, we multiply by 1, using $\frac{a}{a}$:

$$\frac{9}{8} \cdot \frac{a}{a} = \frac{9a}{8a}.$$

Do Exercises 5–9.

b Simplifying Fraction Notation

All of the following are names for eight-ninths:

$$\frac{8}{9}, \ \frac{-8}{-9}, \ \frac{16}{18}, \ \frac{80}{90}, \ \frac{-24}{-27}.$$

We say that $\frac{8}{9}$ is **simplest** because it has the smallest positive denominator. Note that 8 and 9 have no factor in common other than 1.

To simplify, we reverse the process of multiplying by 1. This is accomplished by removing any factors (other than 1) that the numerator and the denominator have in common.

$$\frac{12}{18} = \frac{2 \cdot 6}{3 \cdot 6} \longleftarrow \text{Factoring the numerator}$$
$$\longleftarrow \text{Factoring the denominator}$$

$$= \frac{2}{3} \cdot \frac{6}{6} \qquad \text{Factoring the fraction}$$

$$= \frac{2}{3} \cdot 1 \qquad \frac{6}{6} = 1$$

$$= \frac{2}{3} \qquad \text{Removing the factor 1: } \frac{2}{3} \cdot 1 = \frac{2}{3}$$

EXAMPLES Simplify.

4. $\dfrac{-8}{20} = \dfrac{-2 \cdot 4}{5 \cdot 4} = \dfrac{-2}{5} \cdot \dfrac{4}{4} = \dfrac{-2}{5}$ Removing a factor equal to 1: $\frac{4}{4} = 1$

5. $\dfrac{2}{6} = \dfrac{1 \cdot 2}{3 \cdot 2} = \dfrac{1}{3} \cdot \dfrac{2}{2} = \dfrac{1}{3}$

> Writing 1 allows for pairing of factors in the numerator and the denominator.

6. $\dfrac{30}{6} = \dfrac{5 \cdot 6}{1 \cdot 6} = \dfrac{5}{1} \cdot \dfrac{6}{6} = \dfrac{5}{1} = 5$

> We could also simplify $\frac{30}{6}$ by doing the division $30 \div 6$. That is, $\frac{30}{6} = 30 \div 6 = 5$.

7. $-\dfrac{15}{10} = -\dfrac{3 \cdot 5}{2 \cdot 5}$

$$= -\dfrac{3}{2} \cdot \dfrac{5}{5}$$

$$= -\dfrac{3}{2}$$

Removing a factor equal to 1: $\frac{5}{5} = 1$

8. $\dfrac{4x}{15x} = \dfrac{4 \cdot x}{15 \cdot x}$

$= \dfrac{4}{15} \cdot \dfrac{x}{x}$

$= \dfrac{4}{15}$ ⎫ Removing a factor equal to 1: $\dfrac{x}{x} = 1$

Do Exercises 10–14.

The tests for divisibility can be very helpful when simplifying.

EXAMPLE 9 Simplify: $\dfrac{105}{135}$.

Since both 105 and 135 end in 5, we know that 5 is a factor of both the numerator and the denominator:

$$\dfrac{105}{135} = \dfrac{21 \cdot 5}{27 \cdot 5} = \dfrac{21}{27} \cdot \dfrac{5}{5} = \dfrac{21}{27}.$$ To find the 21, we divided 105 by 5.
To find the 27, we divided 135 by 5.

A fraction is not "simplified" if common factors of the numerator and the denominator remain. Because 21 and 27 are both divisible by 3, we must simplify further:

$$\dfrac{105}{135} = \dfrac{21}{27} = \dfrac{7 \cdot 3}{9 \cdot 3} = \dfrac{7}{9} \cdot \dfrac{3}{3} = \dfrac{7}{9}.$$ To find the 7, we divided 21 by 3.
To find the 9, we divided 27 by 3.

EXAMPLE 10 Simplify: $\dfrac{322}{434}$.

Since 322 and 434 are both even, we know that 2 is a common factor:

$$\dfrac{322}{434} = \dfrac{2 \cdot 161}{2 \cdot 217} = \dfrac{2}{2} \cdot \dfrac{161}{217} = \dfrac{161}{217}.$$ Removing a factor equal to 1: $\dfrac{2}{2} = 1$

Before stating that $\frac{161}{217}$ is in simplified form, we must check for any common factors. We can choose one of the numbers, say 161, and try to find factors of that number. If we find a factor of 161, we check to see if that is also a factor of 217. We need consider only prime numbers as factors. Since 161 does not pass the divisibility test for 3 or for 5, we divide 161 by 7 and find that the remainder is 0. Thus 7 is a factor of 161. Then we divide 217 by 7, and find that 7 is also a factor of 217. Thus, 7 is a common factor and we can simplify further:

$$\dfrac{161}{217} = \dfrac{7 \cdot 23}{7 \cdot 31} = \dfrac{7}{7} \cdot \dfrac{23}{31} = \dfrac{23}{31}.$$ To find the 23, we divided 161 by 7.
To find the 31, we divided 217 by 7.

Thus $\dfrac{322}{434}$ simplifies to $\dfrac{23}{31}$.

Do Exercises 15–19.

Simplify.

10. $\dfrac{8}{14}$ **11.** $\dfrac{-10}{12}$

12. $\dfrac{40}{8}$ **13.** $\dfrac{4a}{3a}$

14. $-\dfrac{50}{30}$

Simplify.

15. $\dfrac{-35}{40}$ **16.** $\dfrac{801}{702}$

17. $\dfrac{280}{960}$ **18.** $\dfrac{429}{561}$

19. Simplify each fraction in this circle graph.

Days Spent Shopping for the Holidays

8 to 14 days $\frac{14}{100}$

1 to 2 days $\frac{20}{100}$

15 or more days $\frac{16}{100}$

Less than a day $\frac{8}{100}$

3 to 7 days $\frac{42}{100}$

Answers on page A-8

CANCELING

Canceling is a shortcut that you may have used for removing a factor that equals 1 when working with fraction notation. With concern, we mention it as a possibility for speeding up your work. Canceling may be done only when removing common factors in numerators and denominators. Each common factor allows us to remove a factor equal to 1 in a product.

In effect, slashes are used to indicate factors equal to 1 that have been removed. For instance, Example 10 might have been done faster as follows:

$$\frac{322}{434} = \frac{2 \cdot 161}{2 \cdot 217} \qquad \text{Factoring the numerator and the denominator}$$

$$= \frac{\cancel{2} \cdot 161}{\cancel{2} \cdot 217} \qquad \begin{array}{l}\text{When a factor equal to 1 is noted,}\\ \text{it is "canceled" as shown: } \frac{2}{2} = 1.\end{array}$$

$$= \frac{161}{217} = \frac{\cancel{7} \cdot 23}{\cancel{7} \cdot 31} = \frac{23}{31}.$$

Caution!

The difficulty with canceling is that it is often applied incorrectly in situations like the following:

$$\frac{\cancel{2} + 3}{\cancel{2}} = 3; \qquad \frac{\cancel{4} + 1}{\cancel{4} + 2} = \frac{1}{2}; \qquad \frac{1\cancel{5}}{\cancel{5}4} = \frac{1}{4}.$$

$\qquad$ Wrong! $\qquad\qquad$ Wrong! $\qquad\qquad$ Wrong!

The correct answers are

$$\frac{2 + 3}{2} = \frac{5}{2}; \qquad \frac{4 + 1}{4 + 2} = \frac{5}{6}; \qquad \frac{15}{54} = \frac{\cancel{3} \cdot 5}{\cancel{3} \cdot 18} = \frac{5}{18}.$$

In each case, the numbers canceled did not form a factor equal to 1. Factors are parts of products. For example, in $2 \cdot 3$, the numbers 2 and 3 are factors, but in $2 + 3$, 2 and 3 are terms, not factors.

- **If you cannot factor, do not cancel! If in doubt, do not cancel!**
- **Only factors can be canceled, and factors are never separated by + or − signs.**

C A Test for Equality

When denominators are the same, we say that fractions have a **common denominator.** One way to compare fractions like $\frac{2}{4}$ and $\frac{3}{6}$ is to find a common denominator and compare numerators. One way to do this is to multiply each fraction by 1, using the other denominator to write 1.

The denominator is 6.

$$\frac{3}{6} = \frac{3}{6} \cdot \frac{4}{4} = \frac{3 \cdot 4}{6 \cdot 4} = \frac{12}{24}$$

$$\frac{2}{4} = \frac{2}{4} \cdot \frac{6}{6} = \frac{2 \cdot 6}{4 \cdot 6} = \frac{12}{24}$$ Both denominators are 24.

The denominator is 4.

Because $\dfrac{12}{24} = \dfrac{12}{24}$ is true, it follows that $\dfrac{3}{6} = \dfrac{2}{4}$ is also true.

The "key" to the above work is that $3 \cdot 4$ and $2 \cdot 6$ are equal. Had these products differed, we would have shown that $\frac{3}{6}$ and $\frac{2}{4}$ were *not* equal. Note where the numbers 3, 4, 2, and 6 appear in the fractions $\frac{3}{6}$ and $\frac{2}{4}$.

A TEST FOR EQUALITY

We multiply these two numbers: $3 \cdot 4$.

We multiply these two numbers: $6 \cdot 2$.

$$\frac{3}{6} \; \square \; \frac{2}{4}$$

We call $3 \cdot 4$ and $6 \cdot 2$ **cross products.** Since the cross products are the same ($3 \cdot 4 = 6 \cdot 2$), we know that

$$\frac{3}{6} = \frac{2}{4}.$$

In the examples that follow, the symbol $\neq$, read "is not equal to," denotes two expressions that are *not* equal.

EXAMPLE 11 Use $=$ or $\neq$ for $\square$ to write a true sentence:

$$\frac{6}{7} \; \square \; \frac{7}{8}.$$

We multiply these two numbers: $6 \cdot 8 = 48$.

We multiply these two numbers: $7 \cdot 7 = 49$.

$$\frac{6}{7} \; \square \; \frac{7}{8}$$

Because $48 \neq 49$, $\frac{6}{7}$ and $\frac{7}{8}$ do not name the same number. Thus, $\frac{6}{7} \neq \frac{7}{8}$.

EXAMPLE 12 Use $=$ or $\neq$ for $\square$ to write a true sentence:

$$\frac{6}{10} \; \square \; \frac{3}{5}.$$

We multiply these two numbers: $6 \cdot 5 = 30$.

We multiply these two numbers: $10 \cdot 3 = 30$.

$$\frac{6}{10} \; \square \; \frac{3}{5}$$

Because the cross products are the same, we have $\frac{6}{10} = \frac{3}{5}$.

Remembering that $\frac{-a}{b}$, $\frac{a}{-b}$, and $-\frac{a}{b}$ all represent the same number can be helpful when checking for equality.

EXAMPLE 13 Use $=$ or $\neq$ for $\square$ to write a true sentence:

$$\frac{-6}{8} \; \square \; -\frac{9}{12}.$$

We rewrite $-\frac{9}{12}$ as $\frac{-9}{12}$ and then check cross products:

$$-6 \cdot 12 = -72 \qquad\qquad -9 \cdot 8 = -72$$

$$\frac{-6}{8} \; \square \; \frac{-9}{12}.$$

Because the cross products are the same, we have $\frac{-6}{8} = -\frac{9}{12}$.

Do Exercises 20–22.

Use $=$ or $\neq$ for $\square$ to write a true sentence.

20. $\dfrac{2}{6} \; \square \; \dfrac{3}{9}$

21. $\dfrac{7}{4} \; \square \; \dfrac{3}{2}$

22. $-\dfrac{10}{15} \; \square \; \dfrac{8}{-12}$

Answers on page A-8

a Find an equivalent expression for each number, using the denominator indicated. Use multiplication by 1.

1. $\dfrac{1}{2} = \dfrac{?}{10}$

2. $\dfrac{1}{6} = \dfrac{?}{12}$

3. $\dfrac{3}{4} = \dfrac{?}{-48}$

4. $\dfrac{2}{9} = \dfrac{?}{-18}$

5. $\dfrac{7}{10} = \dfrac{?}{50}$

6. $\dfrac{3}{8} = \dfrac{?}{48}$

7. $\dfrac{11}{5} = \dfrac{?}{5t}$

8. $\dfrac{5}{3} = \dfrac{?}{3a}$

9. $\dfrac{5}{12} = \dfrac{?}{48}$

10. $\dfrac{7}{8} = \dfrac{?}{56}$

11. $-\dfrac{17}{18} = -\dfrac{?}{54}$

12. $-\dfrac{11}{16} = -\dfrac{?}{256}$

13. $\dfrac{3}{-8} = \dfrac{?}{-40}$

14. $\dfrac{7}{-8} = \dfrac{?}{-32}$

15. $\dfrac{-7}{22} = \dfrac{?}{132}$

16. $\dfrac{-10}{21} = \dfrac{?}{126}$

17. $\dfrac{5}{8} = \dfrac{?}{8x}$

18. $\dfrac{2}{3} = \dfrac{?}{3a}$

19. $\dfrac{10}{7} = \dfrac{?}{7a}$

20. $\dfrac{4}{3} = \dfrac{?}{3n}$

21. $\dfrac{4}{9} = \dfrac{?}{9ab}$

22. $\dfrac{8}{11} = \dfrac{?}{11xy}$

23. $\dfrac{4}{9} = \dfrac{?}{27b}$

24. $\dfrac{8}{11} = \dfrac{?}{55y}$

b Simplify.

25. $\dfrac{2}{4}$

26. $\dfrac{3}{6}$

27. $-\dfrac{6}{9}$

28. $\dfrac{-9}{12}$

29. $\dfrac{10}{25}$

30. $\dfrac{8}{10}$

31. $\dfrac{27}{-3}$

32. $\dfrac{28}{-4}$

33. $\dfrac{27}{36}$

34. $\dfrac{30}{40}$

35. $-\dfrac{24}{14}$

36. $-\dfrac{16}{10}$

37. $\dfrac{16n}{48n}$

38. $\dfrac{150a}{25a}$

39. $\dfrac{-17}{51}$

40. $\dfrac{-425}{525}$

41. $\dfrac{420}{480}$

42. $\dfrac{180}{240}$

43. $\dfrac{153}{136}$

44. $\dfrac{117}{91}$

45. $\dfrac{132}{143}$

46. $\dfrac{91}{259}$

47. $\dfrac{221}{247}$

48. $\dfrac{299}{403}$

49. $\dfrac{3ab}{8ab}$

50. $\dfrac{6xy}{7xy}$

51. $\dfrac{9xy}{6x}$

52. $\dfrac{10ab}{15a}$

53. $\dfrac{-18a}{20ab}$

54. $\dfrac{-19x}{38xy}$

C Use $=$ or $\neq$ for ☐ to write a true sentence.

55. $\dfrac{3}{4}$ ☐ $\dfrac{9}{12}$

56. $\dfrac{4}{8}$ ☐ $\dfrac{3}{6}$

57. $\dfrac{1}{5}$ ☐ $\dfrac{2}{9}$

58. $\dfrac{1}{4}$ ☐ $\dfrac{2}{9}$

59. $\dfrac{3}{8}$ ☐ $\dfrac{6}{16}$

60. $\dfrac{2}{6}$ ☐ $\dfrac{6}{18}$

61. $\dfrac{2}{5}$ ☐ $\dfrac{3}{7}$

62. $\dfrac{1}{3}$ ☐ $\dfrac{1}{4}$

63. $\dfrac{-3}{10}$ ☐ $\dfrac{-4}{12}$

64. $\dfrac{-2}{9}$ ☐ $\dfrac{-8}{36}$

65. $-\dfrac{12}{9}$ ☐ $\dfrac{-8}{6}$

66. $\dfrac{-8}{7}$ ☐ $-\dfrac{16}{14}$

67. $\dfrac{5}{-2}$ ☐ $-\dfrac{17}{7}$

68. $-\dfrac{10}{3}$ ☐ $\dfrac{24}{-7}$

69. $\dfrac{305}{145}$ ☐ $\dfrac{122}{58}$

70. $\dfrac{425}{165}$ ☐ $\dfrac{130}{66}$

71. **D$_W$** Explain in your own words when it *is* possible to "cancel" and when it *is not* possible to "cancel."

72. **D$_W$** Can fraction notation be simplified if the numerator and denominator are two different prime numbers? Why or why not?

SKILL MAINTENANCE

Solve.

73. The East Dorchester soccer field is 90 yd long and 40 yd wide. What is its area?   [1.5c], [1.8a]

74. Yardbird Landscaping buys 13 maple saplings and 17 oak saplings for a project. A maple costs $23 and an oak costs $37. How much is spent altogether for the saplings?   [1.8a]

Multiply.   [2.4a]

75. $-12(-5)$

76. $-5(-13)$

77. $-9 \cdot 7$

78. $-8 \cdot 8$

Solve.   [1.7b]

79. $30 \cdot x = 150$

80. $10{,}947 = 123 \cdot y$

81. $5280 = 1760 + t$

82. $x + 2368 = 11{,}369$

SYNTHESIS

83. **D$_W$** Sometimes a fraction can be simplified more than once. What determines whether or not this can occur?

84. **D$_W$** Why is multiplication of fractions (Section 3.4) discussed before simplification of fractions (Section 3.5)?

Simplify. Use the list of prime numbers on p. 170.

85. $\dfrac{391}{667}$

86. $\dfrac{209ab}{247ac}$

87. $-\dfrac{1073x}{555y}$

88. $-\dfrac{187a}{289b}$

89. ▦ $\dfrac{4247}{4619}$

90. ▦ $\dfrac{3473}{3197}$

91. Sociologists have found that 4 of 10 people are shy. Write fraction notation for the part of the population that is shy; the part that is not shy. Simplify.

92. Sociologists estimate that 3 of 20 people are left-handed. In a crowd of 460 people, how many would you expect to be left-handed?

93. ▦ *Batting Averages.*   For the 2005 season, Derrek Lee of the Chicago Cubs won the National League batting title with 199 hits in 594 times at bat. Michael Young of the Texas Rangers won the American League title with 221 hits in 668 times at bat. Did they have the same fraction of hits in times at bat (batting average)? Why or why not?

Source: Major League Baseball

94. ▦ On a test of 82 questions, Penny got 63 correct. On another test of 100 questions, she got 77 correct. Did she get the same portion of each test correct? Why or why not?

Objectives

a Multiply and simplify using fraction notation.

b Solve applied problems involving multiplication.

a Simplifying When Multiplying

We usually want a simplified answer when we multiply (in Chapter 4 there will be times we don't). To make such simplifying easier, it is generally best not to calculate products in the numerator and denominator until we first factor. Consider

$$\frac{5}{6} \cdot \frac{14}{15}.$$

We proceed as follows:

$$\frac{5}{6} \cdot \frac{14}{15} = \frac{5 \cdot 14}{6 \cdot 15}$$

We do not yet carry out the multiplication. Note that 2 is a factor of 6 and 14. Also, note that 5 is a factor of 5 and 15.

$$= \frac{5 \cdot 2 \cdot 7}{2 \cdot 3 \cdot 5 \cdot 3}$$

Factoring and identifying common factors

$$= \frac{5 \cdot 2}{5 \cdot 2} \cdot \frac{7}{3 \cdot 3}$$

Factoring the fraction

$$\left. \begin{array}{l} = 1 \cdot \dfrac{7}{3 \cdot 3} \\[2mm] = \dfrac{7}{3 \cdot 3} \end{array} \right\}$$

Removing a factor equal to 1: $\frac{5 \cdot 2}{5 \cdot 2} = 1$

$$= \frac{7}{9}.$$

To multiply and simplify:

a) Write the products in the numerator and the denominator, but do not calculate the products.

b) Identify any common factors of the numerator and the denominator.

c) Factor the fraction to remove any factors that equal 1.

d) Calculate the remaining products.

EXAMPLES Multiply and simplify.

1. $\dfrac{2}{3} \cdot \dfrac{5}{4} = \dfrac{2 \cdot 5}{3 \cdot 4}$

Note that 2 is a common factor of 2 and 4.

$$= \frac{2 \cdot 5}{3 \cdot 2 \cdot 2}$$

Try to go directly to this step.

$$\left. \begin{array}{l} = \dfrac{2}{2} \cdot \dfrac{5}{3 \cdot 2} \\[2mm] = 1 \cdot \dfrac{5}{3 \cdot 2} = \dfrac{5}{6} \end{array} \right\}$$

Removing a factor equal to 1: $\frac{2}{2} = 1$

2. $\dfrac{6}{7} \cdot \dfrac{-5}{3} = \dfrac{3 \cdot 2 \cdot (-5)}{7 \cdot 3}$

Note that 3 is a common factor of 6 and 3.

$$= \frac{3}{3} \cdot \frac{2(-5)}{7} = \frac{-10}{7}, \text{ or } -\frac{10}{7}$$

Removing a factor equal to 1: $\frac{3}{3} = 1$

Study Tips

FIND A STUDY PARTNER

It is not always a simple matter to find a study partner. Friendships can become strained and tension can develop if one of the study partners feels that he or she is working too much or too little. Try to find a partner with whom you feel compatible and don't take it personally if your first partner is not a good match. With some effort you will be able to locate a suitable partner. Often tutor centers or learning labs are good places to look for one.

3. $\dfrac{10}{21} \cdot \dfrac{14a}{15} = \dfrac{5 \cdot 2 \cdot 7 \cdot 2a}{7 \cdot 3 \cdot 5 \cdot 3}$ Note that 5 is a common factor of 10 and 15.
                                   Note that 7 is a common factor of 21 and 14a.

$\qquad\qquad = \dfrac{5 \cdot 7}{5 \cdot 7} \cdot \dfrac{2 \cdot 2a}{3 \cdot 3}$

$\qquad\qquad = \dfrac{4a}{9}$ Removing a factor equal to 1: $\dfrac{5 \cdot 7}{5 \cdot 7} = 1$

4. $32 \cdot \dfrac{7}{8} = \dfrac{8 \cdot 4 \cdot 7}{8 \cdot 1}$ Note that 8 is a common factor of 32 and 8.

$\qquad\quad = \dfrac{8}{8} \cdot \dfrac{4 \cdot 7}{1} = 28$ Removing a factor equal to 1: $\dfrac{8}{8} = 1$

Caution!

Canceling can be used as follows for these examples.

1. $\dfrac{2}{3} \cdot \dfrac{5}{4} = \dfrac{\not2 \cdot 5}{3 \cdot \not2 \cdot 2} = \dfrac{5}{6}$ Removing a factor equal to 1: $\dfrac{2}{2} = 1$

2. $\dfrac{6}{7} \cdot \dfrac{-5}{3} = \dfrac{\not3 \cdot 2(-5)}{7 \cdot \not3} = \dfrac{-10}{7}$ Removing a factor equal to 1: $\dfrac{3}{3} = 1$

3. $\dfrac{10}{21} \cdot \dfrac{14a}{15} = \dfrac{\not5 \cdot 2 \cdot \not7 \cdot 2a}{\not7 \cdot 3 \cdot \not5 \cdot 3} = \dfrac{4a}{9}$ Removing a factor equal to 1: $\dfrac{5 \cdot 7}{5 \cdot 7} = 1$

4. $32 \cdot \dfrac{7}{8} = \dfrac{\not8 \cdot 4 \cdot 7}{\not8 \cdot 1} = 28$ Removing a factor equal to 1: $\dfrac{8}{8} = 1$

Remember, only factors can be canceled!

Do Exercises 1–4.

b   Solving Problems

Recall that problems involving repeated addition and those with keywords "of," "twice," and "product" translate to multiplication.

EXAMPLE 5   Elite Elegance is preparing souvenir favors for a charity fund-raising dinner. How many pounds of caramels will be needed to fill 235 boxes if each box contains $\frac{2}{5}$ lb?

1. **Familiarize.**   We first make a drawing or at least visualize the situation. Repeated addition will work here.

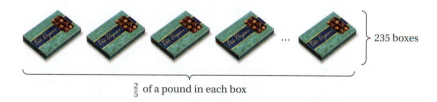

... } 235 boxes

$\frac{2}{5}$ of a pound in each box

    We let $n =$ the number of pounds of caramels.

2. **Translate.**   The problem translates to the following equation:

Rephrase:	Number of pounds of caramels	is	Number of boxes	times	Number of pounds per box
Translate:	n	$=$	235	$\cdot$	$\dfrac{2}{5}$

Multiply and simplify.

1. $\dfrac{2}{3} \cdot \dfrac{7}{8}$

2. $\dfrac{4}{5} \cdot \dfrac{-5}{12}$

3. $16 \cdot \dfrac{3}{8}$

4. $\dfrac{5}{2x} \cdot 6$

Answers on page A-8

5. Rosebud Landscaping uses $\frac{2}{3}$ lb of peat moss for a rosebush. How much will be needed for 21 rosebushes?

3. Solve. To solve the equation, we carry out the multiplication:

$$n = 235 \cdot \frac{2}{5} = \frac{235 \cdot 2}{5} \qquad \text{Multiplying}$$

$$= \frac{5 \cdot 47 \cdot 2}{5 \cdot 1} \qquad \text{Note that 235 is divisible by 5.}$$

$$= \frac{5}{5} \cdot \frac{47 \cdot 2}{1} = 94. \qquad \text{Removing the factor } \frac{5}{5} \text{ and simplifying}$$

4. Check. We could repeat the calculation. We can also think about the reasonableness of the answer. Since we are putting less than a pound of caramels in each box, it makes sense that 235 boxes require fewer than 235 pounds. This provides a partial check of the answer.

A second partial check can be performed using the units:

$$235 \text{ boxes} \cdot \frac{2}{5} \text{ pounds per box}$$

$$= 235 \cdot \frac{2}{5} \cdot \cancel{\text{boxes}} \cdot \frac{\text{pounds}}{\cancel{\text{box}}}$$

$$= 94 \text{ pounds.}$$

Since the resulting unit is pounds, we have another partial check.

5. State. Elite Elegance will need 94 lb of caramels.

Do Exercise 5.

AREA

Multiplication of fractions can arise in geometry problems involving the area of a triangle. Consider a triangle with a base of length b and a height of h.

A rectangle can be formed by splitting and inverting a copy of this triangle.

The rectangle's area, $b \cdot h$, is exactly twice the area of the triangle. We have the following result.

AREA OF A TRIANGLE

The **area A of a triangle** is half the length of the base b times the height h:

$$A = \frac{1}{2} \cdot b \cdot h.$$

EXAMPLE 6 Find the area of this triangle.

$$A = \frac{1}{2} \cdot b \cdot h$$

$$= \frac{1}{2} \cdot 9 \text{ yd} \cdot 6 \text{ yd}$$

$$= \frac{9 \cdot 6}{2} \text{ yd}^2$$

$$= 27 \text{ yd}^2$$

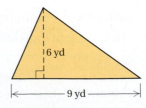

Answer on page A-8

EXAMPLE 7 Find the area of this triangle.

$$A = \frac{1}{2} \cdot b \cdot h$$

$$= \frac{1}{2} \cdot \frac{10}{3} \, \text{cm} \cdot 4 \, \text{cm}$$

$$= \frac{1 \cdot 10 \cdot 4}{2 \cdot 3} \, \text{cm}^2$$

$$= \frac{1 \cdot 2 \cdot 5 \cdot 4}{2 \cdot 3} \, \text{cm}^2 \qquad \text{Removing a factor equal to 1: } \frac{2}{2} = 1$$

$$= \frac{20}{3} \, \text{cm}^2$$

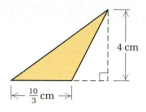

Do Exercises 6 and 7.

EXAMPLE 8 Find the area of this kite.

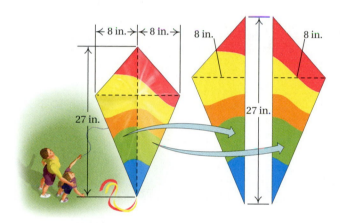

1. **Familiarize.** We look for figures with areas we can calculate using area formulas that we already know. We let K = the kite's area.

2. **Translate.** The kite consists of two triangles, each with a base of 27 in. and a height of 8 in. We can apply the formula $A = \frac{1}{2} \cdot b \cdot h$ for the area of a triangle and then multiply by 2.

Rephrase:	Kite's area	is	twice	Area of long triangle
Translate:	K	$=$	2	$\cdot \quad \frac{1}{2}(27 \text{ in.}) \cdot (8 \text{ in.})$

3. **Solve.** We have

$$K = 2 \cdot \frac{1}{2} \cdot (27 \text{ in.}) \cdot (8 \text{ in.})$$

$$= 1 \cdot 27 \text{ in.} \cdot 8 \text{ in.} = 216 \text{ in}^2.$$

4. **Check.** We can check by repeating the calculations. The unit, in², is appropriate for area.

5. **State.** The area of the kite is 216 in².

Do Exercise 8.

Find the area.

6.

7.

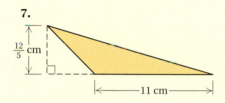

8. Find the area. (*Hint*: The figure is made up of a rectangle and a triangle.)

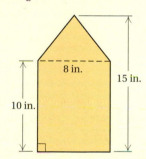

Answers on page A-8

a   Multiply. Don't forget to simplify, if possible.

1. $\dfrac{3}{8} \cdot \dfrac{7}{3}$ **2.** $\dfrac{4}{5} \cdot \dfrac{1}{4}$ **3.** $\dfrac{7}{8} \cdot \dfrac{-1}{7}$ **4.** $\dfrac{5}{6} \cdot \dfrac{-1}{5}$

5. $\dfrac{1}{8} \cdot \dfrac{6}{7}$ **6.** $\dfrac{2}{5} \cdot \dfrac{1}{10}$ **7.** $\dfrac{4}{6} \cdot \dfrac{1}{3}$ **8.** $\dfrac{3}{6} \cdot \dfrac{1}{6}$

9. $\dfrac{9}{-5} \cdot \dfrac{12}{8}$ **10.** $\dfrac{16}{-15} \cdot \dfrac{5}{4}$ **11.** $\dfrac{5x}{9} \cdot \dfrac{4}{5}$ **12.** $\dfrac{25}{4a} \cdot \dfrac{4}{3}$

13. $\dfrac{1}{4} \cdot 12$ **14.** $\dfrac{1}{6} \cdot 12$ **15.** $21 \cdot \dfrac{1}{3}$ **16.** $18 \cdot \dfrac{1}{2}$

17. $-16\left(-\dfrac{3}{4}\right)$ **18.** $-24\left(-\dfrac{5}{6}\right)$ **19.** $\dfrac{3}{8} \cdot 8a$ **20.** $\dfrac{2}{9} \cdot 9x$

21. $\left(-\dfrac{3}{8}\right)\left(-\dfrac{8}{3}\right)$ **22.** $\left(-\dfrac{7}{9}\right)\left(-\dfrac{9}{7}\right)$ **23.** $\dfrac{a}{b} \cdot \dfrac{b}{a}$ **24.** $\dfrac{n}{m} \cdot \dfrac{m}{n}$

25. $\dfrac{1}{26} \cdot 143a$ **26.** $\dfrac{1}{28} \cdot 105n$ **27.** $176\left(\dfrac{1}{-6}\right)$ **28.** $135\left(\dfrac{1}{-10}\right)$

29. $-8x \cdot \dfrac{1}{-8x}$

30. $-5a \cdot \dfrac{1}{-5a}$

31. $\dfrac{2x}{9} \cdot \dfrac{27}{2x}$

32. $\dfrac{10a}{3} \cdot \dfrac{3}{5a}$

33. $\dfrac{7}{10} \cdot \dfrac{34}{150}$

34. $\dfrac{8}{10} \cdot \dfrac{45}{100}$

35. $\dfrac{36}{85} \cdot \dfrac{25}{-99}$

36. $\dfrac{-70}{45} \cdot \dfrac{50}{49}$

37. $\dfrac{-98}{99} \cdot \dfrac{27a}{175a}$

38. $\dfrac{70}{-49} \cdot \dfrac{63}{300x}$

39. $\dfrac{110}{33} \cdot \dfrac{-24}{25x}$

40. $\dfrac{-19}{130} \cdot \dfrac{65}{38x}$

41. $\left(-\dfrac{11}{24}\right)\dfrac{3}{5}$

42. $\left(-\dfrac{15}{22}\right)\dfrac{4}{7}$

43. $\dfrac{10a}{21} \cdot \dfrac{3}{8b}$

44. $\dfrac{17}{21y} \cdot \dfrac{3x}{5}$

b Solve.

The *pitch* of a screw is the distance between its threads. With each complete rotation, the screw goes in or out a distance equal to its pitch. Use this information to answer Exercises 45 and 46.

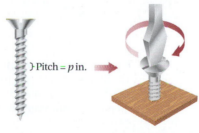

}Pitch = *p* in.

Each rotation moves the screw
in or out *p* in.

45. The pitch of a screw is $\frac{1}{16}$ in. How far will it go into a piece of oak when it is turned 10 complete rotations clockwise?

46. The pitch of a screw is $\frac{3}{32}$ in. How far will it come out of a piece of plywood when it is turned 10 complete rotations counterclockwise?

47. Swimming Speeds. The swimming speed of a killer whale is about 30 mph. The swimming speed of a dolphin is about $\frac{3}{5}$ that of a killer whale. Find the swimming speed of a dolphin.

Source: G. Cafiero and M. Jahoda, *Whales and Dolphins.* New York: Barnes & Noble Books, 1994

48. After Craig completes 60 hr of teacher training at college, he can earn $95 for working a full day as a substitute teacher. How much will he receive for working $\frac{1}{5}$ of a day?

49. Mailing-List Addresses. Analysts have determined that $\frac{1}{4}$ of the addresses on a mailing list will change in one year. A business has a mailing list of 2500 people. After one year, how many addresses on that list will be incorrect?

50. Shy People. Sociologists have determined that $\frac{2}{5}$ of the people in the world are shy. A sales manager is interviewing 650 people for an aggressive sales position. How many of these people might be shy?

51. According to a recent survey, $\frac{2}{3}$ of Americans eat breakfast, and $\frac{3}{4}$ of these eat breakfast at home. What fraction of Americans eat breakfast at home?

Source: *Parade Magazine,* November 16, 2003

52. According to a recent survey, $\frac{7}{10}$ of Americans eat sandwiches for lunch, and $\frac{1}{3}$ of these eat fast food sandwiches. What fraction of Americans eat fast food sandwiches for lunch?

Source: *Parade Magazine,* November 16, 2003

53. A recipe for piecrust calls for $\frac{2}{3}$ cup of flour. A chef is making $\frac{1}{2}$ of the recipe. How much flour should the chef use?

54. Of the students in an entering class, $\frac{2}{5}$ have cameras and $\frac{1}{4}$ of these students will join the college photography club. What fraction of the students in the entering class will join the photography club?

55. Assessed Value. A house worth $154,000 is assessed for $\frac{3}{4}$ of its value. What is the assessed value of the house?

56. Tuition Loan. Christie's tuition was $8400. A loan was obtained for $\frac{3}{4}$ of the tuition. How much was the loan?

57. *Map Scaling.* On a map, 1 in. represents 240 mi. How much does $\frac{2}{3}$ in. represent?

58. *Map Scaling.* On a map, 1 in. represents 120 mi. How much does $\frac{3}{4}$ in. represent?

59. *Household Budgets.* A family has an annual income of $39,600. Of this, $\frac{1}{4}$ is spent for food, $\frac{1}{5}$ for housing, $\frac{1}{10}$ for clothing, $\frac{1}{9}$ for savings, $\frac{1}{4}$ for taxes, and the rest for other expenses. How much is spent for each?

60. *Household Budgets.* A family has an annual income of $37,800. Of this, $\frac{1}{4}$ is spent for food, $\frac{1}{5}$ for housing, $\frac{1}{10}$ for clothing, $\frac{1}{9}$ for savings, $\frac{1}{4}$ for taxes, and the rest for other expenses. How much is spent for each?

Typical Family Income

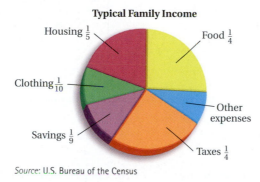

Housing $\frac{1}{5}$ Food $\frac{1}{4}$

Clothing $\frac{1}{10}$

Other expenses

Savings $\frac{1}{9}$ Taxes $\frac{1}{4}$

Source: U.S. Bureau of the Census

Find the area.

61.

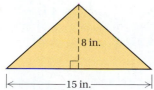

8 in.

15 in.

62.

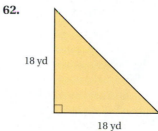

18 yd

18 yd

63.

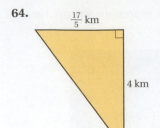

Triangle with height $\frac{7}{2}$ mm and base 5 mm.

64.

$\frac{17}{5}$ km

4 km

65.

Triangle with base $\frac{9}{2}$ m and height $\frac{7}{2}$ m.

66.

$\frac{8}{3}$ yd

$\frac{7}{4}$ yd

67.

10 mi

8 mi

13 mi

68.

15 cm

30 cm

30 cm

69. *Jewelry Making.* A Zuni bolo tie is made by soldering together two identically kite-shaped pieces of sterling silver, as shown below. Determine the surface area of the front of the bolo tie.

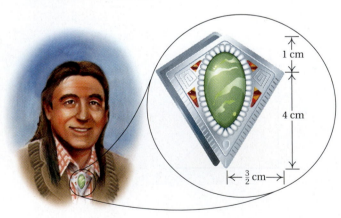

1 cm

4 cm

$\frac{3}{2}$ cm

70. *Construction.* Find the total area of the sides and ends of the town office building shown. Do not subtract for any windows, doors, or steps.

25 ft

11 ft

75 ft

50 ft

Dumont Town Offices

71. $\mathbf{D_W}$ When multiplying using fraction notation, we form products in the numerator and the denominator, but do not immediately calculate the products. Why?

72. $\mathbf{D_W}$ If a fraction's numerator and denominator have no factors (other than 1) in common, can the fraction be simplified? Why or why not?

SKILL MAINTENANCE

Solve.   [1.7b]

73. $48 \cdot t = 1680$

74. $74 \cdot x = 6290$

75. $3125 = 25 \cdot t$

76. $2880 = 24 \cdot y$

77. $t + 28 = 5017$

78. $456 + x = 9002$

79. $8797 = y + 2299$

80. $10{,}000 = 3593 + m$

SYNTHESIS

81. $\mathbf{D_W}$ Why is it useful to remember which numbers are prime when simplifying fractions?

82. $\mathbf{D_W}$ Is the product of two fractions always a fraction? Why or why not?

Simplify. Use the list of prime numbers on p. 170.

83. ▦ $\dfrac{201}{535} \cdot \dfrac{4601}{6499}$

84. ▦ $\dfrac{5767}{3763} \cdot \dfrac{159}{395}$

85. ▦ $\dfrac{667}{899} \cdot \dfrac{558}{621}$

86. *College Profile.*   Of students entering a college, $\frac{7}{8}$ have completed high school and $\frac{2}{3}$ of those are older than 20. If $\frac{1}{7}$ of all students are left-handed, estimate the fraction of students entering the college who are left-handed high school graduates over the age of 20.

87. *College Profile.*   Refer to the information in Exercise 86. If 480 students are entering the college, estimate the number of left-handed high school graduates 20 years old or younger.

88. *College Profile.*   Refer to Exercise 86. What fraction of students entering the college did not graduate high school, are 20 years old or younger, and are left-handed?

89. ▦ *Manufacturing.*   A TriMint candy box is triangular at each end, as shown below. Find the surface area of the box.

30 mm · 30 mm · 26 mm · 140 mm · 30 mm

90. ▦ *Painting.*   Shoreline Painting needs to determine the surface area of an octagonal steeple. Find the total area, if the dimensions are as shown below.

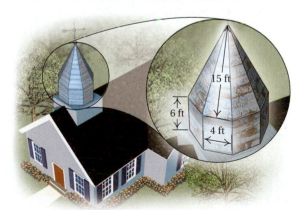

15 ft · 6 ft · 4 ft

91. (See Exercise 70). If both sides and both ends of the town office building are identical, and if the windows are each 3 ft by 4 ft, and the entrances 6 ft by 8 ft, how many square feet of siding does the building require?

Find the reciprocal.

1. $\dfrac{7}{8}$

2. $\dfrac{-6}{x}$

3. 2

4. $\dfrac{1}{5}$

5. $-\dfrac{3}{10}$

Answers on page A-9

210

a ## Reciprocals

Look at these products:

$$8 \cdot \frac{1}{8} = \frac{8}{8} = 1; \qquad \frac{-2}{3} \cdot \frac{3}{-2} = \frac{-6}{-6} = 1.$$

RECIPROCALS

If the product of two numbers is 1, we say that they are **reciprocals** of each other.* To find the reciprocal of a fraction, interchange the numerator and the denominator.

The numbers $\dfrac{3}{4}$ and $\dfrac{4}{3}$ are reciprocals of each other.

EXAMPLES Find the reciprocal.

1. The reciprocal of $\dfrac{4}{5}$ is $\dfrac{5}{4}$. Note that $\dfrac{4}{5} \cdot \dfrac{5}{4} = \dfrac{20}{20} = 1.$

2. The reciprocal of $\dfrac{a}{b}$ is $\dfrac{b}{a}$. Note that $\dfrac{a}{b} \cdot \dfrac{b}{a} = \dfrac{ab}{ba} = 1.$

3. The reciprocal of 8 is $\dfrac{1}{8}$. Think of 8 as $\dfrac{8}{1}$: $\dfrac{8}{1} \cdot \dfrac{1}{8} = \dfrac{8}{8} = 1.$

4. The reciprocal of $\dfrac{1}{3}$ is 3. Note that $\dfrac{1}{3} \cdot 3 = \dfrac{3}{3} = 1.$

5. The reciprocal of $-\dfrac{5}{9}$ is $-\dfrac{9}{5}$. Negative numbers have negative reciprocals: $\left(-\dfrac{5}{9}\right)\left(-\dfrac{9}{5}\right) = \dfrac{45}{45} = 1.$

Do Exercises 1–5.

Does 0 have a reciprocal? If it did, it would have to be a number x such that

$$0 \cdot x = 1.$$

But 0 times any number is 0. Thus, **0 has no reciprocal.**

b ## Division

Consider the division $\frac{3}{4} \div \frac{1}{8}$. This asks how many $\frac{1}{8}$'s are in $\frac{3}{4}$. We can answer this by looking at the figure below.

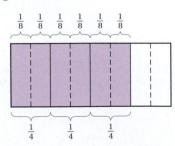

*A reciprocal is also called a multiplicative inverse.

We see that there are six $\frac{1}{8}$'s in $\frac{3}{4}$. Thus,

$$\frac{3}{4} \div \frac{1}{8} = 6.$$

This can be checked by multiplying:

$$6 \cdot \frac{1}{8} = \frac{6}{8} = \frac{3}{4}.$$

Here is a faster way to divide. An explanation of why it works appears on the next page.

DIVISION OF FRACTIONS

To divide by a fraction, multiply by its reciprocal:

$$\frac{a}{b} \div \frac{c}{d} = \frac{a}{b} \cdot \frac{d}{c}. \qquad \text{Multiply by the reciprocal of the divisor.}$$

Recall that when two numbers with unlike signs are multiplied or divided, the result is negative. When both numbers have the same sign, the result is positive.

EXAMPLES Divide and simplify.

6. $\dfrac{5}{6} \div \dfrac{2}{3} = \dfrac{5}{6} \cdot \dfrac{3}{2}$ Multiplying by the reciprocal of the divisor

$\qquad = \dfrac{5 \cdot 3}{3 \cdot 2 \cdot 2}$ Factoring and identifying a common factor

$\qquad = \dfrac{3}{3} \cdot \dfrac{5}{2 \cdot 2}$ Removing a factor equal to 1: $\dfrac{3}{3} = 1$

$\qquad = \dfrac{5}{4}$

7. $\dfrac{-3}{5} \div \dfrac{1}{2} = \dfrac{-3}{5} \cdot 2$ The reciprocal of $\dfrac{1}{2}$ is 2.

$\qquad = \dfrac{-3 \cdot 2}{5} = \dfrac{-6}{5}$

8. $\dfrac{2a}{5} \div 7 = \dfrac{2a}{5} \cdot \dfrac{1}{7}$ The reciprocal of 7 is $\dfrac{1}{7}$.

$\qquad = \dfrac{2a \cdot 1}{5 \cdot 7} = \dfrac{2a}{35}$

9. $\dfrac{\frac{7}{10}}{-\frac{14}{15}} = \dfrac{7}{10} \div \left(-\dfrac{14}{15} \right)$ The fraction bar indicates division.

$\qquad = \dfrac{7}{10} \cdot \left(-\dfrac{15}{14} \right)$ Multiplying by the reciprocal of the divisor

$\qquad = \dfrac{7 \cdot 5(-3)}{2 \cdot 5 \cdot 7 \cdot 2}$ Factoring and identifying common factors

$\qquad = \dfrac{7 \cdot 5}{7 \cdot 5} \cdot \dfrac{-3}{4}$ Removing a factor equal to 1: $\dfrac{7 \cdot 5}{7 \cdot 5} = 1$

$\qquad = -\dfrac{3}{4}$

Divide and simplify.

6. $\dfrac{2}{7} \div \dfrac{6}{5}$

7. $\left(-\dfrac{2}{3} \right) \div \dfrac{1}{4}$

8. $\dfrac{5}{4} \div 10$

9. $60 \div \dfrac{3a}{5}$

10. $\dfrac{\frac{-6}{7}}{\frac{3}{5}}$

Answers on page A-9

11. To remember *why* fractions are divided as they are, multiply by 1 to perform the following division, using the reciprocal of $\frac{4}{5}$ to write 1.

$$\frac{\dfrac{6}{7}}{\dfrac{4}{5}}$$

Answer on page A-9

Multiplying and Dividing Fractions Multiplication and division of fractions on fraction calculators is entered like it is written. To perform the division $\frac{4}{9} \div \frac{6}{5}$ with such a calculator, the following keystrokes can be used:

The simplified fraction notation is $\frac{10}{27}$. On calculators without an $\boxed{a\,b/c}$ key, parentheses must be used around the second fraction when dividing. The following keystrokes can be used to perform the division and convert the result to fraction notation:

Again, the result is $\frac{10}{27}$.

Exercises: Use a calculator to perform the following operations.

1. $\dfrac{5}{8} \cdot \dfrac{4}{15}$ 2. $\dfrac{10}{12} \div \dfrac{3}{8}$

3. $\dfrac{-6}{7} \div \dfrac{2}{3}$ 4. $\dfrac{-9}{10} \div \dfrac{-3}{5}$

Canceling can be used as follows for Examples 6 and 9.

6. $\dfrac{5}{6} \div \dfrac{2}{3} = \dfrac{5}{6} \cdot \dfrac{3}{2} = \dfrac{5 \cdot 3}{6 \cdot 2} = \dfrac{5 \cdot \cancel{3}}{\cancel{3} \cdot 2 \cdot 2} = \dfrac{5}{2 \cdot 2} = \dfrac{5}{4}$

Removing a factor equal to 1: $\frac{3}{3} = 1$

9. $\dfrac{7}{10} \div \left(-\dfrac{14}{15}\right) = \dfrac{7}{10} \cdot \left(-\dfrac{15}{14}\right) = \dfrac{\cancel{7} \cdot \cancel{5}(-3)}{2 \cdot \cancel{5} \cdot \cancel{7} \cdot 2} = \dfrac{-3}{4}$, or $-\dfrac{3}{4}$

Removing a factor equal to 1: $\frac{7 \cdot 5}{7 \cdot 5} = 1$

Remember, if you can't factor, you can't cancel!

Do Exercises 6–10 on the previous page.

Why do we multiply by a reciprocal when dividing? To see why, consider $\frac{2}{3} \div \frac{7}{5}$. We will multiply by 1 to find an equivalent expression. To write 1 we use $(5/7)/(5/7)$, since $\frac{5}{7}$ is the reciprocal of $\frac{7}{5}$.

$$\dfrac{2}{3} \div \dfrac{7}{5} = \dfrac{\dfrac{2}{3}}{\dfrac{7}{5}}$$
Writing fraction notation for the division

$$= \dfrac{\dfrac{2}{3}}{\dfrac{7}{5}} \cdot 1$$
Multiplying by 1

$$= \dfrac{\dfrac{2}{3}}{\dfrac{7}{5}} \cdot \dfrac{\dfrac{5}{7}}{\dfrac{5}{7}}$$
Multiplying by 1; $\frac{5}{7}$ is the reciprocal of $\frac{7}{5}$ and $\frac{\frac{5}{7}}{\frac{5}{7}} = 1$

$$= \dfrac{\dfrac{2}{3} \cdot \dfrac{5}{7}}{\dfrac{7}{5} \cdot \dfrac{5}{7}}$$
Multiplying the numerators and the denominators

$$= \dfrac{\dfrac{2}{3} \cdot \dfrac{5}{7}}{1} = \dfrac{2}{3} \cdot \dfrac{5}{7} = \dfrac{10}{21}$$

After we multiplied, we got 1 for the denominator. The numerator (in color) shows the multiplication by the reciprocal of $\frac{7}{5}$.

Thus,

$$\dfrac{2}{3} \div \dfrac{7}{5} = \dfrac{2}{3} \cdot \dfrac{5}{7} = \dfrac{10}{21}.$$

Expressions of the form $\dfrac{\dfrac{a}{b}}{\dfrac{c}{d}}$ are examples of *complex fractions*:

$$\dfrac{\dfrac{a}{b}}{\dfrac{c}{d}} = \dfrac{a}{b} \div \dfrac{c}{d}.$$

Do Exercise 11.

CHAPTER 3: Fraction Notation: Multiplication and Division

a   Find the reciprocal.

1. $\dfrac{7}{3}$

2. $\dfrac{6}{5}$

3. 9

4. 3

5. $\dfrac{1}{7}$

6. $\dfrac{1}{4}$

7. $-\dfrac{10}{3}$

8. $-\dfrac{12}{5}$

9. $\dfrac{3}{17}$

10. $\dfrac{9}{28}$

11. $\dfrac{-3n}{m}$

12. $\dfrac{8t}{-7r}$

13. $\dfrac{8}{-15}$

14. $\dfrac{-6}{25}$

15. $7m$

16. $5n$

17. $\dfrac{1}{4a}$

18. $\dfrac{1}{9t}$

19. $-\dfrac{1}{3z}$

20. $-\dfrac{1}{2x}$

b   Divide. Don't forget to simplify when possible.

21. $\dfrac{3}{7} \div \dfrac{3}{4}$

22. $\dfrac{2}{3} \div \dfrac{2}{5}$

23. $\dfrac{7}{6} \div \dfrac{5}{-3}$

24. $\dfrac{5}{3} \div \dfrac{4}{-9}$

25. $\dfrac{4}{3} \div \dfrac{1}{3}$

26. $\dfrac{10}{9} \div \dfrac{1}{2}$

27. $\left(-\dfrac{1}{3}\right) \div \dfrac{1}{6}$

28. $\left(-\dfrac{1}{4}\right) \div \dfrac{1}{5}$

29. $\left(-\dfrac{10}{21}\right) \div \left(-\dfrac{2}{15}\right)$

30. $-\dfrac{15}{28} \div \left(-\dfrac{9}{20}\right)$

31. $\dfrac{3}{8} \div 24$

32. $\dfrac{5}{6} \div 45$

33. $\dfrac{12}{7} \div (4x)$

34. $\dfrac{18}{5} \div (2y)$

35. $(-12) \div \dfrac{3}{2}$

36. $(-24) \div \dfrac{3}{8}$

37. $28 \div \dfrac{4}{5a}$

38. $40 \div \dfrac{2}{3m}$

39. $\left(-\dfrac{5}{8}\right) \div \left(-\dfrac{5}{8}\right)$

40. $\left(-\dfrac{2}{5}\right) \div \left(-\dfrac{2}{5}\right)$

41. $\dfrac{-8}{15} \div \dfrac{4}{5}$

42. $\dfrac{6}{-13} \div \dfrac{3}{26}$

43. $\dfrac{77}{64} \div \dfrac{49}{18}$

44. $\dfrac{81}{42} \div \dfrac{33}{56}$

45. $120a \div \dfrac{45}{14}$

46. $360n \div \dfrac{27n}{8}$

47. $\dfrac{\frac{2}{5}}{\frac{3}{7}}$

48. $\dfrac{\frac{5}{6}}{\frac{2}{7}}$

49. $\dfrac{\dfrac{7}{20}}{\dfrac{8}{5}}$

50. $\dfrac{\dfrac{8}{21}}{\dfrac{6}{5}}$

51. $\dfrac{-\dfrac{15}{8}}{\dfrac{9}{10}}$

52. $\dfrac{-\dfrac{27}{10}}{\dfrac{21}{20}}$

53. $\dfrac{-\dfrac{9}{16}}{-\dfrac{6}{5}}$

54. $\dfrac{-\dfrac{35}{18}}{-\dfrac{14}{27}}$

55. **D$_W$** Without performing the division, explain why $5 \div \frac{1}{7}$ is a larger number than $5 \div \frac{2}{3}$.

56. **D$_W$** Carl incorrectly insists that $\frac{2}{5} \div \frac{3}{4}$ is $\frac{15}{8}$. What mistake is he probably making?

SKILL MAINTENANCE

⤶ VOCABULARY REINFORCEMENT

In each of Exercises 57–64, fill in the blank with the correct term from the given list. Some of the choices may not be used and some may be used more than once.

57. The equation $14 + (2 + 30) = (14 + 2) + 30$ illustrates the _____ law of addition.   [1.2a]

58. In the product $10 \cdot \frac{3}{4}$, the numbers 10 and $\frac{3}{4}$ are called _____.   [3.2a]

59. A natural number that has exactly two different factors, itself and 1, is called a _____ number.   [3.2b]

60. In the fraction $\frac{4}{17}$, we call 17 the _____.   [3.3a]

61. Since $a + 0 = a$ for any number a, the number 0 is the _____ identity.   [2.2a]

62. The product of 6 and $\frac{1}{6}$ is 1; we say that 6 and $\frac{1}{6}$ are _____ of each other.   [3.7a]

63. The sum of 6 and -6 is 0; we say that 6 and -6 are _____ of each other.   [2.1d]

64. A sentence with $=$ is called a(n) _____.   [1.7a]

associative

commutative

additive

multiplicative

reciprocals

factors

prime

composite

numerator

denominator

equation

expression

opposites

variables

SYNTHESIS

65. **D$_W$** Is division of fractions associative? Why or why not?

66. **D$_W$** What numbers are their own reciprocals?

Simplify.

67. $\left(\dfrac{4}{15} \div \dfrac{2}{25}\right)^2$

68. $\left(\dfrac{9}{10} \div \dfrac{12}{25}\right)^2$

69. $\left(\dfrac{9}{10} \div \dfrac{2}{5} \div \dfrac{3}{8}\right)^2$

70. $\dfrac{\left(-\dfrac{3}{7}\right)^2 \div \dfrac{12}{5}}{\left(\dfrac{-2}{9}\right)\left(\dfrac{9}{2}\right)}$

71. $\left(\dfrac{14}{15} \div \dfrac{49}{65} \cdot \dfrac{77}{260}\right)^2$

72. $\left(\dfrac{10}{9}\right)^2 \div \dfrac{35}{27} \cdot \dfrac{49}{44}$

Simplify. Use the list of prime numbers on p. 170.

73. ▦ $\dfrac{711}{1957} \div \dfrac{10{,}033}{13{,}081}$

74. ▦ $\dfrac{8633}{7387} \div \dfrac{485}{581}$

75. ▦ $\dfrac{451}{289} \div \dfrac{123}{340}$

76. ▦ $\dfrac{530}{490} \div \dfrac{1060}{980}$

3.8 SOLVING EQUATIONS: THE MULTIPLICATION PRINCIPLE

In Sections 1.7 and 2.8, we learned to solve an equation involving multiplication by dividing on both sides. With fraction notation, we can solve the same type of equation by using multiplication.

a   The Multiplication Principle

We have seen that to divide by a fraction, we multiply by the reciprocal of that fraction. This suggests that we restate the division principle in its more common form—the multiplication principle.

> **THE MULTIPLICATION PRINCIPLE**
>
> For any numbers a, b, and c, with $c \neq 0$,
>
> $\quad a = b \quad$ is equivalent to $\quad a \cdot c = b \cdot c$.

EXAMPLE 1 Solve: $\frac{3}{4}x = 15$.

We can multiply by any nonzero number on both sides to produce an equivalent equation. Since we are looking for an equation of the form $1x = \square$, we multiply both sides of the equation by the reciprocal of $\frac{3}{4}$.

$$\frac{3}{4}x = 15$$

$$\frac{4}{3} \cdot \frac{3}{4}x = \frac{4}{3} \cdot 15 \qquad \text{Using the multiplication principle; note that } \tfrac{4}{3} \text{ is the reciprocal of } \tfrac{3}{4}.$$

$$\left(\frac{4}{3} \cdot \frac{3}{4}\right)x = \frac{4 \cdot 15}{3} \qquad \text{Using an associative law; try to do this mentally.}$$

$$1x = 20 \qquad \text{Multiplying; note that } \frac{4 \cdot 15}{3} = \frac{4 \cdot \cancel{3} \cdot 5}{\cancel{3}}.$$

$$x = 20 \qquad \text{Remember that } 1x \text{ is } x.$$

To confirm that 20 is the solution, we perform a check.

Check:
$$\frac{3}{4}x = 15$$

$$\frac{\dfrac{3}{4} \cdot 20}{} \;?\; 15$$

$$\frac{3 \cdot \cancel{4} \cdot 5}{\cancel{4}}$$

$$3 \cdot 5 \;\Big|\; 15 \quad \text{TRUE}$$

Removing a factor equal to 1: $\frac{4}{4} = 1$

The solution is 20.

Note that using the multiplication principle to multiply by $\frac{4}{3}$ on both sides is the same as using the division principle to divide by $\frac{3}{4}$ on both sides.

Do Exercises 1 and 2.

Solve.

1. $\frac{2}{3}x = 8$

2. $\frac{2}{7}a = -6$

Answers on page A-9

215

Solve.

3. $-\dfrac{9}{8} = 4x$

In an expression like $\frac{3}{4}x$, the constant factor—in this case, $\frac{3}{4}$—is called the **coefficient.** In Example 1, we multiplied on both sides by $\frac{4}{3}$, the reciprocal of the coefficient of x.

■ **EXAMPLE 2** Solve: $5a = -\dfrac{7}{3}$.

We have

$$5a = -\frac{7}{3}$$

$$\frac{1}{5} \cdot 5a = \frac{1}{5} \cdot \left(-\frac{7}{3}\right) \qquad \text{Multiplying both sides by } \tfrac{1}{5}, \text{ the reciprocal of 5}$$

$$1a = -\frac{1 \cdot 7}{5 \cdot 3}$$

$$a = -\frac{7}{15}.$$

Check:

$$5a = -\frac{7}{3}$$

$$\begin{array}{c|c} 5\left(-\dfrac{7}{15}\right) \ ? & -\dfrac{7}{3} \\[2ex] -\dfrac{\cancel{5} \cdot 7}{\cancel{5} \cdot 3} & \\[2ex] -\dfrac{7}{3} & -\dfrac{7}{3} \quad \text{TRUE} \end{array}$$

4. $-\dfrac{6}{7}a = \dfrac{9}{14}$

The solution is $-\dfrac{7}{15}$.

■ **EXAMPLE 3** Solve: $\dfrac{10}{3} = -\dfrac{4}{9}x$.

We have

$$\frac{10}{3} = -\frac{4}{9}x$$

$$-\frac{9}{4} \cdot \frac{10}{3} = -\frac{9}{4} \cdot \left(-\frac{4}{9}\right)x \qquad \text{Multiplying both sides by } -\tfrac{9}{4}, \text{ the reciprocal of } -\tfrac{4}{9}$$

$$-\frac{\cancel{3} \cdot 3 \cdot \cancel{2} \cdot 5}{\cancel{2} \cdot 2 \cdot \cancel{3}} = x$$

$$-\frac{15}{2} = x. \qquad \text{Removing a factor equal to 1: } \frac{3 \cdot 2}{2 \cdot 3} = 1$$

We leave the check to the student. The solution is $-\dfrac{15}{2}$.

Do Exercises 3 and 4.

Answers on page A-9

b   Problem Solving

Equations involving multiplication of fractions arise frequently in problem-solving situations.

EXAMPLE 4 *Herbal Remedies.*   At Sunshine Herbs, Sue needs to fill as many tea bags as possible with $\frac{3}{5}$ g (gram) of chamomile. If she begins with 51 g of chamomile, how many tea bags can she fill?

1. Familiarize.   We first make a drawing or at least visualize the situation.

$\frac{3}{5}$ of a gram in each tea bag

51 grams in all

n tea bags in all

We let $n =$ the number of tea bags that can be filled.

2. Translate.   The problem can be translated to the following equation:

$$\frac{3}{5} \cdot n = 51.$$

3. Solve.   To solve the equation, we use the multiplication principle:

$$\frac{5}{3} \cdot \frac{3}{5} \cdot n = \frac{5}{3} \cdot 51 \qquad \text{Multiplying both sides by } \frac{5}{3}$$

$$1n = \frac{5 \cdot 51}{3}$$

$$n = \frac{5 \cdot 3 \cdot 17}{1 \cdot 3} \Bigg\}$$

$$n = \frac{3}{3} \cdot \frac{5 \cdot 17}{1} \Bigg\} \qquad \text{Identifying a factor equal to 1}$$

$$n = 85. \qquad \text{Simplifying}$$

4. Check.   If each of 85 tea bags took $\frac{3}{5}$ g of chamomile, we would know that

$$85 \cdot \frac{3}{5} = \frac{85 \cdot 3}{5} = \frac{\cancel{5} \cdot 17 \cdot 3}{\cancel{5}} = 17 \cdot 3,$$

or 51 g of chamomile is used. Since the problem states that Sue begins with 51 g, our answer checks. Note too that

$$tea\ bags \cdot \frac{grams}{bag} = grams,$$

so the units also check.

5. State.   Sue can fill 85 tea bags with 51 g of chamomile.

Do Exercises 5 and 6.

---

5. Each loop in a spring uses $\frac{3}{8}$ in. of wire. How many loops can be made from 120 in. of wire?

6. For a party, Jana made an 8-foot submarine sandwich. If one serving is $\frac{2}{3}$ ft, how many servings does Jana's sub contain?

Answers on page A-9

7. Sales Trip. Miles Lanosga sells soybean seeds to seed companies. By lunchtime, he had driven 210 mi. At this point, $\frac{5}{6}$ of his trip was completed. How long was the total trip?

$\frac{5}{6}$ of the trip
210 mi

EXAMPLE 5 *Hand-Knit Scarves.* Nayah knits winter scarves for the Quick-Knit Boutique. She has knitted 63 in. of a scarf, and calculates that she has completed $\frac{7}{8}$ of the scarf. What will be the finished length of the scarf?

1. Familiarize. We ask: "63 in. is $\frac{7}{8}$ of what length?" We make a drawing or at least visualize the problem. We let $s =$ the length of the scarf.

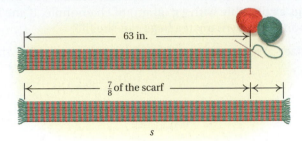

63 in.

$\frac{7}{8}$ of the scarf

s

2. Translate. We translate to an equation.

$$\underbrace{\text{Fraction completed}} \quad \text{of} \quad \underbrace{\begin{matrix}\text{Total} \\ \text{length} \\ \text{of scarf}\end{matrix}} \quad \text{is} \quad \underbrace{\begin{matrix}\text{Amount} \\ \text{already} \\ \text{knitted}\end{matrix}}$$

$$\frac{7}{8} \quad \cdot \quad s \quad = \quad 63$$

3. Solve. The equation that corresponds to the situation is $\frac{7}{8} \cdot s = 63$. To solve, we use the multiplication principle:

$$\frac{7}{8} \cdot s = 63$$

$$\frac{8}{7} \cdot \frac{7}{8} \cdot s = \frac{8}{7} \cdot 63 \qquad \text{Multiplying both sides by } \frac{8}{7}$$

$$1 \cdot s = \frac{8 \cdot 63}{7}$$

$$\left. \begin{matrix} s = \dfrac{8 \cdot 7 \cdot 9}{1 \cdot 7} \\[2mm] s = \dfrac{7}{7} \cdot \dfrac{8 \cdot 9}{1} \end{matrix} \right\} \qquad \text{Identifying a factor equal to 1}$$

$$s = 72.$$

4. Check. If the finished length of the scarf is 72 in., then $\frac{7}{8}$ of the length is

$$\frac{7 \cdot 72}{8} = \frac{7 \cdot \cancel{8} \cdot 9}{1 \cdot \cancel{8}} = 63 \text{ in.}$$

Our answer checks.

5. State. The completed scarf will be 72 in. long.

Do Exercise 7.

Answer on page A-9

Translating for Success

1. *Valentine Boxes.* Jane's Fudge Shop is preparing Valentine boxes. How many pounds of fudge will be needed to fill 80 boxes if each box contains $\frac{5}{16}$ lb?

2. *Gallons of Gasoline.* On the third day of a business trip, a sales representative used $\frac{4}{5}$ of a full tank of gasoline. If the tank is a 20-gal tank, how many gallons were used on the third day?

3. *Purchasing a Shirt.* Tom received $36 for his birthday. If he spends $\frac{3}{4}$ of the gift on a new shirt, what is the cost of the shirt?

4. *Checkbook Balance.* The balance in Pam's checking account is $1456. She writes a check for $28 and makes a deposit of $52. What is the new balance?

5. *Valentine Boxes.* Jane's Fudge Shop prepared 80 lb of fudge for Valentine boxes. If each box contains $\frac{5}{16}$ lb, how many boxes can be filled?

The goal of these matching questions is to practice step (2), *Translate,* of the five-step problem-solving process. Translate each word problem to an equation and select a correct translation from equations A–O.

A. $x = \frac{3}{4} \cdot 36$

B. $28 \cdot x = 52$

C. $x = 80 \cdot \frac{5}{16}$

D. $x = 1456 \div 28$

E. $x = \frac{5}{4} \cdot 20$

F. $20 = \frac{4}{5} \cdot x$

G. $x = 12 \cdot 28$

H. $x = \frac{4}{5} \cdot 20$

I. $\frac{3}{4} \cdot x = 36$

J. $x = 1456 - 52 - 28$

K. $x \div 28 = 1456$

L. $x = 52 - 28$

M. $x = 52 \cdot 28$

N. $x = 1456 - 28 + 52$

O. $\frac{5}{16} \cdot x = 80$

Answers on page A-9

6. *Gasoline Tank.* A gasoline tank contains 20 gal when it is $\frac{4}{5}$ full. How many gallons can it hold when full?

7. *Knitting a Scarf.* It takes Rachel 36 hr to knit a scarf. She can knit only $\frac{3}{4}$ hr per day because she is taking 16 hr of college classes. How many days will it take her to knit the scarf?

8. *Bicycle Trip.* On a recent 52-mi bicycle trip, David stopped to make a cell-phone call after completing 28 mi. How many more miles does he bicycle after the call?

9. *Crème de Menthe Thins.* Andes Candies L.P. makes Crème de Menthe Thins. How many 28-piece packages can be filled with 1456 pieces?

10. *Cereal Donations.* The Williams family donates 28 boxes of cereal weekly to the local Family in Crisis Center. How many boxes does this family donate in one year?

a Use the multiplication principle to solve each equation. Don't forget to check!

1. $\frac{4}{5}x = 12$

2. $\frac{4}{3}x = 20$

3. $\frac{7}{3}a = 21$

4. $\frac{4}{5}a = 24$

5. $\frac{2}{9}x = -10$

6. $\frac{3}{8}x = -21$

7. $6a = \frac{12}{17}$

8. $3a = \frac{15}{14}$

9. $\frac{1}{4}x = \frac{3}{5}$

10. $\frac{1}{6}x = \frac{2}{7}$

11. $\frac{3}{2}t = -\frac{8}{7}$

12. $\frac{4}{3}t = -\frac{5}{2}$

13. $\frac{4}{5} = -10a$

14. $\frac{6}{5} = -12a$

15. $\frac{9}{5}x = \frac{3}{10}$

16. $\frac{10}{3}x = \frac{8}{15}$

17. $-\frac{9}{10}x = 8$

18. $-\frac{2}{11}x = 5$

19. $a \cdot \frac{9}{7} = -\frac{3}{14}$

20. $a \cdot \frac{9}{4} = -\frac{3}{10}$

21. $-x = \frac{7}{13}$

22. $-x = \frac{7}{11}$

23. $-x = -\frac{27}{31}$

24. $-x = -\frac{35}{39}$

25. $7t = 6$

26. $-6t = 1$

27. $-24 = -10a$

28. $-18 = -20a$

29. $-\dfrac{14}{9} = \dfrac{10}{3}t$

30. $-\dfrac{15}{7} = \dfrac{3}{2}t$

31. $n \cdot \dfrac{4}{15} = \dfrac{12}{25}$

32. $n \cdot \dfrac{5}{16} = \dfrac{15}{14}$

33. $-\dfrac{7}{20}x = -\dfrac{21}{10}$

34. $-\dfrac{7}{15}x = -\dfrac{21}{10}$

35. $-\dfrac{25}{17} = -\dfrac{35}{34}a$

36. $-\dfrac{49}{45} = -\dfrac{28}{27}a$

 Solve.

37. Benny uses $\frac{2}{5}$ g of toothpaste each time he brushes his teeth. If Benny buys a 30-g tube, how many times will he be able to brush his teeth?

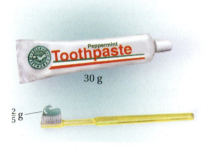

30 g

$\frac{2}{5}$ g

38. A piece of coaxial cable $\frac{4}{5}$ m long is to be cut into 8 pieces of the same length. What is the length of each piece?

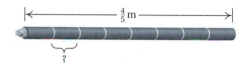

$\frac{4}{5}$ m

?

39. *Gasoline Tanker.* A tanker that delivers gasoline to gas stations had 1400 gal of gasoline when it was $\frac{7}{9}$ full. How much could the tanker hold when it is full?

40. How many $\frac{2}{3}$-cup cereal bowls can be filled from 10 cups of cornflakes?

41. *Honey.* A worker bee will produce $\frac{1}{12}$ tsp (teaspoon) of honey in her lifetime. How many worker bees does it take to produce $\frac{3}{4}$ tsp of honey?

Source: www.pbs.org/wgbh/nova/bees/buzz.html

42. *Syringes.* How many syringes, each containing $\frac{3}{5}$ mL, can a clinical pharmacist fill from a 30-mL vial?

43. *Packaging.* The South Shore Co-op prepackages cheddar cheese in $\frac{3}{4}$-lb packages. How many packages can be made from a 15-lb wheel of cheese?

44. *Meal Planning.* Ian purchased 6 lb of cold cuts for a luncheon. If Ian is to allow $\frac{3}{8}$ lb per person, how many people can attend the luncheon?

45. *Art Supplies.* The Ferristown School District purchased $\frac{3}{4}$ T (ton) of clay. The clay is to be shared equally among the district's 6 art departments. How much will each art department receive?

46. *Gardening.* The Bingham community garden is to be split into 16 equally sized plots. If the garden occupies $\frac{3}{4}$ acre of land, how large will each plot be?

Large quantities of soil, gravel, or loam are normally sold by the *yard* (yd). Although technically the unit for this type of volume should be *cubic yard* (yd^3), in practice only the word *yard* is used. Exercises 47–50 make use of this terminology.

47. *Landscaping.* To cover a walkway at a summer cottage with fresh gravel, $\frac{3}{4}$ yd of gravel is needed. Eric's dump truck has a 6-yd capacity. How many cottage walkways can be covered with one dump truck load?

48. *Landscaping.* To freshly cover a driveway in Surf City with crushed stone requires $\frac{4}{5}$ yd of stone. How many driveways can be freshly covered with one load of a 12-yd dump truck?

49. *Gardening.* Green Season Gardening uses about $\frac{2}{3}$ yd of bark mulch per customer every spring. How many customers can they accommodate with one 30-yd batch of bark mulch?

50. *Gardening.* Bright Moments Greenhouse sells topsoil by the yard. If the typical customer purchases $\frac{2}{5}$ yd, how many customers will it take to use up a 30-yd batch of topsoil?

51. *Sewing.* A pair of basketball shorts requires $\frac{3}{4}$ yd of nylon. How many pairs of shorts can be made from 24 yd of the fabric?

52. *Sewing.* A child's shirt requires $\frac{5}{6}$ yd of cotton fabric. How many shirts can be made from 25 yd of the fabric?

Pitch of a Screw. Refer to Exercises 45 and 46 in Exercise Set 3.6.

53. After a screw has been turned 8 complete rotations, it is extended $\frac{1}{2}$ in. into a piece of wallboard. What is the pitch of the screw?

54. The pitch of a screw is $\frac{3}{32}$ in. How many complete rotations are necessary to drive the screw $\frac{3}{4}$ in. into a piece of pine wood?

CHAPTER 3: Fraction Notation:
Multiplication and Division

55. D_W Does the multiplication principle enable us to solve any questions that could not have been solved with the division principle? Why or why not?

56. D_W Can the multiplication principle be used to solve equations like $7x = 63$? Why or why not?

SKILL MAINTENANCE

Simplify.

57. $-23 + 49$ [2.2a]

58. $-69 + 27$ [2.2a]

59. $-38 - 29$ [2.3a]

60. $-47 - 18$ [2.3a]

61. $36 \div (-3)^2 \times (7 - 2)$ [2.5b]

62. $(-37 - 12 + 1) \div (-2)^3$ [2.5b]

Form an equivalent expression by combining like terms. [2.7a]

63. $13x + 4x$

64. $9a - 5a$

65. $2a + 3 + 5a$

66. $3x - 7 + x$

SYNTHESIS

67. D_W Write a problem for a classmate to solve. Devise the problem so that the solution requires the classmate to divide by a fraction. Arrange for the solution to be "The contents of the barrel will fill 40 bags with $\frac{3}{4}$ lb in each bag."

68. D_W To solve $\frac{3}{4}x = 15$, James multiplies each side of the equation by 4. Is there anything wrong with this? Why or why not?

Solve.

69. $2x - 7x = -\dfrac{10}{9}$

70. $\left(-\dfrac{4}{7}\right)^2 = \left(\dfrac{2^3 - 9}{3}\right)^3 x$

Solve using the five-step problem-solving approach.

71. A package of coffee beans weighed $\frac{21}{32}$ lb when it was $\frac{3}{4}$ full. How much could the package hold when completely filled?

72. After swimming $\frac{2}{7}$ mi, Katie had swum $\frac{3}{4}$ of the race. How long a race was Katie competing in?

73. A block of Swiss cheese is 12 in. long. How many slices will it yield if half of the brick is cut by a slicer set for $\frac{3}{32}$-in. slices and half is cut by a slicer set for $\frac{5}{32}$-in. slices?

74. See Exercise 48. If each driveway required $\frac{3}{5}$ yd of stone and the stone cost $90 a yard, how much would the landscaper receive for a full load?

75. See Exercise 47. If each cottage required $\frac{3}{5}$ yd of gravel and the gravel cost $85 a yard, how much would Eric receive for a full load?

76. ▦ See Exercise 50. If each customer at Bright Moments Greenhouse bought $\frac{7}{8}$ yd of topsoil and the topsoil cost $95 a yard, how much would Bright Moments receive for a 30-yd batch?

77. ▦ See Exercise 49. If each customer at Green Season Gardening used $\frac{3}{4}$ yd of mulch and the mulch cost $65 a yard, how much would Green Season receive for a 25-yd batch?

The review that follows is meant to prepare you for a chapter exam. It consists of two parts. The first part, Concept Reinforcement, is designed to increase understanding of the concepts through true/false exercises. The second part is the Review Exercises. These provide practice exercises for the exam, together with references to section objectives so you can go back and review. Before beginning, stop and look back over the skills you have obtained. What skills in mathematics do you have now that you did not have before studying this chapter?

CONCEPT REINFORCEMENT

Determine whether each statement is true or false. Answers are given at the back of the book.

_____ 1. A number a is divisible by another number b if b is a factor of a.

_____ 2. If a number is not divisible by 6, then it is not divisible by 3.

_____ 3. The number 1 is not prime.

_____ 4. If n is any natural number, $\dfrac{n}{n} > \dfrac{0}{n}$.

_____ 5. A number is divisible by 10 if its ones digit is 0 or 5.

_____ 6. If a number is divisible by 9, then it is also divisible by 3.

_____ 7. The fraction $\dfrac{13}{6}$ is larger than the fraction $\dfrac{11}{6}$.

_____ 8. The fraction $\dfrac{13}{7}$ is larger than the fraction $\dfrac{13}{6}$.

Review Exercises

1. Multiply by 1, 2, 3, and so on, to find ten multiples of 8. [3.1a]

Use the tests for divisibility to answer Exercises 2–6.

2. Determine whether 3920 is divisible by 6. [3.1b]

3. Determine whether 68,537 is divisible by 3. [3.1b]

4. Determine whether 673 is divisible by 5. [3.1b]

5. Determine whether 4936 is divisible by 2. [3.1b]

6. Determine whether 5238 is divisible by 9. [3.1b]

Find all the factors of the number. [3.2a]

7. 60

8. 176

Classify each number as prime, composite, or neither. [3.2b]

9. 37

10. 1

11. 91

Find the prime factorization of each number. [3.2c]

12. 70

13. 72

14. 45

15. 150

16. 648

17. 1200

18. Identify the numerator and the denominator of $\dfrac{9}{7}$. [3.3a]

What part is shaded? [3.3a]

19.

20.

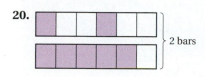

2 bars

21. For a committee in the United States Senate that consists of 3 Democrats and 5 Republicans, what is the ratio of: [3.3a]

a) Democrats to Republicans?
b) Republicans to Democrats?
c) Democrats to the total number of members of the committee?

Simplify, if possible. Assume that all variables are nonzero.

22. $\dfrac{0}{6}$ [3.3b]   **23.** $\dfrac{74}{74}$ [3.3b]   **24.** $\dfrac{48}{1}$ [3.3b]

25. $\dfrac{7x}{7x}$ [3.3b]   **26.** $-\dfrac{10}{15}$ [3.5b]   **27.** $\dfrac{7}{28}$ [3.5b]

28. $\dfrac{-42}{42}$ [3.5b]   **29.** $\dfrac{9m}{12m}$ [3.5b]   **30.** $\dfrac{12}{30}$ [3.5b]

31. $\dfrac{-27}{0}$ [3.3b]   **32.** $\dfrac{6x}{1}$ [3.3b]   **33.** $\dfrac{-9}{-27}$ [3.5b]

Find an equivalent expression for each number, using the denominator indicated. Use multiplication by 1. [3.5a]

34. $\dfrac{5}{7} = \dfrac{?}{21}$ **35.** $\dfrac{-6}{11} = \dfrac{?}{55}$

36. Simplify, if possible, the fractions on this circle graph. [3.5b]

How the Business Travel Dollar Is Spent

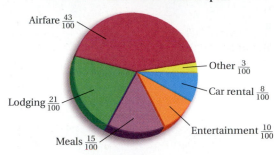

Airfare $\frac{43}{100}$

Other $\frac{3}{100}$

Car rental $\frac{8}{100}$

Lodging $\frac{21}{100}$

Entertainment $\frac{10}{100}$

Meals $\frac{15}{100}$

Use = or ≠ for ☐ to write a true sentence. [3.5c]

37. $\dfrac{3}{5}$ ☐ $\dfrac{4}{6}$ **38.** $\dfrac{4}{7}$ ☐ $\dfrac{8}{14}$

39. $\dfrac{4}{5}$ ☐ $\dfrac{5}{6}$ **40.** $\dfrac{4}{3}$ ☐ $\dfrac{28}{21}$

Find the reciprocal of each number. [3.7a]

41. $\dfrac{2}{13}$ **42.** -7

43. $\dfrac{1}{8}$ **44.** $\dfrac{3x}{5y}$

Perform the indicated operation and, if possible, simplify.

45. $\dfrac{2}{9} \cdot \dfrac{7}{5}$ [3.4b]   **46.** $\dfrac{3}{x} \cdot \dfrac{y}{7}$ [3.4b]

47. $\dfrac{3}{4} \cdot \dfrac{8}{9}$ [3.6a]   **48.** $-\dfrac{5}{7} \cdot \dfrac{1}{10}$ [3.6a]

49. $\dfrac{3a}{10} \cdot \dfrac{2}{15a}$ [3.6a]   **50.** $\dfrac{4a}{7} \cdot \dfrac{7}{4a}$ [3.6a]

51. $9 \div \dfrac{5}{3}$ [3.7b]   **52.** $\dfrac{3}{14} \div \dfrac{6}{7}$ [3.7b]

53. $120 \div \dfrac{3}{5}$ [3.7b]   **54.** $-\dfrac{5}{36} \div \left(-\dfrac{25}{12}\right)$ [3.7b]

225

55. $21 \div \dfrac{7}{2a}$ [3.7b]

56. $-\dfrac{23}{25} \div \dfrac{23}{25}$ [3.7b]

57. $\dfrac{\frac{21}{30}}{\frac{14}{15}}$ [3.7b]

58. $\dfrac{-\frac{3}{40}}{-\frac{54}{35}}$ [3.7b]

Find the area.   [3.6b]

59.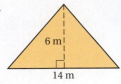
6 m
14 m

60.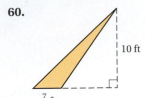
10 ft
$\frac{7}{2}$ ft

Solve.   [3.8a]

61. $\dfrac{2}{3}x = 160$

62. $\dfrac{3}{8} = -\dfrac{5}{4}t$

63. $-\dfrac{1}{7}n = -4$

Solve.

64. A road crew repaves $\frac{1}{12}$ mi of road each day. How long will it take the crew to repave a $\frac{3}{4}$-mi stretch of road?   [3.8b]

65. A chocolate chip cookie recipe calls for $\frac{3}{4}$ cup of sugar. In making $\frac{1}{2}$ of this recipe, how much sugar should be used?   [3.6b]

66. Tanika makes insurance estimates when claims are filed. After driving 180 km (kilometers), she completes $\frac{5}{8}$ of a daily route. How long is the route? How much driving remains?   [3.6b]

67. The Winchester swim team has 4 swimmers in a $\frac{2}{3}$-mi relay race. How far will each person swim?   [3.8b]

68. *Corn Production.* In 2003, the United States produced approximately $\frac{2}{5}$ of the world production of corn. The total world corn production was about 640,000,000 metric tons. How much corn did the United States produce?   [3.6b]

Source: UN Food and Agriculture Organization

69. **D**_W A student claims that "taking $\frac{1}{2}$ of a number is the same as dividing by $\frac{1}{2}$." Explain the error in this reasoning.   [3.7b]

70. **D**_W A student claims that $\frac{2}{8}$ is simplified form of $\frac{20}{80}$. Is the student correct? Why or why not?   [3.5b]

SYNTHESIS

71. Simplify: $\dfrac{15x}{14z} \cdot \dfrac{17yz}{35xy} \div \left(-\dfrac{3}{7}\right)^2$.   [3.6a], [3.7b]

72. What digit(s) could be inserted in the ones place to make 574__ divisible by 6?   [3.1b]

73. A prime number that becomes a prime number when its digits are reversed is called a **palindrome prime.** For example, 17 is a palindrome prime because both 17 and 71 are primes. Which of the following numbers are palindrome primes?   [3.2b]

13, 91, 16, 11, 15, 24, 29, 101, 201, 37

74. ▦ In the division below, find a and b. Assume a/b is in simplified form.   [3.7b]

$$\dfrac{19}{24} \div \dfrac{a}{b} = \dfrac{187,853}{268,224}$$

75. ▦ Use a calculator and the list of prime numbers on p. 170 to find simplified fraction notation for the solution of

$$\dfrac{1751}{267}x = \dfrac{3193}{2759}.$$ [3.8a]

1. Determine whether 5682 is divisible by 3. Do not use long division.

2. Determine whether 7018 is divisible by 5. Do not use long division.

3. Find all the factors of 90.

4. Determine whether 93 is prime, composite, or neither.

Find the prime factorization of each number.

5. 36

6. 60

7. Identify the numerator and the denominator of $\frac{4}{9}$.

8. What part is shaded?

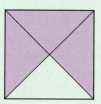

9. What part of the set is shaded?

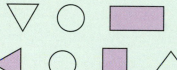

10. *Hockey.* In the 2005–2006 regular season, goaltender Dominik Hasek, of the Ottawa Senators, allowed 90 scores out of 1202 shots on the goal. He had 1112 saves.

   a) What was the ratio of saves to shots on goal?
   b) What was the ratio of scores to shots on goal?

Simplify, if possible. Assume that all variables are nonzero.

11. $\frac{32}{1}$

12. $\frac{-12}{-12}$

13. $\frac{0}{16}$

14. $\frac{-8}{24}$

15. $\frac{9x}{45x}$

16. $\frac{7}{63}$

Use = or ≠ for ☐ to write a true sentence.

17. $\frac{3}{4}$ ☐ $\frac{6}{8}$

18. $\frac{5}{4}$ ☐ $\frac{9}{7}$

19. Find an equivalent expression for $\frac{3}{8}$ with a denominator of 40.

Find the reciprocal.

20. $\frac{a}{42}$

21. -9

Perform the indicated operation. Simplify, if possible.

22. $\dfrac{5}{7} \cdot \dfrac{7}{2}$

23. $\dfrac{2}{11} \div \dfrac{3}{4}$

24. $3 \cdot \dfrac{x}{8}$

25. $\dfrac{\frac{4}{7}}{-\frac{8}{3}}$

26. $12 \div \dfrac{2}{3}$

27. $\dfrac{4a}{13} \cdot \dfrac{9b}{30ab}$

Solve.

28. A $\frac{3}{4}$-lb slab of cheese is shared equally by 5 people. How much does each person receive?

29. Monroe weighs $\frac{5}{7}$ of his dad's weight. If his dad weighs 175 lb, how much does Monroe weigh?

30. $\dfrac{7}{8} \cdot x = 56$

31. $\dfrac{7}{10} = \dfrac{-2}{5} \cdot t$

32. Find the area.

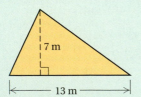

SYNTHESIS

33. A recipe for a batch of buttermilk pancakes calls for $\frac{3}{4}$ tsp of salt. Jacqueline plans to cut the amount of salt in half for each of 5 batches of pancakes. How much salt will she need altogether?

34. Grandma Shelby left $\frac{2}{3}$ of her $\frac{7}{8}$-acre tree farm to Karl. Karl gave $\frac{1}{4}$ of his share to his oldest daughter, Irene. How much land did Irene receive?

35. Simplify: $\left(-\dfrac{3}{8}\right)^2 \div \dfrac{6}{7} \cdot \dfrac{2}{9} \div (-5)$.

36. Solve: $\dfrac{33}{38} \cdot \dfrac{34}{55} = \dfrac{17}{35} \cdot \dfrac{15}{19} x$.

CHAPTER 3: Fraction Notation:
Multiplication and Division

1. Write a word name: 2,056,783.

Add.

2.
$$\begin{array}{r} 2\ 7\ 4\ 3 \\ +\ 8\ 2\ 3\ 9 \\ \hline \end{array}$$

3. $-29 + (-14)$

4. $-45 + 12$

Subtract.

5.
$$\begin{array}{r} 6\ 3\ 2\ 4 \\ -\ 4\ 1\ 9\ 5 \\ \hline \end{array}$$

6. $27 - 50$

7. $-12 - (-4)$

Multiply and, if possible, simplify.

8.
$$\begin{array}{r} 7\ 3\ 5 \\ \times\quad 2\ 3 \\ \hline \end{array}$$

9. $-52 \cdot 6$

10. $\dfrac{6}{7} \cdot (-35x)$

11. $\dfrac{2}{9} \cdot \dfrac{21}{10}$

Divide and, if possible, simplify.

12. $1\ 3\ \overline{)\ 3\ 0\ 5\ 8}$

13. $-85 \div 5$

14. $-16 \div \dfrac{4}{7}$

15. $\dfrac{3}{7} \div \dfrac{9}{14}$

16. Round 4509 to the nearest ten.

17. Estimate the product by rounding to the nearest hundred. Show your work.
$$\begin{array}{r} 9\ 2\ 1 \\ \times\ 4\ 5\ 3 \\ \hline \end{array}$$

18. Find the absolute value: $|-479|$.

19. Simplify: $10^2 \div 5(-2) - 8(2 - 8)$.

20. Determine whether 98 is prime, composite, or neither.

21. Evaluate $a - b^2$ for $a = -5$ and $b = 4$.

Solve.

22. $a + 24 = 49$

23. $7x = 49$

24. $\dfrac{2}{9} \cdot a = -10$

229

25. A 1996 van that gets 25 miles per gallon is traded in toward a 2003 truck that gets 17 miles per gallon. How many more miles per gallon did the older vehicle get?

26. A 48-oz coffee pot is poured into 6 mugs. How much will each mug hold if the coffee is poured out evenly?

Combine like terms.

27. $8 - 4x - 13 + 9x$

28. $-12x + 7y + 15x$

Simplify, if possible.

29. $\dfrac{97}{97}$

30. $\dfrac{0}{81}$

31. $\dfrac{63}{1}$

32. $\dfrac{-10}{54}$

Find the reciprocal.

33. $\dfrac{2}{5}$

34. 57

35. Find an equivalent expression for $\frac{3}{10}$ with a denominator of 70. Use multiplying by 1.

36. There are 7000 students at La Poloma College, and $\frac{5}{8}$ of them live in dorms. How many live in dorms?

37. A thermos of iced tea had 3 qt of tea when it was $\frac{3}{5}$ full. How much tea could it hold when full?

38. Tony has jogged $\frac{2}{3}$ of a course that is $\frac{9}{10}$ of a mile long. How far has Tony gone?

SYNTHESIS

39. Evaluate $\dfrac{ab}{c}$ for $a = -\dfrac{2}{5}$, $b = \dfrac{10}{13}$, and $c = \dfrac{26}{27}$.

40. Evaluate $-|xy|^2$ for $x = -\dfrac{3}{5}$ and $y = \dfrac{1}{2}$.

41. Wayne and Patty each earn $85 a day, while Janet earns $90 a day. They decide to pool their earnings from three days and spend $\frac{2}{5}$ of that on entertainment and save the rest. How much will Wayne, Patty, and Janet end up saving?

Fraction Notation: Addition, Subtraction, and Mixed Numerals

Real-World Application

A guitarist's band is booked for Friday and Saturday nights at a local club. The guitarist's group is a trio on Friday and expands to a quintet on Saturday. Thus the guitarist is paid one-third of one-half the weekend's pay for Friday and one-fifth of one-half the weekend's pay for Saturday. What fractional part of the total pay did the guitarist receive for the weekend's work?

This problem appears as Exercise 99 in Section 4.2.

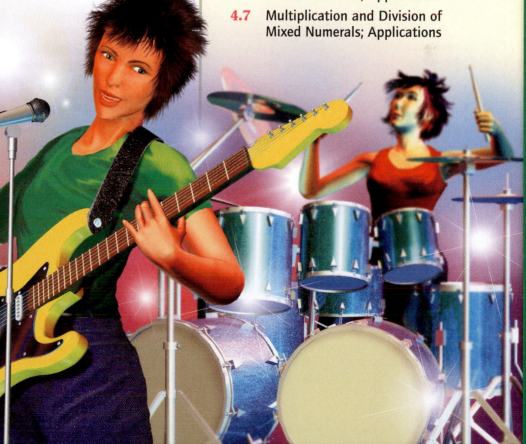

LEAST COMMON MULTIPLES

1. By examining lists of multiples, find the LCM of 9 and 15.

Before discussing addition or subtraction of fractions, it is essential that we be able to find the **least common multiple** of two or more numbers.

a Finding Least Common Multiples

> **LEAST COMMON MULTIPLE, LCM**
>
> The **least common multiple,** or LCM, of two natural numbers is the smallest number that is a multiple of both.

EXAMPLE 1 Find the LCM of 20 and 30.

a) First list some multiples of 20 by multiplying 20 by 1, 2, 3, and so on:

    20, 40, 60, 80, 100, 120, 140, 160, 180, 200, 220, 240, ...

b) Then list some multiples of 30 by multiplying 30 by 1, 2, 3, and so on:

    30, 60, 90, 120, 150, 180, 210, 240, ...

c) Now list the numbers *common* to both lists, the common multiples:

    60, 120, 180, 240, ...

d) These are the common multiples of 20 and 30. Which is the smallest? The LCM of 20 and 30 is 60.

Do Exercises 1 and 2.

2. By examining lists of multiples, find the LCM of 8 and 14.

Next we develop two highly efficient methods for finding LCMs. You may choose to learn only one method (consult your instructor). However, if you intend to study algebra, you should definitely learn method 2.

METHOD 1: FINDING LCMS USING ONE LIST OF MULTIPLES

The first method works especially well when the numbers are relatively small.

> *Method 1.* To find the LCM of two numbers (say, 9 and 12), first determine whether the larger number is a multiple of the other.
>
> 1. If it *is*, it is the LCM.
>
>   (Since 12 is not a multiple of 9, the LCM is not 12.)
>
> 2. If the larger number *is not* a multiple of the other, check each consecutive multiple of the larger number until you find one that *is* a multiple of the other number. That number is the LCM.
>
>   ($2 \cdot 12 = 24$, but 24 *is not* a multiple of 9.
>
> $3 \cdot 12 = 36$, and 36 *is* a multiple of 9, so the LCM of 9 and 12 is 36.)

EXAMPLE 2 Find the LCM of 8 and 10.

1. 10 is the larger number, but it is not a multiple of 8.
2. Check multiples of 10:

   $2 \cdot 10 = 20,$ Not a multiple of 8
   $3 \cdot 10 = 30,$ Not a multiple of 8
   $4 \cdot 10 = 40.$ A multiple of both 8 and 10

The LCM = 40.

EXAMPLE 3 Find the LCM of 4 and 14.

1. 14 is the larger number, but it is not a multiple of 4.
2. Check multiples of 14:

   $2 \cdot 14 = 28.$ A multiple of 4

The LCM = 28.

EXAMPLE 4 Find the LCM of 8 and 32.

1. 32 is the larger number and 32 is a multiple of 8, so it is the LCM.

The LCM = 32.

Do Exercises 3–5.

To find the least common multiple of three numbers, we find the LCM of two of the numbers and then find the LCM of that number and the third number.

EXAMPLE 5 Find the LCM of 4, 10, and 15. We can start by finding the LCM of any two of the numbers. Let's use 10 and 15:

1. 15 is the larger number, but it is not a multiple of 10.
2. Check multiples of 15:

   $2 \cdot 15 = 30.$ A multiple of 10

Note now that any multiple of 30 will automatically be a multiple of 10 and 15. Thus, to find the LCM of 10, 15, and 4, we need only find the LCM of 30 and 4:

1. 30 is not a multiple of 4.
2. Check multiples of 30:

   $2 \cdot 30 = 60.$ Since it is a multiple of 4, we know
                         that 60 is the LCM of 30 and 4.

The LCM of 4, 10, and 15 is 60.

Do Exercise 6.

Find the LCM using one list of multiples.

3. 6, 9

4. 6, 8

5. 7, 14

Find the LCM using lists of multiples.

6. 20, 40, 50

Answers on page A-10

METHOD 2: FINDING LCMS USING PRIME FACTORIZATIONS

A second method for finding LCMs uses prime factorizations and is usually the best method when the numbers are large. Consider again 20 and 30. Their prime factorizations are

$$20 = 2 \cdot 2 \cdot 5 \quad \text{and} \quad 30 = 2 \cdot 3 \cdot 5.$$

The least common multiple must include the factors of each number, so it must include each prime factor the greatest number of times that it appears in either of the factorizations. To find the LCM for 20 and 30, we select one factorization, say,

$$2 \cdot 2 \cdot 5,$$

and observe that since it lacks the factor 3, it does not contain the entire factorization of 30. If we multiply $2 \cdot 2 \cdot 5$ by 3, every prime factor occurs just often enough to have both 20 and 30 as factors.

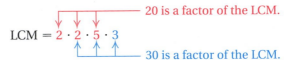

LCM $= 2 \cdot 2 \cdot 5 \cdot 3$ — 20 is a factor of the LCM.
— 30 is a factor of the LCM.

Note that each prime factor is used the greatest number of times that it occurs in either of the individual factorizations.

Method 2. To find the LCM of two numbers (say, 9 and 12):

1. Write the prime factorization of each number.

$$(9 = 3 \cdot 3; \qquad 12 = 2 \cdot 2 \cdot 3)$$

2. Select one of the factorizations and see whether it contains the other.

$$(2 \cdot 2 \cdot 3 \text{ does not contain } 3 \cdot 3.)$$

   a) If it does, it is the LCM.
   b) If it does not, multiply that factorization by those prime factors of the other number that it lacks. The final product is the LCM.

$$(2 \cdot 2 \cdot 3 \cdot 3 \text{ is the LCM.})$$

3. As a check, make sure that the LCM includes each factor the greatest number of times that it occurs in either factorization.

EXAMPLE 6 Find the LCM of 18 and 21.

1. We begin by writing the prime factorization of each number:

$$18 = 2 \cdot 3 \cdot 3 \quad \text{and} \quad 21 = 3 \cdot 7.$$

2. a) We note that $2 \cdot 3 \cdot 3$ does not contain the other factorization, $3 \cdot 7$.

   b) To find the LCM of 18 and 21, we multiply $2 \cdot 3 \cdot 3$ by the factor of 21 that it lacks, 7:

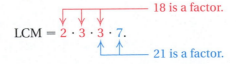

LCM $= 2 \cdot 3 \cdot 3 \cdot 7$. — 18 is a factor.
— 21 is a factor.

3. The greatest number of times that 2 occurs as a factor of 18 or 21 is **one** time; the greatest number of times that 3 occurs as a factor of 18 or 21 is **two** times; and the greatest number of times that 7 occurs as a factor of 18 or 21 is **one** time. To check, note that the LCM has exactly **one** 2, **two** 3's, and **one** 7. The LCM is $2 \cdot 3 \cdot 3 \cdot 7$, or 126.

EXAMPLE 7 Find the LCM of 24 and 36.

1. We begin by writing the prime factorization of each number:

$$24 = 2 \cdot 2 \cdot 2 \cdot 3 \quad \text{and} \quad 36 = 2 \cdot 2 \cdot 3 \cdot 3.$$

2. a) Neither factorization contains the other.

b) To find the LCM of 24 and 36, we multiply the factorization of 24 by any prime factors of 36 that it lacks. We need another factor of 3.

LCM $= 2 \cdot 2 \cdot 2 \cdot 3 \cdot 3$.

— 24 is a factor.
— 36 is a factor.

3. Note that the LCM includes 2 and 3 the greatest number of times that each appears as a factor of either 24 or 36. The LCM is $2 \cdot 2 \cdot 2 \cdot 3 \cdot 3$, or 72.

EXAMPLE 8 Find the LCM of 7 and 21.

1. Because 7 is prime, we think of $7 = 7$ as a "factorization":

$$7 = 7 \quad \text{and} \quad 21 = 3 \cdot 7.$$

2. One factorization, $3 \cdot 7$, contains the other. Thus the LCM is $3 \cdot 7$, or 21.

Do Exercises 7–9.

Exponential notation is often helpful when writing least common multiples. Let's reconsider Example 7:

$$24 = 2 \cdot 2 \cdot 2 \cdot 3 = 2^3 \cdot 3^1 \qquad 2^3 \text{ is the greatest power of 2 and}$$
$$36 = 2 \cdot 2 \cdot 3 \cdot 3 = 2^2 \cdot 3^2 \qquad 3^2 \text{ is the greatest power of 3}$$
$$\text{LCM} = 2 \cdot 2 \cdot 2 \cdot 3 \cdot 3 = 2^3 \cdot 3^2, \text{ or 72.}$$

Note that the greatest power of each factor is used to construct the LCM.

Lining up the different prime numbers in the factorizations can help us construct the LCM. This method also works well when finding the LCM of more than two numbers.

EXAMPLE 9 Find the LCM of 27, 90, and 84.

We find the prime factorization of each number, and write the factorizations in exponential notation.

$$27 = 3 \cdot 3 \cdot 3 = 3^3$$
$$90 = 2 \cdot 3 \cdot 3 \cdot 5 = 2 \cdot 3^2 \cdot 5$$
$$84 = 2 \cdot 2 \cdot 3 \cdot 7 = 2^2 \cdot 3 \cdot 7$$

No one factorization contains the others. The prime numbers 2, 3, 5, and 7 appear as factors.

Use prime factorizations to find the LCM.

7. 8, 10

8. 18, 40

9. 5, 30

Answers on page A-10

Find the LCM.

10. 8, 18, 30

11. 10, 20, 25

Find the LCM.

12. xy, yz

13. $5a^2$, a^3b

14. Find the LCM of $8a^3b^2$ and $10a^2c^4$.

We write the factorizations, lining up all the powers of 2, the powers of 3, and so on.

$$27 = 3^3$$
$$90 = 2 \cdot 3^2 \cdot 5$$
$$84 = 2^2 \cdot 3 \cdot 7$$

The LCM is formed by choosing the greatest power of each factor:

$$2^2 \cdot 3^3 \cdot 5 \cdot 7 = 3780.$$

The LCM of 27, 90, and 84 is 3780.

Do Exercises 10 and 11.

The same method works perfectly with variables.

EXAMPLE 10 Find the LCM of $7a^2b$ and ab^3.

We have the following factorizations:

$$7a^2b = 7 \cdot a \cdot a \cdot b \quad \text{and} \quad ab^3 = a \cdot b \cdot b \cdot b.$$

No one factorization contains the other.

Consider the factorization of $7a^2b$, which is $7 \cdot a \cdot a \cdot b$. Since ab^3 contains two more factors of b, we multiply the factorization of $7a^2b$ by $b \cdot b$.

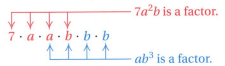

As a second approach, we find the greatest power of each factor using exponential notation.

$$7a^2b = 7 \cdot a^2 \cdot b$$
$$ab^3 = a \cdot b^3$$

The LCM is $7 \cdot a \cdot a \cdot b \cdot b \cdot b$, or $7a^2b^3$.

Do Exercises 12 and 13.

EXAMPLE 11 Find the LCM of $12x^2y^3z$ and $18x^4z^3$.

We write the factorizations using exponential notation:

$$12x^2y^3z = 2^2 \cdot 3 \cdot x^2 \cdot y^3 \cdot z \quad \text{and} \quad 18x^4z^3 = 2 \cdot 3^2 \cdot x^4 \cdot z^3.$$

No one factorization contains the other.

We form the LCM using the greatest power of each factor.

$$12x^2y^3z = 2^2 \cdot 3 \cdot x^2 \cdot y^3 \cdot z$$
$$18x^4z^3 = 2 \cdot 3^2 \cdot x^4 \cdot z^3$$

The LCM is $2^2 \cdot 3^2 \cdot x^4 \cdot y^3 \cdot z^3$, or $36x^4y^3z^3$.

Do Exercise 14.

Answers on page A-10

a   Find the LCM of each set of numbers or expressions.

1. 5, 10

2. 4, 12

3. 10, 25

4. 10, 15

5. 20, 40

6. 8, 12

7. 18, 27

8. 9, 11

9. 30, 50

10. 8, 36

11. 30, 40

12. 21, 27

13. 18, 24

14. 12, 18

15. 60, 70

16. 35, 45

17. 16, 36

18. 24, 32

19. 18, 20

20. 36, 48

21. 2, 3, 7

22. 2, 5, 9

23. 3, 6, 15

24. 6, 12, 18

25. 24, 36, 12

26. 8, 16, 22

27. 5, 12, 15

28. 12, 18, 40

29. 9, 12, 6

30. 8, 16, 12

31. 180, 100, 450

32. 18, 30, 50, 48

33. 75, 100

34. 81, 90

35. ab, bc

36. $7x, xy$

37. $3x, 9x^2$

38. $10x^4, 5x^3$

39. $4x^3, x^2y$

40. $6ab^2, 9a^3b$

41. $6r^3st^4, 8rs^2t$

42. $3m^2n^4p^5, 9mn^2p^4$

43. a^3b, b^2c, ac^2

44. x^2z^3, x^3y, y^2z

Applications of LCMs: Planet Orbits. Earth, Jupiter, Saturn, and Uranus all revolve around the sun. Earth takes 1 yr, Jupiter 12 yr, Saturn 30 yr, and Uranus 84 yr to make a complete revolution. One night, you look at those three distant planets and wonder how many years it will take before they have the same position again. To determine this, you find the LCM of 12, 30, and 84. It will be that number of years.

45. How often will Jupiter and Saturn appear in the same position in the night sky as seen from Earth tonight?

46. How often will Jupiter, Saturn, and Uranus appear in the same position in the night sky as seen from Earth?

47. **D**_W Are the rules for divisibility useful when looking for LCMs? Why or why not?

48. **D**_W What is the difference between a common multiple and a least common multiple?

SKILL MAINTENANCE

Perform the indicated operation and, if possible, simplify.

49. $-38 + 52$ [2.2a]

50. $-18 \div \left(\dfrac{2}{3}\right)$ [3.7b]

51. $23 \cdot 345$ [1.5a]

52. $\dfrac{4}{5} \cdot \dfrac{10}{12}$ [3.6a]

53. $\dfrac{4}{5} \div \left(-\dfrac{7}{10}\right)$ [3.7b]

54. $382 - 549$ [2.3a]

SYNTHESIS

55. **D**_W Under what conditions is the LCM of two composite numbers simply the product of the two numbers?

56. **D**_W Is the LCM of two prime numbers always their product? Why or why not?

57. **D**_W Is the LCM of two numbers always at least twice as large as the larger of the two numbers? Why or why not?

▦ Use a calculator and the multiples method to find the LCM of each pair of numbers.

58. 288; 324

59. 2700; 7800

60. 7719; 18,011

61. 17,385; 24,339

62. The tables at a flea market are either 6 ft long or 8 ft long. If one row is all 6-ft tables and one row is all 8-ft tables, what is the shortest common row length?

238

South African Artistry. In South Africa, the design of every woven handbag, or *gipatsi* (plural *sipatsi*), is created by repeating two or more geometric patterns. Each pattern encircles the bag, sharing the strands of fabric with any pattern above or below. The length, or period, of each pattern is the number of strands required to construct the pattern. For a gipatsi to be considered beautiful, each individual pattern must fit a whole number of times around the bag.

Source: Gerdes, Paulus. *Women, Art and Geometry in Southern Africa.* Asmara, Eritrea: Africa World Press, Inc., p. 5.

63. A weaver is using two patterns to create a gipatsi. Pattern A is 10 strands long, and pattern B is 3 strands long. What is the smallest number of strands that can be used to complete the gipatsi?

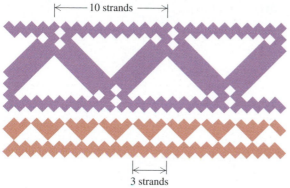

← 10 strands →

← 3 strands →

64. A weaver is using a four-strand pattern, a six-strand pattern, and an eight-strand pattern. What is the smallest number of strands that can be used to complete the gipatsi?

65. *Prescriptions.* A 30-day supply of famatodine and a 14-day supply of pain medication are filled at a pharmacy. Assuming the prescriptions are refilled regularly, how long will it be until they are both refilled on the same day?

66. Consider a^3b^2 and a^2b^5. Determine whether each of the following is the LCM of a^3b^2 and a^2b^5. Tell why or why not.
   a) a^3b^3
   b) a^2b^5
   c) a^3b^5

67. Use Example 9 to help find the LCM of 27, 90, 84, 210, 108, and 50.

68. Use Examples 6 and 7 to help find the LCM of 18, 21, 24, 36, 63, 56, and 20.

69. Find three different pairs of numbers for which 56 is the LCM. Do not use 56 itself in any of the pairs.

70. Find three different pairs of numbers for which 54 is the LCM. Do not use 54 itself in any of the pairs.

4.2 ADDITION, ORDER, AND APPLICATIONS

Objectives

a Add using fraction notation when denominators are the same.

b Add using fraction notation when denominators are different.

c Use < or > to form a true statement using fraction notation.

d Solve problems involving addition with fraction notation.

1. Find $\dfrac{1}{5} + \dfrac{3}{5}$.

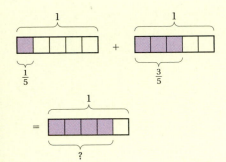

Add and, if possible, simplify.

2. $\dfrac{1}{3} + \dfrac{2}{3}$

3. $\dfrac{5}{12} + \dfrac{1}{12}$

4. $\dfrac{-9}{16} + \dfrac{3}{16}$

5. $\dfrac{3}{x} + \dfrac{-7}{x}$

Answers on page A-10

CHAPTER 4: Fraction Notation: Addition, Subtraction, and Mixed Numerals

---

a Like Denominators

Addition using fraction notation corresponds to combining or putting like things together, just as when we combined like terms in Section 2.7. For example,

We combine two sets, each of which consists of equally sized parts of one object.

This is the resulting set.

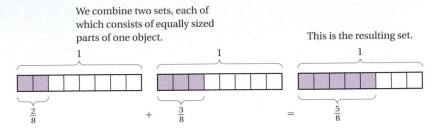

2 eighths + 3 eighths = 5 eighths,

or $2 \cdot \dfrac{1}{8} + 3 \cdot \dfrac{1}{8} = 5 \cdot \dfrac{1}{8}$,

or $\dfrac{2}{8} + \dfrac{3}{8} = \dfrac{5}{8}$.

Do Exercise 1.

> To add when denominators are the same,
>
> **a)** add the numerators,
> **b)** keep the denominator, and
> **c)** if possible, simplify.
>
> $\dfrac{2}{6} + \dfrac{5}{6} = \dfrac{2+5}{6} = \dfrac{7}{6}$

EXAMPLES Add and, if possible, simplify.

1. $\dfrac{2}{4} + \dfrac{1}{4} = \dfrac{2+1}{4} = \dfrac{3}{4}$ No simplifying is possible.

2. $\dfrac{3}{12} + \dfrac{5}{12} = \dfrac{3+5}{12} = \dfrac{8}{12}$ Adding numerators; the denominator remains unchanged.

$= \dfrac{4}{4} \cdot \dfrac{2}{3} = \dfrac{2}{3}$ Simplifying by removing a factor equal to 1: $\frac{4}{4} = 1$

3. $\dfrac{-11}{6} + \dfrac{3}{6} = \dfrac{-11+3}{6} = \dfrac{-8}{6}$

$= \dfrac{2}{2} \cdot \dfrac{-4}{3} = \dfrac{-4}{3}$, or $-\dfrac{4}{3}$ Removing a factor equal to 1: $\frac{2}{2} = 1$

4. $-\dfrac{2}{a} + \left(-\dfrac{3}{a}\right) = \dfrac{-2}{a} + \dfrac{-3}{a}$ Recall that $-\dfrac{m}{n} = \dfrac{-m}{n}$.

$= \dfrac{-2+(-3)}{a} = \dfrac{-5}{a}$, or $-\dfrac{5}{a}$

Do Exercises 2–5.

We may need to add fractions when combining like terms.

■ **EXAMPLE 5** Simplify by combining like terms: $\frac{2}{7}x + \frac{3}{7}x$.

$$\frac{2}{7}x + \frac{3}{7}x = \left(\frac{2}{7} + \frac{3}{7}\right)x \qquad \text{Try to do this step mentally.}$$
$$= \frac{5}{7}x.$$

Do Exercises 6 and 7.

b Addition Using the Least Common Denominator

b   Addition Using the Least Common Denominator

Let's now add $\frac{1}{2} + \frac{1}{3}$:

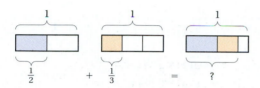

By rewriting $\frac{1}{2}$ as $\frac{1}{2} \cdot \frac{3}{3} = \frac{3}{6}$ and $\frac{1}{3}$ as $\frac{1}{3} \cdot \frac{2}{2} = \frac{2}{6}$, we can determine the sum.

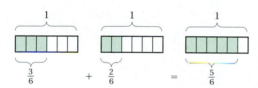

Thus, when denominators differ, before adding we must multiply by a form of 1 to get a common denominator. There is always more than one common denominator that can be used. Consider the addition $\frac{3}{4} + \frac{1}{6}$ using two different denominators.

A. We use 24 as a common denominator:

$$\frac{3}{4} + \frac{1}{6} = \frac{3}{4} \cdot \frac{6}{6} + \frac{1}{6} \cdot \frac{4}{4}$$
$$= \frac{18}{24} + \frac{4}{24} = \frac{22}{24} = \frac{11}{12}.$$

B. We use 12 as a common denominator:

$$\frac{3}{4} + \frac{1}{6} = \frac{3}{4} \cdot \frac{3}{3} + \frac{1}{6} \cdot \frac{2}{2}$$
$$= \frac{9}{12} + \frac{2}{12} = \frac{11}{12}.$$

We had to simplify at the end of (A), but not in (B). In (B), we used the *least* common multiple of the denominators, 12. That number is called the **least common denominator**, or **LCD**.

> **To add when denominators are different:**
>
> **a)** Find the least common multiple of the denominators. That number is the least common denominator, LCD.
> **b)** Multiply by 1, writing 1 in the form of n/n, to find an equivalent sum in which the LCD appears in each fraction.
> **c)** Add and, if possible, simplify.

Simplify by combining like terms.

6. $\frac{3}{10}a + \frac{1}{10}a$

7. $-\frac{3}{4}x + \frac{1}{4}x$

Answers on page A-10

Study Tips

SHOW YOUR WORK

Students sometimes make the mistake of viewing their supporting work as "scrap" work. Most instructors regard your reasoning as more important than your final answer. Try to organize your supporting work so that your instructor (and you as well) can follow your steps.

8. $\dfrac{2}{3} + \dfrac{1}{6}$

9. $\dfrac{3}{8} + \dfrac{5}{6}$

10. Add: $\dfrac{1}{-6} + \dfrac{7}{18}$.

11. Add: $7 + \dfrac{3}{5}$.

Answers on page A-10

■ **EXAMPLE 6** Add: $\dfrac{1}{8} + \dfrac{3}{4}$.

a) Since 4 is a factor of 8, the LCM of 4 and 8 is 8. Thus the LCD is 8.

b) We need to find a fraction equivalent to $\frac{3}{4}$ with a denominator of 8:

$$\frac{1}{8} + \frac{3}{4} = \frac{1}{8} + \frac{3}{4} \cdot \frac{2}{2}.$$

Think: $4 \times \square = 8$. The answer is 2, so we multiply by 1, using $\frac{2}{2}$.

c) We add: $\dfrac{1}{8} + \dfrac{6}{8} = \dfrac{7}{8}.$ $\dfrac{7}{8}$ cannot be simplified.

In Examples 7–10, we follow the same steps without labeling them.

■ **EXAMPLE 7** Add: $\dfrac{5}{6} + \dfrac{1}{9}$.

The LCD is 18. $6 = 2 \cdot 3$ and $9 = 3 \cdot 3$, so the LCM of 6 and 9 is $2 \cdot 3 \cdot 3$, or 18.

$$\frac{5}{6} + \frac{1}{9} = \frac{5}{6} \cdot 1 + \frac{1}{9} \cdot 1$$

$$= \frac{5}{6} \cdot \frac{3}{3} + \frac{1}{9} \cdot \frac{2}{2}$$

Think: $9 \times \square = 18$. The answer is 2, so we multiply by 1, using $\frac{2}{2}$.

Think: $6 \times \square = 18$. The answer is 3, so we multiply by 1, using $\frac{3}{3}$.

$$= \frac{15}{18} + \frac{2}{18} = \frac{17}{18}$$

Do Exercises 8 and 9.

■ **EXAMPLE 8** Add: $\dfrac{3}{-5} + \dfrac{11}{10}$.

$$\frac{3}{-5} + \frac{11}{10} = \frac{-3}{5} + \frac{11}{10}$$

Recall that $\dfrac{m}{-n} = \dfrac{-m}{n}$. We generally avoid negative signs in the denominator. The LCD is 10.

$$= \frac{-3}{5} \cdot \frac{2}{2} + \frac{11}{10}$$

$$= \frac{-6}{10} + \frac{11}{10}$$

$$= \frac{5}{10} = \frac{1}{2}$$

> We may still have to simplify, but simplifying is almost always easier if the LCD has been used.

Do Exercise 10.

■ **EXAMPLE 9** Add: $\dfrac{5}{8} + 2$.

$$\frac{5}{8} + 2 = \frac{5}{8} + \frac{2}{1}$$

Rewriting 2 in fraction notation

$$= \frac{5}{8} + \frac{2}{1} \cdot \frac{8}{8}$$

The LCD is 8.

$$= \frac{5}{8} + \frac{16}{8} = \frac{21}{8}$$

Do Exercise 11.

EXAMPLE 10 Add: $\dfrac{9}{70} + \dfrac{11}{21} + \dfrac{-6}{15}$.

We need to determine the LCM of 70, 21, and 15:

$$70 = 2 \cdot 5 \cdot 7, \qquad 21 = 3 \cdot 7, \qquad 15 = 3 \cdot 5.$$

The LCM is $2 \cdot 3 \cdot 5 \cdot 7$, or 210.

$$\dfrac{9}{70} + \dfrac{11}{21} + \dfrac{-6}{15} = \dfrac{9}{70} \cdot \dfrac{3}{3} + \dfrac{11}{21} \cdot \dfrac{2 \cdot 5}{2 \cdot 5} + \dfrac{-6}{15} \cdot \dfrac{7 \cdot 2}{7 \cdot 2}$$

In each case, we multiply by 1 to obtain the LCD. To form 1, look at the prime factorization of the LCD and use the factor(s) missing from each denominator.

$$= \dfrac{9 \cdot 3}{70 \cdot 3} + \dfrac{11 \cdot 10}{21 \cdot 10} + \dfrac{-6 \cdot 14}{15 \cdot 14}$$

$$= \dfrac{27}{210} + \dfrac{110}{210} + \dfrac{-84}{210}$$

$$= \dfrac{137 + (-84)}{210}$$

$$= \dfrac{53}{210} \qquad \text{Since 53 is prime and not a factor of 210, we cannot simplify.}$$

Do Exercises 12 and 13.

C **Order**

Common denominators are also important for determining the larger of two fractions. When two fractions share a common denominator, the larger number can be found by comparing numerators. For example, 4 is greater than 3, so $\frac{4}{5}$ is greater than $\frac{3}{5}$.

$$\dfrac{4}{5} > \dfrac{3}{5}$$

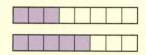

Similarly, because -6 is less than -2, we have

$$\dfrac{-6}{7} < \dfrac{-2}{7}, \quad \text{or} \quad -\dfrac{6}{7} < -\dfrac{2}{7}.$$

Do Exercises 14–16.

EXAMPLE 11 Use $<$ or $>$ for $\square$ to form a true sentence:

$$\dfrac{5}{8} \;\square\; \dfrac{2}{3}.$$

You can confirm that the LCD is 24. We multiply by 1 to find two fractions equivalent to $\frac{5}{8}$ and $\frac{2}{3}$ with denominators the same:

$$\dfrac{5}{8} \cdot \dfrac{3}{3} = \dfrac{15}{24}; \qquad \dfrac{2}{3} \cdot \dfrac{8}{8} = \dfrac{16}{24}.$$

Since $15 < 16$, it follows that $\frac{15}{24} < \frac{16}{24}$. Thus,

$$\dfrac{5}{8} < \dfrac{2}{3}.$$

Add.

12. $\dfrac{4}{10} + \dfrac{1}{100} + \dfrac{3}{1000}$

13. $\dfrac{7}{10} + \dfrac{-2}{21} + \dfrac{1}{7}$

Use $<$ or $>$ for $\square$ to form a true sentence.

14. $\dfrac{3}{8} \;\square\; \dfrac{5}{8}$

15. $\dfrac{7}{10} \;\square\; \dfrac{6}{10}$

16. $\dfrac{-2}{9} \;\square\; \dfrac{-5}{9}$

Answers on page A-10

Use < or > for ☐ to form a true sentence.

17. $\dfrac{2}{3}$ ☐ $\dfrac{3}{4}$

18. $\dfrac{-3}{4}$ ☐ $\dfrac{-8}{12}$

19. $\dfrac{5}{6}$ ☐ $\dfrac{7}{8}$

20. Sally jogs for $\frac{4}{5}$ mi, rests, and then jogs for another $\frac{1}{10}$ mi. How far does she jog in all?

$\frac{4}{5}$ mi

$\frac{1}{10}$ mi

Answers on page A-10

244

CHAPTER 4: Fraction Notation: Addition, Subtraction, and Mixed Numerals

---

EXAMPLE 12 Use < or > for ☐ to form a true sentence:

$$-\dfrac{89}{100} \; \square \; -\dfrac{9}{10}.$$

The LCD is 100.

$$\dfrac{-9}{10} \cdot \dfrac{10}{10} = \dfrac{-90}{100} \qquad \text{We multiply by } \tfrac{10}{10} \text{ to get the LCD.}$$

Since $-89 > -90$, it follows that $-\frac{89}{100} > -\frac{90}{100}$, so

$$-\dfrac{89}{100} \; > \; -\dfrac{9}{10}.$$

Do Exercises 17–19.

d Applications and Problem Solving

EXAMPLE 13 *Subflooring.* Matt Beecher Builders and Developers requires their subcontractors to use two layers of subflooring under a ceramic tile floor. First the contractors install a $\frac{3}{4}$-in. layer of oriented strand board (OSB). Then a $\frac{1}{2}$-in. sheet of cement board is mortared to the OSB. The mortar is $\frac{1}{8}$-in. thick. What is the total thickness of the two installed subfloors?

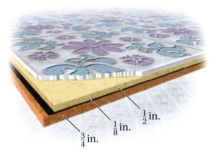

$\frac{1}{2}$ in.

$\frac{1}{8}$ in.

$\frac{3}{4}$ in.

1. **Familiarize.** We first make a drawing. We let $T =$ the total thickness of the subfloors.

2. **Translate.** The problem can be translated to an equation as follows.

OSB	plus	Mortar	plus	Cement board	is	Total thickness
$\dfrac{3}{4}$	$+$	$\dfrac{1}{8}$	$+$	$\dfrac{1}{2}$	$=$	T

3. **Solve.** To solve the equation, we carry out the addition. The LCM of the denominators is 8 because 2 and 4 are factors of 8. We multiply by 1 in order to obtain the LCD:

$$\dfrac{3}{4} + \dfrac{1}{8} + \dfrac{1}{2} = T$$

$$\dfrac{3}{4} \cdot \dfrac{2}{2} + \dfrac{1}{8} + \dfrac{1}{2} \cdot \dfrac{4}{4} = T$$

$$\dfrac{6}{8} + \dfrac{1}{8} + \dfrac{4}{8} = T$$

$$\dfrac{11}{8} = T.$$

4. **Check.** We check by repeating the calculation. We also note that the sum should be larger than any of the individual measurements, which it is. This tells us that the answer is reasonable.

5. **State.** The total thickness of the installed subfloors is $\frac{11}{8}$ in.

Do Exercise 20.

a , **b**   Add and, if possible, simplify.

1. $\dfrac{4}{9} + \dfrac{1}{9}$

2. $\dfrac{1}{4} + \dfrac{1}{4}$

3. $\dfrac{4}{7} + \dfrac{3}{7}$

4. $\dfrac{7}{8} + \dfrac{1}{8}$

5. $\dfrac{7}{10} + \dfrac{3}{-10}$

6. $\dfrac{1}{-6} + \dfrac{5}{6}$

7. $\dfrac{9}{a} + \dfrac{4}{a}$

8. $\dfrac{2}{t} + \dfrac{3}{t}$

9. $\dfrac{-7}{11} + \dfrac{3}{11}$

10. $\dfrac{7}{12} + \dfrac{-5}{12}$

11. $\dfrac{2}{9}x + \dfrac{5}{9}x$

12. $\dfrac{3}{11}a + \dfrac{2}{11}a$

13. $\dfrac{3}{32}t + \dfrac{13}{32}t$

14. $\dfrac{3}{25}x + \dfrac{12}{25}x$

15. $-\dfrac{2}{x} + \left(-\dfrac{7}{x}\right)$

16. $-\dfrac{7}{a} + \dfrac{5}{a}$

17. $\dfrac{1}{8} + \dfrac{1}{6}$

18. $\dfrac{1}{9} + \dfrac{1}{6}$

19. $\dfrac{-4}{5} + \dfrac{7}{10}$

20. $\dfrac{-3}{4} + \dfrac{1}{12}$

21. $\dfrac{7}{12} + \dfrac{3}{8}$

22. $\dfrac{7}{8} + \dfrac{1}{16}$

23. $\dfrac{3}{20} + 4$

24. $\dfrac{2}{15} + 3$

25. $\dfrac{5}{-8} + \dfrac{5}{6}$

26. $\dfrac{5}{-6} + \dfrac{7}{9}$

27. $\dfrac{3}{10}x + \dfrac{7}{100}x$

28. $\dfrac{9}{20}a + \dfrac{3}{40}a$

29. $\dfrac{5}{12} + \dfrac{8}{15}$

30. $\dfrac{3}{16} + \dfrac{1}{12}$

31. $\dfrac{-7}{10} + \dfrac{-29}{100}$

32. $\dfrac{-3}{10} + \dfrac{-27}{100}$

33. $-\dfrac{1}{10}x + \dfrac{1}{15}x$

34. $-\dfrac{1}{6}x + \dfrac{1}{4}x$

35. $-5t + \dfrac{2}{7}t$

36. $-4x + \dfrac{3}{5}x$

37. $-\dfrac{5}{12} + \dfrac{7}{-24}$

38. $-\dfrac{1}{18} + \dfrac{5}{-12}$

39. $\dfrac{4}{10} + \dfrac{3}{100} + \dfrac{7}{1000}$

40. $\dfrac{7}{10} + \dfrac{2}{100} + \dfrac{9}{1000}$

41. $\dfrac{3}{10} + \dfrac{5}{12} + \dfrac{8}{15}$

42. $\dfrac{1}{2} + \dfrac{3}{8} + \dfrac{1}{4}$

43. $\dfrac{5}{6} + \dfrac{25}{52} + \dfrac{7}{4}$

44. $\dfrac{15}{24} + \dfrac{7}{36} + \dfrac{91}{48}$

45. $\dfrac{2}{9} + \dfrac{7}{10} + \dfrac{-4}{15}$

46. $\dfrac{5}{12} + \dfrac{-3}{8} + \dfrac{1}{10}$

47. $-\dfrac{3}{4} + \dfrac{1}{5} + \dfrac{-7}{10}$

48. $\dfrac{1}{3} + \dfrac{-7}{9} + \dfrac{-1}{2}$

c   Use < or > for ☐ to form a true sentence.

49. $\dfrac{3}{8}$ ☐ $\dfrac{2}{8}$ **50.** $\dfrac{7}{9}$ ☐ $\dfrac{5}{9}$ **51.** $\dfrac{2}{3}$ ☐ $\dfrac{5}{6}$ **52.** $\dfrac{11}{18}$ ☐ $\dfrac{5}{9}$

53. $\dfrac{-2}{3}$ ☐ $\dfrac{-5}{7}$ **54.** $\dfrac{-3}{5}$ ☐ $\dfrac{-4}{7}$ **55.** $\dfrac{9}{15}$ ☐ $\dfrac{7}{10}$ **56.** $\dfrac{5}{14}$ ☐ $\dfrac{8}{21}$

57. $\dfrac{3}{4}$ ☐ $-\dfrac{1}{5}$ **58.** $\dfrac{3}{8}$ ☐ $-\dfrac{13}{16}$ **59.** $\dfrac{-7}{20}$ ☐ $\dfrac{-6}{15}$ **60.** $\dfrac{-7}{12}$ ☐ $\dfrac{-9}{16}$

Arrange each group of fractions from smallest to largest.

61. $\dfrac{3}{10}, \dfrac{5}{12}, \dfrac{4}{15}$ **62.** $\dfrac{5}{6}, \dfrac{19}{21}, \dfrac{11}{14}$

d   Solve.

63. Todd bought $\frac{1}{4}$ lb of gumdrops and $\frac{1}{2}$ lb of caramels. How many pounds of candy did Todd buy altogether?

64. Dona bought $\frac{1}{3}$ lb of orange pekoe tea and $\frac{1}{2}$ lb of English cinnamon tea. How many pounds of tea did Dona buy altogether?

65. Kate walked $\frac{7}{8}$ mi to the student union, and then $\frac{2}{5}$ mi to class. How far did Kate walk?

66. Damon walked $\frac{3}{8}$ mi to LeRoy's dormitory, and then $\frac{3}{4}$ mi to class. How far did Damon walk?

67. *Baking.*   A baker used $\frac{1}{2}$ lb of flour for rolls, $\frac{1}{4}$ lb for donuts, and $\frac{1}{3}$ lb for cookies. How much flour was used?

68. *Baking.*   A recipe for muffins calls for $\frac{1}{2}$ qt (quart) of buttermilk, $\frac{1}{3}$ qt of skim milk, and $\frac{1}{16}$ qt of oil. How many quarts of liquid ingredients does the recipe call for?

69. *Meteorology.*   On Monday, April 15, it rained $\frac{1}{2}$ in. in the morning and $\frac{3}{8}$ in. in the afternoon. How much did it rain altogether?

70. *Nursing.*   Janine took $\frac{1}{5}$ g of ibuprofen before lunch and $\frac{1}{2}$ g after lunch. How much did she take altogether?

71. A park naturalist hikes $\frac{3}{5}$ mi to a lookout, another $\frac{3}{10}$ mi to an osprey's nest, and finally, $\frac{3}{4}$ mi to a campsite. How far does the naturalist hike?

72. A triathlete runs $\frac{7}{8}$ mi, canoes $\frac{1}{3}$ mi, and swims $\frac{1}{6}$ mi. How many miles does the triathlete cover?

73. *Punch Recipe.* A recipe for strawberry punch calls for $\frac{1}{5}$ qt of ginger ale and $\frac{3}{5}$ qt of strawberry soda. How much liquid is needed? If the recipe is doubled, how much liquid is needed? If the recipe is halved, how much liquid is needed?

74. *Iced Cookies.* A chef prepared cookies for the freshman orientation reception. He frosted the $\frac{5}{16}$" (in.) cookies with a $\frac{5}{32}$" layer of icing. What is the thickness of the iced cookie?

75. *Masonry.* A tile $\frac{5}{8}$ in. thick is cemented to subflooring that is $\frac{7}{8}$ in. thick. The cement is $\frac{3}{32}$ in. thick. How thick is the result?

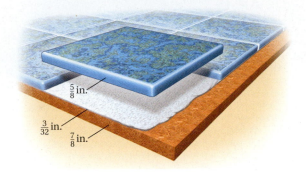

$\frac{5}{8}$ in.

$\frac{3}{32}$ in.

$\frac{7}{8}$ in.

76. *Concrete Mix.* A cubic meter of concrete mix contains 420 kg (kilograms) of cement, 150 kg of stone, and 120 kg of sand. What is the total weight of a cubic meter of the mix? What fractional part is cement? stone? sand? Add these fractional amounts. What is the result?

77. **D**W Explain why $\frac{3}{2000}$ is greater than $\frac{3}{2002}$. Are common denominators required for an explanation? Why or why not?

78. **D**W To add numbers with different denominators, Chris is consistently using the product of the denominators as a common denominator. Is this correct? Why or why not?

SKILL MAINTENANCE

Subtract. [2.3a]

79. $-7 - 6$

80. $-5 - (-9)$

81. $9 - 17$

82. $-8 - 23$

Evaluate. [2.6a]

83. $\frac{x - y}{3}$, for $x = 7$ and $y = -3$

84. $3(x + y)$ and $3x + 3y$, for $x = 5$ and $y = 9$

College Expenditures. The following table shows planned average expenditures by college freshmen to prepare for school in the fall of 2004 and the fall of 2005. Use these data for Exercises 85–90. [1.8a]

BACK TO COLLEGE EXPENSES	2004	2005
Textbooks	$238	$304
Electronics	$760	$540
Clothing and accessories	$58	$92
Dorm furnishings	$83	$128
School supplies	$32	$52
Shoes	$93	$36
Total	?	?

Source: The National Retail Federation

85. How much more did freshmen plan to spend on textbooks in 2005 than in 2004?

86. How much more did freshmen plan to spend on electronics in 2004 than in 2005?

87. How much more did freshmen plan to spend on shoes in 2004 than in 2005?

88. How much more did freshmen plan to spend on dorm furnishings in 2005 than in 2004?

89. What was the total expenditure in 2004?

90. What was the total expenditure in 2005?

SYNTHESIS

91. **D_W** Suppose that a classmate believes, incorrectly, that $\frac{2}{5} + \frac{4}{5} = \frac{6}{10}$. How could you convince the classmate that he or she is mistaken?

92. **D_W** Explain how pictures could be used to convince someone that $\frac{5}{7}$ is larger than $\frac{13}{21}$.

Add and, if possible, simplify.

93. $\frac{3}{10}t + \frac{2}{7} + \frac{2}{15}t + \frac{3}{5}$

94. $\frac{2}{9} + \frac{4}{21}x + \frac{4}{15} + \frac{3}{14}x$

95. $5t^2 + \frac{6}{a}t + 2t^2 + \frac{3}{a}t$

Use $<$, $>$, or $=$ for $\square$ to form a true sentence.

96. ▦ $\frac{10}{97} + \frac{67}{137} \ \square \ \frac{8123}{13{,}289}$

97. ▦ $\frac{12}{169} + \frac{53}{103} \ \square \ \frac{10{,}192}{17{,}407}$

98. ▦ $\frac{37}{157} + \frac{20}{107} \ \square \ \frac{6942}{16{,}799}$

99. A guitarist's band is booked for Friday and Saturday nights at a local club. The guitarist's group is a trio on Friday and expands to a quintet on Saturday. Thus the guitarist is paid one-third of one-half the weekend's pay for Friday and one-fifth of one-half the weekend's pay for Saturday. What fractional part of the total pay did the guitarist receive for the weekend's work? If the band was paid $1200, how much did the guitarist receive?

100. ▦ In the sum below, a and b are digits (so $1b$ is a two-digit number and $35a$ is a three-digit number). Find a and b. (*Hint*: $a < 4$ and $b > 6$.)

$$\frac{a}{17} + \frac{1b}{23} = \frac{35a}{391}$$

101. ▦ Consider only the numbers 3, 4, 5, and 6. Assume each can be placed in only one blank in the following.

$$\square + \frac{\square}{\square} \cdot \square = ?$$

What placement of the numbers in the blanks yields the largest number?

102. ▦ Consider only the numbers 2, 3, 4, and 5. Assume each is placed in a blank in the following.

$$\frac{\square}{\square} + \frac{\square}{\square} = ?$$

What placement of the numbers in the blanks yields the largest sum?

103. ▦ Use a standard calculator. Arrange the following in order from smallest to largest.

$$\frac{3}{4}, \frac{17}{21}, \frac{13}{15}, \frac{7}{9}, \frac{15}{17}, \frac{13}{12}, \frac{19}{22}$$

4.3 SUBTRACTION, EQUATIONS, AND APPLICATIONS

Objectives

a Subtract using fraction notation.

b Solve equations of the type $x + a = b$ and $a + x = b$, where a and b may be fractions.

c Solve applied problems involving subtraction with fraction notation.

Subtract and simplify.

1. $\dfrac{7}{8} - \dfrac{3}{8}$

2. $\dfrac{5}{9a} - \dfrac{1}{9a}$

3. $\dfrac{7}{10} - \dfrac{13}{10}$

4. $-\dfrac{2}{x} - \dfrac{4}{x}$

Answers on page A-10

a Subtraction

LIKE DENOMINATORS

We can consider the difference $\frac{4}{8} - \frac{3}{8}$ as we did before, as either "take away" or "how much more." Let's consider "take away."

We start with $\frac{4}{8}$

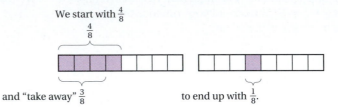

and "take away" $\frac{3}{8}$ to end up with $\frac{1}{8}$.

We start with 4 eighths and take away 3 eighths:

$$4 \text{ eighths} - 3 \text{ eighths} = 1 \text{ eighth},$$

or $\quad 4 \cdot \dfrac{1}{8} - 3 \cdot \dfrac{1}{8} = \dfrac{1}{8}, \quad$ or $\quad \dfrac{4}{8} - \dfrac{3}{8} = \dfrac{1}{8}.$

> To subtract when denominators are the same,
>
> **a)** subtract the numerators,
>
> **b)** keep the denominator, and
>
> **c)** simplify, if possible.
>
> $$\dfrac{7}{10} - \dfrac{4}{10} = \dfrac{7-4}{10} = \dfrac{3}{10}$$

EXAMPLES Subtract and, if possible, simplify.

1. $\dfrac{8}{13} - \dfrac{3}{13} = \dfrac{8-3}{13} = \dfrac{5}{13}$

2. $\dfrac{3}{35} - \dfrac{13}{35} = \dfrac{3-13}{35} = \dfrac{-10}{35} = \dfrac{5}{5} \cdot \dfrac{-2}{7} = \dfrac{-2}{7},$ or $-\dfrac{2}{7}$ Removing a factor equal to 1: $\frac{5}{5} = 1$

3. $\dfrac{13}{2a} - \dfrac{5}{2a} = \dfrac{13-5}{2a} = \dfrac{8}{2a} = \dfrac{2}{2} \cdot \dfrac{4}{a} = \dfrac{4}{a}$ Removing a factor equal to 1: $\frac{2}{2} = 1$

4. $-\dfrac{7}{t} - \dfrac{2}{t} = \dfrac{-7-2}{t} = \dfrac{-9}{t},$ or $-\dfrac{9}{t}$

Do Exercises 1–4.

DIFFERENT DENOMINATORS

> To subtract when denominators are different:
>
> **a)** Find the least common multiple of the denominators. That number is the least common denominator, LCD.
>
> **b)** Multiply by 1, using an appropriate notation, n/n, to express each fraction in an equivalent form that contains the LCD.
>
> **c)** Subtract and, if possible, simplify.

EXAMPLE 5 Subtract: $\dfrac{2}{5} - \dfrac{3}{8}$.

a) The LCM of 5 and 8 is 40, so the LCD is 40.

b) We need to find numbers equivalent to $\frac{2}{5}$ and $\frac{3}{8}$ with denominators of 40:

$$\dfrac{2}{5} - \dfrac{3}{8} = \dfrac{2}{5} \cdot \dfrac{8}{8} - \dfrac{3}{8} \cdot \dfrac{5}{5} \longleftarrow$$

Think: $8 \times ? = 40$. The answer is 5, so we multiply by 1, using $\frac{5}{5}$.

Think: $5 \times ? = 40$. The answer is 8, so we multiply by 1, using $\frac{8}{8}$.

c) We subtract: $\dfrac{16}{40} - \dfrac{15}{40} = \dfrac{16 - 15}{40} = \dfrac{1}{40}$.

d) Since $\frac{1}{40}$ cannot be simplified, we are finished. The answer is $\frac{1}{40}$.

Do Exercise 5.

EXAMPLE 6 Subtract: $\dfrac{7}{12} - \dfrac{5}{6}$.

Since 6 is a factor of 12, the LCM of 6 and 12 is 12. The LCD is 12.

$$\dfrac{7}{12} - \dfrac{5}{6} = \dfrac{7}{12} - \dfrac{5}{6} \cdot \dfrac{2}{2}$$

Think: $6 \times ? = 12$. The answer is 2, so we multiply by 1, using $\frac{2}{2}$.

$$= \dfrac{7}{12} - \dfrac{10}{12}$$

$$= \dfrac{7 - 10}{12} = \dfrac{-3}{12}$$

If we prefer, we can add the opposite: $7 + (-10)$.

$$= \dfrac{3}{3} \cdot \dfrac{-1}{4} = \dfrac{-1}{4}, \text{ or } -\dfrac{1}{4}$$

Simplifying by removing a factor equal to 1: $\frac{3}{3} = 1$

EXAMPLE 7 Subtract: $\dfrac{17}{24} - \dfrac{4}{15}$.

We need to find the LCM of 24 and 15:

$$\left. \begin{array}{l} 24 = 2 \cdot 2 \cdot 2 \cdot 3, \\ 15 = 3 \cdot 5 \end{array} \right\} \quad \text{The LCM is } 2 \cdot 2 \cdot 2 \cdot 3 \cdot 5, \text{ or } 120.$$

$$\dfrac{17}{24} - \dfrac{4}{15} = \dfrac{17}{24} \cdot \dfrac{5}{5} - \dfrac{4}{15} \cdot \dfrac{8}{8}$$

Multiplying by 1 to obtain the LCD. To form 1, use the factors of the LCM that each denominator lacks. Note that $2 \cdot 2 \cdot 2 = 8$.

$$= \dfrac{85}{120} - \dfrac{32}{120} = \dfrac{85 - 32}{120} = \dfrac{53}{120}.$$

Do Exercises 6–9.

EXAMPLE 8 Simplify by combining like terms: $\dfrac{7}{8}x - \dfrac{3}{4}x$.

$$\dfrac{7}{8}x - \dfrac{3}{4}x = \left(\dfrac{7}{8} - \dfrac{3}{4} \right)x$$

Try to do this step mentally.

$$= \left(\dfrac{7}{8} - \dfrac{6}{8} \right)x = \dfrac{1}{8}x$$

Multiplying $\frac{3}{4}$ by $\frac{2}{2}$ and subtracting

Do Exercise 10.

---

5. Subtract: $\dfrac{3}{4} - \dfrac{2}{3}$.

Subtract.

6. $\dfrac{5}{6} - \dfrac{2}{3}$

7. $\dfrac{2}{5} - \dfrac{7}{10}$

8. $\dfrac{2}{3} - \dfrac{5}{6}$

9. $\dfrac{11}{28} - \dfrac{5}{16}$

10. Simplify: $\dfrac{9}{10}x - \dfrac{3}{5}x$.

Answers on page A-10

Solve.

11. $x - \dfrac{2}{5} = \dfrac{1}{5}$

12. $x + \dfrac{2}{3} = \dfrac{5}{6}$

13. $\dfrac{3}{5} + t = -\dfrac{7}{8}$

Answers on page A-10

CHAPTER 4: Fraction Notation: Addition, Subtraction, and Mixed Numerals

---

b Solving Equations

In Section 2.8, we introduced the addition principle as one way to form equivalent equations. We can use that principle here to solve equations containing fractions.

EXAMPLE 9 Solve: $x - \dfrac{1}{3} = -\dfrac{2}{7}$.

$$x - \frac{1}{3} = -\frac{2}{7}$$

$$x - \frac{1}{3} + \frac{1}{3} = -\frac{2}{7} + \frac{1}{3}$$
Using the addition principle: adding $\frac{1}{3}$ to both sides

$$x + 0 = -\frac{2}{7} + \frac{1}{3}$$
Adding $\frac{1}{3}$ "undid" the subtraction of $\frac{1}{3}$ in the previous step.

$$x = -\frac{2}{7} \cdot \frac{3}{3} + \frac{1}{3} \cdot \frac{7}{7}$$
Multiplying by 1 to obtain the LCD, 21

$$x = -\frac{6}{21} + \frac{7}{21} = \frac{1}{21}$$
The solution appears to be $\frac{1}{21}$.

Check:

$$\begin{array}{c|c}
x - \dfrac{1}{3} = -\dfrac{2}{7} \\
\hline
\dfrac{1}{21} - \dfrac{1}{3} \;?\; -\dfrac{2}{7} \\
\dfrac{1}{21} - \dfrac{1}{3} \cdot \dfrac{7}{7} \\
\dfrac{1}{21} - \dfrac{7}{21} \\
\dfrac{-6}{21} \\
-\dfrac{2}{7} \cdot \dfrac{3}{3} \quad \Big| \quad -\dfrac{2}{7} \;\; \text{TRUE}
\end{array}$$

Our answer checks. The solution is $\frac{1}{21}$.

Recall that since subtraction can be regarded as adding the opposite of the number being subtracted, the addition principle allows us to subtract the same number on both sides of an equation.

EXAMPLE 10 Solve: $x + \dfrac{1}{4} = \dfrac{3}{5}$.

$$x + \frac{1}{4} - \frac{1}{4} = \frac{3}{5} - \frac{1}{4}$$
Using the addition principle: adding $-\frac{1}{4}$ to, or subtracting $\frac{1}{4}$ from, both sides

$$x + 0 = \frac{3}{5} \cdot \frac{4}{4} - \frac{1}{4} \cdot \frac{5}{5}$$
The LCD is 20. We multiply by 1 to get the LCD.

$$x = \frac{12}{20} - \frac{5}{20} = \frac{7}{20}$$

The solution is $\frac{7}{20}$. We leave the check to the student.

Do Exercises 11–13.

C Applications and Problem Solving

EXAMPLE 11 *Jewelry Making.* Coldwater Creek offers the pendant necklace illustrated at right. The sterling silver capping at the top measures $\frac{11}{32}$ in. $\left(\frac{11}{32}"\right)$ and the total length of the pendant is $\frac{7}{8}$ in. $\left(\frac{7}{8}"\right)$. Find the length, or diameter, of the pearl ball on the pendant.

1. **Familiarize.** Using a sketch—or a photo, if one is available—we write in the measurements. We let d represent the pearl's diameter, in inches.

2. **Translate.** We see that this is a "how much more" situation. We can translate to an equation.

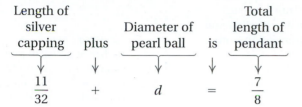

Length of silver capping	plus	Diameter of pearl ball	is	Total length of pendant
↓	↓	↓	↓	↓
$\frac{11}{32}$	$+$	d	$=$	$\frac{7}{8}$

3. **Solve.** To solve the equation, we subtract $\frac{11}{32}$ from both sides:

$$\frac{11}{32} + d = \frac{7}{8}$$

$$\frac{11}{32} + d - \frac{11}{32} = \frac{7}{8} - \frac{11}{32} \qquad \text{Subtracting } \frac{11}{32} \text{ from both sides}$$

$$d + 0 = \frac{7}{8} \cdot \frac{4}{4} - \frac{11}{32} \qquad \text{The LCD is 32. We multiply by 1 to obtain the LCD.}$$

$$d = \frac{28}{32} - \frac{11}{32} = \frac{17}{32}.$$

4. **Check.** To check, we return to the original problem and add:

$$\frac{11}{32} + \frac{17}{32} = \frac{28}{32} = \frac{7}{8} \cdot \frac{4}{4} = \frac{7}{8}.$$

Since the overall length of the pendant is $\frac{7}{8}"$, our answer checks.

5. **State.** The diameter of the pearl is $\frac{17}{32}$ in.

Do Exercise 14.

Source: ©Coldwater Creek Inc. www.coldwatercreek.com

14. Teri has run for $\frac{2}{3}$ mi and will stop when she has run for $\frac{7}{8}$ mi. How much farther does she have to run?

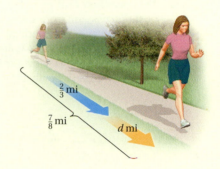

Answer on page A-10

CALCULATOR CORNER

Adding and Subtracting Fractions   To calculate $\frac{2}{15} + \frac{5}{12}$ with a fraction calculator, the following keystrokes can be used (note that the $\boxed{a^{b/c}}$ key usually doubles as the $\boxed{d/c}$ key): $\boxed{2}$ $\boxed{a^{b/c}}$ $\boxed{1}$ $\boxed{5}$ $\boxed{+}$ $\boxed{5}$ $\boxed{a^{b/c}}$ $\boxed{1}$ $\boxed{2}$ $\boxed{\text{ENTER} =}$. The display that appears, $\boxed{\quad 11 \lrcorner 20 \quad}$ or $\boxed{\quad 11/20 \quad}$, represents the fraction $\frac{11}{20}$. Fractions with numerators greater than denominators may be given in mixed numeral notation (see Section 4.5). To convert to fraction notation, press $\boxed{\text{2nd}}$ $\boxed{a^{b/c}}$ $\boxed{\text{ENTER} =}$. Most graphing calculators do not have an $\boxed{a^{b/c}}$ key. To perform the above addition on such a calculator, we use the $\boxed{\text{MATH}}$ key as follows:

$$\boxed{2} \boxed{÷} \boxed{1} \boxed{5} \boxed{+} \boxed{5} \boxed{÷} \boxed{1} \boxed{2} \boxed{\text{MATH}} \boxed{1} \boxed{\text{ENTER}}.$$

Exercises: Calculate.

1. $\frac{3}{8} + \frac{1}{4}$

2. $\frac{5}{12} + \frac{3}{10}$

3. $\frac{8}{7} + \frac{-1}{3}$

4. $\frac{19}{20} - \frac{17}{35}$

5. $\frac{-9}{30} - \frac{1}{25}$

6. $\frac{7}{23} - \left(-\frac{9}{29}\right)$

Translating for Success

1. *Bubble Wrap.* One-Stop Postal Center orders bubble wrap in 64-yd rolls. On the average, $\frac{3}{4}$ yd is used per small package. How many small packages can be prepared with 2 rolls of bubble wrap?

2. *Distance from College.* The post office is $\frac{7}{9}$ mi from the community college. The medical clinic is $\frac{2}{5}$ as far from the college as the post office is. How far is the clinic from the college?

3. *Swimming.* Andrew swims $\frac{7}{9}$ mi every day. One day, he swims $\frac{2}{5}$ mi by 11:00 A.M. How much farther must Andrew swim to reach his daily goal?

4. *Tuition.* The average tuition at Waterside University is $12,000. If a loan is obtained for $\frac{1}{3}$ of the tuition, how much is the loan?

5. *Thermos Bottle Capacity.* A thermos bottle holds $\frac{11}{12}$ gal. How much is in the bottle when it is $\frac{4}{7}$ full?

The goal of these matching questions is to practice step (2), *Translate*, of the five-step problem-solving process. Translate each word problem to an equation and select a correct translation from equations A–O.

A. $\frac{3}{4} \cdot 64 = x$

B. $\frac{1}{3} \cdot 12{,}000 = x$

C. $\frac{1}{3} + \frac{2}{5} = x$

D. $\frac{2}{5} + x = \frac{7}{9}$

E. $\frac{2}{5} \cdot \frac{7}{9} = x$

F. $\frac{3}{4} \cdot x = 64$

G. $\frac{4}{7} = x + \frac{11}{12}$

H. $\frac{2}{5} = x + \frac{7}{9}$

I. $\frac{4}{7} \cdot \frac{11}{12} = x$

J. $\frac{3}{4} \cdot x = 128$

K. $\frac{1}{3} \cdot x = 12{,}000$

L. $\frac{1}{3} + \frac{2}{5} + x = 1$

M. $\frac{4}{3} \cdot x = 64$

N. $\frac{4}{7} + x = \frac{11}{12}$

O. $\frac{1}{3} + x = \frac{2}{5}$

Answers on page A-10

6. *Cutting Rope.* A piece of rope $\frac{11}{12}$ yd long is cut into two pieces. One piece is $\frac{4}{7}$ yd long. How long is the other piece?

7. *Planting Corn.* Each year, Prairie State Farm plants 64 acres of corn. With good weather, $\frac{3}{4}$ of the planting can be completed by April 20. How many acres can be planted with good weather?

8. *Painting Trim.* A painter used $\frac{1}{3}$ gal of white paint for the trim in the library and $\frac{2}{5}$ gal in the family room. How much paint was used for the trim in the two rooms?

9. *Lottery Winnings.* Sally won $12,000 in a state lottery and decides to give the net amount after taxes to three charities. One received $\frac{1}{3}$ of the money, and a second received $\frac{2}{5}$. What fractional part did the third charity receive?

10. *Reading Assignment.* Lowell read 64 pages of his political science assignment. This completed $\frac{3}{4}$ of his required reading. How many total pages were assigned?

a   Subtract and, if possible, simplify.

1. $\dfrac{5}{6} - \dfrac{1}{6}$

2. $\dfrac{7}{5} - \dfrac{2}{5}$

3. $\dfrac{9}{16} - \dfrac{13}{16}$

4. $\dfrac{5}{12} - \dfrac{7}{12}$

5. $\dfrac{8}{a} - \dfrac{6}{a}$

6. $\dfrac{4}{t} - \dfrac{9}{t}$

7. $-\dfrac{2}{9} - \dfrac{5}{9}$

8. $-\dfrac{3}{11} - \dfrac{4}{11}$

9. $-\dfrac{3}{8} - \dfrac{1}{8}$

10. $-\dfrac{3}{10} - \dfrac{1}{10}$

11. $\dfrac{10}{3t} - \dfrac{4}{3t}$

12. $\dfrac{9}{2a} - \dfrac{5}{2a}$

13. $\dfrac{3}{5a} - \dfrac{7}{5a}$

14. $\dfrac{2}{7t} - \dfrac{10}{7t}$

15. $\dfrac{7}{8} - \dfrac{1}{16}$

16. $\dfrac{4}{3} - \dfrac{5}{6}$

17. $\dfrac{7}{15} - \dfrac{4}{5}$

18. $\dfrac{3}{4} - \dfrac{3}{28}$

19. $\dfrac{3}{4} - \dfrac{1}{20}$

20. $\dfrac{3}{4} - \dfrac{4}{16}$

21. $\dfrac{2}{15} - \dfrac{5}{12}$

22. $\dfrac{11}{16} - \dfrac{9}{10}$

23. $\dfrac{7}{10} - \dfrac{23}{100}$

24. $\dfrac{9}{10} - \dfrac{3}{100}$

25. $\dfrac{7}{15} - \dfrac{3}{25}$

26. $\dfrac{18}{25} - \dfrac{4}{35}$

27. $\dfrac{69}{100} - \dfrac{9}{10}$

28. $\dfrac{42}{100} - \dfrac{11}{20}$

29. $\dfrac{1}{8} - \dfrac{2}{3}$

30. $\dfrac{3}{4} - \dfrac{1}{2}$

31. $-\dfrac{3}{10} - \dfrac{7}{25}$

32. $-\dfrac{5}{18} - \dfrac{2}{27}$

33. $\dfrac{2}{3} - \dfrac{4}{5}$

34. $\dfrac{1}{2} - \dfrac{3}{5}$

35. $\dfrac{-5}{18} - \dfrac{7}{24}$

36. $\dfrac{-7}{25} - \dfrac{2}{15}$

37. $\dfrac{13}{90} - \dfrac{17}{120}$

38. $\dfrac{8}{25} - \dfrac{29}{150}$

39. $\dfrac{2}{3}x - \dfrac{4}{9}x$

40. $\dfrac{7}{4}x - \dfrac{5}{12}x$

41. $\dfrac{2}{5}a - \dfrac{3}{4}a$

42. $\dfrac{4}{7}a - \dfrac{1}{3}a$

b Solve.

43. $x - \dfrac{4}{9} = \dfrac{3}{9}$

44. $x - \dfrac{3}{11} = \dfrac{7}{11}$

45. $a + \dfrac{2}{11} = \dfrac{6}{11}$

46. $a + \dfrac{4}{15} = \dfrac{13}{15}$

47. $x + \dfrac{1}{3} = \dfrac{7}{9}$

48. $x + \dfrac{1}{2} = \dfrac{7}{8}$

49. $a - \dfrac{3}{8} = \dfrac{3}{4}$

50. $x - \dfrac{3}{10} = \dfrac{2}{5}$

51. $\dfrac{2}{3} + x = \dfrac{4}{5}$

52. $\dfrac{4}{5} + x = \dfrac{6}{7}$

53. $\dfrac{3}{8} + a = \dfrac{1}{12}$

54. $\dfrac{5}{6} + a = \dfrac{2}{9}$

55. $n - \dfrac{3}{10} = -\dfrac{1}{6}$

56. $n - \dfrac{3}{4} = -\dfrac{5}{12}$

57. $x + \dfrac{3}{4} = -\dfrac{1}{2}$

58. $x + \dfrac{5}{6} = -\dfrac{11}{12}$

c Solve.

59. *Fitness.* As part of a fitness program, Jaci swims $\frac{1}{2}$ mi every day. She has already swum $\frac{1}{5}$ mi today. How much farther should Jaci swim?

60. *Exercise.* As part of an exercise program, Eric walks $\frac{7}{8}$ mi each day. He has already walked $\frac{1}{3}$ mi today. How much farther should Eric walk?

61. A server has a bowl containing $\frac{11}{12}$ cup of grated Parmesan cheese and puts $\frac{1}{8}$ cup on a customer's spaghetti and meatballs. How much remains in the bowl?

62. *Tire Tread.* A long-life tire has a tread depth of $\frac{3}{8}$ in. instead of a more typical depth of $\frac{11}{32}$ in. How much deeper is the long-life tread depth?

Source: *Popular Science*

$\frac{3}{8}$ in.

$\frac{11}{32}$ in.

63. *Woodworking.* Celeste is replacing a $\frac{3}{4}$-in. thick shelf in her bookcase. If her replacement board is $\frac{15}{16}$ in. thick, how much should it be planed down before the repair can be completed?

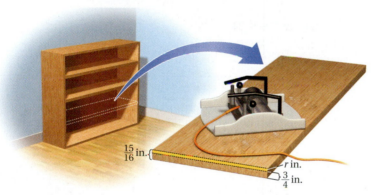

$\frac{15}{16}$ in.

r in.

$\frac{3}{4}$ in.

64. *Furniture Cleaner.* A $\frac{2}{3}$-cup mixture of lemon juice and olive oil makes an excellent cleaner for wood furniture. If the mixture contains $\frac{1}{4}$ cup of lemon juice, how much olive oil is in the cleaner?

$\frac{1}{4}$ cup $\quad + \quad$? $\quad = \quad$ $\frac{2}{3}$ cup

65. From a $\frac{4}{5}$-lb wheel of cheese, a $\frac{1}{4}$-lb piece was served. How much cheese remained on the wheel?

66. Jovan has an $\frac{11}{10}$-lb mixture of cashews and peanuts that includes $\frac{3}{5}$ lb of cashews. How many pounds of peanuts are in the mixture?

67. Jorge's $\frac{3}{4}$-hr drive to a job was a mixture of city and country driving. If $\frac{2}{5}$ hr was city driving, how much time was spent on country driving?

68. Addie spent $\frac{3}{4}$ hr listening to Maroon 5 and U2. She spent $\frac{1}{3}$ hr listening to Maroon 5. How many hours were spent listening to U2?

69. Blake used $\frac{1}{4}$ cup of maple syrup in preparing the batter for a batch of maple oatbran muffins. Sheila pointed out that the recipe actually calls for $\frac{2}{3}$ cup of syrup. How much more syrup should Blake add to the batter?

70. Amber added $\frac{1}{3}$ qt of two-cycle oil to a fuel mixture for her lawn mower. She then noticed that the owner's manual indicates $\frac{1}{2}$ qt should have been added. How much more two-cycle oil should Amber add to the mixture?

71. D_W To solve $x - \frac{1}{2} = -\frac{7}{8}$, Steffen adds $\frac{7}{8}$ to both sides of the equation. Is this correct? Has he formed an equivalent equation? Explain.

72. D_W Ruwanda incorrectly writes $\frac{7}{12} - \frac{7}{8} = \frac{7}{4}$. How could you convince her that this subtraction is incorrect?

SKILL MAINTENANCE

Divide and, if possible, simplify.   [3.7b]

73. $\dfrac{3}{7} \div \dfrac{9}{4}$

74. $\dfrac{9}{10} \div \dfrac{3}{5}$

75. $7 \div \dfrac{1}{3}$

76. $\dfrac{1}{4} \div 8$

Solve.

77. *Digital Tire Gauges.* A factory produces 3885 digital tire gauges per day. How long will it take to fill an order for 66,045 tire gauges?   [1.8a]

78. A batch of fudge requires $\frac{3}{4}$ cup of sugar. How much sugar is needed to make 12 batches?   [3.6b]

79. $3x - 8 = 25$ [2.8d]

80. $5x + 9 = 24$ [2.8d]

SYNTHESIS

81. D_W If a negative fraction is subtracted from another negative fraction, is the result always negative? Why or why not?

82. D_W Without performing the actual computation, explain how you can tell that $\frac{3}{7} - \frac{5}{9}$ is negative.

Simplify.

83. $\dfrac{7}{8} - \dfrac{3}{4} - \dfrac{1}{16}$

84. $\dfrac{9}{10} - \dfrac{1}{2} - \dfrac{2}{15}$

85. $\dfrac{2}{5} - \dfrac{1}{6}(-3)^2$

86. $\dfrac{7}{8} - \dfrac{1}{10}\left(-\dfrac{5}{6}\right)^2$

87. $-4 \cdot \dfrac{3}{7} - \dfrac{1}{7} \cdot \dfrac{4}{5}$

88. $\left(\dfrac{5}{6}\right)^2 + \left(\dfrac{3}{4}\right)^2$

89. $\left(-\dfrac{2}{5}\right)^3 - \left(-\dfrac{3}{10}\right)^3$

90. $\dfrac{3}{17} - \dfrac{2}{19} - \left(\dfrac{3}{17} - \dfrac{2}{19}\right)$

91. A Quizno's franchise is owned by Becky, Clay, and Trey. Becky owns $\frac{1}{3}$ of the business and Clay owns $\frac{1}{2}$. What part of the business does Trey own?

92. A new Chevrolet Aveo costs \$12,600. Pam will pay $\frac{1}{2}$ of the cost, Sam will pay $\frac{1}{4}$ of the cost, Jan will pay $\frac{1}{6}$ of the cost, and Nan will pay the rest.

a) How much will Nan pay?
b) What fractional part will Nan pay?

CHAPTER 4: Fraction Notation: Addition,
Subtraction, and Mixed Numerals

93. The circle graph below shows how long shoppers stay when visiting a mall. What portion of shoppers stay for 0–2 hr?

Less than 1 hour $\frac{26}{50}$

2–3 hours $\frac{5}{50}$

1–2 hours

More than 3 hours $\frac{2}{50}$

94. As part of a rehabilitation program, an athlete must swim and then walk a total of $\frac{9}{10}$ km each day. If one lap in the swimming pool is $\frac{3}{80}$ km, how far must the athlete walk after swimming 10 laps?

95. A VCR can record up to 6 hr on one tape. It can also fill that same tape in either 4 hr or 2 hr when running at faster speeds. A tape is placed in the machine, which records for $\frac{1}{2}$ hr at the 4-hr speed and $\frac{3}{4}$ hr at the 2-hr speed. How much time is left on the tape to record at the 6-hr speed?

96. Mazzi's meat slicer cut 8 slices of turkey and 3 slices of Vermont cheddar. If each turkey slice was $\frac{1}{16}$-in. thick and each cheddar slice was $\frac{5}{32}$-in. thick, how tall was the pile of meat and cheese?

97. Mark Romano owns $\frac{7}{12}$ of Romano-Chrenka Chevrolet and Lisa Romano owns $\frac{1}{6}$. If Paul and Ella Chrenka own the remaining share of the dealership equally, what fractional piece does Paul own?

98. The Fullerton estate was left to four children. One received $\frac{1}{4}$ of the estate, one received $\frac{3}{8}$, and the twins split the rest. What fractional piece did each twin receive?

Solve.

99. ▦ $x + \dfrac{16}{323} = \dfrac{10}{187}$

100. ▦ $x + \dfrac{7}{253} = \dfrac{12}{299}$

101. ▦ Determine what whole number a must be in order for the following to be true:

$$\frac{10 + a}{23} = \frac{330}{391} - \frac{a}{17}.$$

102. A mountain climber, beginning at sea level, climbs $\frac{3}{5}$ km, descends $\frac{1}{4}$ km, climbs $\frac{1}{3}$ km, and then descends $\frac{1}{7}$ km. At what elevation does the climber finish?

103. *Microsoft Interview.*   The following is a question taken from an employment interview with Microsoft. Try to answer it. "Given a gold bar that can be cut exactly twice and a contractor who must be paid one-seventh of a gold bar every day for seven days, how should the bar be cut?"

Source: *Fortune Magazine,* January 22, 2001

Objectives

a Solve equations that involve fractions and require use of both the addition principle and the multiplication principle.

b Solve equations by using the multiplication principle to clear fractions.

In Section 3.8 we used the multiplication principle to solve equations like

$$\frac{2}{3}x = \frac{5}{6} \quad \text{and} \quad 7 = \frac{5}{4}t.$$

In Section 4.3 we used the addition principle to solve equations like

$$\frac{4}{5} + x = \frac{1}{2} \quad \text{and} \quad \frac{7}{3} = t - \frac{2}{9}.$$

We are now ready to solve equations in which both principles are required. Equations of this sort were first encountered in Section 2.8, but at that time no fractions appeared.

a Using the Principles Together

Recall that equations are most commonly solved by producing a sequence of equivalent equations. Thus, the equations $5x + 2 = 47$ and $x = 9$ are equivalent:

$$5x + 2 = 47$$

$$5x + 2 - 2 = 47 - 2 \quad \text{Using the addition principle}$$

$$5x = 45$$

$$\frac{1}{5} \cdot 5x = \frac{1}{5} \cdot 45 \quad \text{Using the multiplication principle}$$

$$x = 9 \quad \text{It appears that the solution is 9.}$$

As a check, note that $5 \cdot 9 + 2 = 45 + 2 = 47$, as desired. The solution is 9.

EXAMPLE 1 Solve: $\dfrac{3}{4}x - \dfrac{1}{8} = \dfrac{1}{2}.$

We first isolate $\frac{3}{4}x$ by adding $\frac{1}{8}$ to both sides:

$$\frac{3}{4}x - \frac{1}{8} = \frac{1}{2}$$

$$\frac{3}{4}x - \frac{1}{8} + \frac{1}{8} = \frac{1}{2} + \frac{1}{8} \quad \text{Using the addition principle}$$

$$\frac{3}{4}x + 0 = \frac{4}{8} + \frac{1}{8} \quad \text{Writing with a common denominator}$$

$$\frac{3}{4}x = \frac{5}{8}.$$

Next, we isolate x by multiplying both sides by $\frac{4}{3}$:

$$\frac{3}{4}x = \frac{5}{8} \quad \text{Note that the reciprocal of } \tfrac{3}{4} \text{ is } \tfrac{4}{3}.$$

$$\frac{4}{3} \cdot \frac{3}{4}x = \frac{4}{3} \cdot \frac{5}{8} \quad \text{Using the multiplication principle}$$

$$1x = \frac{20}{24}, \text{ or } \frac{5}{6}. \quad \text{Simplifying; the solution appears to be } \tfrac{5}{6}.$$

Study Tips

CHECK YOUR ANSWERS

Every solution of an equation can be checked by substituting that value in the original equation. Take advantage of this opportunity to determine if your answers are correct. Note that substitution will not tell you if your answer is in simplified form, so be certain to check for that also.

Check:

$$\frac{3}{4}x - \frac{1}{8} = \frac{1}{2}$$

$$\frac{3}{4} \cdot \frac{5}{6} - \frac{1}{8} \quad ? \quad \frac{1}{2}$$

$$\frac{3 \cdot 5}{4 \cdot 2 \cdot 3} - \frac{1}{8}$$ Removing a factor equal to 1: $\frac{3}{3} = 1$

$$\frac{5}{8} - \frac{1}{8}$$

$$\frac{1}{2} \quad \bigg| \quad \frac{1}{2} \quad \text{TRUE}$$

The solution is $\frac{5}{6}$.

Do Exercises 1 and 2.

EXAMPLE 2   Solve: $5 + \frac{9}{2}t = -\frac{7}{2}$.

We first isolate $\frac{9}{2}t$ by subtracting 5 from both sides:

$$5 + \frac{9}{2}t = -\frac{7}{2}$$

$$5 + \frac{9}{2}t - 5 = -\frac{7}{2} - 5$$ Subtracting 5 from both sides

$$\frac{9}{2}t = -\frac{7}{2} - \frac{10}{2}$$ Writing 5 as $\frac{10}{2}$ to use the LCD

$$\frac{9}{2}t = -\frac{17}{2}$$ Note that the reciprocal of $\frac{9}{2}$ is $\frac{2}{9}$.

$$\frac{2}{9} \cdot \frac{9}{2}t = \frac{2}{9}\left(-\frac{17}{2}\right)$$ Multiplying both sides by $\frac{2}{9}$

$$1t = -\frac{2 \cdot 17}{9 \cdot 2}$$ Removing a factor equal to 1: $\frac{2}{2} = 1$

$$t = -\frac{17}{9}.$$

Check:

$$5 + \frac{9}{2}t = -\frac{7}{2}$$

$$5 + \frac{9}{2}\left(-\frac{17}{9}\right) \quad ? \quad -\frac{7}{2}$$ Removing a factor equal to 1: $\frac{9}{9} = 1$

$$5 + \left(-\frac{17}{2}\right)$$

$$\frac{10}{2} + \left(\frac{-17}{2}\right)$$

$$\frac{10 - 17}{2}$$

$$\frac{-7}{2} \quad \bigg| \quad -\frac{7}{2} \quad \text{TRUE}$$

The solution is $-\frac{17}{9}$.

Do Exercises 3 and 4.

Sometimes the variable appears on the right side of the equation. The strategy for solving the equation remains the same.

Solve.

1. $\dfrac{3}{5}t - \dfrac{8}{15} = \dfrac{2}{15}$

2. $\dfrac{1}{2}x - \dfrac{1}{5} = \dfrac{7}{10}$

Solve.

3. $3 + \dfrac{14}{5}t = -\dfrac{21}{5}$

4. $2x + 4 = \dfrac{1}{2}$

Answers on page A-10

5. Solve: $9 - \dfrac{3}{4}x = 21$.

EXAMPLE 3 Solve: $20 = 6 - \dfrac{2}{3}x$.

Our plan is to first use the addition principle to isolate $-\frac{2}{3}x$ and then to use the multiplication principle to isolate x.

$$20 = 6 - \dfrac{2}{3}x$$

$$20 - 6 = 6 - \dfrac{2}{3}x - 6 \qquad \text{Subtracting 6 (or adding } -6) \text{ on both sides}$$

$$14 = -\dfrac{2}{3}x$$

$$\left(-\dfrac{3}{2}\right)14 = \left(-\dfrac{3}{2}\right)\left(-\dfrac{2}{3}x\right) \qquad \text{Multiplying both sides by } -\tfrac{3}{2}$$

$$-\dfrac{3 \cdot 14}{2} = 1x$$

$$-\dfrac{3 \cdot 7 \cdot \cancel{2}}{\cancel{2}} = 1x \qquad \text{Removing a factor equal to 1: } \tfrac{2}{2} = 1$$

$$-21 = x$$

Check:
$$\begin{array}{c|c} 20 = 6 - \dfrac{2}{3}x \\ \hline 20 \ ? \ 6 - \dfrac{2}{3}(-21) \\ 6 + \dfrac{42}{3} \\ 20 \ \bigg| \ 6 + 14 \qquad \text{TRUE} \end{array}$$

The solution is -21.

Do Exercise 5.

b Clearing Fractions

We now show an alternative approach for solving Examples 1–3. The advantage of this approach is that it minimizes calculations involving fractions. The disadvantage is that it requires an extra step and careful use of the distributive law. Key to this approach is using the multiplication principle in the *first* step to produce an equivalent equation that is "cleared of fractions."

To "clear fractions" in Example 1, we identify the LCD, 8, and use the multiplication principle. Because the LCD is a common multiple of the denominators, when that number is used to multiply the numerator of each fraction, the resulting fractions can then be simplified. An equivalent equation can then be written without using fractions. We demonstrate this approach by solving Examples 1 and 2 by clearing fractions.

Caution!

We can "clear fractions" in equations, not in expressions. Do not multiply to clear fractions when simplifying an expression.

Answer on page A-10

CHAPTER 4: Fraction Notation: Addition,
Subtraction, and Mixed Numerals

EXAMPLE 4 Solve Example 1 by clearing fractions:

$$\frac{3}{4}x - \frac{1}{8} = \frac{1}{2}.$$

The LCD is 8, so we begin by multiplying both sides of the equation by 8:

$$\frac{3}{4}x - \frac{1}{8} = \frac{1}{2}$$

$$8\left(\frac{3}{4}x - \frac{1}{8}\right) = 8 \cdot \frac{1}{2} \qquad \text{Using the multiplication principle}$$

$$\frac{8 \cdot 3}{4}x - 8 \cdot \frac{1}{8} = \frac{8}{2} \qquad \text{Using the distributive law}$$

$$\left.\begin{array}{c} \dfrac{\cancel{4} \cdot 2 \cdot 3}{\cancel{4}}x - 1 = 4 \\[2mm] 6x - 1 = 4 \end{array}\right\} \quad \begin{array}{l}\text{Factoring and simplifying. The equation}\\\text{is now cleared of fractions.}\end{array}$$

$$6x - 1 + 1 = 4 + 1 \qquad \text{Adding 1 to both sides}$$

$$6x = 5$$

$$\frac{6x}{6} = \frac{5}{6} \qquad \begin{array}{l}\text{Dividing both sides by 6 or}\\\text{multiplying both sides by }\frac{1}{6}\end{array}$$

$$x = \frac{5}{6}. \qquad \text{Simplifying}$$

Since $\frac{5}{6}$ was indeed the solution in Example 1, we have a check.

EXAMPLE 5 Solve Example 2 by clearing fractions:

$$5 + \frac{9}{2}t = -\frac{7}{2}.$$

The LCD is 2, so we begin by multiplying both sides of the equation by 2:

$$2\left(5 + \frac{9}{2}t\right) = 2\left(-\frac{7}{2}\right) \qquad \text{Using the multiplication principle}$$

$$2 \cdot 5 + \frac{2 \cdot 9}{2}t = -\frac{2 \cdot 7}{2} \qquad \text{Using the distributive law}$$

$$10 + 9t = -7 \qquad \begin{array}{l}\text{Simplifying and removing a factor}\\\text{equal to 1: }\frac{2}{2} = 1.\text{ The equation is}\\\text{now cleared of fractions.}\end{array}$$

$$10 + 9t - 10 = -7 - 10 \qquad \text{Subtracting 10 from both sides}$$

$$9t = -17 \qquad \text{Simplifying}$$

$$\frac{9t}{9} = -\frac{17}{9} \qquad \begin{array}{l}\text{Dividing both sides by 9 or multiplying}\\\text{both sides by }\frac{1}{9}\end{array}$$

$$t = -\frac{17}{9}. \qquad \text{Simplifying}$$

Since the solution in Example 2 is also $-\frac{17}{9}$, we have a check. The solution is $-\frac{17}{9}$.

Do Exercises 6 and 7.

Either of the methods discussed in this section can be used to solve equations that contain fractions, but for students planning to continue in algebra, it is important that *both* methods be thoroughly understood.

6. Solve Example 3 by clearing fractions:

$$20 = 6 - \frac{2}{3}x.$$

7. Solve Margin Exercise 1 by clearing fractions:

$$\frac{3}{5}t - \frac{8}{15} = \frac{2}{15}.$$

Answers on page A-10

4.4   Solving Equations: Using
the Principles Together

4.4

EXERCISE SET   For Extra Help

Math XL	MyMathLab	InterAct Math	Tutor Center	Video Lectures on CD Disc 2	Student's Solutions Manual
MathXL	MyMathLab	InterAct Math	Math Tutor Center		

a   Solve using the addition principle and/or the multiplication principle. Don't forget to check!

1. $6x - 3 = 15$

2. $7x - 6 = 22$

3. $5x + 7 = -8$

4. $19 = 2x - 7$

5. $31 = 3x - 5$

6. $2a - 9 = -7$

7. $4x - 5 = \dfrac{1}{3}$

8. $5x - 1 = \dfrac{2}{3}$

9. $\dfrac{3}{2}t - \dfrac{1}{4} = \dfrac{1}{2}$

10. $\dfrac{1}{4}t + \dfrac{1}{8} = \dfrac{1}{2}$

11. $\dfrac{2}{5}x + \dfrac{3}{10} = \dfrac{3}{5}$

12. $\dfrac{4}{3}x - \dfrac{2}{15} = \dfrac{2}{15}$

13. $5 - \dfrac{3}{4}x = 3$

14. $3 - \dfrac{2}{5}x = 6$

15. $-1 + \dfrac{2}{5}t = -\dfrac{4}{5}$

16. $-2 + \dfrac{1}{6}t = -\dfrac{7}{4}$

17. $12 = 8 + \dfrac{7}{2}t$

18. $7 = 5 + \dfrac{3}{2}t$

19. $-4 = \dfrac{2}{3}x - 7$

20. $-3 = \dfrac{2}{5}x - 4$

21. $7 = a + \dfrac{14}{5}$

22. $9 = a + \dfrac{47}{10}$

23. $\dfrac{2}{5}t - 1 = \dfrac{7}{5}$

24. $-\dfrac{53}{4} = \dfrac{3}{2}a + 2$

25. $\dfrac{39}{8} = \dfrac{11}{4} + \dfrac{1}{2}x$

26. $\dfrac{17}{2} = \dfrac{2}{7}t - \dfrac{3}{2}$

27. $\dfrac{13}{3}s + \dfrac{11}{2} = \dfrac{35}{4}$

28. $\dfrac{11}{5}t + \dfrac{36}{5} = \dfrac{7}{2}$

Solve by using the multiplication principle to clear fractions.

29. $\dfrac{1}{2}x - \dfrac{1}{4} = \dfrac{1}{2}$

30. $\dfrac{1}{3}x - \dfrac{1}{6} = \dfrac{2}{3}$

31. $7 = \dfrac{4}{9}t + 5$

32. $5 = \dfrac{4}{7}t + 3$

33. $-3 = \dfrac{3}{4}t - \dfrac{1}{2}$

34. $-2 = \dfrac{4}{3}t - \dfrac{5}{6}$

35. $\dfrac{4}{3} - \dfrac{5}{6}x = \dfrac{3}{2}$

36. $\dfrac{3}{2} - \dfrac{5}{3}x = \dfrac{5}{6}$

37. $-\dfrac{3}{4} = -\dfrac{5}{6} - \dfrac{1}{2}x$

38. $-\dfrac{1}{4} = -\dfrac{2}{3} - \dfrac{1}{6}x$

39. $\dfrac{4}{3} - \dfrac{1}{5}t = \dfrac{3}{4}$

40. $\dfrac{2}{5} - \dfrac{3}{4}t = \dfrac{4}{3}$

41. **D$_W$** Describe a procedure that a classmate could use to solve the equation $\dfrac{a}{b}x + c = d$ for x.

42. **D$_W$** Nathan begins solving the equation $-\dfrac{2}{3}x + 7 = -9$ by adding 9 to both sides. Is this a wise thing to do? Why or why not?

SKILL MAINTENANCE

Divide. [2.5a]

43. $39 \div (-3)$

44. $56 \div (-7)$

45. $(-72) \div (-4)$

46. $(-81) \div (-3)$

Solve. [2.3b]

47. Jeremy withdraws $200 from his bank's ATM (automated teller machine), makes a $90 deposit, and then withdraws another $40. How much has Jeremy's account balance changed?

48. Animal Instinct, a pet supply store, makes a profit of $850 on Friday and $375 on Saturday, but suffers a loss of $45 on Sunday. Find the total profit or loss for the three days.

Divide and simplify. [3.7b]

49. $\dfrac{10}{7} \div 2m$

50. $45n \div \dfrac{9}{4}$

51. **D**_{**W**} Emma begins solving the equation $\frac{2}{3}x - \frac{1}{5} = \frac{4}{7}$ by multiplying both sides by $\frac{3}{2}$. Is this a wise thing to do? Why or why not?

52. **D**_{**W**} Andrew begins solving the equation $\frac{2}{3}x + 1 = \frac{5}{6}$ by multiplying both sides by 24. Is this a wise thing to do? Why or why not?

Solve.

53. 🖩 $\dfrac{553}{2451}a - \dfrac{13}{57} = \dfrac{29}{43}$

54. 🖩 $\dfrac{1081}{3599}x - \dfrac{17}{61} = \dfrac{19}{59}$

55. 🖩 $\dfrac{71}{73} = \dfrac{19}{47} - \dfrac{53}{91}t$

56. 🖩 $\dfrac{23}{79} - \dfrac{41}{67}x = \dfrac{37}{83}$

57. $-\dfrac{a}{5} + \dfrac{31}{4} = \dfrac{16}{3}$

58. $\dfrac{47}{5} - \dfrac{a}{4} = \dfrac{44}{7}$

59. $\dfrac{49}{8} + \dfrac{2x}{9} = 4$

60. The perimeter of the figure shown is 15 cm. Solve for x.

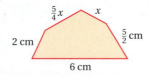

61. The perimeter of the figure is 21 cm. Solve for x.

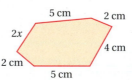

62. The perimeter of the figure is 6 ft. Solve for x.

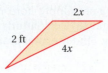

63. The perimeter of the figure is 15 cm. Solve for n.

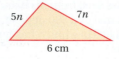

4.5 MIXED NUMERALS

Objectives

a Convert from mixed numerals to fraction notation.

b Convert from fraction notation to mixed numerals.

c Divide, writing a mixed numeral for the quotient.

a What Is a Mixed Numeral?

A symbol like $2\frac{3}{8}$ is called a **mixed numeral.**

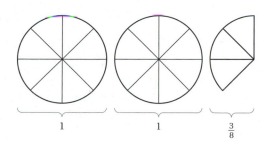

$$2\frac{3}{8} \quad \text{means} \quad 2 + \frac{3}{8}$$

This is a whole number.   This is a fraction less than 1.

The following figure illustrates one use of a mixed numeral in daily life. The bolt shown is $2\frac{3}{8}$ in. long. The length is given as a whole-number part, 2, and a fraction part, $\frac{3}{8}$. We can represent the measurement as $\frac{19}{8}$, but $2\frac{3}{8}$ makes the length easier to visualize and is thus more descriptive.

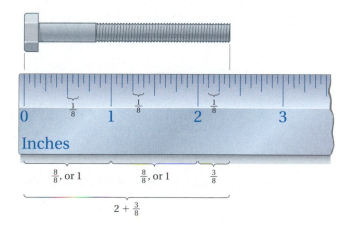

EXAMPLES Convert to a mixed numeral.

1. $7 + \frac{2}{5} = 7\frac{2}{5}$

2. $4 + \frac{3}{10} = 4\frac{3}{10}$

Do Exercises 1–3.

The notation $2\frac{3}{4}$ has a plus sign left out. To aid in understanding, we sometimes write the missing plus sign: $2 + \frac{3}{4}$. Similarly, the notation $-5\frac{2}{3}$ has a minus sign left out since $-5\frac{2}{3} = -\left(5 + \frac{2}{3}\right) = -5 - \frac{2}{3}$.

Mixed numerals can be displayed easily on a number line, as shown here.

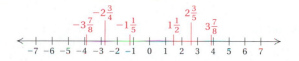

1. $1 + \frac{2}{3} = \boxed{}$ ____ Convert to a mixed numeral.

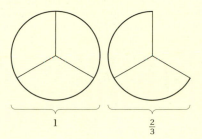

Convert to a mixed numeral.

2. $7 + \frac{1}{4}$ **3.** $15 + \frac{2}{9}$

Answers on page A-11

Study Tips

VISUALIZE

When studying for a quiz or test, don't feel that you need to redo every assigned problem. A more productive use of your time would be simply to reread the assigned problems, making certain that you can visualize the steps that lead to a solution. When you are unsure of how to solve a problem, redo that problem in its entirety, seeking outside help as needed.

Convert to fraction notation.

4. $4\dfrac{2}{5}$ **5.** $6\dfrac{1}{10}$

Convert to fraction notation. Use the faster method.

6. $3\dfrac{2}{7}$

7. $9\dfrac{1}{8}$

8. $20\dfrac{2}{3}$

Convert to fraction notation.

9. $-6\dfrac{2}{5}$

10. $-7\dfrac{2}{9}$

Answers on page A-11

CHAPTER 4: Fraction Notation: Addition, Subtraction, and Mixed Numerals

---

EXAMPLES   Convert to fraction notation.

3. $2\dfrac{3}{4} = 2 + \dfrac{3}{4}$ Inserting the missing plus sign

$\phantom{2\dfrac{3}{4}} = \dfrac{2}{1} + \dfrac{3}{4}$ $2 = \dfrac{2}{1}$

$\phantom{2\dfrac{3}{4}} = \dfrac{2}{1} \cdot \dfrac{4}{4} + \dfrac{3}{4}$ Finding a common denominator

$\phantom{2\dfrac{3}{4}} = \dfrac{8}{4} + \dfrac{3}{4}$

$\phantom{2\dfrac{3}{4}} = \dfrac{11}{4}$ Adding

4. $4\dfrac{3}{10} = 4 + \dfrac{3}{10} = \dfrac{4}{1} + \dfrac{3}{10} = \dfrac{4}{1} \cdot \dfrac{10}{10} + \dfrac{3}{10} = \dfrac{40}{10} + \dfrac{3}{10} = \dfrac{43}{10}$

Do Exercises 4 and 5.

Using Example 4, we can develop a faster way to convert.

> To convert from a mixed numeral like $4\dfrac{3}{10}$ to fraction notation:
>
>   (a) Multiply: $4 \cdot 10 = 40$.
>
>   (b) Add: $40 + 3 = 43$.
>
>   (c) Keep the denominator.
>
> $\quad 4\dfrac{3}{10} = \dfrac{43}{10}$

EXAMPLES   Convert to fraction notation.

5. $6\dfrac{2}{3} = \dfrac{20}{3}$ $6 \cdot 3 = 18;\;\; 18 + 2 = 20;\;\;$ keep the denominator.

6. $8\dfrac{2}{9} = \dfrac{74}{9}$ $9 \cdot 8 = 72;\;\; 72 + 2 = 74;\;\;$ keep the denominator.

7. $10\dfrac{7}{8} = \dfrac{87}{8}$ $8 \cdot 10 = 80;\;\; 80 + 7 = 87;\;\;$ keep the denominator.

Do Exercises 6–8.

To find the opposite of the number in Example 5, we can write either $-6\dfrac{2}{3}$ or $-\dfrac{20}{3}$. Thus, to convert a negative mixed numeral to fraction notation, we remove the negative sign for purposes of computation and then include it in the answer.

EXAMPLES   Convert to fraction notation.

8. $-5\dfrac{1}{3} = -\left(5 + \dfrac{1}{3}\right) = -\dfrac{16}{3}$ $3 \cdot 5 = 15;\;\; 15 + 1 = 16;\;\;$ include the negative sign

9. $-7\dfrac{5}{6} = -\left(7 + \dfrac{5}{6}\right) = -\dfrac{47}{6}$ $6 \cdot 7 = 42;\;\; 42 + 5 = 47$

Do Exercises 9 and 10.

b | Writing Mixed Numerals

We can find a mixed numeral for $\frac{5}{3}$ as follows:

$$\frac{5}{3} = \frac{3}{3} + \frac{2}{3} = 1 + \frac{2}{3} = 1\frac{2}{3}.$$

We can also visualize $\frac{5}{3}$ as one-third of 5 objects, as shown below.

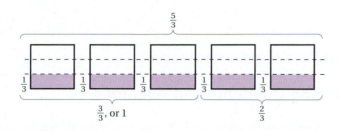

Note that one third of 5 is the same as 5 divided by 3.

> To convert from fraction notation to a mixed numeral, divide.
>
> $$\frac{19}{7}; \qquad 7\overline{)19}$$
> $$\frac{14}{5} \rightarrow \text{The remainder}$$
>
> The quotient $\rightarrow 2\frac{5}{7} \leftarrow$
>
> $$\frac{19}{7} = 2\frac{5}{7}$$

EXAMPLES Convert to a mixed numeral.

10. $\frac{8}{5}$ $5\overline{)8}$ $\frac{8}{5} = 1\frac{3}{5}$
$$\frac{5}{3}$$

> A fraction larger than 1, such as $\frac{8}{5}$, is sometimes referred to as an "improper" fraction. However, the use of notation such as $\frac{8}{5}$, $\frac{69}{10}$, and so on, is quite proper and very common in algebra.

11. $\frac{69}{10}$ $10\overline{)69}$ $\frac{69}{10} = 6\frac{9}{10}$
$$\frac{60}{9}$$

Do Exercises 11 and 12.

Whenever the fraction part of the mixed numeral can be simplified, it is important to do so.

EXAMPLE 12 Convert to a mixed numeral: $\frac{122}{8}$.

$$8\overline{)122} \qquad \frac{122}{8} = 15\frac{2}{8} = 15\frac{1}{4}$$
$$\begin{array}{r} 15 \\ \underline{80} \\ 42 \\ \underline{40} \\ 2 \end{array}$$

Do Exercise 13.

Convert to a mixed numeral.

11. $\frac{7}{3}$

12. $\frac{16}{15}$

Convert to a mixed numeral.

13. $\frac{110}{6}$

Convert to a mixed numeral.

14. $\frac{-12}{5}$

15. $-\frac{134}{12}$

16. Divide. Write a mixed numeral for the answer.

$$6\overline{)4846}$$

17. Over the last 4 yr, Roland Thompson's raspberry patch has yielded 48, 35, 65, and 75 qt of berries. Find the average yield for the four years.

Answers on page A-11

The same procedure also works with negative numbers. Of course, the result will be a negative mixed numeral.

EXAMPLE 13 Convert $\dfrac{-9}{4}$ to a mixed numeral.

$$\text{Since } \quad 4\overline{)9} \atop \underline{8} \atop 1 \quad \text{, } \quad \text{we have} \quad \frac{9}{4} = 2\frac{1}{4}. \quad \text{Thus,} \quad \frac{-9}{4} = -2\frac{1}{4}.$$

Do Exercises 14 and 15 on the preceding page.

C   Finding Quotients and Averages

It is quite common when performing long division to express the quotient as a mixed numeral. As in Examples 10–13, the remainder becomes the numerator of the fraction part of the mixed numeral.

EXAMPLE 14 Divide. Write a mixed numeral for the quotient.

$$7\overline{)6341}$$

We first divide as usual.

$$\begin{array}{r} 905 \\ 7\overline{)6341} \\ \underline{6300} \\ 41 \\ \underline{35} \\ 6 \end{array}$$

The answer is 905 R 6, or $905\frac{6}{7}$. Using fraction notation, we write $\frac{6341}{7} = 905\frac{6}{7}$.

Do Exercise 16 on the preceding page.

EXAMPLE 15 *Nutrition.* The Center for Science in the Public Interest rates the following five meals from fast food restaurants as "Better Bites" based on their nutritional value. This list gives the total fat, in grams, contained in each item. How much fat is contained, on average, in these items?
Source: *Nutrition Action Healthletter,* March 2005

Burger King Chicken or Shrimp Salad	18 g
Wendy's Chili, small	6 g
Wendy's Broccoli & Cheese Potato	15 g
Wendy's Chili, large	9 g
Wendy's Sour Cream & Chives Potato	6 g

Recall from Section 1.9 that to find the *average* of a set of values, we add the values and divide that sum by the number of values being added.

$$\text{Average fat grams} = \frac{18 \text{ g} + 6 \text{ g} + 15 \text{ g} + 9 \text{ g} + 6 \text{ g}}{5} = \frac{54 \text{ g}}{5} = 10\frac{4}{5} \text{ g}$$

On average, these foods contain $10\frac{4}{5}$ g of fat per item.

Do Exercise 17 on the preceding page.

a   Convert to fraction notation.

1. $7\frac{2}{3}$

2. $6\frac{2}{5}$

3. $6\frac{1}{4}$

4. $8\frac{1}{2}$

5. $-20\frac{1}{8}$

6. $-10\frac{1}{3}$

7. $5\frac{1}{10}$

8. $8\frac{1}{10}$

9. $20\frac{3}{5}$

10. $30\frac{4}{5}$

11. $-8\frac{2}{7}$

12. $-8\frac{7}{8}$

13. $6\frac{9}{10}$

14. $1\frac{3}{5}$

15. $-12\frac{3}{4}$

16. $-15\frac{2}{3}$

17. $5\frac{7}{10}$

18. $7\frac{3}{100}$

19. $-5\frac{7}{100}$

20. $-6\frac{4}{15}$

b   Convert to a mixed numeral.

21. $\frac{16}{3}$

22. $\frac{19}{8}$

23. $\frac{45}{6}$

24. $\frac{30}{9}$

25. $\frac{57}{10}$

26. $\frac{-89}{10}$

27. $\frac{65}{9}$

28. $\frac{65}{8}$

29. $\frac{-33}{6}$

30. $\frac{-50}{8}$

31. $\frac{46}{4}$

32. $\frac{39}{9}$

33. $\frac{-12}{8}$

34. $-\frac{57}{6}$

35. $\frac{307}{5}$

36. $\frac{227}{4}$

37. $-\frac{413}{50}$

38. $\frac{467}{100}$

c   Divide. Write a mixed numeral for the answer.

39. $8\overline{)869}$

40. $3\overline{)2126}$

41. $7\overline{)6345}$

42. $9\overline{)9110}$

43. $21\overline{)852}$

44. $85\overline{)7672}$

45. $-302 \div 15$

46. $-475 \div 13$

47. $471 \div (-21)$

48. $542 \div (-25)$

Charities. One way to rate the efficiency of charitable organizations is to calculate how much of the donations are spent on fundraising. The table at right lists some charitable organizations, and how much of every $100 donated is spent on fundraising.

Source: www.forbes.com/lists/2004/11/23/04charityland.html

TYPE	CHARITY	FUNDRAISING EXPENSES per $100
Arts	Nashville Symphony	$1
Health	American Kidney Foundation	$4
Arts	National Public Radio	$5
Health	Shriners Hospital for Children	$6
Health	National Mental Health Association	$7
Children	Marine Toys for Tots Foundation	$8
Museum	American Museum of Natural History	$9
Children	Feed the Children	$9
Arts	Los Angeles Philharmonic Association	$9
Children	Boys & Girls Clubs of America	$10
Health	Children's Hospital	$10
Environment	Nature Conservancy	$11
Museum	Philadelphia Museum of Art	$13
Health	National Kidney Foundation	$14
Environment	World Wildlife Fund	$14
Environment	National Audubon Society	$22
Environment	National Wildlife Federation	$23

49. What is the average fundraising expense per $100 of the health organizations in the list?

50. What is the average fundraising expense per $100 of the environment organizations in the list?

51. What is the average fundraising expense per $100 of the first 6 organizations in the list?

52. What is the average fundraising expense per $100 of the last 8 organizations in the list?

53. $\mathbf{D_W}$ Describe in your own words a method for rewriting a mixed numeral as a fraction.

54. $\mathbf{D_W}$ Describe in your own words a method for rewriting a fraction as a mixed numeral.

Multiply and simplify. [3.6a]

55. $\dfrac{7}{9} \cdot \dfrac{24}{21}$

56. $\dfrac{5}{12} \cdot \dfrac{9}{10}$

57. $\dfrac{7}{10} \cdot \dfrac{5}{14}$

58. $\dfrac{21}{35} \cdot \dfrac{25}{12}$

59. $-\dfrac{17}{25} \cdot \dfrac{15}{34}$

60. $\dfrac{7}{20} \cdot \left(-\dfrac{45}{49}\right)$

61. $\mathbf{D_W}$ Toni claims that $3\frac{1}{5}$ is the reciprocal of $5\frac{1}{3}$. How can you convince her that she is mistaken?

62. $\mathbf{D_W}$ Are the numbers $2\frac{1}{3}$ and $2 \cdot \frac{1}{3}$ equal? Why or why not?

Write a mixed numeral for each number or sum listed.

63. 🖩 $\dfrac{128,236}{541}$

64. 🖩 $\dfrac{103,676}{349}$

65. $\dfrac{56}{7} + \dfrac{2}{3}$

66. $\dfrac{72}{12} + \dfrac{5}{6}$

67. $\dfrac{12}{5} + \dfrac{19}{15}$

68. How many weeks are in a leap year?

69. How many weeks are in a year?

70. _Athletics._ At a track and field meet, the hammer that is thrown has a wire length ranging from 3 ft $10\frac{1}{4}$ in. to 3 ft $11\frac{3}{4}$ in., a $4\frac{1}{8}$-in. grip, and a 16-lb ball with a diameter of $4\frac{3}{8}$ in. to $5\frac{1}{8}$ in. Give specifications for the wire length and diameter of an "average" hammer.

CHAPTER 4: Fraction Notation: Addition, Subtraction, and Mixed Numerals

ADDITION AND SUBTRACTION OF MIXED NUMERALS; APPLICATIONS

Objectives

a Add using mixed numerals.

b Subtract using mixed numerals.

c Solve applied problems involving addition and subtraction with mixed numerals.

d Add and subtract using negative mixed numerals.

a Addition Using Mixed Numerals

To find the sum $1\frac{5}{8} + 3\frac{1}{8}$, we first add the fractions. Then we add the whole numbers and, if possible, simplify the fraction part.

$$
\begin{array}{r}
1\frac{5}{8} = \\
+\ 3\frac{1}{8} = \\
\hline
\frac{6}{8}
\end{array}
\qquad
\begin{array}{r}
1\frac{5}{8} \\
+\ 3\frac{1}{8} \\
\hline
4\frac{6}{8} = 4\frac{3}{4}
\end{array}
$$

↑ Add the fractions.   ↑ Add the whole numbers.   └─ Simplifying

Do Exercises 1 and 2.

Recall that the fraction part of a mixed numeral should always be less than 1.

EXAMPLE 1   Add: $5\frac{2}{3} + 3\frac{5}{6}$. Write a mixed numeral for the answer.

The LCD is 6.

$$
\begin{array}{r}
5\frac{2}{3}\cdot\frac{2}{2} = 5\frac{4}{6} \\
+\ 3\frac{5}{6} = +\ 3\frac{5}{6} \\
\hline
8\frac{9}{6} = 8 + \frac{9}{6} \\
= 8 + 1\frac{1}{2} \\
= 9\frac{1}{2}
\end{array}
$$

To find a mixed numeral for $\frac{9}{6}$, we divide:

$$
\begin{array}{r}
1 \\
6\overline{)9} \\
6 \\
\hline
3
\end{array}
\qquad \frac{9}{6} = 1\frac{3}{6} = 1\frac{1}{2}
$$

$\frac{19}{2}$ is also a correct answer, but it is not a mixed numeral, which is what we are working with in Sections 4.5, 4.6, and 4.7.

Do Exercise 3.

EXAMPLE 2   Add: $10\frac{5}{6} + 7\frac{3}{8}$.

The LCD is 24.

$$
\begin{array}{r}
10\frac{5}{6}\cdot\frac{4}{4} = 10\frac{20}{24} \\
+\ 7\frac{3}{8}\cdot\frac{3}{3} = +\ 7\frac{9}{24} \\
\hline
17\frac{29}{24} = 18\frac{5}{24}
\end{array}
$$

Do Exercise 4.

1. Add.

$$
\begin{array}{r}
4 \\
+\ 5\frac{1}{3} \\
\hline
\end{array}
$$

2. Add.

$$
\begin{array}{r}
2\frac{3}{10} \\
+\ 7\frac{1}{10} \\
\hline
\end{array}
$$

3. Add.

$$
\begin{array}{r}
8\frac{2}{5} \\
+\ 3\frac{7}{10} \\
\hline
\end{array}
$$

4. Add.

$$
\begin{array}{r}
5\frac{3}{4} \\
+\ 8\frac{5}{6} \\
\hline
\end{array}
$$

Answers on page A-11

5. $10\dfrac{7}{8}$

$-\ \ 9\dfrac{3}{8}$

6. $8\dfrac{2}{3}$

$-\ 5\dfrac{1}{2}$

Subtract.

7. 5

$-\ 1\dfrac{1}{3}$

8. $8\dfrac{1}{9}$

$-\ 4\dfrac{5}{6}$

Answers on page A-11

b   Subtraction Using Mixed Numerals

EXAMPLE 3   Subtract: $7\dfrac{3}{4} - 2\dfrac{1}{4}$.

$$
\begin{array}{ll}
7\ \dfrac{3}{4} = & 7\ \dfrac{3}{4} \\[2mm]
-\ 2\ \dfrac{1}{4} = & -\ 2\ \dfrac{1}{4} \\[2mm]
\hline
\dfrac{2}{4} & 5\ \dfrac{2}{4} = 5\dfrac{1}{2} \leftarrow \text{Simplify.}
\end{array}
$$

Subtract the   Subtract the
fractions.   whole numbers.

EXAMPLE 4   Subtract: $9\dfrac{4}{5} - 3\dfrac{1}{2}$.

The LCD is 10.

$$
\begin{array}{ll}
9\ \dfrac{4}{5} \cdot \dfrac{2}{2} = & 9\dfrac{8}{10} \\[3mm]
-\ 3\ \dfrac{1}{2} \cdot \dfrac{5}{5} = & -\ 3\dfrac{5}{10} \\[2mm]
\hline
& 6\dfrac{3}{10}
\end{array}
$$

Do Exercises 5 and 6.

EXAMPLE 5   Subtract: $13 - 9\dfrac{3}{8}$.

$$
\begin{array}{ll}
13\ \ = & 12\dfrac{8}{8} \\[3mm]
-\ 9\dfrac{3}{8} = & -\ 9\dfrac{3}{8} \\[2mm]
\hline
& 3\dfrac{5}{8}
\end{array}
$$

We "borrow" 1, or $\frac{8}{8}$, from 13:
$13 = 12 + 1 = 12 + \frac{8}{8} = 12\frac{8}{8}$.

EXAMPLE 6   Subtract: $7\dfrac{1}{6} - 2\dfrac{1}{4}$.

The LCD is 12.

$$
\begin{array}{ll}
7\ \dfrac{1}{6} \cdot \dfrac{2}{2} = & 7\dfrac{2}{12} \\[3mm]
-\ 2\ \dfrac{1}{4} \cdot \dfrac{3}{3} = & -\ 2\dfrac{3}{12}
\end{array}
$$

To subtract $\frac{3}{12}$ from $\frac{2}{12}$,
we borrow 1, or $\frac{12}{12}$, from 7:
$7\frac{2}{12} = 6 + 1 + \frac{2}{12} = 6 + \frac{12}{12} + \frac{2}{12} = 6\frac{14}{12}$.

We can write this as

$$
\begin{array}{ll}
7\dfrac{2}{12} = & 6\dfrac{14}{12} \\[3mm]
-\ 2\dfrac{3}{12} = & -\ 2\dfrac{3}{12} \\[2mm]
\hline
& 4\dfrac{11}{12}.
\end{array}
$$

Do Exercises 7 and 8.

To combine like terms, we use the distributive law and add or subtract.

EXAMPLE 7 Combine like terms: **(a)** $9\frac{3}{4}x - 4\frac{1}{2}x$; **(b)** $4\frac{5}{6}t + 2\frac{7}{9}t$.

a) $9\dfrac{3}{4}x - 4\dfrac{1}{2}x = \left(9\dfrac{3}{4} - 4\dfrac{1}{2}\right)x$ Using the distributive law; this is often done mentally.

$\qquad\qquad\qquad = \left(9\dfrac{3}{4} - 4\dfrac{2}{4}\right)x$ The LCD is 4.

$\qquad\qquad\qquad = 5\dfrac{1}{4}x$ Subtracting

b) $4\dfrac{5}{6}t + 2\dfrac{7}{9}t = \left(4\dfrac{5}{6} + 2\dfrac{7}{9}\right)t$ This step is often performed mentally.

$\qquad\qquad\qquad = \left(4\dfrac{15}{18} + 2\dfrac{14}{18}\right)t$ The LCD is 18.

$\qquad\qquad\qquad = 6\dfrac{29}{18}t = 7\dfrac{11}{18}t$

Do Exercises 9–11.

C   Applications and Problem Solving

EXAMPLE 8 *Widening a Driveway.* Sherry and Woody are widening their existing $17\frac{1}{4}$-ft driveway by adding $5\frac{9}{10}$ ft on one side. What is the width of the new driveway?

1. **Familiarize.** We let $w =$ the width of the new driveway.
2. **Translate.** We translate as follows.

Existing width	$+$	Additional width	$=$	New width
$\downarrow$	$\downarrow$	$\downarrow$	$\downarrow$	$\downarrow$
$17\frac{1}{4}$	$+$	$5\frac{9}{10}$	$=$	w

3. **Solve.** The translation tells us what to do. We add. The LCD is 20.

$$17\frac{1}{4} = \quad 17\,\frac{1}{4}\cdot\frac{5}{5} = \quad 17\frac{5}{20}$$

$$+\;5\frac{9}{10} = +\;5\,\frac{9}{10}\cdot\frac{2}{2} = +\;5\frac{18}{20}$$

$$\overline{\qquad\qquad\qquad\qquad\qquad\quad 22\frac{23}{20} = 23\frac{3}{20}}$$

Thus, $w = 23\frac{3}{20}$.

Combine like terms.

9. $7\dfrac{1}{6}t + 5\dfrac{2}{3}t$

10. $7\dfrac{11}{12}x - 5\dfrac{2}{3}x$

11. $5\dfrac{11}{15}x + 8\dfrac{3}{10}x$

Answers on page A-11

Study Tips

USE YOUR E-MAIL

Many students overlook an excellent opportunity to get questions cleared up—e-mail. If your instructor makes his or her e-mail address available, consider using it to get help. You can also contact the AW Tutor Center by e-mail. Often, just the act of writing out your question brings some clarity. If you do use e-mail, allow some time for your instructor to reply.

12. Travel Distance. Chrissy Jenkins is a college textbook sales representative for Addison-Wesley. On two business days, Chrissy drove $144\frac{9}{10}$ mi and $87\frac{1}{4}$ mi. What was the total distance Chrissy drove?

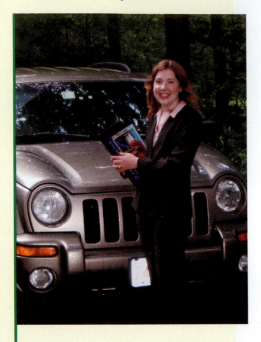

13. There are $20\frac{1}{3}$ gal of water in a rainbarrel; $5\frac{3}{4}$ gal are poured out and $8\frac{2}{3}$ gal are returned after a heavy rainfall. How many gallons of water are then in the barrel?

4. Check. We check by repeating the calculation. We also note that the answer is larger than either of the widths, which means that the answer is reasonable.

5. State. The width of the new driveway is $23\frac{3}{20}$ ft.

Do Exercise 12.

■ **EXAMPLE 9** *Carpentry.* The following illustration shows the layout for the construction of a desk drawer. Find a, the width of the slot.

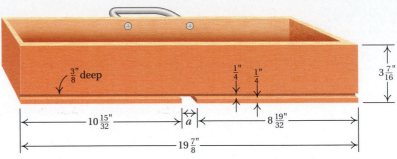

Middle Drawer / Back Layout

1. **Familiarize.** Measurement a is shown in the drawing. It is part of a length totaling $19\frac{7}{8}$".

2. **Translate.** We rephrase and translate as follows.

Rephrase: The total length is the sum of three measurements

Translate: $19\frac{7}{8}$ = $10\frac{15}{32} + a + 8\frac{19}{32}$

3. **Solve.** We solve for a:

$$19\frac{7}{8} = 10\frac{15}{32} + a + 8\frac{19}{32}$$

$$19\frac{7}{8} = a + 10\frac{15}{32} + 8\frac{19}{32} \qquad \text{\color{red}Using a commutative law}$$

$$19\frac{7}{8} = a + 18\frac{34}{32}$$

$$19\frac{7}{8} = a + 19\frac{1}{16} \qquad \text{\color{red}$\frac{34}{32} = 1\frac{2}{32} = 1\frac{1}{16}$}$$

$$19\frac{7}{8} - 19\frac{1}{16} = a + 19\frac{1}{16} - 19\frac{1}{16} \qquad \text{\color{red}Subtracting $19\frac{1}{16}$ from both sides}$$

$$\frac{13}{16} = a. \qquad \text{\color{red}$\frac{7}{8} - \frac{1}{16} = \frac{14}{16} - \frac{1}{16} = \frac{13}{16}$}$$

4. **Check.** We check by repeating the calculation, or by adding the three measures:

$$10\frac{15}{32} + 8\frac{19}{32} + \frac{13}{16} = 19\frac{7}{8}.$$

5. **State.** The length a in the diagram is $\frac{13}{16}$".

Do Exercise 13.

d Negative Mixed Numerals

Consider the numbers $5\frac{3}{4}$ and $-5\frac{3}{4}$ on a number line.

Note that just as $5\frac{3}{4}$ means $5 + \frac{3}{4}$, we can regard $-5\frac{3}{4}$ as $-5 - \frac{3}{4}$.

To subtract a larger number from a smaller number, we must modify the approach of Examples 3–6. To see why, consider the subtraction $4 - 4\frac{1}{2}$. We know that if we have \$4 and make a \$$4\frac{1}{2}$ purchase, we will owe half a dollar. Thus,

$$4 - 4\frac{1}{2} = -\frac{1}{2}.$$

The correct answer, $-\frac{1}{2}$, can be obtained by rewriting the subtraction as addition (see Section 2.3):

$$4 - 4\frac{1}{2} = 4 + \left(-4\frac{1}{2}\right).$$

Because $-4\frac{1}{2}$ has the greater absolute value, the answer will be negative. The difference in absolute value is $4\frac{1}{2} - 4 = \frac{1}{2}$, so

$$4 - 4\frac{1}{2} = -\frac{1}{2}.$$

Another way to see this is to convert to fraction notation and subtract:

$$4 - 4\frac{1}{2} = \frac{8}{2} - \frac{9}{2} = \frac{8}{2} + \left(-\frac{9}{2}\right) = -\frac{1}{2}.$$

Do Exercise 14.

EXAMPLE 10 Subtract: $3\frac{2}{7} - 4\frac{2}{5}$.

Since $4\frac{2}{5}$ is greater than $3\frac{2}{7}$, the answer will be negative. We can also see this by rewriting the subtraction as $3\frac{2}{7} + \left(-4\frac{2}{5}\right)$. The difference in absolute values is

$$
\begin{array}{rcl}
4\frac{2}{5} = & 4\ \dfrac{2}{5} \cdot \dfrac{7}{7} = & 4\dfrac{14}{35} \\[2ex]
-\ 3\frac{2}{7} = & -\ 3\ \dfrac{2}{7} \cdot \dfrac{5}{5} = & -\ 3\dfrac{10}{35} \\[2ex]
\hline
& & 1\dfrac{4}{35}.
\end{array}
$$

Because $-4\frac{2}{5}$ has the larger absolute value, we make the answer negative.

Thus, $3\frac{2}{7} - 4\frac{2}{5} = -1\frac{4}{35}$. ←

Do Exercises 15 and 16.

---

14. Subtract: $7 - 7\frac{3}{4}$.

Subtract.

15. $5\frac{1}{2} - 9\frac{3}{4}$

16. $4\frac{2}{3} - 7\frac{1}{6}$

Answers on page A-11

Subtract.

17. $-7\frac{1}{3} - \left(-5\frac{1}{2}\right)$

18. $-4\frac{1}{10} - \left(-7\frac{2}{5}\right)$

19. Subtract: $-7\frac{1}{10} - 6\frac{2}{15}$.

EXAMPLE 11 Subtract: $-6\frac{4}{5} - \left(-9\frac{3}{10}\right)$.

We write the subtraction as addition:

$$-6\frac{4}{5} - \left(-9\frac{3}{10}\right) = -6\frac{4}{5} + 9\frac{3}{10}.$$ Instead of subtracting, we add the opposite.

Since $9\frac{3}{10}$ has the greater absolute value, the answer will be positive. The difference in absolute values is

$$
\begin{array}{rcccccc}
9\dfrac{3}{10} = & 9\dfrac{3}{10} & = & 9\dfrac{3}{10} & = & 8\dfrac{13}{10} \\[2ex]
-6\dfrac{4}{5} = & -6\,\dfrac{4}{5}\cdot\dfrac{2}{2} & = & -6\dfrac{8}{10} & = & -6\dfrac{8}{10} \\[2ex]
\hline
 & & & & & 2\dfrac{5}{10} = 2\dfrac{1}{2}.
\end{array}
$$

Thus, $-6\frac{4}{5} - \left(-9\frac{3}{10}\right) = 2\frac{1}{2}$.

We can check by converting to fraction notation and redoing the calculation:

$$
\begin{aligned}
-6\frac{4}{5} - \left(-9\frac{3}{10}\right) &= -\frac{34}{5} - \left(-\frac{93}{10}\right) \\[1ex]
&= -\frac{34}{5} + \frac{93}{10} & \text{Adding the opposite} \\[1ex]
&= -\frac{68}{10} + \frac{93}{10} & \text{Writing with a common denominator} \\[1ex]
&= \frac{25}{10} = \frac{5}{2} = 2\frac{1}{2}. & -68 + 93 = 25
\end{aligned}
$$

Do Exercises 17 and 18.

In Section 2.2, we saw that to add two negative numbers we add absolute values and make the answer negative. The same approach is used with mixed numerals.

EXAMPLE 12 Subtract: $-4\frac{1}{6} - 5\frac{2}{9}$.

We rewrite the subtraction as addition:

$$
\begin{aligned}
-4\frac{1}{6} - 5\frac{2}{9} &= -4\frac{1}{6} + \left(-5\frac{2}{9}\right) & \text{Instead of subtracting, we add the opposite.} \\[1ex]
&= -\left(4\frac{1}{6} + 5\frac{2}{9}\right) & \text{The LCD is 18.} \\[1ex]
&= -\left(4\frac{3}{18} + 5\frac{4}{18}\right) & \frac{1}{6}\cdot\frac{3}{3} = \frac{3}{18};\ \ \frac{2}{9}\cdot\frac{2}{2} = \frac{4}{18} \\[1ex]
&= -9\frac{7}{18}.
\end{aligned}
$$

Thus, $-4\frac{1}{6} - 5\frac{2}{9} = -9\frac{7}{18}$.

Do Exercise 19.

Answers on page A-11

CALCULATOR CORNER

Adding or Subtracting Mixed Numerals To add or subtract mixed numerals, we use the $\boxed{a^{b/c}}$ key on a scientific calculator or the FRAC option of the $\boxed{\text{MATH}}$ key on a graphing calculator.

To find $4\frac{2}{5} + 8\frac{1}{3}$ on a scientific calculator, we press

$\boxed{4}\ \boxed{a^{b/c}}\ \boxed{2}\ \boxed{a^{b/c}}\ \boxed{5}\ \boxed{+}\ \boxed{8}\ \boxed{a^{b/c}}\ \boxed{1}\ \boxed{a^{b/c}}\ \boxed{3}\ \boxed{\text{ENTER} =}$.

The display

$\boxed{12\ \lrcorner 11\ \lrcorner 15}$ or $\boxed{12\ \sqcup 11/15}$

indicates that the sum is $12\frac{11}{15}$.

Some calculators are capable of displaying mixed numerals in the way in which we write them, as shown below.

On calculators without an $\boxed{a^{b/c}}$ key, mixed numerals can be entered as sums. For example, $1\frac{1}{3}$ is entered by pressing

$\boxed{1}\ \boxed{+}\ \boxed{1}\ \boxed{\div}\ \boxed{3}$.

When subtracting a mixed numeral written as a sum, parentheses are essential. To be safe, you can enclose every mixed numeral with parentheses. Thus, to find $1\frac{1}{3} - 4\frac{5}{8}$, we press

$\boxed{(}\ \boxed{1}\ \boxed{+}\ \boxed{1}\ \boxed{\div}\ \boxed{3}\ \boxed{)}\ \boxed{-}\ \boxed{(}\ \boxed{4}\ \boxed{+}\ \boxed{5}\ \boxed{\div}\ \boxed{8}\ \boxed{)}\ \boxed{\text{ENTER}}$.

The answer is negative and is written as a decimal number. Convert to fraction notation by using the FRAC option of the $\boxed{\text{MATH}}$ key, then to a mixed numeral as described in Section 4.5. The difference is $-3\frac{7}{24}$. Consult your owner's manual or an instructor if you need help with your particular model.

Exercises: Perform each calculation. Give the answer as a mixed numeral.

1. $4\frac{1}{3} + 5\frac{4}{5}$ 4. $5\frac{3}{20} + 2\frac{11}{12}$

2. $9\frac{2}{7} - 8\frac{1}{4}$ 5. $8\frac{17}{19} - 9\frac{2}{11}$

3. $7\frac{2}{9} - 5\frac{1}{7}$ 6. $6\frac{13}{15} - 9\frac{2}{17}$

a , **b**   Perform the indicated operation. Write a mixed numeral for each answer.

1. $\begin{array}{r} 6 \\ + 5\frac{2}{5} \\ \hline \end{array}$

2. $\begin{array}{r} 3 \\ + 6\frac{5}{7} \\ \hline \end{array}$

3. $\begin{array}{r} 2\frac{7}{8} \\ + 6\frac{5}{8} \\ \hline \end{array}$

4. $\begin{array}{r} 2\frac{5}{6} \\ + 5\frac{5}{6} \\ \hline \end{array}$

5. $\begin{array}{r} 4\frac{1}{4} \\ + 1\frac{2}{3} \\ \hline \end{array}$

6. $\begin{array}{r} 4\frac{1}{3} \\ + 5\frac{2}{9} \\ \hline \end{array}$

7. $\begin{array}{r} 7\frac{3}{4} \\ + 5\frac{5}{6} \\ \hline \end{array}$

8. $\begin{array}{r} 4\frac{3}{8} \\ + 6\frac{5}{12} \\ \hline \end{array}$

9. $\begin{array}{r} 3\frac{2}{5} \\ + 8\frac{7}{10} \\ \hline \end{array}$

10. $\begin{array}{r} 5\frac{1}{2} \\ + 3\frac{7}{10} \\ \hline \end{array}$

11. $\begin{array}{r} 6\frac{3}{8} \\ + 10\frac{5}{6} \\ \hline \end{array}$

12. $\begin{array}{r} \frac{5}{8} \\ + 1\frac{5}{6} \\ \hline \end{array}$

13. $\begin{array}{r} 18\frac{4}{5} \\ + 2\frac{7}{10} \\ \hline \end{array}$

14. $\begin{array}{r} 15\frac{5}{8} \\ + 11\frac{3}{4} \\ \hline \end{array}$

15. $\begin{array}{r} 14\frac{5}{8} \\ + 13\frac{1}{4} \\ \hline \end{array}$

16. $\begin{array}{r} 16\frac{1}{4} \\ + 15\frac{7}{8} \\ \hline \end{array}$

17. $\begin{array}{r} 4\frac{1}{5} \\ - 2\frac{3}{5} \\ \hline \end{array}$

18. $\begin{array}{r} 5\frac{1}{8} \\ - 2\frac{3}{8} \\ \hline \end{array}$

19. $\begin{array}{r} 9\frac{3}{5} \\ - 3\frac{1}{2} \\ \hline \end{array}$

20. $\begin{array}{r} 8\frac{2}{3} \\ - 7\frac{1}{2} \\ \hline \end{array}$

21. $\begin{array}{r} 34\frac{1}{3} \\ - 12\frac{5}{8} \\ \hline \end{array}$

22. $\begin{array}{r} 23\frac{5}{16} \\ - 16\frac{3}{4} \\ \hline \end{array}$

23. $\begin{array}{r} 19 \\ - 5\frac{3}{4} \\ \hline \end{array}$

24. $\begin{array}{r} 17 \\ - 3\frac{7}{8} \\ \hline \end{array}$

25. $\begin{array}{r} 34 \\ - 18\frac{5}{8} \\ \hline \end{array}$

26. $\begin{array}{r} 23 \\ - 19\frac{3}{4} \\ \hline \end{array}$

27. $\begin{array}{r} 21\frac{1}{6} \\ - 13\frac{3}{4} \\ \hline \end{array}$

28. $\begin{array}{r} 42\frac{1}{10} \\ - 23\frac{7}{12} \\ \hline \end{array}$

29. $\begin{array}{r} 25\frac{1}{9} \\ - 13\frac{5}{6} \\ \hline \end{array}$

30. $\begin{array}{r} 23\frac{5}{16} \\ - 14\frac{7}{12} \\ \hline \end{array}$

Combine like terms.

31. $1\frac{3}{14}t + 7\frac{2}{21}t$

32. $6\frac{1}{2}x + 8\frac{3}{4}x$

33. $9\frac{1}{2}x - 7\frac{3}{8}x$

34. $7\frac{3}{4}x - 2\frac{3}{8}x$

35. $5\frac{9}{10}t + 2\frac{7}{8}t$

36. $9\frac{2}{7}x + 2\frac{3}{8}x$

37. $37\frac{5}{9}t - 25\frac{4}{5}t$

38. $23\frac{1}{6}t - 19\frac{2}{5}t$

39. $2\frac{5}{6}x + 3\frac{1}{3}x$

40. $7\frac{3}{20}t + 1\frac{2}{15}t$

41. $1\frac{3}{11}x + 8\frac{2}{3}x$

42. $6\frac{11}{12}t + 3\frac{7}{10}t$

Solve.

43. *Sewing from a Pattern.* Regan wants to make an outfit in size 8. Using 45-in. fabric, she needs $1\frac{3}{8}$ yd for the dress, $\frac{5}{8}$ yd of contrasting fabric for the band at the bottom, and $3\frac{3}{8}$ yd for the jacket. How many yards in all of 45-in. fabric are needed to make the outfit?

44. *Sewing from a Pattern.* Gloria wants to make an outfit in size 12. Using 45-in. fabric, she needs $2\frac{3}{4}$ yd for the dress and $3\frac{1}{2}$ yd for the jacket. How many yards in all of 45-in. fabric are needed to make the outfit?

45. For a family barbecue, Kayla bought packages of hamburger weighing $1\frac{2}{3}$ lb and $5\frac{3}{4}$ lb. What was the total weight of the meat?

46. Marsha's Butcher Shop sold packages of sliced turkey breast weighing $1\frac{1}{3}$ lb and $4\frac{3}{5}$ lb. What was the total weight of the meat?

47. *Heights.* Juan is $187\frac{1}{10}$ cm tall and his daughter is $180\frac{3}{4}$ cm tall. How much taller is Juan?

48. *Heights.* Aunt Louise is $168\frac{1}{4}$ cm tall and her son is $150\frac{7}{10}$ cm tall. How much taller is Aunt Louise?

49. *Plumbing.* A plumber uses pipes of lengths $10\frac{5}{16}$ in. and $8\frac{3}{4}$ in. when installing a sink. How much pipe is used?

50. *Writing Supplies.* The standard pencil is $6\frac{7}{8}$ in. wood and $\frac{1}{2}$ in. eraser. What is the total length of the standard pencil?

Source: Eberhard Faber American

51. *Upholstery Fabric.* Executive Car Care sells 45-in. upholstery fabric for car restoration. Art buys $9\frac{1}{4}$ yd and $10\frac{5}{6}$ yd for two car projects. How many total yards did Art buy?

52. *Carpentry.* When cutting wood with a saw, a carpenter must take into account the thickness of the saw blade. Suppose that from a piece of wood 36 in. long, a carpenter cuts a $15\frac{3}{4}$-in. length with a saw blade that is $\frac{1}{8}$ in. in thickness. How long is the piece that remains?

53. *Liquid Fertilizer.* There are $283\frac{5}{8}$ gal of liquid fertilizer in a fertilizer application tank. After applying $178\frac{2}{3}$ gal to a soybean field, the farmer requests that Braden's Farm Supply deliver an additional 250 gal. How many gallons of fertilizer are in the tank after the delivery?

54. *NCAA Football Goalposts.* In college football, the distance between goalposts was reduced from $23\frac{1}{3}$ ft to $18\frac{1}{2}$ ft. By how much was it reduced?

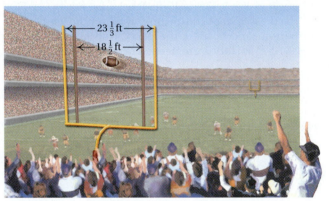

Source: NCAA

55. *Running.* Angela is preparing to run the San Diego marathon. Recently she ran a 10-km "Fun Run." A marathon is $26\frac{7}{32}$ mi and 10 km is $6\frac{1}{5}$ mi. How much farther will Angela run in the marathon?

56. *Running.* Harvey successfully ran both the Spring Lake 5-mi race and the Boys & Girls Club 10-km "Dog Day Race." Given that 5 mi is about $8\frac{1}{20}$ km, how much farther did Harvey run in the Dog Day Race?

57. *Cooking.* Among the ingredients in a recipe for black bean and corn salsa are the following:

Source: Based on *Jane Butel's Southwestern Grill,* HP Books, 1996, p. 206

$1\frac{1}{2}$ cups onion

$1\frac{1}{2}$ cups diced jalapeño

$2\frac{1}{2}$ cups cooked black beans

$1\frac{1}{2}$ cups cooked whole-kernel corn

$\frac{3}{4}$ cup chopped cilantro

How many cups of ingredients are listed?

58. *Cooking.* A recipe for boniato bread includes the following ingredients:

Source: Based on *Miami Spice* by Steven Raichlen, Workman Publishing, 1993, p. 93

$\frac{1}{4}$ cup water

$\frac{1}{3}$ cup sugar

$5\frac{1}{2}$ cups unbleached all-purpose flour

How many cups of ingredients are listed?

59. *Interior Design.* Sue, an interior designer, worked $10\frac{1}{2}$ hr over a three-day period. If Sue worked $2\frac{1}{2}$ hr on the first day and $4\frac{1}{5}$ hr on the second, how many hours did she work on the third day?

60. *Painting.* Geri had $3\frac{1}{2}$ gal of paint. It took $2\frac{3}{4}$ gal to paint the family room. She estimated that it would take $2\frac{1}{4}$ gal to paint the living room. How much more paint was needed?

Find the perimeter of (distance around) the figure.

61.

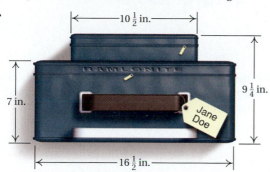

10 $\frac{1}{2}$ in.

9 $\frac{1}{4}$ in.

7 in.

Jane Doe

16 $\frac{1}{2}$ in.

62.

3 $\frac{7}{16}$ ft

3 $\frac{7}{16}$ ft

6 $\frac{7}{8}$ ft

6 $\frac{7}{8}$ ft

63.

4 ft

3 $\frac{3}{4}$ ft

5 $\frac{1}{2}$ ft

6 $\frac{3}{4}$ ft

3 $\frac{3}{4}$ ft

4 ft

64.

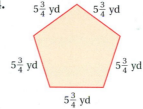

5 $\frac{3}{4}$ yd   5 $\frac{3}{4}$ yd

5 $\frac{3}{4}$ yd   5 $\frac{3}{4}$ yd

5 $\frac{3}{4}$ yd

65.

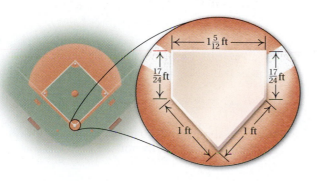

1 $\frac{5}{12}$ ft

$\frac{17}{24}$ ft $\frac{17}{24}$ ft

1 ft   1 ft

66.

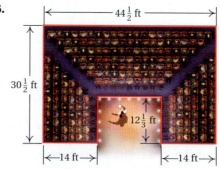

44 $\frac{1}{2}$ ft

3 $0\frac{1}{2}$ ft

12 $\frac{1}{3}$ ft

14 ft   14 ft

Find the length d in the figure.

67.

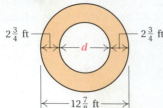

$2\frac{3}{4}$ ft — $\leftarrow d \rightarrow$ — $2\frac{3}{4}$ ft

$\leftarrow 12\frac{7}{8}$ ft $\rightarrow$

68.

$2\frac{1}{5}$ in. — $\leftarrow d \rightarrow$ — $2\frac{1}{5}$ in.

$\leftarrow 10\frac{1}{2}$ in. $\rightarrow$

69. Find the smallest length of a bolt that will pass through a piece of tubing with an outside diameter of $\frac{1}{2}$ in., a washer $\frac{1}{16}$ in. thick, a piece of tubing with an outside diameter of $\frac{3}{4}$ in., another washer $\frac{1}{16}$ in. thick, and a nut $\frac{3}{16}$ in. thick.

70. The front of the stage at the Steel Pony Coffeehouse is $6\frac{1}{2}$ yd long. If renovation work succeeds in adding $2\frac{3}{4}$ yd in length, how long is the renovated stage?

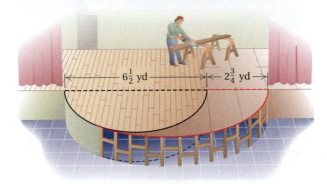

$\leftarrow 6\frac{1}{2}$ yd $\rightarrow\!\!\leftarrow 2\frac{3}{4}$ yd $\rightarrow$

d Subtract.

71. $8\frac{3}{5} - 9\frac{2}{5}$

72. $4\frac{5}{7} - 8\frac{3}{7}$

73. $3\frac{1}{2} - 6\frac{3}{4}$

74. $5\frac{1}{2} - 7\frac{3}{4}$

75. $3\frac{4}{5} - 7\frac{2}{3}$

76. $2\frac{3}{7} - 5\frac{1}{2}$

77. $-3\frac{1}{5} - 4\frac{2}{5}$

78. $-5\frac{3}{8} - 4\frac{1}{8}$

79. $-4\frac{2}{5} - 6\frac{3}{7}$

80. $-2\frac{3}{4} - 3\frac{3}{8}$

81. $-6\frac{1}{9} - \left(-4\frac{2}{9}\right)$

82. $-2\frac{3}{5} - \left(-1\frac{1}{5}\right)$

83. **D𝐰** Write a problem for a classmate to solve. Design the problem so the solution is "The larger package holds $4\frac{1}{2}$ oz more than the smaller package."

84. **D𝐰** Is the sum of two mixed numerals always a mixed numeral? Why or why not?

Solve.

85. Rick's Market prepackages Swiss cheese in $\frac{3}{4}$-lb packages. How many packages can be made from a 12-lb slab of cheese? [3.8b]

86. The Thompson Dairy produced 4578 oz of milk one morning. How many 16-oz cartons would this have filled? How much milk would have been left over? [1.8a]

Determine whether the first number is divisible by the second.

87. 9993 by 3 [3.1b]

88. 9993 by 9 [3.1b]

89. 2345 by 9 [3.1b]

90. 2345 by 5 [3.1b]

91. 2335 by 10 [3.1b]

92. 7764 by 6 [3.1b]

93. 18,888 by 6 [3.1b]

94. 18,888 by 4 [3.1a]

95. Multiply and simplify: $\dfrac{15}{9} \cdot \dfrac{18}{39}$. [3.6a]

96. Divide and simplify: $\dfrac{12}{25} \div \dfrac{24}{5}$. [3.7b]

97. $\mathbf{D_W}$ Explain how the "borrowing" that is used in this section compares with the borrowing used in Section 1.3.

98. $\mathbf{D_W}$ Ryan insists that since $\frac{5}{7} - \frac{2}{7}$ is $\frac{3}{7}$, and $4 - 5$ is -1, it follows that $4\frac{5}{7} - 5\frac{2}{7}$ is $-1\frac{3}{7}$. How could you convince him that he is mistaken?

Calculate each of the following. Write the result as a mixed numeral.

99. ▦ $3289\frac{1047}{1189} + 5278\frac{32}{41}$

100. ▦ $5798\frac{17}{53} - 3909\frac{1957}{2279}$

101. ▦ $4230\frac{19}{73} - 5848\frac{17}{29}$

102. ▦ $57{,}825\frac{13}{79} - 64{,}200\frac{1}{43}$

Solve.

103. $35\dfrac{2}{3} + n = 46\dfrac{1}{4}$

104. $42\dfrac{7}{9} = x - 13\dfrac{2}{5}$

105. $-15\dfrac{7}{8} = 12\dfrac{1}{2} + t$

106. A post for a pier is 29 ft long. Half of the post extends above the water's surface and $8\frac{3}{4}$ ft of the post is buried in mud. How deep is the water at that location?

107. An algebra text is $1\frac{1}{8}$ in. thick, $9\frac{3}{4}$ in. long, and $8\frac{1}{2}$ in. wide. If the front, back, and spine of the book were unfolded, they would form a rectangle. What would the perimeter of that rectangle be?

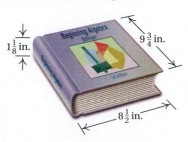

285

Objectives

a Multiply using mixed numerals.

b Divide using mixed numerals.

c Evaluate expressions using mixed numerals.

d Solve problems involving multiplication and division with mixed numerals.

1. Multiply: $8 \cdot 3\frac{1}{2}$.

2. Multiply: $5\frac{1}{2} \cdot \frac{3}{7}$.

Multiply.

3. $-2 \cdot 6\frac{2}{5}$

4. $3\frac{1}{3} \cdot 2\frac{1}{2}$

Answers on page A-11

Carrying out addition and subtraction with mixed numerals is usually easier if the numbers are left as mixed numerals. With multiplication and division, however, it is easier to convert the numbers to fraction notation first.

a Multiplication

> **MULTIPLICATION USING MIXED NUMERALS**
>
> To multiply using mixed numerals, first convert to fraction notation. Then multiply with fraction notation and, if appropriate, rewrite the answer as an equivalent mixed numeral.

EXAMPLE 1 Multiply: $6 \cdot 2\frac{1}{2}$.

$$6 \cdot 2\frac{1}{2} = \frac{6}{1} \cdot \frac{5}{2} = \frac{6 \cdot 5}{1 \cdot 2} = \frac{\cancel{2} \cdot 3 \cdot 5}{\cancel{2} \cdot 1} = 15 \qquad \text{Removing a factor equal to 1: } \frac{2}{2} = 1$$

Here we write fraction notation.

Do Exercise 1.

EXAMPLE 2 Multiply: $3\frac{1}{2} \cdot \frac{3}{4}$.

$$3\frac{1}{2} \cdot \frac{3}{4} = \frac{7}{2} \cdot \frac{3}{4} = \frac{21}{8} = 2\frac{5}{8}$$

> Although fraction notation is needed, *common denominators are not required.*

Do Exercise 2.

EXAMPLE 3 Multiply: $-10 \cdot 5\frac{2}{3}$.

$$-10 \cdot 5\frac{2}{3} = -\frac{10}{1} \cdot \frac{17}{3} = -\frac{170}{3} = -56\frac{2}{3}$$

EXAMPLE 4 Multiply: $2\frac{1}{4} \cdot 5\frac{2}{3}$.

$$2\frac{1}{4} \cdot 5\frac{2}{3} = \frac{9}{4} \cdot \frac{17}{3} = \frac{9 \cdot 17}{4 \cdot 3} = \frac{\cancel{3} \cdot 3 \cdot 17}{2 \cdot 2 \cdot \cancel{3}} = \frac{51}{4} = 12\frac{3}{4}$$

> **Caution!**
>
> $2\frac{1}{4} \cdot 5\frac{2}{3} \neq 10\frac{2}{12}$. A common error is to multiply the whole numbers and then the fractions. The correct answer, $12\frac{3}{4}$, is found only after converting to fraction notation.

Do Exercises 3 and 4.

b Division

The division $1\frac{1}{2} \div \frac{1}{6}$ is shown here. This division means "How many $\frac{1}{6}$'s are in $1\frac{1}{2}$?"

We see that $\frac{1}{6}$ goes into $1\frac{1}{2}$ nine times.

$$1\frac{1}{2} \div \frac{1}{6} = \frac{3}{2} \div \frac{1}{6} \qquad \text{We convert } 1\frac{1}{2} \text{ to fraction notation.}$$

$$= \frac{3}{2} \cdot 6 = \frac{3 \cdot 6}{2} = \frac{3 \cdot 3 \cdot 2}{2 \cdot 1} = \frac{3 \cdot 3}{1} \cdot \frac{2}{2} = \frac{3 \cdot 3}{1} \cdot 1 = 9$$

> ### DIVISION USING MIXED NUMERALS
>
> To divide using mixed numerals, first write an equivalent expression using fraction notation. Then divide and, if appropriate, convert the answer back to an equivalent mixed numeral.

EXAMPLE 5 Divide: $32 \div 3\frac{1}{5}$.

$$32 \div 3\frac{1}{5} = \frac{32}{1} \div \frac{16}{5} \qquad \text{Converting to fraction notation}$$

$$= \frac{32}{1} \cdot \frac{5}{16} = \frac{32 \cdot 5}{1 \cdot 16} = \frac{2 \cdot \cancel{16} \cdot 5}{1 \cdot \cancel{16}} = 10 \qquad \begin{array}{l}\text{Removing a factor} \\ \text{equal to 1: } \frac{16}{16} = 1\end{array}$$

↑ Remember to multiply by the reciprocal of the divisor.

Caution!

The reciprocal of $3\frac{1}{5}$ is neither $5\frac{1}{3}$ nor $3\frac{5}{1}$!

Do Exercise 5.

EXAMPLE 6 Divide: $2\frac{1}{3} \div 1\frac{3}{4}$.

$$2\frac{1}{3} \div 1\frac{3}{4} = \frac{7}{3} \div \frac{7}{4} = \frac{7}{3} \cdot \frac{4}{7} = \frac{\cancel{7} \cdot 4}{\cancel{7} \cdot 3} = \frac{4}{3} = 1\frac{1}{3} \qquad \begin{array}{l}\text{Removing a factor} \\ \text{equal to 1: } \frac{7}{7} = 1\end{array}$$

EXAMPLE 7 Divide: $-1\frac{3}{5} \div \left(-3\frac{1}{3}\right)$.

$$-1\frac{3}{5} \div \left(-3\frac{1}{3}\right) = -\frac{8}{5} \div \left(-\frac{10}{3}\right) = \frac{8}{5} \cdot \frac{3}{10} \qquad \begin{array}{l}\text{The product or quotient of} \\ \text{two negatives is positive.}\end{array}$$

$$= \frac{\cancel{2} \cdot 4 \cdot 3}{5 \cdot \cancel{2} \cdot 5} = \frac{12}{25} \qquad \begin{array}{l}\text{Removing a factor} \\ \text{equal to 1: } \frac{2}{2} = 1\end{array}$$

Do Exercises 6 and 7.

---

5. Divide: $63 \div 5\frac{1}{4}$.

Divide.

6. $2\frac{1}{4} \div 1\frac{1}{5}$

7. $1\frac{3}{4} \div \left(-2\frac{1}{2}\right)$

Answers on page A-11

Evaluate.

8. rt, for $r = 78$ and $t = 2\frac{1}{4}$

C Evaluating Expressions

Mixed numerals can appear in algebraic expressions just as the integers of Section 2.6 did.

EXAMPLE 8 A train traveling r miles per hour for t hours travels a total of rt miles. (*Remember*: Distance = Rate · Time.)

a) Find the distance traveled by a 60-mph train in $2\frac{3}{4}$ hr.

b) Find the distance traveled if the speed of the train is $26\frac{1}{2}$ mph and the time is $2\frac{2}{3}$ hr.

a) We evaluate rt for $r = 60$ and $t = 2\frac{3}{4}$:

$$rt = 60 \cdot 2\frac{3}{4}$$

$$= \frac{60}{1} \cdot \frac{11}{4}$$

$$= \frac{15 \cdot 4 \cdot 11}{1 \cdot 4} = 165. \qquad \text{Removing a factor equal to 1: } \frac{4}{4} = 1$$

In $2\frac{3}{4}$ hr, a 60-mph train travels 165 mi.

b) We evaluate rt for $r = 26\frac{1}{2}$ and $t = 2\frac{2}{3}$:

$$rt = 26\frac{1}{2} \cdot 2\frac{2}{3}$$

$$= \frac{53}{2} \cdot \frac{8}{3} = \frac{53 \cdot 2 \cdot 4}{2 \cdot 3} \qquad \text{Removing a factor equal to 1: } \frac{2}{2} = 1$$

$$= \frac{212}{3} = 70\frac{2}{3}.$$

In $2\frac{2}{3}$ hr, a $26\frac{1}{2}$-mph train travels $70\frac{2}{3}$ mi.

9. $7xy$, for $x = 9\frac{2}{5}$ and $y = 2\frac{3}{7}$

EXAMPLE 9 Evaluate $x + yz$ for $x = 7\frac{1}{3}$, $y = \frac{1}{3}$, and $z = 5$.

We substitute and follow the rules for order of operations:

$$x + yz = 7\frac{1}{3} + \frac{1}{3} \cdot 5$$

$$= 7\frac{1}{3} + \frac{1}{3} \cdot \frac{5}{1} \qquad \text{Multiply first; then add.}$$

$$= 7\frac{1}{3} + \frac{5}{3}$$

$$= 7\frac{1}{3} + 1\frac{2}{3} \qquad \text{Adding mixed numerals}$$

$$= 8\frac{3}{3} = 9.$$

10. $x - y \div z$, for $x = 5\frac{7}{8}$, $y = \frac{1}{4}$, and $z = 2$

Do Exercises 8–10.

Answers on page A-11

d Applications and Problem Solving

EXAMPLE 10 *Average Speed in Indianapolis 500.* Arie Luyendyk won the Indianapolis 500 in 1990 with the highest average speed of about 186 mph. This record high through 2005 is about $2\frac{12}{25}$ times the average speed of the first winner, Ray Harroun, in 1911. What was the average speed of the winner in the first Indianapolis 500?

Source: Indianapolis Motor Speedway

1. **Familiarize.** We ask the question, "186 is $2\frac{12}{25}$ times what number?" We let $s =$ the average speed in 1911. Then the average speed in 1990 was $2\frac{12}{25} \cdot s$.

2. **Translate.** The problem can be translated to an equation as follows.

$$
\underbrace{\text{Average speed in 1990}} \quad \text{is} \quad 2\frac{12}{25} \quad \text{times} \quad \underbrace{\text{Average speed in 1911}}
$$

$$
186 \quad = \quad 2\frac{12}{25} \quad \cdot \quad s
$$

3. **Solve.** To solve the equation, we divide on both sides.

$186 = \dfrac{62}{25} \cdot s$ Converting $2\frac{12}{25}$ to fraction notation

$\dfrac{25}{62} \cdot 186 = \dfrac{25}{62} \cdot \dfrac{62}{25}s$ Using the multiplication principle

$\dfrac{25 \cdot 186}{62} = 1 \cdot s$ Multiplying

$\dfrac{25 \cdot 3 \cdot \cancel{62}}{\cancel{62} \cdot 1} = 1 \cdot s$ Since 62 is in the denominator, we check to see if it is a factor of 186. Since it is, we can remove a factor equal to 1: $\frac{62}{62} = 1$.

$\dfrac{25 \cdot 3}{1} = 1 \cdot s$

$75 = s$ Simplifying

4. **Check.** If the average speed in 1911 was about 75 mph, we find the average speed in 1990 by multiplying 75 by $2\frac{12}{25}$:

$$
2\frac{12}{25} \cdot 75 = \frac{62}{25} \cdot 75 = \frac{62 \cdot 75}{25} = \frac{62 \cdot 25 \cdot 3}{25} = 62 \cdot 3 = 186.
$$

The answer checks.

5. **State.** The average speed of the winner in the first Indianapolis 500 was about 75 mph.

Do Exercises 11 and 12.

11. Kyle's pickup truck travels on an interstate highway at 65 mph for $3\frac{1}{2}$ hr. How far does it travel?

12. Holly's minivan travels 302 mi on $15\frac{1}{10}$ gal of gas. How many miles per gallon did it get?

Answers on page A-11

EXAMPLE 11 *Mirror Area.* The mirror-backed candle shelf, shown below with a carpenter's diagram, was designed and built by Harry Cooper. Such shelves were popular in Colonial times because the mirror provided extra lighting from the candle. A rectangular walnut board is used to make the back of the shelf. Find the area of the original board and the amount left over after the opening for the mirror has been cut out.

Source: Popular Science Woodworking Projects

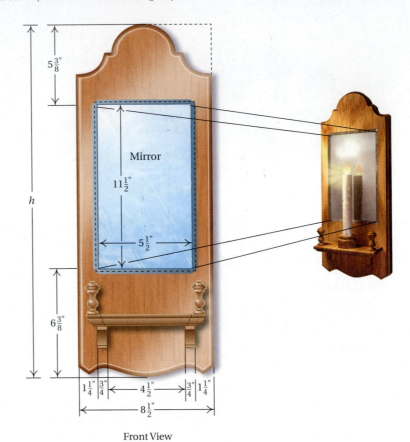

Front View

1. **Familiarize.** Note that there are two rectangles in the diagram: one representing the original board and one representing the mirror. Note too that we can determine the area of the mirror from the given information, but to determine the area of the original board, we must first find its height in inches. This height can be found by adding the vertical measurements, $6\frac{3}{8}''$, $11\frac{1}{2}''$, and $5\frac{3}{8}''$. We let R = the area, in square inches, of the original board after the mirror opening has been removed.

2. **Translate.** We rephrase and translate as follows:

Rephrase: Area of board after opening is cut is Area of original board minus Area of mirror

Translate: $R = 8\frac{1}{2}\left(6\frac{3}{8} + 11\frac{1}{2} + 5\frac{3}{8}\right) - 5\frac{1}{2} \cdot 11\frac{1}{2}$

3. Solve. We calculate as follows:

$$R = 8\frac{1}{2}\left(6\frac{3}{8} + 11\frac{1}{2} + 5\frac{3}{8}\right) - 5\frac{1}{2} \cdot 11\frac{1}{2}$$

$$= 8\frac{1}{2}\left(6\frac{3}{8} + 11\frac{4}{8} + 5\frac{3}{8}\right) - 5\frac{1}{2} \cdot 11\frac{1}{2} \qquad \text{Finding a common denominator}$$

$$= 8\frac{1}{2} \cdot 22\frac{10}{8} - 5\frac{1}{2} \cdot 11\frac{1}{2}$$

$$= 8\frac{1}{2} \cdot 23\frac{1}{4} - 5\frac{1}{2} \cdot 11\frac{1}{2} \qquad 22\frac{10}{8} = 22\frac{5}{4} = 23\frac{1}{4}$$

$$= \frac{17}{2} \cdot \frac{93}{4} - \frac{11}{2} \cdot \frac{23}{2} \qquad \text{Writing fraction notation}$$

$$= \frac{1581}{8} - \frac{253}{4} \qquad \text{Multiplying}$$

This is the area of the mirror.

$$= 197\frac{5}{8} - 63\frac{1}{4}$$

This is the area of the original board.

$$= 197\frac{5}{8} - 63\frac{2}{8} \qquad \Bigg\} \text{ Subtracting}$$

$$= 134\frac{3}{8}.$$

4. Check. To check, we can separately calculate the two rectangular areas and then subtract. This is almost the same as repeating the calculations. We leave this for the student.

5. State. The area of the original board is $197\frac{5}{8}$ in². The area left over is $134\frac{3}{8}$ in².

Do Exercise 13.

---

13. A room measures $22\frac{1}{2}$ ft by $15\frac{1}{2}$ ft. A 9-ft by 12-ft oriental rug is placed in the center of the room. How much area is not covered by the rug?

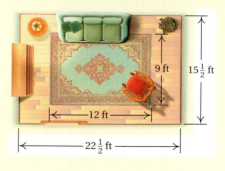

9 ft $15\frac{1}{2}$ ft

12 ft

$22\frac{1}{2}$ ft

CALCULATOR CORNER

Multiplying and Dividing Mixed Numerals   Mixed numerals are multiplied or divided on a fraction calculator or a graphing calculator in much the same way that addition or subtraction is performed. For example, to find $3\frac{2}{3} \cdot 4\frac{1}{5}$, we press

$$\boxed{3}\;\boxed{a^{b/c}}\;\boxed{2}\;\boxed{a^{b/c}}\;\boxed{3}\;\boxed{\times}\;\boxed{4}\;\boxed{a^{b/c}}\;\boxed{1}\;\boxed{a^{b/c}}\;\boxed{5}\;\boxed{\text{ENTER} =}.$$

The calculator displays $\boxed{\quad 15 \; \lrcorner 2 \; \lrcorner 5 \quad}$, so the product is $15\frac{2}{5}$.

For calculators without an $\boxed{a^{b/c}}$ key, use parentheses around mixed numerals written as sums, or convert first to fraction notation.

Exercises: Perform each calculation. Give the answer as a mixed numeral.

1. $2\frac{1}{3} \cdot 4\frac{3}{5}$ 4. $-7\frac{9}{16} \cdot 3\frac{4}{7}$

2. $10\frac{7}{10} \div 3\frac{5}{6}$ 5. $4\frac{12}{13} \cdot 6\frac{4}{11}$

3. $-7\frac{2}{9} \div 4\frac{1}{5}$ 6. $2\frac{10}{17} \cdot 9\frac{11}{13}$

Answer on page A-11

14. After two weeks, Kurt's tomato seedlings measure $9\frac{1}{2}$ in., $10\frac{3}{4}$ in., $10\frac{1}{4}$ in., and 9 in. tall. Find their average height.

EXAMPLE 12 Melody has had three children. Their birth weights were $7\frac{1}{2}$ lb, $7\frac{3}{4}$ lb, and $6\frac{3}{4}$ lb. What was the average weight of her babies?

1. **Familiarize.** Recall that to compute an *average*, we add the values and then divide the sum by the number of values. We let w = the average weight, in pounds.

2. **Translate.** We have
$$w = \frac{7\frac{1}{2} + 7\frac{3}{4} + 6\frac{3}{4}}{3}.$$

3. **Solve.** We first add:

$$7\frac{1}{2} + 7\frac{3}{4} + 6\frac{3}{4} = 7\frac{2}{4} + 7\frac{3}{4} + 6\frac{3}{4} \qquad \text{Finding the LCD}$$

$$= 20\frac{8}{4}$$

$$= 22. \qquad 20\frac{8}{4} = 20 + \frac{8}{4} = 20 + 2$$

Then we divide:

$$w = \frac{7\frac{1}{2} + 7\frac{3}{4} + 6\frac{3}{4}}{3} = \frac{22}{3} = 7\frac{1}{3}. \qquad \text{Dividing by 3}$$

4. **Check.** As a partial check, we note that the average is smaller than the largest individual value and larger than the smallest individual value. We could also repeat our calculations.

5. **State.** The average weight of the three babies was $7\frac{1}{3}$ lb.

Do Exercise 14.

Answer on page A-11

CHAPTER 4: Fraction Notation: Addition, Subtraction, and Mixed Numerals

Translating for Success

1. **Raffle Tickets.** At the Happy Hollow Camp Fall Festival, Rico and Becca, together, spent $270 on raffle tickets that sell for $\frac{9}{20}$ each. How many tickets did they buy?

2. **Irrigation Pipe.** Chad uses two pipes, one of which measures $5\frac{1}{3}$ ft, to repair the irrigation system in the Buxtons' lawn. The total length of the two pipes is $8\frac{7}{12}$ ft. How long is the other pipe?

3. **Vacation Days.** Together, Helmut and Claire have 36 vacation days a year. Helmut has 22 vacation days per year. How many does Claire have?

4. **Enrollment in Japanese Classes.** Last year at the Lakeside Community College, 225 students enrolled in basic mathematics. This number is $4\frac{1}{2}$ times as many as the number who enrolled in Japanese. How many enrolled in Japanese?

5. **Bicycling.** Cole rode his bicycle $5\frac{1}{3}$ mi on Saturday and $8\frac{7}{12}$ mi on Sunday. How far did he ride on the weekend?

The goal of these matching questions is to practice step (2), *Translate*, of the five-step problem-solving process. Translate each word problem to an equation and select a correct translation from equations A–O.

A. $13\frac{11}{12} = x + 5\frac{1}{3}$

B. $\frac{3}{4} \cdot x = 1\frac{2}{3}$

C. $\frac{9}{20} + x = 270$

D. $225 = 4\frac{1}{2} \cdot x$

E. $98 \div 2\frac{1}{3} = x$

F. $22 + x = 36$

G. $x = 4\frac{1}{2} \cdot 225$

H. $x = 5\frac{1}{3} + 8\frac{7}{12}$

I. $22 \cdot x = 36$

J. $x = \frac{3}{4} \cdot 1\frac{2}{3}$

K. $5\frac{1}{3} + x = 8\frac{7}{12}$

L. $\frac{9}{20} \cdot 270 = x$

M. $1\frac{2}{3} + \frac{3}{4} = x$

N. $98 - 2\frac{1}{3} = x$

O. $\frac{9}{20} \cdot x = 270$

Answers on page A-11

6. **Deli Order.** For a promotional open house for contractors last year, the Bayside Builders Association ordered 225 turkey sandwiches. Due to increased registrations this year, $4\frac{1}{2}$ times as many sandwiches will be needed. How many sandwiches are ordered?

7. **Dog Ownership.** In Sam's community, $\frac{9}{20}$ of the households own at least one dog. There are 270 households. How many own dogs?

8. **Magic Tricks.** A magic trick requires a piece of rope $2\frac{1}{3}$ ft long. Gerry, a magician, has 98 ft of rope and needs to divide it into $2\frac{1}{3}$-ft pieces. How many pieces can be cut from the rope?

9. **Painting.** Laura needs $1\frac{2}{3}$ gal of paint to paint the ceiling of the exercise room and $\frac{3}{4}$ gal of the same paint for the bathroom. How much paint does Laura need?

10. **Chocolate Fudge Bars.** A recipe for chocolate fudge bars that serves 16 includes $1\frac{2}{3}$ cups of sugar. How much sugar is needed for $\frac{3}{4}$ of this recipe?

a   Multiply. Write a mixed numeral for each answer.

1. $16 \cdot 1\frac{2}{5}$

2. $5 \cdot 3\frac{3}{4}$

3. $6\frac{2}{3} \cdot \frac{1}{4}$

4. $-9 \cdot 2\frac{3}{5}$

5. $20\left(-2\frac{5}{6}\right)$

6. $7\frac{3}{8} \cdot 4\frac{1}{3}$

7. $3\frac{1}{2} \cdot 4\frac{2}{3}$

8. $4\frac{1}{5} \cdot 5\frac{1}{4}$

9. $-2\frac{3}{10} \cdot 4\frac{2}{5}$

10. $4\frac{7}{10} \cdot 5\frac{3}{10}$

11. $\left(-6\frac{3}{10}\right)\left(-5\frac{7}{10}\right)$

12. $-20\frac{1}{2} \cdot \left(-10\frac{1}{5}\right)$

b   Divide. Write a mixed numeral for each answer whenever possible.

13. $30 \div 2\frac{3}{5}$

14. $18 \div 2\frac{1}{4}$

15. $8\frac{2}{5} \div 7$

16. $3\frac{3}{8} \div 3$

17. $5\frac{1}{4} \div 2\frac{3}{5}$

18. $5\frac{4}{5} \div 2\frac{1}{2}$

19. $-5\frac{1}{4} \div 2\frac{3}{7}$

20. $-4\frac{3}{8} \div 2\frac{5}{6}$

21. $5\frac{1}{10} \div 4\frac{3}{10}$

22. $4\frac{1}{10} \div 2\frac{1}{10}$

23. $20\frac{1}{4} \div (-90)$

24. $12\frac{1}{2} \div (-50)$

c   Evaluate.

25. lw, for $l = 2\frac{3}{5}$ and $w = 9$

26. mv, for $m = 7$ and $v = 3\frac{2}{5}$

27. rs, for $r = 5$ and $s = 3\frac{1}{7}$

28. rt, for $r = 5\frac{2}{3}$ and $t = -2\frac{3}{8}$

29. mt, for $m = 6\frac{2}{9}$ and $t = -4\frac{3}{5}$

30. $M \div NP$, for $M = 2\frac{1}{4}$, $N = -5$, and $P = 2\frac{1}{3}$

31. $R \cdot S \div T$, for $R = 4\frac{2}{3}$, $S = 1\frac{3}{7}$, and $T = -5$

32. $a - bc$, for $a = 18$, $b = 2\frac{1}{5}$, and $c = 3\frac{3}{4}$

33. $r + ps$, for $r = 5\frac{1}{2}$, $p = 3$, and $s = 2\frac{1}{4}$

34. $s + rt$, for $s = 3\frac{1}{2}$, $r = 5\frac{1}{2}$, and $t = 7\frac{1}{2}$

35. $m + n \div p$, for $m = 7\frac{2}{5}$, $n = 4\frac{1}{2}$, and $p = 6$

36. $x - y \div z$, for $x = 9$, $y = 2\frac{1}{2}$, and $z = 3\frac{3}{4}$

d Solve.

37. *Beagles.*   There are about 155,000 Labrador retrievers registered with The American Kennel Club. This is $3\frac{4}{9}$ times the number of beagles registered. How many beagles are registered?

Source: The American Kennel Club

38. *Exercise.*   At one point during a spinning class, Kea's bicycle wheel was completing $76\frac{2}{3}$ revolutions per minute. How many revolutions did the wheel complete in 6 min?

39. *Sodium Consumption.*   The average American woman consumes $1\frac{1}{3}$ tsp of sodium each day. How much sodium do 10 average American women consume in one day?

Source: *Nutrition Action Health Letter,* March 1994, p. 6.
1875 Connecticut Ave., N.W., Washington, DC 20009-5728.

40. *Aeronautics.*   Most space shuttles orbit the earth once every $1\frac{1}{2}$ hr. How many orbits are made every 24 hr?

41. Grape-Nuts Bread. Listed below is the recipe for grape-nuts bread. What are the ingredients for $\frac{1}{2}$ recipe? for 3 recipes?

Source: Reprinted with permission from Taste of Home, Greendale, WI. www.tasteofhome.com

Grape-Nuts Bread

$1\frac{2}{3}$ cups water (70° to 80°)

3 tablespoons canola oil

$4\frac{1}{2}$ teaspoons sugar

1 teaspoon salt

$3\frac{3}{4}$ cups bread flour

$\frac{3}{4}$ cup Grape-Nuts cereal

$1\frac{1}{2}$ teaspoons active dry yeast

42. Sweet 'n' Crunchy Mix. Listed below is the recipe for sweet and crunchy mix. What are the ingredients for $\frac{1}{2}$ recipe? for 5 recipes?

Source: Reprinted with permission from Taste of Home, Greendale, WI. www.tasteofhome.com

Sweet 'n' Crunchy Mix

$2\frac{1}{2}$ cups Rice Chex

$2\frac{1}{2}$ cups Honey-Nut Cheerios

1 package (10 ounces) honey-flavored bear-shaped graham crackers

2 cups miniature pretzels

$\frac{1}{2}$ cup butter, melted

$\frac{1}{3}$ cup packed brown sugar

$4\frac{1}{2}$ teaspoons ground cinnamon

43. Media Usage. In 2005, on the average, a person spent about $9\frac{1}{8}$ times as many hours watching television as in shopping on the Internet. The average person spent about 1825 hours watching television in 2005. About how many hours were spent shopping on the Internet?

Source: Communications Industry Forecast & Report

44. Foreign Students. In U.S. colleges in 2004, the number of foreign students from South Korea was about $8\frac{2}{3}$ times the number of students from Nigeria. About 52,000 students were from South Korea. About how many were from Nigeria?

Source: Institute of International Education

45. Temperatures. Fahrenheit temperature can be obtained from Celsius (centigrade) temperature by multiplying by $1\frac{4}{5}$ and adding 32°. What Fahrenheit temperature corresponds to a Celsius temperature of 20°?

46. Temperature. Fahrenheit temperature can be obtained from Celsius (centigrade) temperature by multiplying by $1\frac{4}{5}$ and adding 32°. What Fahrenheit temperature corresponds to the Celsius temperature of boiling water, which is 100°?

47. Word Processing. Kelly wants to create a table using Microsoft® Word software for word processing. She needs to have two columns, each $1\frac{1}{2}$ in. wide, and five columns, each $\frac{3}{4}$ in. wide. Will this table fit on a piece of standard paper that is $8\frac{1}{2}$ in. wide? If so, how wide will each margin be if her margins on each side are to be of equal width?

48. Construction. A rectangular lot has dimensions of $302\frac{1}{2}$ ft by $205\frac{1}{4}$ ft. A building with dimensions of 100 ft by $25\frac{1}{2}$ ft is built on the lot. How much area is left over?

49. *Servings of Salmon.* A serving of filleted fish is generally considered to be about $\frac{1}{3}$ lb. How many servings can be prepared from $5\frac{1}{2}$ lb of salmon fillet?

50. *Servings of Tuna.* A serving of fish steak (cross section) is generally $\frac{1}{2}$ lb. How many tuna steak servings can be prepared from a cleaned $18\frac{3}{4}$-lb tuna?

51. *Weight of Water.* The weight of water is $62\frac{1}{2}$ lb per cubic foot. What is the weight of $5\frac{1}{2}$ cubic feet of water?

52. *Weight of Water.* The weight of water is $62\frac{1}{2}$ lb per cubic foot. What is the weight of $2\frac{1}{4}$ cubic feet of water?

53. *Video Recording.* The tape in a VCR operating in the short-play mode travels at a rate of $1\frac{3}{8}$ in. per second. How many inches of tape are used to record for 60 sec in the short-play mode?

54. *Audio Recording.* The tape in an audio cassette is played at the rate of $1\frac{7}{8}$ in. per second. How many inches of tape are used when a cassette is played for $5\frac{1}{2}$ sec?

55. A car traveled 213 mi on $14\frac{2}{10}$ gal of gas. How many miles per gallon did it get?

56. A car traveled 385 mi on $15\frac{4}{10}$ gal of gas. How many miles per gallon did it get?

57. *Birth Weights.* The Piper quadruplets of Great Britain weighed $2\frac{9}{16}$ lb, $2\frac{9}{32}$ lb, $2\frac{1}{8}$ lb, and $2\frac{5}{16}$ lb at birth. Find their average birth weight.

Source: *The Guinness Book of Records,* 1998

58. *Vertical Leaps.* Eight-year-old Zachary registered vertical leaps of $12\frac{3}{4}$ in., $13\frac{3}{4}$ in., $13\frac{1}{2}$ in., and 14 in. Find his average vertical leap.

59. *Acceleration.* The results of a *Motor Trend* road acceleration test for five cars are given in the graph below. The test measures the time in seconds required to go from 0 mph to 60 mph. What was the average time?

Source: *Motor Trend,* March 2005, pp. 134–142

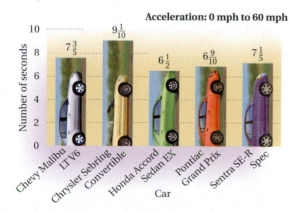

60. *Manufacturing.* A test of five light bulbs showed that they burned for the lengths of time given on the graph below. For how many days, on average, did the bulbs burn?

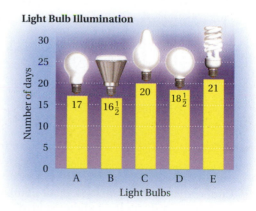

61. *Landscaping.* The previous owners of Ashley's new home had a large L-shaped vegetable garden consisting of a rectangle that is $15\frac{1}{2}$ ft by 20 ft adjacent to one that is $10\frac{1}{2}$ ft by $12\frac{1}{2}$ ft. Ashley wants to cover the garden with sod. What is the total area of the sod she must purchase?

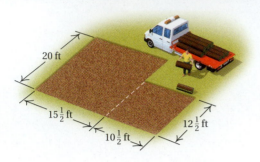

62. *Home Furnishings.* An L-shaped sunroom consists of a rectangle that is $9\frac{1}{2}$ ft by 12 ft adjacent to one that is $9\frac{1}{2}$ ft by 8 ft. What is the total area of a carpet that covers the floor?

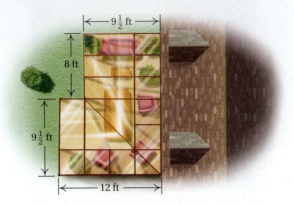

Find the area of each shaded region.

63.

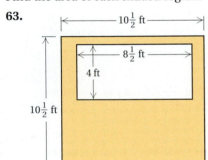

64.

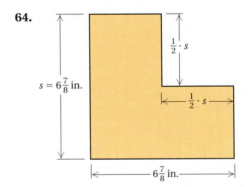

65.

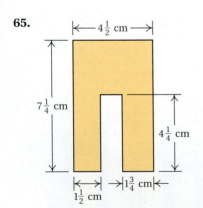

66.

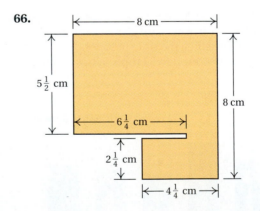

67. **D**_W Under what circumstances is a pair of mixed numerals more easily multiplied than added?

68. **D**_W Under what circumstances is a pair of mixed numerals more easily added than multiplied?

✏ VOCABULARY REINFORCEMENT

In each of Exercises 69–76, fill in the blank with the correct term from the given list. Some of the choices may not be used and some may be used more than once.

69. The set $\{\ldots, -3, -2, -1, 0, 1, 2, 3, \ldots\}$ is the set of
_____ . [2.1a]

70. When denominators are the same, we say that fractions have a _____ denominator. [3.5c]

71. The numbers 91, 95, and 111 are examples of
_____ numbers. [3.2b]

72. The number 22,223,133 is _____ by 9 because the sum of its digits is _____ by 9. [3.1b]

73. To add fractions with different denominators, we must first find the _____ of the denominators. [4.2b]

74. In the equation $2 + 3 = 5$, the numbers 2 and 3 are called
_____ . [1.2a]

75. In the expression $\dfrac{c}{d}$, we call c the _____ . [3.3a]

76. The number 0 has no _____ . [3.7a]

identity	common
reciprocal	prime
least common multiple	composite
	product
irrational numbers	divisible
	digits
integers	factors
numerator	addends
denominator	
equal	

SYNTHESIS

77. **D_W** A turntable for long-playing records (vinyl) typically turns $33\frac{1}{3}$ revolutions per minute. Write a problem that involves a turntable and has "The turntable spins $181\frac{1}{3}$ times" as the solution.

78. **D_W** If Kate and Jessie are both less than 5 ft $6\frac{1}{2}$ in. tall, but Dot is over 5 ft $6\frac{1}{2}$ in. tall, is it possible that the average height of the three exceeds 5 ft $6\frac{1}{2}$ in.? Why or why not?

Simplify. Write each answer as a mixed numeral whenever possible.

79. $-8 \div \dfrac{1}{2} + \dfrac{3}{4} + \left(-5 - \dfrac{5}{8}\right)^2$

80. $\left(\dfrac{5}{9} - \dfrac{1}{4}\right)(-12) + \left(-4 - \dfrac{3}{4}\right)^2$

81. $\dfrac{1}{3} \div \left(\dfrac{1}{2} - \dfrac{1}{5}\right) \times \dfrac{1}{4} + \dfrac{1}{6}$

82. $\dfrac{7}{8} - 1\dfrac{1}{8} \times \dfrac{2}{3} + \dfrac{9}{10} \div \dfrac{3}{5}$

83. Find r if

$$\frac{1}{r} = \frac{1}{40} + \frac{1}{60} + \frac{1}{80}.$$

84. *Heights.* Find the average height of the following NBA players:

Kevin Garnett	6 ft 11 in.
Ray Allen	6 ft 5 in.
Travis Best	5 ft 11 in.
Kobe Bryant	6 ft 7 in.
Shaquille O'Neal	7 ft 1 in.

85. *Water Consumption.* According to the *Consumer Guide to Home Energy Savings* (7th edition, by Alex Wilson et al., 1999, published by American Council for an Energy-Efficient America), washing one load of clothes uses $1\frac{3}{5}$ times the amount of hot water required for the average shower. If the average shower uses 20 gal of hot water, how much hot water will two showers and two loads of wash require?

The review that follows is meant to prepare you for a chapter exam. It consists of two parts. The first part, Concept Reinforcement, is designed to increase understanding of the concepts through true/false exercises. The second part is the Review Exercises. These provide practice exercises for the exam, together with references to section objectives so you can go back and review. Before beginning, stop and look back over the skills you have obtained. What skills in mathematics do you have now that you did not have before studying this chapter?

↪ CONCEPT REINFORCEMENT

Determine whether each statement is true or false. Answers are given at the back of the book.

_____ **1.** If $\dfrac{a}{b} > \dfrac{c}{b}$, $b \neq 0$, then $a > c$.

_____ **2.** All mixed numerals represent numbers larger than 1.

_____ **3.** The least common multiple of two natural numbers is the smallest number that is a factor of both.

_____ **4.** The mixed numeral $5\dfrac{2}{3}$ can be represented by the sum $5 \cdot \dfrac{3}{3} + \dfrac{2}{3}$.

_____ **5.** The least common multiple of two numbers is always larger than or equal to the larger number.

_____ **6.** To add fractions when denominators are the same, we keep the numerator and add the denominators.

Review Exercises

Find the LCM. [4.1a]

1. 16 and 20 **2.** 18 and 45 **3.** 3, 6, and 30

Perform the indicated operation and, if possible, simplify. [4.2a, b], [4.3a]

4. $\dfrac{2}{9} + \dfrac{5}{9}$

5. $\dfrac{7}{x} + \dfrac{2}{x}$

6. $-\dfrac{6}{5} + \dfrac{11}{15}$

7. $\dfrac{5}{16} + \dfrac{3}{24}$

8. $\dfrac{7}{9} - \dfrac{5}{9}$

9. $\dfrac{1}{4} - \dfrac{3}{8}$

10. $\dfrac{10}{27} - \dfrac{2}{9}$

11. $\dfrac{5}{6} - \dfrac{2}{9}$

Use < or > for ☐ to form a true sentence. [4.2c]

12. $\dfrac{4}{7}$ ☐ $\dfrac{5}{9}$

13. $-\dfrac{8}{9}$ ☐ $-\dfrac{11}{13}$

Solve. [4.3b], [4.4a]

14. $x + \dfrac{2}{5} = \dfrac{7}{8}$

15. $7a - 3 = 25$

16. $5 + \dfrac{16}{3}x = \dfrac{5}{9}$

17. $\dfrac{22}{5} = \dfrac{16}{5} + \dfrac{5}{2}x$

Solve by using the multiplication principle to clear fractions. [4.4b]

18. $\dfrac{5}{3}x + \dfrac{5}{6} = \dfrac{3}{2}$

Convert to fraction notation. [4.5a]

19. $7\dfrac{1}{2}$ **20.** $30\dfrac{4}{9}$ **21.** $-9\dfrac{2}{7}$

Convert to a mixed numeral. [4.5b]

22. $\dfrac{13}{5}$ **23.** $\dfrac{-27}{4}$

24. $\dfrac{57}{8}$ **25.** $\dfrac{7}{2}$

26. Divide. Write a mixed numeral for the answer. [4.5c]

$$7896 \div (-9)$$

27. Gina's golf scores were 80, 82, and 85. What was her average score? [4.5c]

Perform the indicated operation. Write a mixed numeral for each answer. [4.6a, b, d]

28. $\quad 7\dfrac{3}{5}$
$\underline{+\ 2\dfrac{4}{5}}$

29. $\quad 6\dfrac{1}{3}$
$\underline{+\ 5\dfrac{2}{5}}$

30. $-3\dfrac{5}{6} + \left(-5\dfrac{1}{6}\right)$ **31.** $-2\dfrac{3}{4} + 4\dfrac{1}{2}$

32. $\quad 14$
$\underline{-\ 6\dfrac{2}{9}}$

33. $\quad 9\dfrac{3}{5}$
$\underline{-\ 4\dfrac{13}{15}}$

34. $4\dfrac{5}{8} - 9\dfrac{3}{4}$ **35.** $-7\dfrac{1}{2} - 6\dfrac{3}{4}$

Combine like terms. [4.2a], [4.6b]

36. $\dfrac{4}{9}x + \dfrac{1}{3}x$ **37.** $8\dfrac{3}{10}a - 5\dfrac{1}{8}a$

Perform the indicated operation. Write a mixed numeral or integer for each answer. [4.7a, b]

38. $6 \cdot 2\dfrac{2}{3}$ **39.** $-5\dfrac{1}{4} \cdot \dfrac{2}{3}$

40. $2\dfrac{1}{5} \cdot 1\dfrac{1}{10}$ **41.** $2\dfrac{2}{5} \cdot 2\dfrac{1}{2}$

42. $-54 \div 2\frac{1}{4}$ **43.** $2\frac{2}{5} \div \left(-1\frac{7}{10}\right)$

44. $3\frac{1}{4} \div 26$ **45.** $4\frac{1}{5} \div 4\frac{2}{3}$

Evaluate.   [4.7c]

46. $5x - y$, for $x = 3\frac{1}{5}$ and $y = 2\frac{2}{7}$

47. $2a \div b$, for $a = 5\frac{2}{11}$ and $b = 3\frac{4}{5}$

Solve.

48. *Home Furnishings.*   Each shelf in June's entertainment center is 27 in. long. A videocassette is $1\frac{1}{8}$ in. thick. How many cassettes can she place on each shelf?   [4.7d]

49. The San Diaz drama club had $\frac{3}{8}$ of a vegetarian pizza, $1\frac{1}{2}$ cheese pizzas, and $1\frac{1}{4}$ pepperoni pizzas remaining after a cast party. How many pizzas remained altogether?   [4.6c]

50. *Bicycling.*   Mica pedals up a $\frac{1}{10}$-mi hill and then coasts for $\frac{1}{2}$ mi down the other side. How far has she traveled?   [4.2d]

51. *Communication Cable.*   Celebrity Cable has two crews who install Internet communication cable. Crew A can install $38\frac{1}{8}$ ft per hour. Crew B can install $31\frac{2}{3}$ ft per hour. How many fewer feet can Crew B install per hour than Crew A?   [4.6c]

52. A wedding-cake recipe requires 12 cups of shortening. Being calorie-conscious, the wedding couple decides to reduce the shortening by $3\frac{5}{8}$ cups and replace it with prune purée. How many cups of shortening are used in their new recipe?   [4.6c]

53. *Book Size.*   One standard book size is $8\frac{1}{2}$ in. by $9\frac{3}{4}$ in. What is the total distance around (perimeter of) the front cover of such a book?   [4.6c]

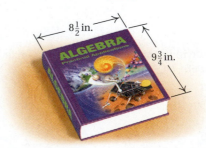

54. *Population.* The population of Louisiana is $2\frac{1}{2}$ times the population of West Virginia. The population of West Virginia is approximately 1,800,000. What is the population of Louisiana? [4.7d]

Source: U.S. Bureau of the Census

55. What is the sum of the areas in the figure below? [4.6c], [4.7d]

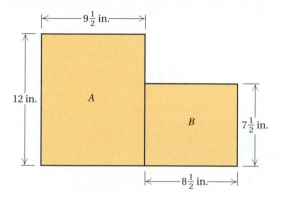

56. In the figure in Exercise 55, how much larger is the area of rectangle *A* than the area of rectangle *B*? [4.6c], [4.7d]

SYNTHESIS

57. **D**_W Rachel insists that $3\frac{2}{5} \cdot 1\frac{3}{7} = 3\frac{6}{35}$. What mistake is she probably making and how should she have proceeded instead? [4.7a]

58. **D**_W Do least common multiples play any role in the addition or subtraction of mixed numerals? Why or why not? [4.6a, b]

59. 🖩 Find the LCM of 141, 2419, and 1357. [4.1a]

60. Find *r* if

$$\frac{1}{r} = \frac{1}{100} + \frac{1}{150} + \frac{1}{200}.$$ [4.2b], [4.4b]

61. Find the smallest integer for which each fraction is greater than $\frac{1}{2}$. [4.2c]

a) $\dfrac{\square}{11}$ **b)** $\dfrac{\square}{8}$

c) $\dfrac{\square}{23}$ **d)** $\dfrac{\square}{35}$

e) $\dfrac{-51}{\square}$ **f)** $\dfrac{-78}{\square}$

g) $\dfrac{-2}{\square}$ **h)** $\dfrac{-1}{\square}$

62. Find the largest integer for which each fraction is greater than 1. [4.2c]

a) $\dfrac{7}{\square}$ **b)** $\dfrac{11}{\square}$

c) $\dfrac{47}{\square}$ **d)** $\dfrac{\frac{9}{8}}{\square}$

e) $\dfrac{\square}{-13}$ **f)** $\dfrac{\square}{-27}$

g) $\dfrac{\square}{-1}$ **h)** $\dfrac{\square}{-\frac{1}{2}}$

1. Find the LCM of 12 and 16.

Perform the indicated operation and, if possible, simplify.

2. $\dfrac{1}{2} + \dfrac{5}{2}$

3. $-\dfrac{7}{8} + \dfrac{2}{3}$

4. $\dfrac{5}{t} - \dfrac{3}{t}$

5. $\dfrac{5}{6} - \dfrac{3}{4}$

6. $\dfrac{5}{8} - \dfrac{17}{24}$

Solve.

7. $x + \dfrac{2}{3} = \dfrac{11}{12}$

8. $-5x - 3 = 9$

9. $\dfrac{3}{4} = \dfrac{1}{2} + \dfrac{5}{3}x$

10. Use $<$ or $>$ for $\square$ to form a true sentence.

$$\dfrac{6}{7} \;\square\; \dfrac{21}{25}$$

Convert to fraction notation.

11. $3\dfrac{1}{2}$

12. $-9\dfrac{3}{8}$

13. Convert to a mixed numeral:

$$-\dfrac{74}{9}.$$

14. Divide. Write a mixed numeral for the answer.

$$1\,1\,\overline{)\,1\,7\,8\,9}$$

Perform the indicated operation. Write a mixed numeral for each answer.

15. $\begin{array}{r} 6\dfrac{2}{5} \\ + \ 7\dfrac{4}{5} \\ \hline \end{array}$

16. $\begin{array}{r} 3\dfrac{1}{4} \\ + \ 9\dfrac{1}{6} \\ \hline \end{array}$

17. $\begin{array}{r} 10\dfrac{1}{6} \\ - \ 5\dfrac{7}{8} \\ \hline \end{array}$

18. $14 + \left(-5\dfrac{3}{7}\right)$

19. $3\dfrac{4}{5} - 9\dfrac{1}{2}$

Combine like terms.

20. $\dfrac{3}{8}x - \dfrac{1}{2}x$

21. $5\dfrac{2}{11}a - 3\dfrac{1}{5}a$

Perform the indicated operation. Write a mixed numeral for each answer.

22. $9 \cdot 4\dfrac{1}{3}$

23. $6\dfrac{3}{4} \cdot \left(-2\dfrac{2}{3}\right)$

24. $33 \div 5\dfrac{1}{2}$

25. $2\dfrac{1}{3} \div 1\dfrac{1}{6}$

Evaluate.

26. $\frac{2}{3}ab$, for $a = 7$ and $b = 4\frac{1}{5}$

27. $4 + mn$, for $m = 7\frac{2}{5}$ and $n = 3\frac{1}{4}$

Solve.

28. *Turkey Chili.* One batch of low-cholesterol turkey chili calls for $1\frac{1}{2}$ lb of roasted turkey breast. How much turkey is needed for 5 batches?

29. *Book Order.* An order of books for a math course weighs 220 lb. Each book weighs $2\frac{3}{4}$ lb. How many books are in the order?

30. *Carpentry.* The following diagram shows a middle drawer support guide for a cabinet drawer. Find each of the following.

  a) The short length a across the top
  b) The length b across the bottom

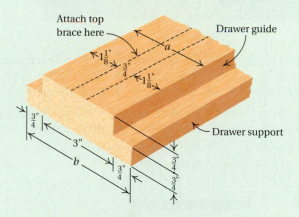

31. *Carpentry.* In carpentry, some pieces of plywood that are called "$\frac{3}{4}$-inch" plywood are actually $\frac{11}{16}$ in. thick. How much thinner is such a piece than its name indicates?

32. *Women's Dunks.* The first three women in the history of college basketball able to dunk a basketball are listed below. Their names, heights, and universities are:

  Michelle Snow, $6\frac{5}{12}$ ft, Tennessee;

  Charlotte Smith, $5\frac{11}{12}$ ft, North Carolina;

  Georgeann Wells, $6\frac{7}{12}$ ft, West Virginia.

Find the average height of these women.

Source: *USA Today,* 11/30/00, p. 3C

SYNTHESIS

33. Yuri and Olga are orangutans who perform in a circus by riding bicycles around a circular track. It takes Yuri $\frac{6}{25}$ min and Olga $\frac{8}{25}$ min to complete one lap. They start their act together at one point and complete their act when they are next together at that point. How long does the act last?

34. Cheri runs 17 laps at her health club. Trent runs 17 laps at his health club. If the track at Cheri's health club is $\frac{1}{7}$ mi long, and the track at Trent's is $\frac{1}{8}$ mi long, who runs farther? How much farther?

35. The students in a math class can be organized into study groups of 8 each such that no students are left out. The same class of students can also be organized into groups of 6 such that no students are left out.

  a) Find some class sizes for which this will work.
  b) Find the smallest such class size.

36. Simplify each of the following, using fraction notation. Try to answer part (e) by recognizing a pattern in parts (a) through (d).

  a) $\dfrac{1}{1 \cdot 2}$

  b) $\dfrac{1}{1 \cdot 2} + \dfrac{1}{2 \cdot 3}$

  c) $\dfrac{1}{1 \cdot 2} + \dfrac{1}{2 \cdot 3} + \dfrac{1}{3 \cdot 4}$

  d) $\dfrac{1}{1 \cdot 2} + \dfrac{1}{2 \cdot 3} + \dfrac{1}{3 \cdot 4} + \dfrac{1}{4 \cdot 5}$

  e) $\dfrac{1}{1 \cdot 2} + \dfrac{1}{2 \cdot 3} + \dfrac{1}{3 \cdot 4} + \dfrac{1}{4 \cdot 5} + \dfrac{1}{5 \cdot 6} + \dfrac{1}{6 \cdot 7} + \dfrac{1}{7 \cdot 8} +$
  $\dfrac{1}{8 \cdot 9} + \dfrac{1}{9 \cdot 10}$

Solve.

1. *Excelsior Made from Aspen.* Excelsior consists of slender, curved wood shavings and is often used for packing. Shown at right are examples of excelsior and the saw blades used to cut it, as made by Western Excelsior Corporation of Mancos, CO. The width of strips for craft decoration is either $\frac{1}{16}$ in. or $\frac{1}{8}$ in. The width for erosion control mats used for stabilizing soil and nourishing young crops is $\frac{1}{24}$ in.

a) How much wider is the $\frac{1}{16}$-in. craft decoration excelsior than the erosion control excelsior?

b) How much wider is the $\frac{1}{8}$-in. craft decoration excelsior than the erosion control excelsior?

Source: Western Excelsior Corporation

2. *DVD Storage.* Gregory is making a home entertainment center. He is planning a 27-in. shelf that holds DVDs that are each $\frac{7}{16}$ in. thick. How many DVDs will the shelf hold?

3. *Cross-Country Skiing.* During a three-day holiday weekend trip, David and Sally Jean cross-country skied $3\frac{2}{3}$ mi on Friday, $6\frac{1}{8}$ mi on Saturday, and $4\frac{3}{4}$ mi on Sunday.

a) Find the total number of miles they skied.

b) Find the average number of miles they skied per day. Express your answer as a mixed numeral.

4. *Room Carpeting.* The Chandlers are carpeting an L-shaped family room consisting of a rectangle that is $8\frac{1}{2}$ ft by 11 ft and one that is $6\frac{1}{2}$ ft by $7\frac{1}{2}$ ft.

a) Find the area of the carpet.

b) Find the perimeter of the carpet.

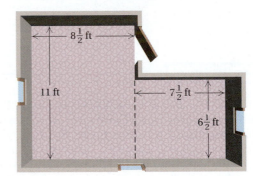

5. How many people can get equal $16 shares from a total of $496?

6. An emergency food pantry fund contains $423. From this fund, $148 and $167 are withdrawn for expenses. How much is left in the fund?

7. A recipe calls for $\frac{4}{5}$ tsp of salt. How much salt should be used for $\frac{1}{2}$ recipe? for 5 recipes?

8. A book weighs $2\frac{3}{5}$ lb. How much do 15 books weigh?

9. How many pieces, each $2\frac{3}{8}$ ft long, can be cut from a piece of wire 38 ft long?

10. In a walkathon, Jermaine walked $\frac{9}{10}$ mi and Oleta walked $\frac{75}{100}$ mi. What was the total distance they walked?

11. In the number 2753, what digit names tens?

12. Write expanded notation for 6075.

13. Write a word name for the number in the following sentence: The diameter of Uranus is 29,500 miles.

14. What part is shaded?

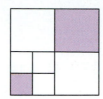

Calculate and, if possible, simplify.

15.
$$\begin{array}{r} 3\ 7\ 5 \\ +\ 2\ 4\ 8 \\ \hline \end{array}$$

16. $29 + (-37)$

17. $\dfrac{3}{8} + \dfrac{1}{24}$

18.
$$\begin{array}{r} 2\dfrac{3}{4} \\ +\ 5\dfrac{1}{2} \\ \hline \end{array}$$

19.
$$\begin{array}{r} 7\ 4\ 6\ 9 \\ -\ 2\ 3\ 4\ 5 \\ \hline \end{array}$$

20. $-9 - (-25)$

21. $\dfrac{4}{t} - \dfrac{9}{t}$

22.
$$\begin{array}{r} 2\dfrac{1}{3} \\ -\ 1\dfrac{1}{6} \\ \hline \end{array}$$

23.
$$\begin{array}{r} 2\ 7\ 8 \\ \times\ \ \ 1\ 8 \\ \hline \end{array}$$

24. $29(-5)$

25. $\dfrac{9}{10} \cdot \dfrac{5}{3}$

26. $18\left(-\dfrac{5}{6}\right)$

27. $2\dfrac{1}{3} \cdot 3\dfrac{1}{7}$

Divide. Write the answer with the remainder in the form 34 R 7.

28. $731 \div 15$

29. $4\ 5\ )\overline{\ 2\ 5\ 3\ 1\ }$

30. In Question 29, write a mixed numeral for the answer.

Divide and, if possible, simplify.

31. $\dfrac{2}{5} \div \left(-\dfrac{7}{10}\right)$

32. $2\dfrac{1}{5} \div \dfrac{3}{10}$

33. Round 38,478 to the nearest hundred.

34. Find the LCM of 24 and 36.

35. Without performing the division, determine whether 4296 is divisible by 6.

36. Find all factors of 16.

Use $<$, $>$, or $=$ for ☐ to form a true sentence.

37. $\dfrac{4}{5}$ ☐ $\dfrac{4}{6}$

38. $-\dfrac{3}{7}$ ☐ $-\dfrac{5}{12}$

Simplify.

39. $\dfrac{36}{45}$

40. $-\dfrac{420}{30}$

41. Convert to fraction notation: $7\dfrac{3}{10}$.

42. Convert to a mixed numeral: $-\dfrac{17}{3}$.

Solve.

43. $x + 37 = 92$

44. $x + \dfrac{7}{9} = \dfrac{4}{3}$

45. $\dfrac{7}{9} \cdot t = -\dfrac{4}{3}$

46. $\dfrac{5}{7} = \dfrac{1}{3} + 4a$

47. Evaluate $\dfrac{t + p}{3}$ for $t = -4$ and $p = 16$.

48. Multiply: $7(b - 5)$.

49. Use the distributive law to write an equivalent expression: $-3(x - 2 + z)$.

50. Combine like terms: $x - 5 - 7x - 4$.

51. *Copier Paper.* A standard sheet of copier paper is $8\frac{1}{2}$ in. by 11 in. What is the perimeter of the paper?

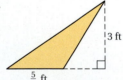

11 in. $8\frac{1}{2}$ in.

52. *Gardening.* Ragheda built a square raised bed for her herb garden. If the length of each side of the square was 12 ft, what was the perimeter of the garden?

Find the area of each figure.

53.

3 ft

$\frac{5}{2}$ ft

54.

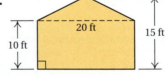

20 ft   15 ft

10 ft

55. Solve: $7x - \dfrac{2}{3}(x - 6) = 6\dfrac{5}{7}$.

56. Each floor of a seven-story office building is 25 m by $22\frac{1}{2}$ m, with a 5-m by $4\frac{1}{2}$-m elevator/stairwell. How many square meters of office space are in the building?

Decimal Notation

5

Real-World Application

Apple sells iPhoto books that users of the iPhoto application can create. The price of a large, softcover 20-page book is $19.99. Additional pages are 69 cents per page. Marta has $35 to spend on a book. What is the greatest number of pages she can put in the book?

Source: www.apple.com

This problem appears as Example 6 in Section 5.8.

DECIMAL NOTATION

Objectives

a Given decimal notation, write a word name, and write a word name for an amount of money.

b Convert between fraction notation and decimal notation.

c Given a pair of numbers in decimal notation, tell which is larger.

d Round decimal notation to the nearest thousandth, hundredth, tenth, one, ten, hundred, or thousand.

249.98

Study Tips

MATH WEB SITES

If for some reason you feel uneasy about asking an instructor or tutor for help, and have already considered the Addison-Wesley Math Tutor Center (see the Preface and page 40), you may benefit from visiting the Web sites www.hotmath.com, www.mathforum.org/dr.math, www.sosmath.com, and http://school.discovery.com/homeworkhelp/webmath/how.html. These sites are not specifically for adult learners, but can be very helpful nonetheless.

The set of **rational numbers** consists of the **integers**

$$\ldots, -3, -2, -1, 0, 1, 2, 3, \ldots,$$

and fractions like

$$\frac{1}{2}, \frac{2}{3}, \frac{-7}{8}, \frac{17}{-10}, \text{ and so on.}$$

We used fraction notation for rational numbers in Chapters 3 and 4. Here in Chapter 5, we will use *decimal notation* to represent the set of rational numbers. For example, $\frac{3}{4}$ will be written as 0.75, and $9\frac{1}{2}$ will be written as 9.5. A number written in decimal notation is often simply referred to as a *decimal*.

The word *decimal* comes from the Latin word *decima*, meaning a *tenth part*. Since our usual counting system is based on tens, we will find decimal notation to be a natural extension of a system with which we are already familiar.

a Decimal Notation and Word Names

A portable DVD player sells for $249.98. The dot in $249.98 is called a **decimal point.** Since $0.98, or 98¢, is $\frac{98}{100}$ of a dollar, it follows that

$$\$249.98 = 249 + \frac{98}{100} \text{ dollars.}$$

Also, since $0.98, or 98¢, has the same value as

$$9 \text{ dimes} + 8 \text{ cents}$$

and 1 dime is $\frac{1}{10}$ of a dollar and 1 cent is $\frac{1}{100}$ of a dollar, we can write

$$249.98 = 2 \cdot 100 + 4 \cdot 10 + 9 \cdot 1 + 9 \cdot \frac{1}{10} + 8 \cdot \frac{1}{100}.$$

This is an extension of the expanded notation for whole numbers that we used in Chapter 1. The place values are 100, 10, 1, $\frac{1}{10}$, $\frac{1}{100}$, and so on. We can see this on a **place-value chart.** The value of each place is $\frac{1}{10}$ as large as the one to its left.

Let's consider decimal notation using a place-value chart to represent 26.3385 min, the men's 10,000-m run record held by Kenenisa Bekele from Ethiopia.

PLACE-VALUE CHART							
Hundreds	Tens	Ones	Ten*ths*	Hundred*ths*	Thousand*ths*	Ten-Thousand*ths*	Hundred-Thousand*ths*
100	10	1	$\frac{1}{10}$	$\frac{1}{100}$	$\frac{1}{1000}$	$\frac{1}{10,000}$	$\frac{1}{100,000}$
	2	6 . 3	3	8	5		

The decimal notation 26.3385 means

2 tens + 6 ones + 3 tenths + 3 hundredths + 8 thousandths + 5 ten-thousandths

or $20 + 6 + \dfrac{3}{10} + \dfrac{3}{100} + \dfrac{8}{1000} + \dfrac{5}{10,000}$.

Using 10,000 as the least common denominator, we have

$$26.3385 = 26 + \frac{3000}{10,000} + \frac{300}{10,000} + \frac{80}{10,000} + \frac{5}{10,000} = 26\frac{3385}{10,000}.$$

We read both 26.3385 and $26\frac{3385}{10,000}$ as

"Twenty-six and three thousand three hundred eighty-five ten-thousandths."

We read the decimal point as "and." Note that the word names to the right of the decimal point always end in *th*. We can also read 26.3385 as "Two six *point* three three eight five," or "Twenty-six point three three eight five."

To write a word name from decimal notation,

a) write a word name for the whole number (the number named to the left of the decimal point),

    397.685 ⟶ Three hundred ninety-seven

b) write the word "and" for the decimal point, and

    397.685 ⟶ Three hundred ninety-seven and

c) write a word name for the number named to the right of the decimal point, followed by the place value of the last digit.

    397.685 ⟶ Three hundred ninety-seven and six hundred eighty-five thousandths

---

EXAMPLE 1 *Ice Cream.* Write a word name for the number in this sentence: The average person eats 26.3 servings of ice cream a year.

Source: NPD Group; J. M. Hirsch, Associated Press

    Twenty-six and three tenths

Write a word name for the decimal numbers.

1. Life Expectancy. In 2005, the life expectancy at birth in Switzerland was 80.39. In the United States, it was 77.71.

Source: *CIA World Factbook,* 2005

2. Kentucky Derby. In 2005, racehorse Giacomo won the Kentucky Derby in a time of 2.04583 min.

Source: msnbc.msn.com

Answers on page A-12

3. 245.89

4. 34.00647

5. 31,079.764

Write in words, as on a check.

6. $4217.56

7. $13.98

Answers on page A-12

EXAMPLE 2 Write a word name for the number in this sentence: The world record in the women's pole vault is 5.01 m, set by Yelena Isinbayeva of Russia.

Source: International Association of Athletics Federations

Five and one hundredth

EXAMPLE 3 Write a word name for the number in this sentence: The world record in the men's 800-m run is 1.6852 min, held by Wilson Kipketer of Denmark.

Source: International Association of Athletics Federations

One and six thousand eight hundred fifty-two ten-thousandths

EXAMPLE 4 Write a word name for the number in this sentence: The current one-mile land speed record of 763.035 mph is held by Andy Green.

Source: www.castrol.com

Seven hundred sixty-three and thirty-five thousandths

Do Exercises 1–5 (Exercises 1 and 2 are on the preceding page.)

Decimal notation is also used with money. It is common on a check to write "and ninety-five cents" as "and $\frac{95}{100}$ dollars."

EXAMPLE 5 Write $5876.95 in words, as on a check.

Five thousand, eight hundred seventy-six and $\frac{95}{100}$ dollars

Do Exercises 6 and 7.

b | Converting Between Decimal Notation and Fraction Notation

We can find fraction notation as follows.

$$9.875 = 9 + \frac{875}{1000} = \frac{9000}{1000} + \frac{875}{1000} = \frac{9875}{1000}$$

Decimal notation ———— Fraction notation

$$9.875 \qquad \frac{9875}{1000}$$

3 decimal places   3 zeros

To convert from decimal to fraction notation,

a) count the number of decimal places,

$$4.98$$

┗━ 2 places

b) move the decimal point that many places to the right, and

4.98. Move 2 places.

c) write the answer over a denominator of 1 followed by that number of zeros.

$$\frac{498}{100}$$ 2 zeros

For a number like 0.876, we write a 0 to call attention to the presence of a decimal point.

EXAMPLE 6 Write fraction notation for 0.876. Do not simplify.

$$0.876 \qquad 0.876. \qquad 0.876 = \frac{876}{1000}$$

3 places   3 zeros

Decimals greater than 1 or less than -1 can be written either as fractions or as mixed numerals.

EXAMPLE 7 Write 56.23 as a fraction and as a mixed numeral.

$$56.23 \qquad 56.23. \qquad 56.23 = \frac{5623}{100}, \qquad \text{and} \qquad 56.23 = 56\frac{23}{100}$$

2 places   2 zeros

As a check, note that both 56.23 and $56\frac{23}{100}$ are read as "fifty-six and twenty-three hundredths."

EXAMPLE 8 Write -2.6073 as a fraction and as a mixed numeral.

We have

$$-2.6073. = -\frac{26{,}073}{10{,}000}, \qquad \text{and} \qquad -2.6073 = -2\frac{6073}{10{,}000}.$$

4 places   4 zeros

Do Exercises 8–10.

To write $\frac{5328}{10}$ as a decimal we can first divide to find an equivalent mixed numeral.

$$\frac{5328}{10} = 532\frac{8}{10}$$

Next note that

$$532\frac{8}{10} = 532 + \frac{8}{10}$$

$$= 532.8$$

$$\begin{array}{r} 5\ 3\ 2 \\ 1\ 0\ )\overline{5\ 3\ 2\ 8} \\ 5\ 0 \\ \hline 3\ 2 \\ 3\ 0 \\ \hline 2\ 8 \\ 2\ 0 \\ \hline 8 \end{array}$$

Write fraction notation. Do not simplify.

8. 0.549

Write as a fraction and as a mixed numeral.

9. 75.069

10. -312.9

Answers on page A-12

Write decimal notation for each number.

11. $\dfrac{743}{100}$

12. $-\dfrac{73}{1000}$

13. $\dfrac{67,089}{10,000}$

14. $-\dfrac{9}{10}$

Write decimal notation for each number.

15. $-7\dfrac{3}{100}$

16. $23\dfrac{47}{1000}$

This procedure can be generalized. It is the reverse of the procedure used in Examples 6–8.

To convert from fraction notation to decimal notation when the denominator is 10, 100, 1000, and so on,

a) count the number of zeros, and

$$\dfrac{8679}{1000}$$

└── 3 zeros

b) move the decimal point that number of places to the left. Leave off the denominator.

8.679. Move 3 places.

$$\dfrac{8679}{1000} = 8.679$$

EXAMPLE 9 Write decimal notation for $\dfrac{47}{10}$.

$$\dfrac{47}{10}$$

└── 1 zero

4.7. $\dfrac{47}{10} = 4.7$ The decimal point is moved 1 place.

EXAMPLE 10 Write decimal notation for $\dfrac{123,067}{10,000}$.

$$\dfrac{123,067}{10,000}$$

└── 4 zeros

12.3067. $\dfrac{123,067}{10,000} = 12.3067$ The decimal point is moved 4 places.

To move the decimal point to the left, we may need to write extra 0's.

EXAMPLE 11 Write decimal notation for $-\dfrac{9}{100}$.

$$-\dfrac{9}{100}$$

└── 2 zeros

−0.09. $-\dfrac{9}{100} = -0.09$ The decimal point is moved 2 places.

Do Exercises 11–14.

For denominators other than 10, 100, and so on, we will usually perform long division. This is examined in Section 5.5.

If a mixed numeral has a fraction part with a denominator that is a power of ten, such as 10, 100, or 1000, and so on, we first write the mixed numeral as a sum of a whole number and a fraction. Then we convert to decimal notation.

EXAMPLE 12 Write decimal notation for $23\dfrac{59}{100}$.

$$23\dfrac{59}{100} = 23 + \dfrac{59}{100} = 23 \text{ and } \dfrac{59}{100} = 23.59$$

Do Exercises 15 and 16.

Answers on page A-12

C Order

To compare numbers in decimal notation, consider 0.85 and 0.9. First note that $0.9 = 0.90$ because $\frac{9}{10} = \frac{90}{100}$. Since $0.85 = \frac{85}{100}$, it follows that $\frac{85}{100} < \frac{90}{100}$ and $0.85 < 0.9$. This leads us to a quick way to compare two numbers in decimal notation.

> To compare two positive numbers in decimal notation, start at the left and compare corresponding digits. When two digits differ, the number with the larger digit is the larger of the two numbers. To ease the comparison, extra zeros can be written to the right of the last decimal place.

EXAMPLE 13 Which is larger: 2.109 or 2.1?

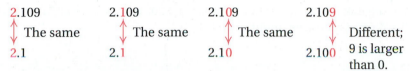

Thus, 2.109 is larger. In symbols, $2.109 > 2.1$.

EXAMPLE 14 Which is larger: 0.09 or 0.108?

0.09   The same   0.09   Different; 1 is larger than 0.
0.108   0.108

Thus, 0.108 is larger. In symbols, $0.108 > 0.09$.

Do Exercises 17–20.

As before, we can use a number line to visualize order. We illustrate Examples 13 and 14 below. Larger numbers are always to the right.

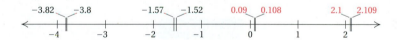

Note from the number line that $-2 < -1$. Similarly, $-1.57 < -1.52$.

> To compare two negative numbers in decimal notation, start at the left and compare corresponding digits. When two digits differ, the number with the smaller digit is the larger of the two numbers.

EXAMPLE 15 Which is larger: −3.8 or −3.82?

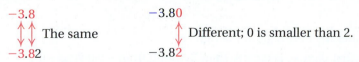

Thus, −3.8 is larger. In symbols, $-3.8 > -3.82$ (see the number line above).

Do Exercises 21–24.

Which number is larger?

17. 2.04, 2.039

18. 0.06, 0.008

19. 0.5, 0.58

20. 1, 0.9999

Which number is larger?

21. 0.8989, 0.09898

22. 21.006, 21.05

23. −34.01, −34.008

24. −9.12, −8.98

Answers on page A-12

Round to the nearest tenth.

25. 2.76   **26.** 13.85

27. −234.448   **28.** 7.009

Round to the nearest hundredth.

29. 0.6362   **30.** −7.8348

31. 34.69514   **32.** −0.02521

Round to the nearest thousandth.

33. 0.94347   **34.** −8.00382

35. −43.111943   **36.** 37.400526

Round 7459.35981 to the nearest:

37. Thousandth.

38. Hundredth.

39. Tenth.

40. One.

41. Ten. (*Caution*: "Tens" are not "tenths.")

42. Hundred.

43. Thousand.

Answers on page A-12

CHAPTER 5: Decimal Notation

---

d Rounding

We round decimals in much the same way that we round whole numbers. To see how, we use a number line.

EXAMPLE 16 Round 0.37 to the nearest tenth.

Here is part of a number line, magnified.

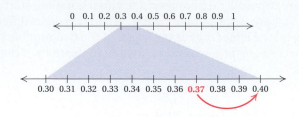

We see that 0.37 is closer to 0.40 than to 0.30. Thus, when 0.37 is rounded to the nearest tenth, we round *up* to 0.4.

> To round to a certain place:
>
> **a)** Locate the digit in that place.
> **b)** Consider the next digit to the right.
> **c)** If the digit to the right is 5 or greater, add one to the original digit. If the digit to the right is 4 or less, the original digit does not change. In either case, drop all numbers to the right of the original digit.

EXAMPLE 17 Round 72.3846 to the nearest hundredth.

a) Locate the digit in the hundredths place.

$$7\ 2.3\ 8\ 4\ 6$$

b) Consider the next digit to the right.

c) Since that digit, 4, is less than 5, we round *down* from 72.3846 to 72.38.

Caution!

72.39 is not a correct answer to Example 17. It is incorrect to round sequentially from right to left as follows: 72.3846, 72.385, 72.39.

72.3846 is closer to 72.38 than to 72.39.

EXAMPLE 18 Round −0.064 to the nearest tenth.

a) Locate the digit in the tenths place.

$$-0.0\ 6\ 4$$

b) Consider the next digit to the right.

c) Since that digit, 6, is greater than 5, round from −0.064 to −0.1.

The answer is −0.1. Since −0.1 < −0.064, we actually rounded *down*.

Do Exercises 25–43.

5.1

EXERCISE SET

For Extra Help

MathXL   MyMathLab   InterAct   Math Tutor   Video   Student's
                        Math   Center   Lectures   Solutions
                                                 on CD   Manual
                                                 Disc 3

a   Write a word name for the number in the sentence.

NYSE INDEX
MOST ACTIVE: SHARE VOLUME

	VOL. (000s)	LAST	CHANGE
Elan	97,202	6.65	−1.29
Lucent	46,527	3.08	−.08
NewsCpA n	32,126	17.28	+.48
BostonSci	24,330	30.90	−.66
Pfizer	21,024	26.59	−.02
ExxonMbl	20,149	63.05	+.37
WalMart	19,358	52.86	+.91
GenElec	17,824	35.55	−.05
Motorola	17,355	15.20	−.38
Texinst	17,234	26.58	−.39

1. *ExxonMbl.* Recently, the stock of Exxon Mobil sold for $63.05 per share.

2. *Motorola.* Recently, the stock of Motorola sold for $15.20 per share.

3. *Pfizer.* Recently, the stock of Pfizer sold for $26.59 per share.

4. *Wal-Mart.* Recently, the stock of Wal-Mart sold for $52.86 per share.

5. *Water Weight.* One gallon of water weighs 8.35 lb.

6. One gallon of paint is equal to 3.785 liters of paint.

7. *Coffee Consumption.* Finland has the largest per capita coffee consumption in the world, with each person using, on average, 24.6875 oz of ground coffee a year.
Source: International Coffee Organization

8. *Theater Admissions.* Of all countries, the United States has the largest per capita movie theater admissions, with each person attending a movie, on average, 5.53 times a year.
Source: Screen Digest

9. The average loss of daylight in October in Anchorage, Alaska, is 5.63 min per day.

10. Recently, one British pound was worth about $0.57790 in U.S. currency.
Source: www.x-rates.com

Write in words, as on a check.

11. $524.95

12. $149.99

13. $36.72

14. $0.67

b   Write each number as a fraction and, if possible, as a mixed numeral. Do not simplify.

15. 7.3

16. 4.9

17. 203.6

18. −57.32

19. −2.703

20. 0.00013

21. 0.0109

22. 1.0008

23. −4.0003

24. −9.012

25. −0.0207

26. −0.00104

27. 70.00105

28. 60.0403

Write decimal notation for each number.

29. $\dfrac{3}{10}$ **30.** $\dfrac{73}{10}$ **31.** $-\dfrac{59}{100}$ **32.** $-\dfrac{67}{100}$ **33.** $\dfrac{3798}{1000}$

34. $\dfrac{780}{1000}$ **35.** $\dfrac{78}{10,000}$ **36.** $\dfrac{56,788}{100,000}$ **37.** $\dfrac{-18}{100,000}$ **38.** $\dfrac{-2347}{100}$

39. $\dfrac{486,197}{1,000,000}$ **40.** $\dfrac{8,953,074}{1,000,000}$ **41.** $7\dfrac{13}{1000}$ **42.** $4\dfrac{909}{1000}$ **43.** $-8\dfrac{431}{1000}$

44. $-49\dfrac{32}{1000}$ **45.** $2\dfrac{1739}{10,000}$ **46.** $9243\dfrac{1}{10}$ **47.** $8\dfrac{953,073}{1,000,000}$ **48.** $2256\dfrac{3059}{10,000}$

c   Which number is larger?

49. 0.06,   0.58   **50.** 0.008,   0.8   **51.** 0.403,   0.410   **52.** 42.06,   42.1

53. −5.046,   −5.043   **54.** −324.19,   −325.19   **55.** 234.07,   235.07   **56.** 0.99999,   1

57. 0.007, $\dfrac{7}{100}$ **58.** $\dfrac{73}{10}$,   0.73   **59.** −0.872,   −0.873   **60.** −0.8437,   −0.84384

d   Round to the nearest tenth.

61. 0.23   **62.** 0.85   **63.** −0.372   **64.** −0.261

65. 2.951   **66.** 7.532   **67.** −327.2347   **68.** −8.7493

Round to the nearest hundredth.

69. 0.893   **70.** 0.675   **71.** −0.6666   **72.** −7.5252

73. 0.9952   **74.** 207.9976   **75.** −0.03488   **76.** −9.27481

Round to the nearest thousandth.

77. 0.5724   **78.** 0.6666   **79.** 17.0015   **80.** 123.4562

81. −20.20202   **82.** −0.10346   **83.** 9.98487   **84.** 67.100602

Round 809.47321 to the nearest:

85. Tenth.   **86.** Thousandth.   **87.** Hundredth.   **88.** One.

89. **D_W** Brian rounds 536.447 to the nearest one and, incorrectly, gets 537. How might he have made this mistake?

90. **D_W** Describe in your own words a procedure for converting from decimal notation to fraction notation.

CHAPTER 5: Decimal Notation

Add or subtract, as indicated.

91.
```
  6 8 1
+ 1 4 9
```
[1.2b]

92. $\dfrac{681}{1000} + \dfrac{149}{1000}$ [4.2a]

93.
```
  2 6 7
−   8 5
```
[1.3d]

94. $\dfrac{267}{100} − \dfrac{85}{100}$ [4.3a]

95. $\dfrac{37}{55} − \dfrac{49}{55}$ [4.3a]

96. $−\dfrac{29}{34} + \dfrac{14}{34}$ [4.2a]

97.
```
  3 4,9 0 3
−    1,9 4 5
```
[1.3d]

98.
```
  4 9 3 7
+ 5 7 8 9
```
[1.2b]

99. **D**W Describe a series of steps that could be used to write fractions like $\frac{3}{4}$, $\frac{2}{5}$, $\frac{7}{20}$, or $\frac{19}{25}$ in decimal form.

100. **D**W Why is it preferable to *not* use the word "and" when reading a number like 457?

101. Arrange the following numbers in order from smallest to largest.

$$−0.989,\ −0.898,\ −1.009,\ −1.09,\ −0.098$$

102. Arrange the following numbers in order from smallest to largest.

$$−2.018,\ −2.1,\ −2.109,\ −2.0119,\ −2.108,$$
$$−2.000001$$

Truncating. There are other methods of rounding decimal notation. A computer often uses a method called **truncating.** To truncate we drop off decimal places right of the rounding place, which is the same as changing all digits to the right of the rounding place to zeros. For example, rounding 6.78093456285102 to the ninth decimal place, using truncating, gives us 6.780934562. Use truncating to round each of the following to the fifth decimal place, that is, the hundred-thousandth.

103. 6.78346123

104. 6.783461902

105. 99.999999999

106. 0.030303030303

Global Warming. The graph below is based on the average global temperatures for January through May of 1910 through 2004. Each bar indicates, in Fahrenheit degrees, how much above or below average the temperature was for the year.

107. For what year(s) was the yearly temperature more than 0.4 degree above average?

108. What was the last year in which the yearly temperature was more than 0.6 degree below average?

A Warming Trend

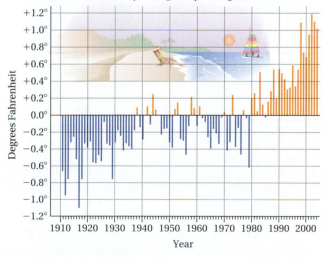

Degrees above or below the average global temperature between 1910 and 2004 for January through May. In degrees Fahrenheit.

Degrees Fahrenheit

+1.2°
+1.0°
+0.8°
+0.6°
+0.4°
+0.2°
0.0°
−0.2°
−0.4°
−0.6°
−0.8°
−1.0°
−1.2°

1910 1920 1930 1940 1950 1960 1970 1980 1990 2000

Year

Sources: Council on Environmental Quality, *The New York Times*, and NASA Goddard Institute for Space Studies

109. What was the last year in which the yearly temperature was below average?

110. For what year(s) was the yearly temperature more than 1.0 degree above average?

319

Objectives

a Add using decimal notation.

b Subtract using decimal notation.

c Add and subtract negative decimals.

d Combine like terms with decimal coefficients.

Add.

1.
```
      0.8 4 7
  + 1 0.0 7
```

2.
```
      2.1
      0.7 3 9
  + 3 1.3 6 8 9
```

Add.

3. $0.02 + 4.3 + 0.649$

4. $0.37 + 6.291 + 0.1372$

5. $0.7438 + 9.10864 + 0.3519$

Answers on page A-12

a Addition

Adding with decimal notation is similar to adding whole numbers. First we line up the decimal points so that we can add corresponding place-value digits. Then we add digits from the right. For example, we add the thousandths, then the hundredths, and so on, carrying if necessary. If desired, we can write extra zeros to the far right of the decimal point so that the number of places is the same in all of the addends.

EXAMPLE 1 Add: $56.314 + 17.78$

```
    5 6 . 3 1 4        Lining up the decimal points in order to add
  + 1 7 . 7 8 0        Writing an extra zero to the far right
                        of the decimal point
```

```
    5 6 . 3 1 4        Adding thousandths
  + 1 7 . 7 8 0
              4
```

```
    5 6 . 3 1 4        Adding hundredths
  + 1 7 . 7 8 0
            9 4
```

```
          1
    5 6 . 3 1 4        Adding tenths
  + 1 7 . 7 8 0        Write a decimal point in the answer.
        . 0 9 4        We get 10 tenths = 1 one + 0 tenths,
                        so we carry the 1 to the ones column.
```

```
    1   1
    5 6 . 3 1 4        Adding ones
  + 1 7 . 7 8 0
      4 . 0 9 4        We get 14 ones = 1 ten + 4 ones,
                        so we carry the 1 to the tens column.
```

```
  1   1
    5 6 . 3 1 4        Adding tens
  + 1 7 . 7 8 0
  7 4 . 0 9 4
```

Do Exercises 1 and 2.

EXAMPLE 2 Add: $3.42 + 0.237 + 14.1$

```
      3.4 2 0        Lining up the decimal points
      0.2 3 7        and writing extra zeros
  + 1 4.1 0 0
    1 7.7 5 7        Adding
```

Do Exercises 3–5.

Consider the addition 3456 + 19.347. Keep in mind that any whole number has an "unwritten" decimal point at the far right, with 0 fractional parts. For example, 3456 can also be written 3456.000. When adding, it is often helpful to write in that decimal point and extra zeros.

EXAMPLE 3 Add: 3456 + 19.347

$$
\begin{array}{r}
\overset{1}{3\,4\,5\,6.0\,0\,0} \\
+\quad\ \ 1\,9.3\,4\,7 \\
\hline
3\,4\,7\,5.3\,4\,7
\end{array}
$$
   Writing in the decimal point and extra zeros
   Lining up the decimal points
   Adding

Do Exercises 6 and 7.

b Subtraction

Subtracting with decimal notation is similar to subtracting whole numbers. First we line up the decimal points so that we can subtract corresponding place-value digits. Then we subtract digits from the right. In the example below, we first subtract the thousandths, then the hundredths, the tenths, and so on, borrowing if necessary.

EXAMPLE 4 Subtract: 56.314 − 17.78

$$
\begin{array}{r}
5\,6.3\,1\,4 \\
-\,1\,7.7\,8\,0 \\
\end{array}
$$
Lining up the decimal points in order to subtract
Writing an extra 0

$$
\begin{array}{r}
5\,6.3\,1\,4 \\
-\,1\,7.7\,8\,0 \\
\hline
4
\end{array}
$$
Subtracting thousandths

$$
\begin{array}{r}
\overset{\ \ \ \ \ 2\ 11}{5\,6.3\,1\,4} \\
-\,1\,7.7\,8\,0 \\
\hline
3\,4
\end{array}
$$
Borrowing a tenth to subtract hundredths

$$
\begin{array}{r}
\overset{\ \ \ 12}{\overset{\ \ 5\ 2\ 11}{5\,6.3\,1\,4}} \\
-\,1\,7.7\,8\,0 \\
\hline
.5\,3\,4
\end{array}
$$
Borrowing a one to subtract tenths

Writing a decimal point

$$
\begin{array}{r}
\overset{\ \ 15\ 12}{\overset{4\ 5\ 2\ 11}{5\,6.3\,1\,4}} \\
-\,1\,7.7\,8\,0 \\
\hline
8.5\,3\,4
\end{array}
$$
Borrowing a ten to subtract ones

$$
\begin{array}{r}
\overset{\ \ 15\ 12}{\overset{4\ 5\ 2\ 11}{5\,6.3\,1\,4}} \\
-\,1\,7.7\,8\,0 \\
\hline
3\,8.5\,3\,4
\end{array}
$$
Subtracting tens

Check:
$$
\begin{array}{r}
\overset{1\ \ 1\ \ 1}{3\,8.5\,3\,4} \\
+\,1\,7.7\,8\,0 \\
\hline
5\,6.3\,1\,4
\end{array}
$$

Do Exercises 8 and 9.

Add.

6. 789 + 123.67

7. 45.78 + 2467 + 1.993

Subtract.

8. 37.428 − 26.674

9.
$$
\begin{array}{r}
0.3\,4\,7 \\
-\,0.0\,0\,8
\end{array}
$$

Answers on page A-12

Study Tips

NEATNESS COUNTS

When working with decimals, make an extra effort to write neatly. Lining up decimal places and clearly distinguishing decimal points from commas will help avert mistakes that could result from sloppiness.

321

5.2 Addition and Subtraction of Decimals

Subtract.

10. $2.9 - 0.36$

11. $0.43 - 0.18762$

12. $5.27 - 0.00008$

Subtract.

13. $1277 - 82.78$

14. $5 - 0.0089$

Add.

15. $7.42 + (-9.38)$

16. $-4.201 + 7.36$

17. Add: $-7.49 + (-5.8)$

Answers on pages A-12 and A-13

EXAMPLE 5   Subtract: $23.08 - 5.0053$

$$
\begin{array}{r}
\overset{1}{2}\,\overset{13}{3}.0\,\overset{7}{8}\,\overset{9}{\cancel{0}}\,\overset{10}{\cancel{0}} \\
-\quad 5.0\,0\,5\,3 \\
\hline
1\,8.0\,7\,4\,7
\end{array}
$$ Writing two extra zeros to the right of the last digit

Subtracting

Do Exercises 10–12.

When subtraction involves an integer, the "unwritten" decimal point can be written in. Extra zeros can then be written in to the right of the decimal point.

EXAMPLE 6   Subtract: $456 - 2.467$

$$
\begin{array}{r}
4\,5\,\overset{5}{6}.\overset{9}{\cancel{0}}\,\overset{9}{\cancel{0}}\,\overset{10}{\cancel{0}} \\
-\quad\quad 2.4\,6\,7 \\
\hline
4\,5\,3.5\,3\,3
\end{array}
$$ Writing in the decimal point and extra zeros

Subtracting

Do Exercises 13 and 14.

C   Adding and Subtracting with Negatives

Negative decimals are added or subtracted just like integers.

> To add a negative number and a positive number:
>
> **a)** Determine the sign of the number with the greater absolute value.
> **b)** Subtract the smaller absolute value from the larger one.
> **c)** The answer is the difference from part (b) with the sign from part (a).

EXAMPLE 7   Add: $-13.82 + 4.69$

a) Since $|-13.82| > |4.69|$, the sign of the number with the greater absolute value is negative.

b)
$$
\begin{array}{r}
1\,3.\overset{7}{\cancel{8}}\,\overset{12}{\cancel{2}} \\
-\quad 4.6\,9 \\
\hline
9.1\,3
\end{array}
$$ Finding the difference of the absolute values

c) Finally, we combine the results of steps (a) and (b): $-13.82 + 4.69 = -9.13$.

Do Exercises 15 and 16.

> To add two negative numbers:
>
> **a)** Add the absolute values.
> **b)** Make the answer negative.

EXAMPLE 8   Add: $-2.306 + (-3.125)$

a)
$$
\left.\begin{array}{r}
2.3\,0\,6 \\
+\;3.1\,2\,5
\end{array}\right\}
$$ $|-2.306| = 2.306$ and $|-3.125| = 3.125$

$$5.4\,3\,1$$ Adding the absolute values

b) $-2.306 + (-3.125) = -5.431$ The sum of two negatives is negative.

Do Exercise 17.

To subtract, we add the opposite of the number being subtracted.

EXAMPLE 9 Subtract: $-3.1 - 4.8$

$$-3.1 - 4.8 = -3.1 + (-4.8) \qquad \text{Adding the opposite of 4.8}$$
$$= -7.9 \qquad \text{The sum of two negatives is negative.}$$

EXAMPLE 10 Subtract: $-7.9 - (-8.5)$

$$-7.9 - (-8.5) = -7.9 + 8.5 \qquad \text{Adding the opposite of } -8.5$$
$$= 0.6 \qquad \text{Subtracting absolute values. The answer is positive since 8.5 has the larger absolute value.}$$

Do Exercises 18–21.

d Combining Like Terms

Recall that like, or similar, terms have exactly the same variable factors. To combine like terms, we add or subtract coefficients to form an equivalent expression.

EXAMPLE 11 Combine like terms: $3.2x + 4.6x$

These are the coefficients.

$$3.2x + 4.6x = (3.2 + 4.6)x \qquad \text{Using the distributive law—try to do this step mentally.}$$
$$= 7.8x \qquad \text{Adding}$$

A similar procedure is used when subtracting like terms.

EXAMPLE 12 Combine like terms: $4.13a - 7.56a$

$$4.13a - 7.56a = (4.13 - 7.56)a \qquad \text{Using the distributive law}$$
$$= (4.13 + (-7.56))a \qquad \text{Adding the opposite of 7.56}$$
$$= -3.43a \qquad \text{Subtracting absolute values. The coefficient is negative since } |-7.56| > |4.13|.$$

When more than one pair of like terms is present, we can rearrange the terms and then simplify.

EXAMPLE 13 Combine like terms: $5.7x - 3.9y - 2.4x + 4.5y$

$$5.7x - 3.9y - 2.4x + 4.5y$$
$$= 5.7x + (-3.9y) + (-2.4x) + 4.5y \qquad \text{Rewriting as addition}$$
$$= 5.7x + (-2.4x) + 4.5y + (-3.9y) \qquad \text{Using the commutative law to rearrange}$$
$$= 3.3x + 0.6y \qquad \text{Combining like terms}$$

With practice, you will be able to perform many of the above steps mentally.

Do Exercises 22–24.

Subtract.

18. $9.25 - 13.41$

19. $-5.72 - 4.19$

20. $9.8 - (-2.6)$

21. $-5.9 - (-3.2)$

Combine like terms.

22. $5.8x - 2.1x$

23. $-5.9a + 7.6a$

24. $-4.8y + 7.5 + 2.1y - 2.1$

Answers on page A-13

a Add.

1.   4 2 6.2 5
     +   3 8.1 2

2.   4 1.8 2 3
     + 6 1 4.9 1 5

3.   6 5 9.4 0 3
     + 9 1 6.8 1 2

4.   8 7 5.7 9 5
     + 3 2 4.8 6 2

5.   9.1 0 4
     + 1 2 3.4 5 6

6.   4.1 5 2 3
     + 3.2 7 7 8

7. 2.006 + 5.817

8. 0.8096 + 0.7856

9. 20.0124 + 30.0124

10. 0.263 + 0.8

11. 0.83 + 0.005

12. 0.347 + 10.04

13. 0.34 + 3.5 + 0.127 + 768

14. 2.3 + 0.729 + 23

15. 17 + 3.24 + 0.256 + 0.3689

16.   4 7.8
      2 1 9.8 5 2
      4 3.5 9
   + 6 6 6.7 1 3

17.   2.7 0 3
       7 8.3 3
      2 8.0 0 0 9
   + 1 1 8.4 3 4 1

18.   1 3.7 2
          9.1 1 2
     6 5 4 2.7 9 0 8
   +   2 3.9 0 1

b Subtract.

19.   4 7.5 9 6
     −   6.2 1 5

20.   1 1.3 4 5
     −   2.1 0 5

21.   5 1.3 1
     −   2.2 9

22.   3 7.4 5
     −   6.3 2

23.
```
    3.6
 − 0.0 3 6
```

24.
```
  2 8.0
 −  0.2 8
```

25.
```
  9 2.3 4 1
 −   6.4 2
```

26.
```
  0.3 4 6
 − 0.0 3 4 6
```

27.
```
    3.0 0 7 4
 −  1.3 4 0 8
```

28.
```
  3 2.7 9 7 8
 −   0.0 5 9 2
```

29.
```
    6.0 7
 −  2.0 0 7 8
```

30.
```
    1.0
 −  0.9 9 9 9
```

31. $30.24 - 0.241$

32. $100.12 - 0.112$

33. $34.07 - 30.7$

34. $36.2 - 16.28$

35. $8.45 - 7.405$

36. $3.801 - 2.81$

37. $6.003 - 2.3$

38. $1 - 0.0098$

39. $2 - 1.0908$

40. $100 - 0.34$

41. $624 - 18.79$

42. $7.48 - 2.6$

43. $57.803 - 4.6$

44. $25.008 - 12.4$

45. $263.7 - 102.08$

46. $19 - 1.198$

47. $45 - 0.999$

48. $10.056 - 0.392$

C   Add or subtract, as indicated.

49. $-5.02 + 1.73$

50. $-4.31 + 7.66$

51. $12.9 - 15.4$

52. $27.2 - 31.9$

53. $-2.9 + (-4.3)$

54. $-7.49 - 1.82$

55. $-4.301 + 7.68$

56. $-5.952 + 7.98$

57. $-12.9 - 3.7$

58. $-8.7 - 12.4$

59. $-2.1 - (-4.6)$

60. $-4.3 - (-2.5)$

61. $14.301 + (-17.82)$

62. $13.45 + (-18.701)$

63. $7.201 - (-2.4)$

64. $2.901 - (-5.7)$

65. $96.9 + (-21.4)$

66. $43.2 + (-10.9)$

67. $-8.9 - (-12.7)$

68. $-4.5 - (-7.3)$

69. $-4.9 - 5.392$

70. $89.3 - 92.1$

71. $14.7 - 23.5$

72. $-7.201 - 1.9$

d Combine like terms.

73. $1.8x + 3.9x$

74. $7.9x + 1.3x$

75. $17.59a - 12.73a$

76. $23.28a - 15.79a$

77. $15.2t + 7.9 + 5.9t$

78. $29.5t - 4.8 + 7.6t$

79. $5.217x - 8.134x$

80. $6.317t - 9.429t$

81. $4.906y - 7.1 + 3.2y$

82. $9.108y + 4.2 + 3.7y$

83. $4.8x + 1.9y - 5.7x + 1.2y$

84. $3.2r - 4.1t + 5.6t + 1.9r$

85. $4.9 - 3.9t + 2.3 - 4.5t$

86. $5.8 + 9.7x - 7.2 - 12.8x$

87. **Dw** Boris claims he can add negative numbers but not subtract them. What advice would you give him?

88. **Dw** Explain the error in the following: Subtract.

$$\begin{array}{r} 7\,3.0\,8\,9 \\ -5.0\,0\,6\,1 \\ \hline 2.3\,0\,2\,8 \end{array}$$

Multiply. [3.4b]

89. $\dfrac{3}{5} \cdot \dfrac{4}{7}$

90. $\dfrac{2}{9} \cdot \dfrac{7}{5}$

91. $\dfrac{3}{10} \cdot \dfrac{21}{100}$

Evaluate. [2.6a]

92. $8 - 2x^2$, for $x = 3$

93. $5 - 3x^2$, for $x = 2$

94. $7 + 2x^2 \div 3$, for $x = 6$

95. **D_W** In what sense is balancing a checkbook (see Exercise 102 below) or determining the perimeter of a shape similar to combining like terms?

96. **D_W** Although the step in which it is used may not always be written out, the commutative law is often used when combining like terms. Under what circumstances would the commutative law *not* be needed for combining like terms?

Combine like terms.

97. ▦ $-3.928 - 4.39a + 7.4b - 8.073 + 2.0001a - 9.931b - 9.8799a + 12.897b$

98. ▦ $79.02x + 0.0093y - 53.14z - 0.02001y - 37.987z - 97.203x - 0.00987y$

99. ▦ $39.123a - 42.458b - 72.457a + 31.462b - 59.491 + 37.927a$

100. Fred presses the wrong key when using a calculator and adds 235.7 instead of subtracting it. The incorrect answer is 817.2. What is the correct answer?

101. Millie presses the wrong key when using a calculator and subtracts 349.2 instead of adding it. The incorrect answer is −836.9. What is the correct answer?

102. ▦ Find the errors, if any, in the balances in this checkbook.

20___		RECORD ALL CHARGES OR CREDITS THAT AFFECT YOUR ACCOUNT					
DATE	CHECK NUMBER	TRANSACTION DESCRIPTION	√ T	(−) PAYMENT/ DEBIT	(+ OR −) OTHER	(+) DEPOSIT/ CREDIT	BALANCE FORWARD
							2767 73
8/16	432	Burch Laundry		23 56			2744 16
8/19	433	Rogers TV		20 49			2764 65
8/20		Deposit				85 00	2848 65
8/21	434	Galaxy Records		48 60			2801 05
8/22	435	Electric Works		267 95			2533 09

Find *a*.

103.
```
   9 3.a 4 3
 − 8 7.9 6 9
 ─────────────
       5.2 7 4
```

104.
```
   4 8 1.a 2 4
 −     7 2.9 7 8
 ─────────────────
     4 0 8.3 4 6
```

Objectives

a Multiply using decimal notation.

b Convert from dollars to cents and cents to dollars, and from notation like 45.7 million to standard notation.

c Evaluate algebraic expressions using decimal notation.

a Multiplication

To develop an understanding of how decimals are multiplied, consider

$$2.3 \times 1.12$$

One way to find this product is to convert each factor to fraction notation, multiply the fractions, and then return to decimal notation.

$$2.3 \times 1.12 = \frac{23}{10} \times \frac{112}{100} = \frac{2576}{1000} = 2.576$$

Note that the number of decimal places in the product is the sum of the number of decimal places in the factors.

$$
\begin{array}{rl}
1.1\,2 & \text{(2 decimal places)} \\
\times\quad 2.3 & \text{(1 decimal place)} \\
\hline
2.5\,7\,6 & \text{(3 decimal places)}
\end{array}
$$

Now consider 0.02×3.412:

$$0.02 \times 3.412 = \frac{2}{100} \times \frac{3412}{1000} = \frac{6824}{100,000} = 0.06824$$

Again, note that the number of decimal places in the product is the sum of the number of decimal places in the factors.

$$
\begin{array}{rl}
3.4\,1\,2 & \text{(3 decimal places)} \\
\times\quad 0.0\,2 & \text{(2 decimal places)} \\
\hline
0.0\,6\,8\,2\,4 & \text{(5 decimal places)}
\end{array}
$$

The 0 after the decimal point is necessary.

> To multiply using decimal notation: 0.8×0.43
>
> **a)** Ignore the decimal points, for the moment, and multiply as though both factors are integers.
>
> $$
> \begin{array}{r}
> \overset{2}{0.4}\,3 \\
> \times\quad 0.8 \\
> \hline
> 3\,4\,4
> \end{array}
> $$
> Ignore the decimal points for now.
>
> **b)** Locate the decimal point so that the number of decimal places in the product is the sum of the number of places in the factors.
>
> $$
> \begin{array}{rl}
> 0.4\,3 & \text{(2 decimal places)} \\
> \times\quad 0.8 & \text{(1 decimal place)} \\
> \hline
> 0.3\,4\,4 & \text{(3 decimal places)}
> \end{array}
> $$
>
> Count off the number of decimal places by starting at the far right and moving the decimal point to the left.

EXAMPLE 1 Multiply: 8.3×74.6

a) Ignore the decimal points and multiply as if both factors are integers.

$$
\begin{array}{r}
\overset{3}{}\overset{4}{} \\
\overset{1}{}\overset{1}{} \\
7\,4.6 \\
\times\quad 8.3 \\
\hline
2\,2\,3\,8 \\
5\,9\,6\,8\,0 \\
\hline
6\,1\,9\,1\,8
\end{array}
$$

We are not yet finished.

Study Tips

DON'T GET STUCK ON A QUESTION

Unless you know the material "cold," don't be surprised if a quiz or test includes a question for which you feel unprepared. Should this happen, do not get rattled—simply skip the question and continue with the quiz or test, returning to the trouble spot after the other questions have been answered.

b) Place the decimal point in the result. The number of decimal places in the product is the sum, $1 + 1$, of the number of decimal places in the factors.

```
        7 4.6      (1 decimal place)
    ×      8.3      (1 decimal place)
    ─────────
      2 2 3 8
    5 9 6 8 0
    ─────────
    6 1 9.1 8      (2 decimal places)
```

Do Exercise 1.

As we catch on to the skill, we can combine the two steps.

EXAMPLE 2   Multiply: 0.0032×2148

```
        2 1 4 8      (0 decimal places)
    × 0.0 0 3 2      (4 decimal places)
    ───────────
        4 2 9 6
      6 4 4 4 0
    ───────────
      6.8 7 3 6      (4 decimal places)
```

EXAMPLE 3   Multiply: -0.14×0.867

Multiplying the absolute values, we have

```
        0.8 6 7      (3 decimal places)
    ×       0.1 4    (2 decimal places)
    ───────────
        3 4 6 8
        8 6 7 0
    ───────────
    0.1 2 1 3 8      (5 decimal places)
```

Since the product of a negative and a positive is negative, the answer is -0.12138.

Do Exercises 2–4.

MULTIPLYING BY 0.1, 0.01, 0.001, AND SO ON

Suppose that a product involves multiplying by a tenth, hundredth, thousandth, or ten-thousandth. From the following products, a pattern emerges.

```
      4 5.6        4 5.6          4 5.6            4 5.6
    ×    0.1     × 0.0 1       × 0.0 0 1        × 0.0 0 0 1
    ────────     ───────       ─────────        ───────────
      4.5 6       0.4 5 6       0.0 4 5 6        0.0 0 4 5 6
```

Note in each case that the product is *smaller* than 45.6. That is, the decimal point in each product is farther to the left than it is in 45.6. Note also that each product can be obtained from 45.6 by moving the decimal point.

---

1. Multiply.

```
      7 6.3
    ×    8.2
```

Multiply.

2.
```
        4 2 1 3
    × 0.0 0 5 1
```

3. 2.3×0.0041

4. $5.2014 \times (-2.41)$

Answers on page A-13

Multiply.

5. 0.1×746

6. 0.001×732.4

7. $(-0.01) \times 6.2$

8. 0.0001×723.6

Multiply.

9. 10×53.917

10. $100 \times (-62.417)$

11. 1000×83.9

12. $10,000 \times 57.04$

To multiply any number by a tenth, hundredth, or thousandth, and so on,

a) count the number of decimal places in the tenth, hundredth, or thousandth, and

0.001×34.45678

$\longrightarrow 3$ places

b) move the decimal point that many places to the left. Use zeros as placeholders when necessary.

$0.001 \times 34.45678 = 0.034.45678$

Move 3 places to the left.

$0.001 \times 34.45678 = 0.03445678$

EXAMPLES Multiply.

4. $0.1 \times 45 = 4.5$ Moving the decimal point one place to the left

5. $0.01 \times 243.7 = 2.437$ Moving the decimal point two places to the left

6. $0.001 \times (-8.2) = -0.0082$ Moving the decimal point three places to the left. This requires writing two extra zeros.

7. $0.0001 \times 536.9 = 0.05369$ Moving the decimal point four places to the left. This requires writing one extra zero.

Do Exercises 5–8.

MULTIPLYING BY 10, 100, 1000, AND SO ON

Next we consider multiplying by 10, 100, 1000, and so on. From the following products, a pattern emerges.

```
    5.2 3 7            5.2 3 7              5.2 3 7
  ×     1 0          ×   1 0 0          ×     1 0 0 0
  ─────────          ─────────          ───────────────
    0 0 0 0            0 0 0 0              0 0 0 0
    5 2 3 7            0 0 0 0              0 0 0 0
  ─────────            5 2 3 7              0 0 0 0
    5 2.3 7 0        ─────────              5 2 3 7
                      5 2 3.7 0 0        ───────────────
                                          5 2 3 7.0 0 0
```

Note in each case that the product is *larger* than 5.237. That is, the decimal point in each product is farther to the right than the decimal point in 5.237. Also, each product can be obtained from 5.237 by moving the decimal point.

To multiply any number by 10, 100, 1000, and so on,

a) count the number of zeros, and

1000×34.45678

$\longrightarrow 3$ zeros

b) move the decimal point that many places to the right. Use zeros as placeholders when necessary.

$1000 \times 34.45678 = 34.456.78$

Move 3 places to the right.

$1000 \times 34.45678 = 34,456.78$

Answers on page A-13

EXAMPLES Multiply.

8. $10 \times 32.98 = 329.8$ Moving the decimal point one place to the right

9. $100 \times 4.7 = 470$ Moving the decimal point two places to the right. The 0 in 470 is a placeholder.

10. $1000 \times (-2.4167) = -2416.7$ Moving the decimal point three places to the right

11. $10,000 \times 7.52 = 75,200$ Moving the decimal point four places to the right and using two zeros as placeholders

Do Exercises 9–12 on the previous page.

b   Applications Using Multiplication with Decimal Notation

NAMING LARGE NUMBERS

We often see notation like the following in newspapers and magazines and on television.

- The largest building in the world is the Pentagon, which has 3.7 million square feet of floor space.
- In 2004, consumers spent $65.1 billion on online retail products. Total spending in specific categories is listed in the following table.

ONLINE CONSUMER SPENDING
(in billions of dollars)

CATEGORY	2004	CATEGORY	2004
1. PCs	$9.8	6. Event Tickets	$3.3
2. Peripherals	$2.7	7. Apparel	$7.5
3. Software	$3.3	8. Jewelry	$1.6
4. Consumer Electronics	$3.3	9. Toys	$0.8
5. Books	$3.5	10. Housewares/ Small Appliances	$2.9

Source: Jupiter Media Metrix, Inc.

To understand such notation, it helps to consider the following table.

NAMING LARGE NUMBERS

1 hundred $= 100 = 10^2$
→ 2 zeros

1 thousand $= 1000 = 10^3$
→ 3 zeros

1 million $= 1,000,000 = 10^6$
→ 6 zeros

1 billion $= 1,000,000,000 = 10^9$
→ 9 zeros

1 trillion $= 1,000,000,000,000 = 10^{12}$
→ 12 zeros

331

Convert the number in the sentence to standard notation.

13. The largest building in the world is the Pentagon, which has 3.7 million square feet of floor space.

14. Online Jewelry Sales. In 2004, consumers spent $1.6 billion online for jewelry.
Source: Jupiter Media Metrix, Inc.

15. Online Ticket Sales. In 2004, consumers spent $3.3 billion online for event tickets.
Source: Jupiter Media Metrix, Inc.

Convert from dollars to cents.

16. $15.69

17. $0.17

Convert from cents to dollars.

18. 35¢

19. 577¢

Answers on page A-13

To convert a large number to standard notation, we proceed as follows.

EXAMPLE 12 Convert the number in this sentence to standard notation: In 1999, the U.S. Mint produced 11.6 billion pennies.
Source: U.S. Mint

$$11.6 \text{ billion} = 11.6 \times 1 \text{ billion}$$
$$= 11.6 \times 1,\underline{000,000,000}$$

9 zeros

$$= 11,600,000,000 \qquad \text{Moving the decimal point 9 places to the right}$$

Do Exercises 13–15.

MONEY CONVERSION

Converting from dollars to cents is like multiplying by 100. To see why, consider $19.43.

$$\$19.43 = 19.43 \times \$1 \qquad \text{We think of \$19.43 as } 19.43 \times 1 \text{ dollar, or } 19.43 \times \$1.$$
$$= 19.43 \times 100¢ \qquad \text{Substituting } 100¢ \text{ for } \$1\text{: } \$1 = 100¢$$
$$= 1943¢ \qquad \text{Multiplying}$$

DOLLARS TO CENTS

To convert from dollars to cents, move the decimal point two places to the right and change the $ sign in front to a ¢ sign at the end.

EXAMPLES Convert from dollars to cents.

13. $189.64 = 18,964¢

14. $0.75 = 75¢

Do Exercises 16 and 17.

Converting from cents to dollars is like multiplying by 0.01. To see why, consider 65¢.

$$65¢ = 65 \times 1¢ \qquad \text{We think of } 65¢ \text{ as } 65 \times 1 \text{ cent, or } 65 \times 1¢.$$
$$= 65 \times \$0.01 \qquad \text{Substituting } \$0.01 \text{ for } 1¢\text{: } 1¢ = \$0.01$$
$$= \$0.65 \qquad \text{Multiplying}$$

CENTS TO DOLLARS

To convert from cents to dollars, move the decimal point two places to the left and change the ¢ sign at the end to a $ sign in front.

EXAMPLES Convert from cents to dollars.

15. 395¢ = $3.95

16. 8503¢ = $85.03

Do Exercises 18 and 19.

C   Evaluating

Algebraic expressions are often evaluated using numbers written in decimal notation.

EXAMPLE 17   Evaluate Prt for $P = 780$, $r = 0.12$, and $t = 0.5$.

We will see in Chapter 8 that this product could be used to determine the interest paid on $780, borrowed at 12 percent simple interest, for half a year. We substitute as follows.

$$Prt = 780 \cdot 0.12 \cdot 0.5 = 780 \cdot 0.06 = 46.8 \qquad \text{This would represent } \$46.80.$$

Do Exercise 20.

EXAMPLE 18   Find the perimeter of a stamp that is 3.25 cm long and 2.5 cm wide.

Recall that the perimeter, P, of a rectangle of length l and width w is given by the formula

$$P = 2l + 2w.$$

Thus, we evaluate $2l + 2w$ for $l = 3.25$ and $w = 2.5$.

$$
\begin{aligned}
2l + 2w &= 2 \cdot 3.25 + 2 \cdot 2.5 \\
&= 6.5 + 5.0 &&\text{Remember the rules for order} \\
&= 11.5 &&\text{of operations.}
\end{aligned}
$$

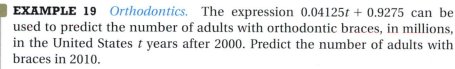

The perimeter is 11.5 cm.

Do Exercise 21.

EXAMPLE 19   *Orthodontics.* The expression $0.04125t + 0.9275$ can be used to predict the number of adults with orthodontic braces, in millions, in the United States t years after 2000. Predict the number of adults with braces in 2010.

Source: American Association of Orthodontists

2010 is 10 years after 2000, so we evaluate $0.04125t + 0.9275$ for $t = 10$.

$$
\begin{aligned}
0.04125t + 0.9275 &= 0.04125 \cdot 10 + 0.9275 \\
&= 0.4125 + 0.9275 \\
&= 1.34
\end{aligned}
$$

In 2010, there will be approximately 1.34 million adults with braces.

Do Exercise 22.

---

20. Evaluate lwh for $l = 3.2$, $w = 2.6$, and $h = 0.8$. (This is the formula for the volume of a rectangular box.)

21. Find the area of the stamp in Example 18.

22. Evaluate $6.28rh + 3.14r^2$ for $r = 1.5$ and $h = 5.1$. (This is the formula for the area of an open can.)

Answers on page A-13

a   Multiply.

1. $\begin{array}{r} 6.8 \\ \times\ \ 7 \\ \hline \end{array}$

2. $\begin{array}{r} 5.7 \\ \times\ 0.9 \\ \hline \end{array}$

3. $\begin{array}{r} 0.8\,4 \\ \times\ \ \ \ 8 \\ \hline \end{array}$

4. $\begin{array}{r} 7.3 \\ \times\ 0.6 \\ \hline \end{array}$

5. $\begin{array}{r} 6.3 \\ \times\ 0.0\,4 \\ \hline \end{array}$

6. $\begin{array}{r} 7.8 \\ \times\ 0.0\,9 \\ \hline \end{array}$

7. $\begin{array}{r} 2\,8.6 \\ \times\ 0.0\,9 \\ \hline \end{array}$

8. $\begin{array}{r} 2\,5.9 \\ \times\ 0.0\,7 \\ \hline \end{array}$

9. 10×42.63

10. 100×2.8793

11. -1000×783.686852

12. -0.34×1000

13. -7.8×100

14. $0.00238 \times (-10)$

15. 0.1×79.18

16. 0.01×789.235

17. 0.001×97.68

18. 8976.23×0.001

19. $28.7 \times (-0.01)$

20. $0.0325 \times (-0.1)$

21. $\begin{array}{r} 2.7\,3 \\ \times\ \ \ 1\,6 \\ \hline \end{array}$

22. $\begin{array}{r} 8.2\,7 \\ \times\ \ \ 5.4 \\ \hline \end{array}$

23. $\begin{array}{r} 0.9\,8\,4 \\ \times\ \ \ \ \ 3.3 \\ \hline \end{array}$

24. $\begin{array}{r} 7.4\,8\,9 \\ \times\ \ \ \ \ 8.2 \\ \hline \end{array}$

25. $(-37.4)(-2.4)$

26. $569(-1.05)$

27. $749(-0.43)$

28. $(-876)(-20.4)$

29. $\begin{array}{r} 0.8\,7 \\ \times\ \ \ 6\,4 \\ \hline \end{array}$

30. $\begin{array}{r} 7.2\,5 \\ \times\ \ \ 6\,0 \\ \hline \end{array}$

31. $\begin{array}{r} 4\,6.5\,0 \\ \times\ \ \ \ \ 7\,5 \\ \hline \end{array}$

32. $\begin{array}{r} 8.2\,4 \\ \times\ 7\,0\,3 \\ \hline \end{array}$

33. $(-0.231)(-0.5)$

34. $(-12.3)(-1.08)$

35. $9.42 \times (-1000)$

36. $-7.6 \times (-1000)$

37. $-95.3 \times (-0.0001)$

38. $-4.23 \times (-0.001)$

b   Convert from dollars to cents.

39. $57.06

40. $49.85

41. $0.95

42. $0.49

43. $0.01

44. $0.09

Convert from cents to dollars.

45. 72¢

46. 52¢

47. 2¢

48. 5¢

49. 6399¢

50. 5238¢

Convert the number in each sentence to standard notation.

51. A century is approximately 3.156 billion seconds.

52. Stonehenge was completed approximately 2.104 billion minutes ago.

Source: Based on information in *The Cambridge Factfinder*, 4th ed.

53. Approximately 63.1 trillion seconds have passed since the continents assumed their present shapes.

Source: Based on information in *The Cambridge Factfinder*, 4th ed.

54. *Rail Transportation.* People in the United States traveled 13.6 billion miles on city rail transportation in 2003.

Source: American Public Transportation Association, NASA

55. *Broadway.* Paid attendance for Broadway productions reached 11.98 million in 2005.

Source: League of American Theaters and Producers

56. The total surface area of Earth is 196.8 million square miles.

C Evaluate.

57. $P + Prt$, for $P = 10{,}000$, $r = 0.04$, and $t = 2.5$
(*A formula for adding interest*)

58. $6.28r(h + r)$, for $r = 10$ and $h = 17.2$
(*Surface area of a cylinder*)

59. $vt + 0.5at^2$, for $v = 10$, $t = 1.5$, and $a = 9.8$
(*A physics formula*)

60. $4lh + 2h^2$, for $l = 3.5$ and $h = 1.2$
(*Surface area of a rectangular prism*)

Find (a) the perimeter and (b) the area of a rectangular room with the given dimensions.

61. 12.5 ft long, 9.5 ft wide

62. 10.25 ft long, 8 ft wide

63. 8.4 m wide, 10.5 m long

64. 8.2 yd long, 6.4 yd wide

e-mail. The expression $5.06t + 9.7$ can be used to predict the average number of e-mail messages, in billions, each day in North America t years after 2000. Predict the average daily number of e-mails in North America in the year indicated.
Source: Pitney Bowes

65. 2010

66. 2009

67. **D**_W If two rectangles have the same perimeter, will they also have the same area? Why or why not?

68. **D**_W You may have seen a sign advertising an article for sale at .99¢. What does this notation really mean? How could a sale price of ninety-nine cents be written correctly?

SKILL MAINTENANCE

Divide. [2.5a]

69. $-162 \div 6$

70. $-216 \div (-6)$

71. $-1035 \div (-15)$

72. $-423 \div 3$

73. $-525 \div (25)$

74. $675 \div (-25)$

75. $-7050 \div 50$

76. $575 \div (-25)$

77. $\mathbf{D_W}$ In your own words, explain why the decimal point in a product is located by adding the number of decimal places in the numbers being multiplied.

78. $\mathbf{D_W}$ Is it possible for the product of two numbers to contain fewer decimal places than either of the numbers being multiplied?

79. One light-year (LY) is 9.46×10^{12} km. The star Regulus is 85 LY from Earth. How many billions of kilometers (km) from Earth is Regulus?
Source: *The Cambridge Factfinder*, 4th ed.

80. The star Deneb is 1600 LY from Earth. How many billions of kilometers from Earth is Deneb (see Exercise 79)?
Source: *The Cambridge Factfinder*, 4th ed.

Evaluate using a calculator.

81. $d + vt + at^2$, for $d = 79.2$, $v = 3.029$, $t = 7.355$, and $a = 4.9$ (*A physics formula for distance traveled*)

82. $3.14r^2 + 6.28rh$, for $r = 5.756$ in. and $h = 9.047$ in. (*Surface area of a toy silo, including bottom*)

83. $0.5(b_1 + b_2)h$, for $b_1 = 9.7$ cm, $b_2 = 13.4$ cm, and $h = 6.32$ cm (*A geometry formula for the area of a trapezoid*)

84. $0.5bh$, for $b = 12.59$ cm and $h = 13.72$ cm (*A formula for the area of a triangle*)

Express as a power of 10.

85. (1 trillion) $\cdot$ (1 billion)

86. (1 million) $\cdot$ (1 billion)

87. In Great Britain, France, and Germany, a billion means a million millions. Write standard notation for the British number 6.6 billion.

88. A quadrillion is 10^{15}. Write standard notation for 5.2 quadrillion.

Electric Bills. Recently, electric bills from the Central Vermont Public Service Corporation consisted of a "customer charge" of $0.374 per day plus an "energy charge" of $0.1174 per kilowatt-hour (kWh) for the first 250 kWh used and $0.09079 per kilowatt-hour for each kilowatt-hour in excess of 250.

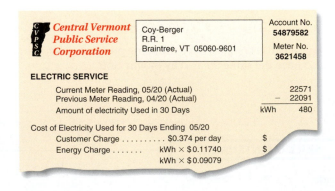

89. From April 20 to May 20, the Coy-Bergers used 480 kWh of electricity. What was their bill for the period?

90. From June 20 to July 20, the D'Amicos used 430 kWh of electricity. What was their bill for the period?

5.4 DIVISION OF DECIMALS

Objectives

a Divide using decimal notation.

b Simplify expressions using the rules for order of operations.

a Division

WHOLE-NUMBER DIVISORS

Now that we have studied multiplication of decimals, we can develop a procedure for division. The following divisions are justified by the multiplication in each *check*:

This is the dividend. ⟶ $\dfrac{651}{7} = 93$ *Check*: $7 \cdot 93 = 651$
This is the divisor. ⟶

This is the quotient.

$\dfrac{65.1}{7} = 9.3$ *Check*: $7 \cdot 9.3 = 65.1$

$\dfrac{6.51}{7} = 0.93$ *Check*: $7 \cdot 0.93 = 6.51$

$\dfrac{0.651}{7} = 0.093$ *Check*: $7 \cdot 0.093 = 0.651$

Note that the number of decimal places in each quotient is the same as the number of decimal places in the dividend.

To perform long division by a whole number,

a) place the decimal point directly above the decimal point in the dividend, and

b) divide as though dividing whole numbers.

$$\begin{array}{r} 0.8\,4 \leftarrow \text{Quotient} \\ 7\,)\,\overline{5.8\,8} \leftarrow \text{Dividend} \\ 5\,6\,0 \\ \hline 2\,8 \\ 2\,8 \\ \hline 0 \leftarrow \text{Remainder} \end{array}$$

Divisor

EXAMPLE 1 Divide: $82.08 \div 24$.

We have

Place the decimal point.

$$\begin{array}{r} 3.4\,2 \\ 24\,)\,\overline{8\,2.0\,8} \\ 7\,2\,0\,0 \\ \hline 1\,0\,0\,8 \\ 9\,6\,0 \\ \hline 4\,8 \\ 4\,8 \\ \hline 0 \end{array}$$

Divide as though dividing whole numbers.

Since $(3.42)(24) = 82.08$, the answer checks.

Do Exercises 1–3.

Divide.

1. $9\,)\,\overline{5.4}$

2. $1\,5\,)\,\overline{2\,5.5}$

3. $8\,2\,)\,\overline{3\,8.5\,4}$

Answers on page A-13

Sometimes we need to write some extra zeros to the right of the dividend's right-most decimal place. This doesn't change the value of the number since adding 0 tenths, or hundredths, or thousandths, and so on, does not change a number.

EXAMPLE 2 Divide: $30 \div 8$.

```
        3.
  8 ) 3 0.        Place the decimal point and divide to find how many ones.
      2 4         Subtracting 8 × 3, or 24, from 30
      ───
        6
```

```
        3.
  8 ) 3 0.0       Write an extra zero. This does not change the number.
      2 4
      ───
        6 0       We can bring down one digit at a time.
```

```
        3.7
  8 ) 3 0.0       Divide to find how many tenths.
      2 4
      ───
        6 0
        5 6       Subtracting 8 × 7, or 56, from 60
        ───
          4
```

```
        3.7
  8 ) 3 0.0 0     Write an extra zero.
      2 4
      ───
        6 0
        5 6
        ───
          4 0     We can bring down one digit at a time.
```

```
        3.7 5
  8 ) 3 0.0 0     Repeat the procedure:  Divide to find how
      2 4         many hundredths are in the quotient.
      ───
        6 0
        5 6
        ───
          4 0
          4 0     Subtracting 8 × 5, or 40, from 40
          ───
            0     Since the remainder is 0, we are finished.
```

To check, the student can confirm that $3.75 \cdot 8 = 30$.
We have $30 \div 8 = 3.75$.

EXAMPLE 3 Divide: $-4 \div 25$.

We first consider $4 \div 25$.

```
         0.1 6
  2 5 ) 4.0 0     ←We can write as many extra zeros as needed.
        2 5
        ───
        1 5 0
        1 5 0
        ─────
            0     ←Since the remainder is 0, we are finished.
```

Since a negative number divided by a positive number is negative, the answer is -0.16. To check, note that $(-0.16)25 = -4$.

Do Exercises 4–6.

CALCULATOR CORNER

Finding Remainders   It is possible to use a calculator to find whole-number remainders when doing division. To see how one method works, consider the quotient 17 ÷ 8.

$$17 \div 8 = 2.125$$

To check, we can multiply:

$$8 \times 2.125 = 17,$$

or, using the distributive law, we can write

$$8 \times (2 + 0.125) = 8 \times 2 + 8 \times 0.125$$
$$= 16 + 1 = 17.$$

Note that 17 ÷ 8 = 2 R 1. Thus, we can find a whole-number remainder by multiplying the decimal portion of a quotient by the divisor.

To find the quotient and the whole-number remainder for 567 ÷ 13, we can use a calculator to find that

$$567 \div 13 \approx 43.61538462 \qquad \text{To isolate the portion to the right of the decimal point, we can subtract 43.}$$

When the decimal part of the quotient is multiplied by the divisor, we have

$$0.61538462 \times 13 = 8.00000006$$

The rounding error in the result may vary, depending on the calculator used. We see that 567 ÷ 13 = 43 R 8.

Exercises: Find the quotient and the whole-number remainder for each of the following.

1. 478 ÷ 17
2. 815 ÷ 7
3. 824 ÷ 11
4. 7888 ÷ 19

DIVISORS THAT ARE NOT WHOLE NUMBERS

Consider the following division.

$$0.24 \overline{)8.208}$$

To understand how to divide, we write the division as $\dfrac{8.208}{0.24}$. Then we multiply by 1 to change to a whole-number divisor.

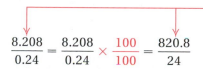

The division $0.24\overline{)8.208}$ is the same as $24\overline{)820.8}$.

$$\frac{8.208}{0.24} = \frac{8.208}{0.24} \times \frac{100}{100} = \frac{820.8}{24}$$

The divisor is now a whole number.

> To divide when the divisor is not a whole number,
>
> **a)** move the decimal point (multiply by 10, 100, and so on) to make the divisor a whole number;
>
> $$0.24 \overline{)8.208}$$
> Move 2 places to the right.
>
> **b)** move the decimal point the same number of places (multiply the same way) in the dividend; and
>
> $$0.24 \overline{)8.208}$$
> Move 2 places to the right.
>
> **c)** place the decimal point for the answer directly above the new decimal point in the dividend and divide as if dividing by a whole number.
>
> ```
>        3 4.2
>  2 4 ) 8 2 0.8
>        7 2
>        1 0 0
>          9 6
>          ───
>            4 8
>            4 8
>            ───
>              0
> ```

7. a) Complete.

$$\frac{3.75}{0.25} = \frac{3.75}{0.25} \times \frac{100}{100}$$

$$= \frac{()}{25}$$

b) Divide.

$$0.2\,5\,\overline{)\,3.7\,5}$$

Divide.

8. $4.067 \div (-0.83)$

9. $-44.8 \div (-3.5)$

Divide.

10. $1.6\,\overline{)\,2\,5}$

11. $-36 \div 0.75$

Answers on page A-13

■ **EXAMPLE 4** Divide: $6.708 \div (-8.6)$.

We first consider $6.708 \div 8.6$:

$$8.6\,\overline{)\,6.7\,_\wedge 0\,8}$$

Multiply the divisor by 10 (move the decimal point 1 place). Multiply the same way in the dividend (move the decimal point 1 place). The caret ^ indicates the new position of the decimal point.

$$
\begin{array}{r}
.7\,8 \\
8\,6\,\overline{)\,6\,7.0\,8} \\
6\,0\,2 \\
\hline
6\,8\,8 \\
6\,8\,8 \\
\hline
0
\end{array}
$$

Then divide.
Note: $\frac{6.708}{8.6} = \frac{6.708}{8.6} \cdot \frac{10}{10} = \frac{67.08}{86}$.

Check: $(0.78)8.6 = 6.708$

Since a positive number divided by a negative number is negative, we have $6.708 \div (-8.6) = -0.78$.

Do Exercises 7–9.

■ **EXAMPLE 5** Divide: $-12 \div (-0.64)$.

Note first that a negative number divided by a negative number is positive. To find the quotient, we consider $12 \div 0.64$:

$$0.6\,4\,\overline{)\,1\,2.0\,0_\wedge}$$

Write a decimal point at the end of the whole number. Multiply the divisor by 100 (move the decimal point 2 places). Multiply the same way in the dividend (move 2 places). Write additional zeros as needed.

$$
\begin{array}{r}
1\,8.7\,5 \\
6\,4\,\overline{)\,1\,2\,0\,0.0\,0} \\
6\,4\,0 \\
\hline
5\,6\,0 \\
5\,1\,2 \\
\hline
4\,8\,0 \\
4\,4\,8 \\
\hline
3\,2\,0 \\
3\,2\,0 \\
\hline
0
\end{array}
$$

Then divide.

Since the remainder is 0, we are finished.
Check: $18.75(-0.64) = -12$

We have $-12 \div (-0.64) = 18.75$.

Do Exercises 10 and 11.

DIVIDING BY 10, 100, 1000, 0.1, 0.01, 0.001, AND SO ON

We can divide quickly by a thousandth, hundredth, tenth, ten, hundred, and so on. Consider $43.9 \div 0.001$.

$$
\begin{array}{r}
4\,3\,9\,0\,0. \\
0.0\,0\,1\,\overline{)\,4\,3.9\,0\,0_\wedge}
\end{array}
$$

We move the decimal point in both the divisor and the dividend 3 places to the right. Note that the divisor is now 1.

To divide by a number like 10, 100, 1000, 0.1, 0.01, or 0.001,

a) move the decimal point in the divisor until the divisor is 1, and
b) move the decimal point in the dividend the same number of places and the same direction.

Examples:

1. Divide 213.745 by 100.

   a) Move the decimal point in the divisor two places to the left to make the divisor 1.

   100. We move 2 places and there are 2 zeros in 100.

   b) Move the decimal point in the dividend two places to the left.

   213.745

   $$\frac{213.745}{100.} = \frac{2.13745}{1.00} = 2.13745$$

2. Divide 69.81 by 0.001.

   a) Move the decimal point in the divisor three places to the right to make the divisor 1.

   0.001 We move 3 places and there are 3 decimal places in 0.001.

   b) Move the decimal point in the dividend three places to the right.

   69.810

   $$\frac{69.810}{0.001} = \frac{69{,}810}{1.} = 69{,}810$$

EXAMPLE 6 Divide: $\dfrac{0.0732}{10}$.

$$\frac{0.0732}{10} = \frac{0.0732}{10.} = \frac{0.00732}{1.0} = 0.00732$$

1 zero   1 place to the left to change 10 to 1

The answer is 0.00732.

EXAMPLE 7 Divide: $\dfrac{-23.738}{0.001}$.

$$\frac{-23.738}{0.001} = \frac{-23.738}{0.001} = \frac{-23{,}738}{1.} = -23{,}738$$

3 places   3 places to the right to change 0.001 to 1

The answer is $-23{,}738$.

Do Exercises 12–15.

b Order of Operations: Decimal Notation

The rules for order of operations apply when simplifying expressions involving decimal notation.

Divide.

12. $\dfrac{0.1278}{0.01}$

13. $\dfrac{0.1278}{100}$

14. $\dfrac{98.47}{1000}$

15. $\dfrac{6.7832}{-0.1}$

Answers on page A-13

Simplify.

16. $0.25 \cdot (1 + 0.08) - 0.0274$

17. $[(19.7 - 17.2)^2 + 3] \div (-1.25)$

18. Tourism. Spain is the world's second most popular tourism destination. The graph below shows the number of international tourist arrivals, in millions, from 1999 to 2003. Find the average number of tourist arrivals in Spain for this period.

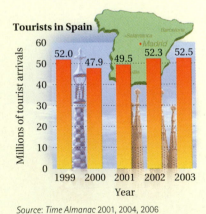

Source: Time Almanac 2001, 2004, 2006

Answers on page A-13

RULES FOR ORDER OF OPERATIONS

1. Do all calculations within grouping symbols first.
2. Evaluate all exponential expressions.
3. Do all multiplications and divisions in order from left to right.
4. Do all additions and subtractions in order from left to right.

EXAMPLE 8 Simplify: $(5 - 0.06) \div 2 + 3.42 \times 0.1$.

$(5 - 0.06) \div 2 + 3.42 \times 0.1 = 4.94 \div 2 + 3.42 \times 0.1$ Working inside the parentheses

$= 2.47 + 0.342$ Multiplying and dividing in order from left to right

$= 2.812$

EXAMPLE 9 Simplify: $13 - [5.4(1.3^2 + 0.21) \div 0.6]$.

$13 - [5.4(1.3^2 + 0.21) \div 0.6]$

$= 13 - [5.4(1.69 + 0.21) \div 0.6]$ Working in the innermost parentheses first

$= 13 - [5.4 \times 1.9 \div 0.6]$

$= 13 - [10.26 \div 0.6]$ Multiplying

$= 13 - 17.1$ Dividing

$= -4.1$

Do Exercises 16 and 17.

EXAMPLE 10 *Tourism.* France is the world's top tourism destination. The graph below shows the number of international tourist arrivals, in millions, from 1999 to 2003. Find the average number of tourist arrivals in France for this period.

Source: Time Almanac 2001, 2004, 2006

To find the average of a set of numbers, we add and then divide by the number of addends. The average of 71.4, 75.6, 76.5, 77.0, and 75.0 is given by

$$(71.4 + 75.6 + 76.5 + 77.0 + 75.0) \div 5 = 375.5 \div 5 = 75.1$$

The average number of tourist arrivals in France from 1999 to 2003 was 75.1 million.

Do Exercise 18.

a   Divide.

1. $5 \overline{)63}$

2. $5 \overline{)62}$

3. $4 \overline{)95.12}$

4. $8 \overline{)25.92}$

5. $12 \overline{)89.76}$

6. $23 \overline{)25.07}$

7. $33 \overline{)237.6}$

8. $12.4 \div (-4)$

9. $5.4 \div (-6)$

10. $3.6 \div 4$

11. $-9.144 \div 8$

12. $-7.254 \div 6$

13. $0.06 \overline{)8.4}$

14. $0.04 \overline{)1.68}$

15. $2.6 \overline{)104}$

16. $6 \div (-15)$

17. $1.8 \div (-12)$

18. $36 \overline{)14.76}$

19. $2.7 \overline{)129.6}$

20. $6.2 \overline{)46.5}$

21. $8.5 \overline{)27.2}$

22. $39.06 \div (-4.2)$

23. $-5 \div (-8)$

24. $-7 \div (-8)$

25. $0.47 \overline{)0.1222}$

26. $0.54 \overline{)0.27}$

27. $0.032 \overline{)0.07488}$

28. $0.017 \overline{)1.581}$

29. $-24.969 \div 82$

30. $-25.221 \div 42$

31. $\dfrac{-213.4567}{100}$ **32.** $\dfrac{-213.4567}{10}$ **33.** $\dfrac{1.0237}{0.001}$ **34.** $\dfrac{1.0237}{-0.01}$ **35.** $\dfrac{92.36}{-0.01}$ **36.** $\dfrac{56.78}{-0.001}$

37. $\dfrac{0.8172}{10}$ **38.** $\dfrac{0.5678}{1000}$ **39.** $\dfrac{0.97}{0.1}$ **40.** $\dfrac{0.97}{0.001}$ **41.** $\dfrac{52.7}{-1000}$ **42.** $\dfrac{8.9}{-100}$

43. $\dfrac{75.3}{-0.001}$ **44.** $\dfrac{63.47}{-0.1}$ **45.** $\dfrac{-75.3}{1000}$ **46.** $\dfrac{23,001}{100}$

b Simplify.

47. $14 \times (82.6 + 67.9)$

48. $(26.2 - 14.8) \times 12$

49. $0.003 + 3.03 \div (-0.01)$

50. $42 \times (10.6 + 0.024)$

51. $(4.9 - 18.6) \times 13$

52. $4.2 \times 5.7 + 0.7 \div 3.5$

53. $210.3 - 4.24 \times 1.01$

54. $-7.32 + 0.04 \div 0.1^2$

55. $12 \div (-0.03) - 12 \times 0.03^2$

56. $(5 - 0.04)^2 \div 4 + 8.7 \times 0.4$

57. $(4 - 2.5)^2 \div 100 + 0.1 \times 6.5$

58. $4 \div 0.4 - 0.1 \times 5 + 0.1^2$

59. $6 \times 0.9 - 0.1 \div 4 + 0.2^3$

60. $5.5^2 \times [(6 - 7.8) \div 0.06 + 0.12]$

61. $12^2 \div (12 + 2.4) - [(2 - 2.4) \div 0.8]$

62. $0.01 \times \{[(4 - 0.25) \div 2.5] - (4.5 - 4.025)\}$

63. Income Taxes. The amount paid in United States individual income taxes in a 5-yr period is shown in the bar graph below. Find the average amount paid per year in individual income tax over the 5-yr period.

United States Individual Income Taxes

2003 — $131.8
2004 — $168.7
2005* — $230.2
2006* — $250.0
2007* — $251.0

Amount paid (in billions)

*Figures are estimates.

Sources: Department of the Treasury; Office of Management and Budget

64. Herbal Supplement. Biologically based therapies in complementary and alternative medicine use substances found in nature such as herbs, food, and vitamins. The graph below shows spending on the herbal supplement echinacea, also known as purple coneflower, which is thought to strengthen the immune system. Find the average amount spent per year from 1999 to 2003.

Spending on Herbal Supplement Echinacea

$220 (1999), $202 (2000), $214 (2001), $188 (2002), $180 (2003)

Source: Nutrition Business Journal

65. Life Expectancy. The information in the following bar graph shows the 5 highest life expectancies in the world. Determine the average life expectancy for these countries.

Life Expectancy at Birth, 2006

Andorra: 83.51, San Marino: 81.71, Singapore: 81.71, Hong Kong: 81.59, Japan: 81.25

Source: U.S. Bureau of the Census, International Data Base

66. Gasoline Consumption. The information in the graph below shows the average number of miles per gallon for U.S. vehicles. Find the average for the entire period.

U.S. Vehicle Fuel Consumption Rates

'96: 16.9, '97: 17.0, '98: 16.9, '99: 16.7, '00: 16.9, '01: 17.1, '02: 16.9, '03: 17.0

Source: U.S. Federal Highway Administration

The following table lists the lengths of the longest tunnels in the world. Use the table for Exercises 67 and 68.

Tunnel	Seikan Japan	Channel Tunnel England–France	Iwate Ichinohe Japan	Daishimizu Japan	Simplon I, II Switzerland-Italy
Length, in miles	33.5	31.1	16.0	13.8	12.3
Length, in kilometers	53.9	50.0	25.8	22.2	19.8

Source: Time Almanac 2006

67. Find the average length of the tunnels, in miles.

68. Find the average length of the tunnels, in kilometers.

69. **D_W** Mel insists that $53 \div 0.1$ is 5.3. Give a "common sense" argument that could convince him that he is mistaken.

70. **D_W** Gilda insists that $7.9 \div 10$ is 79. Give a "common sense" argument that could convince her that she is mistaken.

SKILL MAINTENANCE

Simplify to form an equivalent expression.   [3.5b]

71. $\dfrac{33}{44}$

72. $\dfrac{49}{56}$

73. $-\dfrac{27}{18}$

74. $-\dfrac{18}{60}$

75. $\dfrac{9a}{27}$

76. $\dfrac{12x}{30}$

77. $\dfrac{4r}{20r}$

78. $\dfrac{10t}{15t}$

SYNTHESIS

79. **D_W** Which is easier and why: Dividing a decimal by a decimal or dividing a fraction by a fraction?

80. **D_W** In Exercise 66, the number 17.0 was used. Why do you think the number 17 was not used instead?

Calculate each of the following.

81. 🔲 $7.434 \div (-1.2) \times 9.5 + 1.47^2$

82. 🔲 $-9.46 \times 2.1^2 \div 3.5 + 4.36$

83. 🔲 $9.0534 - 2.041^2 \times 0.731 \div 1.043^2$

84. 🔲 $23.042(7 - 4.037 \times 1.46 - 0.932^2)$

Solve.

85. $439.57 \times 0.01 \div 1000 \cdot x = 4.3957$

86. $5.2738 \div 0.01 \times 1000 \div t = 52.738$

87. $0.0329 \div 0.001 \times 10^4 \div x = 3290$

88. $-4.302 \times 0.1^2 \div 0.001 \cdot t = -430.2$

89. *Television Ratings.*   A television rating point represents 980,000 households. The 2005 NASCAR Nextel Cup Series Dodge/Save Mart 350 was viewed in approximately 5.6 million households, a record for a NASCAR road-course race. How many rating points did the race receive? Round to the nearest tenth.
Source: www.infineonraceway.com

90. *Size of Country.*   The largest country, Russia, has an area of approximately 6.6 million square miles. The smallest country, Vatican City, has an area of approximately 0.2 square miles. How many times larger is Russia?
Source: U.S. Bureau of the Census, International Data Base

🔲 *Electric Bills.*   Recently, electric bills from the Central Vermont Public Service Corporation consisted of a "customer charge" of $0.374 per day plus an "energy charge" of $0.1174 per kilowatt-hour (kWh) for the first 250 kWh used and $0.09079 per kilowatt-hour for each kilowatt-hour in excess of 250.

91. From August 20 to September 20, the Kaufmans' bill was $59.10. How many kilowatt-hours of electricity did they use (round to the nearest kilowatt-hour)?

92. From July 20 to August 20, the McGuires' bill was $70. How many kilowatt-hours of electricity did they use (round to the nearest kilowatt-hour)?

5.5 MORE WITH FRACTION NOTATION AND DECIMAL NOTATION

Objectives

a Use division to convert fraction notation to decimal notation.

b Round numbers named by repeating decimals.

c Convert certain fractions to decimal notation by using equivalent fractions.

d Simplify expressions that contain both fraction and decimal notation.

Now that we know how to divide using decimal notation, we can express *any* fraction as a decimal. This means that any *rational* number (ratio of integers) can be written as a decimal.

a   Using Division to Find Decimal Notation

Recall that $\frac{a}{b}$ means $a \div b$. This gives us one way of converting fraction notation to decimal notation.

EXAMPLE 1   Find decimal notation for $\frac{3}{20}$.

Because $\frac{3}{20}$ means $3 \div 20$, we can perform long division.

$$
\begin{array}{r}
0.1\,5 \\
20\,\overline{)\,3.0\,0} \\
\underline{2\,0} \\
1\,0\,0 \\
\underline{1\,0\,0} \\
0
\end{array}
$$

We are finished when the remainder is 0.

We have $\frac{3}{20} = 0.15$.

EXAMPLE 2   Find decimal notation for $\frac{-7}{8}$.

Since $\frac{-7}{8}$ means $-7 \div 8$ and a negative divided by a positive is negative, we know that the decimal will be negative. We divide 7 by 8 and make the result negative.

$$
\begin{array}{r}
0.8\,7\,5 \\
8\,\overline{)\,7.0\,0\,0} \\
\underline{6\,4} \\
6\,0 \\
\underline{5\,6} \\
4\,0 \\
\underline{4\,0} \\
0
\end{array}
$$

Thus $\frac{-7}{8} = -0.875$.

Do Exercises 1 and 2.

When division with decimals ends, or *terminates*, as in Examples 1 and 2, the result is called a *terminating decimal*. If the division does *not* lead to a remainder of 0, but instead leads to a repeating pattern of nonzero remainders, we have what is called a *repeating decimal*.

Find decimal notation.

1. $\dfrac{2}{5}$

2. $\dfrac{-5}{8}$

Answers on page A-13

Study Tips

IDENTIFY THE HIGHLIGHTS

If you haven't already tried one, consider using a highlighter as you read. By highlighting sentences or phrases that you find especially important, you will make it easier to review important material in the future. Highlighters are only helpful when used, so be sure to keep your highlighter with you whenever you study.

Find decimal notation for each number.

3. $\dfrac{1}{6}$

4. $\dfrac{2}{3}$

Find decimal notation for each number.

5. $\dfrac{5}{11}$

6. $-\dfrac{12}{11}$

7. Find decimal notation for $\dfrac{5}{7}$.

■ **EXAMPLE 3** Find decimal notation for $\frac{5}{6}$.

Since $\frac{5}{6}$ means $5 \div 6$, we have

$$
\begin{array}{r}
0.8\ 3\ 3 \\
6\ \overline{)\ 5.0\ 0\ 0} \\
4\ 8 \\
\hline
2\ 0 \\
1\ 8 \\
\hline
2\ 0 \\
1\ 8 \\
\hline
2
\end{array}
$$

Since 2 keeps reappearing as a remainder, the digits repeat and will continue to do so; therefore,

$$
\frac{5}{6} = 0.83333\ldots.
$$

The dots indicate an endless sequence of digits in the quotient. When there is a repeating pattern, the dots are often replaced by a bar to indicate the repeating part, in this case, only the 3.

$$
\frac{5}{6} = 0.8\overline{3}
$$

Do Exercises 3 and 4.

■ **EXAMPLE 4** Find decimal notation for $-\frac{4}{11}$.

Since $-\frac{4}{11}$ is negative, we divide 4 by 11 and make the result negative.

$$
\begin{array}{r}
0.3\ 6\ 3\ 6 \\
1\ 1\ \overline{)\ 4.0\ 0\ 0\ 0} \\
3\ 3 \\
\hline
7\ 0 \\
6\ 6 \\
\hline
4\ 0 \\
3\ 3 \\
\hline
7\ 0 \\
6\ 6 \\
\hline
4
\end{array}
$$

Since 7 and 4 keep reappearing as remainders, the sequence of digits "36" repeats in the quotient, and

$$
\frac{4}{11} = 0.363636\ldots, \quad \text{or} \quad 0.\overline{36}.
$$

Thus, $-\frac{4}{11} = -0.\overline{36}$.

Do Exercises 5 and 6.

Answers on page A-13

Is there a way to know which fractions represent terminating decimals and which represent repeating decimals? The answer is "yes, provided the fraction is written in simplified form." As illustrated in Examples 1 and 2, when the denominator of a simplified fraction has no prime factor other than 2 or 5, the decimal terminates. When, as in Examples 3 and 4, the denominator of the simplified fraction has a prime factor other than 2 or 5, the decimal repeats.

EXAMPLE 5 Find decimal notation for $\frac{3}{7}$.

Because 7 is not a product of 2's and/or 5's, we expect a repeating decimal.

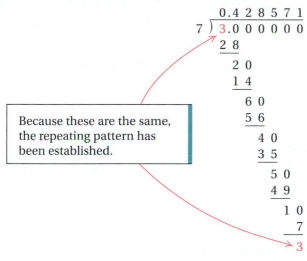

```
        0.4 2 8 5 7 1
  7 ) 3.0 0 0 0 0 0
      2 8
        2 0
        1 4
          6 0
          5 6
            4 0
            3 5
              5 0
              4 9
                1 0
                 7
                 3
```

Because these are the same, the repeating pattern has been established.

Since we have already divided 7 into 3, the sequence of digits "428571" repeats in the quotient.

$$\frac{3}{7} = 0.428571428571\ldots, \quad \text{or} \quad 0.\overline{428571}$$

Do Exercise 7 on the preceding page.

b Rounding Repeating Decimals

The repeating part of a decimal can be so long that it will not fit on a calculator. For example, when $\frac{5}{97}$ is written as a decimal, its repeating part is 96 digits long! Most calculators round off repeating decimals to 9 or 10 decimal places. In applied problems, repeating decimals are generally rounded to a predetermined degree of accuracy.

Round each to the nearest tenth, hundredth, and thousandth.

8. $0.\overline{6}$

9. $0.6\overline{08}$

10. $-7.3\overline{49}$

11. $2.6\overline{891}$

EXAMPLE 6 Round $4.\overline{27}$ to the nearest thousandth.

We first rewrite the decimal without the bar. The repeating part is rewritten until we have passed the thousandths place.

$$4.\overline{27} = 4.2727\ldots.$$

Now we round as in Section 5.1.

a) Locate the digit in the thousandths place. $4.2\,7\,\underset{\uparrow}{2}\,7\ldots$

b) Consider the next digit to the right. $4.2\,7\,2\,\underset{\uparrow}{7}\ldots$

c) Since that digit, 7, is greater than or equal to 5, round up.

     4.273   This is the answer.

EXAMPLES   Round each to the nearest tenth, hundredth, and thousandth.

	Nearest tenth	Nearest hundredth	Nearest thousandth
7. $0.8\overline{3} = 0.83333\ldots$	0.8	0.83	0.833
8. $3.\overline{09} = 3.090909\ldots$	3.1	3.09	3.091
9. $-4.1\overline{763} = -4.1763763\ldots$	-4.2	-4.18	-4.176

Do Exercises 8–11.

c   More with Conversions

Recall that fractions like $\frac{3}{10}$ or $-\frac{71}{1000}$ can be converted to decimal notation, without using long division. When a denominator is a factor of 10, 100, and so on, we can convert to decimal notation by finding (perhaps mentally) an equivalent fraction in which the denominator is a power of 10.

EXAMPLE 10 Find decimal notation for $-\frac{7}{500}$.

Since $500 \cdot 2 = 1000$, and 1000 is a power of 10, we use $\frac{2}{2}$ as an expression for 1.

$$-\frac{7}{500} = -\frac{7}{500} \cdot \frac{2}{2} = -\frac{14}{1000} = -0.014 \qquad \textit{Think}: 1000 \div 500 = 2,$$
$$\text{and } 7 \cdot 2 = 14.$$

EXAMPLE 11 Find decimal notation for $\frac{9}{25}$.

$$\frac{9}{25} = \frac{9}{25} \cdot \frac{4}{4} = \frac{36}{100} = 0.36 \qquad \text{Using } \tfrac{4}{4} \text{ for 1 to get a denominator of 100}$$

As a check, we can divide.

```
       0.3 6
2 5 ) 9.0 0
       7 5          Note that multiplication by 1 is much faster.
       1 5 0
       1 5 0
           0
```

EXAMPLE 12 Find decimal notation for $\frac{7}{4}$.

$$\frac{7}{4} = \frac{7}{4} \cdot \frac{25}{25} = \frac{175}{100} = 1.75$$

Using $\frac{25}{25}$ for 1 to get a denominator of 100. You might also note that 7 quarters is $1.75.

Do Exercises 12–15.

d | Calculations with Fraction and Decimal Notation Together

In certain kinds of calculations, fraction and decimal notation might occur together. In such cases, there are at least three ways in which we can proceed.

EXAMPLE 13 Calculate: $\frac{2}{3} \times 0.576$

METHOD 1 Perhaps the quickest method is to treat 0.576 as $\frac{0.576}{1}$. Then we multiply 0.576 by 2, and divide the result by 3.

$$\begin{array}{r} 0.3\ 8\ 4 \\ 3\overline{)\ 1.1\ 5\ 2} \\ \underline{9} \\ 2\ 5 \\ \underline{2\ 4} \\ 1\ 2 \\ \underline{1\ 2} \\ 0 \end{array}$$

$$\frac{2}{3} \times 0.576 = \frac{2}{3} \times \frac{0.576}{1}$$

$$= \frac{2 \times 0.576}{3} = \frac{1.152}{3}$$

$$= 0.384$$

METHOD 2 A second way to do this calculation is to convert the fraction to decimal notation so that both numbers are decimals. Since $\frac{2}{3}$ is a repeating decimal, it is first rounded to some chosen decimal place. We choose three decimal places. Then, using decimal notation, we multiply. Note that the answer is not as accurate as that found by method 1, due to the rounding.

$$\frac{2}{3} \times 0.576 = 0.\overline{6} \times 0.576$$

$$\approx 0.667 \times 0.576$$

$$= 0.384192 \qquad \text{This is \textit{less} accurate than the result in method 1.}$$

METHOD 3 A third method is to convert the decimal to a fraction so that both numbers are in fraction notation. The answer can be left in fraction notation and simplified, or we can convert back to decimal notation and, if appropriate, round.

$$\frac{2}{3} \times 0.576 = \frac{2}{3} \cdot \frac{576}{1000} = \frac{2 \cdot 576}{3 \cdot 1000}$$

$$= \frac{2 \cdot 2 \cdot 2 \cdot 2 \cdot 2 \cdot 2 \cdot 2 \cdot 3 \cdot 3}{2 \cdot 2 \cdot 2 \cdot 3 \cdot 5 \cdot 5 \cdot 5} \qquad \text{Factoring}$$

$$= \frac{2 \cdot 2 \cdot 2 \cdot 3}{2 \cdot 2 \cdot 2 \cdot 3} \cdot \frac{2 \cdot 2 \cdot 2 \cdot 2 \cdot 3}{5 \cdot 5 \cdot 5} \qquad \text{Removing a factor equal to 1: } \frac{2 \cdot 2 \cdot 2 \cdot 3}{2 \cdot 2 \cdot 2 \cdot 3} = 1$$

$$= \frac{2 \cdot 2 \cdot 2 \cdot 2 \cdot 3}{5 \cdot 5 \cdot 5} = \frac{48}{125}, \text{ or } 0.384$$

Note that we get an exact answer with methods 1 and 3, but method 2 gives an approximation. Method 2 will be exact only if the fraction is a terminating decimal.

Do Exercises 16 and 17.

Multiply by a form of 1 to find decimal notation for each number.

12. $\frac{4}{5}$

13. $-\frac{9}{20}$

14. $\frac{7}{200}$

15. $\frac{33}{25}$

Calculate.

16. $\frac{5}{6} \times 0.864$

17. $\frac{1}{3} \times 0.384 + \frac{5}{8} \times 0.6784$

Answers on page A-13

18. Find the area of a triangular window that is 3.25 ft wide and 2.6 ft tall.

2.6 ft

3.25 ft

Answer on page A-13

■ **EXAMPLE 14**   *Boating.*   A triangular sail from a single-sail day cruiser is 3.4 m wide and 4.2 m tall. Find the area of the sail.

1. **Familiarize.**   We first make a drawing and recall that the formula for the area, A, of a triangle with base b and height h is $A = \frac{1}{2}bh$.

2. **Translate.**   We substitute 3.4 for b and 4.2 for h.

$$A = \frac{1}{2}bh = \frac{1}{2}(3.4)(4.2) \qquad \text{Evaluating}$$

3. **Solve.**   We simplify as follows.

$$A = \frac{1}{2}(3.4)(4.2)$$

$$= \frac{3.4}{2}(4.2) \qquad \text{Multiplying } \tfrac{1}{2} \text{ and } \tfrac{3.4}{1}$$

$$= 1.7(4.2) \qquad \text{Dividing}$$

$$= 7.14 \qquad \text{Multiplying}$$

4. **Check.**   To check, we repeat the calculations using the commutative law to multiply in a different order. We also rewrite $\frac{1}{2}$ as 0.5.

$$\frac{1}{2}(4.2)(3.4) = 0.5(4.2)(3.4)$$

$$= (2.1)(3.4) = 7.14$$

   Our answer checks.

5. **State.**   The area of the sail is 7.14 m² (square meters).

Do Exercise 18.

a , c   Find decimal notation for each number.

1. $\dfrac{3}{8}$

2. $\dfrac{3}{5}$

3. $\dfrac{-1}{2}$

4. $\dfrac{-1}{4}$

5. $\dfrac{3}{25}$

6. $\dfrac{7}{20}$

7. $\dfrac{9}{40}$

8. $\dfrac{3}{40}$

9. $\dfrac{13}{25}$

10. $\dfrac{17}{25}$

11. $\dfrac{-17}{20}$

12. $\dfrac{-13}{20}$

13. $-\dfrac{9}{16}$

14. $-\dfrac{5}{16}$

15. $\dfrac{7}{5}$

16. $\dfrac{3}{2}$

17. $\dfrac{28}{25}$

18. $\dfrac{31}{20}$

19. $\dfrac{11}{-8}$

20. $\dfrac{17}{-10}$

21. $-\dfrac{39}{40}$

22. $-\dfrac{17}{40}$

23. $\dfrac{121}{200}$

24. $\dfrac{32}{125}$

25. $\dfrac{8}{15}$

26. $\dfrac{7}{9}$

27. $\dfrac{1}{3}$

28. $\dfrac{1}{9}$

29. $\dfrac{-4}{3}$

30. $\dfrac{-8}{9}$

31. $\dfrac{7}{6}$

32. $\dfrac{7}{11}$

33. $-\dfrac{14}{11}$ **34.** $-\dfrac{7}{11}$ **35.** $\dfrac{-5}{12}$ **36.** $\dfrac{-11}{12}$

37. $\dfrac{127}{500}$ **38.** $\dfrac{83}{500}$ **39.** $\dfrac{4}{33}$ **40.** $\dfrac{5}{33}$

41. $\dfrac{-12}{55}$ **42.** $\dfrac{-5}{22}$ **43.** $\dfrac{35}{111}$ **44.** $\dfrac{27}{111}$

45. $\dfrac{4}{7}$ **46.** $\dfrac{2}{7}$ **47.** $\dfrac{-37}{25}$ **48.** $\dfrac{-31}{250}$

b   For Exercises 49–60, round the decimal notation for each number to the nearest tenth, hundredth, and thousandth.

49. $\dfrac{4}{11}$ **50.** $\dfrac{3}{11}$ **51.** $-\dfrac{5}{3}$ **52.** $-\dfrac{19}{16}$

53. $\dfrac{-8}{17}$ **54.** $\dfrac{-7}{13}$ **55.** $\dfrac{7}{12}$ **56.** $\dfrac{2}{15}$

57. $\dfrac{29}{-150}$ **58.** $\dfrac{37}{-150}$ **59.** $\dfrac{7}{-9}$ **60.** $\dfrac{5}{-13}$

d Calculate and write the result as a decimal.

61. $\dfrac{7}{8}(10.84)$

62. $\dfrac{4}{5}(264.8)$

63. $\dfrac{47}{9}(-79.95)$

64. $\dfrac{7}{11}(-2.7873)$

65. $\left(\dfrac{1}{6}\right)0.0765 + \left(\dfrac{3}{4}\right)0.1124$

66. $\left(\dfrac{2}{5}\right)6384.1 - \left(\dfrac{5}{8}\right)156.56$

67. $\dfrac{3}{4} \times 2.56 - \dfrac{7}{8} \times 3.94$

68. $\dfrac{2}{5} \times 3.91 - \dfrac{7}{10} \times 4.15$

69. $5.2 \times 1\dfrac{7}{8} \div 0.4$

70. $4\dfrac{3}{4} \times 0.5 \div 0.1$

Solve.

71. Find the area of a triangular shawl that is 1.8 m long and 1.2 m wide.

72. Find the area of a triangular sign that is 1.5 m wide and 1.5 m tall.

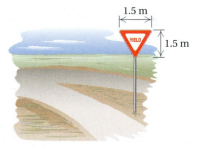

73. Find the area of a triangular stamp that is 3.4 cm wide and 3.4 cm tall.

74. Find the area of a triangular reflector that is 7.4 cm wide and 9.1 cm tall.

75. Find the area of the kite shown on the left.

76. Find the area of the kite shown on the right.

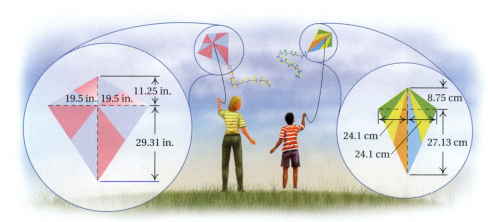

77. **Dw** When is long division *not* the fastest way of converting a fraction to decimal notation?

78. **Dw** Use decimal notation to explain why $\dfrac{-2}{9} = \dfrac{2}{-9} = -\dfrac{2}{9}$.

SKILL MAINTENANCE

79. Round 3572 to the nearest ten.   [1.4a]

80. Round 3572 to the nearest thousand.   [1.4a]

81. Round 78,951 to the nearest hundred.   [1.4a]

82. Round 19,829,996 to the nearest ten.   [1.4a]

83. Simplify: $\dfrac{95}{-1}$.   [3.3b]

84. Solve: $5x - 9 = 7x + 11$.   [4.4a]

85. Simplify: $9 - 4 + 2 \div (-1) \cdot 6$.   [2.5b]

86. Simplify: $\dfrac{-9}{-9}$.   [3.3b]

SYNTHESIS

87. **Dw** Are the numbers $5.1\overline{47}$ and $5.14\overline{747}$ equal? Why or why not?

88. **Dw** ▦ A scientific calculator indicates that

$$\frac{5}{6} = 0.833333333 \quad \text{and} \quad \frac{4{,}999{,}999{,}998}{6{,}000{,}000{,}000} = 0.833333333$$

a) Is it true that $\frac{5}{6} = \frac{4{,}999{,}999{,}998}{6{,}000{,}000{,}000}$? Why or why not?
b) Should decimal notation for $\frac{4{,}999{,}999{,}998}{6{,}000{,}000{,}000}$ repeat? Why or why not?

▦ Find decimal notation. Save the answers for Exercise 94.

89. $\dfrac{1}{7}$

90. $\dfrac{2}{7}$

91. $\dfrac{3}{7}$

92. $\dfrac{4}{7}$

93. $\dfrac{5}{7}$

94. ▦ From the pattern of answers to Exercises 89–93, predict the decimal notation for $\frac{6}{7}$. Check your answer on a calculator.

Find decimal notation. Save the answers for Exercise 98.

95. $\dfrac{1}{9}$

96. $\dfrac{1}{99}$

97. $\dfrac{1}{999}$

98. ▦ From the pattern of Exercises 95–97, predict the decimal notation for $\frac{1}{9999}$. Check your answer on a calculator.

The formula $A = \pi r^2$ is used to find the area, A, of a circle with radius r. For Exercises 99 and 100 find the area of a circle with the given radius, using $\frac{22}{7}$ for π. For Exercises 101 and 102, use 3.14 for π or a calculator with a π key.

99. $r = 2.1$ cm

100. $r = 1.4$ cm

101. ▦ $r = \dfrac{3}{4}$ ft

102. ▦ $r = 4\dfrac{1}{2}$ yd

103. **Dw** Classify each equation as True or False and provide an explanation for each answer.

a) $0.3333 = \dfrac{1}{3}$ **b)** $0.\overline{3} = \dfrac{1}{3}$ **c)** $0.\overline{6} = \dfrac{2}{3}$ **d)** $0.\overline{9} = 1$

5.6 ESTIMATING

a Estimating Sums, Differences, Products, and Quotients

Estimating has many uses. It can be done before a problem is even attempted and it can be done afterward as a check, even when we are using a calculator. Often, an estimate is all we need. We usually estimate by rounding the numbers so that there are one or two nonzero digits. Consider the following advertisements while reading Examples 1–3.

EXAMPLE 1 Estimate by rounding to the nearest ten the total cost of one all-in-one printer/copier and one TV/DVD combo.

We are estimating the sum

$199.98 + $141.99 = Total cost.

The estimate found by rounding the addends to the nearest ten is

$200 + $140 = $340.   (Estimated total cost)

Do Exercise 1.

EXAMPLE 2 About how much more does the refrigerator cost than the TV/DVD combo? Estimate by rounding to the nearest ten.

We are estimating the difference

$309.95 − $141.99 = Price difference.

The estimate to the nearest ten is

$310 − $140 = $170.   (Estimated price difference)

Do Exercise 2.

Objective

a Estimate sums, differences, products, and quotients.

1. Estimate by rounding to the nearest ten the total cost of one refrigerator and one printer/copier.

2. About how much more does the printer/copier cost than the TV/DVD combo? Estimate by rounding to the nearest ten.

Answers on page A-14

Study Tips

PUT MATH TO USE

One excellent way to study math is to use it in your everyday life. The concepts of this section can be easily reinforced if you use estimating when you next go shopping.

357

3. Estimate the total cost of 6 refrigerators by rounding to the nearest ten.

4. About how many Play Station Portable™ systems can be purchased for $4530?

Estimate each product. Do not find the actual product.

5. 2.1 × 8.02

6. 36 × 0.54

7. 0.93 × 472

8. 0.72 × 0.1

9. 0.12 × 180.3

10. 24.359 × 5.2

Answers on page A-14

EXAMPLE 3 Estimate the total cost of 4 TV/DVD combos. (See p. 357).

We are estimating the product

4 × $141.99 = Total cost.

The estimate is found by rounding 141.99 to the nearest ten:

4 × $140 = $560.

Do Exercise 3.

EXAMPLE 4 About how many Play Station Portable™ gaming systems at $249.99 each can be purchased for $1480?

We are estimating the quotient

$1480 ÷ $249.99.

Since we want a whole-number estimate, we choose our rounding appropriately. Rounding $249.99 to the nearest one, we get $250. Since $1480 is close to $1500, which is a multiple of 250, we estimate

$1500 ÷ $250,

so the answer is 6.

Do Exercise 4.

When estimating, we usually look for numbers that are easy to work with. For example, if multiplying, we might round 0.43 to 0.5 and 8.9 to 10, because 0.5 and 10 are convenient numbers to multiply.

EXAMPLE 5 Estimate: 4.8 × 62. Do not find the actual product.

We round 4.8 to the nearest one and 62 to the nearest ten. This gives us two easy numbers with which to work, 5 and 60. Since

4.8 × 62 ≈ 5 × 60

and

5 × 60 = 300,

the estimated product is 300.

Compare these estimates for the product 4.94 × 38:

5 × 40 = 200,   5 × 38 = 190,   4.9 × 40 = 196,   4.9 × 38 = 186.2.

The first estimate was the easiest. You could probably do it mentally. The others had more nonzero digits and were more accurate, but required more work.

Do Exercises 5–10.

EXAMPLE 6   Which of the following is the best estimate of 82.08 ÷ 24?

a) 400   **b)** 16   **c)** 40   **d)** 4

    This is about 80 ÷ 20, so the answer is about 4. We could also estimate the division as 75 ÷ 25, or 3. In any case, of the choices listed, (d) is the most appropriate.

EXAMPLE 7   Which of the following is the best estimate of 94.18 ÷ 3.2?

a) 30   **b)** 300   **c)** 3   **d)** 60

    This is about 90 ÷ 3, so the answer is about 30. Thus, the most appropriate choice is (a).

EXAMPLE 8   Which of the following is the best estimate of 0.0156 ÷ 1.3?

a) 0.2   **b)** 0.002   **c)** 0.02   **d)** 20

    This is about 0.02 ÷ 1, so the answer is about 0.02. Thus, the most appropriate choice is (c).

Do Exercises 11–13.

    In some cases, it is easier to estimate a quotient by checking products than by rounding the divisor and the dividend.

EXAMPLE 9   Which of the following is the best estimate of 0.0074 ÷ 0.23?

a) 0.3   **b)** 0.03   **c)** 300   **d)** 3

    Note that 0.23 is close to 0.25 and that it is easier to multiply than divide by 0.25. Thus we use 0.25 to check some products.
    We first try 3.

$$0.23 \times 3 \approx 0.25 \times 3 = 0.75 \qquad \text{This is too large.}$$

We try a smaller estimate, 0.3.

$$0.23 \times 0.3 \approx 0.25 \times 0.3 = 0.075 \qquad \text{This is also too large.}$$

We make the estimate smaller still, 0.03.

$$0.23 \times 0.03 \approx 0.25 \times 0.03 = 0.0075$$

This is close to 0.0074, so the quotient is close to 0.03. Thus, the most appropriate choice is (b).

Do Exercise 14.

Select the most appropriate estimate for each quotient.

11. 59.78 ÷ 29.1

  a) 200   **b)** 20
  c) 2   **d)** 0.2

12. 82.08 ÷ 2.4

  a) 40   **b)** 4.0
  c) 400   **d)** 0.4

13. 0.1768 ÷ 0.08

  a) 8   **b)** 10
  c) 2   **d)** 20

14. Which of the following is an appropriate estimate of 0.0069 ÷ 0.15?

  a) 0.5   **b)** 50
  c) 0.05   **d)** 23.4

Answers on page A-14

a Consider the following advertisements for Exercises 1–8. Estimate the sums, differences, products, or quotients involved in these problems. Answers will vary, so show all steps.

Weatherband Radio X4402
Rechargeable Two-Way Radios with weather band and up to a five-mile range.
$79 95
For Two

Satellite Radio
Receiver and car dock bundle. Over 120 digital channels of music, sports, entertainment and news including 65 channels of 100% commercial-free music.
$149 99

Upright Vacuum Cleaner
Features a lifetime HEPA filter and 17 ft quick draw hose.
$279

1. Estimate the total cost of one vacuum cleaner and one satellite radio.

2. Estimate the total cost of one satellite radio and one set of two-way radios.

3. About how much more does the vacuum cleaner cost than the satellite radio?

4. About how much more does the satellite radio cost than the two-way radios?

5. Estimate the total cost of 6 sets of two-way radios.

6. Estimate the total cost of 4 vacuum cleaners.

7. About how many sets of two-way radios can be purchased for $830?

8. About how many vacuum cleaners can be purchased for $5627?

Estimate by rounding as directed.

9. $0.02 + 1.31 + 0.34$; nearest tenth

10. $0.88 + 2.07 + 1.54$; nearest one

11. $6.03 + 0.007 + 0.214$; nearest one

12. 1.11 + 8.888 + 99.94;
nearest one

13. 52.367 + 1.307 + 7.324;
nearest one

14. 12.9882 + 1.0115;
nearest tenth

15. 2.678 − 0.445; nearest tenth

16. 12.9882 − 1.0115; nearest one

17. 198.67432 − 24.5007; nearest ten

Choose a rounding digit that gives one or two nonzero digits and select the most appropriate estimate.

18. 234.12321 − 200.3223

    a) 600   **b)** 60
    c) 300   **d)** 30

19. 49 × 7.89

    a) 400   **b)** 40
    c) 4   **d)** 0.4

20. 7.4 × 8.9

    a) 95   **b)** 63
    c) 124   **d)** 6

21. 98.4 × 0.083

    a) 80   **b)** 12
    c) 8   **d)** 0.8

22. 78 × 5.3

    a) 400   **b)** 800
    c) 40   **d)** 8

23. 3.6 ÷ 4

    a) 10   **b)** 1
    c) 0.1   **d)** 0.01

24. 0.0713 ÷ 1.94

    a) 4   **b)** 0.4
    c) 0.04   **d)** 40

25. 74.68 ÷ 24.7

    a) 9   **b)** 3
    c) 12   **d)** 120

26. 914 ÷ 0.921

    a) 10   **b)** 100
    c) 1000   **d)** 1

27. *Fence Posts.* A zoo plans to construct a fence around its proposed African wildlife exhibit. The perimeter of the area to be fenced is 1760 ft. Estimate the number of wooden fence posts needed if the posts are placed 8.625 ft apart.

28. *Paint.* Recently the Home Depot sold Behr semigloss exterior house paint for $22.97 a gallon. Estimate how many gallons could be bought with $4500.

29. $\mathbf{D_W}$ Describe a situation in which an estimate is made by rounding to the nearest 10,000 and then multiplying.

30. $\mathbf{D_W}$ A roll of fiberglass insulation costs $21.95. Describe two situations involving estimating and the cost of fiberglass insulation. Devise one situation so that $21.95 is rounded to $22. Devise the other situation so that $21.95 is rounded to $20.

SKILL MAINTENANCE

🖎 VOCABULARY REINFORCEMENT

In each of Exercises 31–38, fill in the blank with the correct term from the given list. Some of the choices may not be used and some may be used more than once.

31. The decimal $0.57\overline{3}$ is an example of a _____ decimal. [5.5a]

32. The least common _____ of two natural numbers is the smallest number that is a multiple of both. [4.1a]

33. The sentence $5(3 + 8) = 5 \cdot 3 + 5 \cdot 8$ illustrates the _____ law. [1.5a]

34. A(n) _____ of an equation is a replacement for the variable that makes the equation true. [1.7a]

35. The number 1 is the _____ identity. [1.5a]

36. The sentence $13 + 7 = 7 + 13$ illustrates the _____ law of addition. [1.2b]

37. The least common _____ of two or more fractions is the least common _____ of their denominators. [4.2b]

38. The number 3728 is _____ by 9 if the sum of the digits is _____ by 9. [3.1b]

additive

multiplicative

numerator

denominator

commutative

associative

distributive

solution

divisible

terminating

repeating

multiple

factor

SYNTHESIS

39. $\mathbf{D_W}$ Rather than charge $30 for an item, stores often price the item at $29.99, or $29.95, or sometimes even $29.97. Why does this practice exist?

40. $\mathbf{D_W}$ Rick rounded -305.281 to -305.3. Did he round up or down? Explain.

The following were done on a calculator and then rewritten. Estimate to determine whether the decimal point was placed correctly.

41. $178.9462 \times 61.78 = 11,055.29624$

42. $14,973.35 \div 298.75 = 501.2$

43. $19.7236 - 1.4738 \times 4.1097 = 1.366672414$

44. $28.46901 \div 4.9187 - 2.5081 = 3.279813473$

45. 🖩 Use one of $+, -, \times, \div$ in each blank to make a true sentence.

a) $(0.37 \ \boxed{} \ 18.78) \ \boxed{} \ 2^{13} = 156,876.8$

b) $2.56 \ \boxed{} \ 6.4 \ \boxed{} \ 51.2 \ \boxed{} \ 17.4 = 312.84$

46. 🖩 In the subtraction below, a and b are digits. Find a and b.

$$\begin{array}{r} b876.a4321 \\ -\ 1234.a678b \\ \hline 8641.b7a32 \end{array}$$

5.7 SOLVING EQUATIONS

Objectives

a Solve equations containing decimals and one variable term.

b Solve equations containing decimals and two or more variable terms.

In Section 4.4, we used a combination of the addition and multiplication principles to solve equations like $5x + 7 = -3$. We now use those same properties to solve similar equations involving decimals.

a Equations with One Variable Term

Recall that equations like $5x + 7 = -3$ are normally solved by first "undoing" the addition and then "undoing" the multiplication. This reverses the order of operations in which we add last and multiply first.

EXAMPLE 1 Solve: $0.5x + 5 = 8$.

$$0.5x + 5 = 8$$
$$0.5x + 5 - 5 = 8 - 5 \qquad \text{Subtracting 5 from both sides}$$
$$0.5x = 3 \qquad \text{Simplifying}$$
$$\frac{0.5x}{0.5} = \frac{3}{0.5} \qquad \text{Dividing both sides by 0.5}$$
$$x = 6 \qquad \text{Simplifying}$$

Check:

$$\begin{array}{c|c} 0.5x + 5 = 8 \\ \hline 0.5(6) + 5 \; ? \; 8 \\ 3 + 5 \; \\ 8 \; \big| \; 8 \quad \text{TRUE} \end{array}$$

The solution is 6.

Do Exercises 1 and 2.

EXAMPLE 2 Solve: $4.2x + 3.7 = -26.12$.

$$4.2x + 3.7 = -26.12$$
$$4.2x + 3.7 - 3.7 = -26.12 - 3.7 \qquad \text{Subtracting 3.7 from both sides}$$
$$4.2x = -29.82 \qquad \text{Simplifying}$$
$$\frac{4.2x}{4.2} = \frac{-29.82}{4.2} \qquad \text{Dividing both sides by 4.2}$$
$$x = -7.1 \qquad \text{Simplifying}$$

Check:

$$\begin{array}{c|c} 4.2x + 3.7 = -26.12 \\ \hline 4.2(-7.1) + 3.7 \; ? \; -26.12 \\ -29.82 + 3.7 \; \\ -26.12 \; \big| \; -26.12 \quad \text{TRUE} \end{array}$$

The solution is -7.1.

Do Exercises 3 and 4.

Solve.

1. $6x + 7.4 = 11$

2. $0.2 - 0.1x = 1.4$

Solve.

3. $7.4t + 1.25 = 27.89$

4. $-5.7 + 4.8x = -14.82$

Answers on page A-14

Study Tips

DOUBLE-CHECK THE NUMBERS

Solving problems is challenging enough, without miscopying information. Always double-check that you have accurately transferred numbers from the correct exercise in the exercise set.

363

b   Equations with Two or More Variable Terms

Some equations have variable terms on both sides. To solve such an equation, we use the addition principle to get all variable terms on one side of the equation and all constant terms on the other side.

EXAMPLE 3   Solve: $10x - 7 = 2x + 13$.

We begin by subtracting $2x$ from (or adding $-2x$ to) each side. This will group all variable terms on one side of the equation.

$$10x - 7 - 2x = 2x + 13 - 2x \qquad \text{Adding } -2x \text{ to both sides}$$
$$8x - 7 = 13 \qquad\qquad\qquad \text{Combining like terms}$$

We use the addition principle to isolate all constant terms on one side.

$$8x - 7 = 13$$
$$8x - 7 + 7 = 13 + 7 \qquad \text{Adding 7 to both sides}$$
$$8x = 20 \qquad\qquad\quad \text{Simplifying (combining like terms)}$$
$$\frac{8x}{8} = \frac{20}{8} \qquad\qquad \text{Dividing both sides by 8}$$
$$x = 2.5$$

Check:
$$\begin{array}{c|c}
\multicolumn{2}{c}{10x - 7 = 2x + 13} \\
\hline
10(2.5) - 7 \ ? \ 2(2.5) + 13 & \\
25 - 7 & 5 + 13 \\
18 & 18 \qquad \text{TRUE}
\end{array}$$

The solution is 2.5.

Sometimes it may be easier to combine all variable terms on the right side and all constant terms on the left side.

EXAMPLE 4   Solve: $11 - 3t = 7t + 8$.

We can combine all variable terms on the right side by adding $3t$ to both sides.

$$11 - 3t = 7t + 8$$
$$11 - 3t + 3t = 7t + 8 + 3t \qquad \text{Adding } 3t \text{ to both sides}$$
$$11 = 10t + 8 \qquad\qquad\quad \text{Combining like terms}$$
$$11 - 8 = 10t + 8 - 8 \qquad \text{Subtracting 8 from both sides}$$
$$3 = 10t$$
$$\frac{3}{10} = \frac{10t}{10} \qquad\qquad\qquad \text{Dividing both sides by 10}$$
$$0.3 = t$$

Check:
$$\begin{array}{c|c}
\multicolumn{2}{c}{11 - 3t = 7t + 8} \\
\hline
11 - 3(0.3) \ ? \ 7(0.3) + 8 & \\
11 - 0.9 & 2.1 + 8 \\
10.1 & 10.1 \qquad \text{TRUE}
\end{array}$$

The solution is 0.3.

Note that in Example 4 the variable appears on the right side of the last equation. It does not matter whether the variable is isolated on the right or left side. What is important is that you have a clear direction to your work as you proceed from step to step.

Do Exercises 5–7.

EXAMPLE 5 Solve: $5(x + 1) = 3x + 12$.

$$5(x + 1) = 3x + 12$$
$$5 \cdot x + 5 \cdot 1 = 3x + 12 \qquad \text{Using the distributive law to remove parentheses}$$
$$5x + 5 = 3x + 12 \qquad \text{Simplifying}$$
$$5x + 5 - 3x = 3x + 12 - 3x \qquad \text{Subtracting } 3x \text{ from both sides}$$
$$2x + 5 = 12 \qquad \text{Simplifying}$$
$$2x + 5 - 5 = 12 - 5 \qquad \text{Subtracting 5 from both sides}$$
$$2x = 7$$
$$\frac{2x}{2} = \frac{7}{2} \qquad \text{Dividing both sides by 2}$$
$$x = 3.5$$

Check:
$$\begin{array}{c|c} \hline 5(x + 1) = 3x + 12 \\ \hline 5(3.5 + 1) \ ? \ 3(3.5) + 12 \\ 5(4.5) \ \big| \ 10.5 + 12 \\ 22.5 \ \big| \ 22.5 \qquad \text{TRUE} \end{array}$$

The solution is 3.5.

Do Exercise 8.

EXAMPLE 6 Solve: $9(x - 3) + 7 = 5x - 47$.

We use the distributive law and combine like terms before using the addition and multiplication principles.

$$9(x - 3) + 7 = 5x - 47$$
$$9x - 27 + 7 = 5x - 47 \qquad \text{Using the distributive law}$$
$$9x - 20 = 5x - 47 \qquad \text{Simplifying}$$
$$9x - 20 - 5x = 5x - 47 - 5x \qquad \text{Subtracting } 5x \text{ from both sides}$$
$$4x - 20 = -47 \qquad \text{Simplifying}$$
$$4x - 20 + 20 = -47 + 20 \qquad \text{Adding 20 to both sides}$$
$$4x = -27 \qquad \text{Simplifying}$$
$$\frac{4x}{4} = -\frac{27}{4} \qquad \text{Dividing both sides by 4}$$
$$x = -6.75$$

Check:
$$\begin{array}{c|c} \hline 9(x - 3) + 7 = 5x - 47 \\ \hline 9(-6.75 - 3) + 7 \ ? \ 5(-6.75) - 47 \\ 9(-9.75) + 7 \ \big| \ -33.75 - 47 \\ -87.75 + 7 \ \big| \ -80.75 \\ -80.75 \ \big| \ -80.75 \qquad \text{TRUE} \end{array}$$

The solution is -6.75.

Do Exercise 9.

Solve.

5. $10t - 3 = 4t + 18$

6. $8 + 4x = 9x - 3$

7. $2.1x - 45.3 = 17.3x + 23.1$

8. Solve: $3(x + 5) = 20 - x$.

9. Solve: $8(x - 2) - 15 = 4x + 2$.

Answers on page A-14

a Solve. Remember to check.

1. $5x = 27$

2. $36 \cdot y = 14.76$

3. $x + 15.7 = 3.1$

4. $x + 13.9 = 4.2$

5. $5x - 8 = 22$

6. $4x - 7 = 13$

7. $6.9x - 8.4 = 4.02$

8. $7.1x - 9.3 = 8.45$

9. $21.6 + 4.1t = 6.43$

10. $12.4 + 3.7t = 2.04$

11. $-26.05 = 7.5x + 9.2$

12. $-43.42 = 8.7x + 5.3$

13. $-4.2x + 3.04 = -4.1$

14. $-2.9x - 2.24 = -17.9$

15. $-3.05 = 7.24 - 3.5t$

16. $-4.62 = 5.68 - 2.5t$

b Solve. Remember to check.

17. $9x - 2 = 5x + 34$

18. $8x - 5 = 6x + 9$

19. $2x + 6 = 7x - 10$

20. $3x + 4 = 11x - 6$

21. $5y - 3 = 4 + 9y$

22. $6y - 5 = 8 + 10y$

23. $5.9x + 67 = 7.6x + 16$

24. $2.1x + 42 = 5.2x - 20$

25. $7.8a + 2 = 2.4a + 19.28$

26. $7.5a - 5.16 = 3.1a + 12$

27. $6(x + 2) = 4x + 30$

28. $5(x + 3) = 3x + 23$

29. $5(x + 3) = 15x - 6$

30. $2(x + 3) = 4x - 11$

31. $7a - 9 = 15(a - 3)$

32. $2a - 7 = 12(a - 3)$

33. $2.9(x + 8.1) = 7.8x - 3.95$

34. $2(x + 7.3) = 6x - 0.83$

35. $-6.21 - 4.3t = 9.8(t + 2.1)$

36. $-7.37 - 3.2t = 4.9(t + 6.1)$

37. $4(x - 2) - 9 = 2x + 9$

38. $9(x - 4) + 13 = 4x + 12$

39. $43(7 - 2x) + 34 = 50(x - 4.1) + 744$

40. $34(5 - 3.5x) = 12(3x - 8) + 653.5$

41. $\mathbf{D_W}$ Which equation do you consider more challenging to solve and why:

$$4.2x + 3.7 = -26.12 \quad \text{or} \quad 10x - 7 = 2x + 13?$$

42. $\mathbf{D_W}$ Is it "incorrect" to begin solving $5x + 9 = x + 12$ by dividing both sides by 5? Why or why not?

Find the area of each figure.   [3.6b]

43.
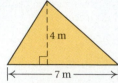
4 m
7 m

44.
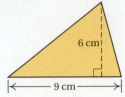
6 cm
9 cm

45.
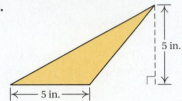
5 in.
5 in.

46.
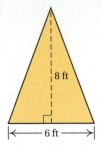
8 ft
6 ft

47.
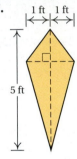
1 ft   1 ft
5 ft

48.
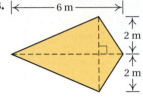
6 m
2 m
2 m

49. Subtract: $\dfrac{3}{25} - \dfrac{7}{10}$.   [4.3a]

50. Simplify: $\dfrac{0}{-18}$.   [3.3b]

51. Add: $-17 + 24 + (-9)$.   [2.2a]

52. Solve: $3x - 10 = 14$.   [2.8d]

53. **D**_W Is it possible for an equation like $x + 3 = x + 5$ to have a solution? Why or why not?

54. **D**_W Is it possible for an equation like $4x - 1 = 4(x - 2) + 4$ to have a solution? Why or why not?

Solve.

55. 🖩 $7.035(4.91x - 8.21) + 17.401 = 23.902x - 7.372815$

56. 🖩 $8.701(3.4 - 5.1x) - 89.321 = 5.401x + 74.65787$

57. $5(x - 4.2) + 3[2x - 5(x + 7)] = 39 + 2(7.5 - 6x) + 3x$

58. $14(2.5x - 3) + 9x + 5 = 4(3.25 - x) + 2[5x - 3(x + 1)]$

59. 🖩 $3.5(4.8x - 2.9) + 4.5 = 9.4x - 3.4(x - 1.9)$

60. 🖩 $4.19 - 1.8(4.5x - 6.4) = 3.1(9.8 + x)$

a Solving Applied Problems

Solving applied problems with decimals is like solving applied problems with integers. We translate first to an equation that corresponds to the situation. Then we solve the equation.

EXAMPLE 1 *Canals.* The Panama Canal in Panama is 50.7 mi long. The Suez Canal in Egypt is 119.9 mi long. How much longer is the Suez Canal?

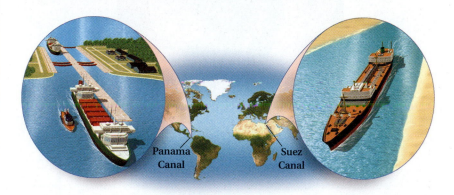

Panama Canal
Suez Canal

1. Familiarize. This is a "how much more" situation. We let l = the distance in miles that the length of the longer canal differs from the length of the shorter canal.

2. Translate. We translate as follows, using the given information.

Length of Panama Canal, the shorter canal	plus	Additional length	is	Length of Suez Canal, the longer canal
50.7 mi	+	l	=	119.9 mi

3. Solve. We solve the equation by subtracting 50.7 mi from both sides.

$$50.7 + l = 119.9$$
$$50.7 + l - 50.7 = 119.9 - 50.7$$
$$l = 69.2$$

4. Check. We can check by adding.

$$50.7 + 69.2 = 119.9$$

5. State. The Suez Canal is 69.2 mi longer than the Panama Canal.

Do Exercise 1.

1. Photofinishing Market. Because an increasing number of people are printing their own digital photographs at home, the commercial photofinishing industry's revenue is declining. Between 1994 and 2005, the total revenue from film processing ranged in value from a low of $3.7 billion to a high of $6.2 billion. By how much did the high value differ from the low value?

Photofinishing Market

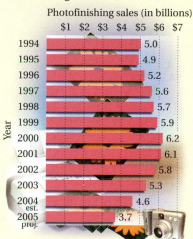

Photofinishing sales (in billions)

Year	Sales
1994	5.0
1995	4.9
1996	5.2
1997	5.6
1998	5.7
1999	5.9
2000	6.2
2001	6.1
2002	5.8
2003	5.3
2004 est.	4.6
2005 proj.	3.7

Source: PMA Marketing Research

2. At Copylot Printing, the cost of copying is 8 cents per page. How much, in dollars, would it cost to make 466 copies?

EXAMPLE 2 A 100-unit syringe is often used by diabetics and nurses to administer insulin. Each unit on the syringe represents 0.01 cc (cubic centimeter). Each day, Wendy averages 42 units of insulin. How many cc's will Wendy use in a typical week?

1. **Familiarize.** We make a drawing or at least visualize the situation. We let a = the amount of insulin used.

2. **Translate.** We translate as follows.

Amount used each day	times	Number of days in a week	is	Total amount injected
↓	↓	↓	↓	↓
0.42 cc	×	7	=	a

3. **Solve.** To solve the equation, we carry out the multiplication.

$$\begin{array}{r} 0.4\,2 \\ \times \quad\;\; 7 \\ \hline 2.9\,4 \end{array}$$

Thus, a = 2.94 cc.

4. **Check.** We can check using approximation.

0.42 cc × 7 ≈ 0.4 cc × 7 = 2.8 cc ≈ 2.94 cc

Note too that 2.94 is bigger than 0.42, so our answer makes sense.

5. **State.** Wendy uses 2.94 cc's of insulin in a typical week.

Do Exercise 2.

Answer on page A-14

MULTISTEP PROBLEMS

EXAMPLE 3 *Student Loans.*
Upon graduation from college,
Aviva must repay a Stafford loan
that totals $23,334. The loan is to
be paid back over 10 yr in equal
monthly payments. Find the
amount of each payment.

1. **Familiarize.** We assume
that the borrowed money
is repaid in monthly checks
that are always the same size.
Since we are not told how many checks there will be, part of solving
will be to determine how many months there are in 10 yr. We let
m = the size of each monthly payment.

2. **Translate.** To find the amount of the monthly payment, we note that the
amount owed is split up, or *divided*, into payments of equal size. The size
of each payment will depend on how many payments there are. To find
the number of payments, we first determine that in 10 yr there are

$$10 \cdot 12 = 120 \text{ months.} \qquad \text{There are 12 months in a year.}$$

We have

Amount of monthly payment	is	Total amount owed	divided by	Number of payments
m	$=$	$\$23,334$	$\div$	120

3. **Solve.** To solve, we carry out the division.

$$
\begin{array}{r}
1\,9\,4.4\,5 \\
1\,2\,0\,)\overline{\,2\,3,3\,3\,4.0\,0\,} \\
\underline{1\,2\,0} \\
1\,1\,3\,3 \\
\underline{1\,0\,8\,0} \\
5\,3\,4 \\
\underline{4\,8\,0} \\
5\,4\,0 \\
\underline{4\,8\,0} \\
6\,0\,0 \\
\underline{6\,0\,0} \\
0
\end{array}
\qquad m = 194.45
$$

4. **Check.** To check, we first verify that there are 120 months in 10 years.

$$120 \text{ months} \div 12 \text{ months per year} = 10 \text{ years}$$

To check that the amount of the monthly payment is correct, we can
estimate the product.

$$\$194.45 \cdot 120 \approx \$200 \cdot 120 = \$24,000 \approx \$23,334$$

5. **State.** Aviva's monthly payments will be $194.45.

Do Exercise 3.

3. **Car Payments.** Kevin's car loan
totals $11,370 and is to be paid
over 5 yr in monthly payments
of equal size. Find the amount
of each payment.

Answer on page A-14

4. Gas Mileage. Fidelis filled his Ford Focus and noted that the odometer read 38,320.8. After the next filling, the odometer read 38,735.5. It took 14.5 gal to fill the tank. How many miles per gallon did the Ford Focus get?

EXAMPLE 4 *Gas Mileage.* Olivia filled her Chevrolet Cobalt with gas and noted that the odometer read 67,507.8. After the next filling, the odometer read 68,006.1. It took 16.5 gal to fill the tank. How many miles per gallon did Olivia's Cobalt get?

1. Familiarize. We first make a drawing.

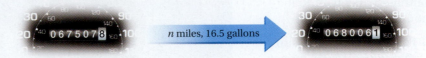

n miles, 16.5 gallons

This is a two-step problem. First, we find the number of miles that have been driven between fillups. We let n = the number of miles driven.

2., 3. Translate and Solve. This is a "how many more" situation. We translate and solve as follows.

First odometer reading	plus	Number of miles driven	is	Second odometer reading
67,507.8	+	n	=	68,006.1

To solve the equation, we subtract 67,507.8 from both sides.

$$n = 68,006.1 - 67,507.8$$
$$= 498.3$$

$$\begin{array}{r} 6\ 8,0\ 0\ 6.1 \\ -\ 6\ 7,5\ 0\ 7.8 \\ \hline 4\ 9\ 8.3 \end{array}$$

Second, we divide the total number of miles driven by the number of gallons. This gives us m = the number of miles per gallon—that is, the mileage. The division that corresponds to the situation is

$$498.3 \div 16.5 = m.$$

To find the number m, we divide.

$$\begin{array}{r} 3\ 0.2 \\ 1\ 6.5\ )\ \overline{4\ 9\ 8.3_\wedge 0} \\ 4\ 9\ 5\ 0 \\ \hline 3\ 3\ 0 \\ 3\ 3\ 0 \\ \hline 0 \end{array}$$

Thus, $m = 30.2$.

4. Check. To check, we first multiply the number of miles per gallon times the number of gallons.

$$16.5 \times 30.2 = 498.3$$

Then we add 498.3 to 67,507.8.

$$67,507.8 + 498.3 = 68,006.1$$

The mileage 30.2 checks.

5. State. Olivia's Cobalt got 30.2 miles per gallon.

Do Exercise 4.

Answer on page A-14

Some problems may require us to recall important formulas. Example 5 involves a formula from geometry that is worth remembering.

5. Suppose that an 8-in.-wide disc is punched out of an 8-in. by 8-in. sheet of metal. How much material is left over?

In any circle, a **diameter** is a segment that passes through the center of the circle with endpoints on the circle. A **radius** is a segment with one endpoint on the center and the other endpoint on the circle. The area, A, of a circle with radius of length r is given by

$$A = \pi \cdot r^2,$$

where $\pi \approx 3.14$.

EXAMPLE 5 The Northfield Tap and Die Company stamps 6-cm-wide discs out of metal squares that are 6 cm by 6 cm. How much metal remains after the disc has been punched out?

1. **Familiarize.** We make, and label, a drawing. The question deals with discs, squares, and leftover material, so we list the relevant area formulas.
   For a square with sides of length s,

   $$Area = s^2.$$

   For a circle with radius of length r,

   $$Area = \pi \cdot r^2,$$

   where $\pi \approx 3.14$.

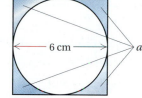

2. **Translate.** To find the amount left over, we subtract the area of the disc from the area of the square. Note that a circle's radius is half of its diameter, or width.

Area of square in square centimeters	minus	Area of disc in square centimeters	is	Area left over in square centimeters
6^2	$-$	$3.14 \times \left(\dfrac{6}{2}\right)^2$	$=$	a

3. **Solve.** We simplify as follows.

$$6^2 - 3.14\left(\frac{6}{2}\right)^2 = a$$
$$36 - 3.14(3)^2 = a$$
$$36 - 3.14 \cdot 9 = a$$
$$36 - 28.26 = a$$
$$7.74 = a$$

4. **Check.** We can repeat our calculation as a check. Note that 7.74 is less than the area of the disc, which in turn is less than the area of the square. This agrees with the impression given by our drawing.

5. **State.** The amount of material left over is 7.74 cm².

Do Exercise 5.

Answer on page A-14

6. Bike Rentals. Mike's Bikes rents mountain bikes. The shop charges $4.00 insurance for each rental plus $6.00 per hour. For how many hours can a person rent a bike with $25.00?

EXAMPLE 6 *Photo Books.* Apple sells iPhoto books that users of the iPhoto application can create. The price of a large, softcover, 20-page book is $19.99. Additional pages are 69 cents per page. Marta has $35 to spend on a book. What is the greatest number of pages she can put in the book?
Source: www.apple.com

1. **Familiarize.** Suppose that Marta puts 30 pages in the book. She would have to pay an additional price per page for $30 - 20$, or 10 pages. The price would be

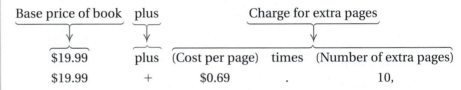

Base price of book — plus — Charge for extra pages

$19.99 — plus — (Cost per page) — times — (Number of extra pages)
$19.99 — $+$ — $0.69 — . — 10,

which is $19.99 + $6.90, or $26.89. This familiarizes us with the way in which price is calculated. Note that we convert 69 cents to $0.69 so that only one unit, dollars, is used. Note also that Marta can put more than 30 pages in the book. To see just how many pages she can add, we could make and check more guesses, but this would be very time-consuming. Instead, we let p = the number of extra pages Marta can add to the 20-page book. Note that the total number of pages in the book will then be $20 + p$.

2. **Translate.** The problem can be rephrased and translated as follows.

Base price of book — plus — Cost per page — times — Number of extra pages — is — Cost of book

$19.99 — $+$ — $0.69 — . — p — $=$ — $35

3. **Solve.** We solve the equation.

$$19.99 + 0.69p = 35$$
$$0.69p = 15.01 \qquad \text{Subtracting 19.99 from both sides}$$
$$p = \frac{15.01}{0.69} \qquad \text{Dividing both sides by 0.69}$$
$$p \approx 21.75 \qquad \text{Rounding}$$

4. **Check.** We check in the original problem. Since Marta cannot pay for parts of a page, we must round the answer *down* to 21 in order to keep the cost under $35. This would make the total length of the book $20 + 21$, or 41 pages. If Marta adds 21 pages to the book, the additional cost will be 21 times $0.69, or $14.49. If we add $14.49 to the base price of $19.99, we get $34.48, which is just under the $35 Marta has to spend.

5. **State.** With $35, Marta can make a 41-page book.

Answer on page A-14

Do Exercise 6.

EXAMPLE 7 *Promotional Buttons.* A sorority has $135 to spend on promotional buttons. They design one-color square buttons that are 2 in. by 2 in. The buttons cost 80 cents each for the first 100 buttons, and 60 cents for each additional button. The shipping cost is $10 per order. How many buttons can the sorority order?

1. **Familiarize.** First of all, note that prices in the problem are given in both dollars and cents. We convert 80 cents to $0.80 and 60 cents to $0.60 so that all prices are in dollars.

   Since the price per button changes after 100 buttons, we see whether the sorority can order more than 100 buttons. The price for 100 buttons will be 100 · $0.80 = $80. With shipping, 100 buttons will cost $80 + $10, or $90. Since the sorority can spend $135, they can order more than 100 buttons. The additional buttons will cost 60 cents each.

   We let *x* represent the number of buttons the sorority can order. Then *x* − 100 is the number of buttons ordered at 60 cents per button.

2. **Translate.** The problem can be rephrased and translated as follows.

Price of 100 buttons	plus	Price of additional buttons	plus	Shipping charges	is	Total cost
$0.80 × 100	+	$0.60($x$ − 100)	+	$10	=	$135

3. **Solve.** We solve the equation.

$$0.80 \times 100 + 0.60(x - 100) + 10 = 135$$

   $0.80 \times 100 + 0.60x - 0.60 \times 100 + 10 = 135$ Using the distributive property

   $80 + 0.60x - 60 + 10 = 135$ Multiplying

   $30 + 0.60x = 135$ Adding

   $0.60x = 105$ Subtracting 30 from both sides

   $x = \dfrac{105}{0.60}$ Dividing both sides by 0.60

   $x = 175$

4. **Check.** We check in the original problem. If the sorority orders 175 buttons, they will pay 80 cents per button for 100 buttons and 60 cents per button for 175 − 100, or 75 buttons. The bill before shipping will be 100($0.80) + 75($0.60) = $80 + $45, or $125. The total with shipping is then $125 + $10 = $135. This is the amount the sorority can spend, so the answer checks.

5. **State.** The sorority can order 175 buttons.

Do Exercise 7.

7. **Promotional Buttons.** Another design considered by the sorority in Example 7 uses several colors. Full-color buttons cost an additional 30 cents per button for the first 100 buttons, and an extra 20 cents per button for additional buttons. The shipping remains the same. How many full-color buttons can the sorority order?

Answer on page A-14

Translating for Success

1. **Gas Mileage.** Art filled his SUV's gas tank and noted that the odometer read 38,271.8 mi. At the next filling, the odometer read 38,677.92 mi. It took 28.4 gal to fill the tank. How many miles per gallon did the SUV get?

2. **Dimensions of a Parking Lot.** Seals' parking lot is a rectangle that measures 85.2 ft by 52.3 ft. What is the area of the parking lot?

3. **Game Snacks.** Three students pay $18.40 for snacks at a football game. What is each person's share?

4. **Electrical Wiring.** An electrician needs 1314 ft of wiring cut into $2\frac{1}{2}$-ft pieces. How many pieces will she have?

5. **College Tuition.** Wayne needs $4638 for the fall semester's tuition. On the day of registration, he had only $3092. How much does he need to borrow?

The goal of these matching questions is to practice step (2), *Translate*, of the five-step problem-solving process. Translate each word problem to an equation and select a correct translation from equations A–O.

A. $2\frac{1}{2} \cdot n = 1314$

B. $18.4 \times 1.87 = n$

C. $n = 85.2 \times 52.3$

D. $19 - (-4) = n$

E. $3 \times 18.40 = n$

F. $2\frac{1}{2} \cdot 1314 = n$

G. $3092 + n = 4638$

H. $18.4 \cdot n = 1.87$

I. $\dfrac{406.12}{28.4} = n$

J. $52.3 \cdot n = 85.2$

K. $n = 19 + (-4)$

L. $52.3 + n = 85.2$

M. $3092 + 4638 = n$

N. $3 \cdot n = 18.40$

O. $85.2 + 52.3 = n$

Answers on page A-14

6. **Cost of Gasoline.** What is the cost, in dollars, of 18.4 gal of gasoline at $1.87 per gallon?

7. **Temperature.** At noon, the temperature in Pierre was 19°F. At midnight, the temperature had fallen to −4°F. By how many degrees had the temperature fallen?

8. **Acres Planted.** This season Sam planted 85.2 acres of corn and 52.3 acres of soybeans. Find the total number of acres that he planted.

9. **Amount Inherited.** Tara inherited $2\frac{1}{2}$ times as much as her cousin. Her cousin received $1314. How much did Tara receive?

10. **Travel Funds.** The athletic department needs travel funds of $4638 for the tennis team and $3092 for the golf team. What is the total amount needed for travel?

a   Solve using the five-step problem-solving procedure.

1. What is the cost of 7 jackets at $32.98 each?

2. What is the cost of 8 pairs of socks at $4.95 each?

3. *Gasoline Cost.* What is the cost, in dollars, of 20.4 gal of gasoline at 224.9 cents per gallon? (224.9 cents = $2.249) Round the answer to the nearest cent.

4. *Gasoline Cost.* What is the cost, in dollars, of 15.3 gal of gasoline at 213.9 cents per gallon? (213.9 cents = $2.139) Round the answer to the nearest cent.

5. *Body Temperature.* Normal body temperature is 98.6°F. During an illness, a patient's temperature rose 4.2°. What was the new temperature?

6. *Gasoline Cost.* What is the cost, in dollars, of 13.8 gal of gasoline at 252.9 cents per gallon? Round the answer to the nearest cent.

7. *Lottery Winnings.* In Texas, one of the state lotteries is called "Cash 5." In a recent weekly game, the lottery prize of $127,315 was shared equally by 6 winners. How much was each winner's share? Round to the nearest cent.

Source: Texas Lottery

8. *Lunch Costs.* A group of 5 students pays $37.45 for lunch and splits the cost equally. What is each person's share?

9. *Stamp.* Find the area and the perimeter of the stamp shown here.

2.5 cm

3.25 cm

10. *Pole Vault Pit.* Find the area and the perimeter of the landing area of the pole vault pit shown here.

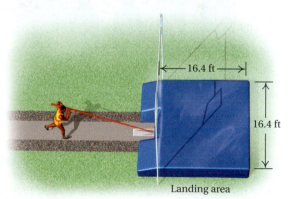

16.4 ft

16.4 ft

Landing area

11. *Odometer Reading.* The Levines checked the odometer before starting a trip. It read 22,456.8 mi and they know that they will be driving 234.7 mi. What will the odometer read at the end of the trip?

12. *Miles Driven.* Petra bought gasoline when the odometer read 14,296.3 mi. At the next gasoline purchase, the odometer read 14,515.8 mi. How many miles had been driven?

13. Andrew bought a DVD of the complete ninth season of the TV show "Friends" for $29.24 plus $1.61 sales tax. He paid for it with a $50 bill. How much change did he receive?

14. Jaden bought the CD "Feels Like Home" by Norah Jones for $13.49 plus $0.81 sales tax. She paid for it with a $20 bill. How much change did she receive?

15. *Medicine.* After taking aspirin, Wanda's temperature dropped from 103.2°F to 99.7°F. How much did her temperature drop?

16. *Nursing.* A nurse draws 17.85 mg of blood and uses 9.68 mg in a blood test. How much is left?

17. *Culinary Arts.* One pound of crabmeat makes three servings at the Key West Seafood Restaurant. If the crabmeat costs $16.95 per pound, what is the cost per serving?

18. *Finance.* A car loan totaling $4425 is to be paid off in 12 monthly payments of equal size. How much is each payment?

19. *Beverage Consumption.* Each year, the average U.S. citizen drinks about 49.0 gal of soft drinks, 41.2 gal of water, 25.3 gal of milk, 24.8 gal of coffee, and 7.8 gal of fruit juice. What is the total amount that the average U.S. citizen drinks?

20. *Medicine.* After being tested for allergies, Mike was given allergy shots of 0.25 mL, 0.4 mL, 0.5 mL, and 0.5 mL over a 7-week period. What was the total amount of the injections?

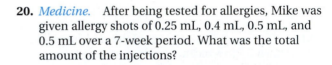

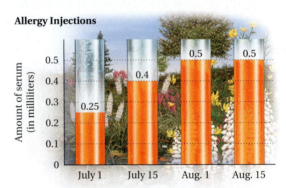

Allergy Injections

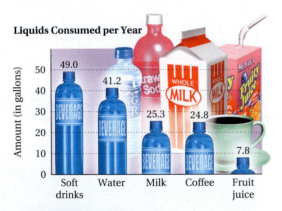

Liquids Consumed per Year

Source: U.S. Department of Agriculture

21. *Gas Mileage.* Peggy filled her van's gas tank and noted that the odometer read 26,342.8 mi. After the next filling, the odometer read 26,736.7 mi. It took 19.5 gal to fill the tank. How many miles per gallon did the van get?

22. *Gas Mileage.* Peter filled his Honda's gas tank and noted that the odometer read 18,943.2 mi. After the next filling, the odometer read 19,306.2 mi. It took 13.2 gal to fill the tank. How many miles per gallon did the car get?

23. *Nursing.* Phil injects a total of 38 units of insulin, each day for a week (see Example 2). How many cc's of insulin does he use in a week?

24. *Nursing.* Carlie averages 49 units of insulin per day, for 15 days (see Example 2). How many cc's of insulin has she used during those 15 days?

25. *Jackie Robinson Poster.* A special limited-edition poster was painted by sports artist Leroy Neiman. Commissioned by Barton L. Kaufman, it commemorates the entrance of the first African-American, Jackie Robinson, into Major League Baseball in 1947. The dimensions of the poster are as shown. How much area is not devoted to the painting?

Source: Barton L. Kaufman, private collection

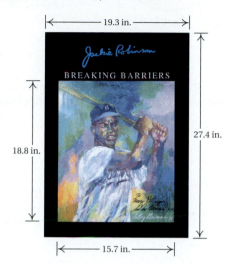

26. *Earth Day Poster.* An Earth Day poster that is 61.8 cm by 73.2 cm includes a 2-cm border. What is the area of the poster inside the border?

27. *Study Cards.* An instructor allows her students to bring one 7.6-cm by 12.7-cm index card to the final exam, with notes of any sort written on the card. If both sides of the card are used, how much area is available for notes?

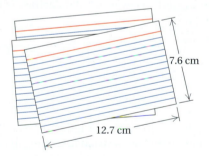

28. *Stamps.* Find the total area of the stamps shown.

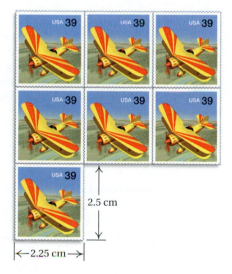

29. *Cost of Video Game.* A certain video game costs 75 cents and runs for 1.5 min. Assuming a player does not win any free games and plays continuously, how much money, in dollars, does it cost to play the video game for 1 hr?

30. *Property Taxes.* The Colavitos own a house with an assessed value of $184,500. For every $1000 of assessed value, they pay $7.68 in taxes. How much do they pay in taxes?

Find the length *d* in each figure.

31.

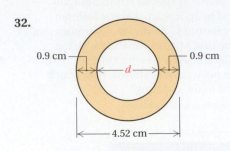

0.8 cm ⎯ | ⎯ 0.8 cm

d

⎯ 3.91 cm ⎯

32.

0.9 cm ⎯ | ⎯ 0.9 cm

d

⎯ 4.52 cm ⎯

33. *Carpentry.* A round, 6-ft-wide hot tub is being built into a 12-ft by 30-ft rectangular deck. How much decking is needed for the surface of the deck?

12 ft   6 ft   30 ft

34. A 4-ft by 4-ft tablecloth is cut from a round tablecloth that is 6 ft wide. Find the area of the cloth left over.

35. *Verizon Wireless.* In 2006, Verizon Wireless offered its America's Choice cellular phone plan with 450 anytime minutes per month for a monthly access fee of $39.99. Minutes in excess of 450 were charged at the rate of $0.45 per minute. In June, Leila used her cell phone for 479 minutes. What was she charged?

Source: vzwshop.com

36. Zachary worked 53 hr during a week one summer. He earned $8.50 per hour for the first 40 hr and $12.75 per hour for overtime. How much did Zachary earn during the week?

37. *Web Space.* Penn State charges a $10 processing fee plus 9 cents per megabyte for Web space. Nikki's bill was $88.75. How many megabytes did she pay for?

Source: aset.its.psu.edu

38. *Service Calls.* JoJo's Service Center charges $30 for a house call plus $37.50 for each hour the job takes. For how long has a repairperson worked on a house call if the bill comes to $123.75?

39. *Verizon Wireless.* In 2006, Verizon Wireless offered its America's Choice cellular phone plan with 900 anytime minutes per month for a monthly access fee of $59.99. Minutes in excess of 900 were charged at the rate of 40 cents per minute. Additional fees of $5.79 are added each month to the bill. One month, Jeff's bill was $89.78. How many minutes did he use that month?

40. *Electric Rates.* Southeast Electric charges 9¢ per kilowatt-hour for the first 200 kWh. The company charges 11¢ per kilowatt-hour for all electrical usage in excess of 200 kWh. How many kilowatt-hours were used if a monthly electric bill was $57.60?

41. *Photo Service.* Consolidated UCLA Photographic Service charges $50 for the first 30 min of video capture and 50 cents for each additional minute. How many minutes of video capture can a sorority buy for $95?

Source: www.uclaphoto.ucla.edu

42. *Overtime Pay.* A construction worker earned $17 per hour for the first 40 hr of work and $25.50 per hour for work in excess of 40 hr. One week she earned $896.75. How much overtime did she work?

43. *Credit Card Processing.* VeriSign Payflow Services processes credit cards for businesses. They charge a setup fee of $179, $19.95 per month for up to 500 transactions, and 10 cents for each additional transaction over 500. Timeless Treasures paid $218.75 for the first month of using VeriSign. How many credit card transactions did they process that month?

Source: www.verisign.com

44. *Limousine Service.* For a chauffeured limousine, Sharp Ride charges a rental fee of $60 plus $5 a mile for the first 10 mi and $3.50 for each additional mile. Jackson paid $195.75 for a limousine ride from his wedding to the reception hall. How many miles was the ride?

45. *Field Dimensions.* The dimensions of a World Cup soccer field are 114.9 yd by 74.4 yd. The dimensions of a standard football field are 120 yd by 53.3 yd. How much greater is the area of a World Cup soccer field?

46. *Technical Support.* Pantek provides technical support for Linux users. They charge $150 an hour, or $1250 for a ten-hour block. The Allen, Benning, & Carmichael law firm paid $2225 for technical support during a recent upgrade to their computer system. How many hours of technical support did they receive?

Source: www.pantek.com

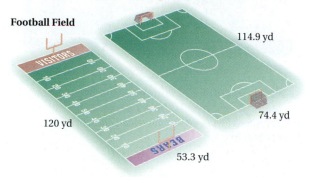

World Cup Soccer Field

Football Field

120 yd

53.3 yd

114.9 yd

74.4 yd

47. Frank has been sent to the store with $40 to purchase 6 lb of cheese at $4.79 a pound and as many bottles of seltzer, at $0.64 a bottle, as possible. How many bottles of seltzer should Frank buy?

48. Janice has been sent to the store with $30 to purchase 5 pt of strawberries at $2.49 a pint and as many bags of chips, at $1.39 a bag, as possible. How many bags of chips should Janice buy?

49. **D**W Write a problem for a classmate to solve. Design the problem so that the solution is "Mona's Buick got 23.5 mpg."

50. **D**W Write a problem for a classmate to solve. Design the problem so that the solution is "The larger field is 200 m^2 bigger."

SKILL MAINTENANCE

51. Simplify: $\dfrac{0}{-13}$. [3.3b]

52. Add: $-\dfrac{4}{5} + \dfrac{7}{10}$. [4.2b]

53. Subtract: $\dfrac{8}{11} - \dfrac{4}{3}$. [4.3a]

54. Solve: $4x - 7 = 9x + 13$. [5.7b]

55. Add: $4\dfrac{1}{3} + 2\dfrac{1}{2}$. [4.6a]

56. Simplify: $\dfrac{-72}{-72}$. [3.3b]

381

57. DW ▦ Which is a better deal and why: a 14-in. pizza that costs $9.95 or a 16-in. pizza that costs $11.95?

58. DW A business received a bill for $2500 for technical support from Pantek. (See Exercise 46.) Can you determine how many hours of technical support the firm received? Why or why not?

59. DW To determine the average amount that each U.S. citizen drinks per year, Alison obtained the following graph and found the average of the numbers listed. Was Alison's approach correct? Why or why not?

Liquids Consumed per Year

Amount (in gallons)

Soft drinks 49.0, Water 41.2, Milk 25.3, Coffee 24.8, Fruit juice 7.8

Source: U.S. Department of Agriculture

60. ▦ A "French Press" coffee pot requires no filters, but costs $34.95. Kenny can also buy a plastic drip cone for $4.49, but the cone requires filters which cost $0.04 per pot dripped. How many pots of coffee must Kenny make for the French Press pot to be the more economical purchase?

61. You can drive from home to work using either of two routes:

> *Route A*: Via interstate highway, 7.6 mi, with a speed limit of 65 mph.
> *Route B*: Via a country road, 5.6 mi, with a speed limit of 50 mph.

Assuming you drive at the posted speed limit, how much time can you save by taking the faster route?

62. ▦ A 25-ft by 30-ft yard contains an 8-ft-wide, round fountain. How many 1-lb bags of grass seed should be purchased to seed the lawn if 1 lb of seed covers 300 ft^2?

63. Find the shaded area. What assumptions must you make?

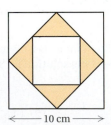

← 10 cm →

64. If the daily rental for a car is $18.90 plus a certain price per mile, and Lindsey must drive 190 mi in one day and still stay within a $55.00 budget, what is the highest price per mile that Lindsey can afford?

65. *Free Agent Pitchers.* In 2003, the average value of a multiyear contract for a free agent Major League Baseball pitcher was $10.3 million; in 2004, it was $12.1 million; in 2005, it was $17.2 million; and in 2006, it was $22.6 million. Determine the average yearly increase in the value of a pitcher's multiyear contract.
Source: *USA Today*, 12/7/05

66. *Free Agent Pitchers.* In 2003, the average length of a multiyear contract for a free agent Major League Baseball pitcher was 2.3 yr; in 2004, it was 2.5 yr; in 2005, it was 2.6 yr; and in 2006, it was 3.2 yr. Determine the average yearly increase in the length of a pitcher's multiyear contract.
Source: *USA Today*, 12/7/05

5 Summary and Review

The review that follows is meant to prepare you for a chapter exam. It consists of two parts. The first part, Concept Reinforcement, is designed to increase understanding of the concepts through true/false exercises. The second part is the Review Exercises. These provide practice exercises for the exam, together with references to section objectives so you can go back and review. Before beginning, stop and look back over the skills you have obtained. What skills in mathematics do you have now that you did not have before studying this chapter?

✋ CONCEPT REINFORCEMENT

Determine whether the statement is true or false. Answers are given at the back of the book.

_____ **1.** In the number 308.00567, the digit 6 names the hundreds place.

_____ **2.** To multiply any number by a multiple of 10, count the number of zeros and move the decimal point that many places to the right.

_____ **3.** One thousand billions is one trillion.

_____ **4.** The number of decimal places in the product of two numbers is the product of the number of places in the factors.

_____ **5.** When writing a word name for decimal notation, we write the word "and" for the decimal point.

Review Exercises

Convert the number in the sentence to standard notation.

1. Russia has the largest total area of any country in the world, at 6.59 million square miles. [5.3b]

2. The total weight of the turkeys consumed by Americans during the Thanksgiving holidays is about 6.9 million pounds. [5.3b]

Write a word name. [5.1a]

3. 3.47

4. 0.031

Write fraction notation and, if possible, as a mixed numeral. [5.1b]

5. 0.09

6. −4.561

7. −0.089

8. 3.0227

Write decimal notation. [5.1b]

9. $-\dfrac{34}{1000}$

10. $\dfrac{42{,}603}{10{,}000}$

11. $27\dfrac{91}{100}$

12. $-867\dfrac{6}{1000}$

Which number is larger? [5.1c]

13. 0.034, 0.0185

14. −0.91, −0.19

Round 17.4287 to the nearest: [5.1d]

15. Tenth.

16. Thousandth.

Perform the indicated operation.

17.
$$
\begin{array}{r}
2\ 3\ 6.2\ 3\ 1 \\
2\ 6\ 3.4 \\
+\qquad 0.1\ 9\ 8 \\
\end{array}
\quad [5.2a]
$$

18.
$$
\begin{array}{r}
3\ 7.6\ 4\ 5 \\
-\qquad 8.4\ 9\ 7 \\
\end{array}
\quad [5.2b]
$$

19. $219.3 + 2.8 + 7$ [5.2a]

20. $745.0109 - 59.959$ [5.2b]

21. $-37.8 + (-19.5)$ [5.2c]

22. $-7.52 - (-9.89)$ [5.2c]

23.
$$
\begin{array}{r}
4\ 8 \\
\times\ 0.2\ 7 \\
\end{array}
\quad [5.3a]
$$

24. $-3.7(0.29)$ [5.3a]

25.
$$
\begin{array}{r}
2\ 4.6\ 8 \\
\times\ 1\ 0\ 0\ 0 \\
\end{array}
\quad [5.3a]
$$

26. $2\ 5\ \overline{)\ 8\ 0}$ [5.4a]

27. $11.52 \div (-7.2)$ [5.4a]

28. $\dfrac{276.3}{1000}$ [5.4a]

Combine like terms.   [5.2d]

29. $3.7x - 5.2y - 1.5x - 3.9y$

30. $7.94 - 3.89a + 4.63 + 1.05a$

31. Evaluate: $P - Prt$ for $P = 1000$, $r = 0.05$, and $t = 1.5$
(*A formula for depreciation*)   [5.3c]

32. Simplify: $9 - 3.2(-1.5) + 5.2^2$.   [5.4b]

33. Estimate the sum $7.298 + 3.961$ by rounding to the nearest tenth.   [5.6a]

34. About how many videotapes, at $2.45 each, can be purchased with $49.95?   [5.6a]

35. Convert 1549 cents to dollars.   [5.3b]

36. Round $248.\overline{27}$ to the nearest hundredth.   [5.5b]

Multiply by a form of 1 to find decimal notation for each number.   [5.5c]

37. $\dfrac{13}{5}$

38. $\dfrac{32}{25}$

Use division to find decimal notation for each number. [5.5a]

39. $\dfrac{13}{4}$

40. $-\dfrac{7}{6}$

41. Calculate: $\dfrac{4}{15} \times 79.05$ [5.5d]

Solve. Remember to check.

42. $t - 4.3 = -7.5$ [5.7a]

43. $4.1x + 5.6 = -6.7$ [5.7a]

44. $6x - 11 = 8x + 4$ [5.7b]

45. $3(x + 2) = 5x - 7$ [5.7b]

Solve.   [5.8a]

46. In the United States, there are 51.81 telephone poles for every 100 people. In Canada, there are 40.65 poles for every 100 people. How many more telephone poles for every 100 people are there in the United States?

47. Stacia, a coronary intensive care nurse, earned $620.74 during a recent 40-hr week. What was her hourly wage? Round to the nearest cent.

48. *Landscaping.* A rectangular yard is 20 ft by 15 ft. The yard is covered with grass except for a circular flower garden with an 8-ft diameter. Find the area of grass in the yard.

49. Derek had $6274.35 in his checking account. He used $485.79 to buy a Personal Digital Assistant with his debit card. How much was left in his account?

50. *Verizon Wireless.* In 2006, Verizon Wireless offered its America's Choice cellular phone plan with 900 anytime minutes per month for a monthly access fee of $59.99. Minutes in excess of 900 were charged at the rate of $0.40 per minute. One month, Jeff used his cell phone for 946 min. What was the charge?

Source: www.vzwshop.com

51. *Storage Space.* Penn State charges a $10 processing fee plus 2 cents per megabyte for information storage. Cody's bill was $46.60. How many megabytes of storage did he pay for?

Source: aset.its.psu.edu

52. *Gas Mileage.* Inge wants to estimate gas mileage per gallon. With an odometer reading 36,057.1, she fills up. At 36,217.6 mi, the tank is refilled with 11.1 gal. Find the mileage per gallon. Round to the nearest tenth.

53. *Seafood Consumption.* The following graph shows the annual consumption, in pounds, of seafood per person in the United States in recent years.

a) Find the total per capita consumption for the seven years.
b) Find the average per capita consumption.

Seafood Consumption

Average number of pounds per person

1980	1985	1990	1995	2000	2001	2002
12.4	15.0	14.9	14.8	15.2	14.7	15.6

Year

54. A taxi driver charges $7.25 plus 95 cents a mile for out-of-town fares. How far can an out-of-towner travel on $15.23?

55. One pound of lean boneless ham contains 4.5 servings. It costs $5.99 per pound. What is the cost per serving? Round to the nearest cent.

56. *Construction.* A rectangular room measures 14.5 ft by 16.25 ft. How many feet of crown molding is needed to go around the top of the room? How many square feet of bamboo tiles are needed for the floor of the room?

57. **Dw** Stacy claims that to convert $\frac{3}{20}$ to decimal notation, she thinks of money—nickels in particular. How do you think she does this?

58. **Dw** Why is $\frac{1}{3} \cdot \frac{1}{6}$ more easily computed in fraction notation than decimal notation? What would be the best way to express this product as a decimal?

SYNTHESIS

59. 🖩 In each of the following, use +, −, ×, or ÷ in each blank to make a true sentence. [5.4b]

a) $2.56 - 6.4 \ \square \ 51.2 - 17.4 + 89.7 = 119.66$
b) $(11.12 \ \square \ 0.29) \ 3^4 = 877.23$

60. Arrange from smallest to largest: [5.1c], [5.5a]

$$-\frac{2}{3}, \ -\frac{15}{19}, \ -\frac{11}{13}, \ \frac{-5}{7}, \ \frac{-13}{15}, \ \frac{-17}{20}.$$

61. *Automobile Leases.* Quentin can lease a BMW 325i for $396 a month. He must pay an additional 20 cents per mile for all miles over 10,000 in one year. In 2006, his total bill for leasing the car was $5952. How many miles did he drive the car in 2006? [5.8a]

Source: www.edmunds.com

62. **Dw** 🖩 Sal's sells Sicilian pizza as a 17-in. by 20-in. pie for $15 or as an 18-in.-diameter round pie for $14. Which is a better buy and why? [5.8a]

Convert the number in the sentence to standard notation.

1. The annual sales of antibiotics in the United States is $8.9 billion.
   Source: IMS Health

2. There are 3.756 million people enrolled in bowling organizations in the United States.
   Source: *Bowler's Journal International,* December 2000

Write a word name.

3. 2.34

4. 105.0005

Write fraction notation.

5. −0.3

6. 2.769

Write decimal notation.

7. $\dfrac{74}{1000}$

8. $-\dfrac{37,047}{10,000}$

9. $756\dfrac{9}{100}$

10. $91\dfrac{703}{1000}$

Which number is larger?

11. 0.07,   0.162

12. −0.173,   −0.25

Round 9.4523 to the nearest:

13. Tenth.

14. Thousandth.

Perform the indicated operation.

15.
```
   4 0 2.3
       2.8 1
+      0.1 0 9
```

16.
```
     0.1 2 5
×    0.2 4
```

17.
```
   2 1 3.4 5
×    0.0 0 1
```

18.
```
   5 2.0 9 1
−      7.3 4 5
```

19. 342.9 + 8.1 + 5.37

20. −9.5 + 7.3

21. 2 − 0.0054

22. 1000 × 73.962

23. $4\overline{)\,1\,9}$

24. $3.3\overline{)\,1\,0\,0.3\,2}$

25. $\dfrac{-346.82}{1000}$

26. $\dfrac{346.82}{0.01}$

27. Convert $179.82 to cents.

28. Combine like terms:
$$4.1x + 5.2 - 3.9y + 5.7x - 9.8$$

29. Evaluate: $2l + 4w + 2h$ for $l = 2.4$, $w = 1.3$, and $h = 0.8$ (*The total girth of a postal package*)

30. Simplify: $20 \div 5(-2)^2 - 8.4$

31. About how many gallons of gasoline, at $2.749 per gallon, can be bought with $20? Round to the nearest gallon.

32. Round $48.\overline{74}$ to the nearest tenth.

Multiply by a form of 1 to find decimal notation for each number.

33. $\dfrac{8}{5}$

34. $\dfrac{21}{4}$

Use division to find decimal notation for each number.

35. $-\dfrac{7}{16}$

36. $\dfrac{14}{9}$

37. Round the answer in Question 36 to the nearest hundredth.

Estimate each of the following.

38. The product 8.91×22.457 by rounding to the nearest one

39. The quotient $78.2209 \div 16.09$ by rounding to the nearest ten

40. Calculate: $\dfrac{3}{8} \times 45.6 - \dfrac{1}{5} \times 36.9$

Solve. Remember to check.

41. $17y - 3.12 = -58.2$

42. $9t - 4 = 6t + 26$

43. $4 + 2(x - 3) = 7x - 9$

44. *Airport Passengers.*   The following graph shows the number of passengers in 2003 who traveled through the country's busiest airports. Find the average number of passengers through these airports.

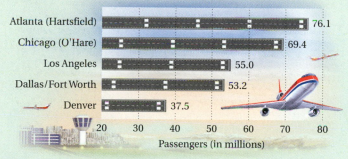

Busiest Airports in the United States

Atlanta (Hartsfield) 76.1
Chicago (O'Hare) 69.4
Los Angeles 55.0
Dallas/Fort Worth 53.2
Denver 37.5

Passengers (in millions)

Source: Airports Council International, www.airports.org

45. *Verizon Wireless.* In 2006, Verizon Wireless offered its America's Choice cellular phone plan with 2000 anytime minutes per month for a monthly access fee of $99.99. Minutes in excess of 2000 were charged $0.25 per minute. One month Trey's bill was $314.99. How many minutes did he use?

Source: www.vzwshop.com

46. *Gas Mileage.* Tina decided to estimate the gas mileage per gallon in her economy car. At 76,843 mi, she filled her gas tank. At 77,310 mi, she filled the tank again, with 16.5 gal of gasoline. Find the mileage per gallon. Round to the nearest tenth.

47. *Checking Account Balance.* Nicholas has a balance of $10,200 in his checking account before making purchases of $123.89, $56.68, and $3446.98 with his debit card. What was the balance after making the purchases?

48. The Drake, Smith, and Nicholas law firm buys 7 cases of copy paper at $25.99 per case. What is the total cost?

SYNTHESIS

49. Use one of the words *sometimes*, *never*, or *always* to complete each of the following.

   a) The product of two numbers greater than 0 and less than 1 is _____ less than 1.

   b) The product of two numbers greater than 1 is _____ less than 1.

   c) The product of a number greater than 1 and a number less than 1 is _____ equal to 1.

   d) The product of a number greater than 1 and a number less than 1 is _____ equal to 0.

50. *Bottled Water Consumption.* The following graph shows the annual consumption, in gallons, of bottled water per person in the United States in recent years. Determine the average yearly increase in the per person consumption of bottled water.

Bottled Water Consumption

20.7 22.1 23.8 25.0

Average number of gallons per person

2002 2003 2004 2005

Year

Source: Beverage Marketing Corporation

51. *Travel Costs.* Roundtrip airfare between Burlington, VT, and Newark, NJ, often costs $189. The cost of driving (gas and general wear and tear) is about 32¢ per mile. Is it more economical to fly or drive the 320 mi for **(a)** an individual; **(b)** a couple; **(c)** a family of 3?

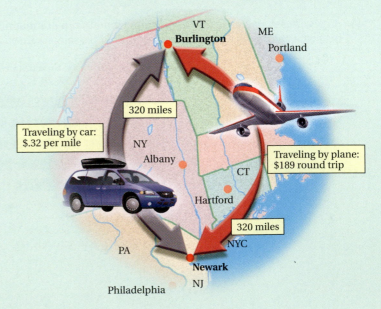

320 miles

Traveling by car: $.32 per mile

Traveling by plane: $189 round trip

320 miles

VT
Burlington
ME
Portland
NY
Albany
CT
Hartford
NYC
Newark
NJ
PA
Philadelphia

1. Write a word name: 207,491.

2. Write standard notation: 6.25 billion.

3. Write fraction notation: 10.09.

4. Convert to fraction notation: $4\frac{3}{8}$.

5. Write decimal notation: $\frac{-35}{1000}$.

6. List all the factors of 66.

7. Find the prime factorization of 154.

8. Find the LCM of 28 and 35.

9. Round 6962.4721 to the nearest hundred.

10. Round 6962.4721 to the nearest hundredth.

Add and, if possible, simplify.

11.
$$\begin{array}{r} 3\frac{2}{3} \\ + 2\frac{5}{9} \\ \hline \end{array}$$

12.
$$\begin{array}{r} 1\ 1\ 0.8\ 6\ 3 \\ 0.7\ 3 \\ 1\ 2\ 1.9 \\ + \quad 1.9\ 0\ 4 \\ \hline \end{array}$$

13.
$$\begin{array}{r} 5\ 2\ 4\ 9 \\ 2\ 1\ 5 \\ + \quad 3\ 1 \\ \hline \end{array}$$

14. $-\frac{4}{15} + \frac{7}{30}$

Subtract.

15. $-23 - 48$

16. $9010 - 563.47$

17. $\frac{8}{9} - \frac{7}{8}$

18. $7\frac{1}{5} - 3\frac{4}{5}$

Multiply and, if possible, simplify.

19.
$$\begin{array}{r} 2\ 3.9 \\ \times \quad 0.2 \\ \hline \end{array}$$

20. $-\frac{3}{5} \times \frac{10}{21}$

21. $3\frac{2}{11} \cdot 4\frac{2}{7}$

22. $5 \cdot \frac{3}{10}$

Divide and, if possible, simplify.

23. $2\frac{4}{5} \div 1\frac{13}{15}$

24. $\frac{6}{5} \div \frac{7}{8}$

25. $-43.795 \div 0.001$

26. $2.1\overline{\smash{)}4\ 3.2\ 6}$

Use <, >, or = for ☐ to write a true sentence.

27. $\frac{2}{3}$ ☐ $\frac{5}{7}$

28. -7 ☐ -4

29. Evaluate $a \div 3 \cdot b$, for $a = 18$ and $b = 2$.

30. Multiply: $4(x - y + 3)$.

Combine like terms.

31. $-4p + 9 + 11p - 17$

32. $x - 9 + 13x - 2$

Solve. Remember to check.

33. $8.32 + x = 9.1$

34. $-75 \cdot x = 2100$

35. $y \cdot 9.47 = 81.6314$

36. $1062 - y = -368{,}313$

37. $t + \dfrac{5}{6} = \dfrac{8}{9}$

38. $\dfrac{7}{8} \cdot t = \dfrac{7}{16}$

39. $2.4x - 7.1 = 2.05$

40. $2(x - 3) = 5x - 13$

Solve.

41. *Hockey.* In December 2005, the active NHL players with the most career penalty minutes were Tie Domi (Toronto) with 3447, Chris Chelios (Detroit) with 2755, Scott Mellanby (Atlanta) with 2382, and Gary Roberts (Florida) with 2361. How many minutes in all have these four players spent in the penalty box?
Source: Internet Hockey Database

42. After making a $450 down payment on a motorcycle, $\frac{3}{10}$ of the total cost was paid. How much did the motorcycle cost?

43. There are 60 seconds in a minute and 60 minutes in an hour. How many seconds are in a day?

44. Claude's college tuition was $4200. A loan was obtained for $\frac{2}{3}$ of the tuition. For how much was the loan?

45. The balance in a checking account is $314.79. After a check is written for $56.02, what is the balance in the account?

46. A clerk in a deli sold $1\frac{1}{2}$ lb of cheese, $2\frac{3}{4}$ lb of turkey, and $2\frac{1}{4}$ lb of roast beef. How many pounds were sold altogether?

47. A triangular sail has a height of 16 ft and a base of 11 ft. Find its area.

48. A 4-in. by 5-in. rectangle is punched from a round piece of steel that is 9 in. wide. How much steel will be left over?

> **SYNTHESIS**

49. A carton of Luna™ Bar boxes weighs about 320 oz. If each box of Luna Bars weighs 25.4 oz, what is the greatest number of boxes that could be inside the carton?

50. In the Newton Market, Brenda used a manufacturer's coupon to buy juice. With the coupon, if 3 cartons of juice were purchased, the fourth carton was free. The price of each carton was $1.89. What was the cost per carton with the coupon? Round to the nearest cent.

51. For the hockey players in Exercise 41, what is the average number of *hours* they have spent in the penalty box?

52. The Fit Fiddle Health Club generally charges a $79 membership fee and $42.50 a month. Alayn has a coupon that will allow her to join the club for $299 for six months. How much will Alayn save if she uses the coupon?

Introduction to Graphing and Statistics

6

Real-World Application

The side-by-side bar graph illustrates the effect of education on earning power for men and women in 2003. A more detailed graph is shown on page 410. How much more did women with bachelor's degrees or higher earn in 2003 than men who ended their education at high school graduation?

This problem appears as Exercise 19 in Section 6.2.

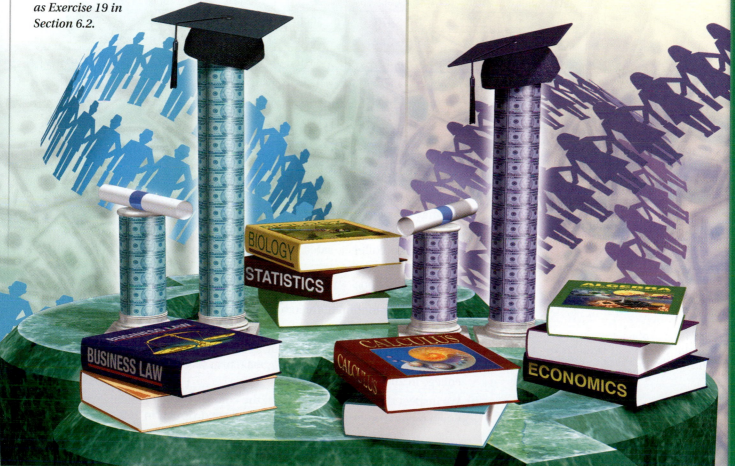

6.1 TABLES AND PICTOGRAPHS

Objectives

a Extract and interpret data from tables.

b Extract and interpret data from pictographs.

c Draw pictographs.

a Reading and Interpreting Tables

A **table** is often used to present data in rows and columns.

EXAMPLE 1 The table below lists the average charges for a full-time student at different types of schools.

INSTITUTIONS OF HIGHER EDUCATION—CHARGES: 2000 TO 2004
[In dollars. Estimated. For the entire academic year ending in year shown. Figures are average charges per full-time equivalent student. Room and board are based on full-time students.]

ACADEMIC CONTROL AND YEAR	TUITION AND REQUIRED FEES[1]		MEAL COSTS		DORMITORY CHARGES	
	2-yr Colleges	4-yr Colleges	2-yr Colleges	4-yr Colleges	2-yr Colleges	4-yr Colleges
Public:						
2000	1338	3349	1834	2406	1549	2519
2001	1333	3501	1906	2499	1600	2654
2002	1380	3735	2036	2645	1722	2816
2003	1483	4046	2164	2712	1954	3029
2004[2]	1670	4630	2224	2877	2092	3212
Private:						
2000	8235	14,588	2922	2881	2808	3237
2001	9067	15,470	3000	2993	2722	3392
2002	10,076	16,211	2633	3109	3116	3576
2003	10,651	16,826	3870	3197	3232	3764
2004[2]	11,635	17,902	4353	3355	3622	3948

[1]For in-state students.
[2]2004 data are preliminary.
Source: U.S. National Center for Education Statistics, *Digest of Education Statistics*, annual

Use the table in Example 1 to answer each of the following.

1. What is the smallest fee on the table and what was it for?

2. In what year did tuition at public 4-yr colleges first exceed $4000?

3. Find the average tuition at a public 4-yr college from 2000 to 2004.

Answers on page A-15

a) What is the most expensive charge on the table? In what year, and at what type of school, did it occur?

b) In what year did meal costs at private 2-yr colleges drop?

c) Mei-Ling went to—and lived at—a private 4-yr college in 2002. Assuming that the college's rates were average for private 4-yr colleges, how much did she pay?

d) From 2000 to 2004, what was the average tuition at a public 2-yr college?

Careful examination of the table will give the answers.

a) To identify the most expensive charge, we inspect the entries in the table. We note that the largest entries occur as tuition for private colleges. Of those entries, the largest charge was $17,902. This was the tuition at a private 4-yr college in 2004.

b) We inspect the "Private" section of the column showing meal costs at 2-yr colleges. Meal costs at a private 2-yr college dropped in 2002.

c) We locate the columns for 4-yr colleges and look down the left-hand column for the year 2002 and "Private" subheading. We find that Mei-Ling paid $16,211 for tuition and fees, $3109 for meals, and $3576 for dormitory charges. Mei-Ling's total charges were $16,211 + $3109 + $3576 = $22,896.

d) To find the average tuition at a public 2-yr college, we locate the 2-yr college column under "Tuition and Required Fees" and use the "Public" figures for 2000–2004:

$$\frac{\$1338 + \$1333 + \$1380 + \$1483 + \$1670}{5} = \$1440.80$$

The average tuition at a public 2-yr college, for the years 2000–2004, was $1440.80.

Do Exercises 1–3 on the previous page.

b Reading and Interpreting Pictographs

Pictographs (or *picture graphs*) are another way to show information. Instead of actually listing the amounts to be considered, a **pictograph** uses symbols to represent the amounts. In addition, a *key* is given telling what each symbol represents.

EXAMPLE 2 *Rhino Population.* The rhinoceros is considered one of the world's most endangered animals. The worldwide total number of rhinoceroses is approximately 17,800. The following pictograph shows the population of the five remaining rhino species. Located in the graph is a key that tells you that each symbol 🦏 represents 300 rhinos.

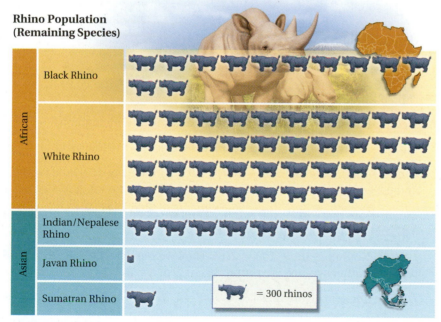

Rhino Population (Remaining Species)

Source: International Rhino Foundation, 2005

On July 5, 2004, at the Cincinnati Zoo and Botanical Gardens, Emi, a critically endangered Sumatran rhinoceros, became the first rhino to produce two calves in captivity.

Use the pictograph in Example 2 to answer Margin Exercises 4–6.

4. Which species has the fewest number of rhinos? Approximately how many are there?

5. How many more Indian/Nepalese rhinos are there than Sumatran rhinos?

6. How does the black rhino population compare with the Sumatran rhino population?

Answers on page A-15

a) Which species has the greatest number of rhinos? Approximately how many are there?

b) Which species has approximately 2400 rhinos?

c) How many more black rhinos are there than Indian/Nepalese rhinos?

We use the information in the pictograph as follows.

a) The species with the most symbols has the greatest number of rhinos: white rhino, represented by about $37\frac{3}{4}$ symbols. Since $37\frac{3}{4} \cdot 300$ is 11,325, there are about 11,325 white rhinos.

b) Since each symbol represents 300 rhinos, a species with 2400 rhinos would have $2400 \div 300$, or 8, symbols. The Indian/Nepalese rhino is shown with 8 symbols, and thus has about 2400 rhinos.

c) From the graph we see that there are $12 \cdot 300$, or 3600, black rhinos, and $8 \cdot 300$, or 2400, Indian/Nepalese rhinos. Thus, there are $3600 - 2400$, or 1200, more black rhinos than Indian/Nepalese rhinos.

Do Exercises 4–6 on the preceding page.

You have probably noticed that, although they seem to be very easy to read, pictographs are difficult to draw accurately because whole symbols reflect loose approximations due to significant rounding. In pictographs, you also need to use some mathematics to find the amounts.

C Drawing Pictographs

EXAMPLE 3 *Movie Production.* The following is a list of the countries that produced the greatest number of films in 2003. Draw a pictograph to represent the data. Let the symbol 🎥 represent 100 movies.

Source: *The Top 10 of Everything 2006,* DK Publishing, New York, 2005

India	1,100
United States	593
Japan	287
France	212
United Kingdom	175

Some computation is necessary before we can draw the pictograph.

India: Note that $1100 \div 100 = 11$. Thus we need 11 whole symbols.

United States: Note that $593 \div 100 = 5.93$. Thus we need 5 whole symbols and 0.93 of another symbol. We estimate 0.93 to be about $\frac{9}{10}$ of a symbol.

Japan: Note that $287 \div 100 = 2.87$. Thus we need 2 whole symbols and 0.87, or about $\frac{9}{10}$, of another symbol.

France: Note that $212 \div 100 = 2.12$. Thus we need 2 whole symbols and 0.12 of another symbol. We round this to $\frac{1}{10}$ of a symbol.

United Kingdom: Note that $175 \div 100 = 1.75$. Thus we need $1\frac{3}{4}$ symbols.

4ᵉ FESTIVAL INTERNATIONAL DU FILM DE MARRAKECH

6-12 DÉCEMBRE 2004

المهرجان الدولي الرابع للفيلم بمراكش

The pictograph can now be drawn as follows. We list the country in one column, draw the number of films using symbols, indicate the key, and title the overall graph "Movie Production." Note that a different choice for the symbol in the key will result in a different pictograph.

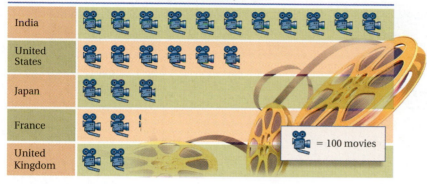

Movie Production, 2003

📽 = 100 movies

Do Exercise 7.

7. Movie Trilogies. The following is a list of the gross revenue (money taken in) worldwide by the five top-grossing movie trilogies through 2004. Draw a pictograph to represent the data. Use a symbol to represent $1,000,000,000.

Source: *The Top 10 of Everything 2006,* DK Publishing, New York, 2005

The Lord of the Rings (2001–2003)
$2,916,544,743

Harry Potter (2001–2004)
$2,652,297,210

Jurassic Park (1993–2002)
$1,902,110,926

The Matrix (1999–2003)
$1,623,924,804

Terminator (1984–2003)
$1,212,019,531

Answer on page A-15

Study Tips

SEEKING HELP?

Using some or all of these resources can make studying easier and more enjoyable.

A variety of resources are available to help you learn math.

- **Textbook supplements.** Are you aware of all the supplements that exist for this textbook? See the Preface for a description of each supplement: the *Student's Solutions Manual,* a complete set of video lectures available on CD-ROM, tutorial software, the Addison-Wesley Math Tutor Center, MathXL, and the Web site www.MyMathLab.com.

- **The Internet.** Our online World Wide Web supplement provides additional practice resources. If you have Internet access, you can reach this site through the address:

  http://www.MyMathLab.com

- **Your college or university.** Your own college or university probably has resources to enhance your math learning.

  1. There may be a learning lab or tutoring center for drop-in tutoring.
  2. There may be study skills workshops or group tutoring sessions tailored for the specific course you are taking.
  3. Often, there is a bulletin board or network where you can locate the names of experienced private tutors.
  4. You might be able to find classmates interested in forming a study group.

- **Your instructor.** Although it may seem obvious, you should consider an often overlooked resource: your instructor. Find out your instructor's office hours and make it a point to visit when you need additional help. Many instructors welcome student e-mail. If you are hesitant to visit your instructor, e-mail may prove quite useful.

a *Nutrition Information.* The following table lists nutrition information for a 1-cup serving of five name-brand cereals (it does not consider the use of milk or sweetener). Use the table for Exercises 1–6.

CEREAL	CALORIES	FAT (in grams)	TOTAL CARBOHYDRATES (in grams)	SODIUM (in milligrams)
Cinnamon Life®	160	1.3	34.7	200
Life® (Regular)	160	2.0	33.3	213.3
Lucky Charms®	120	1.0	25.0	210
Kellogg's Complete®	120	0.7	30.7	280
Wheaties®	110	1.0	24.0	220

Sources: Quaker Oats; General Mills; Kellogg's

1. Which cereal has the least amount of sodium?

2. Which cereal has the greatest amount of fat?

3. Which cereal has the least amount of fat?

4. Find the average total carbohydrates in the cereals listed.

5. Find the average amount of fat in the cereals listed.

6. Find the average amount of sodium in the cereals listed.

Planets. Use the following table, which lists information about the planets, for Exercises 7–12.

PLANET	AVERAGE DISTANCE FROM SUN (in miles)	DIAMETER (in miles)	LENGTH OF PLANET'S DAY IN EARTH TIME (in days)	TIME OF REVOLUTION IN EARTH TIME (in years)
Mercury	35,983,000	3,031	58.82	0.24
Venus	67,237,700	7,520	224.59	0.62
Earth	92,955,900	7,926	1.00	1.00
Mars	141,634,800	4,221	1.03	1.88
Jupiter	483,612,200	88,846	0.41	11.86
Saturn	888,184,000	74,898	0.43	29.46
Uranus	1,782,000,000	31,763	0.45	84.01
Neptune	2,794,000,000	31,329	0.66	164.78
Pluto*	3,666,000,000	1,423	6.41	248.53

*Classified as a planet until 2006
Source: *The Handy Science Answer Book,* Gale Research, Inc.

7. Find the average distance from the sun to Jupiter.

8. How long is a day on Venus?

9. Which planet has a time of revolution of 164.78 yr?

10. Which planet has a diameter of 4221 mi?

11. Which planets have an average distance from the sun that is greater than 500,000,000 mi?

12. Which planets have a diameter that is less than 5000 mi?

Heat Index. In warm weather, a person can feel hotter due to reduced heat loss from the skin caused by higher humidity. The **temperature–humidity index,** or **apparent temperature,** is what the temperature would have to be with no humidity in order to give the same heat effect. The following table lists the apparent temperatures for various actual temperatures and relative humidities. Use this table for Exercises 13–24.

ACTUAL TEMPERATURE (°F)	RELATIVE HUMIDITY									
	10%	20%	30%	40%	50%	60%	70%	80%	90%	100%
	APPARENT TEMPERATURE (°F)									
75°	75	77	79	80	82	84	86	88	90	92
80°	80	82	85	87	90	92	94	97	99	102
85°	85	88	91	94	97	100	103	106	108	111
90°	90	93	97	100	104	107	111	114	118	121
95°	95	99	103	107	111	115	119	123	127	131
100°	100	105	109	114	118	123	127	132	137	141
105°	105	110	115	120	125	131	136	141	146	151

In Exercises 13–16, find the apparent temperature for the given actual temperature and humidity combination.

13. 80°, 60%

14. 90°, 70%

15. 85°, 90%

16. 95°, 80%

17. How many listed temperature–humidity combinations give an apparent temperature of 100°?

18. How many listed temperature–humidity combinations give an apparent temperature of 111°?

19. At a relative humidity of 50%, what actual temperatures give an apparent temperature above 100°?

20. At a relative humidity of 90%, what actual temperatures give an apparent temperature above 100°?

21. At an actual temperature of 95°, what relative humidities give an apparent temperature above 100°?

22. At an actual temperature of 85°, what relative humidities give an apparent temperature above 100°?

23. At an actual temperature of 85°, by how much would the humidity have to increase in order to raise the apparent temperature from 94° to 108°?

24. At an actual temperature of 80°, by how much would the humidity have to increase in order to raise the apparent temperature from 87° to 102°?

Cigarette Consumption. The May 2004 U.S. Surgeon General's report, *The Health Consequences of Smoking,* reports that smoking causes many diseases and reduces the health of smokers. The per capita cigarette consumption in the United States for specific years from 1920 to 2005 is listed in the table below. Use these data for Exercises 25–28.

U.S. CIGARETTE CONSUMPTION (1920–2005) (PER CAPITA AMONG PERSONS 18 AND OLDER)

YEAR	1920	1930	1940	1950	1960	1970	1980	1990	2000	2005
Cigarette Consumption	665	1485	1976	3552	4171	3985	3849	2817	2056	1280

Sources: U.S. Department of Health and Human Services, Centers for Disease Control and Prevention, USDA Economic Research Service, U.S. Bureau of Alcohol, Tobacco, and Firearms; U.S. Bureau of the Census

25. Find the per capita cigarette consumption in 1940 and in 1980. By how much did per capita consumption increase from 1940 to 1980?

26. Find the per capita cigarette consumption in 1960 and in 2000. By how much did per capita consumption decrease from 1960 to 2000?

27. Find the average of the per capita cigarette consumption for the years 1920 to 1950. Find the average of the per capita cigarette consumption for the years 1970 to 2000. By how many cigarettes per capita does the latter average exceed the former?

28. Find the average of the per capita cigarette consumption for the years 2000 and 2005. Find the average of the per capita cigarette consumption for the years 1920 to 2000. By how many cigarettes per capita does the average for 1920 to 2000 exceed the average of the years 2000 and 2005?

World Population Growth. The following pictograph shows world population in various years. Use the pictograph for Exercises 29–36.

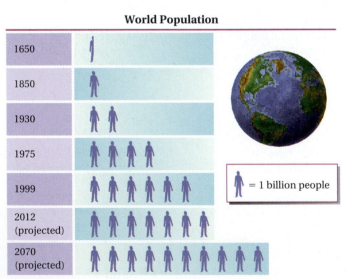

World Population

Source: U.S. Bureau of the Census, International Data Base

29. What was the world population in 1850?

30. What was the world population in 1975?

CHAPTER 6: Introduction to Graphing and Statistics

31. In which year will the population be the greatest?

32. In which year was the population the least?

33. Between which two years was the amount of growth the least?

34. Between which two years was the amount of growth the greatest?

35. How much greater will the world population in 2012 be than in 1975?

36. How much greater was the world population in 1999 than in 1930?

Water Consumption.   The following pictograph shows water consumption, per person, in different regions of the world in a recent year. Water consumption includes agricultural, industrial, and energy use as well as personal use. Use the pictograph for Exercises 37–42.

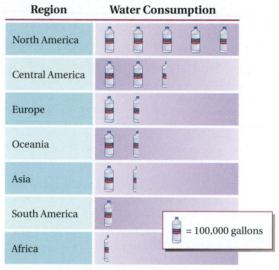

Sources: World Resources Institute; U.S. Energy Information Administration

37. Which region consumes the least water?

38. Which region consumes the most water?

39. About how many gallons are consumed per person in North America?

40. About how many gallons are consumed per person in Europe?

41. Approximately how many more gallons are consumed per person in North America than in Asia?

42. Approximately how many more gallons are consumed per person in Central America than in Africa?

43. *Coffee Consumption.*   The following chart lists approximately how many cups of coffee each person (per capita) drinks annually in several countries. Draw a pictograph that reflects these data. Use the symbol  to represent 100 cups.

COUNTRY	COFFEE CONSUMPTION PER CAPITA
Germany	1113
United States	615
Switzerland	1220
France	790
Italy	730

Source: Based on information from the Beverage Marketing Corporation

Coffee Consumption

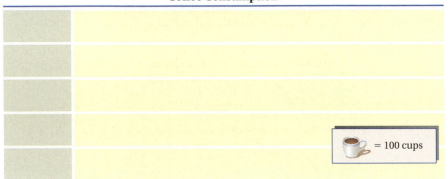

= 100 cups

44. Redo Exercise 43, using the symbol 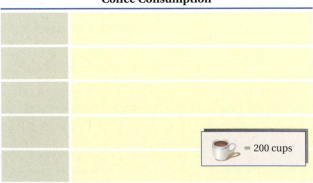 to represent 200 cups.

Coffee Consumption

= 200 cups

CHAPTER 6: Introduction to Graphing
and Statistics

45. *Elephant Population.* The following chart shows the elephant population of various countries in Africa. Draw a pictograph that represents these data. Let the symbol  represent 10,000 elephants.

COUNTRY	ELEPHANT POPULATION
Cameroon	20,050
Zimbabwe	49,800
Sudan	19,800
Zaire	110,120
Tanzania	60,070
Botswana	69,105

Source: Based on information from *National Geographic*

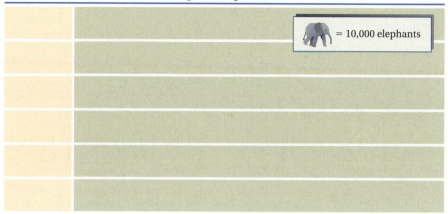

Elephant Population

46. Redo Exercise 45, using the symbol 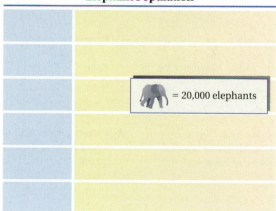 to represent 20,000 elephants.

Elephant Population

47. **Dw** Loreena is drawing a pictograph in which dollar bills are used as symbols to represent the tuition at various private colleges. Should each dollar bill represent $20,000, $2000, or $200? Why?

48. **Dw** What advantage(s) does a table have over a pictograph?

SKILL MAINTENANCE

Perform the indicated operation and, if possible, simplify.

49. $-\dfrac{3}{8} + \dfrac{5}{16}$ [4.2b]

50. $-\dfrac{2}{7} - \dfrac{3}{5}$ [4.3a]

Solve.

51. $9x - 5 = -23$ [2.8d]

52. $3x - 2 = 7x + 10$ [5.7b]

53. $-4x = 3x - 7$ [5.7b]

Convert to decimal notation.   [5.5a]

54. $\dfrac{3}{8}$

55. $\dfrac{29}{25}$

56. $\dfrac{5}{6}$

SYNTHESIS

57. **Dw** Suppose you are drawing a pictograph to represent the number of hours each of three students spends online each month. What would be an appropriate symbol to use and what exactly would the symbol represent? Explain how you arrived at your answer.

58. **Dw** Since information is lost when converting a chart into a pictograph (we lose precision), why would anyone want to draw a pictograph to replace a chart?

The following chart shows phone rates for some long-distance plans offered in California. All rates are in cents per minute. Use the chart for Exercises 59–62.

PLAN	IN-STATE	STATE-TO-STATE, CONTINENTAL U.S.	MONTHLY FEES	GERMANY	SOUTH AFRICA	ANTARCTICA
ECG Easy 2.5	3.90	2.50	$0.59	6	17	21
Pioneer 3.25 TalkCents	3.25	3.25	None	5	13	175
bigredwire 2.7a	3.50	2.70	None	5	13	127
Opex Value Plus 2.7	4.20	3.20	None	5	10	33
Voice Revolution Preferred 3.2	4.20	3.20	None	8	13	58
3U Telecom 4.8	3.40	4.80	None	7	16	115

59. Alan spoke the same number of minutes to a friend in Germany as he did to a friend in South Africa. If his ECG bill was for $16, for how many minutes did he speak to each friend?

60. Patti makes half of her long-distance calls in-state, and half of them state-to-state in the continental United States. If she has $50 budgeted for long-distance, what plan should she select, and for how many minutes can she speak?

61. **Dw** Given that calls are made only to those places shown in the chart, under what circumstances is the Pioneer plan better than the Opex plan?

62. **Dw** Given that calls are made only to those places shown in the chart, under what circumstances is the 3U plan better than the Voice Revolution plan?

CHAPTER 6: Introduction to Graphing and Statistics

b Drawing Bar Graphs

EXAMPLE 3 *Centenarians.* The number of centenarians—that is, people 100 yr or older—is growing rapidly. Projections from the U.S. Bureau of the Census and the National Center for Health Statistics are shown below. Use the projections to form a bar graph.

YEAR	PROJECTED NUMBER OF CENTENARIANS
2000	96,000
2010	129,000
2020	235,000
2030	381,000
2040	551,000
2050	1,095,000

Source: U.S. Bureau of the Census, "National Population Projections—Summary Tables," 1/13/00

First, we draw a horizontal scale with six equally spaced intervals and the different years listed. We title that scale "Year." (See the figure on the left below.) Next, we label the vertical scale "Projected number (in thousands)." Note that the largest number (in thousands) is 1095 and the smallest is 96. If we count by 100's, we can range from 0 to 1100 with 11 marks. Finally, we draw vertical bars to represent the number of centenarians projected for each year and title the graph. (See the figure on the right below.)

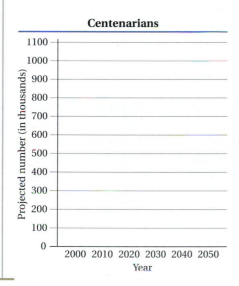

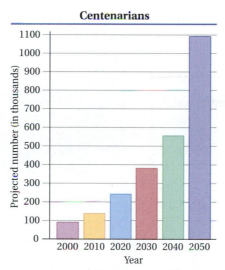

Do Exercise 8.

Bar graphs that are drawn with unmarked scales may be misleading. This is especially true when the scale does not begin with 0, or when an inconsistent scale is used. The graph on the next page creates such a misleading image.

8. Planetary Moons. Make a horizontal bar graph to show the number of moons orbiting the various planets.

PLANET	MOONS
Earth	1
Mars	2
Jupiter	17
Saturn	28
Uranus	21
Neptune	8
Pluto*	1

*Classified as a planet until 2006

Source: National Aeronautics and Space Administration

Answer on page A-15

Study Tips

KEEPING MATH RELEVANT

We have already stated that finding applications of math in your everyday life is a great study aid. Try to extend this idea to the newspapers, periodicals, and books that you read. Look with a critical eye at graphs and their labels. Not only will this help with your math, it will make you a more informed citizen.

9. What is wrong with the following graph?

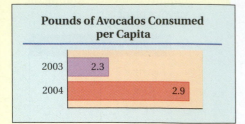

Pounds of Avocados Consumed per Capita

2003	2.3
2004	2.9

Note that the bar representing $894 is nearly twice as long as the bar representing $853, but 894 is not nearly twice 853. As drawn, the graph implies that the cost of books and supplies nearly doubled from 2005 to 2006.

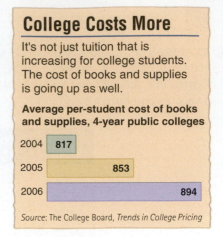

College Costs More

It's not just tuition that is increasing for college students. The cost of books and supplies is going up as well.

Average per-student cost of books and supplies, 4-year public colleges

2004	817
2005	853
2006	894

Source: The College Board, Trends in College Pricing

Do Exercise 9.

C Reading and Interpreting Line Graphs

Line graphs are often used to show a change over time as well as to indicate patterns or trends.

EXAMPLE 4 *New Home Sales.* The following line graph shows the number of new home sales, in thousands, over a recent twelve-month period. The jagged line at the base of the vertical scale indicates an unnecessary portion of the scale. Note that the vertical scale differs from the horizontal scale so that the data can be easily shown.

New Home Sales

Source: U.S. Department of Commerce

a) For which month were new home sales the greatest?

b) Between which months did new home sales increase?

c) For which months were new home sales about 700 thousand?

We look at the graph to answer the questions.

a) The greatest number of new home sales was about 825 thousand in month 1.

b) Reading the graph from left to right, we see that new home sales increased from month 2 to month 3, from month 3 to month 4, from month 5 to month 6, from month 7 to month 8, from month 8 to month 9, from month 9 to month 10, and from month 10 to month 11.

c) We look from left to right along the line at 700.

Answer on page A-15

New Home Sales

Source: U.S. Department of Commerce

Use the line graph in Example 4 to answer each of the following.

10. For which month were new home sales lowest?

11. Between which months did new home sales decrease?

12. For which months were new home sales below 700 thousand?

We see that points are closest to 700 thousand at months 3, 6, 10, 11, and 12.

Do Exercises 10–12.

EXAMPLE 5 *Monthly Loan Payment.* Suppose that you borrow $110,000 at an interest rate of 9% to buy a home. The following graph shows how the size of the monthly payment depends on the length of the loan. (Note that the smaller the monthly payment, the longer the duration of the loan.)

$110,000 Loan Repayment

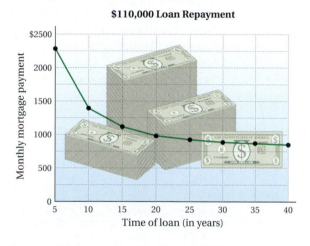

Use the line graph in Example 5 to answer Margin Exercises 13–15.

13. Estimate the monthly payment for a loan of 25 yr.

14. What time period corresponds to a monthly payment of about $850?

15. By how much does the monthly payment decrease when the loan period is increased from 5 yr to 20 yr?

a) Estimate the monthly payment for a loan of 15 yr.

b) What time period corresponds to a monthly payment of about $1400?

c) By how much does the monthly payment decrease when the loan period is increased from 10 yr to 20 yr?

We look at the graph to answer the questions.

a) We find the time period labeled "15" on the bottom scale and move up from that point to the line. We then go straight across to the left and find that the monthly payment is about $1100.

b) We locate $1400 on the vertical axis. Then we move to the right until we hit the line. The point $1400 is on the line at the 10-yr time period.

c) The graph shows that the monthly payment for 10 yr is about $1400; for 20 yr, it is about $990. Thus the monthly payment is decreased by about $1400 − $990, or $410. (Note that you will pay back $990 · 20 · 12 − $1400 · 10 · 12, or $69,600, more in interest for a 20-yr loan.)

Do Exercises 13–15.

Answers on page A-15

16. Traditional SUVs. Listed below are the sales of traditional sport utility vehicles. Make a line graph of the data.

YEAR	TRADITIONAL SUV SALES (in millions)
1998	2.6
1999	2.9
2000	3.0
2001	3.0
2002	3.0
2003	2.8
2004	2.7
2005	2.4
2006*	2.3

Source: USA TODAY, 12/8/05
*Estimated

d Drawing Line Graphs

EXAMPLE 6 *Crossover Utility Vehicles.* Sales of crossover utility vehicles built on car or minivan platforms rose to equal sales of traditional SUVs in 2005. Listed below are the sales of crossover SUVs. Make a line graph of the data.

YEAR	CROSSOVER SUV SALES (in millions)
1998	0.2
1999	0.3
2000	0.5
2001	1.0
2002	1.2
2003	1.6
2004	1.9
2005	2.2
2006*	2.4

Source: USA TODAY, 12/8/05
*Estimated

First, we label the years on the horizontal scale and give it the title "Year." The number of vehicles sold varies from 0.2 million to 2.4 million. Thus we show from 0 to 2.5 on the vertical axis, and scale it by 0.5. The sales are in millions, so we title the vertical scale "Number of vehicles sold (in millions)." We also give the graph the title "Crossover Utility Vehicle Sales."

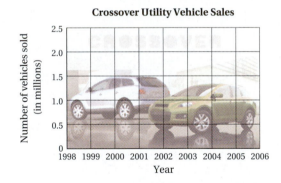

Next, we mark the number of vehicles at the appropriate level above the year. Then we draw line segments connecting the points. The dramatic change over time can now be observed easily from the graph.

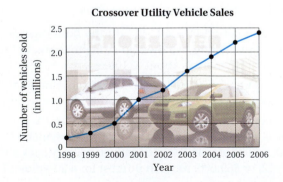

Answer on page A-16

Do Exercise 16.

6.2

EXERCISE SET   For Extra Help

MathXL   MyMathLab   InterAct   Math Tutor   Video   Student's
                        Math   Center   Lectures Solutions
                                                 on CD   Manual
                                                 Disc 3

a *Chocolate Desserts.*   The following horizontal bar graph shows the average caloric content of various kinds of chocolate desserts. Use the bar graph for Exercises 1–12.

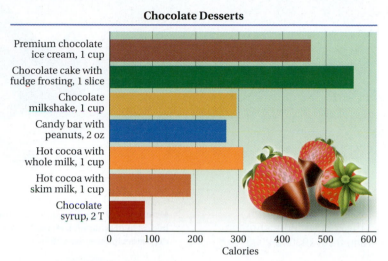

Chocolate Desserts

Source: *Better Homes and Gardens*, December 1996

1. Estimate how many calories there are in 1 cup of hot cocoa with skim milk.

2. Estimate how many calories there are in 1 cup of premium chocolate ice cream.

3. Which dessert has the highest caloric content?

4. Which dessert has the lowest caloric content?

5. Which dessert contains about 460 calories?

6. Which desserts contain about 300 calories?

7. How many more calories are there in 1 cup of hot cocoa made with whole milk than in 1 cup of hot cocoa made with skim milk?

8. Bradey generally drinks a 4-cup chocolate milkshake. How many calories does he consume?

9. Rae likes to eat 2 cups of premium chocolate ice cream at bedtime. How many calories does she consume?

10. Daquan likes to eat a 6-oz chocolate bar with peanuts for lunch. How many calories does he consume?

11. Paul adds a 2-oz chocolate bar with peanuts to his diet each day for 1 yr (365 days) and makes no other changes in his eating or exercise habits. Consumption of 3500 extra calories will add about 1 lb to his body weight. How many pounds will he have gained at the end of the year?

12. Tricia adds one slice of chocolate cake with fudge frosting to her diet each day for 1 yr (365 days) and makes no other changes in her eating or exercise habits. The consumption of 3500 extra calories will add about 1 lb to her body weight. How many pounds will she have gained at the end of the year?

Education and Earnings. The side-by-side bar graph shown at right provides data on the effect of education on earning power for men and women from 1970 to 2003. Use the bar graph in Exercises 13–20.

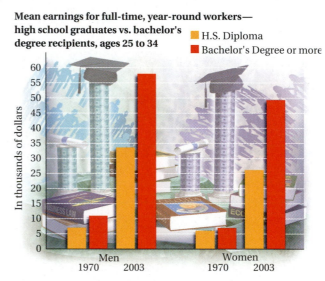

Mean earnings for full-time, year-round workers— high school graduates vs. bachelor's degree recipients, ages 25 to 34

■ H.S. Diploma
■ Bachelor's Degree or more

Sources: USA Group Foundation and U.S. Bureau of the Census

13. How much were the mean earnings for men who earned at least a bachelor's degree in 1970? in 2003? How much had they increased?

14. How much were the mean earnings for women with at least a bachelor's degree in 1970? in 2003? How much had they increased?

15. How much were the mean earnings for women who had ended their education at high school graduation in 1970? in 2003? How much had they increased?

16. How much were the mean earnings for men who had ended their education at high school graduation in 1970? in 2003? How much had they increased?

17. In 1970, how much more did men with at least a bachelor's degree earn than men who ended their education at high school graduation?

18. In 2003, how much more did men with at least a bachelor's degree earn than men who ended their education at high school graduation?

19. In 2003, how much more did women with bachelor's degrees or higher earn than men who ended their education at high school graduation?

20. In 1970, how much more did men with bachelor's degrees or higher earn than women who ended their education at high school graduation?

21. *Commuting Time.* The following table lists the average one-way commuting time in six metropolitan areas with more than 1 million people. Make a vertical bar graph to illustrate the data.

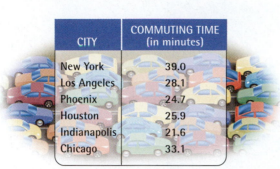

CITY	COMMUTING TIME (in minutes)
New York	39.0
Los Angeles	28.1
Phoenix	24.7
Houston	25.9
Indianapolis	21.6
Chicago	33.1

Source: U.S. Bureau of the Census

Use the data and the bar graph in Exercise 21 to do Exercises 22–25.

22. Which city has the greatest commuting time?

23. Which city has the least commuting time?

24. How much longer is the commuting time in New York than in Los Angeles?

25. What is the average commuting time for the six cities?

26. *Calorie Expenditure.* Use the following information to make a horizontal bar graph showing the number of calories burned during each activity by a person weighing 152 lb.

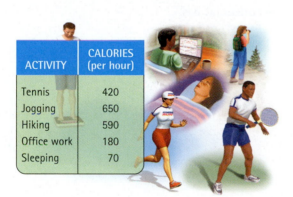

ACTIVITY	CALORIES (per hour)
Tennis	420
Jogging	650
Hiking	590
Office work	180
Sleeping	70

Use the data and the bar graph in Exercise 26 for Exercises 27–30.

27. What is the difference in the number of calories burned per hour between sleeping and jogging?

28. Suppose you were trying to lose weight by exercising and had to choose one of these exercises. If your doctor told you not to jog, what would be the most beneficial exercise?

29. Ryan works at the office for 8 hr and then sleeps for 7 hr. How many calories does Ryan burn doing this?

30. Nancy hiked for 6 hr and then slept for 8 hr. How many calories did she burn doing this?

c *Golf Distances.* In recent years, new equipment and technology have had a tremendous impact on the distance a golfer can hit a golf ball. The line graph below shows the average driving distances for years from 1980 to 2005. Use the graph for Exercises 31–34.

Average Driving Distance on the PGA Tour

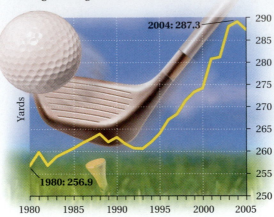

2004: 287.3

1980: 256.9

Yards

1980 1985 1990 1995 2000 2005

Source: PGA of America

31. How much farther was the driving distance in 2004 than in 1980?

32. In which year was the driving distance the highest?

33. In which years was the average driving distance about 264 yd?

34. In which year was the average driving distance about 270 yd?

d

35. *Ozone Level.* Make a line graph of the data, listing years on the horizontal axis.

YEAR	OZONE LEVEL (in Dobson Units)
1996	290
1997	291.1
1998	294.5
1999	293
2000	292.1
2001	290.4
2002	292.6
2003	287.9
2004	284.3

Use the data and the line graph in Exercise 35 for Exercises 36–39.

36. Between which two years was the decrease in the ozone level the greatest?

37. Between which two years was the increase in the ozone level the greatest?

38. What was the average ozone level from 1996 through 2000?

39. What was the average ozone level from 2000 through 2004?

40. *Democratic Governors.* The following table lists the number of state governors affiliated with the Democratic political party for various years. Make a line graph of the data, listing years on the horizontal axis. Use the data and the line graph for Exercises 41–46.

YEAR	NUMBER OF DEMOCRATIC GOVERNORS
1994	29
1995	19
1996	18
1997	17
1998	17
1999	17
2000	17
2001	21
2002	21
2003	24
2004	22

Source: U.S. Bureau of the Census, *Statistical Abstract of the United States*, 2006

41. Between which two years was the increase in the number of Democratic governors the greatest?

42. Between which two years was the decrease in the number of Democratic governors the greatest?

43. In which year did the number of Democratic governors begin to rise?

44. Between which years did the number of Democratic governors remain unchanged?

45. What was the average number of Democratic governors from 1995 to 1999?

46. What was the average number of Democratic governors from 2000 to 2004?

47. D$_W$ Can bar graphs always, sometimes, or never be converted to line graphs? Why?

48. D$_W$ Compare bar graphs and line graphs. Discuss why you might use one rather than the other to graph a particular set of data.

SKILL MAINTENANCE

49. How many 12-oz bottles can be filled from a vat containing 408 oz of catsup?   [1.8a]

50. Managers of pizza restaurants know that if 50 pizzas are ordered in an evening, people will request extra cheese on 9 of them. Find the ratio of pizzas ordered with extra cheese to pizzas ordered.   [3.3a]

51. A can of Cola-Cola contains 12 fl oz (fluid ounces). How many fluid ounces are in a six-pack?   [1.8a]

52. 24 is $\frac{3}{4}$ of what number?   [3.8b]

53. $\frac{2}{3}$ of 75 is what number?   [3.4c]

54. $\frac{3}{5}$ of 30 is what number?   [3.4c]

55. Solve: $-9 = -2x + 3$.   [2.8d]

56. Solve: $17 = -3x - 4$.   [2.8d]

57. **D**_W In his groundbreaking book, *The Visual Display of Quantitative Information,* Edward Tufte cites the following graph as a misleading representation of data. In what way is the graph misleading?

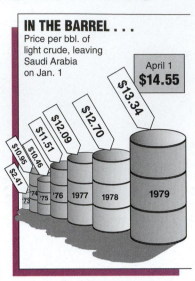

©1979 Time Inc. Reprinted by Permission

58. **D**_W Tufte (see Exercise 57) also cites the following as a misleading presentation of information. In what way is it misleading?

©1979 Los Angeles Times

59. Draw the bar graph shown at the top right of page 406 using an accurate scale.

60. Draw the bar graph in Margin Exercise 9 using an accurate scale.

61. Draw the line graphs from Example 6 and Margin Exercise 16 on the same graph. Extend the graphs to predict the year for which crossover sales will be 1 million more than traditional sales.

62. 🖩 Use the information in Example 2 to approximate the average rate of incidence of breast cancer for all women above the age of 24.

63. **D**_W Consider the graph in Example 4. Sam states that the initial drop shows that sales were nearly cut in half over the first month of the year. What mistake is Sam making?

6.3 ORDERED PAIRS AND EQUATIONS IN TWO VARIABLES

Objectives

a Plot a point, given its coordinates. Find coordinates, given a point.

b Determine the quadrant in which a point lies.

c Determine whether an ordered pair is a solution of an equation with two variables.

Bar graphs and line graphs are used to illustrate relationships between the items or quantities listed along the bottom and the side of the graph. The horizontal and vertical sides of a bar graph or line graph are often called the **axes** (pronounced ăk sēź; singular: **axis**). By using two perpendicular number lines as axes, we will find that we can use points to represent solutions of certain equations. First, however, we must learn to graph points.

a Points and Ordered Pairs

When two number lines are used as axes, a grid can be formed. The grid provides a helpful way of locating any point on the plane. Just as a location in a city might be given as the intersection of an avenue and a side street, a point on a plane can be regarded as the intersection of a vertical line and a horizontal line. In the figure below, these lines pass through 3 on the horizontal axis and 4 on the vertical axis. Thus the **first coordinate** of this point is 3 and the **second coordinate** is 4. **Ordered pair** notation, (3, 4), provides a quick way of stating this.

Caution!

When writing an ordered pair, you should *always* list the coordinate from the horizontal axis first.

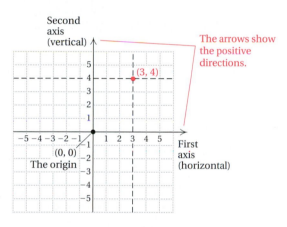

The arrows show the positive directions.

Plot these points on the graph below.

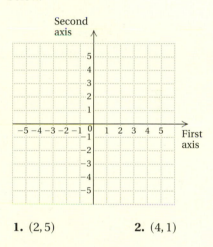

1. (2, 5) **2.** (4, 1)

Answers on page A-16

The point (0, 0), where the axes cross each other, is called the **origin.** To graph, or *plot*, the point (3, 4), we can begin at the origin and move horizontally (along the first axis) to the number 3. From there, we move up 4 units vertically and make a "dot."

It is important to always make sure that the first coordinate matches the number that would be below (or above) the point on the horizontal axis. Similarly, the second coordinate should always match the number that would be to the left (or right) of the point on the vertical axis.

Do Exercises 1 and 2.

Study Tips

MASTER NEW SKILLS

When a brand-new topic, like graphing ordered pairs, arises, try to master that new skill when it first appears. By staying in command of this new material, you will ultimately save yourself time and build a solid foundation that will breed confidence as you move forward.

415

Plot these points on the graph below.

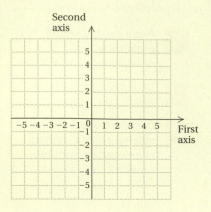

3. $(-2, 5)$

4. $(-3, -4)$

5. $(5, -3)$

6. $(-2, -1)$

7. $(0, -3)$

8. $\left(2\frac{1}{2}, 0\right)$

9. Determine the coordinates of points A, B, C, D, E, F, and G on the graph below.

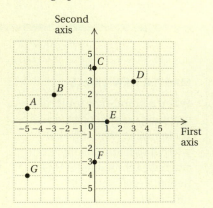

Answers on page A-16

CHAPTER 6: Introduction to Graphing and Statistics

EXAMPLE 1 Plot the points $(-5, 2)$ and $(2, -5)$.

To plot $(-5, 2)$, we locate -5 on the first, or horizontal, axis. From there we go up 2 units and make a dot.

To plot $(2, -5)$, we locate 2 on the first, or horizontal, axis. Then we go down 5 units and make a dot. Note that the order of the numbers within a pair is important: $(2, -5) \neq (-5, 2)$.

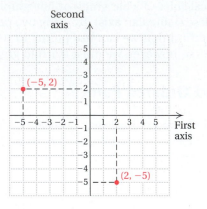

Do Exercises 3–8.

To determine the coordinates of a given point, we first look directly above or below the point to find the point's horizontal coordinate. Then we look to the left (or right) of the point to identify the vertical coordinate.

EXAMPLE 2 Determine the coordinates of points A, B, C, D, E, F, and G.

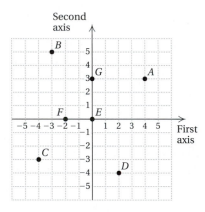

We look below point A to see that its first coordinate is 4. Looking to the left of point A, we find that its second coordinate is 3. Thus the coordinates of point A are $(4, 3)$. The coordinates of the other points are

B: $(-3, 5)$; C: $(-4, -3)$; D: $(2, -4)$;

E: $(0, 0)$; F: $(-2, 0)$; G: $(0, 3)$.

Do Exercise 9.

b Quadrants

The axes divide the plane into four regions, or **quadrants.** For any point in region I (the *first quadrant*), both coordinates are positive. For any point in region II (the *second quadrant*), the first coordinate is negative and the second coordinate is positive. In region III (the *third quadrant*), both coordinates are negative. In region IV (the *fourth quadrant*), the first coordinate is positive and the second coordinate is negative.

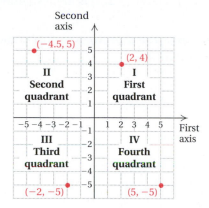

As the figure above illustrates, the point $(2, 4)$ is in the first quadrant, $(-4.5, 5)$ is in the second quadrant, $(-2, -5)$ is in the third quadrant, and $(5, -5)$ is in the fourth quadrant.

Do Exercises 10–15.

c Solutions of Equations

The coordinate system we have just introduced is called the **Cartesian** coordinate system, in honor of the great mathematician and philosopher René Descartes (1596–1650). Legend has it that Descartes hit upon the idea of the coordinate system after watching a fly stop several times on the ceiling over his bed. Descartes used this coordinate system as a method of presenting solutions of equations containing two variables. Equations like $3x + 2y = 8$ have ordered pairs as solutions. In Section 6.4, we will find solutions and graph them. Here we simply practice checking to see if an ordered pair is a solution.

To determine whether an ordered pair is a solution of an equation, we normally substitute the first coordinate for the variable that comes first alphabetically and the second coordinate for the variable that is last alphabetically. The letters x and y are used most often.

EXAMPLE 3 Determine whether the ordered pair $(2, 1)$ is a solution of the equation $3x + 2y = 8$.

$$\frac{3x + 2y = 8}{\begin{array}{c|c} 3 \cdot 2 + 2 \cdot 1\ ?\ 8 \\ 6 + 2 \\ 8 & 8 \quad \text{TRUE} \end{array}}$$

  Substituting 2 for x and 1 for y
  (alphabetical order of variables)

Since the equation becomes true, $(2, 1)$ is a solution.

In a similar manner, we can show that $(0, 4)$ and $(4, -2)$ are also solutions of $3x + 2y = 8$. In fact, there is an infinite number of solutions of $3x + 2y = 8$.

---

10. What can you say about the coordinates of a point in the third quadrant?

11. What can you say about the coordinates of a point in the fourth quadrant?

In which quadrant is each point located?

12. $(5, 3)$

13. $(-6, -4)$

14. $(10, -14)$

15. $\left(-13, 9\frac{1}{2}\right)$

Answers on page A-16

16. Determine whether $(5, 1)$ is a solution of $y = 2x + 3$.

17. Determine whether $(-13.6, 25.4)$ is a solution of $3x + 2y = 10$.

EXAMPLE 4 Determine whether the ordered pair $(-2, 3)$ is a solution of the equation $2t = 4s - 8$.

We substitute:

$$\frac{2t = 4s - 8}{2 \cdot 3 \;?\; 4(-2) - 8}$$
$$\begin{array}{c|c} 6 & -8 - 8 \\ 6 & -16 \quad \text{FALSE} \end{array}$$

Using alphabetical order, and substituting -2 for s and 3 for t

Since the equation becomes false, $(-2, 3)$ is not a solution.

Unless stated otherwise, the coordinates of an ordered pair correspond alphabetically to the variables.

Do Exercises 16 and 17.

Answers on page A-16

CALCULATOR CORNER

Checking Solutions Solutions of equations in two variables can be easily checked on a calculator. For instance, to demonstrate that $(5.1, -3.65)$ is a solution of $3x + 2y = 8$, we press either

$\boxed{3}\ \boxed{\times}\ \boxed{5}\ \boxed{.}\ \boxed{1}\ \boxed{+}\ \boxed{2}\ \boxed{\times}\ \boxed{3}\ \boxed{.}\ \boxed{6}\ \boxed{5}\ \boxed{+/-}\ \boxed{\text{ENTER} =}$

or $\boxed{3}\ \boxed{\times}\ \boxed{5}\ \boxed{.}\ \boxed{1}\ \boxed{+}\ \boxed{2}\ \boxed{\times}\ \boxed{(-)}\ \boxed{3}\ \boxed{.}\ \boxed{6}\ \boxed{5}\ \boxed{\text{ENTER}}$.

The result, 8, shows that $(5.1, -3.65)$ is a solution.

Most calculators now have memory keys. These keys enable us to store and recall a number as needed. Any number being displayed can be stored by pressing a particular key. On most calculators, this key is labeled $\boxed{\text{STO}}$, $\boxed{\text{M}}$, or $\boxed{\text{Min}}$. Once a number has been stored, we can retrieve or recall the number by pressing a key labeled $\boxed{\text{RCL}}$ or $\boxed{\text{MR}}$.

To show that $(7.35, 10.7)$ is a solution of $2t = 4s - 8$, we can first evaluate and store the right side of the equation:

$\boxed{4}\ \boxed{\times}\ \boxed{7}\ \boxed{.}\ \boxed{3}\ \boxed{5}\ \boxed{-}\ \boxed{8}\ \boxed{=}\ \boxed{\text{STO}}$.

The result, 21.4, has been stored in the calculator's memory, so we need not worry about writing it down. To complete the check, we clear the calculator and evaluate the left side of the equation:

$\boxed{2}\ \boxed{\times}\ \boxed{1}\ \boxed{0}\ \boxed{.}\ \boxed{7}\ \boxed{=}$.

To show that this result matches the number stored earlier, we do not clear the display, but instead subtract the number stored:

$\boxed{-}\ \boxed{\text{RCL}}\ \boxed{=}$.

The result, 0, indicates that 2×10.7 and $4 \times 7.35 - 8$ are equal. A result other than 0 would indicate that the ordered pair in question does not check.

As always, keystrokes may vary, so consult your owner's manual if the above keystrokes do not work for your calculator.

Exercises: Determine whether each point is a solution of the given equation.

1. $(7.9, 3.2)$; $5x + 4y = 52.3$

2. $(1.9, 2.3)$; $7x - 8y = 5.1$

3. $(4.3, 4.75)$; $5y = 6x - 7$

4. $(3.8, -4.3)$; $9a = 17 - 4b$

5. $(9.4, -3.9)$; $3a - 15 = 29 + 4b$

6. $(5.6, 8.8)$; $4y + 23 = 7x + 19$

7. $(-2.4, 8.5625)$; $3.5x + 17.4 = 3.2y - 18.4$

8. $(1.8, 2.6)$; $9.2x - 15.3 = 4.8y - 13.7$

CHAPTER 6: Introduction to Graphing and Statistics

a   Plot each group of points on the given graph below.

1. $(4,4)$ $(-2,4)$ $(5,-3)$ $(-5,-5)$ $(0,4)$ $(0,-4)$
$(3,0)$ $(-4,0)$

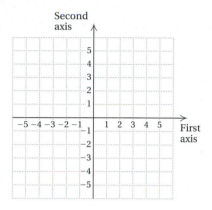

2. $(2,5)$ $(-1,3)$ $(3,-2)$ $(-2,-4)$ $(0,4)$ $(0,-5)$
$(5,0)$ $(-5,0)$

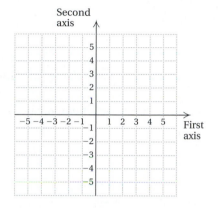

3. $(-2,-4)$ $(5,-4)$ $\left(0,3\frac{1}{2}\right)$ $\left(4,3\frac{1}{2}\right)$ $(-1,-3)$ $(-1,5)$
$(4,-1)$ $(-2,0)$

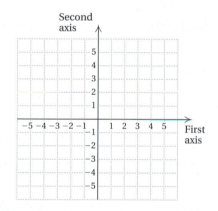

4. $(-3,-1)$ $(5,1)$ $(-1,-5)$ $(0,0)$ $(0,1)$ $(-4,0)$
$\left(2,3\frac{1}{2}\right)$ $\left(4\frac{1}{2},-2\right)$

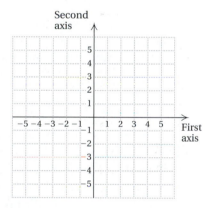

Determine the coordinates of points A, B, C, D, E, and F.

5.

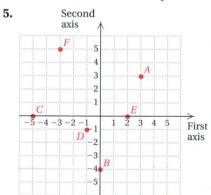

6.

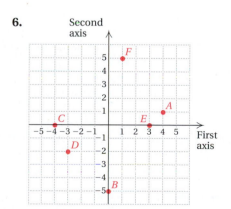

7.

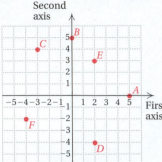

8.

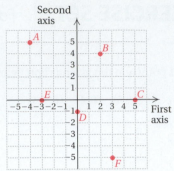

b In which quadrant is each point located?

9. $(-5, 3)$

10. $(-12, 1)$

11. $(100, -1)$

12. $\left(35\frac{1}{2}, -2\frac{1}{2}\right)$

13. $(-6.5, -1.9)$

14. $(-3.4, -5.9)$

15. $\left(3\frac{7}{10}, 9\frac{1}{11}\right)$

16. $(1895, 1492)$

Complete each sentence using the words *positive* or *negative* or the numerals I, II, III, or IV.

17. In quadrant IV, first coordinates are always _____ and second coordinates are always _____.

18. In quadrant III, first coordinates are always _____ and second coordinates are always _____.

19. In quadrant _____, both coordinates are always negative.

20. In quadrant _____, both coordinates are always positive.

21. In quadrants I and _____, the first coordinate is always _____.

22. In quadrants II and _____, the second coordinate is always _____.

c Determine whether each ordered pair is a solution of the given equation.

23. $(4, 3)$; $\quad y = 2x - 5$

24. $(1, 7)$; $\quad y = 2x + 5$

25. $(2, -3)$; $\quad 3x - y = 4$

26. $(-1, 4)$; $\quad 2x + y = 6$

27. $(-2, -1)$; $\quad 3c + 2d = -8$

28. $(0, -4)$; $\quad 4p + 2q = -9$

29. $(5, -4)$; $3x + y = 19$

30. $(-1, 7)$; $x - y = -8$

31. $\left(2\dfrac{1}{3}, 6\right)$; $2q - 3p = 3$

32. $\left(3, 1\dfrac{1}{4}\right)$; $2p - 4q = 1$

33. $(2.4, 0.7)$; $y = 5x - 11.3$

34. $(1.8, 7.4)$; $y = 3x + 2$

35. $\mathbf{D_W}$ Under what conditions will the points (a, b) and (b, a) be in the same quadrant?

36. $\mathbf{D_W}$ Describe in your own words how to plot the point (a, b).

SKILL MAINTENANCE

Solve.

37. $3x - 4 = 17$ [2.8d]

38. $7 + 2x = 25$ [2.8d]

39. $5(x - 2) = 3x - 4$ [5.7b]

40. $\dfrac{3}{7}t - 4 = 2$ [4.4a]

41. $-\dfrac{1}{9}t = \dfrac{2}{3}t$ [5.7b]

42. Simplify: $\dfrac{90}{51}$.   [3.5b]

43. Combine like terms: [4.6b]
$$7\dfrac{2}{11}a - 5\dfrac{1}{3}a.$$

44. Simplify: [2.7a]
$$3(x - 5) + 4x - 9.$$

SYNTHESIS

45. $\mathbf{D_W}$ Write an equation for which $(-2, 5)$ is a solution and explain how you found the equation.

46. $\mathbf{D_W}$ In which quadrant, if any, is the point $(5, 0)$? Why?

Determine whether each ordered pair is a solution of the given equation.

47. ▦ $(-2.37, 1.23)$; $5.2x + 6.1y = -4.821$

48. ▦ $(4.16, -9.35)$; $6.5x - 7.2y = -94.35$

421

In Exercises 49–52, determine the quadrant(s) in which the point could be located.

49. The first coordinate is positive.

50. The second coordinate is negative.

51. The first and second coordinates are equal.

52. The first coordinate is the opposite of the second coordinate.

53. The points $(-1, 1)$, $(4, 1)$, and $(4, -5)$ are three vertices of a rectangle. Find the coordinates of the fourth vertex.

54. A parallelogram is a four-sided polygon with two pairs of parallel sides (two examples are shown below). Three parallelograms share the vertices $(-2, -3)$, $(-1, 2)$, and $(4, -3)$. Find the fourth vertex of each parallelogram.

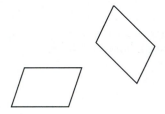

55. Graph eight points such that the sum of the coordinates in each pair is 7.

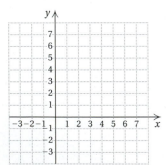

56. Graph eight points such that for each point the first coordinate minus the second coordinate is 3.

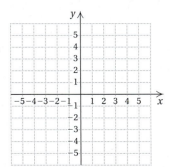

57. Find the perimeter of a rectangle with vertices at $(5, 3)$, $(5, -2)$, $(-3, -2)$, and $(-3, 3)$.

58. Find the area of a rectangle with vertices at $(0, 9)$, $(0, -4)$, $(5, -4)$, and $(5, 9)$.

6.4

GRAPHING LINEAR EQUATIONS

Objectives

a Find solutions of equations in two variables.

b Graph linear equations in two variables.

c Graph equations for horizontal or vertical lines.

In Section 6.3, we saw how to determine whether an ordered pair is a solution of an equation in two variables. We now develop a way of finding such solutions on our own. Once we can find a few ordered pairs that solve an equation, we will be able to graph the equation.

a   Finding Solutions

To solve an equation with one variable, like $3x + 2 = 8$, we isolate the variable, x, on one side of the equation. To solve an equation with two variables, we will first replace one variable with some number choice and then solve the resulting equation.

EXAMPLE 1   Find solutions of $x + y = 7$ for $x = 5$ and for $x = -1$.

If x is 5, then $x + y = 7$ can be rewritten as

$$5 + y = 7.$$

We solve as follows:

$$5 + y = 7$$
$$5 + y - 5 = 7 - 5 \qquad \text{Subtracting 5 from both sides}$$
$$y = 2.$$

The ordered pair $(5, 2)$ is a solution of $x + y = 7$. We leave it for the student to show that $(-1, 8)$ is another solution of $x + y = 7$.

Do Exercise 1.

EXAMPLE 2   Complete these solutions of $2x + 3y = 8$: $(\square, 2)$; $(-2, \square)$.

To complete the pair $(\square, 2)$, we replace y with 2 and solve for x:

$$2x + 3y = 8$$
$$2x + 3 \cdot 2 = 8 \qquad \text{Substituting 2 for } y$$
$$2x + 6 = 8$$
$$2x + 6 - 6 = 8 - 6 \qquad \text{Subtracting 6 from both sides}$$
$$2x = 2$$
$$\tfrac{1}{2} \cdot 2x = \tfrac{1}{2} \cdot 2 \qquad \text{Multiplying both sides by } \tfrac{1}{2}$$
$$x = 1.$$

Thus, $(1, 2)$ is a solution of $2x + 3y = 8$. The check is left for the student.

To complete the pair $(-2, \square)$, we replace x with -2 and solve for y:

$$2x + 3y = 8$$
$$2(-2) + 3y = 8 \qquad \text{Substituting } -2 \text{ for } x$$
$$-4 + 3y = 8$$
$$3y = 12 \qquad \text{Adding 4 to both sides}$$
$$y = 4. \qquad \text{Dividing both sides by 3}$$

Thus, $(-2, 4)$ is also a solution of $2x + 3y = 8$. Again we leave the check for the student.

Do Exercise 2.

1. Find a solution of $x - y = 3$. Let $y = 5$.

2. Complete these solutions of $5x + y = 10$: $(1, \square)$; $(\square, -5)$.

Answers on page A–17

Study Tips

DOES MORE THAN ONE SOLUTION EXIST?

Keep in mind that many problems—in math and elsewhere—have more than one solution. When asked to find *a* solution of an equation containing two variables, there is usually more than one solution.

3. Find three solutions of $x + 2y = 7$. Answers may vary.

EXAMPLE 3 Find three solutions of $2x - y = 5$.

We are free to use *any* number as a replacement for either x or y. To find one solution, we select 1 as a replacement for x. We then solve for y:

$$2x - y = 5$$
$$2 \cdot 1 - y = 5 \qquad \text{Substituting 1 for } x$$
$$2 - y = 5$$
$$-y = 3 \qquad \text{Subtracting 2 from both sides}$$
$$-1y = 3 \qquad \text{Recall that } -a = -1 \cdot a.$$
$$y = -3. \qquad \text{Dividing both sides by } -1$$

Thus, $(1, -3)$ is one solution of $2x - y = 5$.

To find a second solution, we choose to replace y with 0 and solve for x:

$$2x - y = 5$$
$$2x - 0 = 5 \qquad \text{Substituting 0 for } y$$
$$2x = 5 \qquad \text{Simplifying}$$
$$x = 2.5. \qquad \text{Dividing both sides by 2}$$

Thus, $(2.5, 0)$ is a second solution of $2x - y = 5$.

To find a third solution, we can replace x with 0 and solve for y:

$$2x - y = 5$$
$$2 \cdot 0 - y = 5 \qquad \text{Substituting 0 for } x$$
$$0 - y = 5$$
$$-y = 5$$
$$-1y = 5 \qquad \text{Try to do this step mentally.}$$
$$y = -5. \qquad \text{Dividing both sides by } -1$$

The pair $(0, -5)$ is a third solution of $2x - y = 5$.

Note that three different choices for x or y would have given three different solutions. There is an infinite number of ordered pairs that are solutions, so it is unlikely for two students to have solutions that match entirely.

4. Find three solutions of $y = -2x + 7$. Answers may vary.

To find a solution of an equation with two variables:

1) Choose a replacement for one variable.
2) Solve for the other variable.
3) Write the solution as an ordered pair.

Do Exercises 3 and 4.

b   Graphing Equations

Equations like those considered in Examples 1–3 are in the form $Ax + By = C$. All equations that can be written this way are said to be **linear** because the solutions of each equation, when graphed, form a straight line. An equation $Ax + By = C$ is called the **standard form** of a linear equation. When the appropriate line is drawn, we say that we have *graphed* the equation.

EXAMPLE 4 Graph: $2x - y = 5$.

In Example 3, we found that $(1, -3)$, $(2.5, 0)$, and $(0, -5)$ are solutions of $2x - y = 5$. Had we not known that, before graphing we would need to calculate two or three solutions, just as we did in Example 3.

Next, we plot the points and look for a pattern. As expected, the points describe a straight line. We draw the line and label it with the equation as shown on the right below.

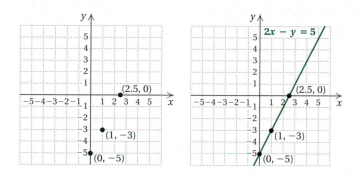

Note that two points are enough to determine a line, but if either point is calculated incorrectly, an incorrect line will be drawn. For this reason, we generally calculate and graph at least three ordered pairs before drawing each line. If the points do not all line up, we know that a mistake has been made.

Do Exercise 5.

Linear equations are not always written in standard form. Equations like $y = 2x$ or $y = x + 2$ are also linear. When a linear equation is written in standard form, as in Example 4, it is convenient to calculate two of the points by choosing 0 for x, and then 0 for y. To find solutions of equations like $y = 2x$, we usually choose values for x and then calculate y.

EXAMPLE 5 Graph: $y = 2x$.

First, we find some ordered pairs that are solutions. To find three ordered pairs, we can choose any three values for x and then calculate the corresponding values for y. One good choice is 0; we also choose -2 and 3.

If x is 0, then $y = 2x = 2 \cdot 0 = 0$.   Thus, $(0, 0)$ is a solution.

If x is -2, then $y = 2x = 2(-2) = -4$.   Thus, $(-2, -4)$ is a solution.

If x is 3, then $y = 2x = 2 \cdot 3 = 6$.   Thus, $(3, 6)$ is a solution.

We can compute additional pairs if we wish and list the ordered pairs that are solutions in a table.

x	y $y = 2x$	(x, y)
3	6	$(3, 6)$
-2	-4	$(-2, -4)$
0	0	$(0, 0)$
1	2	$(1, 2)$

↑ Substitute for x.
↑ Compute the value of y.
↑ Form the ordered pair (x, y).
Plot the points. ⟶
Draw and label the graph. ⟶

Next, we plot these points. We draw the line, or graph, with a ruler and label it $y = 2x$.

Do Exercises 6 and 7.

5. Graph $x + 2y = 7$. Use the results from Margin Exercise 3.

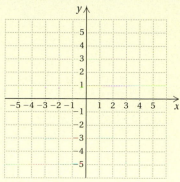

Graph.

6. $y = 3x$

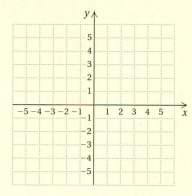

7. $y = \dfrac{1}{2}x$

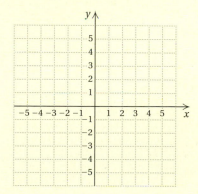

Answers on page A–17

Graph.

8. $y = -x$ (or $y = -1 \cdot x$)

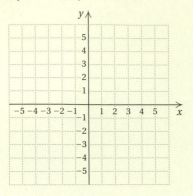

9. $y = -2x$

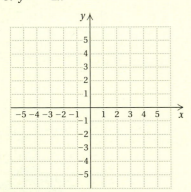

Answers on page A–17

■ **EXAMPLE 6** Graph: $y = -3x$.

We make a table of solutions. Then we plot the points, draw the line with a ruler, and label the line $y = -3x$.

If x is 0, then $y = -3 \cdot 0 = 0$.
If x is 1, then $y = -3 \cdot 1 = -3$.
If x is -2, then $y = -3(-2) = 6$.
If x is 2, then $y = -3 \cdot 2 = -6$.

x	y = −3x	(x, y)
0	0	(0, 0)
1	−3	(1, −3)
−2	6	(−2, 6)
2	−6	(2, −6)

Don't forget the label!

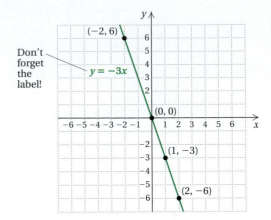

Do Exercises 8 and 9.

■ **EXAMPLE 7** Graph: $y = x + 2$.

We make a table of solutions. Then we plot the points, draw the line with a ruler, and label it.

If x is 0, then $y = 0 + 2 = 2$.
If x is 1, then $y = 1 + 2 = 3$.
If x is -1, then $y = -1 + 2 = 1$.
If x is 3, then $y = 3 + 2 = 5$.

x	y = x + 2	(x, y)
0	2	(0, 2)
1	3	(1, 3)
−1	1	(−1, 1)
3	5	(3, 5)

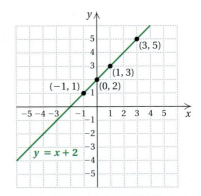

The values of x in these examples were *chosen*. Different choices for x would yield different points, but the same line.

For linear equations, tables can be formed using any numbers for x.

EXAMPLE 8 Graph: $y = \frac{2}{3}x + 1$.

We make a table of solutions, plot the points, and draw and label the line. For this table, we selected multiples of 3 as x-values to avoid fraction values for y.

If x is 6, then $y = \frac{2}{3} \cdot 6 + 1 = 4 + 1 = 5$.

If x is 3, then $y = \frac{2}{3} \cdot 3 + 1 = 2 + 1 = 3$.

If x is 0, then $y = \frac{2}{3} \cdot 0 + 1 = 0 + 1 = 1$.

If x is -3, then $y = \frac{2}{3}(-3) + 1 = -2 + 1 = -1$.

x	$y = \frac{2}{3}x + 1$	(x, y)
6	5	$(6, 5)$
3	3	$(3, 3)$
0	1	$(0, 1)$
-3	-1	$(-3, -1)$

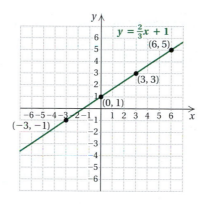

Do Exercises 10–12.

C   Graphing Horizontal or Vertical Lines

We have already stated that any equation in the form $Ax + By = C$ is linear, provided A and B are not both zero. If A is 0 and B is nonzero, there is no x-term and the graph is a horizontal line. If B is 0 and A is nonzero, there is no y-term and the graph is a vertical line. In Examples 9 and 10, we consider both of these possibilities.

EXAMPLE 9 Graph: $y = 3$.

We regard $y = 3$ as $0 \cdot x + y = 3$. No matter what number we choose for x, we find that y must be 3 if the equation is to be solved.

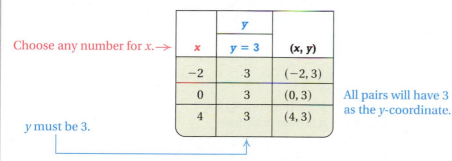

Choose any number for x. →

y must be 3.

All pairs will have 3 as the y-coordinate.

When we plot $(-2, 3)$, $(0, 3)$, and $(4, 3)$ and connect the points, we obtain a horizontal line. Any ordered pair of the form $(x, 3)$ is a solution, so the line is 3 units above the x-axis, as shown in the graph at the top of the next page.

Graph.

10. $y = x + 1$

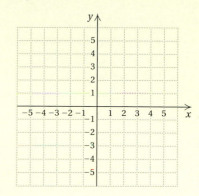

11. $y = -2x + 1$

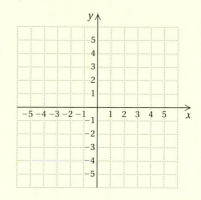

12. $y = \frac{3}{5}x$

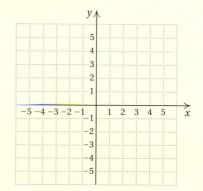

Answers on page A–17

Graph.

13. $y = 4$

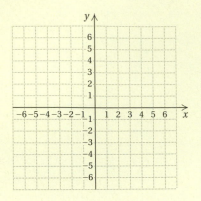

14. $x = 5$

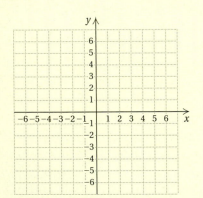

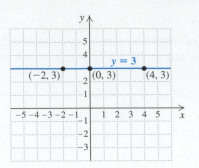

EXAMPLE 10 Graph: $x = -4$.

We regard $x = -4$ as $x + 0 \cdot y = -4$ and make a table with all -4's in the x-column.

x must be -4. →

x		
$x = -4$	y	(x, y)
-4	-5	$(-4, -5)$
-4	1	$(-4, 1)$
-4	3	$(-4, 3)$

All pairs will have -4 as the x-coordinate.

Choose any number for y.

When we plot $(-4, -5)$, $(-4, 1)$, and $(-4, 3)$ and connect them, we obtain a vertical line. Any ordered pair of the form $(-4, y)$ is a solution, so the line is 4 units left of the y-axis.

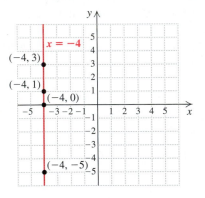

Do Exercises 13 and 14.

The graph of $y = b$ is a horizontal line.
The graph of $x = a$ is a vertical line.

Answers on page A–17

428

Graphing Equations   Calculators or computers with graphing capability have become increasingly common. This technology is generally used for graphing equations that are more complicated than $y = x + 2$ and $y = \frac{2}{3}x + 1$ (Examples 7 and 8) and *in no way decreases the importance of understanding how such equations are graphed by hand.* The purpose of the following discussion is to show how a graphing calculator can be used to check your work and how it might enable you to handle more challenging problems.

All graphing calculators utilize a window, the rectangular portion of the screen in which a graph appears. For now, the "Standard" window extending from -10 to 10 on both the *x*- and *y*-axes will suffice. The standard window is usually selected from the Zoom menu.

To graph $y = x + 2$, we press a key (usually labeled $\boxed{Y=}$) and then

$$\boxed{\text{X, T, } \theta\text{, n}} \ \ \boxed{+} \ \ \boxed{2} \ \ \boxed{\text{GRAPH}}$$

(keystrokes may vary with the calculator used). A graph similar to that shown on the left below should appear. To view some of the ordered pairs that are solutions, a TRACE key can be used to move a cursor along the line. Near the bottom of the window the cursor's coordinates appear (see the graph on the right below).

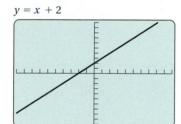

$y = x + 2$

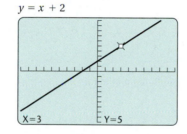

$y = x + 2$

Exercises: Use graphing technology to graph each of the following.

1. $y = \frac{2}{3}x + 1$ (Example 8)

2. $y = x + 1$ (Margin Exercise 10)

3. $y = -2x + 1$ (Margin Exercise 11)

4. $y = \frac{3}{5}x$ (Margin Exercise 12)

a   For each equation, use the indicated value to find an ordered pair that is a solution.

1. $x + y = 8$; let $x = 5$

2. $x + y = 5$; let $x = 4$

3. $2x + y = 7$; let $x = 3$

4. $x + 2y = 9$; let $y = 4$

5. $y = 3x - 1$; let $x = 5$

6. $y = 2x + 7$; let $x = 3$

7. $x + 3y = 1$; let $x = 10$

8. $5x + y = 7$; let $y = -8$

9. $2x + 5y = 17$; let $x = 1$

10. $5x + 2y = 19$; let $x = 1$

11. $3x - 2y = 8$; let $y = -1$

12. $2x - 5y = 12$; let $y = -2$

For each equation, complete the given ordered pairs.

13. $x + y = 4$; ($\square$, 3); (-1, $\square$)

14. $x - y = 6$; ($\square$, 2); (9, $\square$)

15. $x - y = 4$; ($\square$, 3); (10, $\square$)

16. $x + y = 10$; ($\square$, 8); (3, $\square$)

17. $2x + 3y = 15$; (3, $\square$); ($\square$, 1)

18. $3x + 2y = 16$; (4, $\square$); ($\square$, -1)

19. $3x + 5y = 14$; $(3, \square)$; $(\square, 4)$

20. $4x + 3y = 11$; $(5, \square)$; $(\square, 2)$

21. $y = 4x$; $(\square, 4)$; $(-2, \square)$

22. $y = 6x$; $(\square, 6)$; $(-2, \square)$

23. $2x + 5y = 3$; $(0, \square)$; $(\square, 0)$

24. $5x + 7y = 9$; $(0, \square)$; $(\square, 0)$

For each equation, find three solutions. Answers may vary.

25. $x + y = 9$

26. $x + y = 19$

27. $y = 4x$

28. $y = 5x$

29. $3x + y = 13$

30. $x + 5y = 12$

31. $y = 3x - 1$

32. $y = 2x + 5$

33. $y = -7x$

34. $y = -4x$

35. $4 + y = x$

36. $3 + y = x$

37. $3x + 2y = 12$

38. $2x + 3y = 18$

39. $y = \frac{1}{3}x + 2$

40. $y = \frac{1}{2}x + 5$

b Graph each equation.

41. $x + y = 6$

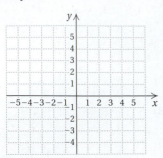

42. $x + y = 4$

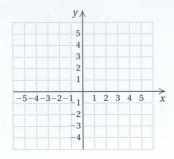

43. $x - 1 = y$

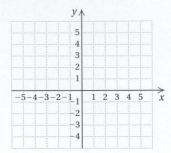

44. $x - 2 = y$

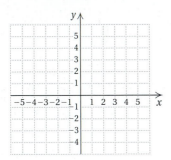

45. $y = x - 4$

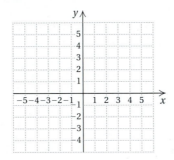

46. $y = x - 5$

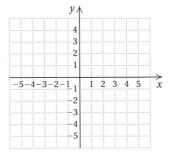

47. $y = \dfrac{1}{3}x$

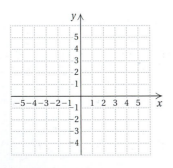

48. $y = -\dfrac{1}{3}x$

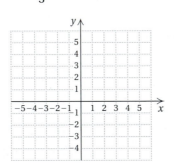

49. $y = x$

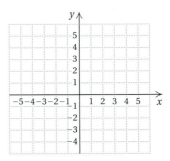

50. $y = x - 3$

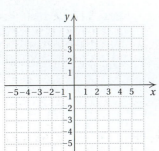

51. $y = 2x - 1$

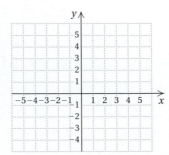

52. $y = 2x - 3$

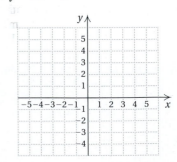

53. $y - 2x + 1$

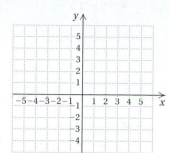

54. $y = 3x + 1$

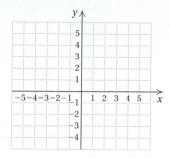

55. $y = \dfrac{2}{5}x$

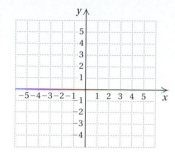

56. $y = \dfrac{3}{4}x$

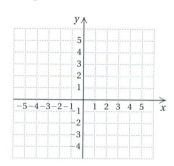

57. $y = -x + 4$

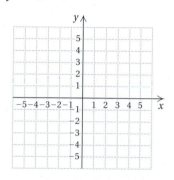

58. $y = -x + 5$

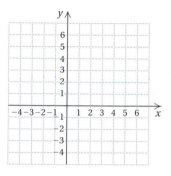

59. $y = \dfrac{2}{3}x + 1$

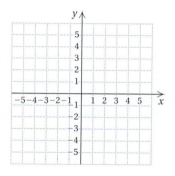

60. $y = \dfrac{2}{5}x - 1$

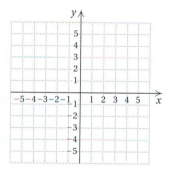

C   Graph.

61. $y = 2$

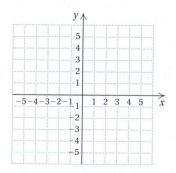

62. $y = 1$

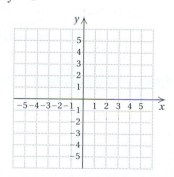

63. $x = 2$

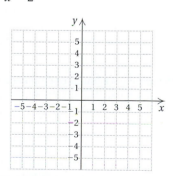

64. $x = 3$

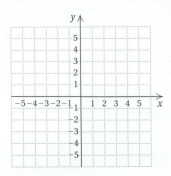

65. $x = -3$

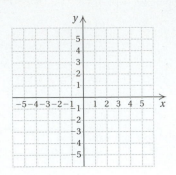

66. $x = -1$

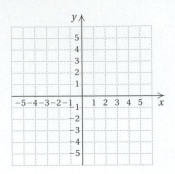

67. $y = -4$

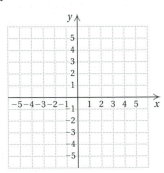

68. $y = -2$

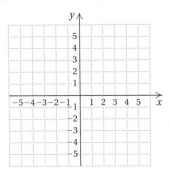

69. ^Dw To graph a linear equation, a student plots three points and discovers that the points do not line up with each other. What should the student do next?

70. ^Dw In Example 8, we found that by choosing multiples of 3 for x, we could avoid fractions. What is the advantage of avoiding fractions? Would it have been incorrect to substitute values for x that are *not* multiples of 3? Why or why not?

SKILL MAINTENANCE

71. The tunes on Miles Davis's classic *Kind of Blue* album are approximately 9 min, $9\frac{1}{2}$ min, $5\frac{1}{2}$ min, $11\frac{1}{2}$ min, and $9\frac{1}{2}$ min long. Find the average length of a tune on that album. [4.5c], [4.6c]

72. The books on Sherry's nightstand are 243, 410, 352, and 274 pages long. What is the average length of a book on the nightstand? [5.4b]

73. A recipe for a batch of chili calls for $\frac{3}{4}$ cup of red wine vinegar. How much vinegar is needed to make $2\frac{1}{2}$ batches of chili? [4.7d]

Simplify.

74. $-\dfrac{49}{77}$ [3.5b]

75. $-8 - 5^2 \cdot 2(3 - 4)$ [2.5b]

76. $\dfrac{3}{10}\left(-\dfrac{25}{12}\right)$ [3.4b]

Solve.

77. $4.8 - 1.5x = 0.9$ [5.7a]

78. $3x - 8 = 5x - 12$ [5.7b]

79. D_W Apart from graphing each equation, how can someone determine that the graphs of $14x + 21y = 63$ and $10x + 15y = 45$ are the same?

80. D_W What is the greatest number of quadrants that a line can pass through? Why?

Find three solutions of each equation. Then graph the equation.

81. ▦ $21x - 70y = -14$

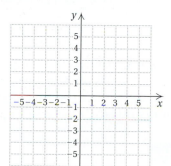

82. ▦ $25x + 80y = 100$

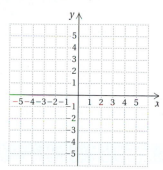

83. ▦ $50x + 75y = 180$

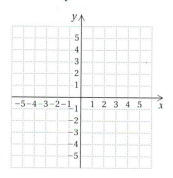

84. Use the graph in Example 4 to find three solutions of $2x - y = 5$. Do not use the ordered pairs already listed.

85. Use the graph in Example 7 to find three solutions of $y = x + 2$. Do not use the ordered pairs already listed.

86. List all solutions of $x + y = 6$ that use only whole numbers.

87. Graph three solutions of $y = |x|$ in the second quadrant and another three solutions in the first quadrant.

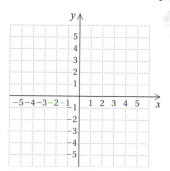

To the Student and the Instructor: Exercises marked with a ⩗ symbol are meant to be solved using a graphing calculator.

⩗ If a graphing calculator is available, use it to graph each of the following.

88. $y = -0.63x + 2.8$

89. $y = 2.3x - 4.1$

MEANS, MEDIANS, AND MODES

Objectives

a Find the mean of a set of numbers and solve applied problems involving means.

b Find the median of a set of numbers and solve applied problems involving medians.

c Find the mode of a set of numbers and solve applied problems involving modes.

d Compare two sets of data using their means.

Find the mean.

1. 12, 15, 27

2. 10.5, 9.5, 8.2, 7.2

3. Wendy scored the following on five tests: 96, 85, 82, 74, 68. What is her mean score?

4. In a five-game series, Antonio scored 26, 21, 13, 14, and 23 points. Find the mean number of points scored per game.

5. Home-Run Batting Average. At the end of the 2005 baseball season, Barry Bonds had the most career home runs of any active player in the major leagues, 708 in 20 seasons. What was his mean number of home runs per season?
Source: Major League Baseball

Answers on page A–18

Pictographs, bar graphs, and line graphs provide three ways of representing a collection of data *visually*. Sometimes it is useful to describe a set of data *numerically,* using *statistics.* A **statistic** is simply a number that is derived from a set of data. There are three statistics used as *center points* or *measures of central tendency.* These are numbers that serve to represent the entire data set. Let's examine all three.

a Means

The most commonly used center point is the *average* of the set of numbers. We have already computed an average several times in this book (see pages 85, 292, and 342). Although the word "average" is often used in everyday speech, in math we generally use the word *mean* instead.

> **MEAN**
>
> To find the **mean** of a set of numbers, add the numbers and then divide by the number of items of data.

EXAMPLE 1   On a 4-day trip, a car was driven the following number of miles each day: 240, 302, 280, 320. What was the mean number of miles per day?

$$\frac{240 + 302 + 280 + 320}{4} = \frac{1142}{4}, \quad \text{or} \quad 285.5$$

The car was driven an average of 285.5 mi per day. Had the car been driven exactly 285.5 mi each day, the same total distance (1142 mi) would have been traveled.

Do Exercises 1–4.

EXAMPLE 2   *Scoring Average.* Kareem Abdul-Jabbar is the all-time leading scorer in the history of the National Basketball Association. He scored 38,387 points in 1560 games. What was the mean number of points scored per game? Round to the nearest tenth.
Source: National Basketball Association

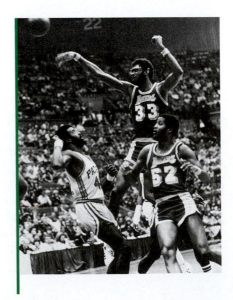

We already know the total number of points, 38,387, and the number of games, 1560. We divide and round to the nearest tenth.

$$\frac{38,387}{1560} = 24.60705\ldots \approx 24.6$$

Abdul-Jabbar's average was 24.6 points per game.

Do Exercise 5.

EXAMPLE 3 *Grade Point Average.* In most colleges, students are assigned grade point values for grades obtained. The **grade point average,** or **GPA,** is the average of the grade point values for each credit hour taken. At most colleges, grade point values are assigned as follows.

A: 4.0   B: 3.0   C: 2.0   D: 1.0   F: 0.0

Meg earned the following grades for one semester. What was her grade point average?

COURSE	GRADE	NUMBER OF CREDIT HOURS IN COURSE
Colonial History	B	3
Basic Mathematics	A	4
English Literature	A	3
French	C	4
Time Management	D	1

Because some of Meg's courses carried more credit than others, the grades in those courses carry more weight mathematically. We could regard Meg's B in history as three B's (since it is a three-credit course), her A in math as four A's, and so on for all 15 credits. Rather than add 15 single-credit grades, we will multiply and then add before finally dividing by the total number of credits:

Colonial History $3.0 \cdot 3 = 9$
Basic Mathematics $4.0 \cdot 4 = 16$
English Literature $4.0 \cdot 3 = 12$ Multiplying grade point values
French $2.0 \cdot 4 = 8$ (in color) by the number of credits
Time Management $1.0 \cdot 1 = 1$ for each course. Meg earned 46
                    ──────   *quality points.*
                    46 (Total)

The total number of credit hours taken is $3 + 4 + 3 + 4 + 1$, or 15. We divide 46 by 15 and round to the nearest tenth.

$$\text{GPA} = \frac{46}{15} \approx 3.1$$

Meg had a 3.1 grade point average.

Do Exercise 6.

EXAMPLE 4 *Grading.* To get a B in math, Geraldo must score a mean of 80 on the tests. On the first four tests, his scores were 79, 88, 64, and 78. What is the lowest score that Geraldo can get on the last test and still get a B?

An average of 80 is equivalent to scoring 80 on each test. Thus Geraldo needs a total score of $5 \cdot 80$, or 400.

The total of the scores on the first four tests is $79 + 88 + 64 + 78 = 309$.

Thus Geraldo needs to get at least $400 - 309$, or 91, in order to get a B. We can check this as follows.

$$\frac{79 + 88 + 64 + 78 + 91}{5} = \frac{400}{5}, \quad \text{or} \quad 80$$

Do Exercise 7.

6. Grade Point Average. Alex earned the following grades one semester.

GRADE	NUMBER OF CREDIT HOURS IN COURSE
B	3
C	4
C	4
A	2

What was Alex's grade point average? Assume that the grade point values are 4.0 for an A, 3.0 for a B, and so on. Round to the nearest tenth.

7. To receive an A in math, Rosa must have a mean test grade of at least 90 on four tests. On the first three tests, her scores were 80, 100, and 86. What is the lowest score that Rosa can get on the last test and still get an A?

Answers on page A–18

Find the median.

8. 17, 13, 18, 14, 19

9. 20, 14, 13, 19, 16, 18, 17

10. 78, 81, 83, 91, 103, 102, 122, 119, 88

Answers on page A–18

438

CALCULATOR CORNER

Computing Means   Means can be easily computed on a calculator if we remember the order in which operations are performed. For example, to calculate $\dfrac{85 + 92 + 79}{3}$ on most calculators, we press

$$\boxed{8}\boxed{5}\boxed{+}\boxed{9}\boxed{2}\boxed{+}\boxed{7}\boxed{9}\boxed{=}\boxed{\div}\boxed{3}\boxed{=},$$

or $\boxed{(}\boxed{8}\boxed{5}\boxed{+}\boxed{9}\boxed{2}\boxed{+}\boxed{7}\boxed{9}\boxed{)}\boxed{\div}\boxed{3}\boxed{=}.$

Exercises:

1. What would the result have been if we had not used parentheses in the latter sequence of keystrokes?

2. Use a calculator to solve Examples 1–4.

b   Medians

Another measure of central tendency is the *median*. Medians are useful when we wish to de-emphasize unusually extreme scores. For example, suppose a small class scored as follows on an exam.

Phil: 78	Pat: 56	Matt: 82
Jill: 81	Olga: 84	

Let's first list the scores in order from smallest to largest.

56, 78, **81**, 82, 84

↑
Middle score

The middle score—in this case, 81—is called the **median.** Note that because of the extremely low score of 56, the average of the scores is 76.2. In this example, the median may be more indicative of how the class as a whole performed.

EXAMPLE 5   What is the median of this set of numbers?

99, 870, 91, 98, 106, 90, 98

We first rearrange the numbers in order from smallest to largest. Then we locate the middle number, 98.

90, 91, 98, **98**, 99, 106, 870

↑
Middle number

The median is 98.

Do Exercises 8–10.

MEDIAN

Once a set of data is listed in order, from smallest to largest, the **median** is the middle number if there is an odd number of values. If there is an even number of values, the median is the number that is the average of the two middle numbers.

EXAMPLE 6 The salaries of the six instructors (one of whom is the owner) of the Belmont Ridge School are as follows.

$35,000, $29,000, $32,000, $31,000, $93,000, $30,000

What is the median salary at the school?

We rearrange the numbers in order from smallest to largest. The two middle numbers are $31,000 and $32,000. Thus the median is halfway between $31,000 and $32,000 (the average of $31,000 and $32,000).

$29,000, $30,000, $31,000, $32,000, $35,000 $93,000

$$\text{Median} = \frac{\$31,000 + \$32,000}{2} = \frac{\$63,000}{2} = \$31,500.$$

The average of the middle numbers is 31,500.

The median salary is $31,500.

Do Exercises 11 and 12.

In Example 6, the mean salary is $41,666.67, whereas the median salary is $31,500. If you were interviewing for a teaching job at Belmont Ridge and given a choice between being told the mean or the median salary, the median would probably give a better indication of what you would likely earn.

C   Modes

The final type of center-point statistic is the **mode**.

> **MODE**
>
> The **mode** of a set of data is the number or numbers that occur most often. If each number occurs the same number of times, there is no mode.

EXAMPLE 7 Find the mode of these data.

13,   14,   17,   17,   18,   19

The number that occurs most often is 17. Thus the mode is 17.

A set of data has just one mean and just one median, but it can have more than one mode. It may also have no mode—when all numbers are equally represented. For example, the set of data 5, 7, 11, 13, 19 has no mode.

EXAMPLE 8 Find the modes of these data.

33,   34,   34,   34,   35,   36,   37,   37,   37,   38,   39,   40

There are two numbers that occur most often, 34 and 37. Thus the modes are 34 and 37.

Do Exercises 13–16.

Which statistic is best for a particular situation? If someone is bowling, the *mean* from several games is a good indicator of that person's ability. If someone is buying a home, the *median* price for a neighborhood is often most indicative of what homes sell for there. Finally, if someone is reordering for a clothing store, the *mode* of the waist sizes sold is probably the most important statistic.

Find the median.

11. $1300, $2000, $3900, $1600, $1800, $1400

12. 68, 34, 67, 69, 58, 70

Find any modes that exist.

13. 23, 45, 45, 45, 78

14. 34, 34, 67, 67, 68, 70

15. 13, 24, 27, 28, 67, 89

16. In a lab, Gina determined the mass, in grams, of each of five eggs.

15 g, 19 g, 19 g, 14 g, 18 g

a) What is the mean?

b) What is the median?

c) What is the mode?

Answers on page A–18

17. Growth of Wheat. Rudy experiments to see which of two kinds of wheat is better. (In this situation, the shorter wheat is considered "better.") He grows both kinds under similar conditions and measures stalk heights, in inches, as follows. Which kind is better?

WHEAT A STALK HEIGHTS (in inches)			
16.2	42.3	19.5	25.7
25.6	18.0	15.6	41.7
22.6	26.4	18.4	12.6
41.5	13.7	42.0	21.6

WHEAT B STALK HEIGHTS (in inches)			
19.7	18.4	19.7	17.2
19.7	14.6	32.0	25.7
14.0	21.6	42.5	32.6
22.6	10.9	26.7	22.8

d Comparing Two Sets of Data

We have seen how to calculate means, medians, and modes from data. A way to analyze two sets of data is to make a determination by comparing the means.

EXAMPLE 9 *Battery Testing.* An experiment is performed to compare battery quality. Two kinds of battery were tested to see how long, in hours, they kept a portable CD player running. On the basis of this test, which battery is better?

BATTERY A: ETERNREADY TIMES (in hours)			BATTERY B: STURDYCELL TIMES (in hours)		
27.9	28.3	27.4	28.3	27.6	27.8
27.6	27.9	28.0	27.4	27.6	27.9
26.8	27.7	28.1	26.9	27.8	28.1
28.2	26.9	27.4	27.9	28.7	27.6

Note that it is difficult to analyze the data at a glance because the numbers are close together. We need a way to compare the two groups. Let's compute the mean of each set of data.

Battery A: Mean

$$= \frac{27.9 + 28.3 + 27.4 + 27.6 + 27.9 + 28.0 + 26.8 + 27.7 + 28.1 + 28.2 + 26.9 + 27.4}{12}$$

$$= \frac{332.2}{12} \approx 27.68$$

Battery B: Mean

$$= \frac{28.3 + 27.6 + 27.8 + 27.4 + 27.6 + 27.9 + 26.9 + 27.8 + 28.1 + 27.9 + 28.7 + 27.6}{12}$$

$$= \frac{333.6}{12} = 27.8$$

We see that the mean time of battery B is higher than that of battery A and thus conclude that battery B is "better." (It should be noted that statisticians might question whether these differences are what they call "significant." The answer to that question belongs to a later math course.)

Do Exercise 17.

Answer on page A–18

Translating for Success

1. *Perimeter.* A rectangular garden is 35 ft long and 27 ft wide. How many feet of fencing is needed to surround the garden?

2. *Craft Sale.* Marissa sold crocheted mittens at a craft sale. She sold 35 pairs on Friday and 27 pairs on Saturday. What was the mean number of pairs of mittens sold each day?

3. *Jogging.* David must jog 2 mi a day for training for basketball. If the alley behind his house is $\frac{1}{5}$ mi long, how many times must he run the length of the alley?

4. *Taxi Fare.* A taxi driver charges $2.25 plus 95 cents a mile. How far can Tonya travel on $11.75?

5. *Fudge.* Amber made $3\frac{1}{2}$ lb of fudge and divided it evenly among 3 packages for gifts. How much fudge was in each package?

The goal of these matching questions is to practice step (2), *Translate*, of the five-step problem-solving process. Translate each word problem to an equation and select a correct translation from equations A–O.

A. $x = 2 \div \frac{1}{5}$

B. $x = 4 + (-8)$

C. $x = 3 \cdot 3\frac{1}{2}$

D. $x = 4 - (-8)$

E. $35 + 27 = x$

F. $x = 2 \cdot 35 + 2 \cdot 27$

G. $x = 3\frac{1}{2} \div 3$

H. $x = 248 \div 15\frac{5}{10}$

I. $x = 3 + 3\frac{1}{2}$

J. $x = 2 \cdot \frac{1}{5}$

K. $35 \cdot 27 = x$

L. $2.25 + 0.95 + x = 11.75$

M. $248 \div x = 15\frac{5}{10}$

N. $x = \dfrac{35 + 27}{2}$

O. $2.25 + 0.95x = 11.75$

Answers on page A-18

6. *Temperature.* The weather forecast for the low temperature in Muncy on December 14 was 4°F. The actual low was −8°F. How many degrees lower than the forecast was the actual temperature?

7. *Pizza Recipe.* A recipe for pizza crust calls for $3\frac{1}{2}$ cups of flour. How many cups of flour are needed for 3 pizza crusts?

8. *Total Purchase.* Oliver spent $11.75 at the Country Store. He purchased candy for 95 cents, nuts for $2.25, and coffee. How much did the coffee cost?

9. *Mileage.* A car traveled 248 mi on $15\frac{5}{10}$ gal of gas. How many miles per gallon did it get?

10. *Cell Phone Minutes.* On Friday, Lathan made phone calls of lengths 35 min and 27 min. How many minutes did he use?

6.5 EXERCISE SET

For Extra Help

MathXL | MyMathLab | InterAct Math | Tutor Center | Video Lectures on CD Disc 3 | Student's Solutions Manual

a, b, c For each set of numbers, find the mean, the median, and any modes that exist.

1. 17, 19, 29, 18, 14, 29

2. 72, 83, 85, 88, 92

3. 5, 37, 20, 20, 35, 5, 25

4. 13, 32, 25, 27, 13

5. 4.3, 7.4, 1.2, 5.7, 7.4

6. 13.4, 13.4, 12.6, 42.9

7. 234, 228, 234, 229, 234, 278

8. $29.95, $28.79, $30.95, $29.95

9. *Tornadoes.* The following bar graph shows the average number of tornado deaths by month for the years 2003–2005. What is the mean number of tornado deaths for the 12 months? the median? the mode?

Average Number of Deaths by Tornado by Month

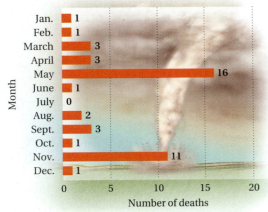

Month	Number of deaths
Jan.	1
Feb.	1
March	3
April	3
May	16
June	1
July	0
Aug.	2
Sept.	3
Oct.	1
Nov.	11
Dec.	1

Source: National Weather Service's Storm Prediction Center

10. *Basketball.* Lisa Leslie of the Los Angeles Sparks once scored 23, 21, 19, 23, and 20 points in consecutive games. What was the mean for the five games? the median? the mode?

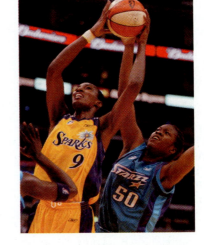

11. *Gas Mileage.* The Acura RSX Type-S gets 279 mi of highway driving on 9 gal of gasoline. What is the average number of miles expected per gallon—that is, what is its gas mileage?

Sources: EPA; *Car and Driver*, September 2005

12. *Gas Mileage.* The Chevrolet Cobalt SS Supercharged gets 322 mi of city driving on 14 gal of gasoline. What is the average number of miles expected per gallon—that is, what is its gas mileage?

Sources: EPA; *Car and Driver*, September 2005

Grade Point Average. The tables in Exercises 13 and 14 show the grades of a student for one semester. In each case, find the grade point average. Assume that the grade point values are 4.0 for an A, 3.0 for a B, and so on. Round to the nearest tenth.

13.

COURSE	GRADE	NUMBER OF CREDIT HOURS IN COURSE
Chemistry	B	4
Prealgebra	A	5
French I	D	3
Pastels	C	4

14.

COURSE	GRADE	NUMBER OF CREDIT HOURS IN COURSE
Botany	A	5
U.S. History	C	4
Drawing I	F	3
Basic Math	B	5

15. *Salmon Prices.* The following prices per pound of Atlantic salmon were found at five fish markets.

$6.99, $8.49, $8.99, $6.99, $9.49

What was the average price per pound? the median price? the mode?

16. *Cheddar Cheese Prices.* The following prices per pound of sharp cheddar cheese were found at five supermarkets.

$5.99, $6.79, $5.99, $6.99, $6.79

What was the average price per pound? the median price? the mode?

17. *Drinking and Driving.* Alcohol-related traffic deaths remain a national problem in the United States. Use the table below to determine the mean number of alcohol-related traffic deaths annually for the years 1995–2002.

YEAR	ALCOHOL-RELATED TRAFFIC DEATHS
1995	17,732
1996	17,673
1997	16,711
1998	16,673
1999	16,572
2000	17,380
2001	17,448
2002	17,419

Source: 2002 FARS Annual Report File, FHWA's Highway Statistics Annual Series, from the National Center for Statistics and Analysis

18. *Commuting Time.* Americans spend more and more time commuting to work. Use the table below to compute the mean commute time for the states listed.

STATE	COMMUTING TIME (in minutes each way)
Connecticut	23.6
Maine	22.6
Massachusetts	26.0
New Hampshire	24.6
Rhode Island	21.8
Vermont	20.3

19. *Grading.* To receive a B in math, Rich must average at least 80 on five tests. Scores on his first four tests were 80, 74, 81, and 75. What is the lowest score that Rich can get on the last test and still receive a B?

20. *Grading.* To receive an A in math, Sybil must average at least 90 on five tests. Scores on her first four tests were 90, 91, 81, and 92. What is the lowest score that Sybil can get on the last test and still receive an A?

21. *Length of Pregnancy.* Marta was pregnant 270 days, 259 days, and 272 days for her first three pregnancies. After her fourth pregnancy, Marta's average pregnancy was exactly the worldwide average of 266 days. How long was her fourth pregnancy?

Source: David Crystal (ed.), *The Cambridge Factfinder.* Cambridge CB2 1RP: Cambridge University Press, 2000, p. 90.

22. *Male Height.* Jason's brothers are 174 cm, 180 cm, 179 cm, and 172 cm tall. The average male is 176.5 cm tall. How tall is Jason if he and his brothers have an average height of 176.5 cm?

d Solve.

23. *Light-Bulb Testing.* An experiment was performed to compare the lives of two types of light bulb. Several bulbs of each type were tested and the results are listed in the following table. On the basis of this test, which bulb is better?

BULB A: HOTLIGHT TIMES (in hours)			BULB B: BRIGHTBULB TIMES (in hours)		
983	964	1214	979	1083	1344
1417	1211	1521	984	1445	975
1084	1075	892	1492	1325	1283
1423	949	1322	1325	1352	1432

24. *Cola Testing.* An experiment was conducted to determine which of two colas tasted better. Students drank each cola and gave it a rating from 1 to 10. The results are given in the following table. On the basis of this test, which cola tastes better?

COLA A: VERVCOLA				COLA B: COLA-COLA			
6	8	10	7	10	9	9	6
7	9	9	8	8	8	10	7
5	10	9	10	8	7	4	3
9	4	7	6	7	8	10	9

25. D_W People fishing in parts of California are forbidden from keeping any salmon that are less than 22 in. long. Kelly's catch of salmon averages 25 in. in length. How is it possible that the law was broken?

26. D_W How is it possible for a sports team to average the most games won per season over a 5-yr span without ever leading the league in games won?

SKILL MAINTENANCE

Multiply.

27. $14 \cdot 14$ [1.5a]

28. $\dfrac{2}{3} \cdot \dfrac{2}{3}$ [3.4b]

29. 1.4×1.4 [5.3a]

30. 1.414×1.414 [5.3a]

31. 12.86×17.5 [5.3a]

32. 222×0.5678 [5.3a]

33. $\dfrac{4}{5} \cdot \dfrac{3}{28}$ [3.6a]

34. $\dfrac{28}{45} \cdot \dfrac{3}{2}$ [3.6a]

Solve.   [5.8a]

35. A disc jockey charges a $40 setup fee and $50 an hour. How long can the disc jockey work for $165?

36. To rent a floor sander costs $15 an hour plus a $10 supply fee. For how long can the machine be rented if $100 has been budgeted for the sander?

SYNTHESIS

37. D_W One way the U.S. Bureau of the Census reports income is by giving a 3-yr mean of median incomes. Why would this statistic be chosen?

38. D_W The following is a list of the number of children in each family in a certain Glen View neighborhood: 0, 2, 3, 0, 5, 2, 2, 0, 0, 2, 0, 0. Explain why the mode might be the most indicative statistic for the number of children in a family.

Bowling Averages.   Bowling averages are always computed by rounding down to the nearest integer. For example, suppose a bowler gets a total of 599 for 3 games. To find the average, we divide 599 by 3 and drop the amount to the right of the decimal point.

$$\frac{599}{3} \approx 199.67 \qquad \text{The bowler's average is 199.}$$

39. ▦ If Frances bowls 4176 in 23 games, what is her average?

40. ▦ If Eric bowls 4621 in 27 games, what is his average?

41. *Hank Aaron.*   Hank Aaron averaged $34\frac{7}{22}$ home runs per year over a 22-yr career. After 21 yr, Aaron had averaged $35\frac{10}{21}$ home runs per year. How many home runs did Aaron hit in his final year?

42. Because of a poor grade on the fifth and final test, Chris's mean test grade fell from 90.5 to 84.0. What did Chris score on the fifth test? Assume that all tests are equally important.

43. *Price Negotiations.*   Amy offers $3200 for a used Ford Taurus advertised at $4000. The first offer from Jim, the car's owner, is to "split the difference" and sell the car for $(3200 + 4000) \div 2$, or $3600. Amy's second offer is to split the difference between Jim's offer and her first offer. Jim's second offer is to split the difference between Amy's second offer and his first offer. If this pattern continues and Amy accepts Jim's third (and final) offer, how much will she pay for the car?

44. The ordered set of data 18, 21, 24, a, 36, 37, b has a median of 30 and a mean of 32. Find a and b.

45. *Gas Mileage.*   The Honda Insight, a gas/electric hybrid car, averages 61 mpg in city driving and has a 10.5-gal gas tank. How much city driving can be done on $\frac{3}{4}$ of a tank of gas?

Source: Based on information from EPA and Honda Motors

46. D_W ▦ After bowling 15 games, Liz had an average of 207. In her 16th game, Liz bowled a 244 and raised her average to 210. Andrew also had a 207 average after 15 games, but needed to bowl a 255 in his 16th game in order to raise his average to 210. How is this possible?

6.6 PREDICTIONS AND PROBABILITY

a Making Predictions

Sometimes we use data to make predictions or estimates of missing data points. One process for doing so is called **interpolation.** Interpolation enables us to estimate missing "in-between values" on the basis of known information.

EXAMPLE 1 *Monthly Mortgage Payments.* When money is borrowed and then repaid in monthly installments, the payment amount increases as the number of payments decreases. The table below lists the size of a monthly payment when $110,000 is borrowed (at 9% interest) for various lengths of time. Use interpolation to estimate the monthly payment on a 35-yr loan.

YEARS	MONTHLY PAYMENT
5	$2283.42
10	1393.43
15	1115.69
20	989.70
25	923.12
30	885.08
35	?
40	848.50

To use interpolation, we first plot the points and look for a trend. It seems reasonable to draw a line between the points corresponding to 30 and 40. We can "zoom-in" to better visualize the situation. To estimate the second coordinate that is paired with 35, we trace a vertical line up from 35 to the graph and then left to the vertical axis. Thus we estimate the value to be 867. We can also estimate this value by averaging $885.08 and $848.50.

$$\frac{\$885.08 + \$848.50}{2} = \$866.79$$

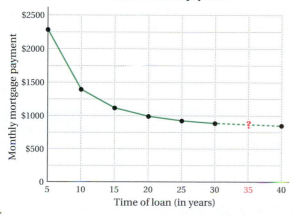

$110,000 Loan Repayment

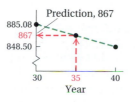

445

When we estimate in this manner to find an in-between value, we are *interpolating*. Real-world information about the data might tell us that an estimate found in this way is unreliable. For example, data from the stock market might be too erratic for interpolation.

Do Exercise 1.

We often analyze data with the intention of going "beyond" the data. One process for doing so is called **extrapolation.**

EXAMPLE 2 *Movies Released.* The data in the following table and graphs show the number of movies released over a period of years. Use extrapolation to estimate the number of movies released in 2006.

YEAR	MOVIES RELEASED
2000	502
2001	475
2002	473
2003	491
2004	558
2005	575

Source: www.the-numbers.com

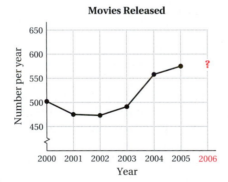 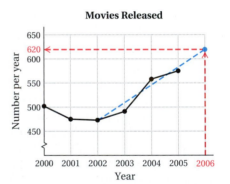

First, we analyze the data and note that they tend to increase from 2002 through 2005. Keeping this trend in mind, we draw a "representative" line through the data and beyond. To estimate a value for 2006, we draw a vertical line up from 2006 until it hits the representative line. We go to the left and read off a value—about 620. When we estimate in this way to find a "go-beyond value," we are *extrapolating*. Estimates found with this method vary depending on the "representative" line chosen.

Do Exercise 2.

In calculus and statistics, other methods of interpolation and extrapolation are developed. The two basic concepts remain unchanged, but more complicated methods of determining what line "best fits" the given data are used. Most of these methods involve use of a graphing calculator or computer software.

Answers on page A-19

b Probability

The predictions made in Examples 1 and 2 have a good chance of being reasonably accurate. A branch of mathematics known as *probability* is used to attach a numerical value to the likelihood that a specific event will occur.

Suppose we were to flip a coin. Because the coin is just as likely to land heads as it is to land tails, we say that the *probability* of it landing heads is $\frac{1}{2}$. Similarly, if we roll a die (plural: dice), we are as likely to roll a ⚀ as we are to roll a ⚁, ⚂, ⚃, ⚄, or ⚅. Because of this, we say that the probability of rolling a ⚄ is $\frac{1}{6}$.

EXAMPLE 3 A die is about to be rolled. Find the probability that a number greater than 4 will be rolled.

Since rolling a ⚀, ⚁, ⚂, ⚃, ⚄, or ⚅ are all equally likely to occur, and since two of these possibilities involve numbers greater than 4, we have

The probability of rolling a number greater than 4 $= \dfrac{2}{6}$ ← Number of ways to roll a 5 or 6
← Number of (equally likely) possible outcomes

$= \dfrac{1}{3}.$

The reasoning shown in Example 3 is used in a variety of applications.

EXAMPLE 4 A cloth bag contains 20 equally sized marbles: 5 are red, 7 are blue, and 8 are yellow. A marble is randomly selected. Find the probability that **(a)** a red marble is selected; **(b)** a blue marble is selected; **(c)** a yellow marble is selected.

a) Since all 20 marbles are equally likely to be selected, we have

The probability of selecting a red marble $= \dfrac{\text{Number of ways to select a red marble}}{\text{Number of ways to select any marble}}$

$= \dfrac{5}{20} = \dfrac{1}{4}$, or 0.25

b) The probability of selecting a blue marble $= \dfrac{\text{Number of ways to select a blue marble}}{\text{Number of ways to select any marble}}$

$= \dfrac{7}{20}$, or 0.35

c) The probability of selecting a yellow marble $= \dfrac{\text{Number of ways to select a yellow marble}}{\text{Number of ways to select any marble}}$

$= \dfrac{8}{20} = \dfrac{2}{5}$, or 0.4

Do Exercise 3.

3. A presentation of *Charlie and the Chocolate Factory* is attended by 100 people: 50 children, 20 seniors, and 30 (nonsenior) adults. After everyone has been seated, one audience member is selected at random. Find the probability of each of the following.

a) A child is selected.
b) A senior is selected.
c) A (nonsenior) adult is selected.

Answer on page A-19

4. A card is randomly selected from a well-shuffled deck of cards. Find the probability of each of the following.

a) The card is a diamond.

b) The card is a king or queen.

Many probability problems involve a standard deck of 52 playing cards. Such a deck is made up as shown below.

A deck of 52 cards

EXAMPLE 5 A card is randomly selected from a well-shuffled (mixed) deck of cards. Find the probability that **(a)** the card is a jack; **(b)** the card is a club.

a) The probability of selecting a jack
$$= \frac{\text{Number of ways to select a jack}}{\text{Number of ways to select any card}}$$
$$= \frac{4}{52} = \frac{1}{13}$$

b) The probability of selecting a club
$$= \frac{\text{Number of ways to select a club}}{\text{Number of ways to select any card}}$$
$$= \frac{13}{52} = \frac{1}{4}$$

Do Exercise 4.

In Examples 3–5, several "events" were discussed: rolling a number greater than 4, selecting a marble of a certain color, and selecting a certain type of playing card. The likelihood of each event occurring was determined by considering the total number of possible outcomes, using the principle formally stated below.

> **THE PRIMARY PRINCIPLE OF PROBABILITY**
>
> If an event E can occur m ways out of n equally likely possible outcomes, then
>
> $$\text{The probability of } E \text{ occurring} = \frac{m}{n}.$$

Answers on page A-19

a   Use interpolation or extrapolation to find the missing data values.

1. *Study Time and Grades.*   A math instructor asked her students to keep track of how much time each spent studying the chapter on decimal notation. They collected the information together with test scores from that chapter's test. The data are given in the following table. Estimate the missing value.

STUDY TIME (in hours)	TEST GRADE
9	75
11	93
13	80
15	85
16	85
17	80
19	?
21	86
23	91

2. *Maximum Heart Rate.*   A person's maximum heart rate depends on his or her gender, age, and resting heart rate. The following table relates resting heart rate and maximum heart rate for a 30-yr-old woman. Estimate the missing value.

RESTING HEART RATE (in beats per minute)	MAXIMUM HEART RATE (in beats per minute)
58	173
65	178
70	?
78	185
85	188

Source: American Heart Association

3. *Motion Picture Revenue.*

YEAR	AVERAGE AMOUNT GROSSED PER MOVIE (in millions of dollars)
2000	15.2
2001	17.0
2002	19.3
2003	18.4
2004	16.9
2005	?

Source: www.the-numbers.com

4. *Ozone Layer.*

YEAR	OZONE LEVEL (in Dobson Units)
2000	292.1
2001	290.4
2002	292.6
2003	287.9
2004	284.3
2005	?

Source: johnstonsarchive.net

5. *Major League Baseball Salaries.*

YEAR	AVERAGE SALARY OF MAJOR LEAGUE BASEBALL PLAYERS (in millions of dollars)
2000	2.0
2001	2.7
2002	2.4
2003	2.3
2004	2.5
2005	2.6
2006	?

Source: Based on information from http://www.baseball-almanac.com

6. *Women's Basketball.*

YEAR	NUMBER OF NCAA WOMEN'S COLLEGE BASKETBALL TEAMS
1999	864
2000	956
2001	958
2002	975
2003	1009
2004	1008
2005	?

Source: National College Athletic Association, Indianapolis, IN

449

7. *Farm Income.*

YEAR	NET FARM INCOME (in billions of dollars)
1998	45.6
1999	46.3
2000	47.9
2001	50.6
2002	?
2003	59.2

Source: Statistical Abstract of the United States, 2006

8. *Passenger Car Travel.*

YEAR	AVERAGE MILES TRAVELED PER VEHICLE (in thousands of miles)
1996	11.3
1997	11.6
1998	11.7
1999	?
2000	12.0
2001	11.8
2002	12.2
2003	12.3
2004	12.5

Source: U.S. Department of Transportation

b Find each of the following probabilities.

Rolling a Die. In Exercises 9–12, assume that a die is about to be rolled.

9. Find the probability that a ⚁ is rolled.

10. Find the probability that a ⚃ is rolled.

11. Find the probability that an odd number is rolled.

12. Find the probability that a number greater than 2 is rolled.

Playing Cards. In Exercises 13–18, assume that one card is randomly selected from a well-shuffled deck (see p. 448).

13. Find the probability that the card is the jack of spades.

14. Find the probability that the card is a picture card (jack, queen, or king).

15. Find the probability that an 8 or a 6 is selected.

16. Find the probability that a black 5 is selected.

17. Find the probability that a red picture card (jack, queen, or king) is selected.

18. Find the probability that a 10 is selected.

Candy Colors. Made by the Tootsie Industries of Chicago, Illinois, Mason Dots® is a gumdrop candy. A box was opened by the authors and found to contain the following number of gumdrops.

Strawberry	7
Lemon	8
Orange	9
Cherry	4
Lime	5
Grape	6

In Exercises 19–22, assume that one of the Dots is randomly chosen from the box.

19. Find the probability that a cherry Dot is selected.

20. Find the probability that an orange Dot is selected.

21. Find the probability that the Dot is *not* lime.

22. Find the probability that the Dot is *not* lemon.

23. **D_W** Would a company considering expansion be more interested in interpolation or extrapolation? Why?

24. **D_W** Would a bookkeeper who is lacking records from a firm's third year of operation be more interested in interpolation or extrapolation? Why?

SKILL MAINTENANCE

 VOCABULARY REINFORCEMENT

In each of Exercises 25–32, fill in the blank with the correct term from the given list. Some of the choices may not be used and some may be used more than once.

25. The set $\{1, 2, 3, 4, 5, \ldots\}$ is called the set of _____ numbers.   [1.1b]

26. The number $\frac{2}{5}$ is written in _____ notation, and the equivalent number 0.4 is written in _____ notation.   [3.3a], [5.1a]

27. To find the _____ of a set of numbers, add the numbers and then divide by the number of items of data.   [6.5a]

28. The decimal $0.\overline{1518}$ is an example of a(n) _____ decimal.   [5.5a]

29. Values in between known values can be estimated using _____.   [6.6a]

30. The statement $a(b + c) = ab + ac$ illustrates the _____ law.   [1.5a]

31. The statement $x + t = t + x$ illustrates the _____ law.   [1.2b]

32. Two perpendicular number lines used to graph ordered pairs are called _____.   [6.3a]

decimal

fraction

terminating

repeating

axes

quadrants

commutative

associative

distributive

mean

median

mode

natural

whole

interpolation

extrapolation

SYNTHESIS

33. **D_W** The answer given for Exercise 7 does not match the real-world figure, which was actually 37.3. What might account for such a discrepancy?

34. **D_W** Bluebird Building, Inc., had 23 employees in 1997, but only 18 employees in 1998, and 11 employees in 2000. Can extrapolation be used to predict the number of employees at Bluebird in 2010? Why or why not?

35. A coin is flipped twice. What is the probability that two heads will occur?

36. A coin is flipped twice. What is the probability that one head and one tail will occur?

37. A die is rolled twice. What is the probability that a ⚁ is rolled twice?

38. A day is chosen randomly during a leap year. What is the probability that the day is in July?

39. **D_W** Is it possible for the probability of an event occurring to exceed 1? Why or why not?

The review that follows is meant to prepare you for a chapter exam. It consists of two parts. The first part, Concept Reinforcement, is designed to increase understanding of the concepts through true/false exercises. The second part is the Review Exercises. These provide practice exercises for the exam, together with references to section objectives so you can go back and review. Before beginning, stop and look back over the skills you have obtained. What skills in mathematics do you have now that you did not have before studying this chapter?

✎ CONCEPT REINFORCEMENT

Determine whether the statement is true or false. Answers are given at the back of the book.

_____ **1.** A set of data has just one mean and just one median, but it can have more than one mode.

_____ **2.** To find the average of a set of numbers, add the numbers and then multiply by the number of items of data.

_____ **3.** It is possible for the mean, median, and mode of a set of data to be the same number.

_____ **4.** If each number in a set of data occurs the same number of times, there is no mode.

Review Exercises

FedEx Mailing Costs. There are three types of Federal Express delivery service for packages of various weights within Zone 5 (shipments from 601 to 1000 mi from origin), as shown in the following table. Use this table for Exercises 1–6. [6.1a]

1. Find the cost of a 3-lb FedEx Priority Overnight delivery.

2. Find the cost of a 10-lb FedEx Standard Overnight delivery.

3. How much would you save by sending the package listed in Exercise 1 by FedEx 2Day delivery?

4. How much would you save by sending the package in Exercise 2 by FedEx 2Day delivery?

Delivery by 10:00 a.m. next business day

Delivery by 3:00 p.m. next business day

Delivery by 4:30 p.m. second business day

All other packaging / weight in lbs. FEDEX LETTER	FEDEX PRIORITY OVERNIGHT®	FEDEX STANDARD OVERNIGHT®	FEDEX 2DAY®
up to 8 oz.	$ 23.04	$ 21.17	$ 13.17
1 lb.	$ 34.28	$ 30.59	$ 13.17
2 lbs.	38.08	33.99	14.02
3	41.71	37.11	15.44
4	45.46	40.18	17.54
5	49.60	42.22	19.24
6	53.12	46.65	20.77
7	56.69	50.28	22.53
8	59.93	53.12	24.29
9	63.50	56.30	25.94
10	67.31	59.64	27.58
11	70.54	62.48	29.28

Source: Federal Express Corporation

5. Is there any difference in price between sending a 5-oz package FedEx Priority Overnight and sending an 8-oz package in the same way?

6. An author has a 4-lb manuscript to send by FedEx Standard Overnight delivery to her publisher. She calls and the package is picked up. Later that day she completes work on another part of her manuscript that weighs 5 lb. She calls and sends it by FedEx Standard Overnight delivery to the same address. How much could she have saved if she had waited and sent both packages as one?

U.S. Police Forces. This pictograph shows the number of officers in the largest U.S. police forces. Use the graph for Exercises 7–10.

America's Largest Police Forces

Source: International Association of Chiefs of Police

7. About how many officers are in the Chicago police force? [6.1b]

8. Which city has about 9000 officers on its force? [6.1b]

9. Of the cities listed, which has the smallest police force? [6.1b]

10. Estimate the average size of these six police forces. [6.1b], [6.5a]

Major League Sports Fans. As NASCAR racing increases in popularity, its fan base is growing rapidly. The following horizontal bar graph shows the number of fans of various major league sports. Use the graph for Exercises 11–16. [6.2a]

Sports Fans

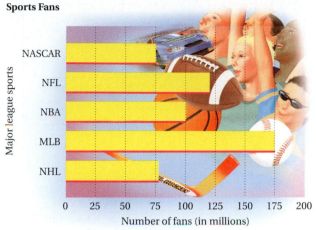

Sources: NASCAR; the individual leagues; ESPN

11. Which of the major league sports shown in the graph has the greatest number of fans?

12. About how many fans does the NFL (National Football League) have?

13. Which of the major league sports has about 100 million fans?

14. Which sports have 100 million or more fans?

15. About how many more MLB fans are there than NASCAR fans?

16. True or false? There are more MLB fans than NASCAR and NHL fans combined.

The following line graph shows the number of accidents per 100 drivers, by age. Use the graph for Exercises 17–22. [6.2c]

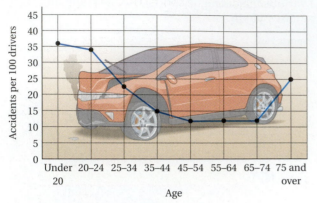

Source: Based on information in *The U.S. Statistical Abstract*, 2002

17. Which age group has the most accidents per 100 drivers?

18. What is the fewest number of accidents per 100 drivers in any age group?

19. How many more accidents per 100 drivers do people over 75 yr of age have than those in the age range of 65–74?

20. Between what ages does the number of accidents per 100 drivers stay basically the same?

21. About how many fewer accidents per 100 drivers do people 25–34 yr of age have than those 20–24 yr of age?

22. Which age group has accidents about three times as often as people 55–64 yr of age?

First-Class Postage. The following table shows the cost of first-class postage in various years. Use the table for Exercises 23 and 24.

YEAR	FIRST-CLASS POSTAGE
1989	25¢
1991	29¢
1995	32¢
1999	33¢
2001	34¢
2002	37¢
2006	39¢

Source: U.S. Postal Service

23. Make a vertical bar graph of the data. [6.2b]

24. Make a line graph of the data. [6.2d]

Determine the coordinates for each point. [6.3a]

25. *A*

26. *B*

27. *C*

28. *D*

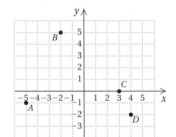

Plot each point on the graph below. [6.3a]

29. (2, 5)   **30.** (0, −3)   **31.** (−4, −2)

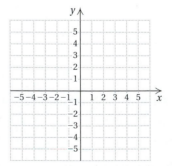

In which quadrant is each point located?   [6.3b]

32. $(3, -8)$ **33.** $(-20, -14)$ **34.** $\left(4\dfrac{9}{10}, 1\dfrac{3}{10}\right)$

35. Complete these solutions of $2x + 4y = 10$:
$(1, \square)$; $(\square, -2)$.   [6.4a]

Graph on a plane.   [6.4b, c]

36. $y = 2x - 5$

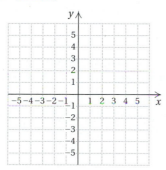

37. $y = -\dfrac{3}{4}x$

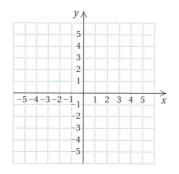

38. $x + y = 4$

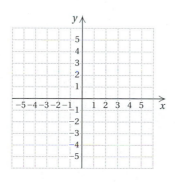

39. $x = -5$

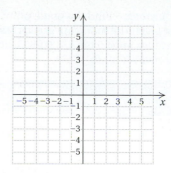

40. $y = 6$

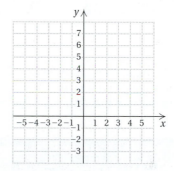

In Exercises 41–45 find **(a)** the mean, **(b)** the median, and **(c)** any modes that exist.   [6.5a, b, c]

41. 26, 51, 34, 26, 43   **42.** 11, 14, 17, 17, 21, 7, 11

43. 500, 25, 470, 190, 470, 280

44. 700, 700, 1900, 2700, 3000

45. $30,000, $75,000, $20,000, $25,000

46. To get an A in math, Sasha must average at least 90 on four tests. Scores on her first three tests were 94, 78, and 92. What is the lowest score that she can receive on the last test and still get an A?   [6.5a]

47. *Grade Point Average.* Find the grade point average for one semester given the following grades. Assume the grade point values are 4.0 for A, 3.0 for B, and so on. Round to the nearest tenth.   [6.5a]

COURSE	GRADE	NUMBER OF CREDIT HOURS IN COURSE
Basic Math	A	5
English	B	3
Computer Applications	C	4
Russian	B	3
College Skills	B	1

48. *Battery Testing.* An experiment was performed to compare battery quality. Two kinds of battery were tested to see how long, in hours, they kept a hand radio running. On the basis of this test, which battery is better?   [6.5d]

BATTERY A: TIMES (in hours)			BATTERY B: TIMES (in hours)		
38.9	39.3	40.4	39.3	38.6	38.8
53.1	41.7	38.0	37.4	47.6	37.9
36.8	47.7	48.1	46.9	37.8	38.1
38.2	46.9	47.4	47.9	50.1	38.2

49. Use interpolation and the graph in Exercises 17–22 to estimate the number of accidents per 100 drivers that are 30 to 40 yr old.   [6.6a]

A deck of 52 playing cards is thoroughly shuffled and a card is randomly selected.   [6.6b]

50. Find the probability that the 5 of clubs was selected.

51. Find the probability that a red card was selected.

52. **D_W** Is it possible for the mean of a set of numbers to be greater than all but one of the numbers in the set? Why or why not?   [6.5a]

53. **D_W** Is it possible for the median of a set of four numbers to be one of the numbers in the set? Why or why not?   [6.5b]

54. ▦ Find three solutions and then graph $34x + 47y = 100$.   [6.4a, b]

55. A typing pool consists of four senior typists who earn $12.35 per hour and nine other typists who earn $11.15 per hour. Find the mean hourly wage.   [6.5a]

56. The ordered set of data 298, 301, 305, a, 323, b, 390 has a median of 316 and a mean of 326. Find a and b.   [6.5a, b]

Graph on a plane.   [6.4b]

57. $1\frac{2}{3}x + \frac{3}{4}y = 2$

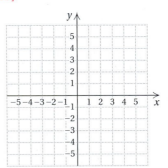

58. $\frac{3}{4}x - 2\frac{1}{2}y = 3$

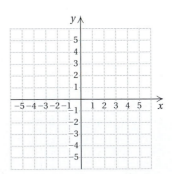

This table lists the number of calories burned during various walking activities. Use it for Questions 1 and 2.

WALKING ACTIVITY	CALORIES BURNED IN 30 MIN		
	110 lb	132 lb	154 lb
Walking			
Fitness (5 mph)	183	213	246
Mildly energetic (3.5 mph)	111	132	159
Strolling (2 mph)	69	84	99
Hiking			
3 mph with 20-lb load	210	249	285
3 mph with 10-lb load	195	228	264
3 mph with no load	183	213	246

1. Which activity provides the greatest benefit in burned calories for a person who weighs 132 lb?

2. What is the least strenuous activity for someone weighing 154 lb who wants to burn at least 250 calories every 30 min?

Waste Generated. The number of pounds of waste generated per person per year varies greatly among countries around the world. In the pictograph at right, each symbol represents approximately 100 lb of waste. Use the pictograph for Questions 3–6.

3. In which country does each person generate 600 lb of waste per year?

4. In which country does each person generate 1000 lb of waste per year?

5. How many pounds of waste per person per year are generated in France?

6. How many pounds of waste per person per year are generated in Finland?

**Amount of Waste Generated 2000
(per person per year)**

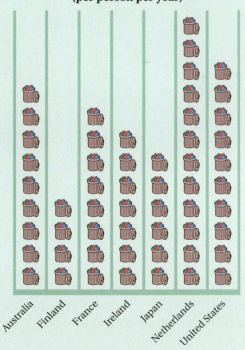

Source: Data from OECD Environmental Data Compendium: 2002

7. *Animal Speeds.* The following table lists maximum speeds of movement for various animals, in miles per hour. Make a vertical bar graph of the data.

ANIMAL	SPEED (in miles per hour)
Antelope	61
Peregrine falcon	225
Cheetah	70
Fastest human	28
Greyhound	42
Golden eagle	150
Grant's gazelle	47

Source: Barbara Ann Kipfer, *The Order of Things.*
New York: Random House, 1998

Refer to the table and the graph in Question 7 for Questions 8 and 9.

8. By how much does the fastest speed exceed the slowest speed?

9. How many times faster can a greyhound run than the fastest human?

Maple Syrup Prices. The line graph below shows the price of a gallon of Vermont maple syrup.

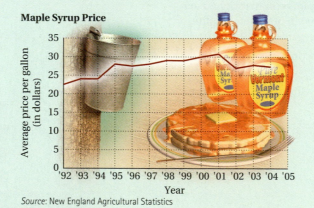

Maple Syrup Price

Source: New England Agricultural Statistics

10. In which year was the average price the highest?

11. Between which years did the price fall the most?

12. For which year was the price about $30?

13. Use extrapolation to estimate the price of a gallon of Vermont maple syrup in 2005.

In which quadrant is each point located?

14. $\left(-\frac{1}{2}, 7\right)$

15. $(-5, -6)$

Determine the coordinates of each point.

16. A

17. B

18. C

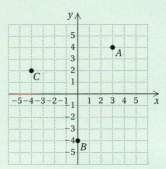

19. Complete the following solution of the equation $y - 3x = -10$: $(\square, 2)$.

Graph.

20. $y = 2x - 2$

21. $y = -\dfrac{3}{2}x$

22. $x = -2$

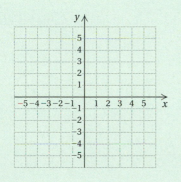

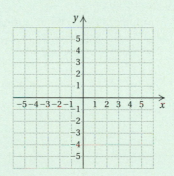

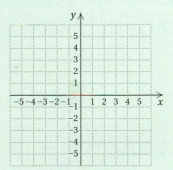

Find the mean.

23. 45, 49, 52, 54

24. 1, 2, 3, 4, 5

25. 3, 17, 17, 18, 18, 20

Find the median and any modes that exist.

26. 45, 47, 54, 54

27. 1, 2, 3, 4, 5

28. 20, 17, 17, 18, 3, 18

29. Gas Mileage. A 2006 Mitsubishi Eclipse GT V-6 gets 432 mi of highway driving on 16 gal of gasoline. What is the gas mileage?

Source: Mitsubishi Motors

30. To get a C in chemistry, Ernie must score an average of 70 on four tests. Scores on his first three tests were 68, 71, and 65. What is the lowest score that Ernie can receive on the last test and still get a C?

31. Chocolate Bars. An experiment is performed to compare the quality of new Swiss chocolate bars being introduced in the United States. People were asked to taste the candies and rate them on a scale of 1 to 10. On the basis of this test, which chocolate bar is better?

BAR A: SWISS PECAN			BAR B: SWISS HAZELNUT		
9	10	8	10	6	8
10	9	7	9	10	10
6	9	10	8	7	6
7	8	8	9	10	8

32. Grade Point Average. Find the grade point average for one semester given the following grades. Assume the grade point values are 4.0 for A, 3.0 for B, and so on. Round to the nearest tenth.

COURSE	GRADE	NUMBER OF CREDIT HOURS IN COURSE
Introductory Algebra	B	3
English	A	3
Business	C	4
Spanish	B	3
Typing	B	2

33. A month of the year is randomly selected for a company's party. What is the probability that a month whose name begins with J is chosen?

Graph.

34. $\frac{1}{4}x + 3\frac{1}{2}y = 1$

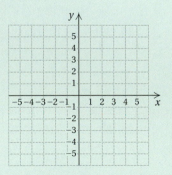

35. $\frac{5}{6}x - 2\frac{1}{3}y = 1$

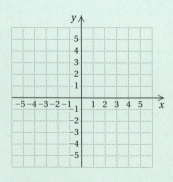

36. Find the area of a rectangle whose vertices are $(-3, 1)$, $(5, 1)$, $(5, 8)$, and $(-3, 8)$.

1. Write exponential notation: $7 \cdot 7 \cdot 7 \cdot 7$.

2. Write standard notation for the number in this sentence: Experts predict the global population to surpass 8 billion by the year 2030.

3. *Peanut Products.* In any given year, the average American eats 2.7 lb of peanut butter, 1.5 lb of salted peanuts, 1.2 lb of peanut candy, 0.7 lb of in-shell peanuts, and 0.1 lb of peanuts in other forms. How many pounds of peanuts and products containing peanuts does the average American eat in one year?

4. Find the perimeter and the area of the rectangle.

   7 cm

   4 cm

5. *Energy Consumption.* In a recent year, American utility companies generated 1464 billion kWh of electricity using coal, 455 billion using nuclear power, 273 billion using natural gas, 250 billion using hydroelectric plants, 118 billion using petroleum, and 12 billion using geothermal technology and other methods. How many kilowatt-hours of electricity were produced that year?

6. A recipe calls for $\frac{3}{4}$ cup of sugar. How much sugar should be used for $\frac{1}{2}$ of the recipe?

7. Tell which integers correspond to this situation: Monique won 8 cases and Jacques lost 7 cases.

Use either $<$ or $>$ for $\square$ to form a true statement.

8. $1 \ \square \ -7$

9. $\dfrac{4}{9} \ \square \ \dfrac{3}{7}$

10. $-4.8 \ \square \ -4.09$

11. Find $-x$ when $x = -9$.

12. Find $-(-x)$ when $x = 17$.

13. Evaluate $2x - y$ for $x = 3$ and $y = 8$.

14. Combine like terms: $6x + 4y - 8x - 3y$.

15. List all factors of 36.

16. Determine whether 732 is divisible by 6.

17. Write two different equivalent expressions for $\dfrac{-7}{x}$ with negative signs in different places.

18. Multiply: $5(2a - 3b + 1)$.

19. What fraction of the figure is shaded?

    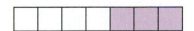

20. Find another name for the given number but with the denominator indicated.

    $$\frac{2}{7} = \frac{?}{35}$$

Perform the indicated operation and, if possible, simplify. Assume that all variables are nonzero.

21. $536 - 398$

22. $17 \cdot 28$

23. $63 \div (-7)$

24. $-18 + (-21)$

25. $\dfrac{3}{7} + \dfrac{2}{7}$

26. $\dfrac{3}{7} \div \dfrac{9}{5}$

27. $\dfrac{5}{6} - \dfrac{1}{9}$

28. $\dfrac{-2}{15} + \dfrac{3}{10}$

29. $\dfrac{8}{11} \cdot \dfrac{11}{8}$

30. $3\dfrac{1}{4} + 5\dfrac{7}{8}$

31. $7\dfrac{2}{3}x - 5\dfrac{1}{4}x$

32. $4\dfrac{1}{5} \cdot 3\dfrac{1}{7}$

33. $39.72 + 43.56$

34. $1334.183 \div 21.4$

35. $17.4(-2.43)$

36. $\dfrac{8t}{8t}$

37. $\dfrac{4x}{1}$

38. $\dfrac{0}{7x}$

Solve.

39. $x + \dfrac{2}{3} = -\dfrac{1}{5}$

40. $\dfrac{3}{8}x + 2 = 11$

41. $3(x - 5) = 7x + 2$

42. In which quadrant is the point $(-4, 9)$ located?

43. Graph: $y = \dfrac{1}{2}x - 4$.

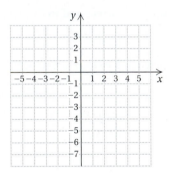

44. Find the mean:

19, 39, 34, 52.

45. Find the median:

7, 9, 12, 35.

46. Find the mode:

43, 56, 56, 43, 49, 49, 49.

47. *Gas Mileage.* A 2005 Subaru Impreza WRX gets 324 mi of highway driving on 12 gal of gasoline. What is the gas mileage?

Source: EPA, *Car and Driver*, September 2005

SYNTHESIS

48. Simplify:

$$\left(\dfrac{3}{4}\right)^2 - \dfrac{1}{8} \cdot \left(3 - 1\dfrac{1}{2}\right)^2.$$

49. Add and write the answer as a mixed numeral:

$$-5\dfrac{42}{100} + \dfrac{355}{100} + \dfrac{89}{10} + \dfrac{17}{1000}.$$

50. A square with sides parallel to the axes has the point $(2, 3)$ at its center. Find the coordinates of the square's vertices if each side is 8 units long.

462

Ratio and Proportion

7

Real-World Application

The number of women attending Purdue University's veterinary school of medicine has grown to surpass the number of men in the past three decades. In 1971, 53 men and 12 women were enrolled. In 1979, 36 men and 36 women were enrolled, and in 2004, there were 58 women and 12 men. What was the ratio of women to men in 1971, in 1979, and in 2004? What was the ratio of men to total enrollment in 2004?

Sources: Purdue University School of Veterinary Medicine, *Indianapolis Star*

This problem appears as Example 7 in Section 7.1.

Objectives

a Find fraction notation for ratios.

b Simplify ratios.

1. Write the ratio of 5 to 11.

2. Write the ratio of 57.3 to 86.1.

3. Write the ratio of $6\frac{3}{4}$ to $7\frac{2}{5}$.

4. Rainfall. The greatest amount of rainfall ever recorded for a 12-month period was 739 in. in Kukui, Maui, Hawaii, from December 1981 to December 1982. Find the ratio of rainfall to time in months.

Source: *The Handy Science Answer Book*

Answers on page A-20

a Ratios

> **RATIO**
>
> A **ratio** is the quotient of two quantities.

In the 2004–2005 season, the Detroit Pistons basketball team averaged 93.3 points per game and allowed their opponents an average of 89.5 points per game. The *ratio* of points earned to points allowed is given by the fraction notation

Points earned → $\dfrac{93.3}{89.5}$ or by the colon notation $93.3 : 89.5$
Points allowed →

We read both forms of notation as "the ratio of 93.3 to 89.5," listing the numerator first and the denominator second.

> **RATIO NOTATION**
>
> The **ratio** of a to b is given by the fraction notation $\dfrac{a}{b}$, where a is the numerator and b is the denominator, or by the colon notation $a : b$.

EXAMPLE 1 Write the ratio of 7 to 8.

The ratio is $\dfrac{7}{8}$,   or $7 : 8$.

EXAMPLE 2 Write the ratio of 31.4 to 100.

The ratio is $\dfrac{31.4}{100}$,   or $31.4 : 100$.

EXAMPLE 3 Write the ratio of $4\frac{2}{3}$ to $5\frac{7}{8}$. You need not simplify.

The ratio is $\dfrac{4\frac{2}{3}}{5\frac{7}{8}}$,   or $4\frac{2}{3} : 5\frac{7}{8}$.

Do Exercises 1–3.

In most of our work, we will use fraction notation for ratios.

EXAMPLE 4 *Recycling.* For every 17 lb of waste produced in the United States, about 5 lb are recycled. Find the ratio of the amount of waste recycled to the amount of waste produced.

Source: U.S. Bureau of the Census, *Statistical Abstract of the United States,* 2006

The ratio is $\dfrac{5}{17}$.

EXAMPLE 5 *Batting.* In the 2005 season, Derrek Lee of the Chicago Cubs got 199 hits in 594 at-bats. What was the ratio of hits to at-bats? of at-bats to hits?

Source: Major League Baseball

The ratio of hits to at-bats is

$$\frac{199}{594}.$$

The ratio of at-bats to hits is

$$\frac{594}{199}.$$

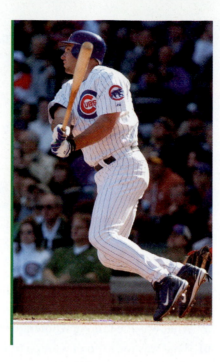

Do Exercises 4–6. (Exercise 4 is on the preceding page.)

EXAMPLE 6 Refer to the triangle below.

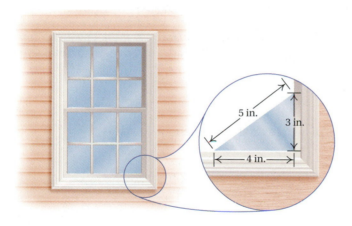

5 in. 3 in. 4 in.

a) What is the ratio of the length of the longest side to the length of the shortest side?

$$\frac{5}{3}.$$

b) What is the ratio of the length of the shortest side to the length of the longest side?

$$\frac{3}{5}.$$

Do Exercise 7.

5. Fat Grams. In one serving $\left(\frac{1}{2}\text{-cup}\right)$ of fried scallops, there are 12 g of fat. In one serving $\left(\frac{1}{2}\text{-cup}\right)$ of fried oysters, there are 14 g of fat. What is the ratio of grams of fat in one serving of scallops to grams of fat in one serving of oysters?

Source: *Better Homes and Gardens: A New Cook Book*

6. Earned Runs. In the 2005 season, Roger Clemens of the Houston Astros gave up 51 earned runs in 211.1 innings pitched. What was the ratio of earned runs to innings pitched? of innings pitched to earned runs?

Source: Major League Baseball

7. In the triangle below, what is the ratio of the length of the shortest side to the length of the longest side?

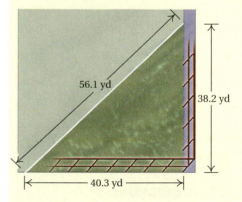

56.1 yd 38.2 yd 40.3 yd

Answers on page A-20

8. **Soap Box Derby.** Participation in the All-American Soap Box Derby World Championship has increased by more than 300 competitors since 1985. In 2004, there were 278 boys and 205 girls participating. What was the ratio of girls to boys? of boys to girls? of boys to total number of participants?

Source: All-American Soap Box Derby

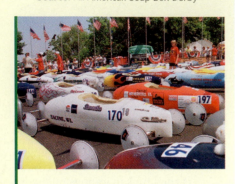

Answer on page A-20

Study Tips

ASKING QUESTIONS

Don't be afraid to ask questions in class. Most instructors welcome this and encourage students to ask them. Other students probably have the same questions you do.

EXAMPLE 7 *Veterinary Medicine.* The number of women attending Purdue University's veterinary school of medicine has grown to surpass the number of men in the past three decades.

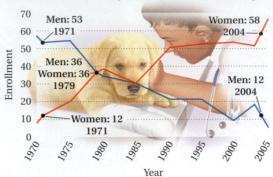

Enrollment in Veterinary Medicine: Purdue University

Sources: Purdue University School of Veterinary Medicine; *Indianapolis Star*

a) What was the ratio of women to men in 1971? in 1979? in 2004?

b) What was the ratio of women to total enrollment in 2004?

c) What was the ratio of men to total enrollment in 2004?

a) The ratio of women to men

$$\text{in 1971: } \frac{12}{53}; \quad \text{in 1979: } \frac{36}{36}; \quad \text{in 2004: } \frac{58}{12}.$$

b) In 2004, there were 58 women and 12 men enrolled, for a total of $58 + 12 = 70$ people. Thus the ratio of women to total enrollment in 2004 was

$$\frac{58}{70}.$$

c) The ratio of men to total enrollment in 2004 was

$$\frac{12}{70}.$$

Do Exercise 8.

b Simplifying Notation for Ratios

Sometimes a ratio can be simplified. This provides a means of finding other numbers with the same ratio.

EXAMPLE 8 Find the ratio of 6 to 8. Then simplify and find two other numbers in the same ratio.

We write the ratio in fraction notation and then simplify:

$$\frac{6}{8} = \frac{2 \cdot 3}{2 \cdot 4} = \frac{2}{2} \cdot \frac{3}{4} = 1 \cdot \frac{3}{4} = \frac{3}{4}.$$

Thus, 3 and 4 have the same ratio as 6 and 8. We can express this by saying "6 is to 8" as "3 is to 4."

EXAMPLE 9 Find the ratio of 50 to 10. Then simplify and find two other numbers in the same ratio.

We write the ratio in fraction notation and then simplify. The simplified form must also be in fraction notation.

$$\frac{50}{10} = \frac{10 \cdot 5}{10 \cdot 1} = \frac{10}{10} \cdot \frac{5}{1} = \frac{5}{1}$$ Since this is a ratio, leave the 1 in the denominator.

Thus, 50 is to 10 as 5 is to 1.

Do Exercises 9 and 10.

EXAMPLE 10 Find the ratio of 2.4 to 10. Then simplify and find two other numbers in the same ratio.

We first write the ratio in fraction notation. Next, we multiply by 1 to clear the decimal from the numerator. Then we simplify.

$$\frac{2.4}{10} = \frac{2.4}{10} \cdot \frac{10}{10} = \frac{24}{100} = \frac{4 \cdot 6}{4 \cdot 25} = \frac{4}{4} \cdot \frac{6}{25} = \frac{6}{25}$$

Thus, 2.4 is to 10 as 6 is to 25.

Do Exercises 11 and 12.

EXAMPLE 11 A standard HDTV screen with a width of 40 in. has a height of $22\frac{1}{2}$ in. Find the ratio of width to height and simplify.

We first write the ratio. Then we rewrite the mixed numeral in fraction notation. Finally, we perform the division.

$$\frac{40}{22\frac{1}{2}} = \frac{40}{\frac{45}{2}} = \frac{40}{1} \div \frac{45}{2}$$

$$= \frac{40}{1} \cdot \frac{2}{45}$$ Multiplying by the reciprocal of $\frac{45}{2}$

$$= \frac{5 \cdot 8 \cdot 2}{1 \cdot 5 \cdot 9}$$

$$= \frac{16}{9}$$ Removing a factor equal to 1: $\frac{5}{5} = 1$

Thus we can say the ratio of width to height is 16 to 9, which can also be expressed as $16:9$.

Do Exercise 13.

CALCULATOR CORNER

Ratios with Mixed Numerals   If you are permitted use of a calculator to solve exercises like Example 11, be careful to enter the mixed numerals using the $\boxed{a^{b/c}}$ key. The final ratio—if it's greater than 1—will also appear as a mixed numeral. This can be converted to fraction notation either by hand or by using the $\boxed{d/c}$ key.

Exercise:

1. Use a calculator to find the ratio of $5\frac{1}{7}$ to $2\frac{3}{4}$.

---

9. Find the ratio of 18 to 27. Then simplify and find two other numbers in the same ratio.

10. Find the ratio of 18 to 3. Then simplify and find two other numbers in the same ratio.

11. Find the ratio of 3.6 to 12. Then simplify and find two other numbers in the same ratio.

12. Find the ratio of 1.2 to 1.5. Then simplify and find two other numbers in the same ratio.

13. In Example 11, find the ratio of the height of the HDTV screen to the width and simplify.

Answers on page A-20

a Find fraction notation for the ratio. You need not simplify.

1. 4 to 5

2. 3 to 2

3. 178 to 572

4. 329 to 967

5. 0.4 to 12

6. 2.3 to 22

7. 3.8 to 7.4

8. 0.6 to 0.7

9. 56.78 to 98.35

10. 456.2 to 333.1

11. $8\frac{3}{4}$ to $9\frac{5}{6}$

12. $10\frac{1}{2}$ to $43\frac{1}{4}$

13. *Corvette Accidents.*   Of every 5 fatal accidents involving a Corvette, 4 do not involve another vehicle. Find the ratio of fatal accidents involving just a Corvette to those involving a Corvette and at least one other vehicle.

Source: *Harper's Magazine*

14. *Price of a Book.*   The paperback book *A Short History of Nearly Everything* by Bill Bryson had a list price of $15.95 but was sold by Amazon.com for $10.85. What was the ratio of the sale price to the list price? of the list price to the sale price?

Source: www.amazon.com

15. *Physicians.*   In 2003, there were 362 physicians in Connecticut per 100,000 residents. In Wyoming, there were 192 physicians per 100,000 residents. Find the ratio of the number of physicians to residents in Connecticut and in Wyoming.

Sources: U.S. Bureau of the Census, American Medical Association

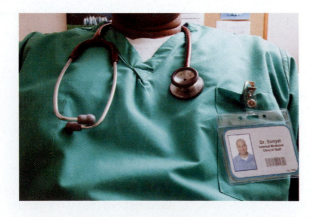

16. *Silicon in the Earth's Crust.*   There are about 28 tons of silicon in every 100 tons of the earth's crust. What is the ratio of silicon to the weight of crust? of the weight of crust to the weight of silicon?

Source: *The Handy Science Answer Book*

17. *Heart Disease.*   In the state of Minnesota, of every 1000 people, 93.2 will die of heart disease. Find the ratio of those who die of heart disease to all people.

Source: "Reforming the Health Care System; State Profiles 1999," AARP

18. *Cancer Deaths.*   In the state of Texas, of every 1000 people, 122.8 will die of cancer. Find the ratio of those who die of cancer to all people.

Source: "Reforming the Health Care System; State Profiles 1999," AARP

19. *Batting.* In the 2005 season, Todd Helton of the Colorado Rockies got 163 hits in 509 at-bats. What was the ratio of hits to at-bats? of at-bats to hits?

Source: Major League Baseball

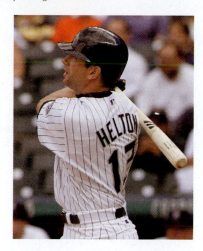

20. *Batting.* In the 2005 season, Manny Ramirez of the Boston Red Sox got 162 hits in 554 at-bats. What was the ratio of hits to at-bats? of at-bats to hits?

Source: Major League Baseball

21. *Field Hockey.* A diagram of the playing area for field hockey is shown below. What is the ratio of width to length? of length to width?

Source: *Sports: The Complete Visual Reference*

Shooting circle

Sideline

Officials' table

Penalty spot

100 yd

60 yd

22. *The Leaning Tower of Pisa.* At the time of this writing, the Leaning Tower of Pisa is still standing. It is 184.5 ft tall but leans about 17 ft out from its base. What is the ratio of the distance it leans to its height? its height to the distance it leans?

Source: *The Handy Science Answer Book*

184.5 ft

17 ft

b   Find the ratio of the first number to the second and simplify.

23. 4 to 6

24. 6 to 10

25. 18 to 24

26. 28 to 36

27. 4.8 to 10

28. 5.6 to 10

29. 2.8 to 3.6

30. 4.8 to 6.4

31. 20 to 30

32. 40 to 60

33. 56 to 100

34. 42 to 100

35. 128 to 256

36. 232 to 116

37. 0.48 to 0.64

38. 0.32 to 0.96

39. In this rectangle, find the ratios of length to width and of width to length.

478 ft

213 ft

40. In this right triangle, find the ratios of shortest length to longest length and of longest length to shortest length.

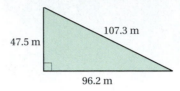

47.5 m

107.3 m

96.2 m

41. The ratio of females to all people worldwide is 51 to 100. Write the ratio of females to males as a simplified fraction.

42. The ratio of Americans ages 18–24 living with their parents to all Americans aged 18–24 is 54 to 100. Write the ratio of those living with parents to those living apart from parents as a simplified fraction.

Write each ratio in simplified form.

43. $3\frac{1}{5}$ to $4\frac{1}{10}$

44. $2\frac{1}{4}$ to $5\frac{1}{2}$

45. $7\frac{1}{8}$ to $2\frac{3}{4}$

46. $6\frac{3}{10}$ to $1\frac{4}{5}$

47. $8\frac{2}{9}$ to $7\frac{1}{6}$

48. $6\frac{3}{8}$ to $5\frac{1}{12}$

49. **D**_W Van Zandt College has a student-to-faculty ratio of 12 to 1 while Townes College has a faculty-to-student ratio of 2 to 26. On the basis of this information, which school would you rather attend? Why?

50. **D**_W The width-to-height ratio of HDTV screens exceeds the width-to-height ratio of conventional TVs. Do HDTV screens always have more area than conventional screens? Why or why not?

Use < or > for ☐ to write a true sentence.   [4.2c]

51. $-\dfrac{5}{6}$ ☐ $-\dfrac{3}{4}$ **52.** $\dfrac{12}{8}$ ☐ $\dfrac{7}{4}$ **53.** $\dfrac{5}{9}$ ☐ $\dfrac{6}{11}$ **54.** $-\dfrac{3}{4}$ ☐ $-\dfrac{2}{3}$

Divide, using decimal notation for the answer.   [5.4a]

55. $200 \div 4$ **56.** $95 \div 10$ **57.** $232 \div 16$ **58.** $342 \div 2.25$

Solve.   [4.6c]

59. Rocky is $187\frac{1}{10}$ cm tall and his daughter is $180\frac{3}{4}$ cm tall. How much taller is Rocky?

60. Aunt Louise is $168\frac{1}{4}$ cm tall and her son is $150\frac{7}{10}$ cm tall. How much taller is Aunt Louise?

61. **D**_W Can every ratio be written as the ratio of some number to 1? Why or why not?

62. **D**_W What can be concluded about the width of a rectangle if the ratio of length to perimeter is 1 to 3? Make some sketches and explain your reasoning.

63. ▦ In 2005, the total payroll of Major League Baseball teams was \$2,191,886,898. The New York Yankees payroll was the highest at \$208,306,817. Find the ratio, as a ratio to 1, of the Yankees payroll to the overall payroll.
Source: *USA Today* Baseball Salaries Database

64. ▦ See Exercise 63. In 2004, the total payroll of Major League Baseball teams was \$2,071,265,943. Find the ratio, as a ratio to 1, of the payroll in 2005 to the payroll in 2004.
Source: *USA Today* Baseball Salaries Database

65. ▦ See Exercises 59 and 60. Find the ratio of the parents' total height to their children's total height and simplify.

Exercises 66 and 67 refer to a common fertilizer known as "5-10-15." This mixture contains 5 parts of potassium for every 10 parts of phosphorus and 15 parts of nitrogen (this is often denoted 5:10:15).

66. Find the ratio of potassium to nitrogen and of nitrogen to phosphorus.

67. Simplify the ratio $5:10:15$.

Objectives

a Give the ratio of two different measures as a rate.

b Find unit prices and use them to compare purchases.

a Rates

A 2005 Kia Sportage EX 4WD can go 414 mi on 18 gal of gasoline. Let's consider the ratio of miles to gallons:

Source: Kia Motors America, Inc.

$$\frac{414 \text{ mi}}{18 \text{ gal}} = \frac{414}{18} \frac{\text{miles}}{\text{gallon}} = \frac{23}{1} \frac{\text{miles}}{\text{gallon}}$$

$$= 23 \text{ miles per gallon} = 23 \text{ mpg.}$$

Miles per gallon is abbreviated mpg.

"per" means "division," or "for each."

The ratio

$$\frac{414 \text{ mi}}{18 \text{ gal}}, \quad \text{or} \quad \frac{414}{18} \frac{\text{mi}}{\text{gal}}, \quad \text{or 23 mpg}$$

is called a **rate.**

> **RATE**
>
> When a ratio is used to compare two different kinds of measure, we call it a **rate.**

Sierra's Ford Focus goes 392.4 mi on 16.8 gal of gasoline. Is the mpg (mileage) of her car better than that of the Kia Sportage above? To determine this, it helps to convert the ratio to decimal notation and perhaps round. Then we have

$$\frac{392.4 \text{ mi}}{16.8 \text{ gal}} = \frac{392.4}{16.8} \text{ mpg} \approx 23.357 \text{ mpg.}$$

Since 23.357 > 23, Sierra's car gets better mileage than the Kia Sportage does.

EXAMPLE 1 It takes 60 oz of grass seed to seed 3000 sq ft of lawn. What is the rate in ounces per square foot?

$$\frac{60 \text{ oz}}{3000 \text{ sq ft}} = \frac{1}{50} \frac{\text{oz}}{\text{sq ft}}, \quad \text{or} \quad 0.02 \frac{\text{oz}}{\text{sq ft}}$$

EXAMPLE 2 A cook buys 10 lb of potatoes for $3.69. What is the rate in cents per pound?

$$\frac{\$3.69}{10 \text{ lb}} = \frac{369 \text{ cents}}{10 \text{ lb}}, \quad \text{or} \quad 36.9 \frac{\text{cents}}{\text{lb}}$$

Rates use two units. Paying attention to units can be helpful in solving certain rate problems. For example, if we expect an answer to be a rate in dollars per month, we look for a calculation that causes an amount of money (in dollars) to be divided by a length of time (in months). In this manner, the rate, dollars/month, will be formed.

Study Tips

RECORDING YOUR LECTURES

Consider recording your lectures and playing them back when convenient, say, while commuting to campus. (Be sure to get permission from your instructor before doing so, however.) Some important points are emphasized verbally.

EXAMPLE 3 A pharmacy student working as a pharmacist's assistant earned $3690 for working 3 months one summer. What was the rate of pay per month?

The rate of pay is the ratio of money earned per length of time worked, or

$$\frac{\$3690}{3 \text{ mo}} = 1230 \frac{\text{dollars}}{\text{month}}, \quad \text{or}$$

$1230 per month.

EXAMPLE 4 *Ratio of Strikeouts to Home Runs.* In the 2005 season, Vladimir Guerrero of the Los Angeles Angels had 48 strikeouts and 32 home runs. What was his strikeout to home-run rate?

Source: Major League Baseball

$$\frac{48 \text{ strikeouts}}{32 \text{ home runs}}$$

$$= \frac{48}{32} \frac{\text{strikeouts}}{\text{home runs}}$$

$$= \frac{48}{32} \text{ strikeouts per home run}$$

$$= 1.5 \text{ strikeouts per home run}$$

Do Exercises 1–8. (Exercise 8 is on the following page.)

b Unit Pricing

> **UNIT PRICE**
>
> A **unit price,** or **unit rate,** is the ratio of price to the number of units.

EXAMPLE 5 *Unit Price of Pears.* A consumer bought a $15\frac{1}{4}$-oz can of pears for $1.07. What is the unit price in cents per ounce?

Often it is helpful to change the cost to cents so we can compare unit prices more easily:

$$\text{Unit price} = \frac{\text{Price}}{\text{Number of units}}$$

$$= \frac{\$1.07}{15\frac{1}{4} \text{ oz}} = \frac{107 \text{ cents}}{15.25 \text{ oz}} = \frac{107}{15.25} \frac{\text{cents}}{\text{oz}}$$

$$\approx 7.016 \text{ cents per ounce.}$$

Do Exercise 9 on the following page.

A ratio of distance traveled to time is called *speed.* What is the rate, or speed, in miles per hour or in kilometers per hour?

1. 45 mi, 9 hr

2. 120 mi, 10 hr

3. 89 km, 13 hr (Round to the nearest hundredth.)

What is the rate, or speed, in feet per second?

4. 2200 ft, 2 sec

5. 52 ft, 13 sec

6. 242 ft, 16 sec

7. Ratio of Home Runs to Strikeouts. Referring to Example 4, determine Guerrero's home-run to strikeout rate.

Source: Major League Baseball

Answers on page A-21

8. Babe Ruth. In his entire career, Babe Ruth had 1330 strikeouts and 714 home runs. What was his home-run to strikeout rate? How does it compare to Guerrero's? (See Margin Exercise 7.)

Source: Major League Baseball

9. Unit Price of Mustard. A consumer bought a 20-oz container of French's yellow mustard for $1.49. What is the unit price in cents per ounce?

10. Meijer Brand Olives. Complete the following table of unit prices for Meijer brand olives. Which package has the better unit price?

Source: Meijer Stores

PACKAGE	PRICE	UNIT PRICE
7 oz	$1.69	
10 oz	$2.59	
$5\frac{3}{4}$ oz	$1.39	

To do comparison shopping, it helps to compare unit prices.

EXAMPLE 6 *Unit Price of Heinz Ketchup.* Many factors can contribute to determining unit pricing in food, such as variations in store pricing and special discounts. Heinz produces ketchup in containers of various sizes. The table below lists several examples of pricing for these packages from a Meijer store. Starting with the price given for each package, compute the unit prices and decide which is the best purchase on the basis of unit price per ounce alone.

Source: Meijer Stores

PACKAGE	PRICE	UNIT PRICE
14 oz	$1.29	9.214¢/oz
24 oz	$1.47	6.125¢/oz
36 oz	$2.49	6.917¢/oz
64 oz	$4.45	6.953¢/oz
101-oz twin pack (two $50\frac{1}{2}$-oz packages)	$5.69	5.634¢/oz Lowest unit price

We compute the unit price for the 24-oz package and leave the remaining prices to the student to check. The unit price for the 24-oz, $1.47 package is given by

$$\frac{\$1.47}{24 \text{ oz}} = \frac{147 \text{ cents}}{24 \text{ oz}} = \frac{147}{24}\frac{\text{cents}}{\text{oz}} = 6.125 \text{ cents per ounce} = 6.125¢/\text{oz}.$$

On the basis of unit price alone, the 101-oz twin pack is the best buy.

Sometimes, as you will see in Margin Exercise 10, a larger size may not have the lower unit price. It is also worth noting that "bigger" is not always "cheaper." (For example, you may not have room for larger packages or the food may spoil before it is used.)

Do Exercise 10.

CALCULATOR CORNER

Reciprocals of Rates If, in Example 1, we wished to determine the rate of lawn coverage in sq ft/oz, we could simply find the reciprocal of 0.02 oz/sq ft by pressing ⎡0⎤⎡.⎤⎡0⎤⎡2⎤⎡1/x⎤.

Exercise:

1. Use the solution of Example 4 to find Guerrero's rate of home runs per strikeout.

7.2 EXERCISE SET

For Extra Help

MathXL MyMathLab InterAct Math Tutor Center Video Lectures on CD Disc 4 Student's Solutions Manual

a In Exercises 1–4, find the rate, or speed, as a ratio of distance to time. Round to the nearest hundredth where appropriate.

1. 120 mi, 3 hr (miles per hour is abbreviated mph)

2. 18 mi, 9 hr

3. 217 mi, 29 sec

4. 443 m, 48 sec

5. *Chevrolet Cobalt LS—City Driving.*   A 2005 Chevrolet Cobalt LS will go 300 mi on 12.5 gal of gasoline in city driving. What is the rate in miles per gallon?
Source: *Car and Driver*, April 2005, p. 104

6. *Audi A6 4.2 Quattro—City Driving.*   A 2005 Audi A6 4.2 Quattro will go 246.5 mi on 14.5 gal of gasoline in city driving. What is the rate in miles per gallon?
Source: *Car and Driver*, May 2005, p. 52

7. *Audi A6 4.2 Quattro—Highway Driving.*   A 2005 Audi A6 4.2 Quattro will go 448.5 mi on 19.5 gal of gasoline in highway driving. What is the rate in miles per gallon?
Source: *Car and Driver*, May 2005, p. 52

8. *Chevrolet Cobalt LS—Highway Driving.*   A 2005 Chevrolet Cobalt LS will go 432 mi on 13.5 gal of gasoline in highway driving. What is the rate in miles per gallon?
Source: *Car and Driver*, April 2005, p. 104

9. *Population Density of Monaco.*   Monaco is a tiny country on the Mediterranean coast of France. It has an area of 0.75 sq mi and a population of 32,270 people. What is the rate of number of people per square mile? The rate per square mile is called the *population density.* Monaco has the highest population density in the world.
Source: *Time Almanac, 2005*

10. *Population Density of Australia.*   The continent of Australia, with the island state of Tasmania, has an area of 2,967,893 sq mi and a population of 19,913,144 people. What is the rate of number of people per square mile? The rate per square mile is called the *population density.* Australia has one of the lowest population densities in the world.
Source: *Time Almanac, 2005*

11. *Lawn Watering.*   To water a lawn adequately requires 623 gal of water for every 1000 ft^2. What is the rate in gallons per square foot?

12. A car is driven 200 km on 40 L of gasoline. What is the rate in kilometers per liter?

13. *Speed of Light.* Light travels 186,000 mi in 1 sec. What is its rate, or speed, in miles per second?
Source: *The Handy Science Answer Book*

14. *Speed of Sound.* Sound travels 1100 ft in 1 sec. What is its rate, or speed, in feet per second?
Source: *The Handy Science Answer Book*

15. Impulses in nerve fibers travel 310 km in 2.5 hr. What is the rate, or speed, in kilometers per hour?

16. A black racer snake can travel 4.6 km in 2 hr. What is its rate, or speed, in kilometers per hour?

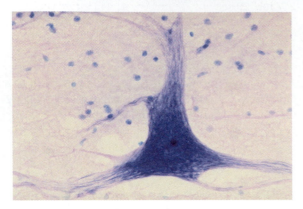

17. A car is driven 500 mi in 20 hr. What is the rate in miles per hour? in hours per mile?

18. A student eats 3 hamburgers in 15 min. What is the rate in hamburgers per minute? in minutes per hamburger?

19. *Points per Game.* In the 2004–2005 season, Yao Ming of the Houston Rockets scored 1465 points in 80 games. What was the rate in points per game?
Source: National Basketball Association

20. *Points per Game.* In the 2004–2005 season, Tayshaun Prince of the Detroit Pistons scored 1206 points in 82 games. What was the rate in points per game?
Source: National Basketball Association

21. *Elephant Heartbeat.* The heart of an elephant, at rest, will beat an average of 1500 beats in 60 min. What is the rate in beats per minute?
Source: *The Handy Science Answer Book*

22. *Human Heartbeat.* The heart of a human, at rest, will beat an average of 4200 beats in 60 min. What is the rate in beats per minute?
Source: *The Handy Science Answer Book*

b Find each unit price in each of Exercises 23–32. Then determine which size has the lowest unit price.

23. *Pert Plus® Shampoo.*

PACKAGE	PRICE	UNIT PRICE
13.5 oz	$2.59	
25.4 oz	$3.99	

24. *Roll-on Deodorant.*

PACKAGE	PRICE	UNIT PRICE
2.25 oz	$2.49	
3.5 oz	$3.98	

25. *Miracle Whip®.*

PACKAGE	PRICE	UNIT PRICE
16-oz jar	$1.84	
18-oz squeezable	$2.49	

26. *Bush's Homestyle Baked Beans®.*

PACKAGE	PRICE	UNIT PRICE
16 oz	$0.99	
28 oz	$1.44	

27. *Meijer® Coffee.*

PACKAGE	PRICE	UNIT PRICE
11.5 oz	$2.09	
34.5 oz	$5.27	

28. *Maxwell House® Coffee.*

PACKAGE	PRICE	UNIT PRICE
13 oz	$3.74	
34.5 oz	$6.36	

29. *Jif® Creamy Peanut Butter.*

PACKAGE	PRICE	UNIT PRICE
18 oz	$1.89	
28 oz	$3.25	
40 oz	$4.99	
64 oz	$7.99	

30. *Downy® Fabric Softener.*

PACKAGE	PRICE	UNIT PRICE
40 oz	$3.23	
60 oz	$4.79	
80 oz	$7.99	
120 oz	$10.69	

31. *Tide® Liquid Laundry Detergent.*

PACKAGE	PRICE	UNIT PRICE
50 fl oz	$4.29	
100 fl oz	$5.29	
200 fl oz	$10.49	
300 fl oz	$15.79	

32. *Del Monte® Green Beans.*

PACKAGE	PRICE	UNIT PRICE
8 oz	$0.59	
14.5 oz	$0.69	
28 oz	$1.19	

33. **D****W** The unit price of an item generally drops when larger packages of that item are purchased. Why do you think manufacturers do this?

34. **D****W** Compare the prices and unit prices for the 16-oz jar of Miracle Whip and the 18-oz squeezable container of Miracle Whip in Exercise 25. What seems unusual about these prices? Explain why you think this has happened.

Replace the ☐ with < or > to form a true statement.   [4.2c]

35. $\dfrac{3}{11}$ ☐ $\dfrac{5}{13}$

36. $\dfrac{9}{7}$ ☐ $\dfrac{7}{5}$

37. $\dfrac{4}{9}$ ☐ $\dfrac{3}{7}$

38. $\dfrac{3}{19}$ ☐ $\dfrac{2}{17}$

39. $-\dfrac{3}{10}$ ☐ $-\dfrac{2}{7}$

40. $-\dfrac{9}{8}$ ☐ $-\dfrac{15}{13}$

41. There are 20.6 million people in this country who play the piano and 18.9 million who play the guitar. How many more play the piano than the guitar?   [5.8a]

42. A serving of fish steak (cross section) is generally $\frac{1}{2}$ lb. How many servings can be prepared from a cleaned $12\frac{3}{4}$-lb salmon?   [4.7d]

43. **D_W** Suppose that the same type of juice is available in two sizes and that the larger bottle has the lower unit price. If the larger bottle costs $3.79 and contains twice as much juice, what can you conclude about the price of the smaller bottle? Why?

44. **D_W** Manufacturers of laundry detergent often charge a higher unit price for their largest packages. Why do you think they do so?

45. Certain manufacturers shrink the size and price of a product so that the consumer thinks the price of a product has been lowered when, in reality, a higher unit price is being charged.

a) Some aluminum juice cans are now concave (curved in) on the bottom. Suppose the volume of the can in the figure has been reduced from a fluid capacity of 6 oz to 5.5 oz, and the price of each can has been reduced from 65¢ to 60¢. Find the unit price of the original and the modified cans in cents per ounce.

b) Suppose that at one time the cost of a certain kind of paper towel was $0.89 for a roll containing 78 ft^2 of absorbent surface. Later the surface area was changed to 65 ft^2, and the price was decreased to $0.79. Find the unit price of each product in cents per square foot.

46. Suppose that a pasta manufacturer shrinks the size of a box from 1 lb to 14 oz, but keeps the price at 85¢ a box. By how much does the unit price change?

47. In 2002, ice cream manufacturers began to package ice cream in 56-oz cartons instead of 64-oz cartons. At first, the price of a carton dropped from $5 to $4.35. After about a year, the price of a carton had again risen to $5. How did the unit price change for consumers?

48. 🖩 Use the formula for the area of a circle, $A = \pi r^2$, to determine which is a better deal: a 14-in. pizza for $10.50 or a 16-in. pizza for $11.95. Use 3.14 for π.

49. 🖩 Suppose that, 25 mi from where you're standing, a bolt of lightning splits a tree. How long will it take for you to hear the accompanying crack of thunder? How long will it take for you to see the flash of light? (*Hint*: Use the information in Exercises 13 and 14.)

50. 🖩 A six-pack of 12-oz cans of Santa Cruz Ginger Ale was recently on sale for $2.99. Find the unit price in ounces per dollar.

51. 🖩 A case of 12-oz cans of Santa Cruz Ginger Ale (4 six-packs) is on sale for $11. Find the unit price in ounces per dollar.

7.3 PROPORTIONS

a Proportions

When two pairs of numbers (such as 3, 2 and 6, 4) have the same ratio, we say that they are **proportional.** The equation

$$\frac{3}{2} = \frac{6}{4}$$

states that the pairs 3, 2 and 6, 4 are proportional. Such an equation is called a **proportion.** We sometimes read $\frac{3}{2} = \frac{6}{4}$ as "3 is to 2 as 6 is to 4."

To check whether two pairs of numbers are proportional, we can compare the ratios as fractions, using the test for equality considered in Section 3.5. To see why this test works, consider $\frac{a}{b}$ and $\frac{c}{d}$ and assume that neither denominator is 0. We show that if the ratios are equal, the cross products are equal.

$$\frac{a}{b} = \frac{c}{d} \qquad \text{Assuming the ratios are equal}$$

$$bd \cdot \frac{a}{b} = bd \cdot \frac{c}{d} \qquad \text{Multiplying both sides by } bd$$

$$da = bc \qquad \text{Simplifying}$$

If neither b nor d is 0, the above equations are all equivalent, and it is also true that if $da = bc$, then $\frac{a}{b} = \frac{c}{d}$.

This is the basis for the following test for equality.

A TEST FOR EQUALITY

We multiply these two numbers: $3 \cdot 4$.

We multiply these two numbers: $2 \cdot 6$.

$$\frac{3}{2} \overset{?}{=} \frac{6}{4}$$

Since $3 \cdot 4 = 2 \cdot 6$, we know that

$$\frac{3}{2} = \frac{6}{4}.$$

We call $3 \cdot 4$ and $2 \cdot 6$ *cross products.*

479

Determine whether the two pairs of numbers are proportional.

1. 3, 4 and 6, 8

2. 1, 4 and 10, 39

3. 1, 2 and 20, 39

EXAMPLE 1 Determine whether 1, 2 and 3, 6 are proportional.

We can use cross products to check an equivalent equation:

$$1 \cdot 6 = 6 \qquad \frac{1}{2} \stackrel{?}{=} \frac{3}{6} \qquad 2 \cdot 3 = 6$$

$$1 \cdot 6 \stackrel{?}{=} 2 \cdot 3$$

$$6 = 6.$$

Since this last equation is true, we know that the first equation is also true and the numbers 1, 2 and 3, 6 are proportional.

EXAMPLE 2 Determine whether 2, 5 and 4, 7 are proportional.

We can check an equivalent equation using cross products:

$$2 \cdot 7 = 14 \qquad \frac{2}{5} \stackrel{?}{=} \frac{4}{7} \qquad 5 \cdot 4 = 20$$

$$2 \cdot 7 \stackrel{?}{=} 5 \cdot 4$$

$$14 \neq 20.$$

Since $14 \neq 20$, we know that $\frac{2}{5} \neq \frac{4}{7}$, so 2, 5 and 4, 7 are not proportional.

Do Exercises 1–3.

EXAMPLE 3 Determine whether $1\frac{1}{4}, \frac{1}{2}$ and $1, \frac{2}{5}$ are proportional.

We can use cross products:

$$1\frac{1}{4} \cdot \frac{2}{5} = \frac{5}{4} \cdot \frac{2}{5} \qquad \frac{1\frac{1}{4}}{\frac{1}{2}} \stackrel{?}{=} \frac{1}{\frac{2}{5}} \qquad \frac{1}{2} \cdot 1 = \frac{1}{2}$$

$$\frac{5}{4} \cdot \frac{2}{5} \stackrel{?}{=} \frac{1}{2}$$

$$\frac{10}{20} = \frac{1}{2}. \qquad \text{We can also note that } \frac{\cancel{5}}{4} \cdot \frac{2}{\cancel{5}} = \frac{2}{4}.$$

Since $\frac{10}{20} = \frac{1}{2}$, we know that

$$\frac{1\frac{1}{4}}{\frac{1}{2}} = \frac{1}{\frac{2}{5}}.$$

The numbers $1\frac{1}{4}, \frac{1}{2}$ and $1, \frac{2}{5}$ are proportional.

Do Exercises 4 and 5.

Determine whether the two pairs of numbers are proportional.

4. $4\frac{2}{3}, 5\frac{1}{2}$ and $14, 16\frac{1}{2}$

5. 7.4, 6.8 and 4.2, 3.6

b Solving Proportions

Often one of the four numbers in a proportion is unknown. Cross products can be used to find the missing number and "solve" the proportion.

EXAMPLE 4 Solve: $\dfrac{x}{8} = \dfrac{3}{5}$.

We form an equivalent equation by equating cross products. Then we solve for x.

$5 \cdot x = 8 \cdot 3$ Equating cross products

$\dfrac{5x}{5} = \dfrac{24}{5}$ Dividing both sides by 5

$x = \dfrac{24}{5}$ Simplifying

$= 4.8$ Dividing

To check that 4.8 is the solution, we replace x with 4.8 and use cross products:

$4.8 \cdot 5 = 24$ $\dfrac{4.8}{8} \overset{?}{=} \dfrac{3}{5}$ $8 \cdot 3 = 24$.   The cross products are the same.

Since the cross products are the same, it follows that $\frac{4.8}{8} = \frac{3}{5}$. Thus 4.8, 8 and 3, 5 are proportional, and 4.8 is the solution of the equation.

SOLVING PROPORTIONS

To solve $\dfrac{a}{b} = \dfrac{c}{d}$ for a specific variable, equate cross products and then divide on both sides to get that variable alone. (Assume $b, d \neq 0$.)

Do Exercise 6.

EXAMPLE 5 Solve: $\dfrac{x}{7} = \dfrac{5}{3}$. Write a mixed numeral for the answer.

We have

$\dfrac{x}{7} = \dfrac{5}{3}$

$3 \cdot x = 7 \cdot 5$ Equating cross products

$\dfrac{3x}{3} = \dfrac{35}{3}$ Dividing both sides by 3

$x = \dfrac{35}{3}, \text{ or } 11\dfrac{2}{3}$.

Check: $\dfrac{\frac{35}{3}}{7} \overset{?}{=} \dfrac{5}{3}$

$\dfrac{35}{3} \cdot 3 = 35$ and $7 \cdot 5 = 35$.

The cross products are the same, so we have a check. The solution is $11\frac{2}{3}$.

Do Exercise 7.

6. Solve: $\dfrac{x}{63} = \dfrac{2}{9}$.

7. Solve: $\dfrac{x}{9} = \dfrac{5}{4}$.

8. Solve: $\dfrac{2.1}{0.5} = \dfrac{n}{2.5}$.

9. Solve: $\dfrac{2}{3} = \dfrac{6}{x}$.

10. Solve: $\dfrac{\frac{2}{5}}{\frac{9}{10}} = \dfrac{4\frac{4}{5}}{t}$.

Answers on page A-21

Solving Proportions Note in Examples 4–8 that when we solve a proportion, we equate cross products and then we divide on both sides to isolate the variable on one side of the equation. We can use a calculator to do the calculations in this situation. In Example 6, for instance, after equating cross products and dividing by 15.4 on both sides, we have

$$y = \frac{7.7 \times 2.2}{15.4}.$$

To compute y on a calculator, we can press $\boxed{7}\,\boxed{.}\,\boxed{7}\,\boxed{\times}$ $\boxed{2}\,\boxed{.}\,\boxed{2}\,\boxed{\div}\,\boxed{1}\,\boxed{5}\,\boxed{.}\,\boxed{4}$ $\boxed{=}$. The result is 1.1, so $y = 1.1$.

Exercises:

1. Use a calculator to solve each of the proportions in Examples 5–7.

2. Use a calculator to solve each of the proportions in Margin Exercises 6–10.

Solve each proportion.

3. $\dfrac{15.75}{20} = \dfrac{a}{35}$

4. $\dfrac{32}{x} = \dfrac{25}{20}$

5. $\dfrac{t}{57} = \dfrac{17}{64}$

6. $\dfrac{71.2}{a} = \dfrac{42.5}{23.9}$

7. $\dfrac{29.6}{3.15} = \dfrac{x}{4.23}$

8. $\dfrac{a}{3.01} = \dfrac{1.7}{0.043}$

EXAMPLE 6 Solve: $\dfrac{7.7}{15.4} = \dfrac{y}{2.2}$. Write decimal notation for the answer.

We have

$$\frac{7.7}{15.4} = \frac{y}{2.2}$$

$$(7.7)(2.2) = 15.4y \qquad \text{Equating cross products; if you prefer,}$$
$$\text{write } 15.4y = (7.7)(2.2).$$

$$\frac{(7.7)(2.2)}{15.4} = \frac{15.4y}{15.4} \qquad \text{Dividing both sides by 15.4}$$

$$\frac{16.94}{15.4} = y \qquad \text{Simplifying}$$

$$1.1 = y. \qquad \text{Dividing:} \quad 1\,5.4\,\overline{)\,1\,6.9_\wedge 4}$$
$$\phantom{1.1 = y. \qquad \text{Dividing:} \quad} \begin{array}{r} 1.1 \\ \underline{1\,5\,4\,0} \\ 1\,5\,4 \\ \underline{1\,5\,4} \\ 0 \end{array}$$

We leave the check for the student. The solution is 1.1.

Do Exercise 8 on the preceding page.

EXAMPLE 7 Solve: $\dfrac{3}{x} = \dfrac{6}{4}$.

We have

$$\frac{3}{x} = \frac{6}{4}$$

$$3 \cdot 4 = x \cdot 6 \qquad \text{Equating cross products; we could also write } 6 \cdot x = 3 \cdot 4.$$

$$\frac{12}{6} = \frac{6x}{6} \qquad \text{Dividing both sides by 6}$$

$$2 = x. \qquad \text{Simplifying}$$

The solution is 2.

Do Exercise 9 on the preceding page.

EXAMPLE 8 Solve: $\dfrac{\frac{17}{8}}{\frac{3}{4}} = \dfrac{1\frac{1}{2}}{n}$.

We have

$$\frac{\frac{17}{8}}{\frac{3}{4}} = \frac{\frac{3}{2}}{n} \qquad \text{Rewriting } 1\frac{1}{2} \text{ as } \frac{3}{2}$$

$$\frac{17}{8} \cdot n = \frac{3}{4} \cdot \frac{3}{2} \qquad \text{Equating cross products}$$

$$n = \frac{8 \cdot 3 \cdot 3}{17 \cdot 4 \cdot 2} \qquad \text{Multiplying both sides by } \frac{8}{17}$$

$$n = \frac{3 \cdot 3}{17} = \frac{9}{17}. \qquad \text{Removing a factor equal to 1: } \frac{8}{4 \cdot 2} = 1$$

The solution is $\frac{9}{17}$.

Do Exercise 10 on the preceding page.

a   Determine whether the two pairs of numbers are proportional.

1. 5, 6 and 7, 9

2. 7, 5 and 6, 4

3. 1, 2 and 10, 20

4. 7, 3 and 21, 9

5. 2.4, 3.6 and 1.8, 2.7

6. 4.5, 3.8 and 6.7, 5.2

7. $5\frac{1}{3}, 8\frac{1}{4}$ and $2\frac{1}{5}, 9\frac{1}{2}$

8. $2\frac{1}{3}, 3\frac{1}{2}$ and 14, 21

Pass-Completion Rates.   The table below lists the records of four NFL quarterbacks from the 2005 season.

PLAYER	TEAM	NUMBER OF PASSES COMPLETED	NUMBER OF PASSES ATTEMPTED	NUMBER OF COMPLETIONS PER ATTEMPT (completion rate)
Tom Brady	New England Patriots	334	530	
Trent Green	Kansas City Chiefs	317	507	
Brett Favre	Green Bay Packers	372	605	
Carson Palmer	Cincinnati Bengals	345	509	

Source: National Football League

9. Find each pass-completion rate rounded to the nearest hundredth. Are any the same?

10. Use cross products to determine whether any quarterback completion rates are the same.

b   Solve.

11. $\frac{18}{4} = \frac{x}{10}$

12. $\frac{x}{45} = \frac{20}{25}$

13. $\frac{x}{8} = \frac{9}{6}$

14. $\frac{8}{10} = \frac{n}{5}$

15. $\frac{t}{12} = \frac{5}{6}$

16. $\frac{12}{4} = \frac{x}{3}$

17. $\frac{2}{5} = \frac{8}{n}$

18. $\frac{10}{6} = \frac{5}{x}$

19. $\dfrac{n}{15} = \dfrac{10}{30}$

20. $\dfrac{2}{24} = \dfrac{x}{36}$

21. $\dfrac{16}{12} = \dfrac{24}{x}$

22. $\dfrac{7}{11} = \dfrac{2}{x}$

23. $\dfrac{6}{11} = \dfrac{12}{x}$

24. $\dfrac{8}{9} = \dfrac{32}{n}$

25. $\dfrac{20}{7} = \dfrac{80}{x}$

26. $\dfrac{5}{x} = \dfrac{4}{10}$

27. $\dfrac{12}{9} = \dfrac{x}{7}$

28. $\dfrac{x}{20} = \dfrac{16}{15}$

29. $\dfrac{x}{13} = \dfrac{2}{9}$

30. $\dfrac{1.2}{4} = \dfrac{x}{9}$

31. $\dfrac{t}{0.16} = \dfrac{0.15}{0.40}$

32. $\dfrac{x}{11} = \dfrac{7.1}{2}$

33. $\dfrac{100}{25} = \dfrac{20}{n}$

34. $\dfrac{35}{125} = \dfrac{7}{m}$

35. $\dfrac{7}{\frac{1}{4}} = \dfrac{28}{x}$

36. $\dfrac{x}{6} = \dfrac{1}{6}$

37. $\dfrac{\frac{1}{4}}{\frac{1}{2}} = \dfrac{\frac{1}{2}}{x}$

38. $\dfrac{1}{7} = \dfrac{x}{4\frac{1}{2}}$

39. $\dfrac{1}{2} = \dfrac{7}{x}$

40. $\dfrac{x}{3} = \dfrac{0}{9}$

41. $\dfrac{\frac{2}{7}}{\frac{3}{4}} = \dfrac{\frac{5}{6}}{y}$

42. $\dfrac{\frac{5}{4}}{\frac{5}{8}} = \dfrac{\frac{3}{2}}{Q}$

43. $\dfrac{2\frac{1}{2}}{3\frac{1}{3}} = \dfrac{x}{4\frac{1}{4}}$

44. $\dfrac{5\frac{1}{5}}{6\frac{1}{6}} = \dfrac{y}{3\frac{1}{2}}$

45. $\dfrac{1.28}{3.76} = \dfrac{4.28}{y}$

46. $\dfrac{10.4}{12.4} = \dfrac{6.76}{t}$

47. $\dfrac{10\frac{3}{8}}{12\frac{2}{3}} = \dfrac{5\frac{3}{4}}{y}$

48. $\dfrac{12\frac{7}{8}}{20\frac{3}{4}} = \dfrac{5\frac{2}{3}}{y}$

49. D_W Instead of equating cross products, a student solves $\frac{x}{7} = \frac{5}{3}$ (see Example 5) by multiplying on both sides by the least common denominator, 21. Is his approach a good one? Why or why not?

50. D_W An instructor predicts that a student's test grade will be proportional to the amount of time the student spends studying. What is meant by this? Write an example of a proportion that involves the grades of two students and their study times.

SKILL MAINTENANCE

✎ VOCABULARY REINFORCEMENT

In each of Exercises 51–58, fill in the blank with the correct term from the given list. Some of the choices may not be used and some may be used more than once.

51. A ratio is the _____ of two quantities. [7.1a]

52. A number is divisible by 3 if the _____ of the digits is divisible by 3. [3.1b]

53. To compute a(n) _____ of a set of numbers, we add the numbers and then divide by the number of addends. [6.5a]

54. To convert from _____ to _____ , move the decimal point two places to the right and change the $ sign in front to the ¢ sign at the end. [5.3b]

55. The numbers -3 and 3 are called _____ . [2.1d]

56. The decimal 0.125 is an example of a(n) _____ decimal. [5.5a]

57. The sentence $\frac{2}{5} \cdot \frac{4}{9} = \frac{4}{9} \cdot \frac{2}{5}$ illustrates the _____ law of multiplication. [1.5a]

58. To solve $\frac{x}{a} = \frac{c}{d}$, equate the _____ and divide on both sides to get x alone. [7.3b]

cross products
cents
dollars
terminating
repeating
mean
median
mode
opposites
reciprocals
associative
commutative
sum
difference
product
quotient

SYNTHESIS

59. D_W If a true proportion is formed using exactly three different nonzero numbers, what can you conclude about the number that is used twice? Why?

60. D_W Joaquin argues that $\frac{0}{0}$ is equal to $\frac{3}{4}$ because $0 \cdot 3 = 4 \cdot 0$. Is he correct? Why or why not?

Solve.

61. 🖩 $\dfrac{1728}{5643} = \dfrac{836.4}{x}$

62. 🖩 $\dfrac{328.56}{627.48} = \dfrac{y}{127.66}$

63. $\dfrac{x}{4} = \dfrac{x-1}{6}$

64. $\dfrac{x+3}{5} = \dfrac{x}{7}$

65. Use a sequence of steps—each of which can be justified—to show that for $a, b, c, d, \neq 0$,

$$\frac{a}{b} = \frac{c}{d} \quad \text{is equivalent to} \quad \frac{d}{b} = \frac{c}{a}.$$

Objective

a Solve applied problems involving proportions.

1. Calories Burned. The readout on Marv's exercise machine tells him that if he exercises for 24 min, he will burn 356 cal. How many calories will he burn if he exercises for 30 min?

Source: Star Trac Treadmill

a Applications and Problem Solving

Proportions have applications in such diverse fields as business, chemistry, health sciences, and home economics, as well as to many areas of daily life. Proportions are useful in making predictions.

EXAMPLE 1 *Predicting Total Distance.* Donna drives her delivery van 800 mi in 3 days. At this rate, how far will she drive in 15 days?

1. **Familiarize.** We let d = the distance traveled in 15 days.

2. **Translate.** We translate to a proportion. We make each side the ratio of distance to time, with distance in the numerator and time in the denominator.

$$\begin{array}{ccc} \text{Distance in 15 days} \rightarrow & \dfrac{d}{15} = \dfrac{800}{3} & \leftarrow \text{Distance in 3 days} \\ \text{Time} \rightarrow & & \leftarrow \text{Time} \end{array}$$

It may help to verbalize the proportion above as "the unknown distance d is to 15 days as the known distance 800 mi is to 3 days."

3. **Solve.** Next, we solve the proportion:

$$3 \cdot d = 15 \cdot 800 \quad \text{Equating cross products}$$

$$\frac{3 \cdot d}{3} = \frac{15 \cdot 800}{3} \quad \text{Dividing both sides by 3}$$

$$d = \frac{15 \cdot 800}{3}$$

$$d = 4000. \quad \text{Multiplying and dividing}$$

4. **Check.** We substitute into the proportion and check cross products:

$$\frac{4000}{15} = \frac{800}{3};$$

$$4000 \cdot 3 = 12{,}000; \qquad 15 \cdot 800 = 12{,}000.$$

The cross products are the same.

5. **State.** Donna will drive 4000 mi in 15 days.

Do Exercise 1.

Problems involving proportion can be translated in more than one way. In Example 1, any one of the following is a correct translation:

$$\frac{800}{3} = \frac{d}{15}, \qquad \frac{15}{d} = \frac{3}{800}, \qquad \frac{15}{3} = \frac{d}{800}, \qquad \frac{800}{d} = \frac{3}{15}.$$

Equating the cross products in each equation gives us the equation $3 \cdot d = 15 \cdot 800$.

Answer on page A-21

EXAMPLE 2 *Recommended Dosage.* To control a fever, a doctor suggests that a child who weighs 28 kg be given 320 mg of Tylenol. If the dosage is proportional to the child's weight, how much Tylenol is recommended for a child who weighs 35 kg?

1. **Familiarize.** We let t = the number of milligrams of Tylenol.

2. **Translate.** We translate to a proportion, keeping the amount of Tylenol in the numerators.

$$\begin{array}{l} \text{Tylenol suggested} \rightarrow \\ \text{Child's weight} \rightarrow \end{array} \frac{320}{28} = \frac{t}{35} \begin{array}{l} \leftarrow \text{Tylenol suggested} \\ \leftarrow \text{Child's weight} \end{array}$$

3. **Solve.** Next, we solve the proportion:

   $320 \cdot 35 = 28 \cdot t$ Equating cross products

   $\dfrac{320 \cdot 35}{28} = \dfrac{28 \cdot t}{28}$ Dividing both sides by 28

   $\dfrac{320 \cdot 35}{28} = t$

   $400 = t.$ Multiplying and dividing

4. **Check.** We substitute into the proportion and check cross products:

   $\dfrac{320}{28} = \dfrac{400}{35};$

   $320 \cdot 35 = 11{,}200;$ $28 \cdot 400 = 11{,}200.$

   The cross products are the same.

5. **State.** The dosage for a child who weighs 35 kg is 400 mg.

Do Exercise 2.

EXAMPLE 3 *Purchasing Tickets.* Carey bought 8 tickets to an international food festival for $52. How many tickets could she purchase with $90?

1. **Familiarize.** We let n = the number of tickets that can be purchased with $90.

2. **Translate.** We translate to a proportion, keeping the number of tickets in the numerators.

$$\begin{array}{l} \text{Tickets} \rightarrow \\ \text{Cost} \rightarrow \end{array} \frac{8}{52} = \frac{n}{90} \begin{array}{l} \leftarrow \text{Tickets} \\ \leftarrow \text{Cost} \end{array}$$

2. Determining Paint Needs.
Lowell and Chris run a summer painting company to support their college expenses. They can paint 1600 ft² of clapboard with 4 gal of paint. How much paint would be needed for a building with 6000 ft² of clapboard?

Answer on page A-21

Study Tips

FIVE STEPS FOR PROBLEM SOLVING

Are you using the five steps for problem solving that were developed in Section 1.8?

1. **Familiarize** yourself with the situation.

   a) Carefully read and reread until you understand what you are being asked to find.

   b) Draw a diagram or see if there is a formula that applies.

   c) Assign a letter, or *variable*, to the unknown.

2. **Translate** the problem to an equation using the letter or variable.

3. **Solve** the equation.

4. **Check** the answer in the original wording of the problem.

5. **State** the answer to the problem clearly with appropriate units.

3. Purchasing Shirts. If 2 shirts can be bought for $47, how many shirts can be bought with $200?

3. Solve. Next, we solve the proportion.

$$52 \cdot n = 8 \cdot 90 \qquad \text{\color{red}Equating cross products}$$

$$\frac{52 \cdot n}{52} = \frac{8 \cdot 90}{52} \qquad \text{\color{red}Dividing both sides by 52}$$

$$n = \frac{8 \cdot 90}{52}$$

$$n \approx 13.8 \qquad \text{\color{red}Multiplying and dividing}$$

Because it is impossible to buy a fractional part of a ticket, we must round our answer *down* to 13.

4. Check. As a check, we use a different approach: We find the cost per ticket and then divide $90 by that price. Since $52 \div 8 = 6.50$ and $90 \div 6.50 \approx 13.8$, we have a check.

5. State. Carey could purchase 13 tickets with $90.

Do Exercise 3.

EXAMPLE 4 *Women's Hip Measurements.* For improved health, it is recommended that a woman's waist-to-hip ratio be 0.85 (or lower). Marta's hip measurement is 40 in. To meet the recommendation, what should Marta's waist measurement be?

Source: David Schmidt, "Lifting Weight Myths," *Nutrition Action Newsletter* 20, no. 4, October 1993

4. Men's Hip Measurements. It is recommended that a man's waist-to-hip ratio be 0.95 (or lower). Malcolm's hip measurement is 40 in. To meet the recommendation, what should Malcolm's waist measurement be?

Source: David Schmidt, "Lifting Weight Myths," *Nutrition Action Newsletter* 20, no. 4, October 1993

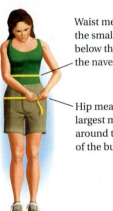

Waist measurement is the smallest measurement below the ribs but above the navel.

Hip measurement is the largest measurement around the widest part of the buttocks.

1. Familiarize. Note that $0.85 = \frac{85}{100}$. We let $w =$ Marta's waist measurement.

2. Translate. We translate to a proportion as follows:

$$\text{Waist measurement} \rightarrow \frac{w}{40} = \frac{85}{100} \quad \text{\color{red}Recommended waist-to-hip ratio}$$
$$\text{Hip measurement} \rightarrow$$

3. Solve. Next, we solve the proportion:

$$100 \cdot w = 40 \cdot 85 \qquad \text{\color{red}Equating cross products}$$

$$\frac{100 \cdot w}{100} = \frac{40 \cdot 85}{100} \qquad \text{\color{red}Dividing both sides by 100}$$

$$w = \frac{40 \cdot 85}{100}$$

$$w = 34. \qquad \text{\color{red}Multiplying and dividing}$$

Answers on page A-21

4. Check. As a check, we divide 34 by 40: $34 \div 40 = 0.85$. This is the desired ratio.

5. State. Marta's recommended waist measurement is 34 in. (or less).

Do Exercise 4 on the preceding page.

5. Construction Plans. In Example 5, the length of the actual deck is 28.5 ft. What is the length of the deck on the blueprints?

EXAMPLE 5 *Construction Plans.* Architects make blueprints of projects being constructed. These are scale drawings in which lengths are in proportion to actual sizes. The Hennesseys are constructing a rectangular deck just outside their house. The architectural blueprints are rendered such that $\frac{3}{4}$ in. on the drawing is actually 2.25 ft on the deck. The width of the deck on the drawing is 4.3 in. How wide is the deck in reality?

1. Familiarize. We let $w =$ the width of the deck.

2. Translate. Then we translate to a proportion, using 0.75 for $\frac{3}{4}$ in.

$$\begin{array}{l} \text{Measure on drawing} \rightarrow \\ \text{Measure on deck} \rightarrow \end{array} \frac{0.75}{2.25} = \frac{4.3}{w} \begin{array}{l} \leftarrow \text{Width of drawing} \\ \leftarrow \text{Width of deck} \end{array}$$

3. Solve. Next, we solve the proportion:

$0.75 \times w = 2.25 \times 4.3$ Equating cross products

$\dfrac{0.75 \times w}{0.75} = \dfrac{2.25 \times 4.3}{0.75}$ Dividing both sides by 0.75

$w = \dfrac{2.25 \times 4.3}{0.75}$

$w = 12.9.$

4. Check. We substitute into the proportion and check cross products:

$\dfrac{0.75}{2.25} = \dfrac{4.3}{12.9};$

$0.75 \times 12.9 = 9.675;$ $2.25 \times 4.3 = 9.675.$

The cross products are the same.

5. State. The width of the deck is 12.9 ft.

Do Exercise 5.

Answer on page A-21

6. **Estimating a Deer Population.**
To determine the number of deer in a forest, a conservationist catches 612 deer, tags them, and releases them. Later, 244 deer are caught, and it is found that 72 of them are tagged. Estimate how many deer are in the forest.

EXAMPLE 6 *Estimating a Wildlife Population.* To determine the number of fish in a lake, a conservationist catches 225 fish, tags them, and throws them back into the lake. Later, 108 fish are caught, and it is found that 15 of them are tagged. Estimate how many fish are in the lake.

1. **Familiarize.** Our strategy is to form two different ratios which can be used to represent the ratio of tagged fish to all fish in the lake. One way to write such a ratio is simply

$$\frac{\text{Number of tagged fish}}{\text{Number of fish in lake}}.$$

A second way to represent this ratio assumes that the tagged fish become uniformly distributed throughout the lake. We then form the ratio

$$\frac{\text{Number of tagged fish caught}}{\text{Total number of fish caught}}.$$

We let F = the total number of fish in the lake.

2. **Translate.** We translate to a proportion as follows:

$$\text{Fish } \textit{tagged} \text{ originally} \rightarrow \frac{225}{F} = \frac{15}{108}. \leftarrow \textit{Tagged} \text{ fish caught later}$$
$$\text{Fish in lake} \rightarrow \qquad\qquad \leftarrow \text{Fish caught later}$$

3. **Solve.** Next, we solve the proportion:

$$225 \cdot 108 = F \cdot 15 \qquad \text{Equating cross products}$$

$$\frac{225 \cdot 108}{15} = \frac{F \cdot 15}{15} \qquad \text{Dividing both sides by 15}$$

$$\frac{225 \cdot 108}{15} = F$$

$$1620 = F. \qquad \text{Multiplying and dividing}$$

4. **Check.** We substitute into the proportion and check cross products:

$$\frac{225}{1620} = \frac{15}{108};$$

$$225 \cdot 108 = 24{,}300; \qquad 1620 \cdot 15 = 24{,}300.$$

The cross products are the same.

5. **State.** We estimate that there are 1620 fish in the lake.

Do Exercise 6.

Answer on page A-21

Translating for Success

1. *Calories in Cereal.* There are 140 calories in a $1\frac{1}{2}$-cup serving of Brand A cereal. How many calories are there in 6 cups of the cereal?

2. *Calories in Cereal.* There are 140 calories in 6 cups of Brand B cereal. How many calories are there in a $1\frac{1}{2}$-cup serving of the cereal?

3. *Gallons of Gasoline.* Nate's SUV traveled 310 mi on 15.5 gal of gasoline. At this rate, how many gallons would be needed to travel 465 mi?

4. *Gallons of Gasoline.* Elizabeth's new fuel-efficient car traveled 465 mi on 15.5 gal of gasoline. At this rate, how many gallons will be needed to travel 310 mi?

5. *Perimeter.* Find the perimeter of a rectangular field that measures 83.7 m by 62.4 m.

The goal of these matching questions is to practice step (2), *Translate*, of the five-step problem-solving process. Translate each word problem to an equation and select a correct translation from equations A–O.

A. $\dfrac{310}{15.5} = \dfrac{465}{x}$

B. $180 = 1\frac{1}{2} \cdot x$

C. $x = 71\frac{1}{8} - 76\frac{1}{2}$

D. $71\frac{1}{8} \cdot x = 74$

E. $74 \cdot 71\frac{1}{8} = x$

F. $x = 83.7 + 62.4$

G. $71\frac{1}{8} + x = 76\frac{1}{2}$

H. $x = 1\frac{2}{3} \cdot 180$

I. $\dfrac{140}{6} = \dfrac{x}{1\frac{1}{2}}$

J. $x = 2(83.7 + 62.4)$

K. $\dfrac{465}{15.5} = \dfrac{310}{x}$

L. $x = 83.7 \cdot 62.4$

M. $x = 180 \div 1\frac{2}{3}$

N. $\dfrac{140}{1\frac{1}{2}} = \dfrac{x}{6}$

O. $x = 1\frac{2}{3} \div 180$

Answers on page A-21

---

6. *Electric Bill.* Last month, Todd's electric bills for his two rentals were $83.70 and $62.40. What was the total electric bill for the two properties?

7. *Package Tape.* A postal service center uses rolls of package tape that each contain 180 ft of tape. If they use an average of $1\frac{2}{3}$ ft per package, how many packages can be taped with one roll?

8. *Online Price.* Jane spent $180 for an area rug in a department store. Later she saw the same rug for sale online and realized she had paid $1\frac{1}{2}$ times the online price. What was the online price?

9. *Heights of Sons.* Henry's three sons play basketball on three different college teams. Jeff's, Jason's, and Jared's heights are 74 in., $71\frac{1}{8}$ in., and $76\frac{1}{2}$ in., respectively. How much taller is Jared than Jason?

10. *Total Investment.* An investor bought 74 shares of stock at $71\frac{1}{8}$ per share. What was the total investment?

a   Solve.

1. *Study Time and Test Grades.* An English instructor asserted that students' test grades are directly proportional to the amount of time spent studying. Lisa studies 9 hr for a particular test and gets a score of 75. At this rate, how many hours would she have had to study to get a score of 92?

2. *Study Time and Test Grades.* A mathematics instructor asserted that students' test grades are directly proportional to the amount of time spent studying. Brent studies 15 hr for a particular test and gets a score of 85. At this rate, what score would he have received if he had studied 16 hr?

3. *Cap'n Crunch's Peanut Butter Crunch® Cereal.* The nutritional chart on the side of a box of Quaker Cap'n Crunch's Peanut Butter Crunch® Cereal states that there are 110 calories in a $\frac{3}{4}$-cup serving. How many calories are there in 6 cups of the cereal?

4. *Rice Krispies® Cereal.* The nutritional chart on the side of a box of Kellogg's Rice Krispies® Cereal states that there are 120 calories in a $1\frac{1}{4}$-cup serving. How many calories are there in 5 cups of the cereal?

Nutrition Facts

Serving Size 3/4 Cup (27g)
Servings Per Container about 15

Amount Per Serving	Cereal Alone	with 1/2 Cup Vitamin A&D Fortified Skim Milk
Calories	110	150
Calories from Fat	25	25
	% Daily Value	
Total Fat 2.5g	4%	4%
Saturated Fat 0.5g	3%	3%
Polyunsaturated Fat 0.5g		
Monounsaturated Fat 1g		
Cholesterol 0mg	0%	0%
Sodium 200mg	8%	11%
Potassium 65mg	2%	8%
Total Carbohydrate 21g	7%	9%
Dietary Fiber 1g	3%	3%
Sugars 9g		
Other Carbohydrate 12g		
Protein 2g		

Nutrition Facts

Serving Size $1\frac{1}{4}$ Cups (33g/1.2oz)
Servings Per Container about 12

Amount Per Serving	Cereal	Cereal with 1/2 Cup Vitamins A&D Fat Free Milk
Calories	120	160
Calories from Fat	0	0
	% Daily Value	
Total Fat 0g	0%	0%
Saturated Fat 0g	0%	0%
Trans Fat 0g		
Cholesterol 0mg	0%	0%
Sodium 320mg	13%	16%
Potassium 40mg	1%	7%
Total Carbohydrate 29g	10%	11%
Dietary Fiber 0g	0%	0%
Sugars 3g		
Other Carbohydrate 26g		
Protein 2g		

5. *Overweight Americans.* A study recently confirmed that of every 100 Americans, 60 are considered overweight. There were 295 million Americans in 2005. How many would be considered overweight?

Source: U.S. Centers for Disease Control

6. *Cancer Death Rate in Illinois.* It is predicted that for every 1000 people in the state of Illinois, 130.9 will die of cancer. The population of Chicago is about 2,721,547. How many of these people will die of cancer?

Source: *2001 New York Times Almanac*

7. *Gasoline Mileage.* Nancy's van traveled 84 mi on 6.5 gal of gasoline. At this rate, how many gallons would be needed to travel 126 mi?

8. *Bicycling.* Roy bicycled 234 mi in 14 days. At this rate, how far would Roy travel in 42 days?

9. Quality Control. A quality-control inspector examined 100 lightbulbs and found 7 of them to be defective. At this rate, how many defective bulbs will there be in a lot of 2500?

10. Grading. A professor must grade 32 essays in a literature class. She can grade 5 essays in 40 min. At this rate, how long will it take her to grade all 32 essays?

11. Sugaring. When 38 gal of maple sap are boiled down, the result is 2 gal of maple syrup. How much sap is needed to produce 9 gal of syrup?

12. Lefties. In a class of 40 students, on average, 6 will be left-handed. If a class includes 9 "lefties," how many students would you estimate are in the class?

13. Painting. Fred uses 3 gal of paint to cover 1275 ft^2 of siding. How much siding can Fred paint with 7 gal of paint?

14. Waterproofing. Bonnie can waterproof 450 ft^2 of decking with 2 gal of sealant. How many gallons should Bonnie buy for a 1200-ft^2 deck?

15. Publishing. Every 6 pages of an author's manuscript corresponds to 5 published pages. How many published pages will a 540-page manuscript become?

16. Turkey Servings. An 8-lb turkey breast contains 36 servings of meat. How many pounds of turkey breast would be needed for 54 servings?

17. Gas Mileage. A 2005 Ford Mustang GT Convertible will go 372 mi on 15.5 gal of gasoline in highway driving.

a) How many gallons of gasoline will it take to drive 2690 mi from Boston to Phoenix?
b) How far can the car be driven on 140 gal of gasoline?
Source: Ford Motor Company

18. Gas Mileage. A 2005 Volkswagen Passat will go 462 mi on 16.5 gal of gasoline in highway driving.

a) How many gallons of gasoline will it take to drive 1650 mi from Pittsburgh to Albuquerque?
b) How far can the car be driven on 130 gal of gasoline?
Source: Volkswagen of America, Inc.

19. Mileage. Jean bought a new car. In the first 8 months, it was driven 9000 mi. At this rate, how many miles will the car be driven in 1 yr?

20. Coffee Production. Coffee beans from 14 trees are required to produce the 17 lb of coffee that the average person in the United States drinks each year. How many trees are required to produce 375 lb of coffee?

21. *Metallurgy.* In a metal alloy, the ratio of zinc to copper is 3 to 13. If there are 520 lb of copper, how many pounds of zinc are there?

22. *Class Size.* Pane College advertises that its student-to-faculty ratio is 14 to 1. If 56 students register for Introductory Spanish, how many sections of the course would you expect to see offered?

23. *Painting.* Helen can paint 950 ft² with 2 gal of paint. How many 1-gal cans does she need in order to paint a 30,000-ft² wall?

24. *Snow to Water.* Under typical conditions, $1\frac{1}{2}$ ft of snow will melt to 2 in. of water. To how many inches of water will $5\frac{1}{2}$ ft of snow melt?

25. *Grass-Seed Coverage.* It takes 60 oz of grass seed to seed 3000 ft² of lawn. At this rate, how much would be needed for 5000 ft² of lawn?

26. *Seating Capacity.* At the Bertocinis' church, two pews can seat 14 people. How many pews will be needed for a wedding party of 44 people?

27. *Estimating a Deer Population.* To determine the number of deer in a game preserve, a forest ranger catches 318 deer, tags them, and releases them. Later, 168 deer are caught, and it is found that 56 of them are tagged. Estimate how many deer are in the game preserve.

28. *Estimating a Trout Population.* To determine the number of trout in a lake, a conservationist catches 112 trout, tags them, and throws them back into the lake. Later, 82 trout are caught, and it is found that 32 of them are tagged. Estimate how many trout there are in the lake.

29. *Map Scaling.* On a road atlas map, 1 in. represents 16.6 mi. If two cities are 3.5 in. apart on the map, how far apart are they in reality?

30. *Map Scaling.* On a map, $\frac{1}{4}$ in. represents 50 mi. If two cities are $3\frac{1}{4}$ in. apart on the map, how far apart are they in reality?

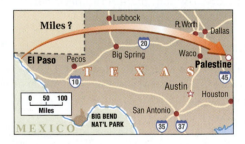

31. **DW** Can unit prices be used to solve proportions that involve money? Why or why not?

32. **DW** Polly solved Example 3 by forming the proportion

$$\frac{90}{52} = \frac{n}{8},$$

whereas Rudy wrote

$$\frac{52}{n} = \frac{90}{8}.$$

Are both approaches valid? Why or why not?

Find the prime factorization of each number. [3.2c]

33. 808
34. 28
35. 866
36. 93
37. 2020

38. Multiply: $-19.3(4.1)$ [5.3a]

39. Divide: $-13.11 \div 5.7$ [5.4a]

40. Divide: $169.36 \div (-23.2)$ [5.4a]

41. Add: $-19.7 + 12.5$ [5.2c]

42. Subtract: $-3.7 - (-1.9)$ [5.2c]

43. **D**w Examine Exercises 27 and 28. Which of the estimates do you think is more reliable? Why?

44. **D**w Rob's waist and hips measure 35 in. and 33 in., respectively (see Margin Exercise 4). Suppose that Rob can either gain or lose 1 in. from one of his measurements. Where should the inch come from or go to? Why?

45. ▦ *Exchanging Money.* On 7 July 2006, 1 U.S. dollar was worth about 0.544174 British pound.
   a) How much would 45 U.S. dollars be worth in British pounds?
   b) How much would a car that costs 8640 British pounds cost in U.S. dollars?

46. ▦ *Exchanging Money.* On 7 July 2006, 1 U.S. dollar was worth about 1.22869 Swiss francs.
   a) How much would 360 U.S. dollars be worth in Swiss francs?
   b) How much would a pair of jeans that costs 80 Swiss francs cost in U.S. dollars?

47. ▦ Carney College is expanding from 850 to 1050 students. To avoid any rise in the student-to-faculty ratio, the faculty of 69 professors must also increase. How many new faculty positions should be created?

48. ▦ In recognition of her outstanding work, Sheri's salary has been increased from $26,000 to $29,380. Tim is earning $23,000 and is requesting a proportional raise. How much more should he ask for?

49. *Baseball Statistics.* Cy Young, one of the greatest baseball pitchers of all time, gave up an average of 2.63 earned runs every 9 innings. Young pitched 7356 innings, more than anyone else in the history of baseball. How many earned runs did he give up?

50. ▦ *Real-Estate Values.* According to Coldwell Banker Real Estate Corporation, a home selling for $189,000 in Austin, Texas, would sell for $665,795 in San Francisco. How much would a $450,000 home in San Francisco sell for in Austin? Round to the nearest $1000.
   Source: Coldwell Banker Real Estate Corporation

51. ▦ The ratio $1:3:2$ is used to estimate the relative costs of a CD player, receiver, and speakers when shopping for a sound system. That is, the receiver should cost three times the amount spent on the CD player and the speakers should cost twice as much as the amount spent on the CD player. If you had $800 to spend, how would you allocate the money, using this ratio?

Objectives

a Find lengths of sides of similar triangles using proportions.

b Use proportions to find lengths in pairs of figures that differ only in size.

1. This pair of triangles is similar. Find the missing length x.

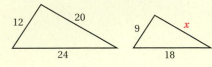

Answer on page A-21

Study Tips

MAKING SKETCHES

One need not be an artist to make highly useful mathematical sketches. That said, it is important to make sure that your sketches are drawn accurately enough to represent the relative sizes within each shape. For example, if one side of a triangle is clearly the longest, make sure your drawing reflects this.

a Proportions and Similar Triangles

Look at the pair of triangles below. Note that they appear to have the same shape, but their sizes are different. These are examples of **similar triangles.** By using a magnifying glass, you could imagine enlarging the smaller triangle to get the larger. This process works because the corresponding sides of each triangle have the same ratio. That is, the following proportion is true.

$$\frac{a}{d} = \frac{b}{e} = \frac{c}{f}$$

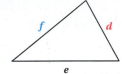

SIMILAR TRIANGLES

Similar triangles have the same shape. The lengths of their corresponding sides have the same ratio—that is, they are proportional.

EXAMPLE 1 The triangles below are similar triangles. Find the missing length x.

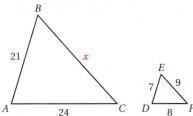

The corresponding sides are x and 9, 21 and 7, and 24 and 8. Thus the ratio of x to 9 is the same as the ratio of 21 to 7 or 24 to 8. One proportion we can write is

$$\text{Side of larger triangle} \rightarrow \frac{x}{9} = \frac{24}{8} \leftarrow \text{Side of larger triangle}$$
$$\text{Side of smaller triangle} \quad \leftarrow \text{Side of smaller triangle}$$

Other proportions could also be used. We could write $\frac{x}{9} = \frac{21}{7}$, or, writing the lengths of the smaller triangle in the numerators, $\frac{9}{x} = \frac{8}{24}$ or $\frac{9}{x} = \frac{7}{21}$. We can solve any of these proportions. We use the first:

$$\frac{x}{9} = \frac{24}{8}$$

$$x \cdot 8 = 24 \cdot 9 \qquad \text{Equating cross products}$$

$$\frac{x \cdot 8}{8} = \frac{24 \cdot 9}{8} \qquad \text{Dividing both sides by 8}$$

$$x = 27. \qquad \text{Simplifying}$$

The missing length x is 27.

Do Exercise 1.

Similar triangles and proportions are often used to compute lengths that would ordinarily be difficult to measure. For example, we could find the height of a flagpole without climbing it or the distance across a river without crossing it.

EXAMPLE 2 *Sculpture.* The works of the great sculptor Auguste Rodin were so popular that he arranged to have enlarged or reduced copies produced of many pieces. Using a Collas machine, Rodin's assistants were able to create a proportionately larger or smaller duplicate of the original sculpture.

Source: www.cantorfoundation.org

2. Rodin's most famous sculpture, *The Thinker*, was originally 28 in. high. If the man's head on that version was 6 in. long, how long would the head be in the more famous 79-in. rendition? Round to the nearest tenth of an inch.

Source: www.cantorfoundation.org

Suppose a Collas machine is used to enlarge a statue that is 35 cm tall into one that is 50 cm tall. If the original statue includes a forearm that is 4 cm long, how long will the forearm be in the enlargement?

We let l = the length of the enlarged forearm, in centimeters. Then we translate to a proportion and solve.

$$\begin{array}{c}\text{Length of}\\\text{enlarged forearm} \rightarrow\end{array} \frac{l}{50} = \frac{4}{35} \begin{array}{c}\leftarrow \text{Length of}\\ \leftarrow \text{original forearm}\end{array}$$

Length of enlarged forearm → $\frac{l}{50}$ = $\frac{4}{35}$ ← Length of original forearm
Height of enlarged statue → ← Height of original statue

$$35 \cdot l = 4 \cdot 50 \qquad \text{Equating cross products}$$

$$l = \frac{4 \cdot 50}{35} \qquad \text{Dividing both sides by 35}$$

$$l \approx 5.7 \qquad \begin{array}{l}\text{Simplifying and rounding to}\\\text{the nearest tenth}\end{array}$$

As a partial check, note that, as expected, the length of the enlarged forearm is longer than the length of the original forearm. The enlarged forearm will be about 5.7 cm long.

Do Exercise 2.

Answer on page A-21

3. Rafters of a House. Referring to Example 3, find the length y in the rafter of the house to the nearest tenth of a foot.

EXAMPLE 3 *Rafters of a House.* Carpenters use similar triangles to determine the lengths of rafters for a house. They first choose the pitch of the roof, or the ratio of the rise over the run. Then using a triangle with that ratio, they calculate the length of the rafter needed for the house. Loren is constructing rafters for a roof with a 6/12 pitch on a house that is 30 ft wide. Using a rafter guide, Loren knows that the rafter length corresponding to the 6/12 pitch is 13.4. Find the length x of the rafter of the house to the nearest tenth of a foot.

13.4
Rise: 6
Run: 12
Pitch: 6/12

We see from the figure that we have the proportion

Length of rafter in 6/12 triangle → $\dfrac{13.4}{x} = \dfrac{12}{15}$. ← Run in 6/12 triangle
Length of rafter on the house → ← Run in similar triangle on the house

Solve: $13.4 \cdot 15 = x \cdot 12$ Equating cross products

$\dfrac{13.4 \cdot 15}{12} = \dfrac{x \cdot 12}{12}$ Dividing both sides by 12

$\dfrac{13.4 \cdot 15}{12} = x$

$16.8 \text{ ft} \approx x$ Rounding to the nearest tenth of a foot

The length x of the rafter of the house is about 16.8 ft.

Do Exercise 3.

b Proportions and Other Geometric Shapes

When one geometric figure is a magnification of another, the figures are similar. Thus the corresponding lengths are proportional.

EXAMPLE 4 The sides in the negative and photograph shown are proportional. Find the width of the photograph.

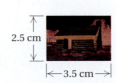

2.5 cm

←3.5 cm→

x

←————— 10.5 cm —————→

We let x = the width of the photograph. Then we translate to a proportion.

Photo width → $\dfrac{x}{2.5} = \dfrac{10.5}{3.5}$ ← Photo length

Negative width →   ← Negative length

Solve: $\quad 3.5 \cdot x = 2.5 \cdot 10.5 \qquad$ Equating cross products

$$\frac{3.5 \cdot x}{3.5} = \frac{2.5 \cdot 10.5}{3.5} \qquad \text{Dividing both sides by 3.5}$$

$$x = \frac{2.5 \cdot 10.5}{3.5} \qquad \text{Simplifying}$$

$$x = 7.5$$

Thus the width of the photograph is 7.5 cm.

Do Exercise 4.

EXAMPLE 5 A scale model of an addition to an athletic facility is 12 cm wide at the base and rises to a height of 15 cm. If the actual base is to be 116 ft, what will be the actual height of the addition?

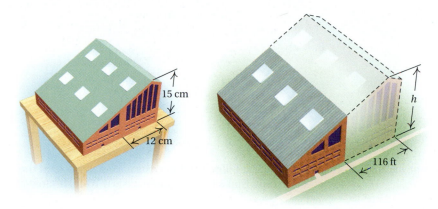

We let h = the height of the addition. Then we translate to a proportion.

Width in model → $\dfrac{12}{116} = \dfrac{15}{h}$ ← Height in model

Actual width →   ← Actual height

Solve: $\quad 12 \cdot h = 116 \cdot 15 \qquad$ Equating cross products

$$\frac{12 \cdot h}{12} = \frac{116 \cdot 15}{12} \qquad \text{Dividing both sides by 12}$$

$$h = \frac{116 \cdot 15}{12} = 145$$

Thus the height of the addition will be 145 ft.

Do Exercise 5.

4. The sides in the photographs below are proportional. Find the width of the larger photograph.

6 cm

←10 cm→

x

35 cm

5. Refer to the figures in Example 5. If a model skylight is 3 cm wide, how wide will the actual skylight be?

Answers on page A-21

7.5 EXERCISE SET For Extra Help

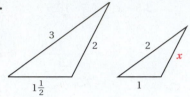

MathXL MyMathLab InterAct Math Math Tutor Center Video Lectures on CD Disc 4 Student's Solutions Manual

a The triangles in each exercise are similar. Find the missing lengths.

1.

2.
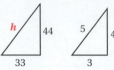

3.

4.
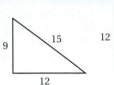

5.

6.

7.
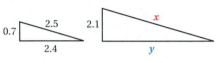

8.

9. How tall is a flagpole that casts a 56-ft shadow at the same time that a 6-ft man casts a 5-ft shadow?

10. How tall is a billboard that casts a 45-ft shadow at the same time that a 5.5-ft woman casts a 10-ft shadow?

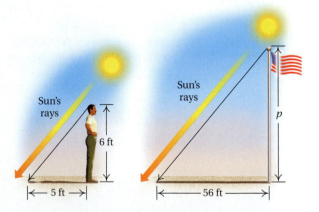

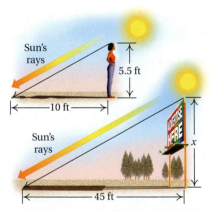

11. How tall is a flagpole that casts a 42-ft shadow at the same time that a $5\frac{1}{2}$-ft woman casts a 7-ft shadow?

12. When a tree 8 m tall casts a shadow 5 m long, how long is the shadow cast by a man who is 2 m tall?

13. How tall is a tree that casts a 32-ft shadow at the same time that an 8-ft light pole casts a 9-ft shadow?

14. How tall is a tree that casts a 27-ft shadow at the same time that a 4-ft fence post casts a 3-ft shadow?

15. Find the height h of the wall.

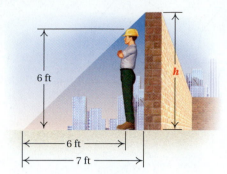

6 ft

6 ft

7 ft

16. Find the length L of the lake. Assume that the ratio of L to 120 yd is the same as the ratio of 720 yd to 30 yd.

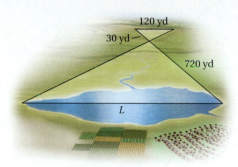

120 yd

30 yd

720 yd

L

17. Find the distance across the river. Assume that the ratio of d to 25 ft is the same as the ratio of 40 ft to 10 ft.

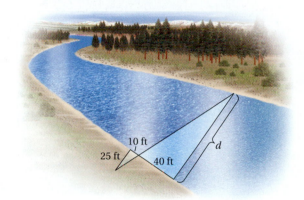

10 ft

25 ft

40 ft

d

18. To measure the height of a hill, a string is drawn tight from level ground to the top of the hill. A 3-ft stick is placed under the string, touching it at point P, a distance of 5 ft from point G, where the string touches the ground. The string is then detached and found to be 120 ft long. How high is the hill?

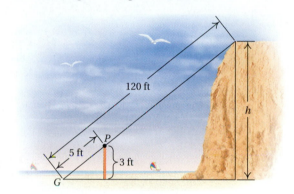

120 ft

P

5 ft

3 ft

h

G

b In each of Exercises 19–28, the sides in each pair of figures are proportional. Find the missing lengths.

19.

6

9

x

6

20.

x

5

7

14

21.

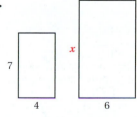

7

x

4

6

22.

11

x

4

3

23.

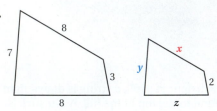

24.

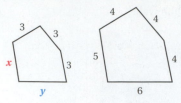

25.

26.

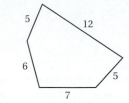

27.

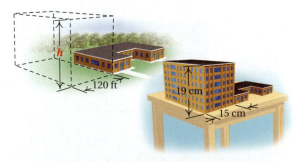

28.

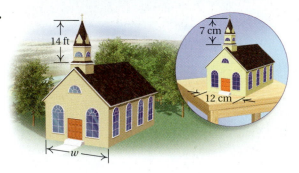

29. A scale model of an addition to an athletic facility is 15 cm wide at the base and rises to a height of 19 cm. If the actual base is to be 120 ft, what will be the height of the addition?

30. Refer to the figures in Exercise 29. If a model skylight is 3 cm wide, how wide will the actual skylight be?

31. **Dw** Is it possible for two triangles to have two pairs of sides that are proportional without the triangles being similar? Why or why not?

32. **Dw** Design a problem for a classmate to solve (see Exercises 9–18) so that the solution is "The height of the sail is 12 ft."

33. *Expense Needs.* Carlos has $34.97 to spend for a book at $49.95, a CD at $14.88, and a sweatshirt at $29.95. How much more money does Carlos need to make these purchases? [5.8a]

34. Divide: $80.892 \div 8.4$ [5.4a]

Multiply. [5.3a]

35. -8.4×80.892

36. 0.01×274.568

37. 100×274.568

38. $-0.002 \times (-274.568)$

Find decimal notation and round to the nearest thousandth, if appropriate. [5.5a, b]

39. $\dfrac{17}{20}$

40. $-\dfrac{73}{40}$

41. $-\dfrac{10}{11}$

42. $\dfrac{43}{51}$

43. **D$_W$** Suppose that all the sides in one triangle are half the size of the corresponding sides in a similar triangle. Does it follow that the area of the smaller triangle is half the area of the larger triangle? Why or why not?

44. **D$_W$** Design for a classmate a problem involving similar triangles for which
$$\frac{18}{128.95} = \frac{x}{789.89}.$$

Hockey Goals. An official hockey goal is 6 ft wide. To make scoring more difficult, goalies often locate themselves far in front of the goal to "cut down the angle." In Exercises 45 and 46, suppose that a slapshot from point A is attempted and that the goalie is 2.7 ft wide. Determine how far from the goal the goalie should be located if point A is the given distance from the goal. (*Hint*: First find how far the goalie should be from point A.)

45. ▦ 25 ft

46. ▦ 35 ft

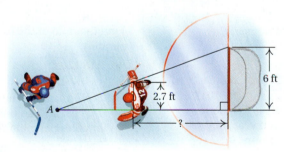

47. A miniature basketball hoop is built for the model referred to in Exercise 29. An actual hoop is 10 ft high. How high should the model hoop be?

▦ The triangles in each exercise are similar triangles. Find the lengths not given.

48.

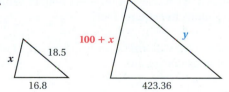

49.

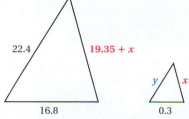

The review that follows is meant to prepare you for a chapter exam. It consists of two parts. The first part, Concept Reinforcement, is designed to increase understanding of the concepts through true/false exercises. The second part is the Review Exercises. These provide practice exercises for the exam, together with references to section objectives so you can go back and review. Before beginning, stop and look back over the skills you have obtained. What skills in mathematics do you have now that you did not have before studying this chapter?

⤴ CONCEPT REINFORCEMENT

Determine whether the statement is true or false. Answers are given at the back of the book.

_____ **1.** The proportion $\dfrac{a}{b} = \dfrac{c}{d}$ can also be written as $\dfrac{c}{a} = \dfrac{d}{b}$.

_____ **2.** Lengths of corresponding sides of similar triangles have the same ratio.

_____ **3.** The larger size of a product always has the lower unit price.

_____ **4.** If $\dfrac{x}{t} = \dfrac{y}{s}$, then $xy = ts$.

_____ **5.** A ratio is a quotient of two quantities.

Review Exercises

Write fraction notation for the ratio. Do not simplify.   [7.1a]

1. 47 to 84

2. 46 to 1.27

3. 83 to 100

4. 0.72 to 197

5. _Kona Jack's Restaurants._   Kona Jack's is a seafood restaurant chain in Indianapolis. Each year they sell 12,480 lb of tuna and 16,640 lb of salmon.   [7.1a]

  a) Write fraction notation for the ratio of tuna sold to salmon sold.

  b) Write fraction notation for the ratio of salmon sold to the total number of pounds of both kinds of fish.

Source: Kona Jack's Restaurants

Find the ratio of the first number to the second number and simplify.   [7.1b]

6. 9 to 12

7. 3.6 to 6.4

8. _Gas Mileage._   The Chrysler PT Cruiser will go 377 mi on 14.5 gal of gasoline in highway driving. What is the rate in miles per gallon?   [7.2a]

Source: DaimlerChrysler Corporation

9. _CD-ROM Spin Rate._   A 12x CD-ROM on a computer will spin 472,500 revolutions if left running for 75 min. What is the rate of its spin in revolutions per minute (rpm)?   [7.2a]

Source: _Electronic Engineering Times,_ June 1997

10. A lawn requires 319 gal of water for every 500 ft². What is the rate in gallons per square foot?   [7.2a]

11. _Turkey Servings._   A 25-lb turkey serves 18 people. Find the rate in servings per pound.   [7.2a]

12. _Calcium Supplement._   The price for a particular calcium supplement is $12.99 for 300 tablets. Find the unit price in cents per tablet.   [7.2b]

13. *Pillsbury Orange Breakfast Rolls.* The price for these breakfast rolls is $1.97 for 13.9 oz. Find the unit price in cents per ounce. [7.2b]

In each of Exercises 14 and 15, find the unit prices. Then determine in each case which has the lowest unit price. [7.2b]

14. *Paper Towels.*

PACKAGE	PRICE	UNIT PRICE PER SHEET
8 rolls, 60 (2 ply) sheets per roll	$6.38	
15 rolls, 60 (2 ply) sheets per roll	$13.99	
6 big rolls, 165 (2 ply) sheets per roll	$10.99	

15. *Crisco Oil.*

PACKAGE	PRICE	UNIT PRICE
32 oz	$2.19	
48 oz	$2.49	
64 oz	$3.59	
128 oz	$7.09	

Determine whether the two pairs of numbers are proportional. [7.3a]

16. 9, 15 and 36, 59

17. 24, 37 and 40, 46.25

Solve. [7.3b]

18. $\dfrac{8}{9} = \dfrac{x}{36}$

19. $\dfrac{6}{x} = \dfrac{48}{56}$

20. $\dfrac{120}{\frac{3}{7}} = \dfrac{7}{x}$

21. $\dfrac{4.5}{120} = \dfrac{0.9}{x}$

Solve. [7.4a]

22. If 3 dozen eggs cost $2.67, how much will 5 dozen eggs cost?

23. *Quality Control.* A factory manufacturing computer circuits found 39 defective circuits in a lot of 65 circuits. At this rate, how many defective circuits can be expected in a lot of 585 circuits?

24. *Exchanging Money.* On 7 July 2006, 1 U.S. dollar was worth about 0.782459 European Monetary Unit (Euro).
   a) How much would 250 U.S. dollars be worth in Euros?
   b) Jamal was traveling in France and saw a sweatshirt that cost 50 Euros. How much would it cost in U.S. dollars?

25. A train travels 448 mi in 7 hr. At this rate, how far will it travel in 13 hr?

26. Fifteen acres are required to produce 54 bushels of tomatoes. At this rate, how many acres are required to produce 97.2 bushels of tomatoes?

27. *Garbage Production.*   It is known that 5 people produce 13 kg of garbage in one day. San Diego, California, has 1,266,753 people. How many kilograms of garbage are produced in San Diego in one day?

28. *Snow to Water.*   Under typical conditions, $1\frac{1}{2}$ ft of snow will melt to 2 in. of water. To how many inches of water will $4\frac{1}{2}$ ft of snow melt?

29. *Lawyers in Michigan.*   In Michigan, there are 2.3 lawyers for every 1000 people. The population of Detroit is 911,402. How many lawyers would you expect there to be in Detroit?

Source: U.S. Bureau of the Census

Each pair of triangles in Exercises 30 and 31 is similar. Find the missing length(s).   [7.5a]

30.

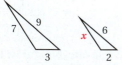

31.

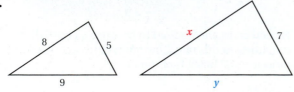

32. How high is a billboard that casts a 25-ft shadow at the same time that an 8-ft sapling casts a 5-ft shadow? [7.5a]

33. The lengths in the figures below are proportional. Find the missing lengths.   [7.5b]

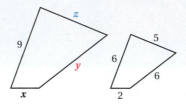

34. **D**_{**W**} If you were a college president, which would you prefer: a low or high faculty-to-student ratio? Why? [7.1a]

35. **D**_{**W**} Write a proportion problem for a classmate to solve. Design the problem so that the solution is "Leslie would need 16 gal of gasoline in order to travel 368 mi." [7.4a]

SYNTHESIS

36. It takes Yancy Martinez 10 min to type two-thirds of a page of his term paper. At this rate, how long will it take him to type a 7-page term paper?   [7.4a]

37. Sari is making beaded bracelets from purple beads and lavender beads. If the ratio of purple to lavender beads is 3 to 5 and each bracelet contains 40 beads, how many bracelets can Sari make from 60 purple beads? How many lavender beads will she need?   [7.4a]

38. A medical insurance claims service analyst can type 8500 keystrokes per hour. At this rate, how many minutes would 34,000 keystrokes require?   [7.4a]

Source: www.or.regence.com

39. Shine-and-Glo Painters uses 2 gal of finishing paint for every 3 gal of primer. Each gallon of finishing paint covers 450 ft². If a surface of 4950 ft² needs both primer and finishing paint, how many gallons of each should be purchased?   [7.4a]

Write fraction notation for the ratio. Do not simplify.

1. 85 to 97

2. 0.34 to 124

3. A diver descends 10 ft in 16 sec. What is the rate of descent in feet per second?

4. *Gas Mileage.* The 2005 Chevrolet Malibu Maxx will go 319 mi on 14.5 gal of gasoline in city driving. What is the rate in miles per gallon?
Source: General Motors Corporation

5. The following table lists prices for various packages of Tide laundry detergent powder. Find the unit price of each package. Then determine which has the lowest unit price.

PACKAGE	PRICE	UNIT PRICE
33 oz	$3.69	
87 oz	$6.22	
131 oz	$10.99	
263 oz	$17.99	

6. Simplify the ratio of length to width in this rectangle.

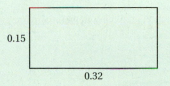

Determine whether the two pairs of numbers are proportional.

7. 7, 8 and 63, 72

8. 1.3, 3.4 and 5.6, 15.2

Solve.

9. $\dfrac{9}{4} = \dfrac{27}{x}$

10. $\dfrac{150}{2.5} = \dfrac{x}{6}$

11. $\dfrac{x}{100} = \dfrac{27}{64}$

12. $\dfrac{68}{y} = \dfrac{17}{25}$

Solve.

13. *Map Scaling.* On a map, 3 in. represents 225 mi. If two cities are 7 in. apart on the map, how far are they apart in reality?

14. *Tower Height.* A birdhouse built on a pole that is 3 m high casts a shadow 5 m long. At the same time, the shadow of a tower is 110 m long. How high is the tower?

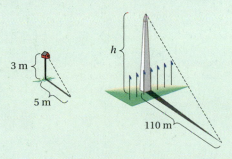

15. *Distance Traveled.* An ocean liner traveled 432 km in 12 hr. At this rate, how far would the boat travel in 42 hr?

16. *Time Loss.* A watch loses 2 min in 10 hr. At this rate, how much will it lose in 24 hr?

17. A drama teacher reserves 9 copies of *The Complete Works of William Shakespeare* for her class. If the ratio of books to students is 3 to 5, how many students are in the class?

18. A city measure in Oregon proposed that homeowners be taxed 39 cents per every $1000 of property value to help maintain parks and to fund recreation programs. How much would the owner of a $135,000 home pay?

The lengths in each pair of figures are proportional. Find the missing lengths.

19.

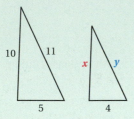

20.

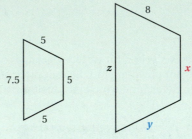

SYNTHESIS

Solve.

21. $\dfrac{x + 3}{4} = \dfrac{5x + 2}{8}$

22. $\dfrac{4 - 6x}{5} = \dfrac{x + 3}{2}$

23. Lara needs 3 balls of burgundy yarn and $2\frac{1}{2}$ balls of black yarn to knit a sweater. She buys 10 balls of the black yarn. How much burgundy yarn does she need and how many sweaters can she knit?

24. The Johnson triplets and the Solomini twins went out to dinner and decided to split the bill of $79.85 proportionately. How much will the Johnsons pay?

25. Find the width of the river in the following diagram.

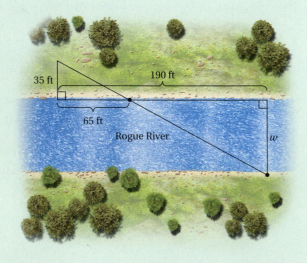

26. A soccer goalie wishing to block an opponent's shot moves toward the shooter to reduce the shooter's view of the goal. If the goalie can only defend a region 10 ft wide, how far in front of the goal should the goalie be? (See the following figure.)

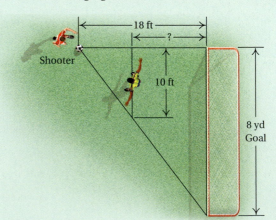

Add and simplify.

1.
```
  1 3 7.1 8 6
    2 3.0 1 9
+ 4 8 3.2 9 7
```

2. $3\dfrac{4}{5}$

$+\ 8\dfrac{7}{10}$

3. $\dfrac{6}{35} + \dfrac{5}{28}$

Perform the indicated operation and, if possible, simplify.

4.
```
  1 9 3 5.0 5
−     6 6.8 3 4
```

5. $-32 - (-15)$

6. $\dfrac{4}{15} - \dfrac{3}{20}$

7.
```
  3 7.6 4
×     5.9
```

8. $-43(15)$

9. $7\dfrac{4}{5} \cdot 3\dfrac{1}{2}$

10. $2.3\,\overline{)\,9\,8.9}$

11. $-306 \div 6$

12. $\dfrac{7}{11} \div \dfrac{14}{33}$

13. Write expanded notation for 30,074.

14. Write a word name for 120.07.

Which number is larger?

15. 0.7,   0.698

16. -0.799, -0.8

17. Find the prime factorization of 144.

18. Find the LCM of 42 and 78.

19. What fractional part is shaded?

20. Simplify: $\dfrac{108}{128}$.

Calculate.

21. $\dfrac{3}{5} \times 9.53$

22. $7.2 \div 0.4(-1.5) + (1.2)^2$

23. Find the mean: 23, 49, 52, 71.

24. Determine whether the pairs 6, 8 and 14, 196 are proportional.

25. Graph on a plane: $y = -2x + 1$.

26. Evaluate $\dfrac{t - 7}{w}$ for $t = -3$ and $w = -2$.

Solve.

27. $\dfrac{14}{25} = \dfrac{x}{54}$

28. $-423 = 16 \cdot t$

29. $9x - 7 = -43$

30. $2(x - 3) + 9 = 5x - 6$

31. $34.56 + n = 67.9$

32. $\dfrac{2}{3} x = \dfrac{16}{27}$

Solve.

33. A truck can reach its destination 143 miles away in 2.6 hr. How far could it travel in 3.8 hr?

34. A machine can stamp out 925 washers in 5 min. The company owning the machine needs 1295 washers by the end of the morning. How long will it take to stamp them out?

35. *Miles Traveled.* A Greyhound tour bus traveled 347.6 mi, 249.8 mi, and 379.5 mi on three separate trips. What was the total mileage of the bus?

36. *Gas Mileage.* The 2005 Volkswagen Jetta 2.5 L will go 319 mi on 14.5 gal of gasoline in city driving. What is the rate in miles per gallon?
Source: Volkswagen of America, Inc.

37. A diamond-shaped insert on a tablecloth consists of two identical triangles, as shown. What is the area of the insert?

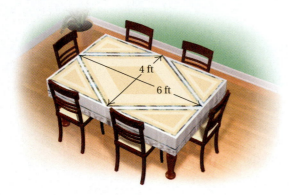

4 ft

6 ft

38. *Airplane Tire Costs.* A Boeing 747-400 jumbo jet has 2 nose tires and 16 rear tires. Each tire costs about $20,000.

   a) What is the total cost of a new set of tires for such a plane?

   b) Suppose an airline has a fleet of 400 such planes. What is the total cost of a new set of tires for all the planes?

   c) Suppose the airline has to change tires every month. What would be the total cost for tires for the airline for an entire year?

Source: *World-Traveler*, October 2000

39. A 46-oz juice can contains $5\frac{3}{4}$ cups of juice. A recipe calls for $3\frac{1}{2}$ cups of juice. How many cups are left?

SYNTHESIS

40. A car travels 88 ft in 1 sec. What is the rate in miles per hour?

41. A 12-oz bag of shredded mozzarella cheese is on sale for $2.07. Blocks of mozzarella cheese are on sale for $2.79 per pound. Which is the better buy?

42. Hans attends a university where the academic year consists of two 16-week semesters. He budgets $1200 for incidental expenses for the academic year. After 3 weeks, Hans has spent $150 for incidental expenses. Assuming he continues to spend at the same rate, will his budget for incidental expenses be adequate? If not, when will the money be exhausted and how much more will be needed to complete the year?

43. A basic sound system consists of a CD player, a receiver, and two speakers. A standard rule of thumb on the relative investment in these components is 1 : 3 : 2. That is, the receiver should cost three times as much as the CD player and the speakers should cost twice as much as the CD player. Eileen has already spent $250 on speakers. If she is to follow this rule of thumb, how much will she spend on her sound system?

44. Connor typically spends his paycheck according to the ratio 1 : 3 : 2, which indicates the ratio of Leisure & Incidental Expenses to Food & Other Necessary Expenses to Debt Payments. If Connor gets paid $4800 a month, how is his paycheck allocated? How much does Connor allot per year to debts? To Leisure & Incidental Expenses?

Percent Notation

8

Real-World Application

There are 1,168,195 people in the United States in active military service. The numbers in the four armed services are as follows: Air Force, 314,477; Army, 391,126; Marine Corps, 135,324; Navy, 327,268. What percent of the total does each branch represent? Round the answers to the nearest tenth of a percent.

Source: U.S. Department of Defense

This problem appears as Exercise 2 in Section 8.4.

Objectives

a Write three kinds of notation for a percent.

b Convert between percent notation and decimal notation.

c Convert between fraction notation and percent notation.

Write three kinds of notation as in Examples 1 and 2.

1. 70%

2. 23.4%

3. 100%

It is thought that the Roman emperor Augustus began percent notation by taxing goods sold at a rate of $\frac{1}{100}$. In time, the symbol "%" evolved by interchanging the parts of the symbol "100" to "0/0" and then to "%."

a Understanding Percent Notation

Of the total surface area of the earth, 70% is covered by water. What does this mean? It means that of every 100 square miles of the earth's surface area, 70 square miles are covered by water. Thus, 70% is a ratio of 70 to 100, or $\frac{70}{100}$.

Source: *The Handy Geography Answer Book*

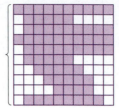

70 of 100 squares are shaded.

70% or $\frac{70}{100}$ or 0.70 of the large square is shaded.

Percent notation is used extensively in our everyday lives. Here are some examples:

18% of household personal vehicles are pickup trucks.

50.3% of all paper used in the United States is recycled.

54% of all kids prefer white bread.

0.08% blood alcohol level is a standard used by some states as the legal limit for drunk driving.

Percent notation is often represented in pie charts to show how the parts of a quantity are related. For example, the circle graph below illustrates the percentage of SUVs, trucks, and vans manufactured in the most popular colors during 2003 in North America.

Most Popular Vehicle Colors, 2003 (SUV/Truck/Van)

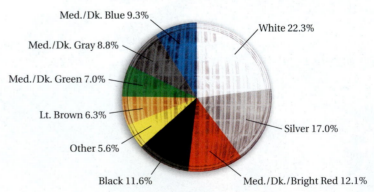

Med./Dk. Blue 9.3%
Med./Dk. Gray 8.8%
Med./Dk. Green 7.0%
Lt. Brown 6.3%
Other 5.6%
Black 11.6%
White 22.3%
Silver 17.0%
Med./Dk./Bright Red 12.1%

Source: DuPont Herberts Automotive Systems, Troy, Michigan, 2003, DuPont Automotive Color Survey Results

Answers on page A-22

PERCENT NOTATION

The notation **n%** means "n per hundred."

This definition leads us to the following equivalent ways of defining percent notation.

NOTATION FOR n%

Percent notation, n%, can be expressed using:

ratio → $n\% = $ the ratio of n to $100 = \dfrac{n}{100}$,

fraction notation → $n\% = n \times \dfrac{1}{100}$,   or

decimal notation → $n\% = n \times 0.01$

From 1998 to 2008, the number of jobs for professional chefs will increase by 13.4%.

Source: *Handbook of U.S. Labor Statistics*

EXAMPLE 1   Write three kinds of notation for 35%.

Using ratio: $35\% = \dfrac{35}{100}$ A ratio of 35 to 100

Using fraction notation: $35\% = 35 \times \dfrac{1}{100}$ Replacing % with $\times \dfrac{1}{100}$

Using decimal notation: $35\% = 35 \times 0.01$ Replacing % with $\times 0.01$

EXAMPLE 2   Write three kinds of notation for 67.8%.

Using ratio: $67.8\% = \dfrac{67.8}{100}$ A ratio of 67.8 to 100

Using fraction notation: $67.8\% = 67.8 \times \dfrac{1}{100}$ Replacing % with $\times \dfrac{1}{100}$

Using decimal notation: $67.8\% = 67.8 \times 0.01$ Replacing % with $\times 0.01$

Do Exercises 1–3 on the preceding page.

b   Converting Between Percent Notation and Decimal Notation

To write decimal notation for a number like 78%, we can replace "%" with " $\times 0.01$" and multiply.

   $78\% = 78 \times 0.01$ Replacing % with $\times 0.01$
   $\quad\;\; = 0.78$ Multiplying

Similarly,

   $4.9\% = 4.9 \times 0.01$ Replacing % with $\times 0.01$
   $\quad\;\;\; = 0.049,$ Multiplying

   $265\% = 265 \times 0.01$ Replacing % with $\times 0.01$
   $\quad\;\;\; = 2.65,$ Multiplying

and

   $100\% = 100 \times 0.01$ Replacing % with $\times 0.01$
   $\quad\;\;\;\; = 1.$ Multiplying

CALCULATOR CORNER

Converting from Percent Notation to Decimal Notation   Many calculators have a ⬜% key that can be used to convert from percent notation to decimal notation. This is often the second operation associated with a particular key and is accessed by first pressing a ⬜2nd or ⬜SHIFT key. To convert 57.6% to decimal notation, for example, you might press ⬜5 ⬜7 ⬜. ⬜6 ⬜2nd ⬜% or ⬜5 ⬜7 ⬜. ⬜6 ⬜SHIFT ⬜%. The display would read ⬜0.576, so 57.6% = 0.576. Read the user's manual to determine whether your calculator can do this conversion.

Exercises: Use a calculator to find decimal notation.

1. 14%   2. 0.069%

3. 43.8%   4. 125%

513

8.1   Percent Notation

Write each percent as an equivalent decimal.

4. 34%

5. 78.9%

6. $6\frac{5}{8}\%$

Find decimal notation for the percent notation in the sentence.

7. Pickup Trucks. Of all household personal vehicles in the United States, 18% are pickup trucks.

Source: U.S. Department of Transportation, 2001 National Household Travel Survey

8. Blood Alcohol Level. A blood alcohol level of 0.08% is a standard used by some states as the legal limit for drunk driving.

Answers on page A-22

To convert from percent notation to decimal notation,	36.5%
a) replace the percent symbol % with × 0.01, and	36.5 × 0.01
b) multiply by 0.01, which means move the decimal point two places to the left.	0.36.5 Move 2 places to the left.
	36.5% = 0.365

EXAMPLE 3   Find equivalent decimal notation for 99.44%.

a) Replace the percent symbol with × 0.01.   99.44 × 0.01

b) Move the decimal point two places to the left.   0.99.44

Thus, 99.44% = 0.9944.

With practice, you will be able to make this conversion mentally.

EXAMPLE 4   The interest rate on a $2\frac{1}{2}$-year certificate of deposit is $6\frac{3}{8}\%$. Find decimal notation for $6\frac{3}{8}\%$.

a) Convert $6\frac{3}{8}$ to decimal notation and replace the percent symbol with × 0.01. $6\frac{3}{8}\%$
6.375 × 0.01

b) Move the decimal point two places to the left.   0.06.375

Thus, $6\frac{3}{8}\% = 0.06375$.

Do Exercises 4–8.

The procedure used in Examples 3 and 4 can be reversed to write a decimal, like 0.38, as an equivalent percent. To see why, note that

$$0.38 = 0.38 \times 100\% = (0.38 \times 100)\% = 38\%.$$

To convert from decimal notation to percent notation, multiply by 100%. That is,	0.675 = 0.675 × 100%
a) move the decimal point two places to the right, and	0.67.5 Move 2 places to the right.
b) write a % symbol.	67.5%
	0.675 = 67.5%

EXAMPLE 5   Find percent notation for 1.27.

a) Move the decimal point two places to the right.   1.27.

b) Write a % symbol.   127%

Thus, 1.27 = 127%.

EXAMPLE 6 Of the time that people declare as sick leave, 0.21 is actually used for family issues. Find percent notation for 0.21.

Source: CCH Inc

a) Move the decimal point two places to the right.

0.21.

b) Write a % symbol.

21%

Thus, 0.21 = 21%.

EXAMPLE 7 Find percent notation for 5.6.

a) Move the decimal point two places to the right, adding an extra zero.

5.60.

b) Write a % symbol.

560%

Thus, 5.6 = 560%.

EXAMPLE 8 Of those who play golf, 0.149 play 8–24 rounds per year. Find percent notation for 0.149.

Source: U.S. Golf Association

0.149 = 14.9%   Moving the decimal point two places to the right and writing a % symbol

Do Exercises 9–13.

C Converting from Fraction Notation to Percent Notation

To convert from fraction notation to percent notation,

$\frac{3}{5}$ Fraction notation

a) find decimal notation by division, and

$$5 \overline{)3.0} \quad \begin{array}{c} 0.6 \\ \underline{3\ 0} \\ 0 \end{array}$$

b) convert the decimal notation to percent notation.

0.6 = 0.60 = 60%   Percent notation

$\frac{3}{5} = 60\%$

EXAMPLE 9 Find percent notation for $\frac{9}{16}$.

a) We first find decimal notation by division.

$$16 \overline{)9.0000} \quad \begin{array}{c} 0.5\ 6\ 2\ 5 \\ \underline{8\ 0} \\ 1\ 0\ 0 \\ \underline{9\ 6} \\ 4\ 0 \\ \underline{3\ 2} \\ 8\ 0 \\ \underline{8\ 0} \\ 0 \end{array}$$

$\frac{9}{16} = 0.5625$

Find percent notation.

9. 0.24

10. 3.47

11. 1

Find percent notation for the decimal notation in the sentence.

12. High School Graduates. The highest level of education for 0.321 of persons 25 and over in the United States is high school graduation.

Source: U.S. Department of Commerce, Bureau of the Census, *Current Population Survey*

13. Golf. Of those who play golf, 0.253 play 25–49 rounds per year.

Source: U.S. Golf Association

Answers on page A-22

Find percent notation.

14. $\dfrac{1}{4}$ **15.** $\dfrac{5}{8}$

Answers on page A-22

CALCULATOR CORNER

Converting Fraction Notation to Percent Notation Calculators are often used to convert fraction notation to percent notation. We simply perform the division on the calculator and then convert the decimal notation to percent notation. For example, percent notation for $\dfrac{17}{40}$ can be found by pressing $\boxed{1}\,\boxed{7}\,\boxed{\div}\,\boxed{4}\,\boxed{0}\,\boxed{=}$ and then converting the result, 0.425, to percent notation, 42.5%. Some calculators have a $\boxed{\%}$ key that will do this conversion automatically.

Exercises: Write each fraction as an equivalent percent. Round to the nearest hundredth of a percent.

1. $\dfrac{13}{25}$ **2.** $\dfrac{5}{13}$

3. $\dfrac{42}{39}$ **4.** $\dfrac{12}{7}$

5. $\dfrac{217}{364}$ **6.** $\dfrac{2378}{8401}$

b) Next, we convert the decimal notation to percent notation. We move the decimal point two places to the right and write a % symbol.

0.56.25

$$\dfrac{9}{16} = 56.25\%, \text{ or } 56\tfrac{1}{4}\%$$

> Don't forget the % symbol.

Do Exercises 14 and 15.

EXAMPLE 10 *Without Health Insurance.* Approximately $\tfrac{1}{6}$ of all people in the United States are without health insurance. Find percent notation for $\tfrac{1}{6}$.

Source: U.S. Bureau of the Census, *Current Population Survey*, March 2003

a) Find decimal notation by division.

```
      0.1 6 6
6 ) 1.0 0 0
    6
    ───
    4 0
    3 6
    ───
      4 0
      3 6
      ───
        4
```

We get a repeating decimal: $0.16\overline{6}$.

b) Convert the answer to percent notation.

0.16.$\overline{6}$

$$\dfrac{1}{6} = 16.\overline{6}\%, \text{ or } 16\tfrac{2}{3}\%$$

> **Caution!**
>
> There are *two* steps for converting fraction notation to percent notation. After you have found decimal notation, don't forget to convert that result to percent notation.

Do Exercises 16 and 17.

In some cases, division is not the fastest way to convert. The following are some optional ways in which conversion might be done.

EXAMPLE 11 Find percent notation for $\frac{69}{100}$.

We use the definition of percent as a ratio.

$$\frac{69}{100} = 69\%$$

EXAMPLE 12 Find percent notation for $\frac{17}{20}$.

We multiply by 1 to get 100 in the denominator. We think of what we have to multiply 20 by in order to get 100. That number is 5, so we multiply by 1 using $\frac{5}{5}$.

$$\frac{17}{20} \cdot \frac{5}{5} = \frac{85}{100} = 85\%$$

Note that this shortcut works only when the denominator of a simplified fraction is a factor of 100.

Do Exercises 18 and 19.

The method used in Example 11 is reversed when we convert from percent notation to fraction notation.

To convert from percent notation to fraction notation,	30% Percent notation
a) use the definition of percent as a ratio, and	$\dfrac{30}{100}$
b) simplify, if possible.	$\dfrac{3}{10}$ Fraction notation
	$30\% = \dfrac{3}{10}$

EXAMPLE 13 Write an equivalent fraction for 75% and simplify.

$$75\% = \frac{75}{100} \qquad \text{Using the definition of percent}$$

$$= \frac{3 \cdot 25}{4 \cdot 25}$$

$$= \frac{3}{4} \cdot \frac{25}{25} \qquad \text{Simplifying by removing a factor equal to 1: } \frac{25}{25} = 1$$

$$= \frac{3}{4}$$

16. Water is the single most abundant chemical in the body. The human body is about $\frac{2}{3}$ water. Find percent notation for $\frac{2}{3}$.

17. Find percent notation: $\dfrac{5}{6}$.

Find percent notation.

18. $\dfrac{57}{100}$

19. $\dfrac{19}{25}$

Answers on page A-22

Find fraction notation.

20. 60%

21. 3.25%

22. $66\dfrac{2}{3}\%$

23. Complete this table.

FRACTION NOTATION	$\dfrac{1}{5}$		
DECIMAL NOTATION		$0.83\overline{3}$	
PERCENT NOTATION			$37\dfrac{1}{2}\%$

Answers on page A-22

EXAMPLE 14   Write an equivalent fraction for 112.5% and simplify.

$$112.5\% = \frac{112.5}{100} \qquad \text{Using the definition of percent}$$

$$= \frac{112.5}{100} \times \frac{10}{10} \qquad \text{Multiplying by 1 to eliminate the decimal point in the numerator}$$

$$= \frac{1125}{1000}$$

$$= \frac{5}{5} \cdot \frac{225}{200}$$

$$= \frac{225}{200}$$

$$= \frac{25}{25} \cdot \frac{9}{8} \qquad \text{Simplifying}$$

$$= \frac{9}{8}$$

Sometimes, rather than divide by 100, it is easier to multiply by $\frac{1}{100}$.

EXAMPLE 15   Write an equivalent fraction for $16\frac{2}{3}\%$ and simplify.

$$16\frac{2}{3}\% = \frac{50}{3}\% \qquad \text{Converting from the mixed numeral to fraction notation}$$

$$= \frac{50}{3} \times \frac{1}{100} \qquad \text{Using the definition of percent}$$

$$= \frac{50 \cdot 1}{3 \cdot 50 \cdot 2}$$

$$= \frac{1}{6} \cdot \frac{50}{50} \qquad \text{Simplifying}$$

$$= \frac{1}{6}$$

Note in Examples 13 and 14 that the definition of percent used was $n\% = \frac{n}{100}$. Generally, this is the definition to use for finding fraction notation when n is a natural number or in decimal notation. In Example 15, the definition used was $n\% = n \cdot \frac{1}{100}$. Use this definition when n is a fraction or mixed numeral.

The table on the inside back cover lists decimal, fraction, and percent equivalents used so often that it would speed up your work if you memorized them. For example, $\frac{1}{3} = 0.\overline{3}$, so we say that the **decimal equivalent** of $\frac{1}{3}$ is $0.\overline{3}$, or that $0.\overline{3}$ has the **fraction equivalent** $\frac{1}{3}$.

EXAMPLE 16   Find fraction notation for $16.\overline{6}\%$.

We can use the table on the inside back cover or recall that $16.\overline{6}\% = 16\frac{2}{3}\% = \frac{1}{6}$. We can also recall from our work with repeating decimals in Chapter 5 that $0.\overline{6} = \frac{2}{3}$. Then we have $16.\overline{6}\% = 16\frac{2}{3}\%$ and can proceed as in Example 15.

Do Exercises 20–23.

CALCULATOR CORNER

Applications of Ratio and Percent: The Price–Earnings Ratio and Stock Yields

The Price–Earnings Ratio If the total earnings of a company one year were $5,000,000 and 100,000 shares of stock were issued, the earnings per share were $50. At one time, the price per share of Coca-Cola was $58.13 and the earnings per share were $0.76. The **price-earnings ratio,** P/E, is the price of the stock divided by the earnings per share. For the Coca-Cola stock, the price–earnings ratio, P/E, is given by

$$\frac{P}{E} = \frac{58.13}{0.76} \approx 76.49 \qquad \text{Dividing, using a calculator, and rounding to the nearest hundredth}$$

Stock Yields At one time, the price per share of Coca-Cola stock was $58.13 and the company was paying a yearly dividend of $0.68 per share. It is helpful to those interested in stocks to know what percent the dividend is of the price of the stock. The percent is called the **yield.** For the Coca-Cola stock, the yield is given by

$$\text{Yield} = \frac{\text{Dividend}}{\text{Price per share}} = \frac{0.68}{58.13} \approx 0.0117 \qquad \text{Dividing and rounding to the nearest ten-thousandth}$$

$$= 1.17\%. \qquad \text{Converting to percent notation}$$

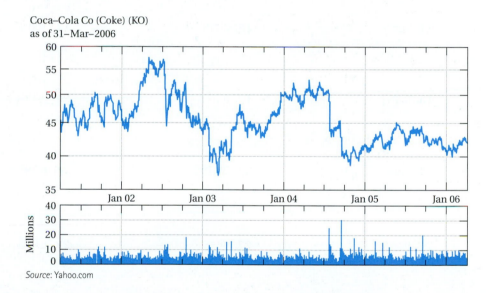

Coca–Cola Co (Coke) (KO)
as of 31–Mar–2006

Source: Yahoo.com

Exercises: Compute the price–earnings ratio and the yield for each stock listed below.

	STOCK	PRICE PER SHARE	EARNINGS	DIVIDEND	P/E	YIELD
1.	Pepsi (PEP)	$42.75	$1.40	$0.56		
2.	Pearson (PSO)	$25.00	$0.78	$0.30		
3.	Quaker Oats (OAT)	$92.38	$2.68	$1.10		
4.	Texas Insts (TEX)	$42.88	$1.62	$0.43		
5.	Ford Motor Co (F)	$27.56	$2.30	$1.19		
6.	Wendy's Intl (WEN)	$25.75	$1.47	$0.23		

a   Write three kinds of notation as in Examples 1 and 2 on p. 513.

1. 90%   **2.** 58.7%   **3.** 12.5%   **4.** 130%

b   Write each percent as an equivalent decimal.

5. 67%   **6.** 17%   **7.** 45.6%   **8.** 76.3%

9. 59.01%   **10.** 30.02%   **11.** 10%   **12.** 80%

13. 1%   **14.** 100%   **15.** 200%   **16.** 300%

17. 0.1%   **18.** 0.4%   **19.** 0.09%   **20.** 0.12%

21. 0.18%   **22.** 5.5%   **23.** 23.19%   **24.** 87.99%

25. $14\frac{7}{8}\%$ **26.** $93\frac{1}{8}\%$ **27.** $56\frac{1}{2}\%$ **28.** $61\frac{3}{4}\%$

Find decimal notation for the percent notation(s) in the sentence.

29. *Pediatricians.*   By 2020, the population of children in the United States is expected to increase by about 9% while the number of pediatricians will leap 58%.

Source: Study by Dr. Scott Shipman, an Oregon Health and Science University pediatrician

30. *Online Sales.*   In 2004, online sales accounted for 1.7% of total U.S. retail sales. The 2004 number was a 23.1% increase in online sales from 2003.

Source: www.computerworld.com

31. *Eating Out.*   On a given day, 44% of all adults eat in a restaurant.

Source: *AARP Bulletin,* November 2004

32. Of those who play golf, 18.6% play 100 or more rounds per year.

Source: U.S. Golf Association

33. *Knitting and Crocheting.* Of all women ages 25 to 34, 36% know how to knit or crochet.

Source: Research Inc. for Craft Yarn Council of America

34. According to a recent survey, 95.1% of those asked to name what sports they participate in said swimming.

Source: Sporting Goods Manufacturers

Write each decimal as an equivalent percent.

35. 0.47 **36.** 0.87 **37.** 0.03 **38.** 0.01 **39.** 8.7

40. 4 **41.** 0.334 **42.** 0.889 **43.** 0.75 **44.** 0.99

45. 0.4 **46.** 0.5 **47.** 0.006 **48.** 0.008 **49.** 0.017

50. 0.024 **51.** 0.2718 **52.** 0.8911 **53.** 0.0239 **54.** 0.00073

Find percent notation for the decimal notation(s) in the sentence.

55. About 0.69 of all newspapers are recycled.

Sources: American Forest and Paper Association; Newspaper Association of America

56. According to a recent survey, 0.526 of those asked to name what sports they participate in said bowling.

Source: Sporting Goods Manufacturers

57. *Bachelor's Degrees.* For 0.177 of the U.S. population 25 and over, the bachelor's degree is the highest level of educational attainment.

Sources: U.S. Department of Commerce, Bureau of the Census, *Current Population Survey;* National Center for Educational Statistics, *Digest of Education Statistics,* 2003

58. *65 and Over.* In Clearwater, Florida, 0.215 of the residents are 65 or over.

Source: U.S. Bureau of the Census

59. *Hours of Sleep.* In 2005, only 0.26 of people get 8 or more hours of sleep a night on weekdays. This rate has declined from 0.38 in 2001.

Source: National Sleep Foundation's 2005 Sleep in America Poll

60. Of those people living in North Carolina, 0.1134 will die of heart disease.

Source: American Association of Retired Persons

C   Write each fraction as an equivalent percent.

61. $\dfrac{41}{100}$ **62.** $\dfrac{36}{100}$ **63.** $\dfrac{5}{100}$ **64.** $\dfrac{1}{100}$ **65.** $\dfrac{2}{10}$ **66.** $\dfrac{7}{10}$

67. $\dfrac{7}{25}$ **68.** $\dfrac{1}{20}$ **69.** $\dfrac{1}{2}$ **70.** $\dfrac{3}{4}$ **71.** $\dfrac{7}{8}$ **72.** $\dfrac{1}{8}$

73. $\dfrac{4}{5}$ **74.** $\dfrac{2}{5}$ **75.** $\dfrac{2}{3}$ **76.** $\dfrac{1}{3}$ **77.** $\dfrac{1}{6}$ **78.** $\dfrac{5}{6}$

79. $\dfrac{3}{16}$ **80.** $\dfrac{11}{16}$ **81.** $\dfrac{3}{20}$ **82.** $\dfrac{31}{50}$ **83.** $\dfrac{29}{50}$ **84.** $\dfrac{13}{25}$

In Exercises 85–90, write percent notation for the fractions in this pie chart.

85. $\dfrac{11}{50}$ **86.** $\dfrac{3}{100}$

87. $\dfrac{1}{20}$ **88.** $\dfrac{59}{100}$

89. $\dfrac{9}{100}$ **90.** $\dfrac{1}{50}$

Time Workers Spend Sorting Unsolicited e-mail and Spam

Less than 5 minutes $\dfrac{59}{100}$

$\dfrac{11}{50}$ 5–15 minutes

$\dfrac{9}{100}$ 15–30 minutes

$\dfrac{1}{20}$ 30–60 minutes

$\dfrac{1}{50}$ More than 1 hour

$\dfrac{3}{100}$ Did not know/ not sure

Source: Data from InsightExpress

Find percent notation for the fraction notation in the sentence.

91. *Driving While Drowsy.*   Almost half of U.S. drivers say they have driven while drowsy. Of this group, $\frac{2}{5}$ fight off sleep by opening a window, and $\frac{9}{50}$ drink a caffeinated beverage.

Source: Harris Interactive for Tylenol PM

92. *Paved Roads.*   About $\frac{13}{20}$ of the roads and streets in the United States are paved.

Source: U.S. Department of Transportation, *Highway Statistics*

For each percent, write an equivalent fraction. When possible, simplify and use the information in the table that appears on the inside back cover.

93. 85%

94. 55%

95. 62.5%

96. 12.5%

97. $33\frac{1}{3}\%$

98. $83\frac{1}{3}\%$

99. $16.\overline{6}\%$

100. $66.\overline{6}\%$

101. 7.25%

102. 4.85%

103. 0.8%

104. 0.2%

105. 150%

106. 110%

107. $33.\overline{3}\%$

108. $83.\overline{3}\%$

Find fraction notation for the percent notation in the following bar graph.

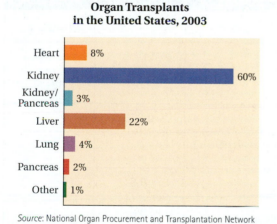

Organ Transplants in the United States, 2003

Heart 8%
Kidney 60%
Kidney/Pancreas 3%
Liver 22%
Lung 4%
Pancreas 2%
Other 1%

Source: National Organ Procurement and Transplantation Network

109. 8%

110. 22%

111. 60%

112. 4%

113. 2%

114. 3%

Find fraction notation for the percent notation in the sentence.

115. A 1.9-oz serving of Raisin Bran Crunch® cereal with $\frac{1}{2}$ cup of skim milk satisfies 35% of the minimum daily requirements for Vitamin B_{12}.

Source: Kellogg's USA, Inc.

116. A 1-cup serving of Wheaties® cereal with $\frac{1}{2}$ cup of skim milk satisfies 15% of the minimum daily requirements for calcium.

Source: General Mills Sales, Inc.

117. Of all those who are 85 or older, 47% have Alzheimer's disease.

Source: Alzheimer's Association

118. In 2003, 24.4% of Americans 18 and older smoked cigarettes.

Source: U.S. Centers for Disease Control and Prevention

Complete the table.

119.

FRACTION NOTATION	DECIMAL NOTATION	PERCENT NOTATION
$\frac{1}{8}$		12.5%, or $12\frac{1}{2}\%$
$\frac{1}{6}$		
		20%
	0.25	
		$33.\overline{3}\%$, or $33\frac{1}{3}\%$
		37.5%, or $37\frac{1}{2}\%$
		40%
$\frac{1}{2}$		

120.

FRACTION NOTATION	DECIMAL NOTATION	PERCENT NOTATION
$\frac{3}{5}$		
	0.625	
$\frac{2}{3}$		
	0.75	75%
$\frac{4}{5}$		
$\frac{5}{6}$		$83.\overline{3}\%$, or $83\frac{1}{3}\%$
$\frac{7}{8}$		87.5%, or $87\frac{1}{2}\%$
		100%

121.

FRACTION NOTATION	DECIMAL NOTATION	PERCENT NOTATION
	0.5	
$\frac{1}{3}$		
		25%
		$16.\overline{6}\%$, or $16\frac{2}{3}\%$
	0.125	
$\frac{3}{4}$		
	$0.8\overline{3}$	
$\frac{3}{8}$		

122.

FRACTION NOTATION	DECIMAL NOTATION	PERCENT NOTATION
		40%
		62.5%, or $62\frac{1}{2}\%$
	0.875	
$\frac{1}{1}$		
	0.6	
	$0.\overline{6}$	
$\frac{1}{5}$		

123. **D_W** *Winning Percentage.* During the 2005 regular baseball season, the Chicago White Sox won 99 of 162 games and went on to win the World Series. Find the ratio of number of wins to total number of games played in the regular season and convert it to decimal notation. Such a rate is often called a "winning percentage." Explain why.

124. **D_W** Athletes sometimes speak of "giving 110%" effort. Does this make sense? Explain.

Solve.

125. $13 \cdot x = 910$ [1.7b]

126. $15 \cdot y = 75$ [1.7b]

127. $0.05 \times b = 20$ [5.7a]

128. $3 = 0.16 \times b$ [5.7a]

129. $\dfrac{24}{37} = \dfrac{15}{x}$ [7.3b]

130. $\dfrac{17}{18} = \dfrac{x}{27}$ [7.3b]

131. $\dfrac{9}{10} = \dfrac{x}{5}$ [7.3b]

132. $\dfrac{7}{x} = \dfrac{4}{5}$ [7.3b]

SYNTHESIS

133. **D_W** Tammy remembers that $\frac{1}{4} = 25\%$. Explain how she can use this to (a) write $\frac{1}{8}$ in percent notation and (b) write $\frac{5}{8}$ in percent notation.

134. **D_W** Is it always best to convert from fraction notation to percent notation by first finding decimal notation? Why or why not?

Write each number as an equivalent percent.

135. ▦ $\dfrac{41}{369}$

136. $2.5\overline{74631}$

Write each percent as an equivalent decimal.

137. $\dfrac{14}{9}\%$

138. $\dfrac{19}{12}\%$

To draw a pie chart, or circle graph, think of a pie cut into 100 equal-sized pieces. Then to represent, say, 20%, draw a wedge equal in size to 20 of these pieces. Use the given information for Exercises 139 and 140 to complete a circle graph.

139. *How Teens Spend Their Money*

Clothing	34%
Entertainment	22%
Food	22%
Other	22%

Source: Rand Youth Poll, eMarketer

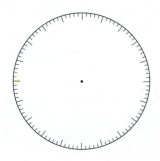

140. *Reasons for Drinking Coffee*

To get going in the morning:	32%
Like the taste:	33%
Not sure:	2%
To relax:	4%
As a pick-me-up:	10%
It's a habit:	19%

Source: LMK Associates survey for Au Bon Pain Co., Inc.

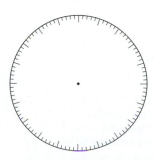

Translate to an equation. Do not solve.

1. 12% of 50 is what?

2. What is 40% of 60?

Translate to an equation. Do not solve.

3. 45 is 20% of what?

4. 120% of what is 60?

Answers on page A-23

*Note: Sections 8.2 and 8.3 present two methods for solving percent problems. You may prefer one method over the other, or your instructor may specify the method to be used.

---

a **Translating to Equations**

To solve a problem involving percents, it is helpful to translate first to an equation. To distinguish the method in Section 8.2 from that of Section 8.3, we will call these *percent equations*.

EXAMPLE 1 Translate:

$$\begin{array}{ccccc} 23\% & \text{of} & 5 & \text{is} & \text{what?} \\ \downarrow & \downarrow & \downarrow & \downarrow & \downarrow \\ 0.23 & \cdot & 5 & = & a \end{array}$$

Note how the key words are translated.

KEY WORDS IN PERCENT TRANSLATIONS	
"**Of**" translates to "·", or "×".	"**Is**" translates to "=".
"**What**" translates to any letter.	"**%**" translates to "×$\frac{1}{100}$" or "× 0.01".

EXAMPLE 2 Translate:

$$\begin{array}{ccccc} \text{What} & \text{is} & 11\% & \text{of} & 49? \\ \downarrow & \downarrow & \downarrow & \downarrow & \downarrow \\ a & = & 0.11 & \cdot & 49 \end{array}$$

Any letter can be used.

Do Exercises 1 and 2.

EXAMPLE 3 Translate:

$$\begin{array}{ccccc} 3 & \text{is} & 10\% & \text{of} & \text{what?} \\ \downarrow & \downarrow & \downarrow & \downarrow & \downarrow \\ 3 & = & 0.10 & \cdot & b \end{array}$$

Caution!

Don't forget to translate percent notation to decimal notation!

EXAMPLE 4 Translate:

$$\begin{array}{ccccc} 45\% & \text{of} & \text{what} & \text{is} & 23? \\ \downarrow & \downarrow & \downarrow & \downarrow & \downarrow \\ 0.45 & \cdot & b & = & 23 \end{array}$$

Do Exercises 3 and 4.

EXAMPLE 5 Translate:

$$\begin{array}{ccccc} 10 & \text{is} & \text{what percent} & \text{of} & 20? \\ \downarrow & \downarrow & \downarrow & \downarrow & \downarrow \\ 10 & = & p & \cdot & 20 \end{array}$$

EXAMPLE 6 Translate:

What percent of 50 is 7?

$$p \cdot 50 = 7$$

Do Exercises 5 and 6.

b | Solving Percent Problems

In solving percent problems, we use the *Translate* and *Solve* steps in the problem-solving strategy used throughout this text.

Percent problems are actually of three different types. Although the method we present does *not* require that you be able to identify which type you are solving, it is helpful to know them.

We know that

15 is 25% of 60,   or $15 = 0.25 \cdot 60$.

We can think of this as:

> Amount = Percent number × Base.

Each of the three types of percent problems depends on which of the three pieces of information is missing.

1. **Finding the *amount* (the result of taking the percent)**

   Example: **What** is 25% of 60?

   Translation: $a = 0.25 \cdot 60$

2. **Finding the *base* (the number you are taking the percent of)**

   Example: 15 is 25% of **what number?**

   Translation: $15 = 0.25 \cdot b$

3. **Finding the *percent number* (the percent itself)**

   Example: 15 is **what percent** of 60?

   Translation: $15 = p \cdot 60$

FINDING THE AMOUNT

EXAMPLE 7 What is 4.6% of 105,000,000?

Translate: $a = 0.046 \cdot 105{,}000{,}000$.

Solve: The letter is by itself. To solve the equation, we multiply:

$$a = (0.046)105{,}000{,}000 = 4{,}830{,}000.$$

Thus, 4,830,000 is 4.6% of 105,000,000. The answer is 4,830,000.

Do Exercise 7.

Translate to an equation. Do not solve.

5. 16 is what percent of 40?

6. What percent of 84 is 10.5?

7. Solve:

What is 12% of 50?

Answers on page A-23

There are 105,000,000 households in the United States and approximately 4.6% own a pet bird. How many households own at least one pet bird?

8. Solve:

64% of $55 is what?

Solve.

9. 20% of what is 45?

10. $60 is 120% of what?

11. Solve:

16 is what percent of 40?

12. Solve:

What percent of $84 is $10.50?

Answers on page A-23

528

EXAMPLE 8   120% of $42 is what?

Translate: $1.20 \cdot 42 = a$.

Solve: The letter is by itself. To solve the equation, we carry out the calculation:

$$a = 1.20 \cdot 42$$
$$= 1.2\,(42) \qquad 1.20 = 1.2$$
$$= 50.4.$$

Thus, 120% of $42 is $50.40. The answer is $50.40.

Do Exercise 8.

FINDING THE BASE

EXAMPLE 9   5% of what is 20?

Translate: $0.05 \cdot b = 20$.

Solve: This time the letter is *not* by itself. To solve the equation, we divide both sides by 0.05:

$$\frac{0.05 \cdot b}{0.05} = \frac{20}{0.05} \qquad \text{Dividing both sides by 0.05}$$

$$b = \frac{20}{0.05}$$

$$b = 400.$$

Thus, 5% of 400 is 20. The answer is 400.

EXAMPLE 10 $3 is 16% of what?

Translate: $3   is   16%   of   what?
$$3 \quad = \quad 0.16 \quad \cdot \quad b.$$

Solve: Again, the variable is not by itself. To solve the equation, we divide both sides by 0.16:

$$\frac{3}{0.16} = \frac{0.16 \cdot b}{0.16} \qquad \text{Dividing both sides by 0.16}$$

$$\frac{3}{0.16} = b$$

$$18.75 = b.$$

$$
\begin{array}{r}
1\,8.7\,5 \\
0.1\,6\,\overline{)\,3.0\,0_{\wedge}0\,0} \\
1\,6 \\
\overline{1\,4\,0} \\
1\,2\,8 \\
\overline{1\,2\,0} \\
1\,1\,2 \\
\overline{8\,0} \\
8\,0 \\
\overline{0}
\end{array}
$$

Thus, $3 is 16% of $18.75. The answer is $18.75.

Do Exercises 9 and 10.

FINDING THE PERCENT NUMBER

In solving these problems, you *must* remember to convert to percent notation after you have solved the equation.

EXAMPLE 11 17 is what percent of 20?

Translate:

17	is	what percent	of	20?
↓	↓	↓	↓	↓
17	=	p	·	20.

Solve: To solve the equation, we divide both sides by 20 and convert the result to percent notation:

$$17 = p \cdot 20$$

$$\frac{17}{20} = \frac{p \cdot 20}{20}$$ Dividing both sides by 20:

$$0.85 = p$$

$$p = 85\%.$$ Converting to percent notation

$$
\begin{array}{r}
.8\,5 \\
20\,\overline{)\,1\,7.0\,0} \\
1\,6\,0 \\
\hline
1\,0\,0 \\
1\,0\,0 \\
\hline
0
\end{array}
$$

Thus, 17 is 85% of 20. The answer is 85%.

Do Exercise 11 on the preceding page.

EXAMPLE 12 What percent of $50 is $16?

Translate:

What percent	of	$50	is	$16?
↓	↓	↓	↓	↓
p	·	50	=	16.

Solve: To solve the equation, we divide both sides by 50 and convert the answer to percent notation:

$$\frac{p \cdot 50}{50} = \frac{16}{50}$$ Dividing both sides by 50

$$p = \frac{16}{50}$$

$$= \frac{16}{50} \cdot \frac{2}{2}$$ Multiplying by $\frac{2}{2}$, or 1, to get a denominator of 100

$$= \frac{32}{100}$$ Had we preferred, we could have divided 16 by 50 to find decimal notation.

$$= 32\%.$$ Converting to percent notation

Thus, 32% of $50 is $16. The answer is 32%.

Do Exercise 12 on the preceding page.

Caution!

When a question asks "what percent?", be sure to give the answer in percent notation.

CALCULATOR CORNER

Using Percents in Computations Many calculators have a [%] key that can be used in computations. (See the Calculator Corner on page 513.) For example, to find 11% of 49, we press [1] [1] [2nd] [%] [×] [4] [9] [=] or [4] [9] [×] [1] [1] [SHIFT] [%]. The display reads [5.39], so 11% of 49 is 5.39.

In Example 9, we divide 20 by 5%. To use the [%] key in this computation, we press [2] [0] [÷] [5] [2nd] [%] [=], or [2] [0] [÷] [5] [SHIFT] [%]. The result is 400.

We can also use the [%] key to find the percent number in a problem. In Example 11, for instance, we answer the question "17 is what percent of 20?" On a calculator, we press [1] [7] [÷] [2] [0] [2nd] [%] [=], or [1] [7] [÷] [2] [0] [SHIFT] [%]. The result is 85, so 17 is 85% of 20.

Exercises: Use a calculator to find each of the following.

1. What is 5% of 24?
2. What is 12.6% of $40?
3. What is 19% of 256?
4. 140% of $16 is what?
5. 0.04% of 28 is what?
6. 33% of $90 is what?
7. Use the percent key to perform the computations in Margin Exercises 10 and 11.

a   Translate to an equation. Do not solve.

1. What is 32% of 78?

2. 98% of 57 is what?

3. 89 is what percent of 99?

4. What percent of 25 is 8?

5. 13 is 25% of what?

6. 21.4% of what is 20?

b   Translate to an equation and solve.

7. What is 85% of 276?

8. What is 74% of 53?

9. 150% of 30 is what?

10. 100% of 13 is what?

11. What is 6% of $300?

12. What is 4% of $45?

13. 3.8% of 50 is what?

14. $33\frac{1}{3}$% of 480 is what?
($Hint:$ $33\frac{1}{3}\% = \frac{1}{3}$.)

15. $39 is what percent of $50?

16. $16 is what percent of $90?

17. 20 is what percent of 10?

18. 60 is what percent of 20?

19. What percent of $300 is $150?

20. What percent of $50 is $40?

21. What percent of 80 is 100?

22. What percent of 60 is 15?

23. 20 is 50% of what?

24. 57 is 20% of what?

25. 40% of what is $32?

26. 100% of what is $74?

27. 56.32 is 64% of what?

28. 71.04 is 96% of what?

29. 70% of what is 14?

30. 70% of what is 35?

31. What is $62\frac{1}{2}$% of 10?

32. What is $35\frac{1}{4}$% of 1200?

33. What is 8.3% of $10,200?

34. What is 9.2% of $5600?

35. $66\frac{2}{3}$% of what is 27.4? $\left(\text{Hint: } 66\frac{2}{3}\% = \frac{2}{3}.\right)$

36. $33\frac{1}{3}$% of what is 17.2?

37. **D**_{**W**} Write a percent problem that could be translated to the equation

$$25 = 0.04 \cdot b.$$

38. **D**_{**W**} Write a percent problem that could be translated to the equation

$$30 = p \cdot 80.$$

SKILL MAINTENANCE

Write fraction notation. [5.1b]

39. 0.623

40. 1.9

41. 2.37

Write decimal notation. [5.1b]

42. $\dfrac{9}{1000}$

43. $\dfrac{39}{100}$

44. $\dfrac{57}{10}$

45. Each mocha served at Razer's Coffee House requires $\frac{2}{3}$ oz of chocolate. How many mochas can be made with one 12-oz can of chocolate?   [3.8b]

46. Each of Sadie's digital photos uses $\frac{3}{5}$ megabytes of computer memory. How many photos can be stored in a file with 12 megabytes available?   [3.8b]

SYNTHESIS

47. **D**_{**W**} Suppose we know that 40% of 92 is 36.8. What is a quick way to find 4% of 92? 400% of 92? Explain.

48. **D**_{**W**} To calculate a 15% tip on a $24 bill, a customer adds $2.40 and half of $2.40, or $1.20, to get $3.60. Is this procedure valid? Why or why not?

Solve.

49. ▦ What is 7.75% of $10,880?

Estimate _____

Calculate _____

50. ▦ 50,951.775 is what percent of 78,995?

Estimate _____

Calculate _____

51. *Recyclables.*   It is estimated that 40% to 50% of all trash is recyclable. If a community produces 270 tons of trash, how much of their trash is recyclable?

52. *Batting.*   An all-star baseball player gets a hit in 30% to 35% of all at-bats. If an all-star had 520 to 580 at-bats, how many hits would he have had?

In the United States, 47% of the labor force are women. In 2003, there were 146,510,000 people in the labor force. This means that 68,859,700 were women.

Sources: U.S. Department of Labor, Bureau of Labor Statistics

*Note: This section presents an alternative method for solving basic percent problems. The method of this section is used as an alternative approach in Section 8.4, but it is not used in other sections of this book.

a Translating to Proportions

A percent is a ratio of some number to 100. For example, 47% is the ratio $\frac{47}{100}$. The numbers 68,859,700 and 146,510,000 have the same ratio as 47 and 100.

$$\frac{47}{100} = \frac{68,859,700}{146,510,000}$$

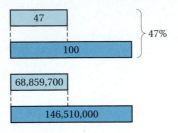

To solve a percent problem using a proportion, we translate as follows:

$$\text{Number} \rightarrow \frac{N}{100} = \frac{a}{b} . \begin{matrix} \leftarrow \text{Amount} \\ \leftarrow \text{Base} \end{matrix}$$

100 ⟶

> You might find it helpful to read this as "part is to whole as part is to whole."

For example, 60% of 25 is 15 translates to

$$\frac{60}{100} = \frac{15}{25} . \begin{matrix} \leftarrow \text{Amount} \\ \leftarrow \text{Base} \end{matrix}$$

A clue in translating is that the base, *b*, corresponds to 100 and usually follows the wording "percent of." Also, *N*% always translates to *N*/100, and the amount *a* is usually written next to the words "is," "are," "was," or "were." Another aid in translating is to make a *comparison drawing*. To do this, we start with the percent side and list 0% at the top and 100% near the bottom. Then we estimate where the specified percent—in this case, 60%—is located. The corresponding quantities are then filled in. The base—in this case, 25—always corresponds to 100% and the amount—in this case, 15—corresponds to the specified percent.

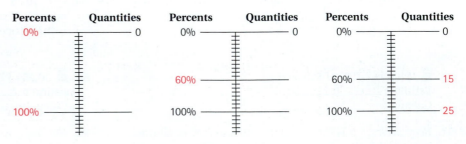

The proportion can then be read easily from the drawing: $\frac{60}{100} = \frac{15}{25}$.

In the examples that follow, we use comparison drawings as an aid for visualization. It is not necessary to always make a drawing to solve problems of this sort.

EXAMPLE 1 Translate to a proportion.

23% of 5 is what?

number of hundredths   base   amount, *a*

$$\frac{23}{100} = \frac{a}{5}$$

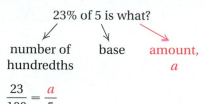

EXAMPLE 2 Translate to a proportion.

What is 124% of 49?

amount, *a*   number of hundredths   base

$$\frac{124}{100} = \frac{a}{49}$$

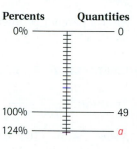

Do Exercises 1–3.

EXAMPLE 3 Translate to a proportion.

3 is 10% of what?

amount   number of hundredths   base, *b*

$$\frac{10}{100} = \frac{3}{b}$$

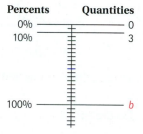

EXAMPLE 4 Translate to a proportion.

45% of what is 23?

number of hundredths   base, *b*   amount

$$\frac{45}{100} = \frac{23}{b}$$

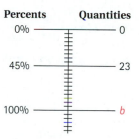

Do Exercises 4 and 5.

EXAMPLE 5 Translate to a proportion.

10 is what percent of 20?

amount   number of hundredths, *N*   base

$$\frac{N}{100} = \frac{10}{20}$$

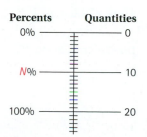

Translate to a proportion. Do not solve.

1. 12% of 50 is what?

2. What is 40% of 60?

3. 130% of 72 is what?

Translate to a proportion. Do not solve.

4. 45 is 20% of what?

5. 120% of what is 60?

Answers on page A-24

Translate to a proportion. Do not solve.

6. 16 is what percent of 40?

7. What percent of 84 is 10.5?

8. Solve:

20% of what is $45?

Solve.

9. 64% of 55 is what?

10. What is 12% of 50?

EXAMPLE 6   Translate to a proportion.

What percent of 50 is 7?

number of hundredths, N base   amount

$$\frac{N}{100} = \frac{7}{50}$$

Percents	Quantities
0%	0
N%	7
100%	50

Do Exercises 6 and 7.

b   Solving Percent Problems

After a percent problem has been translated to a proportion, we solve as in Section 7.3.

EXAMPLE 7   5% of what is $20?

Translate: $\dfrac{5}{100} = \dfrac{20}{b}$

Percents	Quantities
0%	0
5%	20
100%	b

Solve: $5 \cdot b = 100 \cdot 20$ Equating cross products

$$\frac{5 \cdot b}{5} = \frac{100 \cdot 20}{5}$$ Dividing by 5

$$b = \frac{2000}{5}$$

$$b = 400$$ Simplifying

Thus, 5% of $400 is $20. The answer is $400.

Do Exercise 8.

EXAMPLE 8   120% of 42 is what?

Translate: $\dfrac{120}{100} = \dfrac{a}{42}$

Percents	Quantities
0%	0
100%	42
120%	a

Solve: $120 \cdot 42 = 100 \cdot a$ Equating cross products

$$\frac{120 \cdot 42}{100} = \frac{100 \cdot a}{100}$$ Dividing by 100

$$\frac{5040}{100} = a$$

$$50.4 = a$$ Simplifying

Thus, 120% of 42 is 50.4. The answer is 50.4.

Do Exercises 9 and 10.

EXAMPLE 9 3 is 16% of what?

Translate: $\dfrac{3}{b} = \dfrac{16}{100}$

Solve: $3 \cdot 100 = b \cdot 16$ Equating cross products

$\dfrac{3 \cdot 100}{16} = \dfrac{b \cdot 16}{16}$ Dividing by 16

$\dfrac{300}{16} = b$ Multiplying and simplifying

$18.75 = b$ Dividing

Thus, 3 is 16% of 18.75. The answer is 18.75.

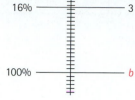

Percents   Quantities

0% — 0
16% — 3
100% — b

Do Exercise 11.

EXAMPLE 10 $10 is what percent of $20?

Translate: $\dfrac{10}{20} = \dfrac{N}{100}$

Solve: $10 \cdot 100 = 20 \cdot N$ Equating cross products

$\dfrac{10 \cdot 100}{20} = \dfrac{20 \cdot N}{20}$ Dividing by 20

$\dfrac{1000}{20} = N$ Multiplying and simplifying

$50 = N$ Dividing

Thus, $10 is 50% of $20. The answer is 50%.

Percents   Quantities

0% — 0
N% — $10
100% — $20

Do Exercise 12.

> Always "look before you leap." Many students can solve this problem mentally: $10 is half, or 50%, of $20.

EXAMPLE 11 What percent of 50 is 16?

Translate: $\dfrac{N}{100} = \dfrac{16}{50}$

Solve: $50 \cdot N = 100 \cdot 16$ Equating cross products

$\dfrac{50 \cdot N}{50} = \dfrac{100 \cdot 16}{50}$ Dividing by 50

$N = \dfrac{1600}{50}$ Multiplying and simplifying

$N = 32$ Dividing

Thus, 32% of 50 is 16. The answer is 32%.

Percents   Quantities

0% — 0
N% — 16
100% — 50

Do Exercise 13.

> **Caution!**
> Don't forget to add the % sign when solving for a percent.

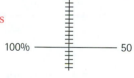

11. Solve:

60 is 120% of what?

12. Solve:

$12 is what percent of $40?

13. Solve:

What percent of 84 is 10.5?

Answers on page A-24

8.3 EXERCISE SET

For Extra Help

a   Translate to a proportion. Do not solve.

1. What is 37% of 74?

2. 66% of 74 is what?

3. 4.3 is what percent of 5.9?

4. What percent of 6.8 is 5.3?

5. 14 is 25% of what?

6. 133% of what is 40?

b   Translate to a proportion and solve.

7. What is 76% of 90?

8. What is 32% of 70?

9. 70% of 660 is what?

10. 80% of 920 is what?

11. What is 4% of 1000?

12. What is 6% of 2000?

13. 4.8% of 60 is what?

14. 63.1% of 80 is what?

15. $24 is what percent of $96?

16. $14 is what percent of $70?

17. 102 is what percent of 100?

18. 103 is what percent of 100?

19. What percent of $480 is $120?

20. What percent of $80 is $60?

21. What percent of 160 is 150?

22. What percent of 33 is 11?

23. $18 is 25% of what?

24. $75 is 20% of what?

25. 60% of what is 54?

26. 80% of what is 96?

27. 65.12 is 74% of what?

28. 63.7 is 65% of what?

29. 80% of what is 16?

30. 80% of what is 10?

31. What is $62\frac{1}{2}\%$ of 40?

32. What is $43\frac{1}{4}\%$ of 2600?

33. What is 9.4% of $8300?

34. What is 8.7% of $76,000?

35. 9.48 is 120% of what?

36. 8.45 is 130% of what?

37. **Dw** In your own words, list steps that a classmate could use to solve any percent problem in this section.

38. **Dw** In solving Example 10, a student simplifies $\frac{10}{20}$ before solving. Is this a good idea? Why or why not?

SKILL MAINTENANCE

Graph. [6.4b]

39. $y = -\frac{1}{2}x$

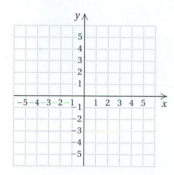

40. $y = 3x$

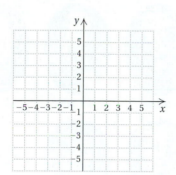

41. $y = 2x - 4$

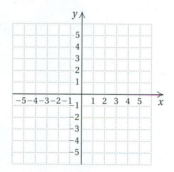

42. $y = \frac{1}{2}x - 3$

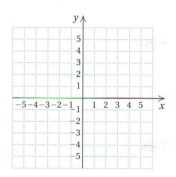

Solve.

43. A recipe for pancakes calls for $\frac{1}{2}$ qt of buttermilk, $\frac{1}{3}$ qt of skim milk, and $\frac{1}{16}$ qt of oil. How many quarts of liquid ingredients does the recipe call for? [4.2d]

44. Guilford Gardeners purchased $\frac{3}{4}$ T (ton) of top soil. If the soil is to be shared equally among 6 gardeners, how much will each gardener receive? [3.8b]

45. $0.05x = 40$ [5.7a]

46. $3 = 0.16t$ [5.7a]

47. $1.3n = 10.4$ [5.7a]

48. $1.2x = 8.4$ [5.7a]

SYNTHESIS

49. **Dw** Does it make sense to talk about a negative percentage? Why or why not?

50. **Dw** Can "comparison drawings," like those used in this section, be used to solve *any* proportion? Why or why not?

Solve.

51. ▦ What is 8.85% of $12,640?
Estimate _____
Calculate _____

52. ▦ 78.8% of what is 9809.024?
Estimate _____
Calculate _____

53. 30% of 80 is what percent of 120?

54. 40% of what is the same as 30% of 200?

55. ▦ What percent of 90 is the same as 26% of 135?

56. ▦ What percent of 80 is the same as 76% of 150?

a Applied Problems Involving Percent

Applied problems involving percent are not always stated in a manner easily translated to an equation. In such cases, it is helpful to rephrase the problem before translating. Sometimes it also helps to make a drawing.

EXAMPLE 1 *Presidential Deaths in Office.* George W. Bush was inaugurated as the 43rd President of the United States in 2001. Because Grover Cleveland was both the 22nd and the 24th presidents, there have been only 42 different presidents. Of the 42 presidents, 8 have died in office: William Henry Harrison, Zachary Taylor, Abraham Lincoln, James A. Garfield, William McKinley, Warren G. Harding, Franklin D. Roosevelt, and John F. Kennedy. What percent have died in office?

Harrison   Taylor   Garfield   McKinley

Harding   Roosevelt   Kennedy

1. **Familiarize.** The question asks for a percent of the presidents who have died in office. We note that 42 is approximately 40 and 8 is $\frac{1}{5}$, or 20%, of 40, so our answer is close to 20%. We let p = the percent who have died in office.

2. **Translate.** There are two ways in which we can translate this problem.

Percent equation (see Section 8.2):

$$
\begin{array}{ccccc}
8 & \text{is} & \text{what percent} & \text{of} & 42? \\
\downarrow & \downarrow & \downarrow & \downarrow & \downarrow \\
8 & = & p & \cdot & 42
\end{array}
$$

Proportion (see Section 8.3):

$$\frac{N}{100} = \frac{8}{42}$$

For proportions, $N\% = p$.

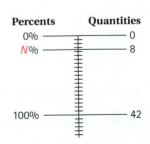

Percents	Quantities
0%	0
$N\%$	8
100%	42

3. Solve. We now have two ways in which to solve this problem.

Percent equation (see Section 8.2):

$$8 = p \cdot 42$$

$$\frac{8}{42} = \frac{p \cdot 42}{42} \qquad \text{Dividing both sides by 42}$$

$$\frac{8}{42} = p$$

$$0.190 \approx p \qquad \text{Finding decimal notation and rounding to the nearest thousandth}$$

$$19.0\% \approx p \qquad \text{Remember to find percent notation.}$$

Note here that the solution, p, includes the % symbol.

Proportion (see Section 8.3):

$$\frac{N}{100} = \frac{8}{42}$$

$$N \cdot 42 = 100 \cdot 8 \qquad \text{Equating cross products}$$

$$\frac{N \cdot 42}{42} = \frac{800}{42} \qquad \text{Dividing both sides by 42}$$

$$N = \frac{800}{42}$$

$$N \approx 19.0 \qquad \text{Dividing and rounding to the nearest tenth}$$

We use the solution of the proportion to express the answer to the problem as 19.0%. Note that in the proportion method, $N\% = p$.

4. Check. To check, we note that the answer 19.0% is close to 20%, as estimated in the *Familiarize* step.

5. State. About 19.0% of the U.S. presidents have died in office.

Do Exercise 1.

EXAMPLE 2 *Transportation to Work.* In the 15 largest cities in the United States, there are about 130,000,000 workers. Approximately 76.3% drive to work alone. How many workers in these 15 cities drive to work alone?

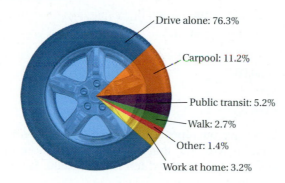

Transportation to Work (in the 15 Largest Cities in the United States)

Drive alone: 76.3%
Carpool: 11.2%
Public transit: 5.2%
Walk: 2.7%
Other: 1.4%
Work at home: 3.2%

Source: U.S. Bureau of the Census

1. Presidential Assassinations in Office. Of the 42 U.S. presidents, 4 have been assassinated in office. These were Garfield, McKinley, Lincoln, and Kennedy. What percent have been assassinated in office?

Answer on page A-24

2. Transportation to Work. There are about 130,000,000 workers in the 15 largest cities in the United States. Approximately 11.2% carpool to work. How many workers in these 15 cities carpool to work?

Source: U.S. Bureau of the Census

1. Familiarize. We can use a simplified form of the pie chart shown to familiarize ourselves with the problem. We let b = the total number of workers who drive to work alone. The chart on the left illustrates percent, and the chart on the right illustrates the actual numbers.

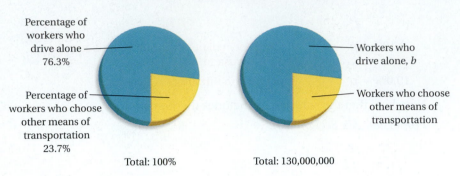

Transportation to Work

Percentage of workers who drive alone 76.3%

Percentage of workers who choose other means of transportation 23.7%

Workers who drive alone, b

Workers who choose other means of transportation

Total: 100%   Total: 130,000,000

2. Translate. There are two ways in which we can translate this problem.

Percent equation:

What number is 76.3% of 130,000,000?

$$b = 0.763 \cdot 130{,}000{,}000$$

Proportion:

$$\frac{76.3}{100} = \frac{b}{130{,}000{,}000}$$

3. Solve. We now have two ways in which to solve this problem.

Percent equation:

$$b = 0.763 \cdot 130{,}000{,}000$$

We multiply:

$$b = (0.763)130{,}000{,}000 = 99{,}190{,}000.$$

Proportion:

$$\frac{76.3}{100} = \frac{b}{130{,}000{,}000}$$

$$76.3 \cdot 130{,}000{,}000 = 100 \cdot b \qquad \text{Equating cross products}$$

$$\frac{76.3 \cdot 130{,}000{,}000}{100} = \frac{100 \cdot b}{100} \qquad \text{Dividing by 100}$$

$$\frac{9{,}919{,}000{,}000}{100} = b$$

$$99{,}190{,}000 = b \qquad \text{Simplifying}$$

Percents	Quantities
0%	0
76.3%	b
100%	130,000,000

4. Check. To check, we can repeat the calculations. We can also do a partial check by estimating. We note that 76.3% is a little more than 75%, or $\frac{3}{4}$. Then we find $\frac{3}{4}$ of 130,000,000, which is 97,500,000. Since 97,500,000 is close to 99,190,000, we have a partial check.

5. State. The number of workers in the 15 largest cities who drive to work alone is 99,190,000.

Answer on page A-24

Do Exercise 2.

b Percent of Increase or Decrease

Percent is often used to state increase or decrease. Let's consider an example of each, using the price of a car as the original number.

PERCENT OF INCREASE

One year a car sold for $20,455. The manufacturer decides to raise the price of the following year's model by 6%. The increase is $0.06 \times$ $20,455, or $1227.30. The new price is $20,455 + $1227.30, or $21,682.30. The *percent of increase* is 6%.

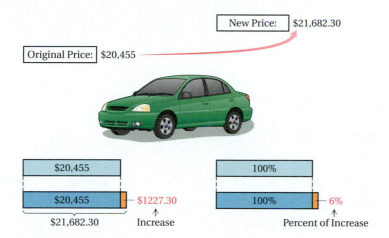

PERCENT OF DECREASE

Lisa buys the car listed above for $20,455. After one year, the car depreciates in value by 25%. This is $0.25 \times$ $20,455, or $5113.75. This depreciation lowers the value of the car to

$$\$20,455 - \$5113.75, \quad \text{or} \quad \$15,341.25$$

Note that the new price is thus 75% of the original price. If Lisa decides to sell the car after a year, $15,341.25 might be the most she could expect to get for it. The *percent of decrease* is 25%, and the decrease is $5113.75.

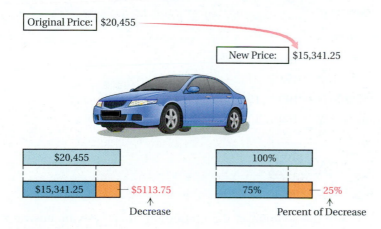

Do Exercises 3 and 4.

3. Percent of Increase. The value of a car is $36,000. The price is increased by 4%.

a) How much is the increase?

b) What is the new price?

4. Percent of Decrease. The value of a car is $36,000. The car depreciates in value by 25% after one year.

a) How much is the decrease?

b) What is the depreciated value of the car?

Answers on page A-24

5. Volume of Mail. The volume of U.S. mail decreased from about 208 billion pieces of mail in 2000 to 202 billion pieces in 2003. What was the percent of decrease?

Source: U.S. Postal Service

When a quantity is decreased by a certain percent, we say we have **percent of decrease.**

EXAMPLE 3 *Chain Link Fence.* For one week only, Sam's Farm Supply had 4 ft × 50 ft rolls of galvanized chain link fence on sale for $39.99. The regular price was $49.99 per roll. What was the percent of decrease?

1. Familiarize. We find the amount of decrease and then make a drawing.

$$
\begin{array}{rl}
4\,9.9\,9 & \text{Retail price} \\
-\,3\,9.9\,9 & \text{Sale price} \\
\hline
1\,0.0\,0 & \text{Decrease}
\end{array}
$$

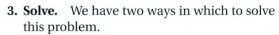

2. Translate. There are two ways in which we can translate this problem.

Percent equation:

$$
\begin{array}{ccccc}
10.00 & \text{is} & \text{what percent} & \text{of} & 49.99? \\
\downarrow & \downarrow & \downarrow & & \downarrow \\
10.00 & = & p & \cdot & 49.99
\end{array}
$$

Proportion:

$$\frac{N}{100} = \frac{10.00}{49.99}$$

For proportions, $N\% = p$.

3. Solve. We have two ways in which to solve this problem.

Percent equation:

$$10.00 = p \cdot 49.99$$

$$\frac{10.00}{49.99} = \frac{p \cdot 49.99}{49.99} \qquad \text{Dividing both sides by 49.99}$$

$$\frac{10.00}{49.99} = p$$

$$0.20 \approx p$$

$$20\% \approx p \qquad \text{Converting to percent notation}$$

Proportion:

$$\frac{N}{100} = \frac{10.00}{49.99}$$

$$49.99 \times N = 100 \times 10 \qquad \text{Equating cross products}$$

$$\frac{49.99 \times N}{49.99} = \frac{100 \times 10}{49.99} \qquad \text{Dividing both sides by 49.99}$$

$$N = \frac{1000}{49.99}$$

$$N \approx 20$$

We use the solution of the proportion to express the answer to the problem as 20%.

Answer on page A-24

4. Check. To check, we note that, with a 20% decrease, the reduced (or sale) price should be 80% of the retail (or original) price. Since

$$80\% \times 49.99 = 0.80 \times 49.99 = 39.992 \approx 39.99,$$

our answer checks.

5. State. The percent of decrease in the price of the roll of fence was 20%.

Do Exercise 5 on the preceding page.

When a quantity is increased by a certain percent, we say we have **percent of increase.**

EXAMPLE 4 *Motor Vehicle Production.* The number of motor vehicles produced worldwide increased from approximately 57.5 million in 2000 to 60.3 million in 2003. What was the percent of increase in motor vehicle production?

Source: Ward's Communications, *Ward's Motor Vehicle Facts & Figures,* 2004

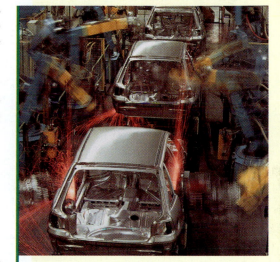

1. Familiarize. We first note that the increase in the number of vehicles produced was $60.3 - 57.5$ million, or 2.8 million. A drawing can help us visualize the situation.

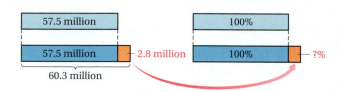

We are asking this question: The increase is what percent of the *original* amount? We let p = the percent of increase.

2. Translate. There are two ways in which we can translate this problem.

Percent equation:

2.8 million   is   what percent   of   57.5 million?

$$2.8 \quad = \quad p \quad \cdot \quad 57.5$$

Proportion:

$$\frac{N}{100} = \frac{2.8}{57.5}$$

For proportions, $N\% = p$.

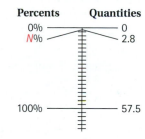

Percents	Quantities
0%	0
N%	2.8
100%	57.5

6. Patents Issued. The number of patents issued per year by the U.S. government increased from 107,332 in 1993 to 189,597 in 2003. What was the percent of increase over the 10-yr period?

Source: U.S. Patent and Trademark Office

Answer on page A-24

Answer on page A-24

CALCULATOR CORNER

Percent of Increase or Decrease On many calculators, there is a fast way to increase or decrease a number by any given percentage. For the price increase on p. 541, the result of taking 6% of $20,455 and adding it to $20,455 can be found by pressing

$$\boxed{2}\boxed{0}\boxed{4}\boxed{5}\boxed{5}\boxed{+}$$
$$\boxed{6}\boxed{\text{SHIFT}}\boxed{\%}\boxed{=}.$$

The displayed result would be

$$\boxed{21682.3}.$$

For the car's depreciation on p. 541, the computation would be

$$\boxed{2}\boxed{0}\boxed{4}\boxed{5}\boxed{5}\boxed{-}$$
$$\boxed{2}\boxed{5}\boxed{\text{SHIFT}}\boxed{\%}\boxed{=}.$$

The result would then be

$$\boxed{15341.25}.$$

Check your manual for other procedures for determining percents.

Exercises:

1. Use a calculator with a $\boxed{\%}$ key to confirm your answer to Margin Exercises 3 and 4.

2. The selling price of Lisa's $87,000 condominium was reduced 8%. Find the new price.

---

3. Solve. We have two ways in which to solve this problem.

Percent equation:

$$2.8 = p \cdot 57.5$$

$$\frac{2.8}{57.5} = \frac{p \cdot 57.5}{57.5} \qquad \text{Dividing both sides by 57.5}$$

$$\frac{2.8}{57.5} = p$$

$$0.049 \approx p$$

$$4.9\% \approx p \qquad \text{Converting to percent notation}$$

Proportion:

$$\frac{N}{100} = \frac{2.8}{57.5}$$

$$57.5 \times N = 100 \times 2.8 \qquad \text{Equating cross products}$$

$$\frac{57.5 \times N}{57.5} = \frac{100 \times 2.8}{57.5} \qquad \text{Dividing both sides by 57.5}$$

$$N = \frac{280}{57.5}$$

$$N \approx 4.9$$

We use the solution of the proportion to express the answer to the problem as 4.9%.

4. Check. To check, we take 4.9% of 57.5:

$$4.9\% \times 57.5 = 0.049 \times 57.5 = 2.8175.$$

Since we rounded the percent, this approximation is close enough to 2.8 to be a good check.

5. State. The percent of increase in the number of motor vehicles produced was about 4.9%.

Do Exercise 6.

The percent of increase or decrease is *always* based on the original amount. To find a percent of increase or decrease, we need to know (1) the amount of increase or decrease and (2) the original amount. The "new" amount after the increase or decrease is not used in the calculation.

$$\text{Percent of increase/decrease} = \frac{\text{Amount of increase/decrease}}{\text{Original amount}}$$

Translating for Success

1. **Distance Walked.** After knee replacement, Alex walked $\frac{1}{8}$ mi each morning and $\frac{1}{5}$ mi each afternoon. How much farther did he walk in the afternoon?

2. **Stock Prices.** A stock sold for $5 per share on Monday and only $2\frac{1}{8}$ per share on Friday. What was the percent of decrease from Monday to Friday?

3. **SAT Score.** After attending a class titled *Improving Your SAT Scores,* Jacob raised his total score from 884 to 1040. What was the percent of increase?

4. **Change in Population.** The population of a small farming community decreased from 1040 to 884. What was the percent of decrease?

5. **Lawn Mowing.** During the summer, brothers Steve and Rob earned money for college by mowing lawns. The largest lawn they mowed was $2\frac{1}{8}$ acres. Steve can mow $\frac{1}{5}$ acre per hour, and Rob can mow only $\frac{1}{8}$ acre per hour. Working together, how many acres did they mow per hour?

The goal of these matching questions is to practice step (2), *Translate,* of the five-step problem-solving process. Translate each word problem to an equation and select a correct translation from equations A–O.

A. $x + \dfrac{1}{5} = \dfrac{1}{8}$

B. $250 = x \cdot 1040$

C. $884 = x \cdot 1040$

D. $\dfrac{250}{16.25} = \dfrac{1000}{x}$

E. $156 = x \cdot 1040$

F. $16.25 = 250 \cdot x$

G. $\dfrac{1}{5} + \dfrac{1}{8} = x$

H. $2\dfrac{1}{8} = x \cdot 5$

I. $5 = 2\dfrac{7}{8} \cdot x$

J. $\dfrac{1}{8} + x = \dfrac{1}{5}$

K. $1040 = x \cdot 884$

L. $\dfrac{250}{16.25} = \dfrac{x}{1000}$

M. $2\dfrac{7}{8} = x \cdot 5$

N. $x \cdot 884 = 156$

O. $x = 16.25 \cdot 250$

Answers on page A-24

6. **Land Sale.** Cole sold $2\frac{1}{8}$ acres from the 5 acres he inherited from his uncle. What percent did he sell?

7. **Travel Expenses.** A magazine photographer is reimbursed 16.25¢ per mile for business travel up to 1000 mi per week. In a recent week, he traveled only 250 mi. What was the total reimbursement for travel?

8. **Trip Expenses.** The total expenses for Claire's recent business trip were $1040. She put $884 on her charge card and paid the balance in cash. What percent did she place on her charge card?

9. **Cost of Copies.** During the first summer session at a community college, the campus copy center advertised 250 copies for $16.25. At this rate, what is the cost of 1000 copies?

10. **Cost of Insurance.** Following a raise in the cost of health insurance, 250 of a company's 1040 employees dropped their health coverage. What percent of the employees canceled their insurance?

a Solve.

1. *Wild Horses.* There are 27,369 wild horses on land managed by the Federal Bureau of Land Management. It is estimated that 48.4% of this total is in Nevada. How many wild horses are in Nevada?

Source: Federal Bureau of Land Management, 2005

2. *U.S. Armed Forces.* There are 1,168,195 people in the United States in active military service. The numbers in the four armed services are listed in the table below. What percent of the total does each branch represent? Round the answers to the nearest tenth of a percent.

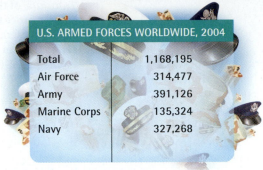

U.S. ARMED FORCES WORLDWIDE, 2004	
Total	1,168,195
Air Force	314,477
Army	391,126
Marine Corps	135,324
Navy	327,268

Source: U.S. Department of Defense

3. *Car Value.* In a recent year, the base price of a Nissan 350Z was $34,000. This vehicle is expected to retain 62% of its value at the end of 3 yr and 52% at the end of 5 yr. What is the value of a Nissan 350Z after 3 yr? after 5 yr?

Source: November/December *Kelley Blue Book Residual Values Guide*

4. *Panda Survival.* Breeding the much-loved panda bear in captivity has been quite difficult for zookeepers.

a) From 1964 to 1997, of 133 panda cubs born in captivity, only 90 lived to be one month old. What percent lived to be one month old?

b) In 1999, Mark Edwards of the San Diego Zoo developed a nutritional formula on which 18 of 20 newborns lived to be one month old. What percent lived to be one month old?

5. *Overweight and Obese.* Of the 294 million people in the United States, 60% are considered overweight and 25% are considered obese. How many are overweight? How many are obese?

Source: U.S. Centers for Disease Control

6. *Smoking and Diabetes.* Of the 294 million people in the United States, 26% are smokers. How many are smokers?

Source: SAMHSA, Office of Applied Studies, National Survey on Drug Use and Health

7. A lab technician has 680 mL of a solution of water and acid; 3% is acid. How many milliliters are acid? water?

8. A lab technician has 540 mL of a solution of alcohol and water; 8% is alcohol. How many milliliters are alcohol? water?

9. *Mississippi River.* The Mississippi River, which extends from Minneapolis, Minnesota, to the Gulf of Mexico, is 2348 mi long. Approximately 77% of the river is navigable. How many miles of the river are navigable?

Source: National Oceanic and Atmospheric Administration

Mississippi River

10. *Immigrants.* In 2003, 705,827 immigrants entered the United States. Of this total, 16.4% were from Mexico and 7.1% were from India. How many immigrants came from Mexico? from India?

Source: U.S. Department of Justice, *2003 Yearbook of Immigration Statistics*

11. *Hispanic Population.* The Hispanic population is growing rapidly in the United States. In 2003, the population of the United States was about 291,000,000 and 13.7% of this total was Hispanic. How many Hispanic people lived in the United States in 2003?

Source: U.S. Bureau of the Census

12. *Age 65 and Over.* By 2010, it is predicted that 13.2% of the U.S. population will be 65 and over. If the population of the United States in 2010 is 307,000,000, how many people will be 65 and over?

Source: U.S. Bureau of the Census

13. *Test Results.* On a test of 40 items, Christina got 91% correct. (There was partial credit on some items.) How many items did she get correct? incorrect?

14. *Test Results.* On a test of 80 items, Pedro got 93% correct. (There was partial credit on some items.) How many items did he get correct? incorrect?

15. *Test Results.* On a test, Maj Ling got 86%, or 81.7, of the items correct. (There was partial credit on some items.) How many items were on the test?

16. *Test Results.* On a test, Juan got 85%, or 119, of the items correct. How many items were on the test?

17. *TV Usage.* Of the 8760 hr in a year, most television sets are on for 2190 hr. What percent is this?

18. *Colds from Kissing.* In a medical study, it was determined that if 800 people kiss someone who has a cold, only 56 will actually catch a cold. What percent is this?

Source: U.S. Centers for Disease Control

19. *Maximum Heart Rate.* Treadmill tests are often administered to diagnose heart ailments. A guideline in such a test is to try to get you to reach your *maximum heart rate,* in beats per minute. The maximum heart rate is found by subtracting your age from 220 and then multiplying by 85%. What is the maximum heart rate of someone whose age is 25? 36? 48? 55? 76? Round to the nearest one.

20. It costs an oil company $40,000 a day to operate two refineries. Refinery A accounts for 37.5% of the cost, and refinery B for the rest of the cost.

a) What percent of the cost does it take to run refinery B?

b) What is the cost of operating refinery A? refinery B?

 Solve.

21. *Savings Increase.* The amount in a savings account increased from $200 to $216. What was the percent of increase?

22. *Population Increase.* The population of a small mountain town increased from 840 to 882. What was the percent of increase?

23. During a sale, a dress decreased in price from $90 to $72. What was the percent of decrease?

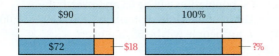

24. A person on a diet goes from a weight of 125 lb to a weight of 110 lb. What is the percent of decrease?

25. *Cooling Costs.* By increasing the thermostat from 72° to 78°, a family can reduce its cooling bill by 50%. If the cooling bill was $106.00, what would the new bill be?

26. *World Population.* World population is increasing by 1.6% each year. In 2006, it was 6.5 billion. Estimate the population in 2007.

Sources: Population Reference Bureau; U.S. Bureau of the Census

27. *Doormat.* A bordered coir doormat that measures 42 in. × 24 in. × $1\frac{1}{2}$ in. has a retail price of $89.95. Over the holidays it is on sale for $65.49. What is the percent of decrease?

28. *Business Tote.* A leather business tote retails for $239.99. An insurance company bought 30 totes for its sales staff at a reduced price of $184.95. What was the percent of decrease in the price?

29. A person earns $28,600 one year and receives a 5% raise in salary. What is the new salary?

30. A person earns $43,200 one year and receives an 8% raise in salary. What is the new salary?

31. *Car Depreciation.* Irwin buys a car for $21,566. It depreciates 25% each year that he owns it. What is the depreciated value of the car after 1 yr? after 2 yr?

32. *Car Depreciation.* Janice buys a car for $22,688. It depreciates 25% each year that she owns it. What is the depreciated value of the car after 1 yr? after 2 yr?

33. *Two-by-Four.* A cross section of a standard or nominal "two-by-four" board actually measures $1\frac{1}{2}$ in. by $3\frac{1}{2}$ in. The rough board is 2 in. by 4 in. but is planed and dried to the finished size. What percent of the wood is removed in planing and drying?

34. *Tipping.* Diners frequently add a 15% tip when charging a meal to a credit card. What is the total amount charged if the cost of the meal, without tip, is $18? $34? $49?

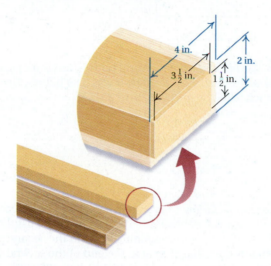

Life Insurance Rates for Smokers and Nonsmokers. The following table provides data showing how yearly rates (premiums) for a $500,000 term life insurance policy are increased for smokers. Complete the missing numbers in the table. Round percents to the nearest percent.

TYPICAL INSURANCE PREMIUMS (DOLLARS)

	AGE	RATE FOR NONSMOKER	RATE FOR SMOKER	PERCENT INCREASE FOR SMOKER
	35	$ 345	$ 630	83%
35.	40	$ 430	$ 735	
36.	45	$ 565		84%
37.	50	$ 780		100%
38.	55	$ 985		117%
39.	60	$1645	$2955	
40.	65	$2943	$5445	

Source: Pacific Life PL Protector Term Life Portfolio, OYT Rates

Population Increase. The following table provides data showing how the populations of various states increased from 1990 to 2003. Complete the missing numbers in the table. Round percents to the nearest tenth of a percent.

	STATE	POPULATION IN 1990	POPULATION IN 2003	CHANGE	PERCENT CHANGE
41.	Alaska	550,043	648,818		
42.	Connecticut	3,287,116		196,256	
43.	Montana		917,621	118,556	
44.	Texas		22,118,509	5,131,999	
45.	Colorado	3,294,394		1,256,294	
46.	Pennsylvania	11,881,643	12,365,455		

Source: U.S. Bureau of the Census

47. ▦ *Population Decrease.* Between 1990 and 2000, the population of Detroit, Michigan, decreased from 1,028,000 to 951,000.

   a) What is the percent of decrease?
   b) If this percent of decrease over a 10-yr period repeated itself in the following decade, what would the population be in 2010?

Source: U.S. Bureau of the Census

48. ▦ *World Population.* World population is increasing by 1.14% each year. In 2004, it was 6.39 billion. How much will it be in 2006? 2010? 2015?

Source: *The World Factbook,* 2004

49. *Car Depreciation.* A car generally depreciates 25% of its original value in the first year. A car is worth $27,300 after the first year. What was its original cost?

50. *Car Depreciation.* Given normal use, an American-made car will depreciate 25% of its original cost the first year and 14% of its remaining value in the second year. What is the value of a car at the end of the second year if its original cost was $36,400? $28,400? $26,800?

51. *Fetal Acoustic Stimulation.* Each year there are about 4 million births in the United States. Of these, about 120,000 births occur in breech position (with the buttocks or feet appearing first). A technique called *fetal acoustic stimulation (FAS)* uses sound directed through a mother's abdomen to stimulate movement of the fetus to a safer position. In a study of this procedure, FAS enabled doctors to turn the baby in 34 of 38 cases.

Source: Johnson and Elliott, "Fetal Acoustic Stimulation, an Adjunct to External Cephalic Versions: A Blinded, Randomized Crossover Study," *American Journal of Obstetrics & Gynecology* **173**, no. 5 (1995): 1369–1372

   a) What percent of U.S. births are breech?
   b) What percent (rounded to the nearest tenth) of cases showed success with FAS?
   c) About how many breech babies yearly might be turned if FAS could be implemented in all breech births in the United States?
   d) Breech position is one reason for performing Caesarean section (or C-section) birth surgery. Researchers expect that FAS alone can eliminate the need for about 2000 C-sections yearly in the United States. Given this information, how many C-sections per year are due to breech position alone?

52. *Drunk Driving Fatalities.* The data in the table show the number of deaths due to drunk drivers from 1990 to 2003.

   a) What is the percent of increase in the number of alcohol-related deaths from 1999 to 2003?
   b) What is the percent of decrease in the number of alcohol-related deaths from 1990 to 2003?

ALCOHOL-RELATED TRAFFIC DEATHS INCREASING OR DECREASING?

YEAR	DEATHS	YEAR	DEATHS
1990	22,587	2000	17,380
1995	17,732	2001	17,400
1997	16,711	2002	17,524
1998	16,673	2003	17,013
1999	16,572		

Source: National Highway Traffic Safety Administration

53. Strike Zone. In baseball, the *strike zone* is normally a 17-in. by 30-in. rectangle. Some batters give the pitcher an advantage by swinging at pitches thrown out of the strike zone. By what percent is the area of the strike zone increased if a 2-in. border is added to the outside?

Source: Major League Baseball

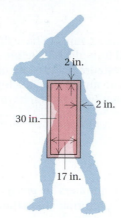

54. Tony has planted grass seed on a 24-ft by 36-ft area in his backyard. He has also installed a 6-ft by 8-ft garden. By what percent does the garden reduce the area he will have to mow?

55. **D**_W What makes better sense (see Exercises 49 and 50): buying a new car, or buying a 1-yr-old car for 25% less? Why?

56. **D**_W See Exercise 25. Does it follow that raising the thermostat from 78° to 84° would reduce the cooling bill to 0?

SKILL MAINTENANCE

Convert to decimal notation. [5.1b], [5.5a]

57. $\dfrac{25}{11}$

58. $\dfrac{11}{25}$

59. $\dfrac{27}{8}$

60. $\dfrac{43}{9}$

61. $\dfrac{23}{25}$

62. $\dfrac{20}{24}$

63. $\dfrac{14}{32}$

64. $\dfrac{2317}{1000}$

65. $\dfrac{34,809}{10,000}$

66. $\dfrac{27}{40}$

Find the perimeter of each polygon. [2.7b]

67.
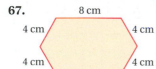

68.
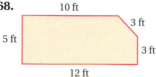

69.
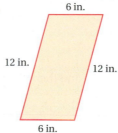

70.
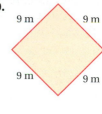

SYNTHESIS

71. **D**_W Exercise 17 reads: "Of the 8760 hr in a year, most television sets are on for 2190 hr. What percent is this?" Will the answer to this change for a leap year? Why or why not?

72. **D**_W Which is better for a wage earner, and why: a 10% raise followed by a 5% raise a year later, or a 5% raise followed by a 10% raise a year later?

73. *Adult Height.* It has been determined that at the age of 10, a girl has reached 84.4% of her final adult height. Cynthia is 4 ft 8 in. at the age of 10. What will be her final adult height?

74. *Adult Height.* It has been determined that at the age of 15, a boy has reached 96.1% of his final adult height. Claude is 6 ft 4 in. at the age of 15. What will be his final adult height?

75. If p is 120% of q, then q is what percent of p?

76. A coupon allows a couple to have dinner and then have $10 subtracted from the bill. Before subtracting $10, however, the restaurant adds a tip of 15%. If the couple is presented with an adjusted bill for $44.05, how much would the dinner (without tip) have cost without the coupon?

The U.S. Bureau of the Census reported the following household incomes by state. The numbers are given as averages of median incomes over 2-yr periods, and have been standardized to 2004 dollars. Use the information in the table to answer Exercises 77 and 78.

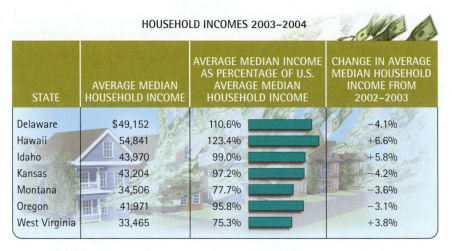

HOUSEHOLD INCOMES 2003–2004

STATE	AVERAGE MEDIAN HOUSEHOLD INCOME	AVERAGE MEDIAN INCOME AS PERCENTAGE OF U.S. AVERAGE MEDIAN HOUSEHOLD INCOME	CHANGE IN AVERAGE MEDIAN HOUSEHOLD INCOME FROM 2002–2003
Delaware	$49,152	110.6%	−4.1%
Hawaii	54,841	123.4%	+6.6%
Idaho	43,970	99.0%	+5.8%
Kansas	43,204	97.2%	−4.2%
Montana	34,506	77.7%	−3.6%
Oregon	41,971	95.8%	−3.1%
West Virginia	33,465	75.3%	+3.8%

77. How much higher was the household income for 2002–2003 in Hawaii than in Montana?

78. Which state had the smallest actual amount of decrease in median household income?

79. 🖩 A worker receives raises of 3%, 6%, and then 9%. By what percent has the original salary increased?

80. **Dᴡ** The former baseball player and manager Yogi Berra once said that "Ninety percent of the game is half mental." If this is true, what percent of the game is mental? Explain your reasoning.

81. **Dᴡ** 🖩 A workers' union is offered either a 5% "across-the-board" raise in which all salaries would increase 5%, or a flat $1650 raise for each worker. If the total payroll for the 123 workers is $4,213,365, which offer should the union select? Why?

8.5 SALES TAX, COMMISSION, AND DISCOUNT

Objectives

a Solve applied problems involving sales tax and percent.

b Solve applied problems involving commission and percent.

c Solve applied problems involving discount and percent.

a Sales Tax

Sales tax computations represent a special type of percent of increase problem. For example, the sales tax rate in Maryland is 5%. This means that the tax is 5% of the purchase price. Suppose the purchase price of a coat is $124.95. The sales tax is then 5% of $124.95, or $0.05 \cdot 124.95$, or 6.2475, or about $6.25.

$124.95
+ 5% sales tax

Baltimore
Annapolis

BILL:

Purchase price	=	$124.95
Sales tax (5% of $124.95)	=	+ 6.25
Total price		$131.20

The total that you pay is the price plus the sales tax.

$124.95 + $6.25, or $131.20

SALES TAX

Sales tax = Sales tax rate × Purchase price

Total price = Purchase price + Sales tax

EXAMPLE 1 *Florida Sales Tax.* The sales tax rate in Florida is 6%. How much tax is charged on the purchase of 4 inflatable rafts at $89.99 each? What is the total price?

a) We first find the cost of the rafts. It is

$4 \cdot \$89.99 = \$359.96$

89^{99} plus 6% each

b) The sales tax on items costing $359.96 is

$$\underbrace{\text{Sales tax rate}}_{6\%} \times \underbrace{\text{Purchase price}}_{\$359.96}$$

or $0.06 \cdot 359.96$, or 21.5976. Thus the tax is $21.60 (rounded to the nearest cent).

c) The total price is given by the purchase price plus the sales tax.

$359.96 + $21.60, or $381.56

To check, note that the total price is the purchase price plus 6% of the purchase price. Thus the total price is 106% of the purchase price. Since $1.06 \cdot 359.96 \approx 381.56$, we have a check. The sales tax is $21.60 and the total price is $381.56.

Do Exercises 1 and 2.

---

1. California Sales Tax. The sales tax rate in California is 7.25%. How much tax is charged on the purchase of a refrigerator that sells for $668.95? What is the total price?

2. Louisiana Sales Tax. Sam buys 5 hardcover copies of Dean Koontz's novel *From the Corner of His Eye* for $26.95 each. The sales tax rate in Louisiana is 4%. How much sales tax will be charged? What is the total price?

Answers on page A-24

3. The sales tax on the purchase of a night table that costs $849 is $50.94. What is the sales tax rate?

4. The sales tax on the purchase of a portable navigation system is $59.94 and the sales tax rate is 6%. Find the purchase price (the price before taxes are added).

Portable Navigation System

$?

Tax at 6% = $59.94

NEW LOW PRICE

$?

$8.93 Tax @ 5%

EXAMPLE 2 The sales tax on the purchase of this wooden play center, which costs $1099, is $43.96. What is the sales tax rate?

Wooden Play Center 27 sq ft play deck with slide and swing set

$1099

+ $43.96 sales tax at ?% sales tax rate

Rephrase: Sales tax   is   what percent   of   purchase price?

Translate: $43.96   = r $\cdot$ $1099

To solve the equation, we divide both sides by 1099:

$$\frac{43.96}{1099} = \frac{r \cdot 1099}{1099}$$

$$\frac{43.96}{1099} = r$$

$$0.04 = r$$

$$4\% = r.$$

The sales tax rate is 4%.

Do Exercise 3.

EXAMPLE 3 The sales tax on the purchase of a Tiffany torchiere lamp is $8.93 and the sales tax rate is 5%. Find the purchase price (the price before taxes are added).

Rephrase: Sales tax   is   5%   of   what?

Translate: $8.93   =   0.05 $\cdot$ b,   or $8.93 = 0.05 \cdot b$

To solve, we divide both sides by 0.05:

$$\frac{8.93}{0.05} = \frac{0.05 \cdot b}{0.05}$$

$$\frac{8.93}{0.05} = b$$

$$\$178.60 = b.$$

The purchase price is $178.60.

Do Exercise 4.

Answers on page A-24

b Commission

When you work for a **salary,** you receive the same amount of money each week or month. When you work for a **commission,** you are paid a percentage of the total sales for which you are responsible.

COMMISSION	
Commission = Commission rate · Sales	

EXAMPLE 4 *Exercise Equipment Sales.* A salesperson's commission rate is 16%. What is the commission from the sale of $9700 worth of exercise equipment?

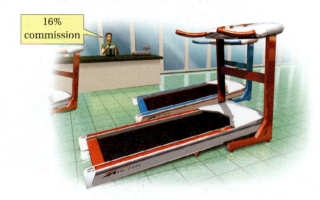

16% commission

$$
\begin{array}{ccccc}
Commission & = & Commission\ rate & \cdot & Sales \\
C & = & 16\% & \cdot & 9700 \\
C & = & 0.16 & \cdot & 9700 \\
C & = & 1552 & &
\end{array}
$$

The commission is $1552.

Do Exercise 5.

EXAMPLE 5 *Farm Machinery Sales.* Dawn earns a commission of $31,000 selling $620,000 worth of farm machinery. What is the commission rate?

FARM SUPPLIES

COMMISSION $31,000
TOTAL SALES $620,000

$$
\begin{array}{ccccc}
Commission & = & Commission\ rate & \cdot & Sales \\
31,000 & = & r & \cdot & 620,000
\end{array}
$$

5. Raul's commission rate is 30%. What is the commission from the sale of $18,760 worth of air conditioners?

Answer on page A-24

Study Tips

WRITING ALL THE STEPS

Take the time to include all the steps when working your homework problems. Doing so will help you organize your thinking and avoid computational errors. If you find a wrong answer, having all the steps allows easier checking of your work. It will also give you complete, step-by-step solutions of the exercises that can be used to study for an exam.

Writing down all the steps and keeping your work organized may also give you a better chance of getting partial credit.

6. Liz earns a commission of $3000 selling $24,000 worth of U2 concert tickets. What is the commission rate?

To solve this equation, we divide both sides by 620,000:

$$\frac{31,000}{620,000} = \frac{r \cdot 620,000}{620,000}$$

$$\frac{1}{20} \cdot \frac{31,000}{31,000} = r \qquad \text{Removing a factor equal to 1: } \frac{31,000}{31,000} = 1$$

$$0.05 = r$$

$$5\% = r.$$

The commission rate is 5%.

Do Exercise 6.

EXAMPLE 6 *Motorcycle Sales.* Joyce's commission rate is 12%. She receives a commission of $936 from the sale of a motorcycle. How much did the motorcycle cost?

Motorcycle Shop Sales Commission Contract

12%

The Salesperson, known as _Joyce Stone_, will receive a commission of 12% of the final price on each motorcycle he/she sells. This contract shall be legal for the extent of the employment of the salesperson.

Employee _____
Date _____

Commission	=	Commission rate	·	Sales		
936	=	12%	·	S,	or	$936 = 0.12 \cdot S$

To solve this equation, we divide both sides by 0.12:

$$\frac{936}{0.12} = \frac{0.12 \cdot S}{0.12}$$

$$\frac{936}{0.12} = S$$

$$7800 = S.$$

The motorcycle cost $7800.

Do Exercise 7.

7. Ben's commission rate is 16%. He receives a commission of $268 from sales of clothing. How many dollars worth of clothing were sold?

C   Discount

Suppose that the regular price of a rug is $60, and the rug is on sale at 25% off. Since 25% of $60 is $15, the sale price is $60 − $15, or $45. We call $60 the **original,** or **marked price,** 25% the **rate of discount,** $15 the **discount,** and $45 the **sale price.** Note that discount problems are a type of percent of decrease problem.

> **DISCOUNT AND SALE PRICE**
>
> **Discount** = Rate of discount × Original price
> **Sale price** = Original price − Discount

EXAMPLE 7 A masonite door marked $1389 is on sale at $33\frac{1}{3}\%$ off. What is the discount? the sale price?

DOOR
$1389 orig.

Save
33$\frac{1}{3}$%

Sale price = ?

a) *Discount = Rate of discount · Original price*

$$D = 33\frac{1}{3}\% \cdot 1389$$

$$D = \frac{1}{3} \cdot 1389 \qquad 33\frac{1}{3}\% = \frac{1}{3}$$

$$D = \frac{1389}{3} = 463$$

b) *Sale price = Original price − Discount*

$$S = 1389 - 463$$

$$S = 926$$

The discount is $463 and the sale price is $926.

Do Exercise 8.

EXAMPLE 8 *Antique Pricing.* The price of an antique table is marked down from $620 to $527. What is the rate of discount?

We first find the discount by subtracting the sale price from the original price:

$$620 - 527 = 93.$$

The discount is $93.

Next, we use the equation for discount:

$$Discount = Rate\ of\ discount \cdot Original\ price$$
$$93 = r \cdot 620.$$

To solve, we divide both sides by 620:

$$\frac{93}{620} = \frac{r \cdot 620}{620}$$

$$\frac{93}{620} = r$$

$$0.15 = r$$

$$15\% = r.$$

The discount rate is 15%.

> To check, note that a 15% discount rate means that 85% of the original price is paid:
> $$0.85 \times 620 = 527.$$

Do Exercise 9.

8. A suit marked $540 is on sale at $33\frac{1}{3}\%$ off. What is the discount? the sale price?

9. The price of a pair of hiking boots is reduced from $75 to $60. Find the rate of discount.

Reduced to
$60
Original price
$75

Answers on page A-24

a   Solve.

1. Tennessee Sales Tax.   The sales tax rate in Tennessee is 7%. How much sales tax would be charged on a lawn mower that costs $279?

2. Arizona Sales Tax.   The sales tax rate in Arizona is 5.6%. How much sales tax would be charged on a lawn mower that costs $279?

3. Kansas Sales Tax.   The sales tax rate in Kansas is 5.3%. How much sales tax would be charged on a video game, Quest of the Planets, which sells for $49.99?

4. New Jersey Sales Tax.   The sales tax rate in New Jersey is 6%. How much sales tax would be charged on a copy of John Grisham's novel *A Painted House*, which sells for $27.95?

Sources: Borders Bookstore; Andrea Sutcliffe, *Numbers*

5. Utah Sales Tax.   The sales tax rate in Utah is 4.75%. How much tax is charged on a purchase of 5 telephones at $69 apiece? What is the total price?

6. New York Sales Tax.   The sales tax rate in New York is 4.25%. How much tax is charged on a purchase of 5 teapots at $37.99 apiece? What is the total price?

7. The sales tax is $48 on the purchase of a dining room set that sells for $960. What is the sales tax rate?

8. The sales tax is $15 on the purchase of a diamond ring that sells for $500. What is the sales tax rate?

9. The sales tax is $35.80 on the purchase of a refrigerator–freezer that sells for $895. What is the sales tax rate?

10. The sales tax is $9.12 on the purchase of a patio set that sells for $456. What is the sales tax rate?

11. The sales tax on the purchase of a used car is $100 and the sales tax rate is 5%. Find the purchase price (the price before taxes are added).

12. The sales tax on the purchase of a new boat is $112 and the sales tax rate is 2%. Find the purchase price.

13. The sales tax on the purchase of a dining room set is $28 and the sales tax rate is 3.5%. Find the purchase price.

14. The sales tax on the purchase of a portable DVD player is $24.75 and the sales tax rate is 5.5%. Find the purchase price.

15. The sales tax rate in Austin is 2% for the city and county and 6.25% for the state. Find the total amount paid for 2 shower units at $332.50 apiece.

16. The sales tax rate in Omaha is 1.5% for the city and 4.5% for the state. Find the total amount paid for 3 air conditioners at $260 apiece.

17. The sales tax on an automobile purchase of $18,400 is $1030.40. What is the sales tax rate?

18. The sales tax on an automobile purchase of $15,800 is $979.60. What is the sales tax rate?

b Solve.

19. Kiersten's commission rate is 6%. What is the commission from the sale of $45,000 worth of furnaces?

20. Jose's commission rate is 32%. What is the commission from the sale of $12,500 worth of sailboards?

21. Mitchell earns $120 selling $2400 worth of television sets in a consignment shop. What is the commission rate?

22. Donna earns $408 selling $3400 worth of shoes. What is the commission rate?

23. An art gallery's commission rate is 40%. They receive a commission of $392. How many dollars worth of artwork were sold?

24. A real estate agent's commission rate is 7%. She receives a commission of $5600 from the sale of a home. How much did the home sell for?

25. A real estate commission is 6%. What is the commission from the sale of a $98,000 home?

26. A real estate commission is 8%. What is the commission from the sale of a piece of land for $68,000?

27. Bonnie earns $280.80 selling $2340 worth of tee shirts. What is the commission rate?

28. Chuck earns $1147.50 selling $7650 worth of ski passes. What is the commission rate?

29. Miguel's commission is increased according to how much he sells. He receives a commission of 5% for the first $2000 and 8% on the amount over $2000. What is the total commission on sales of $6000?

30. Lucinda earns a salary of $500 a month, plus a 2% commission on sales. One month, she sold $990 worth of encyclopedias. What were her wages that month?

C Find what is missing.

	MARKED PRICE	RATE OF DISCOUNT	DISCOUNT	SALE PRICE
31.	$300	10%		
32.	$2000	40%		
33.	$17	15%		
34.	$20	25%		
35.		10%	$12.50	
36.		15%	$65.70	
37.	$600		$240	
38.	$12,800		$1920	

39. Find the discount and the rate of discount for the car seat in this ad.

Sale
Car Seat
$149.99
Was $179.99

40. Find the discount and the rate of discount for the wicker chair in this ad.

Best price of the season!
Now only
$90
Was $125

41. Find the marked price and the rate of discount for the stacked tool storage units in this ad.

CLOSEOUT
$349
buys both
SAVE
$200

42. Find the marked price and the rate of discount for the basketball system in this ad.

SAVE $400
Basketball System
Now
$599⁹⁹

43. ^D**W** Is the following ad mathematically correct? Why or why not?

FAMOUS MAKER WATCHES

$6.95 Regularly $9.95

30% OFF

Choose from men's and ladies' casual or dress designs

Limited time offer

44. ^D**W** Is a 40% discount the same as taking 20% off of a price that reflects a previous 20% discount? Why or why not?

SKILL MAINTENANCE

Solve. [7.3b]

45. $\dfrac{x}{12} = \dfrac{24}{16}$

46. $\dfrac{7}{2} = \dfrac{11}{x}$

Graph. [6.4b]

47. $y = \dfrac{4}{3}x$

48. $y = -\dfrac{4}{3}x + 1$

Write decimal notation. [5.1b]

49. $\dfrac{5}{9}$

50. $\dfrac{23}{11}$

Simplify. [5.4b]

51. $80\left(1 + \dfrac{0.06}{2}\right)^2$

52. $70\left(1 + \dfrac{0.08}{2}\right)^2$

SYNTHESIS

53. ^D**W** Carl's Car Care mistakenly charged Dawn 5% tax for a cleaning job that was all labor. To correct the mistake, they subtracted 5% from what Dawn paid. Was this correct? Why or why not?

54. ^D**W** An item that is no longer on sale at "25% off" receives a price tag that is $33\frac{1}{3}\%$ more than the sale price. Has the item price been restored to its original price? Why or why not?

55. Before taxes are added, the price of a pack of cigarettes is approximately $2.70. State sales tax and federal cigarette tax combine to add another 39¢ (AL, CO, DE, MT, NH, OR) to 69¢ (RI, WA) to the price of a pack. Approximate the range of state and federal taxes (combined) as a percentage of the per-pack price.

Source: Campaign for Tobacco-Free Kids® 2003

56. Tee shirts are being sold at the mall for $5 each, or 3 for $10. If you buy three tee shirts, what is the rate of discount?

57. 🖩 *Magazine Subscriptions.* In a recent subscription drive, *People* offered a subscription of 52 weekly issues for a price of $1.89 per issue. They advertised that this was a savings of 29.7% off the newsstand price. What was the newsstand price?

58. 🖩 Gordon receives a 10% commission on the first $5000 in sales and 15% on all sales beyond $5000. If Gordon receives a commission of $2405, how much did he sell? Use a calculator and trial and error if you wish.

59. 🖩 A real-estate commission rate is 7.5%. A house sells for $98,500. How much does the seller get for the house after paying the commission?

60. ^D**W** Herb collects baseball memorabilia. He bought two autographed plaques, but became short of funds and had to sell them quickly for $200 each. On one, he made a 20% profit and on the other, he lost 20%. Did he make or lose money on the sale? Explain.

SIMPLE AND COMPOUND INTEREST

Objectives

a. Solve applied problems involving simple interest.

b. Solve applied problems involving compound interest.

1. What is the simple interest on $4300 invested at an interest rate of 7% for 1 year?

2. What is the simple interest on a principal of $4300 invested at an interest rate of 7% for $\frac{3}{4}$ year?

Answers on page A-25

a Simple Interest

Suppose you put $1000 into an investment for 1 year. The $1000 is called the **principal.** If the **interest rate** is 8%, in addition to the principal, you get back 8% of the principal, which is

8% of $1000, or 0.08 · 1000, or $80.00

The $80.00 is called the **simple interest.** It is, in effect, the price that a financial institution pays for the use of the money over time.

> **SIMPLE INTEREST FORMULA**
>
> The **simple interest I** on principal P, invested for t years at interest rate r, is given by
>
> $$I = P \cdot r \cdot t.$$

EXAMPLE 1 What is the simple interest on $2500 invested at an interest rate of 6% for 1 year?

We use the formula $I = P \cdot r \cdot t$:

$$I = P \cdot r \cdot t = \$2500 \cdot 6\% \cdot 1$$
$$= \$2500 \cdot 0.06$$
$$= \$150.$$

The simple interest for 1 year is $150.

Do Exercise 1.

EXAMPLE 2 What is the simple interest on a principal of $2500 invested at an interest rate of 6% for $\frac{1}{4}$ year?

We use the formula $I = P \cdot r \cdot t$:

$$I = P \cdot r \cdot t = \$2500 \cdot 6\% \cdot \frac{1}{4}$$
$$= \frac{\$2500 \cdot 0.06}{4}$$
$$= \frac{\$150}{4} \quad \text{Multiplying}$$
$$= \$37.50$$

> We could instead have found $\frac{1}{4}$ of 6% and then multiplied by 2500.

$$\begin{array}{r} 3\,7.5 \\ 4\,)\,\overline{1\,5\,0.0} \\ \underline{1\,2\,0} \\ 3\,0 \\ \underline{2\,8} \\ 2\,0 \\ \underline{2\,0} \\ 0 \end{array}$$

The simple interest for $\frac{1}{4}$ year is $37.50.

Do Exercise 2.

Study Tips

DO EXTRA PROBLEMS

When an exercise gives you difficulty, it is usually wise to practice solving some other exercises that are very similar to the one that gave you trouble. Usually, if the troubling exercise is odd, the next (even) exercise is quite similar. Checking the Chapter Review and Test for similar problems is also a good idea.

When time is given in days, we generally divide it by 365 to express the time as a fractional part of a year.

EXAMPLE 3 To pay for a shipment of tee shirts, New Wave Designs borrows $8000 at $9\frac{3}{4}$% for 60 days. Find **(a)** the amount of simple interest that is due and **(b)** the total amount that must be paid after 60 days.

a) We express 60 days as a fractional part of a year:

$$I = P \cdot r \cdot t = \$8000 \cdot 9\frac{3}{4}\% \cdot \frac{60}{365}$$

$$= \$8000 \cdot 0.0975 \cdot \frac{60}{365}$$

$$\approx \$128.22$$

The interest due for 60 days is $128.22.

b) The total amount to be paid after 60 days is the principal plus the interest:

$$\$8000 + \$128.22 = \$8128.22$$

The total amount due is $8128.22.

Do Exercise 3.

3. The Glass Nook borrows $4800 at $8\frac{1}{2}$% for 30 days. Find **(a)** the amount of simple interest due and **(b)** the total amount that must be paid after 30 days.

b Compound Interest

When interest is paid *on interest,* we call it **compound interest.** This is the type of interest usually paid on investments or loans. Suppose you have $5000 in a savings account at 6%. In 1 year, the account will contain the original $5000 plus 6% of $5000. Thus the total in the account after 1 year will be

106% of $5000, or 1.06 · $5000, or $5300.

Now suppose that the total of $5300 remains in the account for another year. At the end of this second year, the account will contain the $5300 plus 6% of $5300. The total in the account would thus be

106% of $5300, or 1.06 · $5300, or **$5618**.

Note that in the second year, interest is earned on the first year's interest. When this happens, we say that interest is **compounded annually.**

4. Find the amount in an account if $2000 is invested at 9%, compounded annually, for 2 years.

EXAMPLE 4 Find the amount in an account if $2000 is invested at 8%, compounded annually, for 2 years.

a) After 1 year, the account will contain 108% of $2000.

$$1.08 \cdot \$2000 = \$2160$$

b) At the end of the second year, the account will contain 108% of $2160.

$$1.08 \cdot \$2160 = \$2332.80$$

The amount in the account after 2 years is $2332.80.

Do Exercise 4.

Suppose that the interest in Example 4 were **compounded semi-annually**—that is, every half year. Interest would then be calculated twice a year at a rate of 8% ÷ 2, or 4% each time. The approach used in Example 4 can then be adapted, as follows.

After the first $\frac{1}{2}$ year, the account will contain 104% of $2000.

$$1.04 \cdot \$2000 = \$2080$$

After a second $\frac{1}{2}$ year (1 full year), the account will contain 104% of $2080.

$$1.04 \cdot \$2080 = \$2163.20$$

After a third $\frac{1}{2}$ year $\left(1\frac{1}{2} \text{ full years}\right)$, the account will contain 104% of $2163.20.

$$1.04 \cdot \$2163.20 = \$2249.728$$
$$\approx \$2249.73 \qquad \text{Rounding to the nearest cent}$$

Finally, after a fourth $\frac{1}{2}$ year (2 full years), the account will contain 104% of $2249.73.

$$1.04 \cdot \$2249.73 = \$2339.7192$$
$$\approx \$2339.72 \qquad \text{Rounding to the nearest cent}$$

Let's summarize our results and look at them another way:

End of 1st $\frac{1}{2}$ year $\rightarrow 1.04 \cdot 2000 = 2000 \cdot (1.04)^1$;
End of 2nd $\frac{1}{2}$ year $\rightarrow 1.04 \cdot (1.04 \cdot 2000) = 2000 \cdot (1.04)^2$;
End of 3rd $\frac{1}{2}$ year $\rightarrow 1.04 \cdot (1.04 \cdot 1.04 \cdot 2000) = 2000 \cdot (1.04)^3$;
End of 4th $\frac{1}{2}$ year $\rightarrow 1.04 \cdot (1.04 \cdot 1.04 \cdot 1.04 \cdot 2000) = 2000 \cdot (1.04)^4$.

Note that each multiplication was by 1.04 and that

$$\$2000 \cdot 1.04^4 \approx \$2339.72 \qquad \text{Using a calculator and rounding to the nearest cent}$$

We have illustrated the following result.

COMPOUND INTEREST FORMULA

If a principal P has been invested at interest rate r, compounded n times a year, in t years it will grow to an amount A given by

$$A = P \cdot \left(1 + \frac{r}{n}\right)^{n \cdot t}.$$

In the compound interest formula, $n \cdot t$ is the total number of compounding periods, r is written in decimal notation, and $\frac{r}{n}$ is the interest rate for each period.

Answer on page A-25

Let's apply this formula to confirm our preceding discussion, where the amount invested is $P = \$2000$, the number of years is $t = 2$, and the number of compounding periods each year is $n = 2$. Substituting into the compound interest formula, we have

$$A = P \cdot \left(1 + \frac{r}{n}\right)^{n \cdot t} = 2000 \cdot \left(1 + \frac{8\%}{2}\right)^{2 \cdot 2}$$

$$= 2000 \cdot \left(1 + \frac{0.08}{2}\right)^{4} = 2000(1.04)^4$$

$$= 2000 \cdot 1.16985856 \approx \$2339.72$$

If you are using a calculator, you could perform this computation in one step.

EXAMPLE 5 The Ibsens invest $4000 in an account paying $5\frac{5}{8}\%$, compounded quarterly. Find the amount in the account after $2\frac{1}{2}$ years.

The compounding is quarterly, so n is 4. We substitute $4000 for P, $5\frac{5}{8}\%$, or 0.05625, for r, 4 for n, and $2\frac{1}{2}$, or $\frac{5}{2}$, for t and compute A:

$$A = P \cdot \left(1 + \frac{r}{n}\right)^{n \cdot t} = \$4000 \cdot \left(1 + \frac{5\frac{5}{8}\%}{4}\right)^{4 \cdot 5/2}$$

$$= \$4000 \cdot \left(1 + \frac{0.05625}{4}\right)^{10}$$

$$= \$4000(1.0140625)^{10}$$

$$\approx \$4599.46$$

The amount in the account after $2\frac{1}{2}$ years is $4599.46.

Do Exercise 5.

Answer on page A-25

5. A couple invests $7000 in an account paying $6\frac{3}{8}\%$, compounded semiannually. Find the amount in the account after $1\frac{1}{2}$ years.

CALCULATOR CORNER

Compound Interest   A calculator is useful in computing compound interest. Not only does it do computations quickly but it also eliminates the need to round until the computation is completed. This minimizes "round-off errors" that occur when rounding is done at each stage of the computation. We must keep order of operations in mind when computing compound interest.

To find the amount due on a $20,000 loan made for 25 days at 11% interest, compounded daily, we would compute $20{,}000\left(1 + \dfrac{0.11}{365}\right)^{25}$. To do this on a calculator, we press 2 0 0 0 0 × ((1 + · 1 1 ÷

3 6 5)) y^x (or x^y or ∧) 2 5 = . Without parentheses, we would first find $1 + \dfrac{0.11}{365}$, raise this result to the

25th power, and then multiply by 20,000. To do this, we press 1 + · 1 1 ÷ 3 6 5 = y^x (or x^y or ∧) 2 5 = × 2 0 0 0 0 . In either case, the result is 20,151.23, rounded to the nearest cent.

Some calculators have business keys that allow such computations to be done more quickly.

Exercises:

1. Find the amount due on a $16,000 loan made for 62 days at 13% interest, compounded daily.

2. An investment of $12,500 is made for 90 days at 8.5% interest, compounded daily. How much is the investment worth after 90 days?

---

a   Find the simple interest.

	PRINCIPAL	RATE OF INTEREST	TIME	SIMPLE INTEREST
1.	$200	4%	1 year	
2.	$450	2%	1 year	
3.	$2000	8.4%	$\frac{1}{2}$ year	
4.	$200	7.7%	$\frac{1}{2}$ year	
5.	$4300	10.56%	$\frac{1}{4}$ year	
6.	$8000	9.42%	$\frac{1}{6}$ year	
7.	$20,000	$4\frac{5}{8}\%$	1 year	
8.	$100,000	$3\frac{7}{8}\%$	1 year	
9.	$50,000	$5\frac{3}{8}\%$	$\frac{1}{4}$ year	
10.	$80,000	$6\frac{3}{4}\%$	$\frac{1}{12}$ year	

Solve. Assume that simple interest is being calculated in each case.

11. CopiPix, Inc., borrows $10,000 at 9% for 60 days. Find **(a)** the amount of interest due and **(b)** the total amount that must be paid after 60 days.

12. Sal's Laundry borrows $8000 at 10% for 90 days. Find **(a)** the amount of interest due and **(b)** the total amount that must be paid after 90 days.

13. Animal Instinct, a pet supply shop, borrows $6500 at 5% for 90 days. Find **(a)** the amount of interest due and **(b)** the total amount that must be paid after 90 days.

14. Andante's Cafe borrows $4500 at 12% for 60 days. Find **(a)** the amount of interest due and **(b)** the total amount that must be paid after 60 days.

15. Jean's Garage borrows $5600 at 10% for 30 days. Find (a) the amount of interest due and (b) the total amount that must be paid after 30 days.

16. Shear Delights, a hair salon, borrows $3600 at 4% for 30 days. Find (a) the amount of interest due and (b) the total amount that must be paid after 30 days.

b Interest is compounded annually. Find the amount in the account after the given length of time. Round to the nearest cent.

	PRINCIPAL	RATE OF INTEREST	TIME	AMOUNT IN THE ACCOUNT
17.	$400	5%	2 years	
18.	$450	4%	2 years	
19.	$2000	8.8%	4 years	
20.	$4000	7.7%	4 years	
21.	$4300	10.56%	6 years	
22.	$8000	9.42%	6 years	
23.	$20,000	$6\frac{5}{8}\%$	25 years	
24.	$100,000	$5\frac{7}{8}\%$	30 years	

Interest is compounded semiannually. Find the amount in the account after the given length of time. Round to the nearest cent.

	PRINCIPAL	RATE OF INTEREST	TIME	AMOUNT IN THE ACCOUNT
25.	$4000	6%	1 year	
26.	$1000	5%	1 year	
27.	$20,000	8.8%	4 years	
28.	$40,000	7.7%	4 years	
29.	$5000	10.56%	6 years	
30.	$8000	9.42%	8 years	
31.	$20,000	$7\frac{5}{8}\%$	25 years	
32.	$100,000	$4\frac{7}{8}\%$	30 years	

Solve.

33. The Lopez family invests $4000 in an account paying 6%, compounded monthly. How much is in the account after 5 months?

34. Chrissy and Kyle invest $2500 in an account paying 3%, compounded monthly. How much is in the account after 6 months?

35. A couple invests $1200 in an account paying 10%, compounded quarterly. How much is in the account after 1 year?

36. The O'Hares invest $6000 in an account paying 8%, compounded quarterly. How much is in the account after 18 months?

37. ▦ Emilio loans his niece's business $20,000 for 50 days at 6% interest, compounded daily. How much is Emilio owed after 50 days?

38. ▦ Elsa loans her nephew's business $25,000 for 40 days at 5% interest, compounded daily. How much is Elsa owed after 40 days?

39. **D**_{**W**} Which is a better investment and why: $1000 invested at $7\frac{3}{4}$% simple interest for 1 year, or $1000 invested at 7% compounded monthly for 1 year?

40. **D**_{**W**} A firm must choose between borrowing $5000 at 10% for 30 days and borrowing $10,000 at 8% for 60 days. Give arguments in favor of and against each option.

🔖 **VOCABULARY REINFORCEMENT**

In each of Exercises 41–48, fill in the blank with the correct term from the given list. Some of the choices may not be used and some may be used more than once.

41. If the product of two numbers is 1, they are _____ of each other.   [3.7a]

42. A number is _____ if its ones digit is even and the sum of its digits is divisible by 3.   [3.1b]

43. The number 0 is the _____ identity.   [2.2a]

44. A(n) _____ is the ratio of price to the number of units.   [7.2b]

45. The distance around an object is its _____ .   [1.2c]

46. A number is _____ if the sum of its digits is divisible by 3.   [3.1b]

47. A natural number that has exactly two different factors, only itself and 1, is called a _____ number.   [3.2b]

48. When two pairs of numbers have the same ratio, they are _____ .   [7.3a]

divisible by 3

divisible by 4

divisible by 6

divisible by 9

perimeter

area

unit rate

reciprocals

proportional

composite

prime

additive

multiplicative

49. **D**_{**W**} Without performing the multiplications, determine which gives the most interest: $1000 × 8% × $\frac{1}{12}$, or $1000 × 8% × $\frac{30}{365}$? How did you decide?

50. **D**_{**W**} Erik is in the 25% tax bracket. Would he be better off investing in an account earning 4.5% interest, compounded daily, that would be taxed, or a tax-free account earning 3.5% compounded daily? Explain.

Effective Yield.   The *effective yield* is the yearly rate of simple interest that corresponds to a rate for which interest is compounded two or more times a year. For example, if P is invested at 12%, compounded quarterly, we would multiply P by $(1 + 0.12/4)^4$, or 1.03^4. Since $1.03^4 \approx 1.126$, or 112.6%, the 12% compounded quarterly corresponds to an effective yield of approximately 12.6%. In Exercises 51 and 52, find the effective yield for the indicated account.

51. ▦ The account pays 9% compounded monthly.

52. ▦ The account pays 10% compounded daily.

53. ▦ Rather than spend $20,000 on a new car that will lose 30% of its value in 1 year, the Coniglios invest the money at 9%, compounded daily. By not buying the car (which would have *de*creased their net worth), and investing instead (which *in*creased their net worth), how much have the Coniglios saved after 1 year?

568

8.7 INTEREST RATES ON CREDIT CARDS AND LOANS

Objective

a Solve applied problems involving interest rates on credit cards and loans.

a Credit Cards and Loans

Look at the following graphs. They offer good reason for a study of the real-world applications of percent, interest, loans, and credit cards.

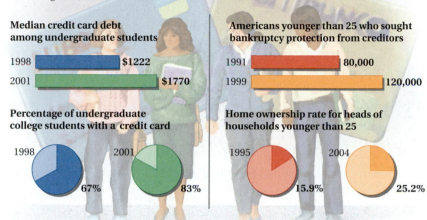

The True Cost of "Free" Money

Many 18–25-year-olds are borrowing and mortgaging their futures, according to debt and consumer advocates.

Median credit card debt among undergraduate students

1998	$1222
2001	$1770

Americans younger than 25 who sought bankruptcy protection from creditors

1991	80,000
1999	120,000

Percentage of undergraduate college students with a credit card

1998 — 67%   2001 — 83%

Home ownership rate for heads of households younger than 25

1995 — 15.9%   2004 — 25.2%

Sources: Nellie Mae, U.S. Bureau of the Census, and Harvard Law School

Comparing interest rates is essential if one is to become financially responsible. A small change in an interest rate can make a *large* difference in the cost of a loan. When you make a payment on a loan, do you know how much of that payment is interest and how much is applied to reducing the principal?

We begin with an example involving credit cards. A balance carried on a credit card is a type of loan. In a recent year in the United States, 100,000 young adults declared bankruptcy because of excessive credit card debt. The money you obtain through the use of a credit card is not "free" money. There is a price (interest) to be paid for the privilege.

EXAMPLE 1 *Credit Cards.*   After the holidays, Sarah has a balance of $3216.28 on a credit card with an annual percentage rate (APR) of 19.7%. She decides not to make additional purchases with this card until she has paid off the balance.

a) Many credit cards require a minimum monthly payment of 2% of the balance. What is Sarah's minimum payment on a balance of $3216.28? Round the answer to the nearest dollar.

b) Find the amount of interest and the amount applied to reduce the principal in the minimum payment found in part (a).

c) If Sarah had transferred her balance to a card with an APR of 12.5%, how much of her first payment would be interest and how much would be applied to reduce the principal?

d) By how much more was the principal reduced in part (c) than in part (b)? How much more interest was paid in part (b) than in part (c)?

1. Credit Cards. After the holidays, Jamal has a balance of $4867.59 on a credit card with an annual percentage rate (APR) of 21.3%. He decides not to make additional purchases with this card until he has paid off the balance.

a) Many credit cards require a minimum monthly payment of 2% of the balance. What is Jamal's minimum payment on a balance of $4867.59? Round the answer to the nearest dollar.

b) Find the amount of interest and the amount applied to reduce the principal in the minimum payment found in part (a).

c) If Jamal had transferred his balance to a card with an APR of 13.6%, how much of his first payment would be interest and how much would be applied to reduce the principal?

d) By how much more was the principal reduced in part (c) than in part (b)? How much less interest is paid in part (c) than in part (b)?

Answers on page A-25

We solve as follows.

a) The minimum payment is 2% of $3216.28.

$$0.02 \cdot \$3216.28 = \$64.3256 \qquad \text{Sarah's minimum payment, rounded to the nearest dollar, is \$64.}$$

b) The amount of interest on $3216.28 at 19.7% for one month* is given by

$$I = P \cdot r \cdot t = \$3216.28 \cdot 0.197 \cdot \frac{1}{12} \approx \$52.80.$$

To find the amount applied to reduce the principal in the first payment, we use the answer from part (a) and subtract.

$$\text{Amount applied to reduce the principal} = \text{Minimum payment} - \text{Interest for the month}$$

$$= \$64 - \$52.80$$

$$= \$11.20$$

Thus the principal of $3216.28 is decreased by only $11.20 with the first payment. (Sarah still owes $3205.08.)

c) The amount of interest on $3216.28 at 12.5% for one month is

$$I = P \cdot r \cdot t = \$3216.28 \cdot 0.125 \cdot \frac{1}{12} \approx \$33.50.$$

To find the amount applied to reduce the principal in the first payment, we use the answer from part (a) and subtract.

$$\text{Amount applied to reduce the principal} = \text{Minimum payment} - \text{Interest for the month}$$

$$= \$64 - \$33.50$$

$$= \$30.50$$

Thus the principal of $3216.28 is decreased by $30.50 with the first payment. (Sarah still owes $3185.78.)

d) Let's organize the information for both rates in the following table.

BALANCE BEFORE FIRST PAYMENT	FIRST MONTH'S PAYMENT	% APR	AMOUNT OF INTEREST	AMOUNT APPLIED TO PRINCIPAL	BALANCE AFTER FIRST PAYMENT
$3216.28	$64	19.7%	$52.80	$11.20	$3205.08
3216.28	64	12.5	33.50	30.50	3185.78

Difference in balance after first payment → $19.30

At 19.7%, the interest is $52.80 and the principal is decreased by $11.20. At 12.5%, the interest is $33.50 and the principal is decreased by $30.50. Thus the principal is decreased by $30.50 − $11.20, or $19.30 more with the 12.5% rate than with the 19.7% rate. The interest at 19.7% is $19.30 greater than the interest at 12.5%.

Do Exercise 1.

---

*Actually, the interest on a credit card is computed daily with a rate called a daily percentage rate (DPR). The DPR for Example 1 would be 19.7%/365 ≈ 0.054%. When no payments or additional purchases are made during the month, the difference in total interest for the month is minimal and we will not deal with it here.

Even though the mathematics of the information in the chart below is beyond the scope of this text, it is interesting to compare how long it takes to pay off the balance of Example 1 if Sarah pays $64 for each payment with how long it takes if she pays double that amount, $128. Financial consultants frequently tell clients that if they want to control their debt, they should double the minimum payment.

RATE	PAYMENT PER MONTH	NUMBER OF PAYMENTS TO PAY OFF DEBT	TOTAL PAID BACK	ADDITIONAL COST OF PURCHASES
19.7%	$64	107, or 8 yr 11 mo	$6848	$3631.72
19.7	128	33, or 2 yr 9 mo	4224	1007.72
12.5	64	72, or 6 yr	4608	1391.72
12.5	128	29, or 2 yr 5 mo	3712	495.72

As with most loans, if you pay an extra amount toward the principal with each payment, the length of the loan can be greatly reduced. Note that at 19.7%, it will take Sarah almost 9 yr to pay off her debt if she pays only $64 per month (and does not make additional purchases). If she transfers her balance to a card with a 12.5% rate and pays $128 per month, she could eliminate her debt in approximately $2\frac{1}{2}$ yr. Debt can quickly grow out of control if you continue to make purchases and pay only the minimum payment. The debt will never be eliminated.

The Federal Stafford Loan program provides educational loans to students at interest rates that are much lower than those on credit cards. Payments on such a loan do not begin until 6 months after graduation. At that time, the student has 10 years, or 120 monthly payments, to pay off the loan.

EXAMPLE 2 *Federal Stafford Loans.* After graduation, the balance on Taylor's Stafford loan is $28,650. If the rate on his loan is 3.37%, he will make 120 payments of approximately $282 each to pay off the loan.

a) Find the amount of interest and the amount of principal in the first payment.

b) If the interest rate were 5.25%, he would make 120 monthly payments of approximately $307 each. How much more of the first payment is interest if the rate is 5.25% rather than 3.37%?

c) Compare the total amount of interest on the loan at 3.37% with the interest on the loan at 5.25%. How much more would Taylor pay at 5.25% than at 3.37%?

We solve as follows.

a) We use the formula $I = P \cdot r \cdot t$.

$$I = \$28{,}650 \cdot 0.0337 \cdot \frac{1}{12} \qquad \text{Substituting}$$

$$\approx \$80.46$$

The amount of interest in the first payment is $80.46, and the payment is $282. To determine the amount applied to the principal, we subtract.

$$\$282 - \$80.46 = \$201.54$$

With the first payment, the principal will be reduced by $201.54.

2. Federal Stafford Loans. After graduation, the balance on Maggie's Stafford loan is $32,680. To pay off the loan at 3.37%, she will make 120 payments of approximately $321 each.

a) Find the amount of interest and the amount applied to the principal in the first payment.

b) If the interest rate were 5.5%, she would make 120 payments of approximately $355 each. How much more of the first payment is interest if the loan is 5.5% rather than 3.37%?

c) Compare the total amount of interest on the loan at 3.37% with the amount of interest on the loan at 5.5%. How much more would Maggie pay in interest on the 5.5% loan than on the 3.37% loan?

Answers on page A-25

---

b) The interest at 5.25% would be

$$I = \$28{,}650 \cdot 0.0525 \cdot \frac{1}{12}$$

$$\approx \$125.34.$$

At 5.25%, the additional interest in the first payment is

$$\$125.34 - \$80.46 = \$44.88.$$

The higher interest rate results in an additional $44.88 in interest in the first payment.

c) For the 3.37% loan, there will be 120 payments of $282 each:

$$120 \cdot \$282 = \$33{,}840.$$

The total amount of interest at this rate is

$$\$33{,}840 - \$28{,}650 = \$5190.$$

For the 5.25% loan, there will be 120 payments of $307 each:

$$120 \cdot \$307 = \$36{,}840.$$

The total amount of interest at this rate is

$$\$36{,}840 - \$28{,}650 = \$8190.$$

At the rate of 5.25%, Taylor would pay

$$\$8190 - \$5190 = \$3000$$

more in interest than at the rate of 3.37%.

Do Exercise 2.

EXAMPLE 3   *Home Loans.* The Sawyers recently purchased their first home. They borrowed $153,000 at $6\frac{5}{8}\%$ for 30 years (360 payments). Their monthly payment (excluding insurance and taxes) is $979.68.

a) How much of the first payment is interest and how much is applied to reduce the principal?

b) If the Sawyers pay the entire 360 payments, how much interest will be paid on the loan?

We solve as follows.

a) To find the amount of interest paid in the first payment, we use the formula $I = P \cdot r \cdot t$:

$$I = P \cdot r \cdot t = \$153{,}000 \cdot 0.06625 \cdot \frac{1}{12} \approx \$844.69.$$

The amount applied to the principal is

$$\$979.68 - \$844.69, \text{ or } \$134.99.$$

b) Over the 30-year period, the total paid will be

$$360 \cdot \$979.68, \text{ or } \$352{,}684.80.$$

The total amount of interest paid over the lifetime of the loan is

$$\$352{,}684.80 - \$153{,}000, \text{ or } \$199{,}684.80.$$

Do Exercises 3 and 4 on the following page.

AMORTIZATION TABLES

If we make 360 calculations as in Example 3(a) and continue with a decreased principal as in Margin Exercise 3, we can create an *amortization table,* part of which is shown below. Such tables are also found in reference books. The beginning, middle, and last part of the loan described are shown. Look over the table and note how little each payment reduces the principal at the beginning of the loan. Do you see again why a loan is not "free"?

MORTGAGE AMORTIZATION PROGRAM

MORTGAGE AMOUNT: $153,000
INTEREST RATE: 6.625%
NUMBER OF YEARS: 30
MONTHLY PAYMENTS ARE: $979.68

PAYMENT	PRINCIPAL	INTEREST	BALANCE	
1	$134.99	$844.69	$152,865.01	← Example 3
2	$135.74	$843.94	$152,729.27	← Margin
3	$136.48	$843.19	$152,592.80	Exercise 3
4	$137.24	$842.44	$152,455.56	
5	$137.99	$841.68	$152,317.56	
6	$138.76	$840.92	$152,178.81	
7	$139.52	$840.15	$152,039.29	
8	$140.29	$839.38	$151,898.99	
9	$141.07	$838.61	$151,757.93	
10	$141.85	$837.83	$151,616.08	
11	$142.63	$837.05	$151,473.45	
12	$143.42	$836.26	$151,330.04	

Interest for 12 periods = $10,086.14

175	$351.84	$627.84	$113,369.72	
176	$353.78	$625.90	$113,015.94	
177	$355.73	$623.94	$112,660.20	
178	$357.70	$621.98	$112,302.51	
179	$359.67	$620.00	$111,942.83	
180	$361.66	$618.02	$111,581.18	
181	$363.65	$616.02	$111,217.52	
182	$365.66	$614.01	$110,851.86	
183	$367.68	$611.99	$110,484.18	
184	$369.71	$609.96	$110,114.47	
185	$371.75	$607.92	$109,742.71	
186	$373.80	$605.87	$109,368.91	

Interest for 12 periods = $7303.45

349	$917.04	$62.63	$10,427.84	
350	$922.11	$57.57	$9,505.74	
351	$927.20	$52.48	$8,578.54	
352	$932.32	$47.36	$7,646.23	
353	$937.46	$42.21	$6,708.76	
354	$942.64	$37.04	$5,766.13	
355	$947.84	$31.83	$4,818.28	
356	$953.07	$26.60	$3,865.21	
357	$958.34	$21.34	$2,906.87	
358	$963.63	$16.05	$1,943.24	
359	$968.95	$10.73	$974.30	
360	$974.30	$5.38	$0.00	

Interest for 12 periods = $411.22

Total interest for 360 periods = $199,684.80

Refer to Example 3 for Margin Exercises 3 and 4.

3. **Home Loans.** Since the principal has been reduced by the first payment, at the time of the second payment of the Sawyers' 30-yr loan, the new principal is the decreased principal

$$\$153,000 - \$134.99,$$

or

$$\$152,865.01.$$

Use $152,865.01 as the principal, and determine how much of the second payment is interest and how much is applied to reduce the principal. (In effect, repeat Example 3(a) using the new principal.)

4. **Home Loans.** The Sawyers decide to change the period of their home loan from 30 years to 15 years. Their monthly payment increases to $1343.33.

   a) How much of the first payment is interest and how much is applied to reduce the principal?

   b) If the Sawyers pay the entire 180 payments, how much interest will be paid on this loan?

   c) How much less interest will the Sawyers pay with the 15-yr loan than with the 30-yr loan?

Answers on page A-25

5. Refinancing a Home Loan.
Consider Example 4 for a 15-yr loan. The new monthly payment is $1250.14.

a) How much of the first payment is interest and how much is applied to reduce the principal?

b) If the Sawyers pay the entire 180 payments, how much interest will be paid on this loan?

c) How much less interest is paid in the course of paying off the 15-yr loan at $5\frac{1}{2}$% than in the course of paying off the 15-yr loan at $6\frac{5}{8}$% in Margin Exercise 4?

Answers on page A-25

Study Tips

PROFESSORS' MISTAKES

Even the best professors sometimes make mistakes. If, as you review your notes, you find that something doesn't make sense, it may be due to your instructor having made a mistake. If, after double-checking, you still perceive a mistake, calmly ask your instructor about it. In doing so you may be helping everyone in your class.

EXAMPLE 4 *Refinancing a Home Loan.* Refer to Example 3. Ten months after the Sawyers buy their home financed at a rate of $6\frac{5}{8}$%, they decide to re-finance at $5\frac{1}{2}$% even though the new loan will cost them refinance charges. They have reduced the principal a small amount in the 10 payments they have made, but they decide to again borrow $153,000 for 30 years at the new rate. Their new monthly payment is $868.72.

a) How much of the first payment is interest and how much is applied to the principal?

b) Compare the amounts at $5\frac{1}{2}$% found in part (a) with the amounts at $6\frac{5}{8}$% found in Example 3(a).

c) If the Sawyers pay the entire 360 payments, how much interest will be paid on this loan?

d) How much less is the total interest at $5\frac{1}{2}$% than at $6\frac{5}{8}$%?

We solve as follows.

a) To find the interest paid in the first payment, we use the formula $I = P \cdot r \cdot t$.

$$I = P \cdot r \cdot t = \$153{,}000 \cdot 0.055 \cdot \frac{1}{12} = \$701.25$$

The amount applied to the principal is

$868.72 − $701.25, or $167.47.

b) We compare the amounts found in part (a) with the amounts found in Example 3(a):

Rate	Monthly payment	Interest in first payment	Amount applied to principal
$6\frac{5}{8}$%	$979.68	$844.69	$134.99
$5\frac{1}{2}$%	$868.72	$701.25	$167.47

At $5\frac{1}{2}$%, the interest in the first payment is

$844.69 − $701.25, or $143.44,

less than at $6\frac{5}{8}$%. The amount applied to the principal is

$167.47 − $134.99, or $32.48,

more than $6\frac{5}{8}$%.

c) Over the 30-year period at $5\frac{1}{2}$%, the total paid will be

360 · $868.72, or $312,739.20.

The total amount of interest paid over the lifetime of the $5\frac{1}{2}$% loan is

$312,739.20 − $153,000, or $159,739.20.

d) The total interest paid at $6\frac{5}{8}$% is $199,684.80 (from Example 3). The difference between the interest paid at the two rates is

$199,684.80 − $159,739.20, or $39,945.60.

Thus the $5\frac{1}{2}$% loan saves the Sawyers approximately $40,000 in interest charges over the 30 years.

Do Exercise 5.

a Solve.

1. *Credit Cards.* At the end of his first year of college, Antonio has a balance of $4876.54 on a credit card with an annual percentage rate (APR) of 21.3%. He decides not to make additional purchases with his card until he has paid off the balance.

  a) Many credit cards require a minimum monthly payment of 2% of the balance. What is Antonio's minimum payment on a balance of $4876.54? Round the answer to the nearest dollar.

  b) Find the amount of interest and the amount applied to reduce the principal in the minimum payment found in part (a).

  c) If Antonio had transferred his balance to a card with an APR of 12.6%, how much of his first payment would be interest and how much would be applied to reduce the principal?

  d) By how much do the amounts for 12.6% from part (c) and the amounts for 21.3% from part (b) differ?

2. *Credit Cards.* At the end of her junior year of college, Becky had a balance of $5328.88 on a credit card with an annual percentage rate (APR) of 18.7%. She decides not to make additional purchases with this card until she has paid off the balance.

  a) Many credit cards require a minimum monthly payment of 2% of the balance. What is Becky's minimum payment on a balance of $5328.88? Round the answer to the nearest dollar.

  b) Find the amount of interest and the amount applied to reduce the principal in the minimum payment found in part (a).

  c) If Becky had transferred her balance to a card with an APR of 13.2%, how much of her first payment would be interest and how much would be applied to reduce the principal?

  d) By how much do the amounts for 13.2% from part (c) and the amounts for 18.7% from part (b) differ?

3. *Federal Stafford Loans.* After graduation, the balance on Grace's Stafford loan is $44,560. To pay off the loan at 3.37%, she will make 120 payments of approximately $437.93 each.

  a) Find the amount of interest and the amount applied to reduce the principal in the first payment.

  b) If the interest rate were 4.75%, she would make 120 monthly payments of approximately $467.20 each. How much more of the first payment is interest if the loan is at 4.75% rather than 3.37%?

  c) Compare the total amount of interest on a loan at 3.37% with the amount on the loan at 4.75%. How much more would Grace pay on the 4.75% loan than on the 3.37% loan?

4. *Federal Stafford Loans.* After graduation, the balance on Ricky's Stafford loan is $38,970. To pay off the loan at 3.37%, he will make 120 payments of approximately $382.99 each.

  a) Find the amount of interest and the amount applied to reduce the principal in the first payment.

  b) If the interest rate were 5.4%, he would make 120 monthly payments of approximately $421 each. How much more of the first payment is interest if the loan is at 5.4% rather than 3.37%?

  c) Compare the total amount of interest on the loan at 3.37% with the amount on the loan at 5.4%. How much more would Ricky pay on the 5.4% loan than on the 3.37% loan?

5. *Home Loan.* The Martinez family recently purchased a home. They borrowed $164,000 at $6\frac{1}{4}$% for 30 years (360 payments). Their monthly payment (excluding insurance and taxes) is $1009.78.

  a) How much of the first payment is interest and how much is applied to reduce the principal?

  b) If this family pays the entire 360 payments, how much interest will be paid on the loan?

  c) Determine the new principal after the first payment. Use that new principal to determine how much of the second payment is interest and how much is applied to reduce the principal.

6. *Home Loan.* The Kaufmans recently purchased a home. They borrowed $136,000 at 5.75% for 30 years (360 payments). Their monthly payment (excluding insurance and taxes) is $793.66.

  a) How much of the first payment is interest and how much is applied to reduce the principal?

  b) If the Kaufmans pay the entire 360 payments, how much interest will be paid on the loan?

  c) Determine the new principal after the first payment. Use that new principal to determine how much of the second payment is interest and how much is applied to reduce the principal.

7. *Refinancing a Home Loan.* Refer to Exercise 5. The Martinez family decides to change the period of their home loan to 15 years. Their monthly payment increases to $1406.17.

a) How much of the first payment is interest and how much is applied to reduce the principal?

b) If the Martinez family pays the entire 180 payments, how much interest will be paid on the loan?

c) How much less interest is paid with the 15-yr loan than with the 30-yr loan?

8. *Refinancing a Home Loan.* Refer to Exercise 6. The Kaufmans decide to change the period of their home loan to 15 years. Their monthly payment increases to $1129.36.

a) How much of the first payment is interest and how much is applied to reduce the principal?

b) If the Kaufmans pay the entire 180 payments, how much interest will be paid on the loan?

c) How much less interest is paid with the 15-yr loan than with the 30-yr loan?

Complete the following table, assuming monthly payments as given.

	INTEREST RATE	HOME MORTGAGE	TIME OF LOAN	MONTHLY PAYMENT	PRINCIPAL AFTER FIRST PAYMENT	PRINCIPAL AFTER SECOND PAYMENT
9.	6.98%	$100,000	360 mos	$663.96		
10.	6.98%	$100,000	180 mos	$897.71		
11.	8.04%	$100,000	180 mos	$957.96		
12.	8.04%	$100,000	360 mos	$736.55		
13.	7.24%	$150,000	360 mos	$1022.25		
14.	7.24%	$75,000	180 mos	$684.22		
15.	7.24%	$200,000	180 mos	$1824.60		
16.	7.24%	$180,000	360 mos	$1226.70		

17. *Dealership's Car Loan Offer.* For a trip to Colorado, Michael and Rebecca buy a Toyota Sienna van whose selling price is $23,950. For financing, they accept the promotion from the manufacturer that offers a 48-month loan at 2.9% with 10% down. Their monthly payment is $454.06.

a) What is the down payment? the amount borrowed?

b) How much of the first payment is interest and how much is applied to reduce the principal?

c) What is the total interest cost of the loan if they pay all of the 48 payments?

18. *New-Car Loan.* After working at her first job for 2 years, Janice buys a new Saturn for $16,385. She makes a down payment of $1385 and finances $15,000 for 4 years at a new-car loan rate of 8.99%. Her monthly payment is $373.20.

a) How much of her first payment is interest and how much is applied to reduce the principal?

b) Find the principal balance at the beginning of the second month and determine how much less interest she will pay in the second payment than in the first.

c) What is the total interest cost of the loan if she pays all of the 48 payments?

19. *Used-Car Loan.* Twin brothers, Jerry and Terry, each take a job at the college cafeteria in order to have the money to make payments on the purchase of a 2004 Chrysler PT Cruiser for $11,900. They make a down payment of 5% and finance the remainder at 9.3% for 3 years. (Used-car loan rates are generally higher than new-car loan rates.) Their monthly payment is $361.08.

a) What is the down payment? the amount borrowed?
b) How much of the first payment is interest and how much is applied to reduce the principal?
c) If they pay all 36 payments, how much interest will they pay for the loan?

20. *Used-Car Loan.* For his construction job, Clint buys a 2003 Dodge Ram 1500 truck for $13,800. He makes a down payment of $1380 and finances the remainder for 4 years at 8.8%. The monthly payment is $307.89.

a) How much is financed?
b) How much of the first payment is interest and how much is applied to reduce the principal?
c) If he pays all 48 payments, how much interest will he pay for the loan?

21. **D**_{**W**} Look over the examples and exercises in this section. What happens to the monthly payment on a loan if the time of payment changes from 30 years to 15 years, assuming the interest rate stays the same? Discuss the pros and cons of both time periods.

22. **D**_{**W**} Examine the information in the graphs at the beginning of this section. Discuss how a knowledge of this section might have been of help to some of these students.

SKILL MAINTENANCE

Find the area of each figure.

23. 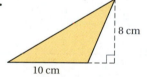 [3.6b]

8 cm
10 cm

24. 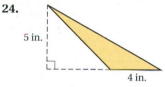 [3.6b]

5 in.
4 in.

25. 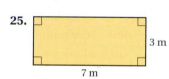 [1.5c]

3 m
7 m

26. 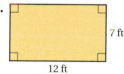 [1.5c]

7 ft
12 ft

27. 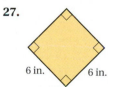 [1.5c]

6 in. 6 in.

28. 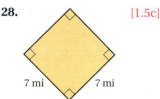 [1.5c]

7 mi 7 mi

SYNTHESIS

29. **D**_{**W**} After receiving her monthly credit card bill, Beverly usually waits about 10 days before mailing her check. Assuming she pays off her entire balance each month, why is this a smart thing for her to do?

30. **D**_{**W**} Suppose a friend owes you $100 and offers to repay you by giving you either $10 a month for 12 months, or $60 every 6 months for 12 months. Which would you select and why?

31. *Refinancing a Home Loan.* Refer to Examples 3 and 4. It cost the Sawyers $1200 to refinance their loan. With the lower house payment, how long will it take the Sawyers to recoup the refinance charge?

The review that follows is meant to prepare you for a chapter exam. It consists of three parts. The first part, Concept Reinforcement, is designed to increase understanding through true/false exercises. The second part is a list of important properties and formulas. The third part is the Review Exercises. These provide practice exercises for the exam, together with references to section objectives so you can go back and review. Before beginning, stop and look back over the skills you have obtained. What skills in mathematics do you have now that you did not have before studying this chapter?

CONCEPT REINFORCEMENT

Determine whether the statement is true or false. Answers are given at the back of the book.

_____ 1. A fixed principal invested for four years will earn more interest when interest is compounded quarterly than when interest is compounded semiannually.

_____ 2. Of the numbers 0.5%, $\frac{5}{1000}\%$, $\frac{1}{2}\%$, $\frac{1}{5}$, and $0.\overline{1}$, the largest number is $0.\overline{1}$.

_____ 3. If principal A equals principal B and principal A is invested for 2 years at 4% compounded quarterly while principal B is invested for 4 years at 2% compounded semiannually, the interest earned from each investment is the same.

_____ 4. The symbol % is equivalent to $\times \frac{1}{10}$.

IMPORTANT PROPERTIES AND FORMULAS

Commission = Commission rate · Sales

Discount = Rate of discount · Original price

Sale price = Original price − Discount

Simple Interest: $I = P \cdot r \cdot t$

Compound Interest: $A = P \cdot \left(1 + \dfrac{r}{n}\right)^{n \cdot t}$

Review Exercises

Find percent notation for the decimal notation in the sentence in Exercises 1 and 2.   [8.1b]

1. Of all the snacks eaten on Super Bowl Sunday, 0.56 of them are chips and salsa.

   Source: Korbel Research and Pace Foods

2. Of all the vehicles in Mexico City, 0.017 of them are taxis.

   Source: *The Handy Geography Answer Book*

Find percent notation. [8.1c]

3. $\dfrac{3}{8}$

4. $\dfrac{1}{3}$

Find decimal notation. [8.1b]

5. 73.5%

6. $6\dfrac{1}{2}\%$

Find fraction notation. [8.1c]

7. 24%

8. 6.3%

Translate to a percent equation. Then solve. [8.2a, b]

9. 30.6 is what percent of 90?

10. 63 is 84 percent of what?

11. What is $38\dfrac{1}{2}\%$ of 168?

Translate to a proportion. Then solve. [8.3a, b]

12. 24 percent of what is 16.8?

13. 42 is what percent of 30?

14. What is 10.5% of 84?

Solve. [8.4a, b]

15. *Favorite Ice Creams.* According to a recent survey, 8.9% of those interviewed chose chocolate as their favorite ice cream flavor and 4.2% chose butter pecan. Of the 2500 students in a freshman class, how many would choose chocolate as their favorite ice cream? butter pecan?

Source: International Ice Cream Association

16. *Prescriptions.* Of the 295 million people in the United States, 123.64 million take at least one kind of prescription drug per day. What percent take at least one kind of prescription drug per day?

Source: American Society of Health-System Pharmacies

17. *Water Output.* The average person expels 200 mL of water per day by sweating. This is 8% of the total output of water from the body. How much is the total output of water?

Source: Elaine N. Marieb, *Essentials of Human Anatomy and Physiology,* 6th ed. Boston: Addison Wesley Longman, Inc., 2000

18. *Test Scores.* Jason got a 75 on a math test. He was allowed to go to the math lab and take a retest. He increased his score to 84. What was the percent of increase?

19. *Test Scores.* Jenny got an 81 on a math test. By taking a retest in the math lab, she increased her score by 15%. What was her new score?

Solve. [8.5a, b, c]

20. A state charges a meals tax of $4\dfrac{1}{2}\%$. What is the meals tax charged on a dinner party costing $320?

21. In a certain state, a sales tax of $378 is collected on the purchase of a used car for $7560. What is the sales tax rate?

22. Kim earns $753.50 selling $6850 worth of televisions. What is the commission rate?

23. An air conditioner has a marked price of $350. It is placed on sale at 12% off. What is the discount and what is the sale price?

24. A fax machine priced at $305 is discounted at the rate of 14%. What is the discount and what is the sale price?

25. An insurance salesperson receives a 7% commission. If $42,000 worth of life insurance is sold, what is the commission?

26. Find the rate of discount.

Our lowest price of the season!

18-ft metal frame

$399⁶⁹

Reg. 489.99

Family-size metal-frame pool

Solve.   [8.6a, b]

27. What is the simple interest on $1800 at 6% for $\frac{1}{3}$ year?

28. The Dress Shack borrows $24,000 at 10% simple interest for 60 days. Find **(a)** the amount of interest due and **(b)** the total amount that must be paid after 60 days.

29. What is the simple interest on $2200 principal at the interest rate of 5.5% for 1 year?

30. The Kleins invest $7500 in an investment account paying an annual interest rate of 12%, compounded monthly. How much is in the account after 3 months?

31. Find the amount in an investment account if $8000 is invested at 9%, compounded annually, for 2 years.

Solve.   [8.7a]

32. *Credit Cards.*   At the end of her junior year of college, Judy has a balance of $6428.74 on a credit card with an annual percentage rate (APR) of 18.7%. She decides not to make additional purchases with this card until she has paid off the balance.

a) Many credit cards require a minimum payment of 2% of the balance. What is Judy's minimum payment on a balance of $6428.74? Round the answer to the nearest dollar.

b) Find the amount of interest and the amount applied to reduce the principal in the minimum payment found in part (a).

c) If Judy had transferred her balance to a card with an APR of 13.2%, how much of her first payment would be interest and how much would be applied to reduce the principal?

d) By how much more is the principal decreased at 13.2% than at 18.7%? How much less interest is paid at 13.2% than at 18.7%?

33. **Dw** Ollie buys a microwave oven during a 10%-off sale. The sale price that Ollie paid was $162. To find the original price, Ollie calculates 10% of $162 and adds that to $162. Is this correct? Why or why not?   [8.5c]

34. **Dw** Which is the better deal for a consumer and why: a discount of 40% or a discount of 20% followed by another of 22%?   [8.5c]

SYNTHESIS

35. 🔢 *Land Area of the United States.*   After Hawaii and Alaska became states, the total land area of the United States increased from 2,963,681 mi² to 3,540,939 mi². What was the percent of increase?   [8.4b]

36. Rhonda's Dress Shop reduces the price of a dress by 40% during a sale. By what percent must the store increase the sale price, after the sale, to get back to the original price?   [8.5c]

37. A $200 coat is marked up 20%. After 30 days, it is marked down 30% and sold. What was the final selling price of the coat?   [8.5c]

1. *Bookmobiles.* Since 1991, the number of bookmobiles has decreased by approximately 6.4%. Find decimal notation for 6.4%.
   Source: American Library Association

2. *Gravity.* The gravity of Mars is 0.38 as strong as Earth's. Find percent notation for 0.38.
   Source: www.marsinstitute.info/epo/mermarsfacts.html

3. Find percent notation for $\frac{11}{8}$.

4. Find fraction notation for 65%.

5. Translate to a percent equation. Then solve.

   What is 40% of 55?

6. Translate to a proportion. Then solve.

   What percent of 80 is 65?

Solve.

7. *Cruise Ship Passengers.* Of the passengers on a typical cruise ship, 16% are in the 25–34 age group and 23% are in the 35–44 age group. A cruise ship has 2500 passengers. How many are in the 25–34 age group? the 35–44 age group?
   Source: Polk

8. *Population Increase.* The population of the state of Utah increased from 1,722,850 in 1990 to 2,547,389 in 2005. What was the percent of increase?
   Source: U.S. Bureau of the Census

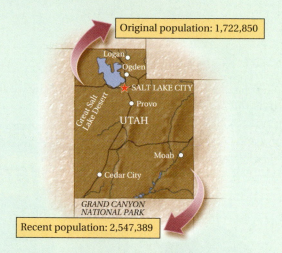

Original population: 1,722,850

UTAH

Logan
Ogden
SALT LAKE CITY
Provo
Moab
Cedar City

Great Salt Lake Desert

GRAND CANYON NATIONAL PARK

Recent population: 2,547,389

9. *Airline Profits.* Profits of the entire U.S. airline industry decreased from $5.5 billion in 1999 to $2.7 billion in 2000. Find the percent of decrease.
   Source: Air Transport Association

10. There arc 6.6 billion people living in the world today. It is estimated that the total number who have ever lived is about 120 billion. What percent of people who have ever lived are alive today?
    Source: *The Handy Geography Answer Book*

11. *Maine Sales Tax.* The sales tax rate in Maine is 5%. How much tax is charged on a purchase of $324? What is the total price?

12. Gwen's commission rate is 15%. What is the commission from the sale of $4200 worth of merchandise?

13. The marked price of a DVD player is $200 and the item is on sale at 20% off. What is the discount and what is the sale price?

14. What is the simple interest on a principal of $120 at the interest rate of 7.1% for 1 year?

15. The Burnham Parents–Teachers Association invests $5200 at 6% simple interest. How much is in the account after $\frac{1}{2}$ year?

16. Find the amount in an account if $1000 is invested at $5\frac{3}{8}\%$, compounded annually, for 2 years.

17. The Suarez family invests $10,000 at an annual interest rate of 4.9%, compounded monthly. How much is in the account after 3 years?

18. *Job Opportunities.*   The table below lists job opportunities, in millions, in 2002 and projected increases to 2012. Find the missing numbers.

OCCUPATION	NUMBER OF JOBS IN 2002 (in millions)	NUMBER OF JOBS IN 2012 (in millions)	CHANGE	PERCENT OF INCREASE
Retail salespersons	4.1	4.8	0.7	17.1%
Registered nurses	2.3		0.6	
Post-secondary teachers	1.6	2.2		
Food preparation and service workers		2.4	0.4	
Restaurant servers		2.5		19.0%

Source: Department of Labor

19. Find the discount and the discount rate of the washer-dryer duet in this ad.

SHORT TIME ONLY

$1675

Was $1950

No interest for 18 months

20. *Home Loan.*   Complete the following table, assuming the monthly payment as given.

Interest Rate	7.4%
Mortgage	$120,000
Time of Loan	360 mos
Monthly Payment	$830.86
Principal after First Payment	
Principal after Second Payment	

21. By selling a home without using a realtor, Juan and Marie can avoid paying a 7.5% commission. They receive an offer of $180,000 from a potential buyer. In order to give a comparable offer, for what price would a realtor need to sell the house? Round to the nearest hundred.

22. Karen's commission rate is 16%. She invests her commission from the sale of $15,000 worth of merchandise at the interest rate of 12%, compounded quarterly. How much is Karen's investment worth after 6 months?

1. Write fraction notation for 0.091.

2. Write decimal notation for $\frac{13}{6}$.

3. Write an equivalent decimal for 3%.

4. Write $\frac{9}{8}$ as an equivalent percent.

5. Write fraction notation for the ratio 0.6 to 8.

6. Write the rate in kilometers per hour.

$$350 \text{ km}, \quad 15 \text{ hr}$$

Use <, >, or = for ☐ to write a true sentence.

7. $\frac{5}{7}$ ☐ $\frac{6}{8}$

8. -3.78 ☐ -37.8

Estimate the sum or difference by rounding to the nearest hundred.

9. $263{,}961 + 32{,}090 + 127.89$

10. $73{,}510 - 23{,}450$

11. Calculate: $46 - [4(6 + 4 \div 2) + 2 \times 3 - 5]$.

12. Combine like terms: $5x - 9 - 7x - 5$.

Perform the indicated operation and, if possible, simplify.

13. $\frac{6}{5} + 1\frac{5}{6}$

14. $-46.9 + 32.7$

15.
$$\begin{array}{r} 4\,8\,7{,}0\,9\,4 \\ 6{,}9\,3\,6 \\ +\quad 2\,1{,}1\,2\,0 \\ \hline \end{array}$$

16. $35 - 34.98$

17. $3\frac{1}{3} - 2\frac{2}{3}$

18. $-\frac{8}{9} - \frac{6}{7}$

19. $\frac{7}{9} \cdot \frac{3}{14}$

20. $(-32)(-4)(-3)$

21.
$$\begin{array}{r} 4\,6.0\,1\,2 \\ \times\quad 0.0\,3 \\ \hline \end{array}$$

22. $6\frac{3}{5} \div 4\frac{2}{5}$

23. $431.2 \div (-35.2)$

24. $15\,\overline{)\,1\,8\,5\,0}$

Solve.

25. $18 \cdot x = 1710$

26. $y + 142.87 = 151$

27. $\frac{3}{7}x - 5 = 16$

28. $\frac{3}{4} + x = \frac{5}{6}$

29. $3(x - 7) + 2 = 12x - 3$

30. $\frac{16}{n} = \frac{21}{11}$

31. In what quadrant does the point $(-3, -1)$ lie?

32. Graph on a plane: $y = -\frac{3}{5}x$.

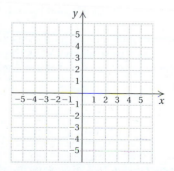

33. Find the mean: 19, 29, 34, 39, 45.

34. Find the median: 7, 7, 12, 15, 21.

35. Find the perimeter of a 15-in. by 15-in. chessboard.

36. Find the area of a 40-yd by 90-yd soccer field.

Solve.

37. A 12-oz box of cereal costs $3.60. Find the unit price in cents per ounce.

38. A bus travels 456 km in 6 hr. At this rate, how far would the bus travel in 8 hr?

39. In December 2004, one U.S. dollar was worth about 103 Japanese yen. In December 2005, the exchange rate was one U.S. dollar to 117 yen. Find the percent of increase.
Source: www.oanda.com

40. *Office Communications.* About 27.4% of office communications are telephone calls, and 11.6% are voice mail. If Alannah receives 150 communications on Monday, how many will be telephone calls? How many will be voice mail?

41. How many pieces of ribbon $1\frac{4}{5}$ yd long can be cut from a length of ribbon 9 yd long?

42. Bobbie walked $\frac{7}{10}$ km to school and then walked $\frac{8}{10}$ km to the library. How far did she walk?

SYNTHESIS

On a trip through the mountains, a Dodge Neon traveled 240 mi on $7\frac{1}{2}$ gal of gasoline. Going across the plains, the same car averaged 36 miles per gallon.

43. What was the percent of increase or decrease in miles per gallon when the car left the mountains for the plains?

44. How many miles per gallon did the Dodge average over the entire trip if it used 5 gal of gas to cross the plains?

45. A housing development is constructed on a dead-end road along a river and ends in a cul-de-sac, as shown in the figure. The property owners agree to share the cost of maintaining the road in the following manner. The cost of the first fifth of the road in front of lot 1 is to be shared equally among all five lot owners. The cost of the second fifth in front of lot 2 is to be shared equally among the owners of lots 2–5, and so on. Assume that all five sections of the road cost the same to maintain.

  a) What fractional part of the cost is paid by each owner?
  b) What percent of the cost is paid by each owner?
  c) If lots 3, 4, and 5 were all owned by the same person, what percent of the cost of maintenance would this person pay?

46. How many successive 10% discounts are necessary to lower the price of an item to below 50% of its original price?

Geometry and Measurement

9

Real-World Application

"The Gates," designed by artists Christo and Jeanne-Claude, is the largest art project in New York City's history. For the 16-day exhibit in February 2005, 7500 gates with fabric panels were placed at 12-ft intervals along 23 miles of pedestrian paths in Central Park. The panels contained 46 miles of saffron-colored fabric. Convert 46 miles to yards.

Sources: *Indianapolis Star,* February 13, 2005; Associated Press, Verena Dobnik

This problem appears as Example 5 in Section 9.1.

Objectives

a Convert from one American unit of length to another.

b Convert from one metric unit of length to another.

c Convert between American and metric units of length.

Use the unit below to measure the length of each segment or object.

1.

2.

3.

4.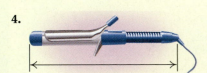

Answers on page A-26

586

CHAPTER 9: Geometry and Measurement

Length, or distance, is one kind of measure. To find lengths, we start with some **unit segment** and assign to it a measure of 1. Suppose $\overline{AB}$ below is a unit segment.

Let's measure segment $\overline{CD}$ below, using $\overline{AB}$ as our unit segment.

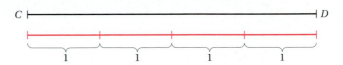

Since 4 unit segments fit end to end along $\overline{CD}$, the measure of $\overline{CD}$ is 4.

Sometimes we have to use parts of units. For example, the measure of the segment $\overline{MN}$ below is $1\frac{1}{2}$.

Do Exercises 1–4.

a   American Measures

American units of length are related as follows.

(Actual size, in inches)

AMERICAN UNITS OF LENGTH	
12 inches (in.) = 1 foot (ft)	3 feet = 1 yard (yd)
36 inches = 1 yard	5280 feet = 1 mile (mi)

The symbolism 13 in. = 13″ and 27 ft = 27′ is also used for inches and feet. American units have also been called "English," or "British–American," because at one time they were used in both North America and Britain. Today, both Canada and England have officially converted to the metric system. However, if you travel in England, you will still see units such as "miles" on road signs.

When we write a given length using a different unit, we say that we are "converting" from one unit to another. For example, we convert 1 yard to inches by saying that

$$1 \text{ yd} = 36 \text{ in.}$$

Conversions are made using either *substitution* or *multiplying by 1*. To use substitution, we replace one length with an equivalent length using another unit.

EXAMPLE 1 Complete: 5 yd = _____ in.

Since we are converting yards to inches, we use the fact that 1 yd = 36 in.

5 yd = 5 · 1 yd	Rewriting to aid in substitution
= 5 · 36 in.	Substituting 36 in. for 1 yd
= 180 in.	Multiplying

EXAMPLE 2 Complete: 2 mi = _____ in.

We want to convert miles to inches, but we do not know from the box on the previous page how many inches are in a mile. Since we do know how many feet are in a mile and how many inches are in a foot, we substitute twice.

2 mi = 2 · 1 mi	
= 2 · 5280 ft	Substituting 5280 ft for 1 mi
= 10,560 · 1 ft	Multiplying
= 10,560 · 12 in.	Substituting 12 in. for 1 ft
= 126,720 in.	The student should check the multiplication.

Do Exercises 5–7.

In Examples 1 and 2, we converted from larger units to smaller ones and used substitution. Sometimes—especially when converting from smaller to larger units—it helps to multiply by 1. For example, 12 in. = 1 ft, so we might choose to write 1 as

$$\frac{12 \text{ in.}}{1 \text{ ft}} \quad \text{or} \quad \frac{1 \text{ ft}}{12 \text{ in.}}.$$

EXAMPLE 3 Complete: 48 in. = _____ ft.

Since we are converting "in." to "ft," we choose a symbol for 1 containing the equivalent lengths of "12 in." and "1 ft." We write "12 in." on the bottom to eliminate the inches unit.

$48 \text{ in.} = \dfrac{48 \text{ in.}}{1} \cdot \dfrac{1 \text{ ft}}{12 \text{ in.}}$	Multiplying by 1 using $\dfrac{1 \text{ ft}}{12 \text{ in.}}$ to eliminate in.
$= \dfrac{48 \text{ in.}}{12 \text{ in.}} \cdot 1 \text{ ft}$	Pay careful attention to the units.
$= \dfrac{48}{12} \cdot \dfrac{\text{in.}}{\text{in.}} \cdot 1 \text{ ft}$	The $\dfrac{\text{in.}}{\text{in.}}$ acts like 1, so we can omit it.
$= 4 \cdot 1 \text{ ft} = 4 \text{ ft}$	Dividing

The conversion can also be regarded as "canceling" units:

$$48 \text{ in.} = \frac{48 \text{ in.}}{1} \cdot \frac{1 \text{ ft}}{12 \text{ in.}} = \frac{48}{12} \cdot 1 \text{ ft} = 4 \text{ ft}.$$

Do Exercises 8 and 9.

Complete.

5. 8 yd = _____ in.

6. 14.5 yd = _____ ft

7. 3.8 mi = _____ in.

Complete.

8. 72 in. = _____ ft

9. 24 ft = _____ yd

Answers on page A-26

Complete.

10. 35 ft = _____ yd

11. 18 yd = _____ ft

Complete.

12. 6 mi = _____ ft

13. 26,400 ft = _____ mi

14. Pedestrian Paths. There are 23 miles of pedestrian paths in Central Park in New York City. Convert 23 miles to yards.

Answers on page A-26

In Examples 4 and 5, we will use only the "canceling" method.

■ **EXAMPLE 4** Complete: 75 ft = _____ yd.

Since we are converting from "ft" to "yd," we choose a symbol for 1 with "yd" on the top and "ft" on the bottom:

$$75 \text{ ft} = 75 \text{ ft} \cdot \frac{1 \text{ yd}}{3 \text{ ft}} \qquad \text{If it helps, write 75 ft as } \frac{75 \text{ ft}}{1}.$$

$$= \frac{75}{3} \cdot 1 \text{ yd} = 25 \text{ yd.} \qquad \text{Multiplying}$$

Do Exercises 10 and 11.

■ **EXAMPLE 5** *Art Project in Central Park.* "The Gates," designed by artists Christo and Jeanne-Claude, is the largest art project in New York City's history. For the 16-day exhibit in February 2005, 7500 gates with fabric panels were placed at 12-ft intervals along 23 miles of pedestrian paths in Central Park. The panels contained 46 miles of saffron-colored fabric. Convert 46 miles to yards.

Sources: *Indianapolis Star,* February 13, 2005; Associated Press, Verena Dobnik

We multiply by 1 twice, to convert mi to ft and then to convert ft to yd. The first symbol for 1 will contain "1 mi" in the denominator to eliminate miles; the second symbol will contain "3 ft" in the denominator to eliminate feet.

$$46 \text{ mi} = 46 \text{ mi} \cdot \frac{5280 \text{ ft}}{1 \text{ mi}} \cdot \frac{1 \text{ yd}}{3 \text{ ft}}$$

$$= \frac{46 \cdot 5280}{1 \cdot 3} \cdot 1 \text{ yd}$$

$$= 80,960 \text{ yd}$$

"The Gates" project contained 80,960 yd of fabric.

Note that when we convert from smaller units to larger units, we divide. When converting from larger units to smaller units, we multiply.

Do Exercises 12–14.

b The Metric System

The **metric system** is gradually replacing traditional systems of measurement. Because it is based on powers of 10, the metric system allows for easy conversion between units. The metric system does not use inches, feet, pounds, and so on, but the units for time and electricity are the same as those you use now.

The basic unit of length is the **meter.** It is just over a yard. In fact, 1 meter ≈ 1.1 yd.

(Comparative sizes are shown.)

| 1 Meter |

| 1 Yard |

The other units of length are multiples of the length of a meter: 10 times a meter, 100 times a meter, 1000 times a meter; or fractions of a meter: $\frac{1}{10}$ of a meter, $\frac{1}{100}$ of a meter, $\frac{1}{1000}$ of a meter.

METRIC UNITS OF LENGTH

1 *kilo*meter (km) = 1000 meters (m)

1 *hecto*meter (hm) = 100 meters (m)

1*deka*meter (dam) = 10 meters (m)

    1 meter (m)

> The units hm, dam, and dm are not used often.

1 *deci*meter (dm) = $\frac{1}{10}$ meter (m)

1 *centi*meter (cm) = $\frac{1}{100}$ meter (m)

1 *milli*meter (mm) = $\frac{1}{1000}$ meter (m)

It is important to remember these names and abbreviations—especially *kilo-* for 1000, *deci-* for $\frac{1}{10}$, *centi-* for $\frac{1}{100}$, and *milli-* for $\frac{1}{1000}$. These prefixes are also used when measuring volume and mass (weight).

To familiarize yourself with metric units, consider the following.

1 kilometer (1000 meters)	is a bit more than $\frac{1}{2}$ mile (≈0.6 mi).
1 meter	is just over a yard (≈1.1 yd).
1 centimeter (0.01 meter)	is a little more than the width of a jumbo paperclip (≈0.4 in.).
1 millimeter	is about the diameter of a paperclip wire.

1 cm

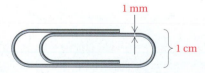

1 mm

1 cm

1 inch is about 2.54 centimeters.

589

Use a centimeter ruler. Measure each object. (Actual sizes are shown.)

15.

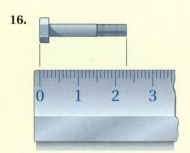

16.

17.

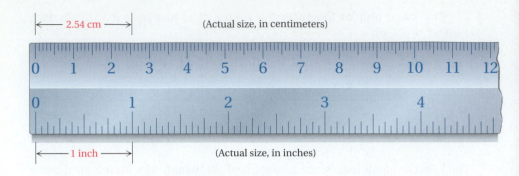

2.54 cm   (Actual size, in centimeters)

1 inch   (Actual size, in inches)

Millimeters (mm) are used for small distances, especially in industry.

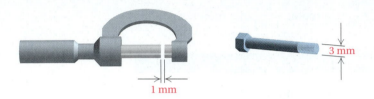

1 mm   3 mm

Centimeters (cm) are used for body dimensions, clothing sizes, and household measurements.

163 cm
(64.2 in.)
5 ft 4 in.

RELAXED FIT
97cm/81cm
(38 in./32 in.)

Do Exercises 15–17.

Meters (m) are used for expressing dimensions of larger objects—say, the height of a diving board—and for shorter distances, like the length of a rug.

10 m
(33 ft)

3.7 m
(12 ft)

2.7 m
(9 ft)

Kilometers (km) are used for longer distances, mostly in cases where miles are now being used.

1 mile is about 1.6 km.

Do Exercises 18–23.

Metric conversions from larger units to smaller ones are often most easily made using substitution, much as we did in Examples 1 and 2.

EXAMPLE 6   Complete: 4 km = _____ m.

$$4 \text{ km} = 4 \cdot 1 \text{ km}$$
$$= 4 \cdot 1000 \text{ m} \qquad \text{Substituting 1000 m for 1 km}$$
$$= 4000 \text{ m} \qquad \text{Multiplying by 1000}$$

Do Exercises 24 and 25.

Since

$$\frac{1}{10} \text{ m} = 1 \text{ dm}, \qquad \frac{1}{100} \text{ m} = 1 \text{ cm}, \quad \text{and} \quad \frac{1}{1000} \text{ m} = 1 \text{ mm},$$

it follows that

METRIC CONVERSIONS
1 m = 10 dm,   1 m = 100 cm,   and   1 m = 1000 mm.

Remembering these equations will help you to write forms of 1 when canceling to make conversions. The procedure is the same as that used in Examples 4 and 5.

EXAMPLE 7   Complete: 93.4 m = _____ cm.

To convert from "m" to "cm," we multiply by 1 using a symbol for 1 with "m" on the bottom and "cm" on the top. This process introduces centimeters and at the same time eliminates meters.

$$93.4 \text{ m} = 93.4 \text{ m} \cdot \frac{100 \text{ cm}}{1 \text{ m}} \qquad \text{Multiplying by 1 using } \frac{100 \text{ cm}}{1 \text{ m}}$$
$$= 93.4 \text{ m} \cdot \frac{100 \text{ cm}}{1 \text{ m}} = 93.4 \cdot 100 \text{ cm} = 9340 \text{ cm}$$

Complete with mm, cm, m, or km.

18. A stick of gum is 7 _____ long.

19. Minneapolis is 3213 _____ from San Francisco.

20. A penny is 1 _____ thick.

21. The halfback ran 7 _____ .

22. The book is 3 _____ thick.

23. The desk is 2 _____ long.

Complete.
24. 37 km = _____ m

25. 5 hm = _____ m

Answers on page A-26

Complete.

26. 1.78 m = _____ cm

27. 9.04 m = _____ mm

Complete.

28. 7814 m = _____ km

29. 7814 m = _____ dam

Complete.

30. 87.2 mm = _____ cm

31. 89 km = _____ cm

Answers on page A-26

EXAMPLE 8   Complete: 0.248 m = _____ mm.

We are converting from "m" to "mm," so we choose a symbol for 1 with "mm" on the top and "m" on the bottom:

$$0.248 \text{ m} = 0.248 \text{ m} \cdot \frac{1000 \text{ mm}}{1 \text{ m}} = 0.248 \cdot 1000 \text{ mm} = 248 \text{ mm}.$$

Do Exercises 26 and 27.

EXAMPLE 9   Complete: 2347 m = _____ km.

We multiply by 1 using $\frac{1 \text{ km}}{1000 \text{ m}}$:

$$2347 \text{ m} = 2347 \text{ m} \cdot \frac{1 \text{ km}}{1000 \text{ m}} = \frac{2347}{1000} \cdot 1 \text{ km} = 2.347 \text{ km}.$$

Do Exercises 28 and 29.

It is helpful to remember that 1000 mm = 100 cm and, more simply, 10 mm = 1 cm.

EXAMPLE 10   Complete: 8.42 mm = _____ cm.

We can multiply by 1 using either $\frac{1 \text{ cm}}{10 \text{ mm}}$ or $\frac{100 \text{ cm}}{1000 \text{ mm}}$. Both expressions for 1 will eliminate mm and leave cm:

$$8.42 \text{ mm} = 8.42 \text{ mm} \cdot \frac{1 \text{ cm}}{10 \text{ mm}} = \frac{8.42}{10} \cdot 1 \text{ cm} = 0.842 \text{ cm}.$$

Do Exercises 30 and 31.

MENTAL CONVERSION

Note in Examples 6–10 that changing from one unit to another in the metric system involves moving a decimal point. This occurs because the metric system is based on 10. To find a faster way to convert, consider these equivalent ways of expressing the width of a standard sheet of paper.

EQUIVALENT WAYS OF EXPRESSING DIMENSIONS
Width of a standard sheet of paper = 216 mm = 21.6 cm = 2.16 dm = 0.216 m = 0.0216 dam = 0.00216 hm = 0.000216 km

Each unit in the box above is ten times as large as the next smaller unit. Thus, converting to the next larger unit means moving the decimal point one place to the left.

EXAMPLE 11   Complete: 35.7 mm = _____ cm.

Think: Centimeters is the next larger unit after millimeters. Thus, we move the decimal point one place to the left.

35.7   3.5.7   35.7 mm = 3.57 cm   Converting to a larger unit shifts the decimal point to the left.

Converting to the next *smaller* unit means moving the decimal point one place to the right.

EXAMPLE 12 Complete: 3 m = _____ cm.

Think: A centimeter is two units smaller than a meter (cm, dm, m), so we move the decimal point two places to the right. To do so, we write two additional zeros.

3   3.00.   3 m = 300 cm   Converting to a smaller unit shifts the decimal point to the right.

The following table can help with mental conversions.

1000 m	100 m	10 m	1 m	0.1 m	0.01 m	0.001 m
1 km	1 hm	1 dam	1 m	1 dm	1 cm	1 mm

Larger ←——————————————— ——————————————→ Smaller

Do Exercises 32–34.

EXAMPLE 13 Complete: 4.37 km = _____ cm.

Think: A kilometer is 1000 times as large as a meter and a meter is 100 times as large as a centimeter. Thus, we move the decimal point 3 + 2, or 5, places to the right. This requires writing three additional zeros.

4.37   4.37000.   4.37 km = 437,000 cm

> The most commonly used metric units of length are km, m, cm, and mm. We have purposely used these more often than the others in the exercises and examples.

Do Exercise 35.

C   Converting Between American and Metric Units

We can make conversions between American and metric units by using the following table. Again, we either make a substitution or multiply by 1 appropriately.

AMERICAN	METRIC
1 in.	2.540 cm
1 ft	0.305 m
1 yd	0.914 m
1 mi	1.609 km
0.621 mi	1 km
1.094 yd	1 m
3.281 ft	1 m
39.370 in.	1 m

Complete. Try to do this mentally.

32. 6780 m = _____ km

33. 9.74 cm = _____ mm

34. 1 mm = _____ cm

35. Complete: 845.1 mm = _____ dm.

Answers on page A-26

Complete. Answers may vary slightly, depending on the conversion used.

36. 100 yd = _____ m
(The length of a football field)

EXAMPLE 14 Complete: 26.2 mi = _____ km (the length of the Olympic marathon).

Since we are given that 1 mi ≈ 1.609 km, we can convert using substitution.

$$26.2 \text{ mi} = 26.2 \cdot 1 \text{ mi}$$
$$\approx 26.2 \cdot 1.609 \text{ km} \qquad \text{Converting from mi to km}$$
$$\approx 42.16 \text{ km}$$

The table also indicates that 1 km ≈ 0.621 mi, so we can also convert by multiplying by 1 using $\dfrac{1 \text{ km}}{0.621 \text{ mi}}$.

$$26.2 \text{ mi} \approx \frac{26.2 \text{ mi}}{1} \cdot \frac{1 \text{ km}}{0.621 \text{ mi}}$$
$$\approx 42.19 \text{ km}$$

Note that, since the conversion factors given in the table are approximations, the answers found by the two methods differ slightly.

37. 500 mi = _____ km
(The Indianapolis 500-mile race)

EXAMPLE 15 Complete: 100 m = _____ yd (the length of a dash in track).

$$100 \text{ m} = 100 \cdot 1 \text{ m}$$
$$\approx 100 \cdot 1.094 \text{ yd} \qquad \text{Converting from m to yd}$$
$$\approx 109.4 \text{ yd}$$

EXAMPLE 16 Complete: 170 cm = _____ in. (a common ski length).

We multiply by 1, using the fact that 1 in. ≈ 2.540 cm, and writing 2.540 cm on the bottom to eliminate the centimeters unit.

$$170 \text{ cm} \approx 170 \text{ cm} \cdot \frac{1 \text{ in.}}{2.540 \text{ cm}} \qquad \text{Converting to inches from cm}$$
$$\approx 66.93 \text{ in.}$$

Do Exercises 36–38.

38. 2383 km = _____ mi
(The distance from St. Louis to Phoenix)

Answers on page A-26

a   Complete.

1. 1 yd = _____ ft

2. 1 ft = _____ in.

3. 1 in. = _____ ft

4. 1 mi = _____ ft

5. 1 mi = _____ yd

6. 1 ft = _____ yd

7. 3 yd = _____ ft

8. 4 yd = _____ in.

9. 84 in. = _____ ft

10. 18 in. = _____ ft

11. 48 ft = _____ yd

12. 29 ft = _____ yd

13. 5 mi = _____ yd

14. 5 mi = _____ ft

15. 48 in. = _____ ft

16. 19 ft = _____ yd

17. 11,616 ft = _____ mi

18. 5.2 yd = _____ ft

19. 15,840 ft = _____ mi

20. 10 mi = _____ ft

21. $7\frac{1}{2}$ ft = _____ yd

22. 360 in. = _____ yd

23. 36 in. = _____ ft

24. 7.2 ft = _____ in.

25. 1760 yd = _____ mi

26. 330 ft = _____ yd

27. 3520 yd = _____ mi

28. 100 yd = _____ ft

29. 25 mi = _____ ft

30. 240 in. = _____ ft

31. 2 mi = _____ in.

32. 63,360 in. = _____ mi

b   Complete. Do as much as possible mentally.

33. a) 1 km = _____ m

    b) 1 m = _____ km

34. a) 1 hm = _____ m

    b) 1 m = _____ hm

35. a) 1 dam = _____ m

    b) 1 m = _____ dam

36. a) 1 dm = _____ m

    b) 1 m = _____ dm

37. a) 1 cm = _____ m

    b) 1 m = _____ cm

38. a) 1 mm = _____ m

    b) 1 m = _____ mm

39. 8.3 km = _____ m

40. 27 km = _____ m

41. 98 cm = _____ m

42. 53 cm = _____ m

43. 8921 m = _____ km

44. 8664 m = _____ km

45. 32.17 m = _____ km

46. 4.733 m = _____ km

47. 289 m = _____ cm

48. 869 m = _____ cm

49. 477 cm = _____ m

50. 6.27 mm = _____ m

51. 6.88 m = _____ cm

52. 6.88 m = _____ dm

53. 1 mm = _____ cm

54. 1 cm = _____ km

55. 1 km = _____ cm

56. 2 km = _____ cm

57. 14.2 cm = _____ mm

58. 25.3 cm = _____ mm

59. 8.2 mm = _____ cm

60. 9.7 mm = _____ cm

61. 4500 mm = _____ cm

62. 8,000,000 m = _____ km

63. 0.024 mm = _____ m

64. 60,000 mm = _____ dam

65. 6.88 m = _____ dam

66. 7.44 m = _____ hm

67. 2.3 dam = _____ dm

68. 9 km = _____ hm

C   Complete. Answers may vary slightly, depending on the conversion used.

69. 10 km = _____ mi
(A common running distance)

70. 5 mi = _____ km
(A common running distance)

71. 14 in. = _____ cm
(A common paper length)

72. 400 m = _____ yd
(A common race distance)

73. 65 mph = _____ km/h
(A common speed limit in the
United States)

74. 100 km/h = _____ mph
(A common speed limit in
Canada)

75. 94 ft = _____ m
(The length of an NCAA basketball court)

76. 165 cm = _____ in.
(A common height for a woman)

77. 180 cm = _____ in.
(A common snowboard length)

78. 450 ft = _____ m
(The length of a long home run
in baseball)

79. 36 yd = _____ m
(A common length for a roll
of tape)

80. 70 in. = _____ cm
(A common height for a man)

81. **D_W** Would you expect the world record for the 100-m
dash to be greater or less than the record for the
100-yd dash? Why?

82. **D_W** A student incorrectly writes the following
conversion:

23 in. = 23 · (12 ft) = 276 ft.

What mistake has been made?

Solve.

83. $-7x - 9x = 24$ [5.7b]

84. $-2a + 9 = 5a + 23$ [5.7b]

85. If 3 calculators cost $43.50, how much would 7 calculators cost?   [7.4a]

86. A principal of $500 is invested at a rate of 8.9% for 1 year. Find the simple interest.   [8.6a]

Convert to percent notation.

87. 0.47   [8.1b]

88. $\dfrac{7}{20}$ [8.1c]

89. A living room is 12 ft by 16 ft. Find the perimeter and area of the room.   [1.2c], [1.5c]

90. A bedroom measures 10 ft by 12 ft. Find the perimeter and area of the room.   [1.2c], [1.5c]

91. **D**$_W$ Why do you think 8-km road races are more common than 8-mi road races?

92. **D**$_W$ The Olympic marathon is 26.2 mi. Do you think that this was originally a metric length? Why or why not?

In Exercises 93–96, each sentence is incorrect. Insert or alter a decimal point to make the sentence correct.

93. When my right arm is extended, the distance from my left shoulder to the end of my right hand is 10 m.

94. The height of the Empire State Building is 38.1 m.

95. A stack of ten quarters is 140 cm high.

96. The width of an adult's hand is 112 cm.

97. *Noah's Ark.* In biblical measures, it is thought that 1 cubit ≈ 18 in. The dimensions of Noah's ark are given as follows: "The length of the ark shall be three hundred cubits, the breadth of it fifty cubits, and the height of it thirty cubits." What were the dimensions of Noah's ark in inches? in feet?
Source: *Holy Bible, King James Version,* Gen. 6:15

98. *Goliath's Height.* In biblical measures, a span was considered to be half of a cubit (1 cubit ≈ 18 in.; see Exercise 97). The giant Goliath's height was "six cubits and a span." What was the height of Goliath in inches? in feet?
Source: *Holy Bible, King James Version,* 1 Sam. 17:4

Complete. Answers may vary, depending on the conversion used.

99. ▦ 2 mi = _____ cm

100. ▦ 10 km = _____ in.

101. ▦ Audio cassettes are generally played at a rate of $1\frac{7}{8}$ in. per second. How many meters of tape are used for a 60-min cassette? (*Note*: A 60-min cassette has 30 min of playing time on each side.)

102. ▦ The current world record for the 100-m dash is 9.78 sec. How fast is this in miles per hour? Round to the nearest tenth of a mile per hour.

103. ▦ *National Debt.* Recently the national debt was $5.103 trillion. To get an idea of this amount, picture that if that many $1 bills were stacked on top of each other, they would reach 1.382 times the distance to the moon. The distance to the moon is 238,866 mi. How thick, in inches, is a $1 bill?

Use < or > to complete the following. Perform only approximate, mental calculations.

104. 59 in. ☐ 59 cm

105. 35 yd ☐ 35 m

106. 7 km ☐ 6 mi

107. 9 mi ☐ 18 km

108. 24 ft ☐ 6 m

109. 30 in. ☐ 90 cm

9.2 CONVERTING UNITS OF AREA

Objectives

a Convert from one American unit of area to another.

b Convert from one metric unit of area to another.

a American Units

It is often necessary to convert units of area. First we will convert from one American unit of area to another.

EXAMPLE 1 Complete: $1 \text{ yd}^2 = $ _____ ft^2.

We recall that 1 yd = 3 ft and make a sketch. Note that $1 \text{ yd}^2 = 9 \text{ ft}^2$. The same result can be found as follows:

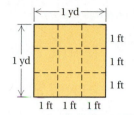

$$1 \text{ yd}^2 = 1 \cdot (3 \text{ ft})^2 \quad \text{Substituting 3 ft for 1 yd}$$
$$= 3 \text{ ft} \cdot 3 \text{ ft}$$
$$= 9 \text{ ft}^2. \quad \text{Note that ft} \cdot \text{ft} = \text{ft}^2.$$

EXAMPLE 2 Complete: $2 \text{ ft}^2 = $ _____ in^2.

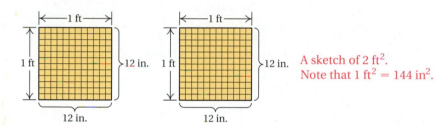

A sketch of 2 ft^2.
Note that $1 \text{ ft}^2 = 144 \text{ in}^2$.

$$2 \text{ ft}^2 = 2 \cdot (12 \text{ in.})^2 \quad \text{Substituting 12 in. for 1 ft}$$
$$= 2 \cdot 12 \text{ in.} \cdot 12 \text{ in.} = 288 \text{ in}^2 \quad \text{Note that in.} \cdot \text{in.} = \text{in}^2.$$

Do Exercises 1–3.

American units of area are related as follows.

AMERICAN UNITS OF AREA

1 square yard (yd^2) = 9 square feet (ft^2)

1 square foot (ft^2) = 144 square inches (in^2)

1 square mile (mi^2) = 640 acres

1 acre = 43,560 ft^2

EXAMPLE 3 Complete: $36 \text{ ft}^2 = $ _____ yd^2.

To convert from "ft^2" to "yd^2" we write 1 with yd^2 on top and ft^2 on the bottom.

$$36 \text{ ft}^2 = 36 \text{ ft}^2 \cdot \frac{1 \text{ yd}^2}{9 \text{ ft}^2} \quad \text{Multiplying by 1 using } \frac{1 \text{ yd}^2}{9 \text{ ft}^2}$$

$$= \frac{36}{9} \cdot \text{yd}^2 = 4 \text{ yd}^2$$

Complete.

1. $1 \text{ ft}^2 = $ _____ in^2

2. $10 \text{ ft}^2 = $ _____ in^2

3. $7 \text{ yd}^2 = $ _____ ft^2

Answers on page A-26

Complete.

4. $360 \text{ in}^2 = \underline{\hspace{1.5cm}} \text{ ft}^2$

5. $5 \text{ mi}^2 = \underline{\hspace{1.5cm}} \text{ acres}$

Complete.

6. $1 \text{ m}^2 = \underline{\hspace{1.5cm}} \text{ mm}^2$

7. $1 \text{ mm}^2 = \underline{\hspace{1.5cm}} \text{ cm}^2$

Answers on page A-26

As in Section 9.1, when converting from larger to smaller units, it is usually easiest to use substitution.

EXAMPLE 4 Complete: $7 \text{ mi}^2 = \underline{\hspace{1.5cm}} \text{ acres}$.

$$7 \text{ mi}^2 = 7 \cdot 640 \text{ acres} \qquad \text{Substituting 640 acres for 1 mi}^2$$
$$= 4480 \text{ acres}$$

Had we used canceling, we could have multiplied 7 mi^2 by $\dfrac{640 \text{ acres}}{1 \text{ mi}^2}$ to find the same answer.

Do Exercises 4 and 5.

b   Metric Units

We next convert from one metric unit of area to another.

EXAMPLE 5 Complete: $1 \text{ cm}^2 = \underline{\hspace{1.5cm}} \text{ mm}^2$.

Since $1 \text{ cm} = 10 \text{ mm}$, a square centimeter will have sides of length 10 mm.

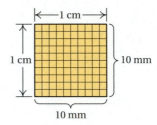

The area of the square can be written

$$(1 \text{ cm})(1 \text{ cm}) = 1 \text{ cm}^2$$

or

$$(10 \text{ mm})(10 \text{ mm}) = 100 \text{ mm}^2.$$

Thus $1 \text{ cm}^2 = 100 \text{ mm}^2$.

EXAMPLE 6 Complete: $1 \text{ km}^2 = \underline{\hspace{1.5cm}} \text{ m}^2$.

$$1 \text{ km}^2 = (1 \text{ km})(1 \text{ km})$$
$$= (1000 \text{ m})(1000 \text{ m}) \qquad \text{Substituting 1000 m for 1 km}$$
$$= 1,000,000 \text{ m}^2 \qquad \text{Note that m} \cdot \text{m} = \text{m}^2.$$

EXAMPLE 7 Complete: $1 \text{ cm}^2 = \underline{\hspace{1.5cm}} \text{ m}^2$.

$$1 \text{ cm}^2 = (1 \text{ cm})(1 \text{ cm})$$
$$= \left(\frac{1}{100} \text{ m}\right)\left(\frac{1}{100} \text{ m}\right) \qquad \text{Substituting } \frac{1}{100} \text{ m for 1 cm}$$
$$= \frac{1}{10,000} \text{ m}^2 \qquad \text{Note that m} \cdot \text{m} = \text{m}^2.$$
$$= 0.0001 \text{ m}^2$$

Do Exercises 6 and 7.

MENTAL CONVERSION

Note in Example 6 that whereas it takes 1000 m to make 1 km, it takes $1,000,000$ m^2 to make 1 km^2. That is, to convert 1 km to m, we move the decimal point 3 places to the right, and to convert 1 km^2 to m^2, we move the decimal point 6 places to the right.

Similarly, in Example 7, we saw that although 1 cm is equivalent to 0.01 m (moving the decimal point 2 places to the left), 1 cm^2 is equivalent to 0.0001 m^2 (moving the decimal point 4 places to the left).

In general, a metric area conversion requires moving the decimal point twice as many places as the corresponding length conversion. For example, below we list four equivalent ways of expressing the area of a standard sheet of paper.

$$\text{Area of a standard sheet of paper} = 60,264.\ mm^2$$
$$= 602.64\ cm^2$$
$$= 0.060264\ m^2$$
$$\approx 0.00000006\ km^2$$

EXAMPLE 8 Complete: $3.48\ km^2 = $ _____ m^2.

Think: A kilometer is 1000 times as big as a meter, so 1 km^2 is 1,000,000 times as big as 1 m^2. To convert from km to m, we shift the decimal point 3 places to the right; to convert from km^2 to m^2, we shift the decimal point 6 places to the right.

3.48 $3.480000.$ $3.48\ km^2 = 3,480,000\ m^2$

EXAMPLE 9 Complete: $586.78\ cm^2 = $ _____ m^2.

Think: To convert from cm to m, we shift the decimal point two places to the left. To convert from cm^2 to m^2, we shift the decimal point *four* places to the left.

586.78 $0.0586.78$ $586.78\ cm^2 = 0.058678\ m^2$

Do Exercises 8–10.

Complete.

8. $2.88\ m^2 = $ _____ cm^2

9. $4.3\ mm^2 = $ _____ cm^2

10. $678,000\ m^2 = $ _____ km^2

Answers on page A-26

a Complete.

1. $5 \text{ yd}^2 = $ _____ ft^2

2. $4 \text{ ft}^2 = $ _____ in^2

3. $7 \text{ ft}^2 = $ _____ in^2

4. $2 \text{ acres} = $ _____ ft^2

5. $432 \text{ in}^2 = $ _____ ft^2

6. $54 \text{ ft}^2 = $ _____ yd^2

7. $22 \text{ yd}^2 = $ _____ ft^2

8. $40 \text{ ft}^2 = $ _____ in^2

9. $15 \text{ ft}^2 = $ _____ in^2

10. $144 \text{ ft}^2 = $ _____ yd^2

11. $20 \text{ mi}^2 = $ _____ acres

12. $576 \text{ in}^2 = $ _____ ft^2

13. $69 \text{ ft}^2 = $ _____ yd^2

14. $1 \text{ mi}^2 = $ _____ yd^2

15. $720 \text{ in}^2 = $ _____ ft^2

16. $27 \text{ ft}^2 = $ _____ yd^2

17. $1 \text{ in}^2 = $ _____ ft^2

18. $72 \text{ in}^2 = $ _____ ft^2

19. $1 \text{ acre} = $ _____ mi^2

20. $4 \text{ acres} = $ _____ ft^2

b Complete.

21. $19 \text{ km}^2 = $ _____ m^2

22. $39 \text{ km}^2 = $ _____ m^2

23. $6.31 \text{ m}^2 = $ _____ cm^2

24. $2.7 \text{m}^2 = $ _____ mm^2

25. $6.5432 \text{ mm}^2 = $ _____ cm^2

26. $8.38 \text{ cm}^2 = $ _____ mm^2

27. $349 \text{ cm}^2 = $ _____ m^2

28. $125 \text{ mm}^2 = $ _____ m^2

29. $250{,}000 \text{ mm}^2 = $ _____ cm^2

30. $5900 \text{ mm}^2 = $ _____ cm^2

31. $472{,}800 \text{ m}^2 = $ _____ km^2

32. $1.37 \text{ cm}^2 = $ _____ mm^2

Find the area of the shaded region of each figure. Give the answer in square feet. (Figures are not drawn to scale.)

33.

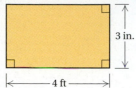

3 in.

|← 4 ft →|

34.

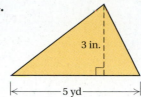

3 in.

|← 5 yd →|

35.

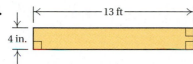

|← 13 ft →|

4 in.

36.

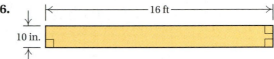

|← 16 ft →|

10 in.

Find the area of the shaded region of each figure.

37.

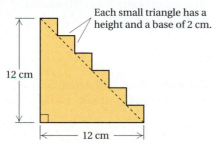

Each small triangle has a height and a base of 2 cm.

12 cm

|← 12 cm →|

38.

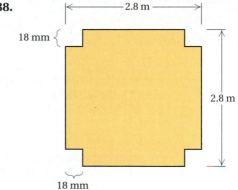

|← 2.8 m →|

18 mm

2.8 m

18 mm

39. **D**w What advantage do metric units offer over American units when we are converting area measurements?

40. **D**w Why might a scientist choose to give area measurements in mm² rather than cm²?

Find the simple interest.   [8.6a]

	Principal	Rate of interest	Time
41.	$700	5%	$\frac{1}{2}$ year
42.	$500	4%	$\frac{3}{4}$ year
43.	$450	6%	$\frac{1}{4}$ year
44.	$1200	8.9%	30 days
45.	$1800	12%	60 days
46.	$2500	14%	90 days

SYNTHESIS

47. D$_W$ Which is larger and why: one square meter or nine square feet?

48. D$_W$ What unit of measurement from this section would you use for measuring the area of a postage stamp? Why?

49. ▦ A 30-ft by 60-ft ballroom is to be turned into a nightclub by placing an 18-ft by 42-ft dance floor in the middle and carpeting the rest of the room. The new dance floor is laid in tiles that are 8-in. by 8-in. squares. How many such tiles are needed? What percent of the area is the dance floor?

Complete the following, using the conversions on page 593. Answers may vary slightly, depending on the conversion used.

50. ▦ $1 \text{ m}^2 = $ _____ ft^2

51. ▦ $1 \text{ in}^2 = $ _____ cm^2

52. ▦ $2 \text{ yd}^2 = $ _____ m^2

53. ▦ $1 \text{ acre} = $ _____ m^2

54. ▦ The president's family has about 20,175 ft^2 of living area in the White House. Estimate the living area in square meters.

55. ▦ A handwoven scarf is 2 m long and 10 in. wide. Find its area in square centimeters.

56. ▦ In order to remodel an office, a carpenter needs to purchase carpeting, at $8.45 a square yard, and molding for the base of the walls, at $0.87 a foot. If the room is 9 ft by 12 ft, with a 3-ft–wide doorway, what will the materials cost?

57. Janie's Rubik's Cube has 54 cm^2 of area. Each side of Norm's cube is twice as wide as Janie's. Find the area of Norm's cube.

9.3 MORE WITH PERIMETER AND AREA

Objectives

a Find the area of a parallelogram or trapezoid.

b Find the circumference, area, radius, or diameter of a circle, given the length of a radius or diameter.

We have already studied how to find the perimeter of polygons and the area of squares, rectangles, and triangles. In this section, we learn how to find the area of *parallelograms*, *trapezoids*, and *circles*. We also learn how to calculate the perimeter, or *circumference*, of a circle.

a Parallelograms and Trapezoids

A **parallelogram** is a four-sided figure with two pairs of parallel sides, as shown below.

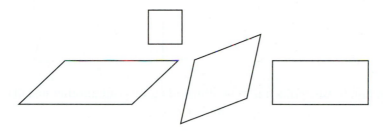

To find the area of a parallelogram, consider the one below.

If we cut off a piece and move it to the other end, we get a rectangle.

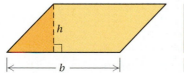

We can find the area by multiplying the length of a **base** *b*, by *h*, the **height.**

AREA OF A PARALLELOGRAM

The **area of a parallelogram** is the product of the length of a base *b* and the height *h*:

$$A = b \cdot h.$$

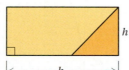

EXAMPLE 1 Find the area of this parallelogram.

$A = b \cdot h$

$= 7 \text{ km} \cdot 5 \text{ km}$

$= 35 \text{ km}^2$

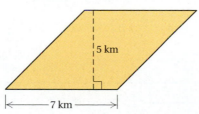

5 km

7 km

Study Tips

SKETCH CAREFULLY

When writing notes, we often use abbreviations and sometimes list only the first letter of a word if we are rushing. If a sketch of a geometric shape is called for, try to make sure the sketch is drawn carefully enough to be realistic.

605

Find the area.

1.

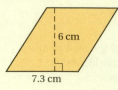

6 cm

7.3 cm

2.

5.5 km

2.25 km

EXAMPLE 2 Find the area of this parallelogram.

$$A = b \cdot h$$
$$= (1.2 \text{ m}) \cdot (6 \text{ m})$$
$$= 7.2 \text{ m}^2$$

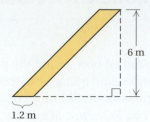

6 m

1.2 m

Do Exercises 1 and 2.

A **trapezoid** is a polygon with four sides, two of which are parallel to each other.* The parallel sides are called the **bases.**

To find the area of a trapezoid, think of cutting out another just like it.

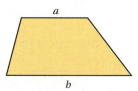

a

b

Then place the second one like this.

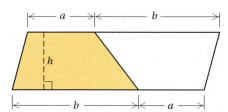

a   b

h

b   a

The resulting figure is a parallelogram with an area of

$$h \cdot (a + b). \qquad \text{The base of the parallelogram has length } a + b.$$

The trapezoid we started with has half the area of the parallelogram, or

$$\frac{1}{2} \cdot h \cdot (a + b).$$

AREA OF A TRAPEZOID

The **area of a trapezoid** is half the product of the height and the sum of the lengths of the parallel sides, or the product of the height and the average length of the bases:

$$A = \frac{1}{2} \cdot h \cdot (a + b) = h \cdot \frac{a + b}{2}.$$

a

h

b

---

*Some definitions of trapezoid specify *exactly* two parallel sides. We refrain from doing so. Thus we consider a parallelogram a special type of trapezoid.

EXAMPLE 3 Find the area of this trapezoid.

$$A = \frac{1}{2} \cdot h \cdot (a + b)$$

$$= \frac{1}{2} \cdot 7 \text{ cm} \cdot (12 + 18) \text{ cm}$$

$$= \frac{7 \cdot 30}{2} \cdot \text{cm}^2 = \frac{7 \cdot 15 \cdot 2}{1 \cdot 2} \text{ cm}^2$$

$$= 105 \text{ cm}^2 \qquad \text{Removing a factor equal to 1} \left(\frac{2}{2} = 1\right) \text{ and multiplying}$$

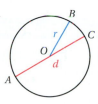

Do Exercises 3 and 4.

b   Circles

RADIUS AND DIAMETER

At right is a circle with center O. Segment $\overline{AC}$ is a *diameter*. A **diameter** is a segment that passes through the center of the circle and has endpoints on the circle. Segment $\overline{OB}$ is called a *radius*. A **radius** is a segment with one endpoint on the center and the other endpoint on the circle. The words *radius* and *diameter* are also used to represent the lengths of a circle's radius and diameter, respectively.

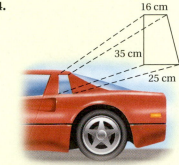

DIAMETER AND RADIUS

Suppose that d is the diameter of a circle and r is the radius. Then

$$d = 2 \cdot r \quad \text{and} \quad r = \frac{d}{2}.$$

EXAMPLE 4 Find the length of a radius of this circle.

$$r = \frac{d}{2}$$

$$= \frac{12 \text{ m}}{2}$$

$$= 6 \text{ m}$$

The radius is 6 m.

EXAMPLE 5 Find the length of a diameter of this circle.

$$d = 2 \cdot r$$

$$= 2 \cdot \frac{1}{4} \text{ ft}$$

$$= \frac{1}{2} \text{ ft}$$

The diameter is $\frac{1}{2}$ ft.

Do Exercises 5 and 6.

Find the area.

3.

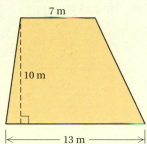

4.

5. Find the length of a radius.

6. Find the length of a diameter.

Answers on page A-26

7. Find the circumference of this circle. Use 3.14 for π.

20 m

CIRCUMFERENCE

The perimeter of a circle is called its **circumference.** Take a 12-oz soda can and measure its circumference C and diameter d. Next, consider the ratio C/d:

$$\frac{C}{d} \approx \frac{7.8 \text{ in.}}{2.5 \text{ in.}} \approx 3.1.$$

$C \approx 7.8$ in.

$d \approx 2.5$ in.

If we did this with cans and circles of several sizes, the result would always be a number close to 3.1. For any circle, if we divide the circumference C by the diameter d, we get the same number. We call this number π (pi). The number π is a nonterminating, nonrepeating decimal. It is impossible to precisely express π as a decimal or a fraction.

CIRCUMFERENCE AND DIAMETER

The circumference C of a circle of diameter d is given by

$$C = \pi \cdot d.$$

The number π is about 3.14, or about $\frac{22}{7}$.

EXAMPLE 6 Find the circumference of this circle. Use 3.14 for π.

$C = \pi \cdot d$

$\approx 3.14 \cdot 6 \text{ cm}$

$\approx 18.84 \text{ cm}$

The circumference is about 18.84 cm.

6 cm

Do Exercise 7.

Since $d = 2 \cdot r$, where r is the length of a radius, it follows that

$$C = \pi \cdot d = \pi \cdot (2 \cdot r).$$

CIRCUMFERENCE AND RADIUS

The circumference C of a circle of radius r is given by

$$C = 2 \cdot \pi \cdot r.$$

EXAMPLE 7 Find the circumference of this circle. Use $\frac{22}{7}$ for π.

$C = 2 \cdot \pi \cdot r$

$\approx 2 \cdot \frac{22}{7} \cdot 70 \text{ in.}$

$\approx 2 \cdot 22 \cdot \frac{70}{7} \text{ in.}$

$\approx 44 \cdot 10 \text{ in.}$

$\approx 440 \text{ in.}$

70 in.

The circumference is about 440 in.

Answer on page A-26

EXAMPLE 8 Find the perimeter of this figure. Use 3.14 for π.

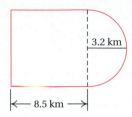

We let $P =$ the perimeter. Note that the figure has three straight edges (line segments) and one curved edge. The curved edge is half the circumference of a circle of radius 3.2 km. The top and bottom straight edges have lengths of 8.5 km. The left edge is the same length as twice the radius of the semicircle on the right, or 6.4 km.

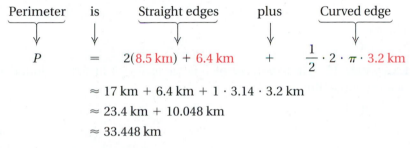

$$\approx 17\ \text{km} + 6.4\ \text{km} + 1 \cdot 3.14 \cdot 3.2\ \text{km}$$

$$\approx 23.4\ \text{km} + 10.048\ \text{km}$$

$$\approx 33.448\ \text{km}$$

The perimeter is about 33.448 km.

Do Exercises 8–10.

AREA

To find the area of a circle with radius r, we cut half a circular region into small slices and arrange them as shown below. Note that half of the circumference is half of $2\pi r$, or simply πr.

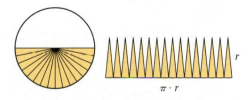

Next we slice the remaining half of the circular region and arrange the pieces in between the others as shown below.

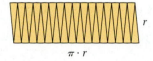

The thinner the slices, the closer this comes to being a rectangle. Its length is πr and its width is r, so its area is

$$(\pi \cdot r) \cdot r.$$

This is the area of a circle.

---

8. Find the circumference of this circle. Use 3.14 for π.

9. Find the circumference of this bicycle wheel. Use $\frac{22}{7}$ for π.

10. Find the perimeter of this figure. Use 3.14 for π.

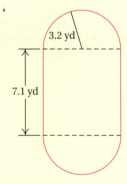

Answers on page A-26

11. Find the area of this circle. Use $\frac{22}{7}$ for π.

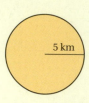

5 km

12. Find the area of this circle. Use 3.14 for π.

10.4 cm

13. Find the area of the shaded region. Use 3.14 for π. (*Hint:* The figure consists of a square and two semicircles, or one complete circle.)

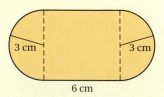

3 cm   3 cm

6 cm

Answers on page A-26

The **area of a circle** with radius of length r is given by

$$A = \pi \cdot r \cdot r, \quad \text{or} \quad A = \pi \cdot r^2.$$

r

EXAMPLE 9 Find the area of this circle. Use $\frac{22}{7}$ for π.

$$A = \pi \cdot r^2 = \pi \cdot r \cdot r$$

$$\approx \frac{22}{7} \cdot 14 \text{ cm} \cdot 14 \text{ cm}$$

$$\approx \frac{22}{7} \cdot \frac{7 \cdot 2}{1} \text{ cm} \cdot 14 \text{ cm}$$

$$\approx 616 \text{ cm}^2 \qquad \textit{Note: } r^2 \neq 2r.$$

14 cm

The area is about 616 cm^2.

EXAMPLE 10 Find the area of this circle. Use 3.14 for π. Round to the nearest hundredth.

$$A = \pi \cdot r \cdot r$$

$$\approx 3.14 \cdot 2.1 \text{ m} \cdot 2.1 \text{ m}$$

$$\approx 3.14 \cdot 4.41 \text{ m}^2$$

$$\approx 13.8474 \text{ m}^2 \approx 13.85 \text{ m}^2$$

2.1 m

The area is about 13.85 m^2.

Do Exercises 11 and 12.

Some regions that appear irregular are actually a combination of familiar shapes. We can add or subtract areas to determine the area of the region.

EXAMPLE 11 Find the area of the shaded region.

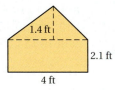

1.4 ft

2.1 ft

4 ft

The region consists of a triangle and a rectangle, as illustrated below. Note that the base of the triangle is the same as the length of the rectangle. We calculate the areas separately and add to find the area of the shaded region.

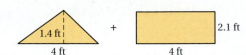

1.4 ft   +   2.1 ft

4 ft   4 ft

$$\text{Area} = \frac{1}{2} \cdot 4 \text{ ft} \cdot 1.4 \text{ ft} + 4 \text{ ft} \cdot 2.1 \text{ ft}$$

$$= 2.8 \text{ ft}^2 + 8.4 \text{ ft}^2$$

$$= 11.2 \text{ ft}^2$$

Do Exercise 13.

EXAMPLE 12 *Art Design.* As an assignment in an art design course at Austin Community College, students design artwork to decorate a CD "jewel-box." Part of the artwork inside the case covers a 12-cm square, over which a CD with a 12-cm diameter rests. How much of the artwork behind the CD will be visible?

1. **Familiarize.** From examining a drawing, we see that the corners of the design will be visible when the CD is in place. We let A = the area of the design visible behind the CD.

2. **Translate.** The problem can be rephrased as follows:

Area of case	minus	Area of CD	is	Visible area.
↓	↓	↓	↓	↓
$s \cdot s$	$-$	$\pi \cdot r \cdot r$	$=$	A

3. **Solve.** Note that the radius of the CD is half the diameter, or 6 cm.

$12 \text{ cm} \cdot 12 \text{ cm} - 3.14 \cdot 6 \text{ cm} \cdot 6 \text{ cm} \approx A$ Substituting 6 cm for r, 12 cm for s, and 3.14 for π

$144 \text{ cm}^2 - 113.04 \text{ cm}^2 \approx A$

$31.0 \text{ cm}^2 \approx A$ Rounding to the nearest tenth

4. **Check.** We can check by repeating our calculations. Note also that the area of the case does exceed the area of the CD, as expected.

5. **State.** When the CD is in place, about 31 cm² of artwork will be visible.

Do Exercise 14.

AREA AND CIRCUMFERENCE FORMULAS

Area of a parallelogram: $A = b \cdot h$

Area of a trapezoid: $A = \dfrac{1}{2} \cdot h \cdot (a + b)$

or $A = h \cdot \dfrac{a + b}{2}$

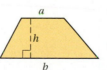

Circumference of a circle: $C = \pi \cdot d$
or $C = 2 \cdot \pi \cdot r$

Area of a circle: $A = \pi \cdot r^2$

Note: $d = 2r$ and $\pi \approx 3.14 \approx \frac{22}{7}$.

---

14. Which is larger and by how much: a 10-ft square flower bed or a 12-ft diameter flower bed?

Answer on page A-26

CALCULATOR CORNER

Pi   On certain calculators, there is a pi key, $\boxed{\pi}$. You can use a $\boxed{\pi}$ key for most computations instead of stopping to round the value of π. Rounding, if necessary, is done at the end.

Exercises:

1. If you have a $\boxed{\pi}$ key on your calculator, to how many decimal places does this key give the value of π?

2. Find the circumference and the area of a circle with a radius of 225.68 in.

3. Find the area of a circle with a diameter of $46\frac{12}{13}$ in.

4. Find the area of a large irrigated farming circle with a diameter of 400 ft.

a   Find the area of each parallelogram or trapezoid.

1.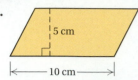
5 cm
10 cm

2.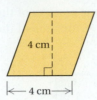
4 cm
4 cm

3.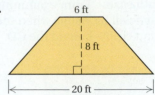
6 ft
8 ft
20 ft

4.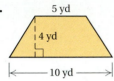
5 yd
4 yd
10 yd

5.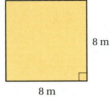
8 m
8 m

6.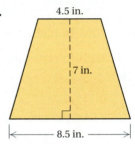
4.5 in.
7 in.
8.5 in.

7.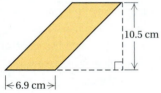
10.5 cm
6.9 cm

8.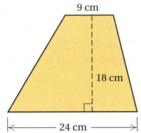
9 cm
18 cm
24 cm

9.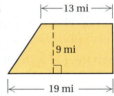
13 mi
9 mi
19 mi

10.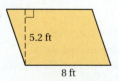
5.2 ft
8 ft

11.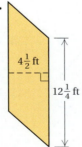
$4\frac{1}{2}$ ft
$12\frac{1}{4}$ ft

12.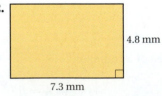
4.8 mm
7.3 mm

13.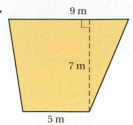

9 m

7 m

5 m

14.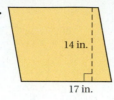

14 in.

17 in.

15.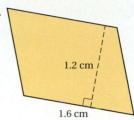

1.2 cm

1.6 cm

16.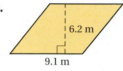

6.2 m

9.1 m

17.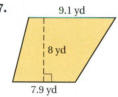

9.1 yd

8 yd

7.9 yd

18.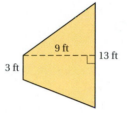

9 ft

13 ft

3 ft

b Find the length of a diameter of each circle.

19.

7 cm

20.

8 m

21.

$\frac{7}{8}$ in.

22.

$8\frac{2}{3}$ mi

Find the length of a radius of each circle.

23.
20 ft

24.
10 in.

25.
1.4 cm

26.
60.9 km

Find the circumference of each circle in Exercises 19–22. Use $\frac{22}{7}$ for π.

27. Exercise 19   **28.** Exercise 20   **29.** Exercise 21   **30.** Exercise 22

Find the circumference of each circle in Exercises 23–26. Use 3.14 for π.

31. Exercise 23   **32.** Exercise 24   **33.** Exercise 25   **34.** Exercise 26

Find the area of each circle in Exercises 19–22. Use $\frac{22}{7}$ for π.

35. Exercise 19   **36.** Exercise 20   **37.** Exercise 21   **38.** Exercise 22

Find the area of each circle in Exercises 23–26. Use 3.14 for π.

39. Exercise 23   **40.** Exercise 24   **41.** Exercise 25   **42.** Exercise 26

Little Caesars® is a national chain of pizzerias that offers both square and round pizzas. The various sizes are shown in the table below. Use the table to answer Exercises 43–50. Use 3.14 for π.

SIZE	ROUND (diameter)	SQUARE (width)
Small	10 in.	8.5 in.
Medium	12 in.	10 in.
Large	14 in.	12 in.
Extra-Large	16 in.	NA

43. Which shape of small pizza has more area? How much more area does it have?

44. Which shape of medium pizza has more area? How much more area does it have?

45. Which shape of large pizza has more area? How much more area does it have?

46. Which has more area: a medium round pizza or a large square pizza?

47. Which has more area: two small round pizzas, or one large square pizza? How much more?

48. Which has more area: two small square pizzas, or one large round pizza? How much more?

49. Which has more area: two medium square pizzas, or one extra-large round pizza? How much more?

50. Which has more area: three small square pizzas or one extra-large round pizza? How much more?

Solve. Use 3.14 for π.

51. *Trampoline.* The standard backyard trampoline has a diameter of 14 ft. What is its area?

Source: International Trampoline Industry Association, Inc.

14 ft

Frame height: 36 in.

52. *Soda-Can Top.* The top of a soda can has a 6-cm diameter. What is its radius? its circumference? its area?

6 cm

53. A college radio station is allowed by the FCC to broadcast over an area with a radius of 7 mi. How much area is this?

54. *Penny.* A penny has a 1-cm radius. What is its diameter? its circumference? its area?

1 cm

55. *Gypsy-Moth Tape.* To protect an elm tree, gypsy-moth caterpillar tape is wrapped around the trunk. It takes 47.1 in. of tape to go around the trunk once. What is the diameter of the tree?

56. *Farming.* The circumference of a silo is 62.8 ft. What is the diameter of the silo?

57. *Masonry.* The Harris-Regency Hotel plans to install a 1-yd–wide walk around a circular swimming pool. The diameter of the pool is 20 yd. What is the area of the walk?

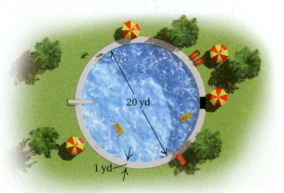

20 yd

1 yd

58. *Roller-Rink Floor.* A roller-rink floor is shown below. Each end is a semicircle. What is the area of the floor? If hardwood flooring costs $32.50 per square meter, how much will the flooring cost?

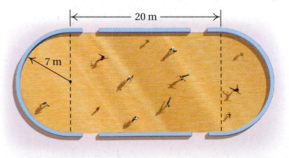

20 m

7 m

Find the perimeter of each figure. Use 3.14 for π.

59.

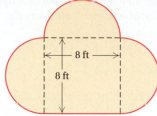

8 ft

8 ft

60.

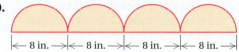

|← 8 in. →|← 8 in. →|← 8 in. →|← 8 in. →|

61.

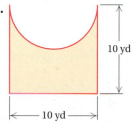

10 yd

10 yd

62.

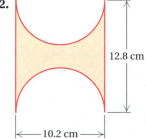

12.8 cm

10.2 cm

Find the area of the shaded region in each figure. Use 3.14 for π.

63.

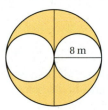

8 m

64.

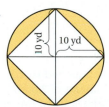

10 yd   10 yd

65.

2.8 cm

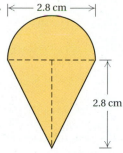

2.8 cm

66.

8 km

8 km

67. **Dw** Devise a problem for a classmate to solve that involves the area of a circular object. Design the problem so that it is most easily solved using the approximation 3.14 for π.

68. **Dw** Devise a problem for a classmate to solve that involves the circumference of a circular object. Design the problem so that it is most easily solved using the approximation $\frac{22}{7}$ for π.

Convert to fraction notation. [8.1c]

69. 9.25%

70. $87\frac{1}{2}$%

Convert to percent notation. [8.1c]

71. $\frac{11}{8}$

72. $\frac{2}{3}$

73. $\frac{5}{4}$

74. $\frac{8}{5}$

---

75. **Dw** The radius of one circle is twice the size of another circle's radius. Is the area of the first circle twice the area of the other circle? Why or why not?

76. **Dw** The radius of one circle is twice the size of another circle's radius. Is the circumference of the first circle twice the circumference of the other circle? Why or why not?

77. 🖩 Calculate the surface area of an unopened steel can that has a height of 3.5 in. and a diameter of 2.5 in. (*Hint*: Make a sketch and "unroll" the sides of the can.) Use 3.14 for π.

78. 🖩 The sides of a cake box are trapezoidal, as shown in the figure. Determine the surface area of the box.

79. **Dw** 🖩 See Exercises 43–50. Little Caesars® recently charged $10.99 for a large pizza. Which shaped pizza is a better value? Why?

80. **Dw** 🖩 *Pricing.* See Exercises 43–50. Little Caesars® recently charged $6.99 for a 10-in. square pizza or a 12-in.–diameter pizza. Which is a better value? Why?

81. *Sports Marketing.* Tennis balls are generally packed vertically, three in a can, one on top of another. Without using a calculator, determine the larger measurement: the can's circumference or the can's height.

82. 🖩 The distance from Kansas City to Indianapolis is 500 mi. A car was driven this distance using tires with a radius of 14 in. How many revolutions of each tire occurred on the trip? Use $\frac{22}{7}$ for π.

83. **D**_W 🖩 *Urban Planning.* Years ago, when a 12-in.–diameter tree was cut down in New York City, new trees with a combined diameter of 12 in. had to be planted. Now, instead of being able to use four 3-in.–diameter trees as replacement, a total of *sixteen* 3-in.–diameter trees must be planted. Consider area and explain why the new replacement calculation is more correct mathematically.

Source: *The New York Times* 7/24/88, p. 6; article by David W. Dunlap

84. *Baseball's Strike Zones.* Between the 2000 and 2001 seasons, Major League Baseball redefined the shape of the strike zone. Over the years before 2001, the strike zone evolved to something resembling the region *AQRST* in the diagram. In 2001, the strike zone was changed to rectangle *ABCD* in the illustration. By what percent has the area of the strike zone been increased by the change?

Sources: *The Cincinnati Enquirer;* Major League Baseball; Gannett News Service; *The Sporting News Official Baseball Rules Book*

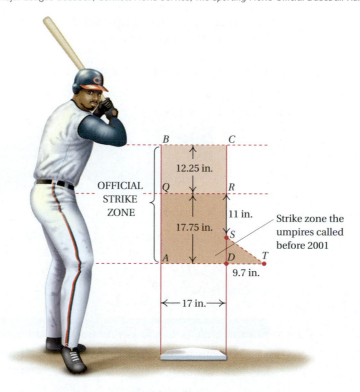

Objectives

a Find the volume of a rectangular solid, a cylinder, and a sphere.

b Convert from one unit of capacity to another.

c Solve applied problems involving volume and capacity.

1. Find the volume.

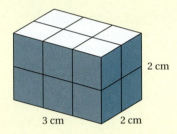

3 cm   2 cm   2 cm

Answer on page A-26

Study Tips

EXTEND PATTERNS

In this section you will learn that for certain solids the volume is the product of the solid's base area and its height. In future work with solids that are not studied in this text, you may want to try extending the base · height approach to these new objects.

a Volume

The **volume** of an object is the number of unit cubes needed to fill it.

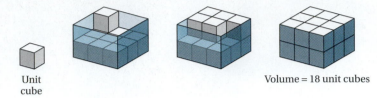

Unit cube   Volume = 18 unit cubes

Two common units are shown below (actual size).

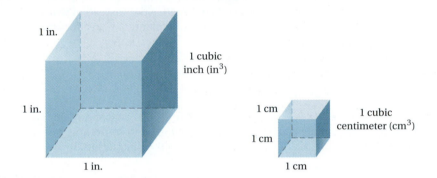

1 in.

1 in.

1 in.

1 cubic inch (in^3)

1 cm

1 cm

1 cm

1 cubic centimeter (cm^3)

The volume of a **rectangular solid** can be found by taking the product of the length, width, and height.

EXAMPLE 1 Find the volume.

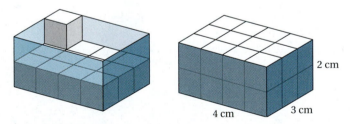

4 cm   3 cm   2 cm

The figure is made up of 2 layers of 12 cubes each, so its volume is 24 cubic centimeters (cm^3). Note that $24 = 4 \cdot 3 \cdot 2$.

Do Exercise 1.

VOLUME OF A RECTANGULAR SOLID

The **volume of a rectangular solid** is found by multiplying length by width by height:

$$V = l \cdot w \cdot h.$$

Area of base   Height

h

w l

EXAMPLE 2 *Carry-on Luggage.* The largest piece of luggage that you can carry on an airplane measures 23 in. by 10 in. by 13 in. Find the volume of this solid.

$$V = l \cdot w \cdot h$$
$$= 23 \text{ in.} \cdot 10 \text{ in.} \cdot 13 \text{ in.}$$
$$= 230 \cdot 13 \text{ in}^3$$
$$= 2990 \text{ in}^3$$

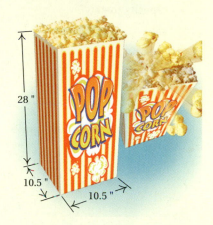

Do Exercises 2 and 3.

Note that volumes are described in units such as cubic centimeters (cm³) and cubic inches (in³). To determine the correct unit for an answer, examine the steps leading up to the answer. If two or more measurements are being added, as in a perimeter problem, a one-dimensional unit of length is used (3 ft + 2 ft + 7 ft = 12 ft). If two measurements of length are multiplied, a two-dimensional unit of area is used (8 ft × 7 ft = 56 ft²). And finally, if three measurements of length are multiplied, a three-dimensional unit of volume is formed (3 m · 2 m · 4 m = 24 m³). Dimensional analysis is an excellent way of figuring out what units are needed for an answer.

A rectangular solid is shown below. Note that we can think of the volume as the product of the area of the base times the height:

$$V = l \cdot w \cdot h$$
$$= (l \cdot w) \cdot h$$
$$= (\text{Area of the base}) \cdot h$$
$$= B \cdot h,$$

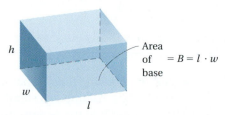

Area of base $= B = l \cdot w$

where *B* represents the area of the base.

Like rectangular solids, **circular cylinders** have bases of equal area that lie in parallel planes. The bases of circular cylinders are circular regions.

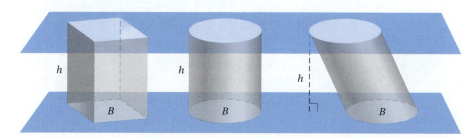

The volume of a circular cylinder is found in a manner similar to finding the volume of a rectangular solid. The volume is the product of the area of the base times the height. The height is always measured perpendicular to the base.

2. Popcorn. The average American eats 54 qt of popcorn annually. This quantity of popcorn would fill a bag measuring 10.5 in. by 10.5 in. by 28 in. Find the volume of such a bag.

Source: The Popcorn Board

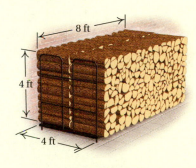

3. Cord of Wood. A cord of wood measures 4 ft by 4 ft by 8 ft. What is the volume of a cord of wood?

Answers on page A-26

4. Find the volume of the cylinder. Use 3.14 for π.

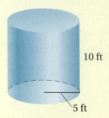

10 ft

5 ft

5. Find the volume of the cylinder. Use $\frac{22}{7}$ for π.

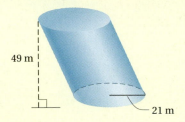

49 m

21 m

6. Find the volume of the sphere. Use $\frac{22}{7}$ for π.

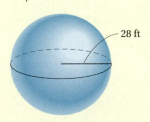

28 ft

7. The radius of a standard-sized golf ball is 2.1 cm. Find the volume of a standard-sized golf ball. Use 3.14 for π.

VOLUME OF A CIRCULAR CYLINDER

The **volume of a circular cylinder** is the product of the area of the base B and the height h:

$$V = B \cdot h, \quad \text{or} \quad V = \pi \cdot r^2 \cdot h.$$

EXAMPLE 3 Find the volume of this circular cylinder. Use 3.14 for π.

12 cm

4 cm

$$V = Bh = \pi \cdot r^2 \cdot h$$
$$\approx 3.14 \cdot 4\text{ cm} \cdot 4\text{ cm} \cdot 12\text{ cm}$$
$$= 602.88 \text{ cm}^3$$

Do Exercises 4 and 5.

A **sphere** is the three-dimensional counterpart of a circle. It is the set of all points in space that are a given distance (the radius) from a given point (the center). The volume of a sphere depends on its radius.

We find the volume of a sphere as follows.

VOLUME OF A SPHERE

The **volume of a sphere** of radius r is given by

$$V = \frac{4}{3} \cdot \pi \cdot r^3.$$

EXAMPLE 4 *Bowling Ball.* The radius of a standard-sized bowling ball is 4.2915 in. Find the volume of a standard-sized bowling ball (disregarding the finger holes). Round to the nearest hundredth of a cubic inch. Use 3.14 for π.

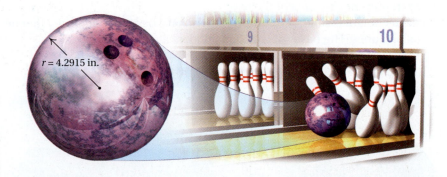

$r = 4.2915$ in.

9   10

We have

$$V = \frac{4}{3} \cdot \pi \cdot r^3 \approx \frac{4}{3} \times 3.14 \times (4.2915 \text{ in.})^3$$
$$\approx 330.90 \text{ in}^3. \qquad \text{Using a calculator}$$

Do Exercises 6 and 7.

Answers on page A-26

622

CHAPTER 9: Geometry and Measurement

b Capacity

To answer a question like "How much soda is in the bottle?" we need measures of **capacity.** American units of capacity are ounces, cups, pints, quarts, and gallons. These units are related as follows.

AMERICAN UNITS OF CAPACITY
1 gallon (gal) = 4 quarts (qt)   1 pt = 2 cups
= 16 fluid ounces (fl oz)
1 qt = 2 pints (pt)   1 cup = 8 fl oz

Fluid ounces, abbreviated fl oz, are often referred to as ounces, or oz.

EXAMPLE 5 Complete: 24 qt = _____ gal.

In this case, we multiply by 1 using 1 gal in the numerator, since we are converting *to* gallons, and 4 qt in the denominator, since we are converting *from* quarts.

$$24 \text{ qt} = \frac{24 \text{ qt}}{1} \cdot \frac{1 \text{ gal}}{4 \text{ qt}} = \frac{24}{4} \cdot 1 \text{ gal} = 6 \text{ gal}$$

To check that our answer is reasonable, note that since we are converting from smaller to larger units, our answer is a smaller number than the one with which we started.

EXAMPLE 6 Complete: 9 gal = _____ oz.

The box above does not list how many ounces are in 1 gal. We convert gallons to quarts, quarts to pints, and pints to ounces, using the relationships given in the box.

$$9 \text{ gal} = \frac{9 \text{ gal}}{1} \cdot \frac{4 \text{ qt}}{1 \text{ gal}} \qquad$$ We write 1 gal on the bottom to eliminate the gallons and 4 qt on the top to convert to quarts.

$$= 9 \cdot 4 \text{ qt}$$

$$= 36 \text{ qt} \qquad$$ We have converted to quarts.

$$= \frac{36 \text{ qt}}{1} \cdot \frac{2 \text{ pt}}{1 \text{ qt}} \qquad$$ We write 1 qt on the bottom to eliminate the quarts and 2 pt on the top to convert to pints.

$$= 36 \cdot 2 \text{ pt}$$

$$= 72 \text{ pt} \qquad$$ We have converted to pints.

$$= \frac{72 \text{ pt}}{1} \cdot \frac{16 \text{ oz}}{1 \text{ pt}} \qquad$$ We write 1 pt on the bottom to eliminate the pints and 16 oz on the top to convert to ounces.

$$= 72 \cdot 16 \text{ oz}$$

$$= 1152 \text{ oz} \qquad$$ We have converted to ounces.

This conversion could have been done in one step had we known that 1 gal = 128 oz. We then would have multiplied 9 gal by $\frac{128 \text{ oz}}{1 \text{ gal}}$.

Do Exercises 8 and 9.

Complete.

8. 80 qt = _____ gal

9. 4 gal = _____ pt

Answers on page A-26

Complete with mL or L.

10. To prevent infection, a patient received an injection of 2 _____ of penicillin.

11. There are 250 _____ in a coffee cup.

12. The gas tank holds 80 _____.

13. Bring home 8 _____ of milk.

The basic unit of capacity for the metric system is the **liter**. A liter is just a bit more than a quart (1 liter = 1.06 quarts). It is defined as follows.

1 liter   1 quart

METRIC UNITS OF CAPACITY

1 liter (L) = 1000 cubic centimeters (1000 cm^3)

The script letter ℓ is also used for "liter."

Metric prefixes are used with liters. The most common is **milli-**. The milliliter (mL) is, then, $\frac{1}{1000}$ liter. Thus,

$$1\ L = 1000\ mL = 1000\ cm^3;$$
$$0.001\ L = 1\ mL = 1\ cm^3.$$

A common unit for drug dosage is the milliliter (mL) or the cubic centimeter (cm^3). The notation "cc" is also used for cubic centimeter, especially in medicine. A milliliter and a cubic centimeter are the same size. Each is about the size of a sugar cube.

1 cm^3 5 mL   3 cm^3

$$1\ mL = 1\ cm^3 = 1\ cc$$

Volumes for which quarts and gallons are used are expressed in liters. Large volumes may be expressed using cubic meters (m^3).

Do Exercises 10–13.

EXAMPLE 7   Complete: 4.5 L = _____ mL.

$$4.5 \text{ L} = 4.5 \text{ L} \cdot \frac{1000 \text{ mL}}{1 \text{ L}}$$

$$= 4.5 \cdot 1000 \text{ mL}$$

$$= 4500 \text{ mL}$$

EXAMPLE 8   Complete: 280 mL = _____ L.

$$280 \text{ mL} = 280 \text{ mL} \cdot \frac{1 \text{ L}}{1000 \text{ mL}}$$

$$= \frac{280}{1000} \text{ L}$$

$$= 0.28 \text{ L}$$

Do Exercises 14 and 15.

C   Solving Problems

EXAMPLE 9   At a self-service gasoline station, 89-octane gasoline sells for 102.6¢ a liter. Estimate the price of 1 gal in dollars.

Since 1 liter is about 1 quart and there are 4 quarts in a gallon, the price of a gallon is about 4 times the price of a liter.

$$4 \cdot 102.6¢ = 410.4¢ = \$4.104$$

Thus 89-octane gasoline sells for about $4.10 a gallon.

Do Exercise 16.

EXAMPLE 10   *Propane Gas Tank.*   A propane gas tank is shaped like a circular cylinder with half of a sphere at each end. Find the volume of the tank if the cylindrical section is 5 ft long with a 4-ft diameter. Use 3.14 for π.

1. Familiarize.   We first make a drawing.

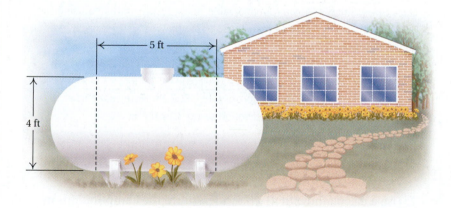

Complete.

14. 0.97 L = _____ mL

15. 8990 mL = _____ L

16. At the same station, the price of 87-octane gasoline is 96.7 cents a liter. Estimate the price of 1 gal in dollars.

Answers on page A-26

17. Medicine Capsule. A cold capsule is 8 mm long and 4 mm in diameter. Find the volume of the capsule. Use 3.14 for π. (*Hint*: First find the length of the cylindrical section.)

2. Translate. This is a two-step problem. We first find the volume of the cylindrical portion. Then we find the volume of the two ends and add. Note that the radius is 2 ft and that together the two ends make a sphere. We let V = the total volume.

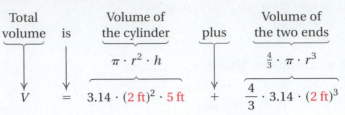

$$V = 3.14 \cdot (2\ \text{ft})^2 \cdot 5\ \text{ft} + \frac{4}{3} \cdot 3.14 \cdot (2\ \text{ft})^3$$

3. Solve. The volume of the cylinder is approximately

$$3.14 \cdot (2\ \text{ft})^2 \cdot 5\ \text{ft} \approx 3.14 \cdot 2\ \text{ft} \cdot 2\ \text{ft} \cdot 5\ \text{ft}$$
$$\approx 62.8\ \text{ft}^3.$$

The volume of the two ends is approximately

$$\frac{4}{3} \cdot 3.14 \cdot (2\ \text{ft})^3 \approx 1.33 \cdot 3.14 \cdot 2\ \text{ft} \cdot 2\ \text{ft} \cdot 2\ \text{ft}$$
$$\approx 33.4\ \text{ft}^3.$$

The total volume is approximately

$$62.8\ \text{ft}^3 + 33.4\ \text{ft}^3 = 96.2\ \text{ft}^3.$$

4. Check. The check is left to the student.

5. State. The volume of the tank is about $96.2\ \text{ft}^3$.

Do Exercise 17.

CALCULATOR CORNER

Volumes using Pi Most calculators have a $\boxed{\pi}$ key that can be used to enter the value of π in a computation (see p. 611). It might be necessary to press a $\boxed{\text{2nd}}$ or $\boxed{\text{SHIFT}}$ key before pressing the $\boxed{\pi}$ key on some calculators. Since 3.14 is a rounded value for π, results obtained using the $\boxed{\pi}$ key will be more precise than those obtained when 3.14 is used in a computation.

To find the volume of the circular cylinder in Example 3, we press $\boxed{\text{2nd}}\ \boxed{\pi}\ \boxed{\times}\ \boxed{4}\ \boxed{\times}\ \boxed{4}\ \boxed{\times}\ \boxed{1}\ \boxed{2}\ \boxed{=}$ or $\boxed{\text{SHIFT}}\ \boxed{\pi}\ \boxed{\times}\ \boxed{4}\ \boxed{\times}\ \boxed{4}\ \boxed{\times}\ \boxed{1}\ \boxed{2}\ \boxed{=}$. The result is approximately 603.19. Note that this is slightly different from the result found using 3.14 for π.

Exercises:

1. Use a calculator with a $\boxed{\pi}$ key to perform the computations in Examples 4 and 10.

2. Use a calculator with a $\boxed{\pi}$ key to perform the computations in Margin Exercises 4–7.

Answer on page A-26

a Find each volume. Use 3.14 for π in Exercises 9–12.

1.

10 cm 5 cm 5 cm

2.

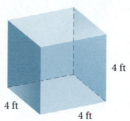

4 ft 4 ft 4 ft

3.

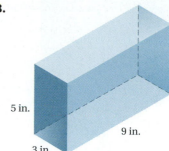

5 in. 9 in. 3 in.

4.

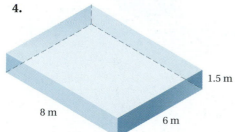

8 m 6 m 1.5 m

5.

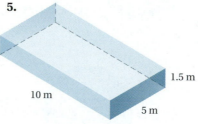

10 m 5 m 1.5 m

6.

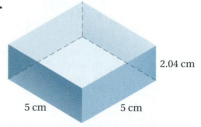

5 cm 5 cm 2.04 cm

7.

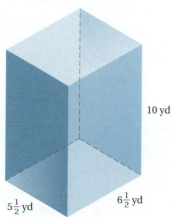

10 yd $5\frac{1}{2}$ yd $6\frac{1}{2}$ yd

8.

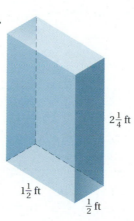

$2\frac{1}{4}$ ft $1\frac{1}{2}$ ft $\frac{1}{2}$ ft

9.

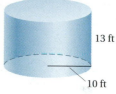

13 ft 10 ft

10.

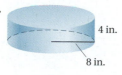

4 in. 8 in.

11.

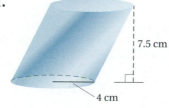

7.5 cm 4 cm

12.

15.1 m 3 m

Find each volume. Use $\frac{22}{7}$ for π in Exercises 13, 14, 19, and 20. Use 3.14 for π in Exercises 15–18.

13.

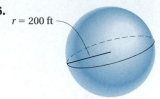

300 yd

210 yd

14.

28 km

4 km

15.

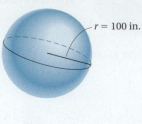

$r = 100$ in.

16.

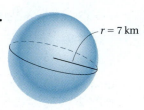

$r = 200$ ft

17.

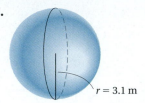

$r = 3.1$ m

18.

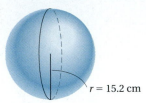

$r = 15.2$ cm

19.

$r = 7$ km

20.

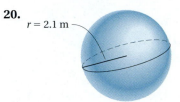

$r = 2.1$ m

---

b   Complete.

21. $1 \text{ L} = $ _____ mL = _____ cm^3

22. _____ L $= 1$ mL $= $ _____ cm^3

23. $59 \text{ L} = $ _____ mL

24. $714 \text{ L} = $ _____ mL

25. $49 \text{ mL} = $ _____ L

26. $43 \text{ mL} = $ _____ L

27. $27.3 \text{ L} = $ _____ cm^3

28. $49.2 \text{ L} = $ _____ cm^3

29. $5 \text{ gal} = $ _____ pt

30. $48 \text{ oz} = $ _____ pt

31. $10 \text{ qt} = $ _____ oz

32. $2 \text{ gal} = $ _____ cups

33. $24 \text{ oz} = $ _____ cups

34. $20 \text{ cups} = $ _____ pt

35. $10 \text{ gal} = $ _____ qt

36. $5 \text{ gal} = $ _____ cups

37. $3 \text{ gal} = $ _____ cups

38. $72 \text{ oz} = $ _____ cups

39. $15 \text{ pt} = $ _____ gal

40. $9 \text{ qt} = $ _____ gal

C   Solve.

41. *Volume of a Trash Can.*   The diameter of the base of a cylindrical trash can is 0.7 yd. The height is 1.1 yd. Find the volume. Use 3.14 for π.

42. *Ladder Rung.*   A rung of a ladder is 2 in. in diameter and 16 in. long. Find the volume. Use 3.14 for π.

43. *Barn Silo.*   A barn silo, excluding the top, is a circular cylinder. The silo is 6 m in diameter and the height is 13 m. Find the volume of the silo. Use 3.14 for π.

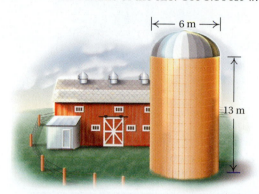

44. *Clouds.*   Find the volume of a spherical cloud with a 1000-m diameter. Use 3.14 for π.

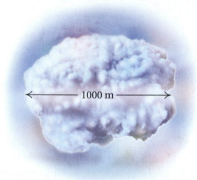

45. *Tennis Ball.*   The diameter of a tennis ball is 6.5 cm. Find the volume. Use 3.14 for π.
Source: D. Ellenbogen

46. *Oak Log.*   An oak log has a diameter of 12 cm and a length (height) of 42 cm. Find the volume. Use 3.14 for π.

47. 🖩 *Volume of Earth.*   The radius of the earth is about 3980 mi. Find the volume of the earth. Use 3.14 for π. Round to the nearest ten thousand cubic miles.
Source: *The Cambridge Factfinder,* 4th ed.

48. 🖩 *Astronomy.*   The radius of Pluto's moon is about 600 km. Find the volume of this satellite. Use $\frac{22}{7}$ for π.
Source: *The Cambridge Factfinder,* 4th ed.

49. *Oceanographic Research.*   The original "yellow submarine," the *Alvin,* is a deep-sea research vessel that has completed thousands of dives, including those in which the remains of the *Titanic* were discovered. The "pressure sphere," occupied by three humans, measures 2 m in diameter. Find the volume of the pressure sphere. Use 3.14 for π.
Source: Woods Hole Oceanographic Institution

50. *Weather Forecasting.*   Every day, the National Weather Service launches spherical weather balloons from 100 locations in the United States. Each balloon can rise over 100,000 ft and measures anywhere from 15 in. to 20 ft in diameter. Find the volume of a 6-ft–diameter weather balloon. Use 3.14 for π.
Source: Kaysam Worldwide, Inc.

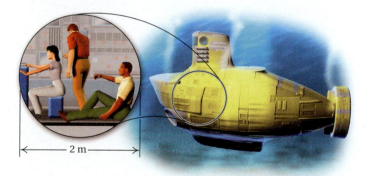

629

51. *Tennis-Ball Packaging.* Tennis balls are generally packaged in circular cylinders that hold 3 balls each. The diameter of a tennis ball is 6.5 cm. Find the volume of an empty can of tennis balls. Use 3.14 for π.

52.  *Golf-Ball Packaging.* The box shown is just big enough to hold 3 golf balls. If the radius of a golf ball is 2.1 cm, how much air surrounds the three balls? Use 3.14 for π.

53. *Conservation.* Many people leave the water running while brushing their teeth. Suppose that one person wastes 32 oz of water in such a manner each day. How much water, in gallons, would that person waste in a week? in 30 days? in a year? If each of 261 million Americans wastes water this way, estimate how much water is wasted in a year.

54. *Water Storage.* A water storage tank is a circular cylinder with a radius of 5 m and a height of 14 m. What is the tank's volume? Use $\frac{22}{7}$ for π.

55. 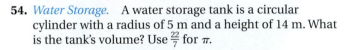 *Truck Rental.* The storage compartment of the U-Haul Mini Mover® is 9.83 ft by 5.67 ft by 5.83 ft, with an "attic" measuring 1.5 ft by 5.67 ft by 2.5 ft. Find the total volume of the compartment.

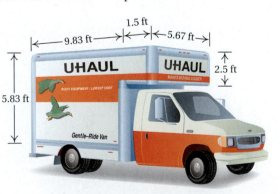

56. *Metallurgy.* If all the gold in the world could be gathered together, it would form a cube 18 yd on a side. Find the volume of the world's gold.

57. **D**_W How could you use the volume formulas given in this section to help estimate the volume of an egg?

58. **D**_W What advantages do metric units of capacity have over American units?

59. Find the simple interest on $600 at 8% for $\frac{1}{2}$ yr. [8.6a]

60. Find the simple interest on $5000 at 7% for $\frac{1}{2}$ yr. [8.6a]

61. If 9 pens cost $8.01, how much would 12 pens cost? [7.4a]

62. Solve: $9(x - 1) = 3x + 5$. [5.7b]

63. Solve: $-5y + 3 = -12y - 4$. [5.7b]

64. A barge travels 320 km in 15 days. At this rate, how far will it travel in 21 days? [7.4a]

65. Evaluate $\frac{9}{5}C + 32$ for $C = 15$. [4.7c]

66. Evaluate $\frac{5}{9}(F - 32)$ for $F = 50$. [4.7c]

67. **D$_W$** Which occupies more volume: two spheres, each with radius r, or one sphere with radius $2r$? Explain why.

68. **D$_W$** Nate reasons that since 1 yard is 3 times the length of a foot, 1 cubic yard is 3 times the volume of a cubic foot. Is his reasoning valid? Why or why not?

69. *Remarkable Feat.* In 1982, Larry Walters captured the world's imagination by riding a lawn chair attached to 42 helium-filled weather balloons to an altitude of 16,000 ft, before using a BB gun to pop a few balloons and safely descend. Walters used balloons measuring approximately 7 ft in diameter. Find the total volume of the balloons used. Use $\frac{22}{7}$ for π.

Source: MarkBarry.com

70. The volume of a ball is 36π cm^3. Find the dimensions of a rectangular box that is just large enough to hold the ball.

71. 🖩 Audio-cassette cases are typically 7 cm by 10.75 cm by 1.5 cm and contain 90 min of music. Compact-disc cases are typically 12.4 cm by 14.1 cm by 1 cm and contain 50 min of music. Which container holds the most music per cubic centimeter?

72. 🖩 A 2-cm–wide stream of water passes through a 30-m–long garden hose. At the instant that the water is turned off, how many liters of water are in the hose? Use 3.141593 for π.

73. 🖩 The volume of a basketball is 2304π cm^3. Find the volume of a cube-shaped box that is just large enough to hold the ball.

74. 🖩 *Moving Trucks.* The U-Haul Easy Loading Mover® has cargo space measuring 14' 1" by 7' 7" by 7' 2" with an attic of 2' 5" by 7' 7" by 2' 7". How much more volume than the Mini Mover (see Exercise 55) does it have?

75. 🖩 The width of a dollar bill is 2.3125 in., the length is 6.0625 in., and the thickness is 0.0041 in. Find the volume occupied by 1 million one-dollar bills.

76. 🖩 *Circumference of Earth.* The circumference of the earth at the equator is about 24,901.55 mi. Due to the irregular shape of the earth, the circumference of a circle of longitude wrapped around the earth between the north and south poles is about 24,859.82 mi. Describe and carry out a procedure for estimating the volume of the earth.

Source: *The Handy Geography Answer Book*

9.5 ANGLES AND TRIANGLES

a Measuring Angles

We see a real-world application of *angles* of various types in the spokes of these bicycles and the different back postures of the riders.

Style of Biking Determines Cycling Posture

Road About 180° flat	Mountain About 45°	Comfort About 90°

Riders prefer a more aerodynamic flat-back position.

Riders prefer a semi-upright position to help lift the front wheel over obstacles.

Riders prefer an upright position that lessens stress on the lower back and neck.

Source: USA TODAY research

An **angle** is a set of points consisting of two **rays,** or half-lines, with a common endpoint. The endpoint is called the **vertex.**

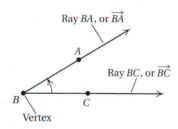

The rays are called the *sides*. The angle above can be named

angle *ABC*,   angle *CBA*,   angle *B*,   ∠*ABC*,   ∠*CBA*,   or   ∠*B*.

Note that the name of the vertex is either in the middle or, if no confusion results, listed by itself.

Do Exercises 1 and 2.

When a figure contains more than one angle, the angles are often numbered. In the following figure, another name for ∠*ABC* is ∠1, and another name for ∠*CBD* is ∠2.

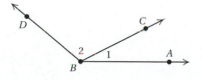

Do Exercise 3.

Objectives

a Name an angle in six different ways and given an angle, measure it with a protractor.

b Classify an angle as right, straight, acute, or obtuse.

c Identify complementary, supplementary, and vertical angles and find the measure of a complement or a supplement of a given angle.

d Classify a triangle as equilateral, isosceles, or scalene, and as right, obtuse, or acute.

e Given two of the angle measures of a triangle, find the third.

Name the angle in six different ways.

1.

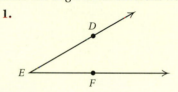

2.

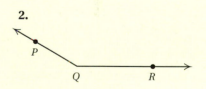

3. Give another name for ∠1.

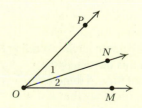

Answers on page A-27

633

4. Use a protractor to measure this angle.

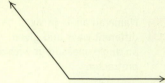

Answer on page A-27

Study Tips

LEARNING VOCABULARY

Occasionally you may encounter a lesson with which you are already familiar. When this occurs, make a special effort to remember the terminology that is used. By learning the proper names for the concepts, you will solidify your understanding of those concepts.

Measuring angles is similar to measuring segments. To measure angles, we start with some predetermined angle and assign to it a measure of 1. We call it a *unit angle*. Suppose that ∠U is a unit angle. Let's measure ∠DEF. If we made 3 copies of ∠U, they would "fill up" ∠DEF. Thus the measure of ∠DEF would be 3 units.

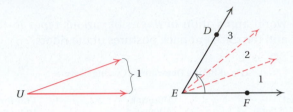

The unit most commonly used for angle measure is the degree. Below is such a unit. Its measure is 1 degree, or 1°.

A 1° angle:

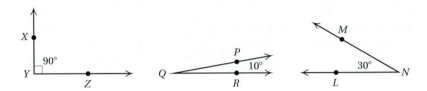

Here are some other angles with their degree measures.

To indicate the *measure* of ∠XYZ, we write $m\angle XYZ = 90°$. The symbol ⌐ is sometimes drawn on a figure to indicate a 90° angle.

A device called a **protractor** is used to measure angles. Protractors have two scales. In the center of the protractor is a vertex indicator such as ▲ or a small hole. To measure an angle like ∠Q below, we place the protractor's ▲ at the vertex and line up one of the angle's sides at 0°. Then we check where the angle's other side crosses the scale. In the figure below, 0° is on the inside scale, so we check where the angle's other side crosses the inside scale. We see that $m\angle Q = 145°$. The notation $m\angle Q$ is read "the measure of angle Q."

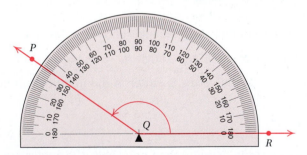

Do Exercise 4.

Let's find the measure of ∠ABC. This time we will use the 0° on the outside scale. We see that m ∠ABC = 68°.

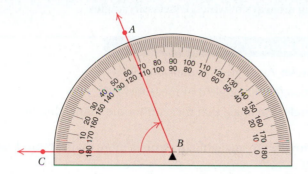

Do Exercise 5.

A protractor can be used to draw a circle graph.

EXAMPLE 1 *Transportation.* According to a recent poll, 45% of adults believe that flying is the safest mode of transportation, 39% believe that cars are the safest, and 16% believe that trains are the safest. Draw a circle graph to represent these figures.

Source: *Marist Institute for Public Opinion*

Every circle graph contains a total of 360°. Thus

45% of the circle is a 0.45(360°), or 162° angle;

39% of the circle is a 0.39(360°), or 140.4° angle;

16% of the circle is a 0.16(360°), or 57.6° angle.

We begin by drawing a 162° angle. Beginning at the center of the circle, we draw a horizontal segment to the circle. That segment is one side of the angle. We use a protractor to mark off a 162° angle. From that mark, we draw a segment to the center of the circle to complete the angle. This section of the circle graph we label with both the percent (45%) and the type of transportation (Airplanes).

From the second segment drawn, we repeat the above procedure to draw a 140.4° angle. Since protractors are marked in units of 1°, we must approximate this angle. This section we label with 39% and Cars.

The remainder of the circle represents Trains and should be a 57.6° angle; we measure to confirm this, and label the section with 16% and Trains.

Finally, we give a title to the graph: Safest Mode of Transportation.

Safest Mode of Transportation

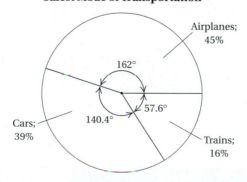

Do Exercise 6.

---

5. Use a protractor to measure this angle.

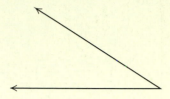

6. Lengths of Engagement of Married Couples. The data below list the percent of married couples who were engaged for a certain time period before marriage. Use this information to draw a circle graph.

Source: Bruskin Goldring Research

Less than 1 yr:	24%
1–2 yr:	21%
More than 2 yr:	35%
Never engaged:	20%

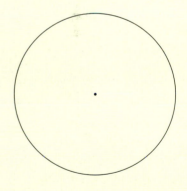

Answers on page A-27

Classify each angle as right, straight, acute, or obtuse. Use a protractor if necessary.

7.

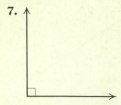

8.

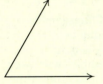

9.

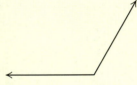

10.

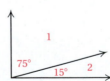

b Classifying Angles

The following are ways in which we classify angles.

> **TYPES OF ANGLES**
>
> **Right angle:** An angle that measures 90°.
>
> **Straight angle:** An angle that measures 180°.
>
> **Acute angle:** An angle that measures more than 0° but less than 90°.
>
> **Obtuse angle:** An angle that measures more than 90° but less than 180°.

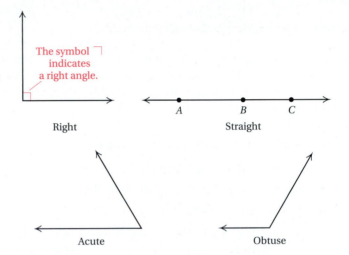

The symbol ⌐ indicates a right angle.

Right   Straight

Acute   Obtuse

Do Exercises 7–10.

c Complementary, Supplementary, and Vertical Angles

Certain pairs of angles share special properties.

COMPLEMENTARY ANGLES

When the sum of the measures of two angles is 90°, the angles are said to be **complementary.** For example, in the figure below, ∠1 and ∠2 are complementary.

$$m\angle 1 + m\angle 2 = 90°$$
$$75° + 15° = 90°$$

> **COMPLEMENTARY ANGLES**
>
> Two angles are **complementary** if the sum of their measures is 90°. Each angle is called a **complement** of the other.

If two angles are complementary, each is an acute angle. When complementary angles are adjacent to each other, that is, they have a side in common, they form a right angle.

EXAMPLE 2 Identify each pair of complementary angles.

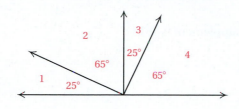

∠1 and ∠2   25° + 65° = 90°   ∠2 and ∠3
∠1 and ∠4   ∠3 and ∠4

EXAMPLE 3 Find the measure of a complement of a 39° angle.

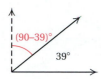   90° − 39° = 51°

The measure of a complement is 51°.

Do Exercises 11–14.

SUPPLEMENTARY ANGLES

Next, consider ∠1 and ∠2 as shown below. Because the sum of their measures is 180°, ∠1 and ∠2 are said to be **supplementary.** Note that when supplementary angles are adjacent, they form a straight angle.

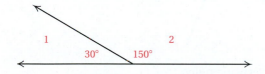 $m∠1 + m∠2 = 180°$
                              $30° + 150° = 180°$

SUPPLEMENTARY ANGLES
Two angles are **supplementary** if the sum of their measures is 180°. Each angle is called a **supplement** of the other.

EXAMPLE 4 Identify each pair of supplementary angles.

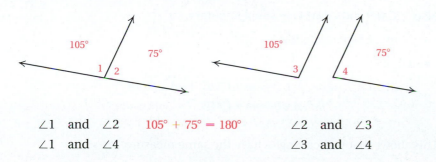

∠1 and ∠2   105° + 75° = 180°   ∠2 and ∠3
∠1 and ∠4   ∠3 and ∠4

11. Identify each pair of complementary angles.

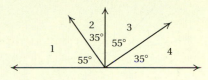

Find the measure of a complement of each angle.

12.

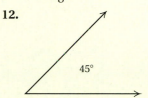

13.

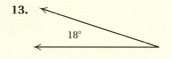

14.

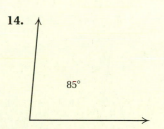

Answers on page A-27

15. Identify each pair of supplementary angles.

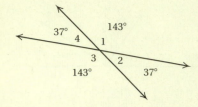

Find the measure of a supplement of an angle with the given measure.

16. 38°

17. 157°

18. 90°

19. Identify each pair of vertical angles.

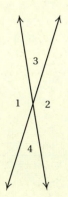

Answers on page A-27

EXAMPLE 5 Find the measure of a supplement of an angle of 112°.

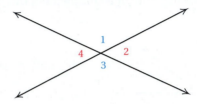

$$180° - 112° = 68°$$

The measure of a supplement is 68°.

Do Exercises 15–18.

VERTICAL ANGLES

When two lines intersect, four angles are formed. The pairs of angles that do not share any side in common are said to be **vertical** (or *opposite*) angles. Thus, in the drawing below, ∠1 and ∠3 are vertical angles, as are ∠4 and ∠2. Note that $m\angle 1 = m\angle 3$ and $m\angle 4 = m\angle 2$.

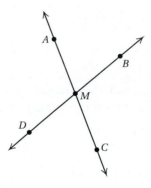

EXAMPLE 6 Identify each pair of vertical angles.

∠AMB and ∠CMD are vertical angles.
∠BMC and ∠DMA are vertical angles.

Do Exercise 19.

Note in the figure in Example 6 that ∠DMA and ∠AMB are supplementary; that is,

$$m\angle DMA + m\angle AMB = 180°.$$

Also, ∠CMD and ∠DMA are supplementary, so

$$m\angle CMD + m\angle DMA = 180°.$$

Therefore,

$$m\angle DMA + m\angle AMB = m\angle CMD + m\angle DMA$$

$$m\angle AMB = m\angle CMD. \quad \text{Subtracting } m\angle DMA \text{ from both sides}$$

This shows that vertical angles have the same measure.

Two angles are **vertical** if they are formed by two intersecting lines and have no side in common. Vertical angles have the same measure.

Do Exercise 20.

If two angles have the same measure, we say that they are **congruent,** denoted by the symbol ≅. In the figure below, the measures of angles *WVZ* and *XVY* are equal, and the angles are congruent:

$$m \angle WVZ = m \angle XVY$$
$$\angle WVZ \cong \angle XVY.$$

Note that we do not write that angles are equal: The *measures are equal* and the *angles are congruent.*

Do Exercise 21.

d Triangles

A **triangle** is a polygon made up of three segments, or sides. Consider these triangles. The triangle with vertices *A*, *B*, and *C* can be named △*ABC*.

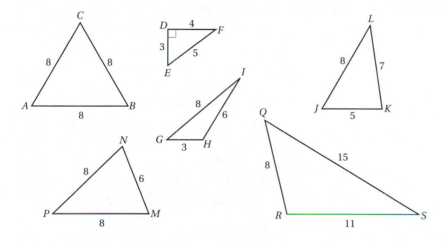

We can classify triangles according to sides and according to angles.

TYPES OF TRIANGLES

Equilateral triangle: All sides are the same length.

Isosceles triangle: Two or more sides are the same length.

Scalene triangle: All sides are of different lengths.

Right triangle: One angle is a right angle.

Obtuse triangle: One angle is an obtuse angle.

Acute triangle: All three angles are acute.

Do Exercises 22–25.

---

20. Complete:

$$m \angle ANC = \underline{\qquad}.$$
$$m \angle ANB = \underline{\qquad}.$$

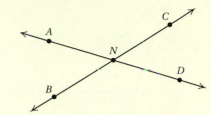

21. Complete:

$$\angle PMQ \cong \underline{\qquad}.$$
$$\angle SMP \cong \underline{\qquad}.$$

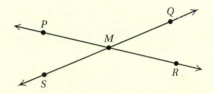

22. Which triangles on this page are:
 a) equilateral?
 b) isosceles?
 c) scalene?

23. Are all equilateral triangles isosceles?

24. Are all isosceles triangles equilateral?

25. Which triangles on this page are:
 a) right triangles?
 b) obtuse triangles?
 c) acute triangles?

Answers on page A-27

26. Find
$$m(\angle P) + m(\angle Q) + m(\angle R).$$

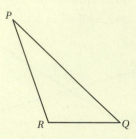

e Sum of the Angle Measures of a Triangle

The sum of the angle measures of every triangle is 180°. To see this, note that we can think of cutting apart a triangle as shown on the left below. If we reassemble the pieces, we see that a straight angle is formed.

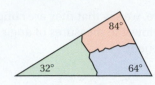
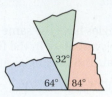

$$64° + 32° + 84° = 180°$$

SUM OF THE ANGLE MEASURES OF A TRIANGLE

In any triangle *ABC*, the sum of the measures of the angles is 180°:
$$m(\angle A) + m(\angle B) + m(\angle C) = 180°.$$

Do Exercise 26.

If we know the measures of two angles of a triangle, we can calculate the measure of the remaining angle.

EXAMPLE 7 Find the missing angle measure.

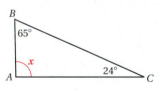

$$m(\angle A) + m(\angle B) + m(\angle C) = 180°$$
$$x + 65° + 24° = 180°$$
$$x + 89° = 180°$$
$$x = 180° - 89° \qquad \text{Subtracting 89° from both sides}$$
$$x = 91°$$

The measure of $\angle A$ is 91°.

Do Exercise 27.

27. Find the missing angle measure.

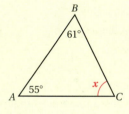

a   Name each angle in six different ways.

1.

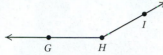

G   H   I

2.

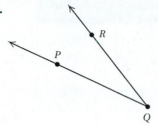

R   P   Q

Give another name for ∠1 in each figure.

3.

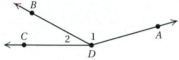

B   C   2   1   A   D

4.

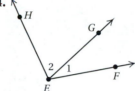

H   G   2   1   F   E

Use a protractor to measure each angle.

5.

6.

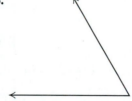

7.

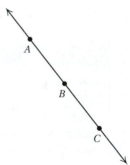

A   B   C

8.

9.

10.

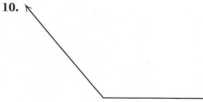

Use the given information and a protractor to draw a circle graph.

11. *Snacking Habits.*   Some adults snack heavily and some not at all. The table below lists the snacking habits of American adults.

FREQUENCY OF SNACKING	PERCENT
Never	10%
Occasionally	45%
Moderately	35%
Heavily	10%

Source: Market Facts for Hershey Foods

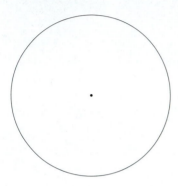

12. *e-mail.*   The table below lists the numbers of e-mails people get per day at work.

NUMBER OF E-MAILS PER DAY	PERCENT
Less than 1	28%
1–5	20%
6–10	12%
11–20	9%
21 or more	31%

Source: John J. Heldrich Center for Workforce Development

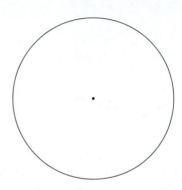

13. *Kids in Foster Care.*   There are approximately one-half million children in foster care in the United States. Most of these children are under the age of 10. The table below lists the percentages by ages of children in foster care.

AGE GROUP	PERCENT
Under 1	3%
1–5	25%
6–10	27%
11–15	27%
16+	18%

Source: The Administration for Children and Families

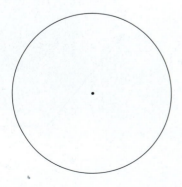

14. *Causes of Spinal Cord Injuries.*   The table below lists the causes of spinal cord injury.

CAUSES	PERCENT
Motor vehicle accidents	44%
Acts of violence	24%
Falls	22%
Sports	8%
Other	2%

Source: National Spinal Cord Injury Association

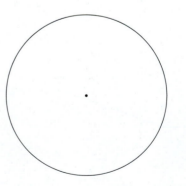

15.–20. Classify each of the angles in Exercises 5–10 as right, straight, acute, or obtuse.

21.–24. Classify each of the angles in Margin Exercises 1, 2, 4 and 5 as right, straight, acute, or obtuse.

 Identify two pairs of vertical angles for each figure.

25.

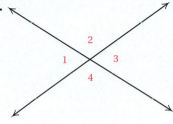

26.

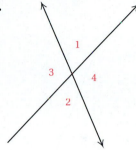

27.

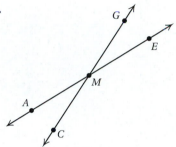

28.

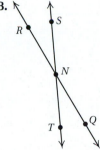

Complete.

29. Refer to Exercise 25.

$m \angle 2 = $ _____

$m \angle 3 = $ _____

30. Refer to Exercise 26.

$m \angle 4 = $ _____

$m \angle 2 = $ _____

31. Refer to Exercise 27.

$\angle AMC \cong$ _____

$\angle AMG \cong$ _____

32. Refer to Exercise 28.

$\angle RNS \cong$ _____

$\angle TNR \cong$ _____

Find the measure of a complement of an angle with the given measure.

33. 11°

34. 83°

35. 67°

36. 5°

37. 58°

38. 32°

39. 29°

40. 54°

Find the measure of a supplement of an angle with the given measure.

41. 3°

42. 54°

43. 139°

44. 13°

45. 75°

46. 128°

47. 104°

48. 49°

Classify each triangle as equilateral, isosceles, or scalene. Then classify it as right, obtuse, or acute.

49.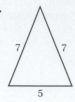

50.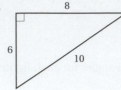

51.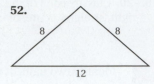

52.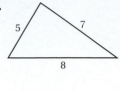

53.

54.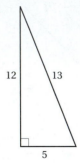

55.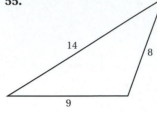

56.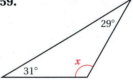

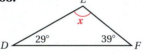 Find each missing angle measure.

57.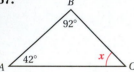

58.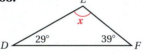

59.

60.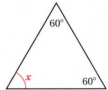

61. **D**_W Can both an angle and its supplement be obtuse? Why or why not?

62. **D**_W Explain a procedure that could be used to determine the measure of an angle's supplement from the measure of the angle's complement.

SKILL MAINTENANCE

Find the simple interest. [8.6a]

	PRINCIPAL	RATE OF INTEREST	TIME	SIMPLE INTEREST
63.	$2000	8%	1 year	
64.	$750	6%	$\frac{1}{2}$ year	
65.	$4000	7.4%	$\frac{1}{2}$ year	
66.	$200,000	6.7%	$\frac{1}{12}$ year	

Interest is compounded semiannually. Find the amount in the account after the given length of time. Round to the nearest cent.   [8.6b]

	PRINCIPAL	RATE OF INTEREST	TIME	AMOUNT IN THE ACCOUNT
67.	$25,000	6%	5 years	
68.	$150,000	$6\frac{7}{8}$%	15 years	
69.	$150,000	7.4%	20 years	
70.	$160,000	7.4%	20 years	

Simplify.   [1.9c]

71. $2^2 + 3^2 + 4^2$

72. $5^2 - 4^2 + 1^2$

SYNTHESIS

73. **D**_W Explain how you might use triangles to find the sum of the angle measures of this figure.

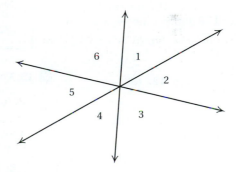

74. **D**_W Do parallelograms always contain two acute and two obtuse angles? Why or why not?

75. ▦ In the figure, $m\angle 1 = 79.8°$ and $m\angle 3 = 33.07°$. Find $m\angle 2$, $m\angle 4$, $m\angle 5$, and $m\angle 6$.

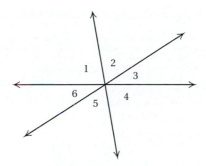

76. ▦ In the figure, $m\angle 2 = 42.17°$ and $m\angle 6 = 81.9°$. Find $m\angle 1$, $m\angle 3$, $m\angle 4$, and $m\angle 5$.

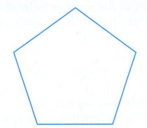

77. Find $m\angle ACB$, $m\angle CAB$, $m\angle EBC$, $m\angle EBA$, $m\angle AEB$, and $m\angle ADB$ in the rectangle shown below.

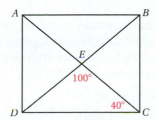

78. The angles in the figure are supplementary. Find the measure of each angle.

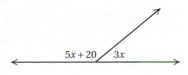

SQUARE ROOTS AND THE PYTHAGOREAN THEOREM

Objectives

a Simplify square roots of squares, such as $\sqrt{25}$.

b Approximate square roots.

c Given the lengths of any two sides of a right triangle, find the length of the third side.

d Solve applied problems involving right triangles.

Find each square.

1. 9^2

2. $(-10)^2$

3. 11^2

4. 12^2

It would be helpful to memorize the squares of numbers from 1 to 25.

5. 13^2

6. 14^2

7. 15^2

8. 16^2

Find all square roots. Use the results of Exercises 1–8 above, if necessary.

9. 100

10. 81

11. 49

12. 196

Simplify. Use the results of Exercises 1–8 above, if necessary.

13. $\sqrt{49}$

14. $\sqrt{16}$

15. $\sqrt{121}$

16. $\sqrt{100}$

17. $\sqrt{81}$

18. $\sqrt{64}$

19. $\sqrt{225}$

20. $\sqrt{169}$

21. $\sqrt{1}$

22. $\sqrt{0}$

Answers on page A-27

a Square Roots

> **SQUARE ROOT**
>
> If a number is a product of a factor times itself, then that factor is a **square root** of the number. (If $c^2 = a$, then c is a square root of a.)

For example, 36 has two square roots, 6 and -6. To see this, note that $6 \cdot 6 = 36$ and $(-6) \cdot (-6) = 36$.

EXAMPLE 1 Find the square roots of 25.

The square roots of 25 are 5 and -5, because $5^2 = 25$ and $(-5)^2 = 25$.

> **Caution!**
>
> To find the *square* of a number, multiply the number by itself. To find a *square root* of a number, find a number that, when squared, gives the original number.

Do Exercises 1–12.

Since every positive number has two square roots, the symbol $\sqrt{}$ (called a *radical* sign) is used to denote the positive square root of the number underneath. Thus, $\sqrt{9}$ means 3, not -3.

> **RADICAL SIGN, $\sqrt{}$**
>
> If n is a positive number, $\sqrt{n}$ means the positive square root of n. *The square root of n means $\sqrt{n}$.*

EXAMPLES Simplify.

2. $\sqrt{36} = 6$ The square root of 36 is 6 because $6^2 = 36$ and 6 is positive.

3. $\sqrt{144} = 12$ Note that $12^2 = 144$.

Do Exercises 13–22.

b Approximating Square Roots

Many square roots can't be written as whole numbers or fractions. For example, $\sqrt{2}$, $\sqrt{3}$, $\sqrt{39}$, and $\sqrt{70}$ cannot be precisely represented in decimal notation. To see this, consider the following decimal approximations for $\sqrt{2}$. Each gives a closer approximation, but none is exactly $\sqrt{2}$:

$$\sqrt{2} \approx 1.4 \qquad \text{because} \quad (1.4)^2 = 1.96;$$
$$\sqrt{2} \approx 1.41 \qquad \text{because} \quad (1.41)^2 = 1.9881;$$
$$\sqrt{2} \approx 1.414 \qquad \text{because} \quad (1.414)^2 = 1.999396.$$

Decimal approximations like these are commonly found by using a calculator.

EXAMPLE 4 Approximate $\sqrt{3}$, $\sqrt{27}$, and $\sqrt{180}$ to the nearest thousandth. Use a calculator.

We use a calculator to find each square root. Since more than three decimal places are given, we round back to three places.

$$\sqrt{3} \approx 1.732,$$
$$\sqrt{27} \approx 5.196,$$
$$\sqrt{180} \approx 13.416$$

As a check, note that $1 \cdot 1 = 1$ and $2 \cdot 2 = 4$, so we expect $\sqrt{3}$ to be between 1 and 2. Similarly, we expect $\sqrt{27}$ to be between 5 and 6 and $\sqrt{180}$ to be between 13 and 14.

If you continue in algebra, you will probably learn techniques for rewriting $\sqrt{27}$ as $3\sqrt{3}$ and $\sqrt{180}$ as $6\sqrt{5}$. Such techniques are beyond the scope of this text.

Do Exercises 23–25 on the following page.

C   The Pythagorean Theorem

A **right triangle** is a triangle with a 90° angle, as shown here.

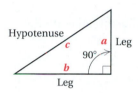

In a right triangle, the longest side is called the **hypotenuse.** It is also the side opposite the right angle. The other two sides are called **legs.** We generally use the letters a and b for the lengths of the legs and c for the length of the hypotenuse. They are related as follows.

THE PYTHAGOREAN THEOREM

In any right triangle, if a and b are the lengths of the legs and c is the length of the hypotenuse, then

$$a^2 + b^2 = c^2, \quad \text{or}$$
$$(\text{Leg})^2 + (\text{Other leg})^2 = (\text{Hypotenuse})^2.$$

The equation $a^2 + b^2 = c^2$ is called the **Pythagorean equation.***

---

*The *converse* of the Pythagorean theorem is also true. That is, if $a^2 + b^2 = c^2$, then the triangle is a right triangle.

Approximate to the nearest thousandth. Use a calculator.

23. $\sqrt{5}$

24. $\sqrt{78}$

25. $\sqrt{168}$

26. Find the length of the hypotenuse of this right triangle.

Answers on page A-27

Answers on page A-27

Study Tips

WORKING WITH A CLASSMATE

If you are finding it difficult to master a particular topic or concept, try talking about it with a classmate. Verbalizing your questions about the material might help clarify it. If your classmate is also finding the material difficult, it is possible that the majority of the people in your class are confused and you can ask your instructor to explain the concept again.

648

The Pythagorean theorem is named for the Greek mathematician Pythagoras (569?–500? B.C.). We can think of this relationship as adding areas.

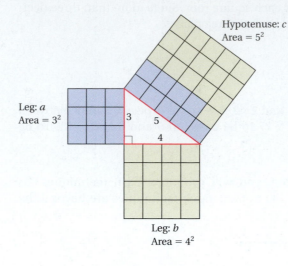

$$a^2 + b^2 = c^2$$
$$3^2 + 4^2 = 5^2$$
$$9 + 16 = 25$$

If we know the lengths of any two sides of a right triangle, we can use the Pythagorean equation to determine the length of the third side.

EXAMPLE 5 Find the length of the hypotenuse of this right triangle.

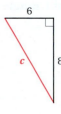

We substitute in the Pythagorean equation:

$$a^2 + b^2 = c^2$$
$$6^2 + 8^2 = c^2 \qquad \text{Substituting}$$
$$36 + 64 = c^2$$
$$100 = c^2.$$

The solution of this equation is the square root of 100, which is 10:

$$c = \sqrt{100} = 10.$$

Do Exercise 26.

EXAMPLE 6 Find the length b for the right triangle shown. Give an exact answer and an approximation to three decimal places.

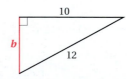

We substitute in the Pythagorean equation:

$$a^2 + b^2 = c^2$$
$$10^2 + b^2 = 12^2 \qquad \text{Substituting}$$
$$100 + b^2 = 144.$$

Next, we solve for b^2 and then b, as follows:

$$100 + b^2 - 100 = 144 - 100 \qquad \text{Subtracting 100 from both sides}$$
$$b^2 = 144 - 100$$
$$b^2 = 44 \qquad \text{Solving for } b^2$$

Exact answer: $\qquad b = \sqrt{44} \qquad$ Solving for b

Approximation: $\quad b \approx 6.633.$ $\qquad$ Using a calculator

Do Exercises 27–29.

d | Applications

EXAMPLE 7 *Height of Ladder.* A 12-ft ladder leans against a building. The bottom of the ladder is 7 ft from the building. How high is the top of the ladder? Give an exact answer and an approximation to the nearest tenth of a foot.

1. **Familiarize.** We first make a drawing. In it we see a right triangle. We let $h =$ the unknown height.

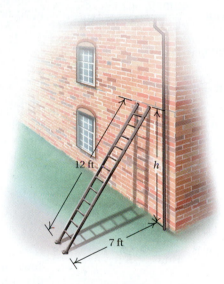

2. **Translate.** We substitute 7 for a, h for b, and 12 for c in the Pythagorean equation:

$$a^2 + b^2 = c^2 \qquad \text{Pythagorean equation}$$
$$7^2 + h^2 = 12^2.$$

3. **Solve.** We solve for h^2 and then h:

$$49 + h^2 = 144 \qquad 7^2 = 49 \text{ and } 12^2 = 144$$
$$49 + h^2 - 49 = 144 - 49 \qquad \text{Subtracting 49 from both sides}$$
$$h^2 = 144 - 49$$
$$h^2 = 95$$

Exact answer: $\qquad h = \sqrt{95} \qquad$ Solving for h

Approximation: $\quad h \approx 9.7$ ft.

4. **Check.** $7^2 + \left(\sqrt{95}\right)^2 = 49 + 95 = 144 = 12^2.$

5. **State.** The top of the ladder is $\sqrt{95}$, or about 9.7 ft from the ground.

Do Exercise 30.

Find the length of the leg of the right triangle. Give an exact answer and an approximation to three decimal places.

27.

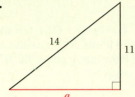

28.

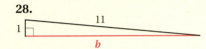

29.

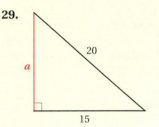

30. How long is a guy wire reaching from the top of an 18-ft pole to a point on the ground 10 ft from the pole? Give an exact answer and an approximation to the nearest tenth of a foot.

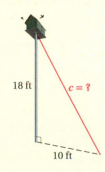

Answers on page A-27

Translating for Success

1. **Test Items.** On a test of 90 items, Sally got 80% correct. How many items did she get correct?

2. **Suspension Bridge.** The San Francisco/Oakland Bay suspension bridge is 0.4375 mi long. Convert this distance to yards.

3. **Population Growth.** Brookdale's growth rate per year is 0.9%. If the population was 1,500,000 in 2005, what is the population in 2006?

4. **Roller Coaster Drop.** The Manhattan Express Roller Coaster at the New York-New York Hotel and Casino, Las Vegas, Nevada, has a 144-ft drop. The California Screamin' Roller Coaster at Disney's California Adventure, Anaheim, California, has a 32.635-m drop. How much larger in meters is the drop of the Manhattan Express than the drop of the California Screamin'?

5. **Driving Distance.** Nate drives the company car 675 mi in 15 days. At this rate, how far will he drive in 20 days?

The goal of these matching questions is to practice step (2), *Translate*, of the five-step problem-solving process. Translate each word problem to an equation and select a correct translation from equations A–O.

A. $32.635 \text{ m} + x = 144 \text{ ft} \cdot \dfrac{0.305 \text{ m}}{1 \text{ ft}}$

B. $x = 0.4375 \text{ mi} \times \dfrac{5280 \text{ ft}}{1 \text{ mi}} \times \dfrac{12 \text{ in.}}{1 \text{ ft}}$

C. $80\% \cdot x = 90$

D. $x = 0.89 \text{ km} \cdot \dfrac{1000 \text{ m}}{1 \text{ km}} \cdot \dfrac{1 \text{ m}}{3.281 \text{ ft}}$

E. $x = 80\% \cdot 90$

F. $x = 420 \text{ m} + 75 \text{ ft} \cdot \dfrac{0.305 \text{ m}}{1 \text{ ft}}$

G. $\dfrac{x}{20} = \dfrac{675}{15}$

H. $x = 0.4375 \text{ mi} \times \dfrac{5280 \text{ ft}}{1 \text{ mi}} \times \dfrac{1 \text{ yd}}{3 \text{ ft}}$

I. $x = 0.89 \text{ km} \cdot \dfrac{0.621 \text{ mi}}{1 \text{ km}} \cdot \dfrac{5280 \text{ ft}}{1 \text{ mi}}$

J. $\dfrac{x}{15} = \dfrac{675}{20}$

K. $x = 420 \text{ m} \cdot \dfrac{3.281 \text{ ft}}{1 \text{ m}} + 75 \text{ ft}$

L. $144 \text{ ft} + x = 32.635 \text{ m} \cdot \dfrac{1 \text{ ft}}{0.305 \text{ m}}$

M. $x = 1{,}500{,}000 - 0.9\%(1{,}500{,}000)$

N. $20 \cdot x = 675$

O. $x = 1{,}500{,}000 + 0.9\%(1{,}500{,}000)$

Answers on page A-27

---

6. **Test Items.** Jason answered correctly 90 items on a recent test. These items represented 80% of the total number of questions. How many items were on the test?

7. **Population Decline.** Flintville's growth rate per year is −0.9%. If the population was 1,500,000 in 2005, what is the population in 2006?

8. **Bridge Length.** The Tatara Bridge in Onomichi-Imabari, Japan, is 0.89 km long. Convert this distance to feet.

9. **Height of Tower.** The Sears Tower in Chicago is 75 ft taller than the Jin Mao Building in Shanghai. The height of the Jin Mao Building is 420 m. What is the height of the Sears Tower in feet?

10. **Gasoline Usage.** Nate's company car gets 20 miles to the gallon in city driving. How many gallons will it use in 675 mi of city driving?

9.6

EXERCISE SET

For Extra Help

MathXL MyMathLab InterAct Math Tutor Video Student's
Math Center Lectures Solutions
on CD Manual
Disc 5

a Find both square roots for each number listed.

1. 16

2. 9

3. 121

4. 49

5. 169

6. 144

7. 2500

8. 3600

Simplify.

9. $\sqrt{64}$

10. $\sqrt{4}$

11. $\sqrt{81}$

12. $\sqrt{49}$

13. $\sqrt{225}$

14. $\sqrt{121}$

15. $\sqrt{625}$

16. $\sqrt{900}$

17. $\sqrt{400}$

18. $\sqrt{169}$

19. $\sqrt{10,000}$

20. $\sqrt{1,000,000}$

b Approximate each number to the nearest thousandth. Use a calculator.

21. $\sqrt{48}$

22. $\sqrt{17}$

23. $\sqrt{8}$

24. $\sqrt{7}$

25. $\sqrt{3}$

26. $\sqrt{6}$

27. $\sqrt{12}$

28. $\sqrt{18}$

29. $\sqrt{19}$

30. $\sqrt{75}$

31. $\sqrt{110}$

32. $\sqrt{10}$

c Find the length of the third side of each right triangle. Give an exact answer and, when appropriate, an approximation to the nearest thousandth.

33.
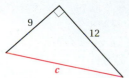

34.

35.

36.

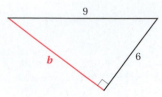

37.

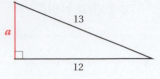

38.

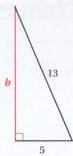

39.

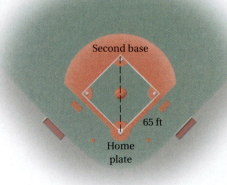

40.

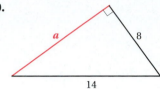

For each right triangle, find the length of the side not given. Assume that *c* represents the length of the hypotenuse. Give an exact answer and, when appropriate, an approximation to the nearest thousandth.

41. $a = 10, b = 24$

42. $a = 5, b = 12$

43. $a = 9, c = 15$

44. $a = 18, c = 30$

45. $a = 4, b = 5$

46. $a = 5, b = 6$

47. $a = 1, c = 32$

48. $b = 1, c = 20$

d In Exercises 49–56, give an exact answer and an approximation to the nearest tenth.

49. How long is a string of lights reaching from the top of a 12-ft pole to a point 8 ft from the base of the pole?

50. How long must a wire be in order to reach from the top of a 13-m telephone pole to a point on the ground 9 m from the base of the pole?

51. *Softball Diamond.* A slow-pitch softball diamond is actually a square 65 ft on a side. How far is it from home plate to second base?

52. *Baseball Diamond.* A baseball diamond is actually a square 90 ft on a side. How far is it from home plate to second base?

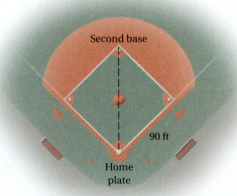

Second base

65 ft

Home plate

Second base

90 ft

Home plate

53. How tall is this tree?

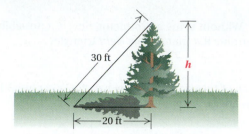

54. How far is the base of the fence post from point *A*?

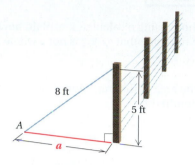

55. An airplane is flying at an altitude of 4100 ft. The slanted distance directly to the airport is 15,100 ft. How far is the airplane horizontally from the airport?

56. A surveyor had poles located at points *P*, *Q*, and *R* around a lake. The distances that the surveyor was able to measure are marked on the drawing. What is the distance from *P* to *R* across the lake?

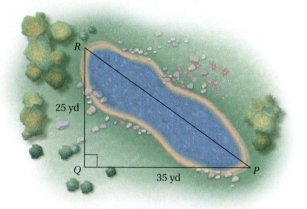

57. $\mathbf{D_W}$ Write a problem similar to Exercises 49–56 for a classmate to solve. Design the problem so that its solution involves the length $\sqrt{58}$ m.

58. $\mathbf{D_W}$ Does every number have two square roots? Why or why not?

SKILL MAINTENANCE

Solve.

59. Food expenses account for 26% of the average family's budget. A family makes $1800 one month. How much do they spend for food? [8.4a]

60. Blakely County has a population that is increasing by 4% each year. This year the population is 180,000. What will it be next year? [8.4b]

61. Dexter College has a student body of 1850 students. Of these, 17.5% are seniors. How many students are seniors? [8.4a]

62. The price of a cellular phone was reduced from $70 to $61.60. Find the percent of decrease in price. [8.4b]

Simplify. [1.9b]

63. 2^3

64. 5^3

65. 4^3

66. 3^3

67. **D**_W Without using a calculator, explain how you could convince someone that $\sqrt{902}$ is not a whole number.

68. **D**_W Without using a protractor, how is it possible to determine if a triangle is a *right* triangle?

69. 🖩 Find the area of the trapezoid shown. Round to the nearest hundredth.

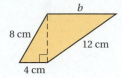

70. Which of the triangles below has the larger area? If the areas are the same, state so.

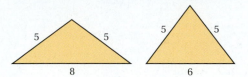

71. 🖩 Caiden's new TV has a screen that measures $31\frac{3}{4}$ in. by $56\frac{1}{2}$ in. Determine the diagonal measurement of the screen. Round your answer to the nearest tenth of an inch.

72. A Philips 42-in. plasma television has a rectangular screen that measures 42 in. diagonally. The ratio of width to height is 16 to 9. Find the width and the height of the screen.

73. A conventional 19-in. television set has a rectangular screen that measures 19 in. diagonally. The ratio of width to height in a conventional television set is 4 to 3. Find the width and the height of the screen.

74. 🖩 A cube is circumscribed by a sphere with a 1-m diameter. How much more volume is in the sphere?

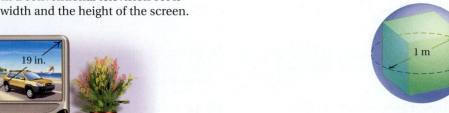

9.7

WEIGHT, MASS, AND TEMPERATURE

Objectives

a Convert from one American unit of weight to another.

b Convert from one metric unit of mass to another.

c Convert temperatures from Celsius to Fahrenheit and from Fahrenheit to Celsius.

a Weight: The American System

The American units of weight are as follows.

AMERICAN UNITS OF WEIGHT
1 lb = 16 ounces (oz)
1 ton (T) = 2000 pounds (lb)

The term "ounce" used here for weight is different from the "ounce" we used for capacity in Section 9.4. Often, however, 1 fluid ounce weighs approximately 1 ounce.

EXAMPLE 1 A well-known hamburger is called a "quarter-pounder." Find its name in ounces: a "_____ ouncer."

$$\frac{1}{4}\,\text{lb} = \frac{1}{4} \cdot 1\,\text{lb}$$

$$= \frac{1}{4} \cdot 16\,\text{oz} \qquad \text{Substituting 16 oz for 1 lb}$$

$$= 4\,\text{oz}$$

A "quarter-pounder" can also be called a "four-ouncer."

EXAMPLE 2 Complete: 15,360 lb = _____ T.

$$15{,}360\,\text{lb} = 15{,}360\,\text{lb} \cdot \frac{1\,\text{T}}{2000\,\text{lb}} \qquad \text{Multiplying by 1}$$

$$\left. \begin{array}{l} = \dfrac{15{,}360}{2000}\,\text{T} \\[2mm] = 7.68\,\text{T} \end{array} \right\} \quad \text{Dividing by 2000}$$

Do Exercises 1–3.

b Mass: The Metric System

There is a difference between **mass** and **weight,** but the terms are often used interchangeably. People sometimes use the word "weight" instead of "mass." Weight is related to the force of the earth's gravity. The farther you are from the center of the earth, the less you weigh. Your mass, on the other hand, stays the same no matter where you are.

The basic unit of mass is the **gram** (g), which is the mass of 1 cubic centimeter (1 cm^3 or 1 mL) of water. Since a cubic centimeter is small, a gram is a small unit of mass.

Complete.

1. 5 lb = _____ oz

2. 8640 lb = _____ T

3. 1 T = _____ oz

Answers on page A-28

Complete with mg, g, kg, or t.

4. A laptop computer has a mass of 2 _____.

5. Evan has a body mass of 76 _____.

6. For her sprained ankle, Roz took 400 _____ of ibuprofen.

7. A pen has a mass of 12 _____.

8. A pickup truck has a mass of 1.5 _____.

Answers on page A-28

Study Tips

BEGINNING TO STUDY FOR THE FINAL EXAM

It is never too soon to begin to study for the final examination. Take a few minutes each week to review the highlighted information, such as formulas, properties, and procedures. Make special use of the Summary and Reviews, Chapter Tests, and Cumulative Reviews, as well as the supplements such as the InterAct Math Tutorial Web site and MathXL. The Cumulative Reviews and Final Examination for Chapters 1–10 can be used for practice.

1 g = 1 gram = the mass of 1 cm^3 (1 mL) of water

The metric prefixes for mass are the same as those used for length and capacity.

METRIC UNITS OF MASS

1 metric ton (t) = 1000 kilograms (kg)

1 *kilo*gram (kg) = 1000 grams (g)

1 *hecto*gram (hg) = 100 grams (g)

1 *deka*gram (dag) = 10 grams (g)

1 gram (g)

1 *deci*gram (dg) = $\frac{1}{10}$ gram (g)

1 *centi*gram (cg) = $\frac{1}{100}$ gram (g)

1 *milli*gram (mg) = $\frac{1}{1000}$ gram (g)

Hectograms, dekagrams, decigrams, and centigrams are only rarely used.

THINKING METRIC

The mass of 1 raisin or 1 paperclip or 1 package of "Splenda" sweetener is approximately 1 gram (g). One metric ton (t) is about 2200 lb, or about 10% more than 1 American ton (T). The metric ton is used for very large masses, such as vehicles; the kilogram is used for masses of people or larger food packages; the gram is used for smaller food packages or objects like a coin or a ring; the milligram is used for even smaller masses like a dosage of medicine.

Each 2.5 mg   2 g   15 g   2 kg   90 kg

Do Exercises 4–8.

CHANGING UNITS MENTALLY

As before, changing from one metric unit to another amounts to only the movement of a decimal point. Consider these equivalent masses.

> **MASS OF A STANDARD SHEET OF PAPER**
>
> 4260 mg = 426 cg = 4.26 g = 0.00426 kg

EXAMPLE 3 Complete: 8 kg = _____ g.

Think: A kilogram is 1000 times the mass of a gram. Thus we move the decimal point three places to the right.

8.0   8.000.   8 kg = 8000 g

EXAMPLE 4 Complete: 4235 g = _____ kg.

Think: There are 1000 grams in 1 kilogram. Thus we move the decimal point three places to the left.

4235.0   4.235.0   4235 g = 4.235 kg

EXAMPLE 5 Complete: 6.98 cg = _____ mg.

Think: One centigram has the mass of 10 milligrams. Thus we move the decimal point one place to the right.

6.98   6.9.8   6.98 cg = 69.8 mg

The following table is helpful in visualizing mental conversions.

1000 g	100 g	10 g	1 g	0.1 g	0.01 g	0.001 g
1 kg	1 hg	1 dag	1 g	1 dg	1 cg	1 mg

Larger ←——————————————————————→ Smaller

> The most commonly used metric units of mass are kg, g, and mg. We have intentionally used those more often than the others in the exercises.

Do Exercises 9–12.

Complete.

9. 6.2 kg = _____ g

10. 93.1 g = _____ kg

11. 7.7 cg = _____ mg

12. 2344 mg = _____ cg

Answers on page A-28

Convert to Celsius. Use a straightedge and the scales shown on this page. Approximate to the nearest ten degrees.

13. 180°F (Brewing coffee)

14. 25°F (Cold day)

15. −10°F (Extremely cold day)

Convert to Fahrenheit. Use a straightedge and the scales shown on this page. Approximate to the nearest ten degrees.

16. 25°C (Warm day)

17. 40°C (Temperature of a patient with a high fever)

18. 10°C (A cold bath)

C **Temperature**

ESTIMATED CONVERSIONS

Below are two temperature scales: **Fahrenheit** for American measure and **Celsius,** used internationally and in science.

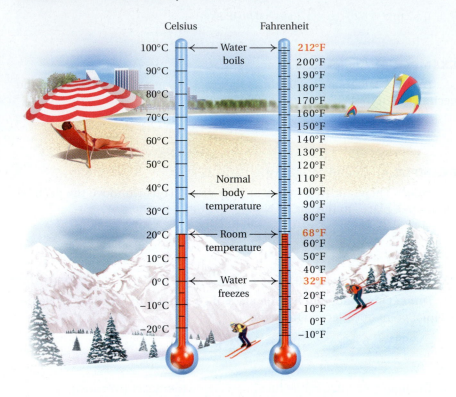

By laying a straightedge horizontally between the scales, we can make an approximate conversion from one measure of temperature to another and get an idea of how the temperature scales compare.

EXAMPLES Convert to Celsius (using the scales shown above). Approximate to the nearest ten degrees.

6. 212°F (Boiling point of water)   100°C   This is exact.
7. 32°F (Freezing point of water)   0°C   This is exact.
8. 105°F   40°C   This is approximate.

Do Exercises 13–15.

EXAMPLES Make an approximate conversion to Fahrenheit.

9. 44°C (Hot bath)   110°F   This is approximate.
10. 20°C (Room temperature)   68°F   This is exact.
11. 83°C   180°F   This is approximate.

Do Exercises 16–18.

EXACT CONVERSIONS

A formula allows us to make exact conversions from Celsius to Fahrenheit.

CELSIUS TO FAHRENHEIT

$$F = \frac{9}{5} \cdot C + 32, \quad \text{or} \quad F = 1.8 \cdot C + 32$$

$\left(\text{Multiply the Celsius temperature by } \dfrac{9}{5}, \text{ or } 1.8, \text{ and add } 32.\right)$

EXAMPLES Convert to Fahrenheit.

12. 0°C (Freezing point of water)

$$F = \frac{9}{5} \cdot C + 32 = \frac{9}{5} \cdot 0 + 32 = 0 + 32 = 32°$$

Thus, 0°C = 32°F.

13. 37°C (Normal body temperature)

$$F = 1.8 \cdot C + 32 = 1.8 \cdot 37 + 32 = 66.6 + 32 = 98.6°$$

Thus, 37°C = 98.6°F.

Check the answers to Examples 12 and 13 using the scales on p. 658.

Do Exercises 19 and 20.

A second formula allows us to make exact conversions from Fahrenheit to Celsius.

FAHRENHEIT TO CELSIUS

$$C = \frac{5}{9} \cdot (F - 32), \quad \text{or} \quad C = \frac{F - 32}{1.8}$$

$\left(\text{Subtract 32 from the Fahrenheit temperature and multiply by } \dfrac{5}{9} \text{ or}\right.$

divide by 1.8. $\Big)$

EXAMPLES Convert to Celsius.

14. 212°F (Boiling point of water)

$$C = \frac{5}{9} \cdot (F - 32)$$

$$= \frac{5}{9} \cdot (212 - 32)$$

$$= \frac{5}{9} \cdot 180 = 100°$$

Thus, 212°F = 100°C.

15. 77°F

$$C = \frac{F - 32}{1.8}$$

$$= \frac{77 - 32}{1.8}$$

$$= \frac{45}{1.8} = 25°$$

Thus, 77°F = 25°C.

Check the answers to Examples 14 and 15 using the scales on p. 658.

Do Exercises 21 and 22.

Convert to Fahrenheit.

19. 80°C

20. 35°C

Convert to Celsius.

21. 95°F

22. 113°F

Answers on page A-28

CALCULATOR CORNER

Temperature Conversions

Temperature conversions can be done quickly using a calculator. To convert 37°C to Fahrenheit, for example, we press $\boxed{1}\;\boxed{.}\;\boxed{8}\;\boxed{\times}\;\boxed{3}\;\boxed{7}$ $\boxed{+}\;\boxed{3}\;\boxed{2}\;\boxed{=}$. The calculator displays $\boxed{98.6}$, so 37°C = 98.6°F. We can convert 212°F to Celsius by pressing $\boxed{(}\;\boxed{2}\;\boxed{1}\;\boxed{2}\;\boxed{-}\;\boxed{3}\;\boxed{2}\;\boxed{)}$ $\boxed{\div}\;\boxed{1}\;\boxed{.}\;\boxed{8}\;\boxed{=}$. The display reads $\boxed{100}$, so 212°F = 100°C. Note that we must use parentheses when converting from Fahrenheit to Celsius in order to get the correct result.

Exercises: Use a calculator to convert each temperature to Fahrenheit.

1. 5°C

2. 50°C

Use a calculator to convert each temperature to Celsius.

3. 68°F

4. 113°F

---

a   Complete.

1. 1 lb = _____ oz

2. 1 T = _____ lb

3. 8000 lb = _____ T

4. 7 T = _____ lb

5. 3 lb = _____ oz

6. 10 lb = _____ oz

7. 4.5 T = _____ lb

8. 2.5 T = _____ lb

9. 4800 lb = _____ T

10. 7500 lb = _____ T

11. 72 oz = _____ lb

12. 960 oz = _____ lb

b   Complete.

13. 4 kg = _____ g

14. 9 kg = _____ g

15. 1 g = _____ kg

16. 1 dg = _____ g

17. 1 cg = _____ g

18. 1 mg = _____ g

19. 1 g = _____ mg

20. 1 g = _____ cg

21. 1 g = _____ dg

22. 57 kg = _____ g

23. 934 kg = _____ g

24. 678 g = _____ kg

25. 6345 g = _____ kg

26. 42.75 kg = _____ g

27. 897 mg = _____ kg

28. 45 cg = _____ g

29. 7.32 kg = _____ g

30. 0.439 cg = _____ mg

31. 9350 g = _____ kg

32. 5640 g = _____ kg

33. 69 mg = _____ cg

34. 76.1 mg = _____ cg

35. 8 kg = _____ cg

36. 0.02 kg = _____ mg

37. 1 t = _____ kg

38. 2 t = _____ kg

39. 3.4 cg = _____ dag

40. 9.34 g = _____ mg

C   Convert to Celsius. Round the answer to the nearest ten degrees. Use the scales on p. 658.

41. 178°F

42. 195°F

43. 140°F

44. 107°F

45. 68°F

46. 45°F

47. 10°F

48. 120°F

Convert to Fahrenheit. Round the answer to the nearest ten degrees. Use the scales on p. 658.

49. 86°C

50. 93°C

51. 58°C

52. 33°C

53. −10°C

54. −5°C

55. 5°C

56. 15°C

Convert to Fahrenheit. Use the formula $F = \frac{9}{5} \cdot C + 32$.

57. 30°C   **58.** 85°C   **59.** 40°C   **60.** 90°C

61. 3000°C (Melting point of iron)   **62.** 1000°C (Melting point of gold)

Convert to Celsius. Use the formula $C = \frac{5}{9} \cdot (F - 32)$.

63. 77°F   **64.** 59°F   **65.** 131°F   **66.** 140°F

67. 98.6°F (Normal body temperature)   **68.** 104°F (High-fevered body temperature)

69. *Highest Temperatures.* The highest temperature ever recorded in the world is 136°F in the desert of Libya in 1922. The highest temperature ever recorded in the United States is $56\frac{2}{3}$°C in California's Death Valley in 1913.

Source: *The Handy Geography Answer Book*

a) Convert each temperature to the other scale.
b) How much higher in degrees Fahrenheit was the world record than the U. S. record?

70. *Boiling Point and Altitude.* The boiling point of water actually changes with altitude. The boiling point is 212°F at sea level, but lowers about 1°F for every 500 ft that the altitude increases above sea level.

Sources: *The Handy Geography Answer Book; The New York Times Almanac*

a) What is the boiling point at an elevation of 1500 ft above sea level?
b) The elevation of Tucson is 2564 ft above sea level and that of Phoenix is 1117 ft. What is the boiling point in each city?
c) How much lower is the boiling point in Denver, whose elevation is 5280 ft, than in Tucson?
d) What is the boiling point at the top of Mt. McKinley in Alaska, the highest point in the United States, at 20,320 ft?

71. **D_W** Describe a situation in which one object weighs 70 kg, another object weighs 3 g, and a third object weighs 125 mg.

72. **D_W** Give at least two reasons why someone might prefer the use of grams to the use of ounces.

↪ **VOCABULARY REINFORCEMENT**

In each of Exercises 73–80, fill in the blank with the correct term from the given list. Some of the choices may not be used and some may be used more than once.

73. When interest is paid on interest, it is called _____ interest.   [8.6b]

74. A _____ is the quotient of two quantities.   [7.1a]

75. The _____ of a set of data is the middle number if there is an odd number of data items.   [6.5b]

76. When you work for a _____, you are paid a percentage of the total sales for which you are responsible. [8.5b]

77. In _____ triangles, the lengths of their corresponding sides have the same ratio.   [7.5a]

78. To find the _____ of a set of data, add the numbers and then divide by the number of items of data. [6.5a]

79. A natural number, other than 1, that is not prime is _____ .   [3.2b]

80. The prefix _____ means 1000.   [9.1b]

| |
| mean |
| median |
| mode |
| complementary |
| composite |
| commission |
| salary |
| ratio |
| centi- |
| kilo- |
| simple |
| compound |
| isosceles |
| similar |

SYNTHESIS

81. **D**_W Near the Canadian border, a radio forecast calls for an overnight low of 60°. Was the temperature given in Celsius or Fahrenheit? Explain how you can tell.

82. **D**_W Which represents a bigger change in temperature: a drop of 5°F or a drop of 5°C? Why?

Complete. Use 453.6 g = 1 lb. Round to four decimal places.

83. ▦ 1 lb = _____ kg

84. ▦ 1 g = _____ lb

85. *Large Diamonds.*   A **carat** (also spelled **karat**) is a unit of weight for precious stones; 1 carat = 200 mg. The Golden Jubilee Diamond weighs 545.67 carats and is the largest cut diamond in the world. The Hope Diamond, located at the Smithsonian Institution Museum of Natural History, weighs 45.52 carats.
   Source: *National Geographic,* February 2001

   a) How many grams does the Golden Jubilee Diamond weigh?
   b) How many grams does the Hope Diamond weigh?
   c) ▦ Given that 1 lb = 453.6 g, how many ounces does each diamond weigh?

86. *Chemistry.* Another temperature scale often used is the **Kelvin** scale. Conversions from Celsius to Kelvin can be carried out using the formula

$$K = C + 273.$$

A chemistry textbook describes an experiment in which a reaction takes place at a temperature of 400° Kelvin. A student wishes to perform the experiment, but has only a Fahrenheit thermometer. At what Fahrenheit temperature will the reaction take place?

87. 🖩 A large egg is about $5\frac{1}{2}$ cm tall with a diameter of 4 cm. Estimate the mass of such an egg by averaging the volumes of two spheres. (*Hint*: 1 cc of water has a mass of 1 g.)

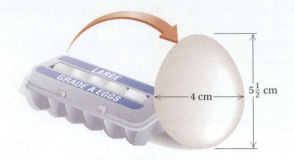

88. Use the formula $F = \frac{9}{5} \cdot C + 32$ to find the temperature that is the same for both the Fahrenheit and Celsius scales.

89. 🖩 *Track and Field.* A woman's shot put weighs 8.8 lb and has a 4.5-in. diameter. Find its mass per cubic centimeter, given that 1 lb = 453.6 g.

Source: National Collegiate Athletic Association

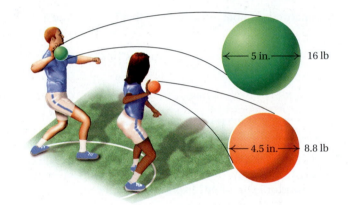

90. 🖩 *Track and Field.* A man's shot put weighs 16 lb and has a 5-in. diameter. Find its mass per cubic centimeter, given that 1 lb = 453.6 g.

91. 🖩 A shrink-wrapped brick of pudding boxes weighs $15\frac{17}{20}$ lb. If each box of pudding weighs $1\frac{3}{4}$ oz, how many boxes are in the brick and how much does the plastic shrink wrap weigh?

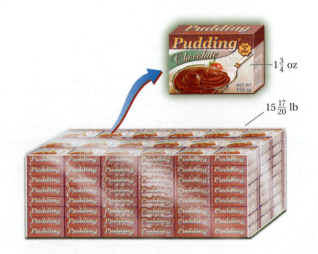

9.8

MEDICAL APPLICATIONS

a Measurements and Medicine

Measurements play a critical role in health care. Doctors, nurses, aides, technicians, and others all need to use the proper units and perform the proper calculations to assure the best possible care of patients.

Because of the ease with which conversions can be made and its extensive use in science—among other reasons—the metric system is the primary system of measurement in medicine.

EXAMPLE 1 *Medical Dosage.* A physician orders 3.5 L of 5% dextrose in water (abbrev. D5W) to be administered over a 24-hr period. How many milliliters were ordered?

We convert 3.5 L to milliliters:

$$3.5\,\text{L} = 3.5 \cdot 1\,\text{L}$$
$$= 3.5 \cdot 1000\,\text{mL} \qquad \text{Substituting}$$
$$= 3500\,\text{mL}.$$

The physician had ordered 3500 mL of D5W.

Do Exercise 1.

Liquids at a pharmacy are often labeled in liters or milliliters. Thus if a physician's prescription is given in ounces, it must be converted. For conversion, a pharmacist knows that 1 oz ≈ 29.57 mL.*

EXAMPLE 2 *Prescription Size.* A prescription calls for 3 oz of theophylline, a drug commonly used for children with asthma. For how many milliliters is the prescription?

We convert as follows:

$$3\,\text{oz} = 3 \cdot 1\,\text{oz}$$
$$\approx 3 \cdot 29.57\,\text{mL} \qquad \text{Substituting}$$
$$= 88.71\,\text{mL}.$$

The prescription calls for 88.71 mL of theophylline.

Do Exercise 2.

---

*In practice, most physicians use 30 mL as an approximation to 1 oz.

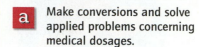

a Make conversions and solve applied problems concerning medical dosages.

1. **Medical Dosage.** A physician orders 2400 mL of 0.9% saline solution to be administered intravenously over a 24-hr period. How many liters were ordered?

2. **Prescription Size.** A prescription calls for 2 oz of theophylline.
   a) For how many milliliters is the prescription?
   b) For how many liters is the prescription?

Answers on page A-28

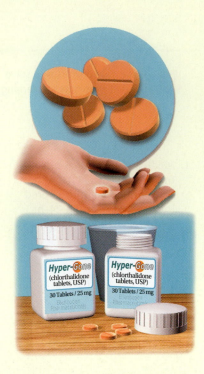

EXAMPLE 3 *Pill Splitting.* Chlorthalidone is a commonly prescribed drug used to treat hypertension. Tania's physician directs her to reduce her dosage from 25 mg to 12.5 mg. Tania's original prescription contained 30 tablets, each 25 mg.

a) How many milligrams of chlorthalidone, in total, were in the original prescription?

b) How many 12.5-mg doses can Tania obtain from the original prescription?

a) The original prescription contained 30 tablets, each containing 25 mg of chlorthalidone. To find the total amount in the prescription, we multiply:

30 tablets · 25 mg/tablet = 750 mg.

The original prescription contained 750 mg of chlorthalidone.

b) Since 12.5 mg is half of 25 mg (25 ÷ 2 = 12.5), Tania's original 30 doses can each be split in half, yielding 30 · 2, or 60, doses at 12.5 mg per dose.

Do Exercise 3.

Another metric unit that is used in medicine is the microgram (mcg). It is defined as follows.

> **MICROGRAM**
>
> $$1 \text{ microgram} = 1 \text{ mcg} = \frac{1}{1,000,000} \text{ g} = 0.000001 \text{ g}$$
>
> $$1,000,000 \text{ mcg} = 1 \text{ g}$$

One microgram is one-millionth of a gram, so one million micrograms is one gram. A microgram is also one-thousandth of a milligram, so one thousand micrograms is one milligram.

EXAMPLE 4 Complete: 1 mg = _____ mcg.

We convert to grams and then to micrograms:

$$1 \text{ mg} = 0.001 \text{ g}$$
$$= 0.001 \cdot 1 \text{ g}$$
$$= 0.001 \cdot 1,000,000 \text{ mcg} \qquad \text{Substituting 1,000,000 mcg for 1 g}$$
$$= 1000 \text{ mcg}.$$

Do Exercise 4.

3. Pill Splitting. If Tania's physician originally prescribed 14 tablets that were each 25 mg, how many milligrams were in the original prescription? How many 12.5-mg doses could be obtained from the original prescription?

4. Complete:
3 mg = _____ mcg.

Answers on page A-28

EXAMPLE 5 *Medical Dosage.* Nitroglycerin sublingual tablets come in 0.4-mg tablets. How many micrograms are in each tablet?

Source: Steven R. Smith, M.D.

We are to complete: 0.4 mg = _____ mcg. Thus,

$$0.4 \text{ mg} = 0.4 \cdot 1 \text{ mg}$$
$$= 0.4 \cdot 1000 \text{ mcg} \qquad \text{Substituting 1000 mcg for 1 mg}$$
$$= 400 \text{ mcg}.$$

We can also do this problem in a manner similar to Example 4.

Do Exercise 5.

5. Medical Dosage. A physician prescribes 500 mcg of alprazolam, an antianxiety medication. How many milligrams is this dosage?

Source: Steven R. Smith, M.D.

Answer on page A-28

Study Tips

Although this does not qualify as a Study Tip, you can use the information to motivate your study of mathematics. The book Best Jobs for the 21st Century, 3rd edition, developed by Michael Farr with database work by Laurence Shatkin, Ph.D, lists 500 jobs, based on overall scores for pay, growth rate through 2010, and number of annual openings. Note that the use of mathematics is significant in most of the 20 top jobs.

THE 20 BEST JOBS

JOB	ANNUAL EARNINGS	GROWTH RATE THROUGH 2010	ANNUAL OPENINGS
1. Computer Software Engineers, Applications	$70,210	100.0%	28,000
2. Computer Systems Analysts	$61,990	59.7%	34,000
3. Computer and Information Systems Managers	$82,480	47.9%	28,000
4. Teachers, Postsecondary	$52,115	23.5%	184,000
5. Management Analysts	$57,970	28.9%	50,000
6. Registered Nurses	$46,670	25.6%	140,000
7. Computer Software Engineers, Systems Software	$73,280	89.7%	23,000
8. Medical and Health Services Managers	$59,220	32.3%	27,000
9. Sales Agents, Financial Services	$59,690	22.3%	55,000
10. Sales Agents, Securities and Commodities	$59,690	22.3%	55,000
11. Securities, Commodities, and Financial Services Sales Agents	$59,690	22.3%	55,000
12. Computer Support Specialists	$38,560	97.0%	40,000
13. Sales Managers	$71,620	32.8%	21,000
14. Computer Security Specialists	$53,770	81.9%	18,000
15. Network and Computer Systems Administrators	$53,770	81.9%	18,000
16. Financial Managers	$70,210	18.5%	53,000
17. Financial Managers, Branch or Department	$70,210	18.5%	53,000
18. Treasurers, Controllers, and Chief Financial Officers	$70,210	18.5%	53,000
19. Accountants	$45,380	18.5%	100,000
20. Accountants and Auditors	$45,380	18.5%	100,000

a *Medical Dosage.* Solve each of the following. (None of these medications should be taken without consulting your own physician.)

1. An emergency-room physician orders 2.0 L of Ringer's lactate to be administered over 2 hr for a patient in shock. How many milliliters is this?

2. Ingrid receives 84 mL per hour of normal saline solution. How many liters did Ingrid receive in a 24-hr period?

3. To battle hypertension and prostate enlargement, Rick is directed to take 4 mg of doxazosin each day for 30 days. How many grams is this?

4. To battle high cholesterol, Kit is directed to take 40 mg of atorvastatin for 60 days. How many grams is this?

5. Cephalexin is an antibiotic that frequently is prescribed in a 500-mg tablet form. Dr. Bouvier prescribes 2 g of cephalexin per day for a patient with a skin abscess. How many 500-mg tablets would have to be taken in order to achieve this daily dosage?

6. Quinidine gluconate is a liquid mixture, part medicine and part water, which is administered intravenously. There are 80 mg of quinidine gluconate in each cubic centimeter (cc) of the liquid mixture. Dr. Nassat orders 500 mg of quinidine gluconate to be administered daily to a patient with malaria. How much of the solution would have to be administered in order to achieve the recommended daily dosage?

7. Albuterol is a medication used for the treatment of asthma. It comes in an inhaler that contains 17 mg albuterol mixed with a liquid. One actuation (inhalation) from the mouthpiece delivers a 90-mcg dose of albuterol.

   a) Dr. Martinez orders 2 inhalations 4 times per day. How many micrograms of albuterol does the patient inhale per day?

   b) How many actuations/inhalations are contained in one inhaler?

   c) Danielle is going away for 4 months of college and wants to take enough albuterol to last for that time. Her physician has prescribed 2 inhalations 4 times per day. Estimate how many inhalers Danielle will need to take with her for the 4-month period.

8. Amoxicillin is a common antibiotic prescribed for children. It is a liquid suspension composed of part amoxicillin and part water. In one formulation of amoxicillin suspension, there are 250 mg of amoxicillin in 5 cc of the liquid suspension. Dr. Scarlotti prescribes 400 mg per day for a 2-year-old child with an ear infection. How much of the amoxicillin liquid suspension would the child's parent need to administer in order to achieve the recommended daily dosage of amoxicillin?

9. Dr. Norris tells a patient to purchase 0.5 L of hydrogen peroxide. Commercially, hydrogen peroxide is found on the shelf in bottles that hold 4 oz, 8 oz, and 16 oz. Which bottle comes closest to filling the prescription?

10. Dr. Lopez wants a patient to receive 3 L of a normal glucose solution in a 24-hr period. How many milliliters per hour should the patient receive?

11. Remeron® is a commonly prescribed antianxiety drug. Joanne has a 14-tablet supply of 30-mg tablets and her physician now directs her to reduce her dosage to 15 mg.

  a) How many grams were originally prescribed?
  b) How many 15-mg doses can Joanne obtain from the original prescription?

12. Serzone® is a commonly prescribed antidepressant. Chad has a 90-tablet supply of 200-mg tablets, when his physician directs him to cut his dosage size to 100 mg.

  a) How many grams are in Chad's current supply?
  b) How many 100-mg doses can Chad obtain from his supply?

13. Amoxicillin is an antibiotic obtainable in a liquid suspension form, part medication and part water, and is frequently used to treat infections in infants. One formulation of the drug contains 125 mg of amoxicillin per 5 mL of liquid. A pediatrician orders 150 mg per day for a 4-month-old child with an ear infection. How much of the amoxicillin suspension would the parent need to administer to the infant in order to achieve the recommended daily dose?

14. Diphenhydramine HCL is an antihistamine available in liquid form, part medication and part water. One formulation contains 25 mg of medication in 5 mL of liquid. An allergist orders 40-mg doses for a high school student. How many milliliters should be in each dose?

Complete.

15. 1 mg = _____ mcg

16. 1 mcg = _____ mg

17. 325 mcg = _____ mg

18. 0.45 mg = _____ mcg

19. Dr. Djihn prescribes 0.25 mg of alprazolam, an antianxiety medication. How many micrograms are in this dose?

20. Dr. Kramer prescribes 0.4 mg of alprazolam, an antianxiety medication. How many micrograms are in this dose?

21. Digoxin is a medication used to treat heart problems. A cardiologist orders 0.125 mg of digoxin to be taken once daily. How many micrograms of digoxin are there in the daily dosage?

22. Digoxin is a medication used to treat heart problems. An internist orders 0.25 mg of digoxin to be taken once a day. How many micrograms of digoxin are there in the daily dosage?

23. Triazolam is a medication used for the short-term treatment of insomnia. A physician advises her patient to take one of the 0.125-mg tablets each night for 7 nights. How many milligrams of triazolam will the patient have ingested over that 7-day period? How many micrograms?

24. Clonidine is a medication used to treat high blood pressure. The usual starting dose of clonidine is one 0.1-mg tablet twice a day. If a patient is started on this dose by his physician, how many total milligrams of clonidine will the patient have taken before he returns to see his physician 14 days later? How many micrograms?

25. **D**_W A nursing student is reprimanded for moving a decimal point one place to the right instead of one place to the left. Why was this considered such a serious mistake?

26. **D**_W Why might a patient want to purchase 200-mg tablets and split them in half rather than simply purchase 100-mg tablets?

SKILL MAINTENANCE

Subtract. [1.3d]

27.   5 7 8 9
       − 2 4 3 1

28.   8 4 2 9
       − 1 0 1 5

29.   4 0 9 7
       − 3 2 4 3

30.   8 3 9 0
       − 2 0 5 6

Simplify. [2.7a]

31. $7x + 9 - 2x - 1$

32. $8x + 12 - 2x - 7$

33. $8t - 5 - t - 4$

34. $9r - 6 - r - 4$

SYNTHESIS

35. **D**_W Describe a situation in which someone would need to convert from kilograms to micrograms.

36. **D**_W Alan's dose is being reduced, yet the size of the tablets in the prescription is being increased. Why do you think this might be?

37. ▦ A patient is directed to take 200 mg of Serzone® 3 times a day for one week, then 200 mg twice a day for a week, and then 100 mg three times a day for a week.

   a) How many grams of medication are used altogether?
   b) What is the average dosage size?

38. ▦ A patient is directed to take 200 mg of Wellbutrin® twice a day for a week, then 200 mg in the evening and 100 mg in the morning for a week, and then 100 mg twice a day for a week.

   a) How many grams of medication are used altogether?
   b) What is the average dosage size?

39. Naproxen sodium, sometimes sold under the brand name Aleve, is a painkiller that lasts approximately 12 hours. A typical dose is one 220-mg tablet. Ibuprofen is a similar painkiller that lasts about 6 hours and has a typical dosage of two 200-mg tablets. What would be a better purchase: a bottle containing 44 g of naproxen sodium costing $11, or a bottle containing 72 g of ibuprofen costing $11.24? Why?

40. **D**_W Why might someone ignore the answer to Exercise 39 when shopping for naproxen sodium or ibuprofen?

The review that follows is meant to prepare you for a chapter exam. It consists of three parts. The first part, Concept Reinforcement, is designed to increase understanding of the concepts through true/false exercises. The second part is a list of important properties and formulas. The third part is the Review Exercises. These provide practice exercises for the exam, together with references to section objectives so you can go back and review. Before beginning, stop and look back over the skills you have obtained. What skills in mathematics do you have now that you did not have before studying this chapter?

↪ CONCEPT REINFORCEMENT

Determine whether the statement is true or false. Answers are given at the back of the book.

_____ **1.** Distances that are measured in miles in the American system would probably be measured in meters in the metric system.

_____ **2.** One meter is slightly more than one yard.

_____ **3.** When converting from meters to centimeters we move the decimal point to the right.

_____ **4.** To convert mm^2 to cm^2, move the decimal point 2 places to the left.

_____ **5.** Since 1 yd = 3 ft, we multiply by 3 to convert square yards to square feet.

_____ **6.** You would probably use your furnace when the temperature is 40°C.

IMPORTANT PROPERTIES AND FORMULAS

Area of a Parallelogram: $\qquad\quad A = b \cdot h$

Area of a Trapezoid: $\qquad\qquad A = \dfrac{1}{2} \cdot h \cdot (a + b)$

Radius and Diameter of a Circle: $\quad d = 2 \cdot r$, or $r = \dfrac{d}{2}$

Circumference of a Circle: $\qquad C = \pi \cdot d$, or $C = 2 \cdot \pi \cdot r$

Area of a Circle: $\qquad\qquad\quad A = \pi \cdot r \cdot r$, or $A = \pi \cdot r^2$

Volume of a Rectangular Solid: $\quad V = l \cdot w \cdot h$

Volume of a Circular Cylinder: $\quad V = \pi \cdot r^2 \cdot h$

Volume of a Sphere: $\qquad\qquad V = \frac{4}{3} \cdot \pi \cdot r^3$

Pythagorean Equation: $\qquad\quad a^2 + b^2 = c^2$

Temperature Conversion: $\qquad F = \frac{9}{5} \cdot C + 32; \quad C = \frac{5}{9} \cdot (F - 32)$

See tables inside chapter for units of length, weight, mass, and capacity.

Review Exercises

Complete.

1. 10 ft = _____ yd
[9.1a]

2. $\frac{5}{6}$ yd = _____ in.
[9.1a]

3. 1.7 mm = _____ cm
[9.1b]

4. 6 m = _____ km
[9.1b]

5. 4 km = _____ cm
[9.1b]

6. 14 in. = _____ ft
[9.1a]

7. 5 lb = _____ oz
[9.7a]

8. 3 g = _____ kg
[9.7b]

9. 50 qt = _____ gal
[9.4b]

10. 28 gal = _____ pt
[9.4b]

11. 60 mL = _____ L
[9.4b]

12. 0.4 L = _____ mL
[9.4b]

13. 0.7 T = _____ lb
[9.7a]

14. 0.2 g = _____ mg
[9.7b]

15. 4.7 kg = _____ g
[9.7b]

16. 4 cg = _____ g
[9.7b]

17. 4 yd^2 = _____ ft^2
[9.2a]

18. 0.7 km^2 = _____ m^2
[9.2b]

19. 1008 in^2 = _____ ft^2
[9.2a]

20. 570 cm^2 = _____ m^2
[9.2b]

21. Find the circumference of a circle of radius 5 m. Use 3.14 for π. [9.3b]

22. Find the length of a radius of the circle. [9.3b]

$\frac{28}{11}$ in.

23. Find the length of a diameter of the circle. [9.3b]

12 m

24. *Track and Field.* Track meets are held on a track similar to the one shown below. Find the shortest distance around the track. [9.3b]

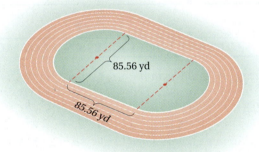

85.56 yd

85.56 yd

Find the area of each figure in Exercises 25–30. [9.3a, b]

25.

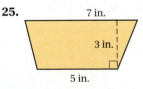

7 in.

3 in.

5 in.

26.

4.2 ft

7.1 ft

27.

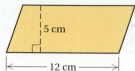

5 cm

12 cm

28.

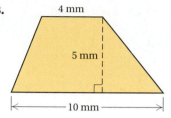

4 mm

5 mm

10 mm

29. Use $\frac{22}{7}$ for π.

7 ft

30. Use 3.14 for π.

10 cm

31. Find the area of the shaded region. Use 3.14 for π. [9.3b]

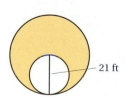

21 ft

32. A "Norman" window is designed with dimensions as shown. Find its area. Use 3.14 for π. [9.3b]

2 ft

5 ft

33. Find the measure of a complement and of a supplement of $\angle BAC$. [9.5c]

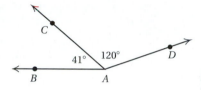

C

41° 120° D

B A

34. List two pairs of vertical angles. [9.5c]

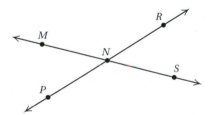

R

M

N

S

P

Use the following triangle for Exercises 35–37.

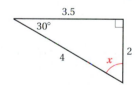

3.5

30°

4

2

x

35. Find the missing angle measure. [9.5e]

36. Classify the triangle as equilateral, isosceles, or scalene. [9.5d]

37. Classify the triangle as right, obtuse, or acute. [9.5d]

Find the volume of each figure. Use 3.14 for π. [9.4a]

38.

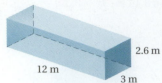

12 m   2.6 m   3 m

39.

14 ft
3 ft   4.6 ft

40.

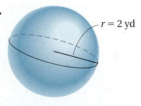

90 cm
10 cm

41.

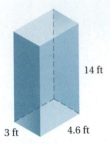

$r = 2$ yd

42.

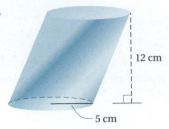

12 cm
5 cm

43. Simplify: $\sqrt{64}$. [9.6a]

For each right triangle, find the length of the side not given. Find an exact answer and an approximation to three decimal places. Assume that c represents the length of the hypotenuse. [9.6c]

44. $a = 15$, $b = 25$ **45.** $a = 4$, $c = 10$

46.

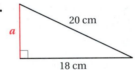

c 5 ft
8 ft

47.

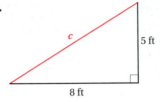

20 cm
a
18 cm

48. How tall is this tree? [9.6d]

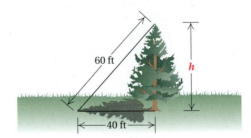

60 ft h
40 ft

49. Convert 35°C to Fahrenheit. [9.7c]

50. Convert 68°F to Celsius. [9.7c]

Medical Dosage. Solve.   [9.8a]

51. Amoxicillin is an antibiotic obtainable in a liquid suspension form, part medication and part water, and is frequently used to treat infections in infants. One formulation of the drug contains 125 mg of amoxicillin per 5 mL of liquid. A pediatrician orders 200 mg per day for a child with an ear infection. How much of the amoxicillin suspension would the parent need to administer to the child in order to achieve the recommended daily dose?

52. An emergency-room physician orders 3 L of Ringer's lactate to be administered over 4 hr for a patient suffering from shock and severe low blood pressure. How many milliliters is this?

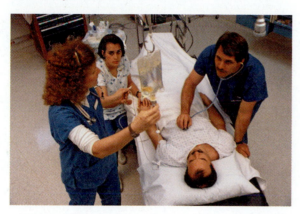

53. Dr. Tanner prescribes 0.5 mg of alprazolam, an antianxiety medication. How many micrograms are in this dose?

54. **D_W** Is a square a special type of parallelogram? Why or why not?   [9.3a]

55. **D_W** Is it possible for a triangle to contain two 90° angles? Why or why not?   [9.5e]

56. **D_W** Which is a larger measure of volume: 1 m^3 or 27 ft^3? Explain how you can tell without using a calculator. [9.1c], [9.4a]

57. **D_W** What weighs more: 32 oz or 1 kg? Explain how you can tell without using a calculator.   [9.7a, b]

58. ▦ Find the area of the largest round pizza that can be baked in a 35-cm by 50-cm pan.   [9.3b]

59. ▦ One lap around a standard running track is 440 yd. A marathon is 26 mi, 385 yd long. How many laps around a track does a marathon require?   [9.1a]

60. Lumber that starts out at a certain measure must be trimmed to take out warps and get boards that are straight. Because of trimming, a "two-by-four" is trimmed to an actual size of $1\frac{1}{2}$ in. by $3\frac{1}{2}$ in. What percent of the wood in an untrimmed 10-ft two-by-four is lost by trimming?   [9.4c]

61. A community center has a rectangular swimming pool that is 50 ft wide, 100 ft long, and 10 ft deep. The center decides to fill the pool with water to a line that is 1 ft from the top. Water costs \$2.25 per 1000 ft^3. How much does it cost to fill the pool?   [9.4c]

675

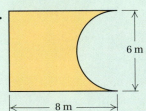
Complete.

1. 8 ft = _____ in.

2. 280 cm = _____ m

3. 2 yd^2 = _____ ft^2

4. 5 km = _____ m

5. 9.1 mm = _____ cm

6. 4520 m^2 = _____ km^2

7. 2983 mL = _____ L

8. 3.8 kg = _____ g

9. 10 gal = _____ oz

10. 0.69 L = _____ mL

11. 9 lb = _____ oz

12. 4.11 T = _____ lb

13. Find the length of a radius of this circle.

16 cm

14. Find the area of a circle of radius 4 m. Use 3.14 for π.

15. Find the circumference of a circle of radius 14 ft. Use $\frac{22}{7}$ for π.

Find each area. Use 3.14 for π.

16.

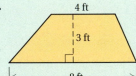

2.5 cm

10 cm

17.

6 m

8 m

18.

4 ft

3 ft

8 ft

19. *Scrapbooking.* A 3-in.–diameter circle is cut from a photo and mounted in the center of a 4-in.–diameter circle of red card stock. What is the area of the red card stock that borders the photo?

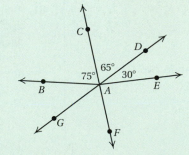

← 3 in. →

← 4 in. →

20. Find the measure of a supplement of $\angle CAD$.

21. Find the measure of $\angle GAF$.

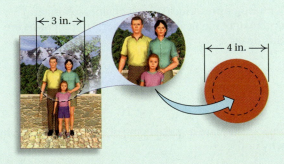

C

D

65°

75°   30°

B   A   E

G

F

22. A twelve-box carton of 12-oz juice boxes comes in a rectangular box $10\frac{1}{2}$ in. by 8 in. by 5 in. What is the volume of the carton?

In Exercises 23–25 find the volume in each figure. Use 3.14 for π.

23.

5 ft

8 ft

24.

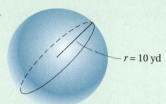

$r = 10$ yd

25.

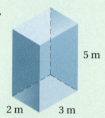

5 m

2 m   3 m

26. Dr. Pietrofiro wants a patient to receive 0.5 L of a dextrose solution every 8 hr. How many milliliters will the patient have received after one 48-hr period?

For each right triangle, find the length of the side not given. Find an exact answer and an approximation to three decimal places.

27.

c

1

1

28.

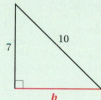

7   10

b

29. Find $\sqrt{121}$.

30. Convert 32°F to Celsius.

31. An antihistamine solution contains 25 mg of medication in 5 mL of liquid. Pat is directed to take a 30-mg dose of the antihistamine. How many mL's of the solution should Pat take?

SYNTHESIS

32. The measure of ∠SMC is three times that of its complement. Find the measure of the supplement of ∠SMC.

33. 🖩 A *board foot* is the amount of wood in a piece 12 in. by 12 in. by 1 in. A carpenter places the following order for a certain kind of lumber:

> 25 pieces: 2 in. by 4 in. by 8 ft;
> 32 pieces: 2 in. by 6 in. by 10 ft;
> 24 pieces: 2 in. by 8 in. by 12 ft.

The price of this type of lumber is $225 per thousand board feet. What is the total cost of the carpenter's order?

Find the volume of the solid. (Note that the solids are not drawn in perfect proportion.) Give the answer in cubic feet. Use 3.14 for π.

34.

12 ft   2.6 in.
3 in.

35.

18 ft

$\frac{3}{4}$ in.

Cumulative Review

Perform the indicated operations and, if possible, simplify.

1. $4\dfrac{2}{3} + 5\dfrac{1}{2}$

2. $\left(\dfrac{1}{4}\right)^2 \div \left(\dfrac{1}{2}\right)^3 \times 2^4 + (10.3)(4)$

3. $120.5 - 32.98$

4. $-27{,}148 \div 22$

5. $14 \div [33 \div 11 + 8 \times 2 - (15 - 3)]$

6. $8^3 + 45 \cdot 24 - 9^2 \div 3$

7. Write 1.2 as an equivalent fraction. Simplify, if possible.

8. Write $\dfrac{9}{20}$ as an equivalent percent.

Use $<$, $>$, or $=$ for $\square$ to write a true sentence.

9. $\dfrac{5}{6} \; \square \; \dfrac{7}{8}$

10. $\dfrac{5}{12} \; \square \; \dfrac{3}{10}$

Complete.

11. $6\,\text{oz} = \underline{\hspace{1cm}} \text{lb}$

12. $100°\text{C} = \underline{\hspace{1cm}} °\text{F}$

13. $0.087\,\text{L} = \underline{\hspace{1cm}} \text{mL}$

14. $2.5\,\text{yd} = \underline{\hspace{1cm}} \text{in.}$

15. $3\,\text{yd}^2 = \underline{\hspace{1cm}} \text{ft}^2$

16. $37\,\text{cm} = \underline{\hspace{1cm}} \text{m}$

17. Find the perimeter and the area.

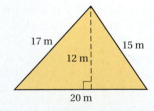

17 m   15 m
12 m
20 m

18. Combine like terms: $12a - 7 - 3a - 9$.

19. Graph: $y = -\dfrac{1}{3}x + 2$.

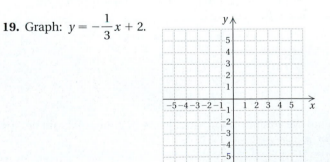

The following graph records Vermont Medicaid spending on clients of the Department of Prevention, Assistance, Transition and Health Access (PATH).

Source: *Burlington Free Press,* 1/5/03

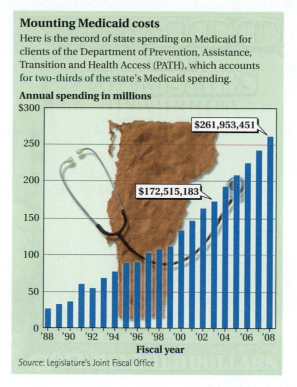

Mounting Medicaid costs

Here is the record of state spending on Medicaid for clients of the Department of Prevention, Assistance, Transition and Health Access (PATH), which accounts for two-thirds of the state's Medicaid spending.

Annual spending in millions

$261,953,451

$172,515,183

Fiscal year

Source: Legislature's Joint Fiscal Office

20. Estimate the Medicaid spent on PATH clients in 1998.

21. In what year was Medicaid spending on PATH clients first projected to exceed $200 million?

Solve.

22. $\dfrac{12}{15} = \dfrac{x}{18}$

23. $1 - 7x = 4 - (x + 9)$

24. $-15x = 280$

25. $x + \dfrac{3}{4} = \dfrac{7}{8}$

26. *Tourist Spending.* Foreign tourists spend $79.3 billion in this country annually. The most money, $18.2 billion, is spent in Florida. What is the ratio of amount spent in Florida to total amount spent? What is the ratio of total amount spent to amount spent in Florida?

Source: Travel Industry Association of America, 2000

27. Find the rate of discount.

Cookware Set **$58⁹⁹**

Lowest Price of the Year
Original price $100

28. Find the mean: 49, 53, 60, 62, 69.

29. What is the simple interest on $800 at 12% for $\frac{1}{4}$ year?

30. How long must a rope be in order to reach from the top of an 8-m tree to a point on the ground 15 m from the bottom of the tree?

31. The sales tax on a purchase of $5.50 is $0.33. What is the sales tax rate?

32. A bolt of fabric in a fabric store has $10\frac{3}{4}$ yd on it. A customer purchases $8\frac{5}{8}$ yd. How many yards remain on the bolt?

33. What is the cost, in dollars, of 15.6 gal of gasoline at 178.9¢ per gallon? Round to the nearest cent.

34. A box of powdered milk that makes 20 qt costs $4.99. A box that makes 8 qt costs $1.99. Which size has the lower unit price?

35. It is $\frac{7}{10}$ km from Ida's dormitory to the library. She starts to walk there, changes her mind after going $\frac{1}{4}$ of the distance, and returns home. How far did Ida walk?

Consider the following figure.

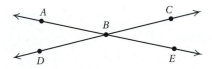

36. If the measure of $\angle ABC$ is 148°, find the measure of $\angle DBE$.

37. If the measure of $\angle ABD$ is 43°, find the measure of $\angle DBE$.

SYNTHESIS

38. A house sits on a lot measuring 75 ft by 200 ft. The lot is a corner lot, and includes sidewalks on two edges of the lot. If the sidewalks are 3 ft wide and 4 in. of snow falls, what volume of snow must be shoveled?

39. The U.S. Postal Service will not ship a box if the sum of the box's lengthwise perimeter and widthwise perimeter exceeds 108 in. Will a 1-ft by 2-ft by 3-ft box be accepted for shipping? Support your answer mathematically.

Polynomials

Real-World Application

The number of degrees granted in foreign languages and literature x years after 1970 can be estimated by the polynomial

$$-x^3 + 80x^2 - 1723x + 22669.$$

Determine the number of degrees in foreign languages and literature in 2005 ($x = 35$).

Source: Based on information from U.S. National Center for Education Statistics, *Digest of Education Statistics,* annual

This problem appears as Exercise 65 in Exercise Set 10.1.

Objectives

a Add polynomials.

b Find the opposite of a polynomial.

c Subtract polynomials.

d Evaluate a polynomial.

ADDITION AND SUBTRACTION OF POLYNOMIALS

In Section 2.7, we defined a *term* as a number, a variable, a product of numbers and/or variables, or a quotient of numbers and/or variables. Thus, expressions like

$$5x^2, \quad -34, \quad \frac{3}{4}ab^2, \quad xy^3z^5, \quad \text{and} \quad \frac{7n}{m}$$

are terms. A term is called a **monomial** if there is no division by a variable expression. Thus all of the terms above, except for $\frac{7n}{m}$, are monomials. Monomials are used to form **polynomials** like the following:

$$a^2b + c^3, \quad 5y + 3, \quad 3x^2 + 2x - 5, \quad -7a^3 + \tfrac{1}{2}a, \quad 37p^4, \quad x, \quad 0.$$

POLYNOMIAL

A **polynomial** is a monomial or a combination of sums and/or differences of monomials.

The following algebraic expressions are *not* polynomials, because each involves division by a variable expression:

$$\frac{x + 3}{x - 4}, \quad 5x^3 - 2x^2 + \frac{1}{x}, \quad \frac{1}{x^3 - 2}.$$

In a monomial, the number multiplied by the variable or variables is called the **coefficient**. The coefficient of $-6a^2b$ is -6. The coefficient of $\frac{1}{3}x^4$ is $\frac{1}{3}$.

a Adding Polynomials

Recall that the commutative and associative laws are often used to make addition easier to perform. For example,

$$(9 + 17) + (1 + 13)$$

can be rewritten as the equivalent expression

$$(9 + 1) + (17 + 13), \quad \text{or} \quad 10 + 30.$$

A similar approach can be used for adding polynomials. Recall that when two terms have the same variable(s) raised to the same power(s), they are "like" terms and can be combined.

EXAMPLE 1 Add: $(5x^3 + 4x^2 + 3x) + (2x^3 + 5x^2 - x)$.

$(5x^3 + 4x^2 + 3x) + (2x^3 + 5x^2 - x)$

$= (5x^3 + 2x^3) + (4x^2 + 5x^2) + (3x - x)$ Using the commutative and associative laws to pair like terms

$= 7x^3 + 9x^2 + 2x$ Combining like terms. Remember that x means $1x$.

Often, the terms do not all form pairs of like terms.

EXAMPLE 2 Add: $(3a^2 + 7a^2b) + (5a^2 - 6ab^2)$.

$$(3a^2 + 7a^2b) + (5a^2 - 6ab^2) = (3a^2 + 5a^2) + 7a^2b - 6ab^2$$
$$= 8a^2 + 7a^2b - 6ab^2 \qquad \text{Combining like terms}$$

EXAMPLE 3 Add: $(7x^2 + 5) + (5x^3 + 4x)$.

$$(7x^2 + 5) + (5x^3 + 4x) = 7x^2 + 5 + 5x^3 + 4x \qquad \text{There are no like terms here.}$$
$$= 5x^3 + 7x^2 + 4x + 5 \qquad \text{Rearranging the order}$$

Note that in Example 3 we wrote the answer so that the powers of x decrease as we read from left to right. This **descending order** is the traditional way of expressing an answer.

Do Exercises 1–3.

b ┃ Opposites of Polynomials

To subtract a number, we can add its opposite. We can similarly subtract a polynomial by adding its opposite. To check if two polynomials are opposites, recall that 5 and -5 are opposites, because $5 + (-5) = 0$.

> **THE OPPOSITE OF A POLYNOMIAL**
>
> Two polynomials are **opposites,** or **additive inverses,** of each other if their sum is zero.

The opposite of $5t^3 - 2$ is $-5t^3 + 2$ because

$$(5t^3 - 2) + (-5t^3 + 2) = 5t^3 + (-5t^3) + (-2) + 2$$
$$= 0.$$

Similarly, the opposite of $-9x^2 + x - 7$ is $9x^2 - x + 7$ because

$$(-9x^2 + x - 7) + (9x^2 - x + 7) = -9x^2 + 9x^2 + x + (\quad x) + (-7) + 7$$
$$= 0.$$

This can be said using algebraic symbolism:

$$\underbrace{\text{The opposite of}} \quad \underbrace{(5t^3 - 2)} \quad \overset{\downarrow}{\text{is}} \quad \underbrace{-5t^3 + 2.}$$
$$\downarrow \qquad \qquad \downarrow \qquad \qquad \downarrow \qquad \qquad \downarrow$$
$$- \qquad \quad (5t^3 - 2) \quad = \quad -5t^3 + 2.$$

Similarly,

$$\underbrace{\text{The opposite of}} \quad \underbrace{(-9x^2 + x - 7)} \quad \overset{\downarrow}{\text{is}} \quad \underbrace{9x^2 - x + 7.}$$
$$\downarrow \qquad \qquad \downarrow \qquad \qquad \downarrow \qquad \qquad \downarrow$$
$$- \qquad \quad (-9x^2 + x - 7) \quad = \quad 9x^2 - x + 7.$$

> **TO FIND THE OPPOSITE OF A POLYNOMIAL**
>
> We can find an equivalent polynomial for the opposite, or additive inverse, of a polynomial by replacing each term with its opposite—that is, *changing the sign of every term.*

Add.

1. $(7a^2 + 2a + 8) + (2a^2 + a - 9)$

2. $(5x^2y + 3x^2 + 4) + (2x^2y + 4x)$

3. $(2a^3 + 17) + (2a^2 - 9a)$

Answers on page A-29

683

Find two equivalent expressions for the opposite of each polynomial.

4. $12x^4 - 3x^2 + 4x$

5. $-4x^4 + 3x^2 - 4x$

6. $-13x^6 + 2x^4 - 3x^2 + x - \frac{5}{13}$

7. $-8a^3b + 5ab^2 - 2ab$

Simplify.

8. $-(4x^3 - 6x + 3)$

9. $-(5x^3y + 3x^2y^2 - 7xy^3)$

10. $-\left(14x^{10} - \frac{1}{2}x^5 + 5x^3 - x^2 + 3x\right)$

Subtract.

11. $(7x^3 + 2x + 4) - (5x^3 - 4)$

12. $(-3x^2 + 5x - 4) - (-4x^2 + 11x - 2)$

13. Subtract:

$(7x^3 + 3x^2 - xy) - (5x^3 + 3xy + 2).$

Answers on page A-29

EXAMPLE 4 Find two equivalent expressions for the opposite of

$$4x^5 - 7x^3 - 8x + \tfrac{5}{6}.$$

a) $-\left(4x^5 - 7x^3 - 8x + \tfrac{5}{6}\right)$ This is one expression for the opposite of $4x^5 - 7x^3 - 8x + \tfrac{5}{6}$.

b) $-4x^5 + 7x^3 + 8x - \tfrac{5}{6}$ Changing the sign of every term

Thus, $-\left(4x^5 - 7x^3 - 8x + \tfrac{5}{6}\right)$ is equivalent to $-4x^5 + 7x^3 + 8x - \tfrac{5}{6}$, and each is the opposite of the original polynomial $4x^5 - 7x^3 - 8x + \tfrac{5}{6}$.

Do Exercises 4–7.

EXAMPLE 5 Simplify: $-\left(-7x^4 - \tfrac{5}{9}x^3 + 8x^2 - x + 67\right)$.

$$-\left(-7x^4 - \tfrac{5}{9}x^3 + 8x^2 - x + 67\right) = 7x^4 + \tfrac{5}{9}x^3 - 8x^2 + x - 67$$

Do Exercises 8–10.

C Subtracting Polynomials

We can now subtract a polynomial by adding the opposite of that polynomial. That is, for any polynomials p and q, $p - q = p + (-q)$.

EXAMPLE 6 Subtract:

$$(9x^5 + x^3 - 2x^2 + 4) - (2x^5 + x^4 - 4x^3 - 3x^2).$$

We have

$(9x^5 + x^3 - 2x^2 + 4) - (2x^5 + x^4 - 4x^3 - 3x^2)$

$= (9x^5 + x^3 - 2x^2 + 4) + [-(2x^5 + x^4 - 4x^3 - 3x^2)]$ Adding the opposite

$= (9x^5 + x^3 - 2x^2 + 4) + [-2x^5 - x^4 + 4x^3 + 3x^2]$ Changing the sign of every term

$= 9x^5 + x^3 - 2x^2 + 4 - 2x^5 - x^4 + 4x^3 + 3x^2$

$= 7x^5 - x^4 + 5x^3 + x^2 + 4.$ Combining like terms

Do Exercises 11 and 12.

To shorten our work, we often begin by changing the sign of each term in the polynomial being subtracted.

EXAMPLE 7 Subtract: $(5a^4 - 7a^3 + 5a^2b) - (-3a^4 + 4a^2b + 6)$.

We have

$(5a^4 - 7a^3 + 5a^2b) - (-3a^4 + 4a^2b + 6)$

$= 5a^4 - 7a^3 + 5a^2b + 3a^4 - 4a^2b - 6$ Removing all parentheses; changing the sign of every term in the second polynomial

$= 8a^4 - 7a^3 + a^2b - 6.$ Combining like terms

Do Exercise 13.

d Evaluating Polynomials and Applications

It is important to keep in mind that when we add or subtract polynomials, we are *not* solving an equation. Rather, we are finding an equivalent expression that is usually more concise. One reason we do this is to make it easier to evaluate.

EXAMPLE 8 Evaluate both $(5x^3 + 4x^2 + 3x) + (2x^3 + 5x^2 - x)$ and $7x^3 + 9x^2 + 2x$ for $x = 2$ (see Example 1).

a) When x is replaced by 2 in $(5x^3 + 4x^2 + 3x) + (2x^3 + 5x^2 - x)$, we have

$$5 \cdot 2^3 + 4 \cdot 2^2 + 3 \cdot 2 + 2 \cdot 2^3 + 5 \cdot 2^2 - 2,$$

or $\quad 5 \cdot 8 + 4 \cdot 4 + 6 + 2 \cdot 8 + 5 \cdot 4 - 2,$

or $\quad 40 + 16 + 6 + 16 + 20 - 2, \quad$ which is 96.

b) Similarly, when x is replaced by 2 in $7x^3 + 9x^2 + 2x$, we have

$$7 \cdot 2^3 + 9 \cdot 2^2 + 2 \cdot 2,$$

or $\quad 7 \cdot 8 + 9 \cdot 4 + 4,$

or $\quad 56 + 36 + 4. \quad$ As expected, this is also 96.

Note how much easier it is to evaluate the simplified sum in part (b) rather than the original expression.

Do Exercise 14.

Polynomials are frequently evaluated in real-world situations.

EXAMPLE 9 *Athletics.* In a sports league of n teams in which all teams play each other twice, the total number of games played is given by the polynomial

$$n^2 - n.$$

A women's softball league has 10 teams. If each team plays every other team twice, what is the total number of games played?

We evaluate the polynomial for $n = 10$:

$$n^2 - n = 10^2 - 10 = 100 - 10 = 90.$$

The league plays 90 games.

Do Exercises 15–17.

14. Evaluate each expression for $a = 2$. (See Margin Exercise 1.)

a) $(7a^2 + 2a + 8) + 2a^2 + a - 9$

b) $9a^2 + 3a - 1$

15. In the situation of Example 9, how many games are played in a league with 12 teams?

Wendy is pedaling down a hill. Her distance from the top of the hill, in meters, can be approximated by

$$\frac{1}{2}t^2 + 3t,$$

where t is the number of seconds she has been pedaling and $t < 30$.

16. How far has Wendy traveled in 4 sec?

17. How far has Wendy traveled in 10 sec?

Answers on page A-29

It is never too early to start studying for your final exam.

Best Scenario: Two Weeks of Study Time

1. **Begin by browsing through each chapter, reviewing the highlighted or boxed information regarding important formulas in both the text and the Summary and Review.** There may be some formulas that you will need to memorize.

2. **Retake all tests that your instructor has returned.** Restudy the objectives in the text that correspond to each question you miss.

3. **Then work the Chapter Tests and Cumulative Reviews in the portion of the text that you covered.** Be careful to avoid any questions corresponding to skipped material. Again, restudy the objectives in the text that correspond to each question you miss.

4. **Attend a final-exam review session if one is available.**

5. **If you are still having difficulty, use the supplements for extra review.** You might try the Video Lectures on CD, the *Student's Solutions Manual*, or the InterAct Math Tutorial Web site.

6. **For any remaining difficulties, see your instructor, go to a tutoring session, or participate in a study group.**

7. **See if previous final exams are available.** If they are, use them for practice, being alert to trouble spots.

8. **Take the Final Examination in the text during the last couple of days before the final.** See how much of the final exam you can complete under test-like conditions.

Moderate Scenario: Three Days to Two Weeks of Study Time

1. **Begin by browsing through each chapter, reviewing the highlighted or boxed information regarding important formulas in both the text and the Summary and Review.** There may be some formulas that you will need to memorize.

2. **Retake all tests that your instructor has returned.** Restudy the objectives in the text that correspond to each question you miss.

3. **Then work the Chapter Tests and Cumulative Reviews in the portion of the text that you covered.** Avoid any questions corresponding to skipped material. Again, restudy the objectives in the text that correspond to each question you miss.

4. **Attend a final-exam review session if one is available.**

5. **For any remaining difficulties, see your instructor, go to a tutoring session, or participate in a study group.**

6. **Take the Final Examination in the text during the last couple of days before the final.** See how much of the final exam you can complete under test-like conditions.

Worst Scenario: One or Two Days of Study Time

1. **Begin by browsing through each chapter, reviewing the highlighted or boxed information regarding important formulas in both the text and the Summary and Review.** There may be some formulas that you will need to memorize.

2. **Then work the last Cumulative Review in the portion of the text that you covered.** Avoid any questions corresponding to skipped material. Restudy the objectives in the text that correspond to each question you miss.

3. **Attend a final-exam review session if one is available.**

4. **Take the Final Examination in the text as preparation for the final.** See how much of the final exam you can complete under test-like conditions.

Promise yourself that next semester you will allow more time for final exam preparation.

a   Add.

1. $(3x + 7) + (-7x + 3)$

2. $(6x + 1) + (-7x + 2)$

3. $(-9x + 7) + (x^2 + x - 2)$

4. $(x^2 - 5x + 4) + (8x - 9)$

5. $(x^2 - 7) + (x^2 + 7)$

6. $(x^3 + x^2) + (2x^3 - 5x^2)$

7. $(6t^4 + 4t^3 - 1) + (5t^2 - t + 1)$

8. $(5t^2 - 3t + 12) + (2t^2 + 8t - 30)$

9. $(2 + 4x + 6x^2 + 7x^3) + (5 - 4x + 6x^2 - 7x^3)$

10. $(3x^4 - 6x - 5x^2 + 5) + (6x^2 - 4x^3 - 1 + 7x)$

11. $(9x^8 - 7x^4 + 2x^2 + 5) + (8x^7 + 4x^4 - 2x)$

12. $(4x^5 - 6x^3 - 9x + 1) + (6x^3 + 9x^2 + 9x)$

13. $(8t^4 + 6t^3 - t^2 + 3t) + (5t^4 - 2t^3 + t - 3)$

14. $(7t^5 - 3t^4 - 2t^2 + 5) + (3t^5 - 2t^4 + 4t^3 - t^2)$

15. $(-5x^4y^3 + 7x^3y^2 - 4xy^2) + (2x^3y^3 - 3x^3y^2 - 5xy)$

16. $(-9a^5b^4 + 7a^3b^3 + 2a^2b^2) + (2a^4b^4 - 5a^3b^3 - a^2b^2)$

17. $(8a^3b^2 + 5a^2b^2 + 6ab^2) + (5a^3b^2 - a^2b^2 - 4a^2b)$

18. $(6x^3y^3 - 4x^2y^2 + 3xy^2) + (x^3y^3 + 7x^3y^2 - 2xy^2)$

19. $(17.5abc^3 + 4.3a^2bc) + (-4.9a^2bc - 5.2abc)$

20. $(23.9x^3yz - 19.7x^2y^2z) + (-14.6x^3yz - 8x^2yz)$

b Find two equivalent expressions for the opposite of each polynomial.

21. $-5x$

22. $x^2 - 3x$

23. $-x^2 + 13x - 7$

24. $-7x^3 - x^2 - x$

25. $12x^4 - 3x^3 + 3$

26. $4x^3 - 6x^2 - 8x + 1$

Simplify.

27. $-(3x - 5)$

28. $-(-2x + 4)$

29. $-(4x^2 - 3x + 2)$

30. $-(-6a^3 + 2a^2 - 9a + 1)$

31. $-\left(-4x^4 + 6x^2 + \frac{3}{4}x - 8\right)$

32. $-(-5x^4 + 4x^3 - x^2 + 0.9)$

c Subtract.

33. $(3x + 2) - (-4x + 3)$

34. $(6x + 1) - (-7x + 2)$

35. $(9t^2 + 7t + 5) - (5t^2 + t - 1)$

36. $(8t^2 - 5t + 7) - (3t^2 - 2t + 1)$

37. $(-8x + 2) - (x^2 + x - 3)$

38. $(x^2 - 5x + 4) - (8x - 9)$

39. $(7a^2 + 5a - 9) - (2a^2 + 7)$

40. $(8a^2 - 6a + 5) - (2a^2 - 19a)$

41. $(8x^4 + 3x^3 - 1) - (4x^2 - 3x + 5)$

42. $(-4x^2 + 2x) - (3x^3 - 5x^2 + 3)$

43. $(1.2x^3 + 4.5x^2 - 3.8x) - (-3.4x^3 - 4.7x^2 + 23)$

44. $(0.5x^4 - 0.6x^2 + 0.7) - (2.3x^4 + 1.8x - 3.9)$

45. $\left(\frac{5}{8}x^3 - \frac{1}{4}x - \frac{1}{3}\right) - \left(-\frac{1}{8}x^3 + \frac{1}{4}x - \frac{1}{3}\right)$

46. $\left(\frac{1}{5}x^3 + 2x^2 - 0.1\right) - \left(-\frac{2}{5}x^3 + 2x^2 + 0.01\right)$

47. $(9x^3y^3 + 8x^2y^2 + 7xy) - (3x^3y^3 - 2x^2y + 3xy)$

48. $(3x^4y + 2x^3y - 7x^2y) - (5x^4y + 2x^2y^2 - 2x^2y)$

CHAPTER 10: Polynomials

d   Evaluate each polynomial for $x = 4$.

49. $-7x + 5$

50. $-3x + 1$

51. $2x^2 - 5x + 7$

52. $3x^2 + x + 7$

53. $x^3 - 5x^2 + x$

54. $7 - x + 3x^2$

Evaluate each polynomial for $x = -1$.

55. $2x + 9$

56. $6 - 2x$

57. $x^2 - 2x + 1$

58. $5x - 6 + x^2$

59. $-3x^3 + 7x^2 - 3x - 2$

60. $-2x^3 - 5x^2 + 4x + 3$

Falling Distance.   The distance, in feet, traveled by a body falling freely from rest in t seconds is approximated by the polynomial $16t^2$.

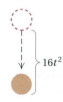

$16t^2$

61. A stone is dropped from a cliff and takes 8 sec to hit the ground. How high is the cliff?

62. A brick falls from the top of a building and takes 3 sec to hit the ground. How high is the building?

Minutes of Daylight.   The number of minutes of daylight in Chicago, on a date n days after December 21, can be approximated by

$$-0.01096n^2 + 4n + 548.$$

63. ▦ Determine the number of minutes of daylight in Chicago 92 days after December 21.

64. ▦ Determine the number of minutes of daylight in Chicago 123 days after December 21.

The number of degrees granted in foreign languages and literature x years after 1970 can be estimated by the polynomial

$$-x^3 + 80x^2 - 1723x + 22669.$$

Source: Based on information from U.S. National Center for Education Statistics, *Digest of Education Statistics,* annual

65. ▦ Determine the number of degrees in foreign languages and literature in 2005 ($x = 35$).

66. ▦ Determine the number of degrees in foreign languages and literature in 1980 ($x = 10$).

Daily Accidents. The average number of accidents per day involving drivers who are a years old is approximated by the polynomial

$$0.4a^2 - 40a + 1039.$$

67. Evaluate the polynomial for $a = 18$ to find the daily number of accidents involving 18-year-old drivers.

68. Evaluate the polynomial for $a = 20$ to find the daily number of accidents involving 20-year-old drivers.

Total Revenue. Cutting Edge Electronics is marketing a new kind of stereo. *Total revenue* is the total amount of money taken in. The firm determines that when it sells x stereos, it takes in

$$280x - 0.4x^2 \text{ dollars.}$$

69. What is the total revenue from the sale of 75 stereos?

70. What is the total revenue from the sale of 100 stereos?

Total Cost. Cutting Edge Electronics determines that the total cost of producing x stereos is given by

$$5000 + 0.6x^2 \text{ dollars.}$$

71. What is the total cost of producing 500 stereos?

72. What is the total cost of producing 650 stereos?

73. **D_W** Is every term a monomial? Why or why not?

74. **D_W** Suppose that two polynomials, each containing 3 terms, are added. Is it possible for the sum to contain more than 3 terms? fewer than 3 terms? exactly 3 terms? Explain.

SKILL MAINTENANCE

75. A 10-lb fish serves 7 people. What is the ratio of servings to pounds? [7.1a]

76. A bicycle salesperson's commission rate is 22%. A commission of $783.20 is received. How many dollars' worth of bicycles were sold? [8.5b]

77. In 2006, the sales tax rate in California was 6%. How much tax would be paid in California for a laptop computer that sold for $1350? [8.5a]

78. Find the area of a rectangle that is 6.5 m by 4 m. [5.8a]

79. Find the area of a circle with radius 20 cm. Use 3.14 for π. [9.3b]

80. Melba earned $4740 for working 12 weeks. What was the rate of pay? [7.2a]

Write the prime factorization for each number. [3.2c]

81. 168

82. 192

83. 735

84. 117

85. **D**_{**W**} 🔳 While using the formulas for number of minutes of daylight in Exercises 63 and 64, Alicia argues that instead of replacing n with 360, she can get nearly the same result using 5 instead. What do you think her reasoning is?

86. **D**_{**W**} Explain, in your own words, how the distributive law is used when subtracting polynomials.

Minutes of Daylight. The number of minutes of daylight in Los Angeles, on a date n days after December 21, can be approximated by

$$-0.0085n^2 + 3.1014n + 593.$$

87. 🔳 Determine the number of minutes of daylight in Los Angeles on "Ground Hog Day" (February 2).

88. 🔳 How much more daylight is available in Chicago than in Los Angeles on July 4? (See Exercises 63 and 64.)

89. 🔳 The polynomial used in Exercises 65 and 66 describes the trend over a number of years of degrees granted in foreign languages and literature. To visualize this trend, make a vertical bar graph, with the number of degrees on the vertical axis and years on the horizontal axis. Show a value for every 5 years from 1970 to 2010.

90. **D**_{**W**} How might the information in Exercise 89 be used by a college?

91. 🔳 *Medicine.* When a person swallows 400 mg of ibuprofen, the number of milligrams in the bloodstream t hours later can be approximated by the polynomial

$$0.5t^4 + 3.45t^3 - 96.65t^2 + 347.7t,$$

with $0 \le t \le 6$. Determine the amount of ibuprofen in the bloodstream **(a)** 1 hr after swallowing 400 mg; **(b)** 2 hr after swallowing 400 mg; **(c)** 6 hr after swallowing 400 mg.

92. 🔳 *Cellular Phone Sales.* The polynomial

$$0.04x^3 - 0.23x^2 + 0.94x - 0.05$$

can be used to estimate the number of cellular phones in use, in millions, x years after 1985. Estimate the number of cellular phones in use in 2004.

Perform the indicated operations and simplify.

93. $(7y^2 - 5y + 6) - (3y^2 + 8y - 12) + (8y^2 - 10y + 3)$

94. $(3x^2 - 4x + 6) - (-2x^2 + 4) + (-5x - 3)$

95. $(-y^4 - 7y^3 + y^2) + (-2y^4 + 5y - 2) - (-6y^3 + y^2)$

96. $(-4 + x^2 + 2x^3) - (-6 - x + 3x^3) - (-x^2 - 5x^3)$

97. Complete: $9x^4 + \underline{\hspace{0.5cm}} + 5x^2 - 7x^3 + \underline{\hspace{0.5cm}} - 9 + \underline{\hspace{0.5cm}} = 12x^4 - 5x^3 + 5x^2 - 16.$

98. Complete: $8t^4 + \underline{\hspace{0.5cm}} - 2t^3 + \underline{\hspace{0.5cm}} - 2t^2 + t - \underline{\hspace{0.5cm}} - 3 + \underline{\hspace{0.5cm}} = 8t^4 + 7t^3 - 3t + 4.$

691

Objectives

a Multiply monomials.

b Multiply a monomial and any polynomial.

c Use the distributive law to factor.

Multiply.

1. $(6a)(3a)$

2. $(-7x)(2x)$

Answers on page A-29

Study Tips

SUMMING IT ALL UP

In preparation for a final exam, many students find it helpful to write up a few pages of notes that represent the most important concepts of the course. After doing so, it is a good idea to try to condense those notes down to just one page. This exercise will help you focus on the most important material.

a **Multiplying Monomials**

Recall that the area of a square with sides of length x is x^2.

Area $= x^2$

If a rectangle is 3 times as long as it is wide, we can represent its width by x and its length by $3x$.

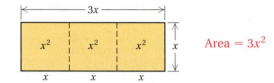

Area $= 3x^2$

The area, $3x^2$, is the product of $3x$ and x. This product can be found using an associative law:

$$(3x)x = 3(xx) = 3x^2.$$

To find other products of monomials, we may need to use a commutative law as well.

EXAMPLE 1 Multiply: $(4x)(5x)$.

$$
\begin{aligned}
(4x)(5x) &= 4 \cdot x \cdot 5 \cdot x && \text{Using an associative law} \\
&= 4 \cdot 5 \cdot x \cdot x && \text{Using a commutative law} \\
&= (4 \cdot 5)(xx) && \text{Using an associative law} \\
&= 20x^2
\end{aligned}
$$

Example 1 can be regarded as finding the area of a rectangle of width $4x$ and length $5x$. Note that the area consists of 20 squares, each of which has area x^2.

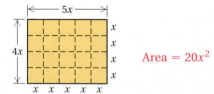

Area $= 20x^2$

Do Exercises 1 and 2.

Usually the steps in Example 1 are combined: We multiply coefficients and we multiply variables.

EXAMPLES   Multiply.

2. $(5x)(6x) = (5 \cdot 6)(x \cdot x)$ Multiplying the coefficients
$= 30x^2$ Simplifying

3. $(3x)(-x) = (3x)(-1x)$ Rewriting $-x$ as $-1x$
$= (3)(-1)(x \cdot x)$
$= -3x^2$

4. $(7x)(4y) = (7 \cdot 4)(x \cdot y)$
$= 28xy$

Do Exercises 3–5.

MULTIPLYING POWERS WITH LIKE BASES

In later courses, you will likely learn several rules for manipulating exponents. The one rule that we develop now is useful when multiplying powers with like bases. Consider the following:

$$a^3 \cdot a^2 = \underbrace{(a \cdot a \cdot a)}_{\text{3 factors}} \underbrace{(a \cdot a)}_{\text{2 factors}} = \underbrace{a \cdot a \cdot a \cdot a \cdot a}_{\text{5 factors}} = a^5.$$

Note that the exponent in a^5 is the sum of those in $a^3 \cdot a^2$. That is, $3 + 2 = 5$.

Adding the exponents gives the correct result.

> **THE PRODUCT RULE FOR EXPONENTS**
>
> For any number a and any positive integers m and n,
> $$a^m \cdot a^n = a^{m+n}.$$
>
> (When multiplying with exponential notation, if the bases are the same, keep the base and add the exponents.)

EXAMPLES   Multiply and simplify.

5. $x^2 \cdot x^5 = x^{2+5}$ Adding exponents
$= x^7$

6. $(3a^4)(5a^2) = (3 \cdot 5)(a^4 \cdot a^2)$ Multiplying coefficients; adding exponents
$= 15a^6$

7. $(-4x^2y^3)(3x^6y^7) = (-4 \cdot 3)(x^2 \cdot x^6)(y^3 \cdot y^7)$
$= -12x^8y^{10}$

Do Exercises 6–9.

We have not yet determined what the number 1 will mean when used as an exponent. Consider the following:

$$m \cdot m^2 = m \cdot m \cdot m = m^3$$
and $\quad x \cdot x^3 = x \cdot x \cdot x \cdot x = x^4.$

Multiply.
3. $(4a)(12a)$

4. $(-m)(5m)$

5. $(-6a)(-7b)$

Multiply.
6. $a^5 \cdot a^4$

7. $(2x^8)(4x^5)$

8. $(-7m^4)(-5m^7)$

9. $(3a^5b^4)(5a^2b^8)$

Answers on page A-29

10. Evaluate 7^1 and -19^1.

Note that if $m = m^1$ and $x = x^1$, the same results can be found using the product rule:

$$m \cdot m^2 = m^1 \cdot m^2 = m^3$$

and $x \cdot x^3 = x^1 \cdot x^3 = x^4$.

This suggests the following definition.

THE EXPONENT 1

$b^1 = b$ for any number b.

EXAMPLE 8   Evaluate 23^1, -23^1, and $(-23)^1$.

$$23^1 = 23;$$
$$-23^1 = -23; \qquad \text{We read } -23^1 \text{ as "the opposite of } 23^1\text{."}$$
$$(-23)^1 = -23. \qquad \text{We read } (-23)^1 \text{ as "negative 23 to the first."}$$

Do Exercise 10.

Multiply.

11. $4x$ and $3x + 5$

b   Multiplying a Monomial and Any Polynomial

When a polynomial contains two terms, it is called a **binomial.** The product of the monomial x and the binomial $x + 2$ can be visualized as the area of a rectangle with width x and length $x + 2$.

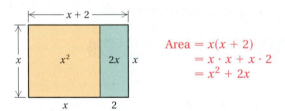

$$\text{Area} = x(x + 2)$$
$$= x \cdot x + x \cdot 2$$
$$= x^2 + 2x$$

12. $3a(2a^2 - 5a + 7)$

The distributive law is used to find products of polynomials algebraically.

EXAMPLE 9   Multiply: $2x$ and $5x + 3$.

$$2x(5x + 3) = 2x \cdot 5x + 2x \cdot 3 \qquad \text{Using the distributive law}$$
$$= 10x^2 + 6x \qquad \qquad \text{Multiplying each pair of monomials}$$

EXAMPLE 10   Multiply: $5x(2x^2 - 3x + 4)$.

13. $4a^3b^2(2a^2 + 5b^4)$

$$5x(2x^2 - 3x + 4) = 5x \cdot 2x^2 - 5x \cdot 3x + 5x \cdot 4$$
$$= 10x^3 - 15x^2 + 20x \qquad \text{Note that } x \cdot x^2 = x^1 \cdot x^2 = x^3.$$

EXAMPLE 11   Multiply: $-3r^2s(2r^3s^2 - 5rs)$.

$$-3r^2s(2r^3s^2 - 5rs) = -3r^2s \cdot 2r^3s^2 - (-3r^2s)5rs$$
$$= -6r^5s^3 + 15r^3s^2$$

Answers on page A-29

Do Exercises 11–13.

C Factoring

Factoring is the reverse of multiplying. We use the distributive law, beginning with a sum or a difference of terms that contain a common factor:

$$ab + ac = a(b + c) \quad \text{and} \quad rs - rt = r(s - t).$$

> **FACTORING**
>
> To **factor** an expression is to find an equivalent expression that is a product.

To *factor* an expression like $10y + 15$, we find an equivalent expression that is a product. To do this, we look to see if both terms have a factor in common. If there *is* a common factor, we can "factor it out" using the distributive law. Note the following:

The prime factorization of $10y$ is $2 \cdot 5 \cdot y$.

The prime factorization of 15 is $3 \cdot 5$.

We factor out the common factor, 5:

$$10y + 15 = 5 \cdot 2y + 5 \cdot 3 \qquad \text{Try to do this step mentally.}$$
$$= 5(2y + 3). \qquad \text{Using the distributive law}$$

We generally factor out the *largest* common factor. This is the product of all factors common to all terms.

EXAMPLE 12 Factor $12a - 30$.

The prime factorization of $12a$ is $2 \cdot 2 \cdot 3 \cdot a$.

The prime factorization of 30 is $2 \cdot 3 \cdot 5$.

Both factorizations include a factor of 2 and a factor of 3. Thus, 2 is a common factor, 3 is a common factor, and $2 \cdot 3$ is a common factor. The largest common factor is $2 \cdot 3$, or 6:

$$12a - 30 = 6(2a - 5). \qquad \text{Try to go directly to this step.}$$

EXAMPLE 13 Factor $9x + 27y - 9$.

The prime factorization of $9x$ is $3 \cdot 3 \cdot x$.

The prime factorization of $27y$ is $3 \cdot 3 \cdot 3 \cdot y$.

The prime factorization of 9 is $3 \cdot 3$.

$$9x + 27y - 9 = 9 \cdot x + 9 \cdot 3y - 9 \cdot 1 \qquad \text{The largest common factor is } 3 \cdot 3,$$
$$\text{or 9.}$$
$$= 9(x + 3y - 1)$$

In Example 13, the 1 in the factorization is necessary. To illustrate this, reverse the factorization process by multiplying. This provides a check for the answer.

$$9(x + 3y - 1) = 9 \cdot x + 9 \cdot 3y - 9 \cdot 1 = 9x + 27y - 9$$

> Factorizations can always be checked by multiplying.

695

Factor.

14. $6z - 12$

15. $3x - 6y + 12$

16. $16a - 36b + 42$

17. $-12x + 32y - 16z$

Factor.

18. $5a^3 + 10a$

19. $14x^3 - 7x^2 + 21x$

20. $9a^2b - 6ab^2$

Answers on page A-29

> **Caution!**
>
> Note in Example 13 that although $3(3x + 9y - 3)$ is also equivalent to $9x + 27y - 9$, it is not factored "completely." However, we can complete the process by factoring out another factor of 3:
>
> $$9x + 27y - 9 = 3(3x + 9y - 3) = 3 \cdot 3(x + 3y - 1) = 9(x + 3y - 1).$$
>
> Remember to factor out the *largest common factor*.

EXAMPLES   Factor. Try to write just the answer.

14. $-3x + 6y - 9z = -3(x - 2y + 3z)$

We generally factor out a negative factor when the first coefficient is negative. We might also factor as $-3x + 6y - 9z = 3(-x + 2y - 3z)$.

15. $18z - 12x - 24 = 6(3z - 2x - 4)$

The largest common factor is $2 \cdot 3$.

$$\begin{cases} 18z = 2 \cdot 3 \cdot 3 \cdot z; \\ 12x = 2 \cdot 2 \cdot 3 \cdot x; \\ 24 = 2 \cdot 2 \cdot 2 \cdot 3 \end{cases}$$

Check: $6(3z - 2x - 4) = 6 \cdot 3z - 6 \cdot 2x - 6 \cdot 4 = 18z - 12x - 24$

> *Remember*: An expression is factored when it is written as a product.

Do Exercises 14–17.

EXAMPLE 16   Factor each of the following:

a) $10x^6 + 15x^2$

b) $8xy^3 - 6xy^2 + 4xy$

a) The prime factorization of $10x^6$ is $2 \cdot 5 \cdot x \cdot x \cdot x \cdot x \cdot x \cdot x$.
The prime factorization of $15x^2$ is $3 \cdot 5 \cdot x \cdot x$.

$$10x^6 + 15x^2 = 5x^2 \cdot 2x^4 + 5x^2 \cdot 3 \qquad \text{The largest common factor is } 5x^2.$$
$$= 5x^2(2x^4 + 3)$$

b) The prime factorization of $8xy^3$ is $2 \cdot 2 \cdot 2 \cdot x \cdot y \cdot y \cdot y$.
The prime factorization of $6xy^2$ is $2 \cdot 3 \cdot x \cdot y \cdot y$.
The prime factorization of $4xy$ is $2 \cdot 2 \cdot x \cdot y$.

$$8xy^3 - 6xy^2 + 4xy = 2xy \cdot 4y^2 - 2xy \cdot 3y + 2xy \cdot 2 \qquad \text{The largest common factor is } 2xy.$$
$$= 2xy(4y^2 - 3y + 2)$$

The checks are left for the student.

The largest common factor can be determined by considering the coefficients and the variables separately. The largest common factor of the coefficients is found using prime factorizations. The largest common variable factors can be found by examining the exponents.

> When a variable appears in every term of a polynomial, the *largest* common factor of that variable is the *smallest* of the powers of that variable in the polynomial.

Do Exercises 18–20.

a Multiply.

1. $(4a)(7a)$

2. $(7x)(6x)$

3. $(-4x)(15x)$

4. $(-9a)(10a)$

5. $(7x^5)(4x^3)$

6. $(10a^2)(3a^2)$

7. $(-0.1x^6)(0.7x^3)$

8. $(0.3x^3)(-0.4x^6)$

9. $(5x^2y^3)(7x^4y^9)$

10. $(9a^5b^4)(2a^4b^7)$

11. $(4a^3b^4c^2)(3a^5b^4)$

12. $(7x^3y^5z^2)(8x^3z^4)$

13. $(3x^2)(-4x^3)(2x^6)$

14. $(-2y^5)(10y^4)(-3y^3)$

b Multiply.

15. $3x(-x + 7)$

16. $2x(4x - 6)$

17. $-3x(x - 2)$

18. $-9x(-x - 1)$

19. $x^2(x^3 + 1)$

20. $-2x^3(x^2 - 1)$

21. $5x(2x^2 - 6x + 1)$

22. $-4x(2x^3 - 6x^2 - 5x + 1)$

23. $4xy(3x^2 + 2y)$

24. $7xy(3x^2 - 6y^2)$

25. $3a^2b(4a^5b^2 - 3a^2b^2)$

26. $4a^2b^2(2a^3b - 5ab^2)$

c Factor. Check by multiplying.

27. $2x + 8$

28. $3x + 12$

29. $7a - 35$

30. $9a - 18$

31. $28x + 21y$

32. $8x - 10y$

33. $9a - 27b + 81$

34. $5x + 10 + 15y$

35. $18 - 6m$

36. $28 - 4y$

37. $-16 - 8x + 40y$

38. $-35 + 14x - 21y$

39. $9x^5 + 9x$

40. $5x^6 + 5x$

41. $a^3 - 8a^2$

42. $a^5 - 9a^2$

43. $8x^3 - 6x^2 + 2x$

44. $9x^4 - 12x^3 + 3x$

45. $12a^4b^3 + 18a^5b^2$

46. $15a^5b^2 + 20a^2b^3$

47. **D_W** If a binomial consists of two like terms, can it be factored? Why or why not?

48. **D_W** If all of a polynomial's coefficients are prime, is it still possible to factor the polynomial? Why or why not?

SKILL MAINTENANCE

49. When used for a singles match, a regulation tennis court is 27 ft by 78 ft. Find its perimeter.   [2.7b]

50. The Floral Doctor's delivery van traveled 147 mi on 10.5 gal of gas. How many miles per gallon did the van get?   [7.2a]

51. Ramon's new truck gets 21 mpg. This is 20% more than the mileage his old truck got. What mileage did the old truck get?   [8.4b]

52. A 5% sales tax is added to the price of a two-speed washing machine. If the machine is priced at $399, find the total amount paid.   [8.5a]

53. Of the 8 fish Mac caught, 3 were trout. What percentage were not trout?   [8.4a]

54. The diameter of a compact disc is 12 cm. What is its circumference? (Use 3.14 for π.)   [9.3b]

55. Multiply: $-57 \cdot 48$.   [2.4a]

56. Multiply: $(-72)(-46)$.   [2.4a]

SYNTHESIS

57. **D_W** Describe a method for creating a binomial that has $5x^2$ as its largest common factor.

58. **D_W** Explain in your own words why the product rule for exponents "works."

Factor.

59. ▦ $391x^{391} + 299x^{299}$

60. ▦ $703a^{437} + 437a^{703}$

61. $84a^7b^9c^{11} - 42a^8b^6c^{10} + 49a^9b^7c^8$

62. Draw a figure similar to those preceding Examples 1 and 9 to show that $2x \cdot 3x = 6x^2$.

10.3 MORE MULTIPLICATION OF POLYNOMIALS

a Multiplying Two Binomials

To find an equivalent expression for the product of two binomials, we use the distributive law more than once. In the example that follows, the distributive law is used three times.

EXAMPLE 1 Multiply: $x + 5$ and $x + 4$.

$$(x + 5)(x + 4) = (x + 5)x + (x + 5)4 \qquad \text{Using the distributive law}$$
$$= x \cdot x + 5 \cdot x + x \cdot 4 + 5 \cdot 4 \qquad \text{Using the distributive law two more times}$$
$$= x^2 + 5x + 4x + 20 \qquad \text{Multiplying monomials}$$
$$= x^2 + 9x + 20 \qquad \text{Combining like terms}$$

Do Exercises 1 and 2.

We can visualize the product $(x + 5)(x + 4)$ as the area of a rectangle with width $x + 4$ and length $x + 5$. Note that the total area is the sum of the four smaller areas.

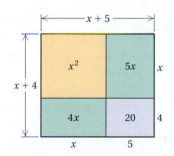

Area $= (x + 5)(x + 4)$
$= x \cdot x + 4x + 5x + 5 \cdot 4$
$= x^2 + 9x + 20$

More complicated products of binomials are not as easily visualized, but can be simplified using the steps of Example 1.

EXAMPLE 2 Multiply: $4x + 3$ and $x - 2$.

$$(4x + 3)(x - 2) = (4x + 3)(x + (-2)) \qquad \text{Rewriting } x - 2 \text{ as } x + (-2)$$
$$= (4x + 3)x + (4x + 3)(-2) \qquad \text{Using the distributive law}$$
$$= 4x \cdot x + 3 \cdot x + 4x(-2) + 3(-2) \qquad \text{Using the distributive law two more times}$$
$$= 4x^2 + 3x + (-8x) + (-6) \qquad \text{Multiplying monomials}$$
$$= 4x^2 - 5x - 6 \qquad \text{Combining like terms}$$

Do Exercises 3 and 4.

Note in Examples 1 and 2 that four products were found. These products are found by multiplying the First terms, the Outer terms, the Inner terms, and the Last terms in the binomials. We use the word FOIL to help remember these products.

Multiply.

1. $x + 8$ and $x + 5$

2. $(x + 5)(x - 4)$

Multiply.

3. $5x + 3$ and $x - 4$

4. $(3x - 2)(5x - 1)$

Answers on page A-29

Study Tips

DON'T GIVE UP NOW!

Often students are tempted to ease up once the end of a course is in sight. Don't let this happen to you! You have invested a great deal of time and energy thus far. Don't tarnish that hard effort by doing anything less than your best work as the course winds down.

699

EXAMPLE 3 Use FOIL to multiply $(x + 3)(x + 7)$.

We have

$$
\begin{array}{l}
\text{First} \quad \text{Last} \\
(x + 3)(x + 7) = \overset{\text{F}}{x \cdot x} + \overset{\text{O}}{7 \cdot x} + \overset{\text{I}}{3 \cdot x} + \overset{\text{L}}{3 \cdot 7} \\
\qquad\qquad\quad\text{Inner} \\
\qquad\qquad = x^2 + 7x + 3x + 21 \\
\qquad\quad\text{Outer} \\
\qquad\qquad = x^2 + 10x + 21.
\end{array}
$$

6. $(x - 3)(x - 8)$

Do Exercises 5 and 6.

b Multiplying Any Polynomials

A polynomial containing three terms is called a **trinomial.** To find the product of a binomial and a trinomial, we again use the distributive law.

EXAMPLE 4 Multiply: $(x^2 + 2x - 3)(x^2 + 4)$.

$$
\begin{aligned}
(x^2 + 2x - 3)(x^2 + 4) &= (x^2 + 2x - 3)x^2 + (x^2 + 2x - 3)4 \\
&= x^2 \cdot x^2 + 2x \cdot x^2 - 3 \cdot x^2 + x^2 \cdot 4 + 2x \cdot 4 - 3 \cdot 4 \\
&= x^4 + 2x^3 - 3x^2 + 4x^2 + 8x - 12 \\
&= x^4 + 2x^3 + x^2 + 8x - 12 \qquad \text{Combining like terms}
\end{aligned}
$$

Multiply.

7. $(x^2 + 3x - 4)(x^2 + 5)$

Do Exercises 7 and 8.

> **MULTIPLYING POLYNOMIALS**
>
> To multiply polynomials P and Q, select one of the polynomials—say, P. Then multiply each term of P by every term of Q and combine any like terms.

8. $(3y^2 - 7)(2y^3 - 2y + 5)$

Columns can be used for long multiplication. To do so, we multiply each term at the top by every term below. We write like terms in columns and add the results. This is like long multiplication of numbers with all the steps done out in detail.

Multiply.

9.
$$
\begin{array}{r}
3x^2 - 2x + 4 \\
x + 5 \\
\hline
\end{array}
$$

$$
\begin{array}{r}
2\ 3\ 1 \\
\times \quad 3\ 2 \\
\hline
4\ 6\ 2 \\
6\ 9\ 3\ 0 \\
\hline
7\ 3\ 9\ 2
\end{array}
\qquad
\begin{array}{r}
2\ 3\ 1 \\
\times \qquad\qquad 3\ 2 \\
\hline
400 + 60 + 2 \\
6000 + \ 900 + 30 \\
\hline
6000 + 1300 + 90 + 2
\end{array}
\qquad
\begin{array}{l}
= 200 + 30 + 1 \\
= \qquad\quad 30 + 2 \\
\\
= 2(231) = 2(200 + 30 + 1) \\
= 30(231) = 30(200 + 30 + 1) \\
\\
= 7392
\end{array}
$$

EXAMPLE 5 Multiply: $(4x^2 - 2x + 3)(x + 2)$.

$$
\begin{array}{r}
4x^2 - 2x + 3 \\
x + 2 \\
\hline
8x^2 - 4x + 6 \\
4x^3 - 2x^2 + 3x \\
\hline
4x^3 + 6x^2 - \quad x + 6
\end{array}
$$

It helps that both polynomials are in descending order.

Multiplying the top row by 2
Multiplying the top row by x

Combining like terms

Line up like terms in columns and then combine.

Do Exercise 9.

a Multiply.

1. $(x + 6)(x + 2)$

2. $(x + 5)(x + 2)$

3. $(x + 5)(x - 2)$

4. $(x + 1)(x - 3)$

5. $(x + 6)(x - 2)$

6. $(x - 4)(x - 3)$

7. $(x - 7)(x - 3)$

8. $(x + 3)(x - 6)$

9. $(x + 5)(x - 5)$

10. $(x - 7)(x + 7)$

11. $(3 + x)(6 + 2x)$

12. $(2x + 5)(2x + 5)$

13. $(3x - 4)(3x - 4)$

14. $(5x - 1)(5x - 1)$

15. $\left(x - \frac{5}{2}\right)\left(x + \frac{2}{5}\right)$

16. $\left(x + \frac{4}{3}\right)\left(x + \frac{3}{2}\right)$

b Multiply.

17. $(x^2 + x - 3)(x + 1)$

18. $(x^2 - x + 1)(x + 2)$

19. $(2x + 1)(2x^2 + 6x + 1)$

20. $(3x - 1)(4x^2 - 2x - 1)$

21. $(y^2 - 3)(3y^2 - 6y + 2)$

22. $(3y^2 - 3)(y^2 + 6y + 1)$

23. $(x^3 + x^2)(x^3 + x^2 - x)$

24. $(x^3 - x^2)(x^3 - x^2 + x)$

25. $(2t^2 - t - 4)(3t^2 + 2t - 1)$

26. $(3a^2 - 5a + 2)(2a^2 - 3a + 4)$

27. $(x - x^3 + x^5)(x^2 - 1 + x^4)$

28. $(x - x^3 + x^5)(3x^2 + 3x^6 + 3x^4)$

29. **D_W** Ron says that since $(xy)^2 = (xy) \cdot (xy) = x^2y^2$, it follows that $(x + y)^2 = x^2 + y^2$. Is he correct? Why or why not?

30. **D_W** Joyce insists that since $x \cdot x$ is x^2 and $5 \cdot 4 = 20$, it follows that $(x + 5)(x + 4) = x^2 + 20$. How could you convince her that this is not correct?

31. A sidewalk of uniform width is built around three sides of a store, as shown in the figure. What is the area of the sidewalk? [1.8a]

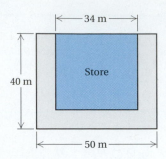

32. A real estate agent's commission rate is 6%. A commission of $7380 is received on the sale of a home. For how much did the home sell? [8.5b]

33. What percent of 24 is 32? [8.3b]

34. 39 is 150% of what number? [8.3b]

35. In 2002, the Oakland Athletics won 103 of 162 games. What percentage of games did they win? [8.4a]

36. The Sanchez's flower garden covers a 14-ft–wide circular region of their yard. Find the garden's area. Use $\frac{22}{7}$ for π. [9.3b]

Add. [2.2a]

37. $-5 + (-12)$

38. $-19 + 12$

39. $17 + (-24)$

40. $-15 + (-2)$

41. **D**_{**W**} Explain how the product of two 2-digit numbers can be regarded as the product of two binomials.

42. **D**_{**W**} Is the product of two binomials always a trinomial? Why or why not?

43. ▦ (See Example 4.) Check that the expressions $(x^2 + 2x - 3)(x^2 + 4)$ and $x^4 + 2x^3 + x^2 + 8x - 12$ are equivalent by evaluating both expressions for $x = 5$, $x = 3.5$, and $x = -1.2$.

44. Simplify: $(x + 2)(x + 3) + (x - 4)^2$.

For each figure below, find a simplified expression for **(a)** the perimeter and **(b)** the area.

45.

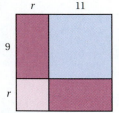

46.

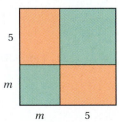

47. Find a polynomial for the shaded area.

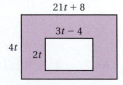

48. A box with a square bottom and no top is to be made from a 12-in. square piece of cardboard. Squares with side x are cut out of the corners and the sides are folded up. Find polynomials for the volume and the outside surface area of the box.

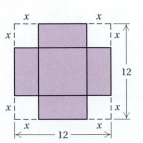

Multiply.

49. $(3x - 5)^2$

50. $(9x + 4)^2$

10.4 INTEGERS AS EXPONENTS

Objectives

a Evaluate algebraic expressions containing whole-number exponents.

b Express exponential expressions involving negative exponents as equivalent expressions containing positive exponents.

We have already used the numbers $1, 2, 3, \ldots$, as exponents. In this section we consider 0, as well as negative integers, as exponents.

a Zero as an Exponent

Look for a pattern in the following:

$$8 \cdot 8 \cdot 8 \cdot 8 = 8^4 \quad \text{We divide by 8 each time.}$$
$$8 \cdot 8 \cdot 8 = 8^3$$
$$8 \cdot 8 = 8^2$$
$$8 = 8^1$$
$$1 = 8^?.$$

The exponents decrease by 1 each time. Continuing the pattern,

$$1 = 8^0.$$

We make the following definition.

THE EXPONENT 0
$b^0 = 1$, for any nonzero number b.

We leave 0^0 undefined.

EXAMPLE 1 Evaluate 3^0, $(-3)^0$, and -3^0.

$$3^0 = 1; \qquad\qquad \text{3 is the base; 0 is the exponent.}$$
$$(-3)^0 = 1; \qquad\qquad \text{-3 is the base; 0 is the exponent.}$$
$$-3^0 = -1 \cdot 3^0 = -1 \cdot 1 = -1. \qquad \text{Note that } -3^0 \neq (-3)^0.$$

Do Exercises 1–3.

EXAMPLE 2 Evaluate $m^0 + 5$ for $m = 9$.

$$m^0 + 5 = 9^0 + 5 = 1 + 5 = 6$$

EXAMPLE 3 Evaluate $(3x + 2)^0$ for $x = -5$.

We substitute -5 for x and follow the rules for order of operations:

$$(3x + 2)^0 = (3(-5) + 2)^0 \qquad \text{Substituting}$$
$$= (-15 + 2)^0 \qquad \text{Multiplying}$$
$$= (-13)^0$$
$$= 1.$$

Caution!
Don't confuse the powers 1 and 0. Be careful: $8^0 = 1$, but $8^1 = 8$.

Do Exercises 4 and 5.

Evaluate.

1. 7^0

2. $(-9)^0$

3. -8^0

4. Evaluate $t^0 - 4$ for $t = 7$.

5. Evaluate $(2x - 9)^0$ for $x = 3$.

Answers on page A-29

Write an equivalent expression with positive exponents. Then simplify.

6. 4^{-3}

7. 5^{-2}

8. 2^{-4}

9. $(-2)^{-3}$

10. $\left(\dfrac{5}{3}\right)^{-2}$

Answers on page A-29

b Negative Integers as Exponents

The pattern used to help define the exponent 0 can also be used to define negative-integer exponents:

$$8 \cdot 8 \cdot 8 = 8^3 \quad \text{We divide by 8 each time.}$$
$$8 \cdot 8 = 8^2$$
$$8 = 8^1$$
$$1 = 8^0$$
$$\frac{1}{8} = 8^{?}$$
$$\frac{1}{8 \cdot 8} = 8^{?}.$$

The exponents decrease by 1 each time. To continue the pattern, we would say that

$$\frac{1}{8} = 8^{-1}$$

and $\quad \dfrac{1}{8 \cdot 8} = 8^{-2}.$

Thus, if we are to preserve the above pattern, we must have

$$\frac{1}{8^1} = 8^{-1} \quad \text{and} \quad \frac{1}{8^2} = 8^{-2}.$$

This leads to our definition of negative exponents:

NEGATIVE EXPONENTS

For any nonzero numbers a and b, and any integer n,

$$a^{-n} = \frac{1}{a^n}, \quad \text{and} \quad \left(\frac{a}{b}\right)^{-n} = \left(\frac{b}{a}\right)^n.$$

(A base raised to a negative exponent is equal to the reciprocal of the base raised to a positive exponent.)

EXAMPLES Write an equivalent expression using positive exponents. Then simplify.

4. $4^{-2} = \dfrac{1}{4^2} = \dfrac{1}{16}$ $\quad$ Note that 4^{-2} represents a *positive* number.

5. $(-3)^{-2} = \dfrac{1}{(-3)^2} = \dfrac{1}{(-3)(-3)} = \dfrac{1}{9}$

6. $m^{-3} = \dfrac{1}{m^3}$

7. $ab^{-1} = a\left(\dfrac{1}{b^1}\right) = a\left(\dfrac{1}{b}\right) = \dfrac{a}{b}$ $\quad$ Think of a as $\dfrac{a}{1}$ if you wish.

8. $\left(\dfrac{5}{6}\right)^{-2} = \left(\dfrac{6}{5}\right)^2$ $\qquad \left(\dfrac{a}{b}\right)^{-n} = \left(\dfrac{b}{a}\right)^n$

$\qquad = \dfrac{6}{5} \cdot \dfrac{6}{5} = \dfrac{36}{25}$

Note in Example 4 that

$$4^{-2} \neq 4(-2) \quad \text{and} \quad \frac{1}{4^2} \neq 4(-2).$$

Similarly, in Example 5,

$$(-3)^{-2} \neq (-3)(-2) \quad \text{and} \quad \frac{1}{(-3)^2} \neq (-3)(-2).$$

In general, $a^{-n} \neq a(-n)$. The negative exponent also does not indicate that a negative number is in the result. That is,

$$4^{-2} = \frac{1}{16}, \quad \textit{not} \quad \frac{1}{-16}.$$

Do Exercises 6–10 on the preceding page.

EXAMPLES Write an equivalent expression using negative exponents.

9. $\dfrac{1}{7^2} = 7^{-2}$ ⟶ Reading $a^{-n} = \dfrac{1}{a^n}$ from right to left: $\dfrac{1}{a^n} = a^{-n}$

10. $\dfrac{5}{x^8} = 5 \cdot \dfrac{1}{x^8} = 5x^{-8}$

Do Exercises 11 and 12.

Consider an expression like

$$\frac{a^2}{b^{-3}},$$

in which the denominator is a negative power. We can simplify as follows:

$$\frac{a^2}{b^{-3}} = \frac{a^2}{\dfrac{1}{b^3}} \qquad \text{Rewriting } b^{-3} \text{ as } \dfrac{1}{b^3}$$

$$= a^2 \cdot \frac{b^3}{1} \qquad \begin{array}{l}\text{To divide by a fraction, we} \\ \text{multiply by its reciprocal.}\end{array}$$

$$= a^2 b^3.$$

Do Exercises 13 and 14.

Our work above indicates that to divide by a base raised to a negative power, we can instead multiply by the opposite power of the same base. This will shorten our work.

EXAMPLES Write an equivalent expression using positive exponents.

11. $\dfrac{x^3}{y^{-2}} = x^3 y^2$ ⟶ Instead of dividing by y^{-2}, multiply by y^2.

12. $\dfrac{a^2 b^5}{c^{-6}} = a^2 b^5 c^6$ ⟶ Instead of dividing by c^{-6}, multiply by c^6.

13. $\dfrac{x^{-2} y}{z^{-3}} = x^{-2} y z^3 = \dfrac{y z^3}{x^2}$

Do Exercises 15–17.

Write an equivalent expression with negative exponents.

11. $\dfrac{1}{9^2}$

12. $\dfrac{7}{x^4}$

Write an equivalent expression with positive exponents.

13. $\dfrac{m^3}{n^{-5}}$

14. $\dfrac{ab}{c^{-1}}$

Write an equivalent expression with positive exponents.

15. $\dfrac{a^4}{b^{-6}}$

16. $\dfrac{x^7 y}{z^{-4}}$

17. $\dfrac{a^4 b^{-7}}{c^{-3}}$

Answers on pages A-29–A-30

Simplify. Use positive powers in the answer.

18. $5^{-2} \cdot 5^4$

19. $x^{-3} \cdot x^{-4}$

20. $(5x^{-3}y)(4x^{12}y^5)$

21. $(a^{-9}b^{-4})(a^2b^7)$

Answers on page A-30

The product rule, developed in Section 10.2, still holds when exponents are zero or negative.

EXAMPLES Simplify. Use positive powers in the answer.

14. $7^{-3} \cdot 7^6 = 7^{-3+6}$ Adding exponents
$$= 7^3$$

15. $x^4 \cdot x^{-3} = x^{4+(-3)} = x^1 = x$

16. $(2a^3b^{-4})(3a^2b^7) = 2 \cdot 3 \cdot a^3 \cdot a^2 \cdot b^{-4} \cdot b^7$ Using the commutative and associative laws
$$= 6a^{3+2}b^{-4+7}$$ Using the product rule
$$= 6a^5b^3$$

17. $(x^{-4}y^5)(x^7y^{-11}) = x^{-4+7}y^{5+(-11)}$
$$= x^3y^{-6}$$
$$= \frac{x^3}{y^6}$$

Do Exercises 18–21.

Study Tips

GET SOME REST

The final exam is probably your most important math test of the semester. Do yourself a favor and see to it that you get a good night's sleep the night before. Being well rested will help guarantee that you put forth your best work.

a   Evaluate.

1. 4^0

2. 17^0

3. 3.14^0

4. 2.67^1

5. $(-19.57)^1$

6. $(-34.6)^0$

7. $(-98.6)^0$

8. $(-98.6)^1$

9. $x^0, x \neq 0$

10. $a^0, a \neq 0$

11. $(3x - 17)^0$, for $x = 10$

12. $(7x - 45)^0$, for $x = 8$

13. $(5x - 3)^1$, for $x = 4$

14. $(35 - 4x)^1$, for $x = 8$

15. $(4m - 19)^0$, for $m = 3$

16. $(9 - 2x)^0$, for $x = 5$

17. $3x^0 + 4$, for $x = -2$

18. $7x^0 + 6$, for $x = -3$

19. $(3x)^0 + 4$, for $x = -2$

20. $(7x)^0 + 6$, for $x = -3$

21. $(5 - 3x^0)^1$, for $x = 19$

22. $(5x^1 - 29)^0$, for $x = 4$

b   Write an equivalent expression with positive exponents. Then simplify, if possible.

23. 3^{-2}

24. 2^{-3}

25. 10^{-4}

26. 5^{-6}

27. t^{-4}

28. x^{-2}

29. $(-5)^{-2}$

30. $(-4)^{-3}$

31. $3x^{-7}$

32. $-6y^{-2}$

33. $\dfrac{x}{y^{-4}}$

34. $\dfrac{r}{t^{-7}}$

35. $\dfrac{r^5}{t^{-3}}$

36. $\dfrac{x^7}{y^{-5}}$

37. $-7a^{-9}$

38. $9p^{-4}$

39. $\left(\dfrac{2}{5}\right)^{-2}$

40. $\left(\dfrac{3}{7}\right)^{-2}$

41. $\left(\dfrac{5}{a}\right)^{-3}$

42. $\left(\dfrac{x}{3}\right)^{-4}$

Write an equivalent expression using negative exponents.

43. $\dfrac{1}{7^3}$

44. $\dfrac{1}{5^2}$

45. $\dfrac{9}{x^3}$

46. $\dfrac{4}{y^2}$

Simplify. Do not use negative exponents in the answer.

47. $x^{-2} \cdot x$

48. $x \cdot x^{-1}$

49. $x^4 \cdot x^{-4}$

50. $x^9 \cdot x^{-9}$

51. $t^{-4} \cdot t^{-11}$

52. $y^{-5} \cdot y^{-8}$

53. $(3a^2b^{-7})(2ab^9)$

54. $(5xy^8)(3x^4y^{-5})$

55. $(-2x^{-3}y^8)(3xy^{-2})$

56. $(5a^{-1}b^{-7})(-2a^4b^2)$

57. $(3a^{-4}bc^2)(2a^{-2}b^{-5}c)$

58. $(5x^2y^{-7}z)(-4xy^{-3}z^{-4})$

59. D$_W$ Consider the expression x^{-3}. When evaluated, will the expression ever be negative? Explain.

60. D$_W$ What number is larger and why: 5^{-8} or 6^{-8}? Do not use a calculator.

↪ **VOCABULARY REINFORCEMENT**

In each of Exercises 61–68, fill in the blank with the correct term from the given list. Some of the choices may not be used and some may be used more than once.

61. A _____ is a polynomial with two terms.   [10.2b]

62. A parallelogram is a four-sided figure with two pairs of _____ sides.   [9.3a]

63. In the metric system, the _____ is the basic unit of mass.   [9.7b]

64. A natural number, other than 1, that is not _____ is composite.   [3.2b]

65. A(n) _____ is a set of points consisting of two rays with a common endpoint.   [9.5a]

66. To convert from _____ to _____ , move the decimal point two places to the left and change the ¢ sign at the end to the $ sign in front.   [5.3b]

67. The _____ of a polygon is the sum of the lengths of its sides.   [2.7b]

68. The number 1 is known as the _____ identity, and the number 0 is known as the _____ identity.   [1.5a], [2.2a]

prime

composite

binomial

trinomial

dollars

cents

perimeter

parallel

perpendicular

additive

multiplicative

meter

gram

vertex

angle

area

> **SYNTHESIS**

69. D$_W$ Explain in your own words why it is that the product rule applies to negative exponents.

70. D$_W$ True or false, and why: If $a > b$, then $a^{-1} < b^{-1}$?

71. D$_W$ Under what circumstances is $a^0 > a^1$?

72. ▦ Evaluate $\dfrac{3^x}{3^{x-1}}$ for $x = -4$ and then for $x = -40$.

73. ▦ Evaluate $\dfrac{5^x}{5^{x+1}}$ for $x = -3$ and then for $x = -30$.

74. D$_W$ How can negative exponents and the product rule be used to answer Exercises 72 and 73 without using a calculator?

Simplify.

75. $(y^{2x})(y^{3x})$

76. $a^{5k} \div a^{3k}$

77. $\dfrac{a^{6t}(a^{7t})}{a^{9t}}$

CHAPTER 10: Polynomials

10.5 SCIENTIFIC NOTATION

a Writing Scientific Notation

Objectives

a Convert between scientific notation and decimal notation.

b Multiply and divide using scientific notation.

c Solve applied problems using scientific notation.

There are many kinds of symbols, or notation, for numbers. You are already familiar with fraction notation, decimal notation, and percent notation. Now we study another, **scientific notation,** which makes use of exponential notation. Scientific notation is especially useful when calculations involve very large or very small numbers. The following are examples of scientific notation:

① *Niagara Falls*: On the Canadian side, during the summer the amount of water that spills over the falls in 1 day is about

$$4.9793 \times 10^{10} \text{ gal} = 49{,}793{,}000{,}000 \text{ gal.}$$

② *The mass of the earth*:

$$6.615 \times 10^{21} \text{ tons} = 6{,}615{,}000{,}000{,}000{,}000{,}000{,}000 \text{ tons.}$$

③ *The mass of a hydrogen atom*:

$$1.7 \times 10^{-24} \text{ g} = 0.0000000000000000000000017 \text{ g.}$$

①

②

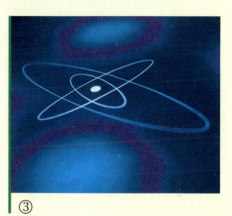

③

SCIENTIFIC NOTATION

Scientific notation for a number is an expression of the type

$$M \times 10^{n},$$

where n is an integer, M is greater than or equal to 1 and less than 10 ($1 \le M < 10$), and M is expressed in decimal notation.

You should try to make conversions to scientific notation mentally as much as possible. Here is a handy mental device.

In scientific notation, a positive power of 10 indicates a large number (greater than or equal to 10) and a negative power of 10 indicates a small number (between 0 and 1).

Convert each number to scientific
notation.

1. 0.000517

2. 523,000,000

Convert each number to decimal
notation.

3. 6.893×10^{11}

4. 5.67×10^{-5}

Answers on page A-30

To convert a number to scientific notation, we write it in the form $M \times 10^n$, where M looks just like the original number but with exactly one nonzero digit to the immediate left of the decimal point. For large numbers, this requires moving the decimal point n places to the left. For small numbers, the decimal point moves $|n|$ places to the right, where n is a negative integer.

EXAMPLES   Convert each number to scientific notation.

1. $78{,}000 = 78{,}000. = 7.8000 \times 10^4 = 7.8 \times 10^4$

4 places

Check: $7.8 \times 10^4 = 7.8 \times 10{,}000 = 78{,}000$

2. $0.0000923 = 000009.23 \times 10^{-5} = 9.23 \times 10^{-5}$

5 places

Check: $9.23 \times 10^{-5} = 9.23 \times 0.00001 = 0.0000923$

Each of the following is *not* scientific notation.

$$12.46 \times 10^7 \qquad\qquad 0.347 \times 10^{-5}$$

This number is greater than 10.   This number is less than 1.

Do Exercises 1 and 2.

The process is reversed when we convert from scientific notation to decimal notation.

EXAMPLES   Convert each number to decimal notation.

3. $7.893 \times 10^5 = 789{,}300$ The decimal point moves right 5 places.

Think: Positive exponent indicates large number.

Check: $7.893 \times 10^5 = 7.893 \times 100{,}000 = 789{,}300$

4. $4.7 \times 10^{-8} = 0.000000047$ The decimal point moves left 8 places.

Think: Negative exponent indicates small number.

Check: $4.7 \times 10^{-8} = 4.7 \times 0.00000001 = 0.000000047$

Do Exercises 3 and 4.

b Multiplying and Dividing Using Scientific Notation

MULTIPLYING

Consider the product

$$400 \cdot 2000 = 800,000.$$

In scientific notation, this is

$$(4 \times 10^2) \cdot (2 \times 10^3) = (4 \cdot 2)(10^2 \cdot 10^3) = 8 \times 10^5.$$

By applying the commutative and associative laws, we found this product by multiplying $4 \cdot 2$, to get 8, and $10^2 \cdot 10^3$, to get 10^5 (adding exponents).

EXAMPLE 5 Multiply: $(1.8 \times 10^6) \cdot (2.3 \times 10^{-4})$.

We apply the commutative and associative laws to get

$$(1.8 \times 10^6) \cdot (2.3 \times 10^{-4}) = (1.8 \cdot 2.3) \times (10^6 \cdot 10^{-4})$$
$$= 4.14 \times 10^{6+(-4)}$$
$$= 4.14 \times 10^2.$$

EXAMPLE 6 Multiply: $(3.1 \times 10^5) \cdot (4.5 \times 10^{-3})$.

We have

$$(3.1 \times 10^5) \cdot (4.5 \times 10^{-3}) = (3.1 \times 4.5)(10^5 \cdot 10^{-3})$$
$$= 13.95 \times 10^2 \qquad \text{Not scientific notation;}$$
$$\qquad\qquad\qquad\qquad \text{13.95 is greater than 10.}$$
$$= (1.395 \times 10^1) \times 10^2 \qquad \text{Substituting } 1.395 \times 10^1$$
$$\qquad\qquad\qquad\qquad \text{for 13.95}$$
$$= 1.395 \times (10^1 \times 10^2) \qquad \text{Using the associative law}$$
$$= 1.395 \times 10^3. \qquad \text{Adding exponents;}$$
$$\qquad\qquad\qquad\qquad \text{the answer is now in}$$
$$\qquad\qquad\qquad\qquad \text{scientific notation.}$$

Do Exercises 5 and 6.

DIVIDING

Consider the quotient

$$800,000 \div 400 = 2000.$$

In scientific notation, this is

$$(8 \times 10^5) \div (4 \times 10^2) = \frac{8 \times 10^5}{4 \times 10^2} = \frac{8}{4} \times 10^5 \cdot 10^{-2} = 2 \times 10^3.$$

EXAMPLE 7 Divide: $(3.41 \times 10^5) \div (1.1 \times 10^{-3})$.

We have

$$(3.41 \times 10^5) \div (1.1 \times 10^{-3}) = \frac{3.41 \times 10^5}{1.1 \times 10^{-3}}$$
$$= \frac{3.41}{1.1} \times \frac{10^5}{10^{-3}}$$
$$= 3.1 \times 10^5 \cdot 10^3$$
$$= 3.1 \times 10^8.$$

Multiply and write scientific notation for the result.

5. $(1.12 \times 10^{-8})(5 \times 10^{-7})$

6. $(9.1 \times 10^{-17})(8.2 \times 10^3)$

Answers on page A-30

CALCULATOR CORNER

Scientific Notation   Most scientific calculators have a key labeled [EE] which is used for scientific notation. We simply enter the decimal portion of the number, press [EE], and then enter the exponent. Addition, subtraction, multiplication, and division are performed as usual. The results will appear in scientific notation.

```
1.789ᴇ−11
              1.789ᴇ−11
```

To enter a number in scientific notation on a graphing calculator, we key in the decimal portion of the number, press [2nd] [EE] (EE is the second operation associated with the [·] key), and then key in the exponent. For example, to enter 1.789×10^{-11}, we press [1] [.] [7] [8] [9] [2nd] [EE] [(−)] [1] [1] [ENTER]. The decimal portion of the number appears before a small E and the exponent follows. To find the product in Example 5 and express the result in scientific notation, we first set the calculator to Scientific mode by pressing [MODE], positioning the cursor over Sci on the first line, and pressing [ENTER]. Then we press [2nd] [QUIT] to go to the home screen and press [1] [.] [8] [2nd] [EE] [6] [×] [2] [.] [3] [2nd] [EE] [(−)] [4] [ENTER].

```
Normal Sci Eng
Float 0123456789
Radian Degree
Func Par Pol Seq
Connected Dot
Sequential Simul
Real a+bi re^θi
Full Horiz G−T
```

Exercises: Multiply or divide and express each answer in scientific notation.

1. $(3.15 \times 10^7)(4.3 \times 10^{-12})$
2. $(4.76 \times 10^{-5})(1.9 \times 10^{10})$
3. $(8 \times 10^9)(4 \times 10^{-5})$
4. $(4 \times 10^4)(9 \times 10^7)$
5. $\dfrac{4.5 \times 10^6}{1.5 \times 10^{12}}$
6. $\dfrac{6.4 \times 10^{-5}}{1.6 \times 10^{-10}}$

```
1.8ᴇ6*2.3ᴇ−4
              4.14ᴇ2
```

7. $\dfrac{4 \times 10^{-9}}{5 \times 10^{16}}$
8. $\dfrac{9 \times 10^{11}}{3 \times 10^{-2}}$

Divide and write scientific notation for each result.

7. $\dfrac{4.2 \times 10^5}{2.1 \times 10^2}$

8. $\dfrac{1.1 \times 10^{-4}}{2.0 \times 10^{-7}}$

EXAMPLE 8   Divide: $(6.4 \times 10^{-7}) \div (8.0 \times 10^6)$.

We have

$$(6.4 \times 10^{-7}) \div (8.0 \times 10^6) = \frac{6.4 \times 10^{-7}}{8.0 \times 10^6}$$

$$= \frac{6.4}{8.0} \times \frac{10^{-7}}{10^6}$$

$$= 0.8 \times 10^{-7} \cdot 10^{-6}$$

$$= 0.8 \times 10^{-13} \qquad \text{Not scientific notation.}$$
$$\qquad\qquad\qquad\qquad\qquad \text{0.8 is less than 1.}$$

$$= (8.0 \times 10^{-1}) \times 10^{-13} \qquad \text{Substituting } 8.0 \times 10^{-1} \text{ for 0.8}$$

$$= 8.0 \times (10^{-1} \times 10^{-13}) \qquad \text{Using the associative law}$$

$$= 8.0 \times 10^{-14}. \qquad \text{Adding exponents}$$

Do Exercises 7 and 8.

Answers on page A-30

C   Applications with Scientific Notation

EXAMPLE 9   *Distance from the Sun to Earth.*   Light from the sun traveling at a rate of 300,000 km/s (kilometers per second) reaches Earth in 499 sec. Find the distance, expressed in scientific notation, from the sun to Earth.

   The time it takes light to reach Earth from the sun is 4.99×10^2 sec. The speed is 3.0×10^5 km/s. Recall that distance can be expressed in terms of speed and time as

   Distance = Speed · Time.

We substitute 3.0×10^5 for the speed and 4.99×10^2 for the time:

Distance $= (3.0 \times 10^5)(4.99 \times 10^2)$ Substituting
   $= 14.97 \times 10^7$ Note that 14.97 is greater than 10.
   $= (1.497 \times 10^1) \times 10^7$
   $= 1.497 \times (10^1 \times 10^7)$ ⎫ Converting to scientific notation
   $= 1.497 \times 10^8$ km.   ⎭

Thus the distance from the sun to Earth is 1.497×10^8 km.

Do Exercise 9.

EXAMPLE 10   *DNA.*   A strand of DNA (deoxyribonucleic acid) is about 150 cm long and 1.3×10^{-10} cm wide. How many times the width of DNA is the length?
Source: Human Genome Project Information

   To determine how many times the width can fit into the length, we divide the length by the width:

$$\frac{150}{1.3 \times 10^{-10}} = \frac{150}{1.3} \times \frac{1}{10^{-10}}$$

$$\approx 115.385 \times 10^{10}$$
$$= (1.15385 \times 10^2) \times 10^{10} \quad \text{Converting to}$$
$$= 1.15385 \times 10^{12}. \quad \text{scientific notation}$$

Thus the length of DNA is about 1.15385×10^{12} times its width.

Do Exercise 10.

---

9. Niagara Falls Water Flow.   On the Canadian side, during the summer the amount of water that spills over the falls in 1 min is about

$$1.3088 \times 10^8 \text{ L}.$$

How many liters of water spill over the falls in one day? Express the answer in scientific notation.

Source: *Collier's Encyclopedia*, 1997, Vol. 17

10. Earth vs. Saturn.   The mass of Earth is about 6×10^{21} metric tons. The mass of Saturn is about 5.7×10^{23} metric tons. About how many times the mass of Earth is the mass of Saturn? Express the answer in scientific notation.

Answers on page A-30

Translating
for Success

1. Servings of Pork. An 8-lb pork roast contains 37 servings of meat. How many pounds of pork would be needed for 55 servings?

2. Height of a Ladder. A 14.5-ft ladder leans against a house. The bottom of the ladder is 9.4 ft from the building. How high is the top of the ladder?

3. Cruise Cost. A group of 6 college students pays $4608 for a spring break cruise. What is each person's share?

4. Sales Tax Rate. The sales tax is $14.95 on the purchase of a new ladder which costs $299. What is the sales tax rate?

5. Elevation. Jamal hiked from an elevation 450 ft above sea level to an elevation 25 ft below sea level. What was the change in his elevation?

The goal of these matching questions is to practice step (2), *Translate*, of the five-step problem-solving process. Translate each word problem to an equation and select a correct translation from equations A–O.

A. $x = 450 - (-25)$

B. $6 \cdot x = \$4608$

C. $x = (7 \times 10^2)(2.9 \times 10^8)$

D. $x = \pi \cdot \left(5\frac{1}{2} \div 2\right)^2 \cdot 7$

E. $x = 6\% \times 5 \times \$14.95$

F. $x = \pi\left(5\frac{1}{2}\right)^2 + 7$

G. $(9.4)^2 + x^2 = (14.5)^2$

H. $\$14.95 = x \cdot \$299$

I. $x = 2(14.5 + 9.4)$

J. $(9.4 + 14.5)^2 = x$

K. $\dfrac{8}{37} = \dfrac{x}{55}$

L. $x = (2.9 \times 10^8) \div (7 \times 10^2)$

M. $x = 6 \cdot \$4608$

N. $8 \cdot 37 = 55 \cdot x$

O. $x = 450 - 25$

Answers on page A-30

---

6. Inheritance. Each of 6 children inherits $4608 from their mother's estate. What is the total inheritance?

7. Sales Tax. Erica buys 5 pairs of earrings at $14.95 each. The sales tax rate in Indiana is 6%. How much sales tax will be charged?

8. Food Consumption. In 2003, Americans ate 700 lb of fruits and vegetables per person. If there were 290 million Americans in 2003, how many pounds of fruits and vegetables were consumed?

9. Volume of a Storage Tank. The diameter of a cylindrical grain-storage tank is $5\frac{1}{2}$ yd. Its height is 7 yd. Find its volume.

10. Perimeter of a Photo. A rectangular photo is 14.5 cm by 9.4 cm. What is the perimeter of the photo?

a   Convert each number to scientific notation.

1. 28,000,000,000

2. 4,900,000,000,000

3. 907,000,000,000,000,000

4. 168,000,000,000,000

5. 0.00000304

6. 0.000000000865

7. 0.000000018

8. 0.00000000002

9. 100,000,000,000

10. 0.0000001

Convert the number in each sentence to scientific notation.

11. *Population of the United States.*   After the 2000 census, the population of the United States was 281 million (1 million $= 10^6$).
Source: U.S. Bureau of the Census

12. *NASCAR.*   Total revenue of NASCAR (National Association of Stock Car Auto Racing) is expected to be $3423 million by 2006.
Source: NASCAR

13. *California Lottery.*   The probability of winning the California state lottery is about 67/1,000,000,000.
Source: James Walsh, *True Odds: How Risk Affects Your Everyday Life,* Silver Lake Publishing, 1996, p. 342

14. *Cancer Death Rate.*   In Michigan, the death rate due to cancer is about 127.1/1000.
Source: AARP

Convert each number to decimal notation.

15. 8.74×10^7

16. 1.85×10^8

17. 5.704×10^{-8}

18. 8.043×10^{-4}

19. 10^7

20. 10^6

21. 10^{-5}

22. 10^{-8}

b   Multiply or divide and write scientific notation for each result.

23. $(3 \times 10^4)(2 \times 10^5)$

24. $(3.9 \times 10^8)(8.4 \times 10^{-3})$

25. $(5.2 \times 10^5)(6.5 \times 10^{-2})$

26. $(7.1 \times 10^{-7})(8.6 \times 10^{-5})$

27. $(9.9 \times 10^{-6})(8.23 \times 10^{-8})$

28. $(1.123 \times 10^4) \times 10^{-9}$

715

29. $\dfrac{8.5 \times 10^8}{3.4 \times 10^{-5}}$

30. $\dfrac{5.6 \times 10^{-2}}{2.5 \times 10^5}$

31. $(3.0 \times 10^6) \div (6.0 \times 10^9)$

32. $(1.5 \times 10^{-3}) \div (1.6 \times 10^{-6})$

33. $\dfrac{7.5 \times 10^{-9}}{2.5 \times 10^{12}}$

34. $\dfrac{4.0 \times 10^{-3}}{8.0 \times 10^{20}}$

C Solve.

35. *River Discharge.* The average discharge at the mouths of the Amazon River is 4,200,000 cubic feet per second. How much water is discharged from the Amazon River in 1 yr? Express the answer in scientific notation.

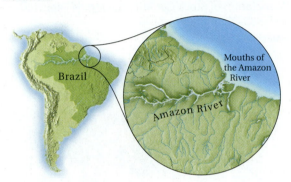

36. *Computers.* A gigabyte is a measure of a computer's storage capacity. One gigabyte holds about 1 billion bytes of information. If a firm's computer network contains 2500 gigabytes of memory, how many bytes are in the network? Express the answer in scientific notation.

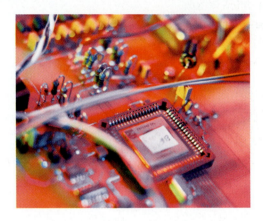

37. *Earth vs. Jupiter.* The mass of Earth is about 6×10^{21} metric tons. The mass of Jupiter is about 1.908×10^{24} metric tons. About how many times the mass of Earth is the mass of Jupiter? Express the answer in scientific notation.

38. *Water Contamination.* In the United States, 200 million gal of used motor oil are improperly disposed of each year. One gallon of used oil can contaminate 1 million gal of drinking water. How many gallons of drinking water can 200 million gal of oil contaminate? Express the answer in scientific notation.

Source: *The Macmillan Visual Almanac*

39. *Stars.*   It is estimated that there are 10 billion trillion stars in the universe. Express the number of stars in scientific notation (1 billion = 10^9; 1 trillion = 10^{12}).

40. *Closest Star.*   Excluding the sun, the closest star to Earth is Proxima Centauri, which is 4.3 light-years away (one light-year = 5.88×10^{12} mi). How far, in miles, is Proxima Centauri from Earth? Express the answer in scientific notation.

41. *Earth vs. Sun.*   The mass of Earth is about 6×10^{21} metric tons. The mass of the sun is about 1.998×10^{27} metric tons. About how many times the mass of Earth is the mass of the sun? Express the answer in scientific notation.

42. *Red Light.*   The wavelength of light is given by the velocity divided by the frequency. The velocity of red light is 300,000,000 m/sec, and its frequency is 400,000,000,000,000 cycles per second. What is the wavelength of red light? Express the answer in scientific notation.

Space Travel.   Use the following information for Exercises 43 and 44.

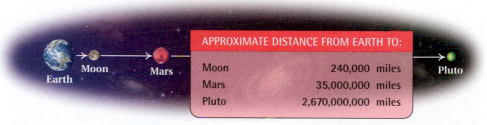

APPROXIMATE DISTANCE FROM EARTH TO:	
Moon	240,000 miles
Mars	35,000,000 miles
Pluto	2,670,000,000 miles

43. *Time to Reach Mars.*   Suppose that it takes about 3 days for a spacecraft to travel from Earth to the moon. About how long would it take the same spacecraft traveling at the same speed to reach Mars? Express the answer in scientific notation.

44. *Time to Reach Pluto.*   Suppose that it takes about 3 days for a spacecraft to travel from Earth to the moon. About how long would it take the same spacecraft traveling at the same speed to reach Pluto? Express the answer in scientific notation.

45. **D**_W Emma can give an answer using the unit km or mm. Which measurement would require the larger exponent if scientific notation were used? Why?

46. **D**_W Why is it important to be able to add negative integers before learning how to use scientific notation?

SKILL MAINTENANCE

Solve. [5.7b]

47. $2x - 4 - 5x + 8 = x - 3$

48. $8x + 7 - 9x = 12 - 6x + 5$

49. $8(2x + 3) - 2(x - 5) = 10$

50. $4(x - 3) + 5 = 6(x + 2) - 8$

Graph. [6.4b]

51. $y = x - 5$

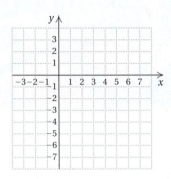

52. $y = -2x + 8$

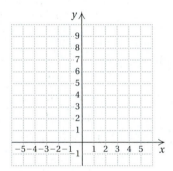

SYNTHESIS

53. **D**_W Why do you think scientific notation was first devised?

54. **D**_W Explain how it is possible to design a problem in which the problem cannot be solved by a calculator, despite the fact that scientific notation is in use.

55. ▦ Carry out the indicated operations. Express the result in scientific notation.

$$\frac{(5.2 \times 10^6)(6.1 \times 10^{-11})}{1.28 \times 10^{-3}}$$

56. Find the reciprocal and express it in scientific notation.

$$6.25 \times 10^{-3}$$

57. Find the LCM for 6.4×10^8 and 1.28×10^4.

10 Summary and Review

The review that follows is meant to prepare you for a chapter exam. It consists of three parts. The first part, Concept Reinforcement, is designed to increase understanding of the concepts through true/false exercises. The second part is a list of important properties and formulas. The third part is the Review Exercises. These provide practice exercises for the exam, together with references to section objectives so you can go back and review. Before beginning, stop and look back over the skills you have obtained. What skills in mathematics do you have now that you did not have before studying this chapter?

✎ CONCEPT REINFORCEMENT

Determine whether each statement is true or false. Answers are given at the back of the book.

_____ **1.** If a polynomial is written as a product, it is factored.

_____ **2.** FOIL stands for First, Outer, Inner, Last.

_____ **3.** If x^7 is divided by x^7, the result is the same as x^0.

_____ **4.** Raising a number to a negative power may not result in a negative number.

_____ **5.** Any number raised to the zero power is zero.

_____ **6.** In scientific notation, if the exponent is negative, the number is very large.

IMPORTANT PROPERTIES AND FORMULAS

Exponents: $\qquad$ $b^1 = b;$ $\quad a^0 = 1$ for $a \neq 0$

The Product Rule: $\quad a^m \cdot a^n = a^{m+n}$

Negative exponents: $\quad a^{-n} = \dfrac{1}{a^n}$ and $\left(\dfrac{a}{b}\right)^{-n} = \left(\dfrac{b}{a}\right)^n$

Review Exercises

Perform the indicated operation. [10.1a, c]

1. $(-4x + 9) + (7x - 15)$

2. $(7x^4 - 5x^3 + 3x - 5) + (x^3 - 4x + 2)$

3. $(9a^5 + 8a^3 + 4a + 7) - (a^5 - 4a^3 + a^2 - 2)$

4. $(8a^3b^3 + 9a^2b^3) - (3a^3b^3 - 2a^2b^3 + 7)$

5. Find two equivalent expressions for the opposite of $12x^3 - 4x^2 + 9x - 3$. [10.1b]

Evaluate.

6. $(-59)^0$ [10.4a]

7. $5t^3 + t$, for $t = -2$ [10.1d]

8. The altitude, in feet, of a falling golf ball t seconds after it reaches the peak of its flight can be estimated by $-16t^2 + 200$. Find the ball's altitude 3 sec after it has reached its peak.   [10.1d]

Multiply.

9. $(5x^3)(6x^4)$ [10.2a]

10. $3x(6x^3 - 4x - 1)$ [10.2b]

11. $2a^4b(7a^3b^3 + 5a^2b^3)$ [10.2b]

12. $(x - 7)(x + 9)$ [10.3a]

13. $(3x - 1)(5x - 2)$ [10.3a]

14. $(a^2 - 1)(a^2 + 2a - 1)$ [10.3b]

Factor.   [10.2c]

15. $45x^3 - 10x$

16. $7a - 35b - 49ac$

17. $6x^3y - 9x^2y^5$

Write an equivalent expression using positive exponents. Then simplify, if possible.   [10.4b]

18. 12^{-2}

19. $8a^{-7}$

20. $\dfrac{x^{-3}}{y^5 z^{-6}}$

21. $\left(\dfrac{4}{5}\right)^{-2}$

22. Write an expression equivalent to $\dfrac{1}{x^7}$ using a negative exponent.   [10.4b]

Simplify. Use positive exponents in the answer.   [10.4b]

23. $x^{-5} \cdot x^{-12}$

24. $(-7x^3y^{-5})(-2x^4y^{-2})$

25. Write scientific notation for 42,700,000.   [10.5a]

26. Write scientific notation for 0.0001924.   [10.5a]

Simplify. Write the answer in scientific notation.   [10.5b]

27. $(5.1 \times 10^6)(2.3 \times 10^4)$

28. The wavelength of a radio wave is given by the velocity divided by the frequency. The velocity of radio waves is approximately 300,000,000 m/sec, and the frequency of Rick's ham radio repeater is 1,200,000,000 cycles per sec. What is the wavelength of Rick's repeater frequency? Express the answer in scientific notation.   [10.5c]

29. A good length for a radio antenna is one fourth of the wavelength of the frequency it is designed to receive. What should the length of such a *quarter-wave* antenna be for the frequency described in Exercise 28?   [10.5c]

30. D$_W$ Adi claims that
$$(3x^{-5})(-4x^{-2}) = -x^{10}.$$
What mistake(s) is she probably making?   [10.4b]

31. D$_W$ Can x^{-2} represent a negative number? Why or why not?   [10.4b]

SYNTHESIS

Simplify.

32. ▦ $(2349x^7 - 357x^2)(493x^{10} + 597x^5)$ [10.3a]

33. $-3x^5 \cdot 3x^3 - x^6(2x)^2 + (3x^4)^2 + (2x^4)^2 - 40x^2(x^3)^2$ [10.2a]

Factor.   [10.2c]

34. $39a^3b^7c^6 - 130a^2b^5c^8 + 52a^4b^6c^5$

35. $w^5x^6y^4z^5 - w^7x^3y^7z^3 + w^6x^2y^5z^6 - w^6x^7y^3z^4$

36. $10a^4b^{-5} + 12a^7b^{-3}$

721

1. Add: $(12a^3 - 9a^2 + 8) + (6a^3 + 4a^2 - a)$.

2. Find two equivalent expressions for the opposite of $-9a^4 + 7b^2 - ab + 3$.

3. Subtract: $(12x^4 + 7x^2 - 6) - (9x^4 + 8x^2 + 5)$.

Evaluate.

4. 193^1

5. $(3x - 7)^0$, for $x = 2$

6. The height, in meters, of a ball t sec after it has been thrown is approximated by $-4.9t^2 + 15t + 2$. How high is the ball 2 sec after it has been thrown?

Multiply.

7. $(-5x^4y^3)(2x^2y^5)$

8. $2a(5a^2 - 4a + 3)$

9. $(x - 5)(x + 9)$

10. $(2a + 1)(a^2 - 3a + 2)$

Factor.

11. $35x^6 - 25x^3 + 15x^2$

12. $6ab - 9bc + 12ac$

Write an equivalent expression with positive exponents. Then simplify, if possible.

13. 5^{-3}

14. $\dfrac{5a^{-3}}{b^{-2}}$

15. $\left(\dfrac{3}{5}\right)^{-3}$

Simplify. Use positive exponents in the answer.

16. $x^{-7} \cdot x^{-9}$

17. $(3a^{-7}b^9)(-2a^{10}b^{-12})$

18. Write scientific notation for 0.00047.

19. Write scientific notation for 8,250,000.

20. Find the product and write the answer using scientific notation:
$(3.2 \times 10^{-8})(5.7 \times 10^{-9})$.

| SYNTHESIS |

21. The polynomial
$$0.041h - 0.018A - 2.69$$
can be used to estimate the lung capacity, in liters, of a female of height h, in centimeters, and age A, in years. Find the lung capacity of a 30-yr-old woman who is 150 cm tall.

22. Write an equivalent expression with positive exponents and then simplify:
$$12a^6(2a^3 - 6a)^{-2}.$$

723

This exam reviews the entire textbook. A question may arise as to what notation to use for a particular problem or exercise. Although there is no hard-and-fast rule, especially as you use mathematics outside the classroom, here is the guideline that we follow: Use the notation given in the problem. That is, if the problem is given using mixed numerals, give the answer as a mixed numeral. If the problem is given in decimal notation, give the answer in decimal notation.

Solve.

1. *Quito, Ecuador.* In Quito, Ecuador, there are 1.5 million people living under threat of ash clouds and mudflows from a giant volcano, Mt. Antisana. Find standard notation for 1.5 million.
   Source: *National Geographic Magazine, February 2001*

2. *Dead Sea.* The lowest point in the world is the Dead Sea on the border of Israel and Jordan. It is 1312 ft below sea level. Convert 1312 ft to yards; to meters.
   Source: *The Handy Geography Answer Book*

Egg Consumption. As shown in the line graph below, egg consumption per person in the United States has been increasing in recent years. Use the graph for Exercises 3–8.

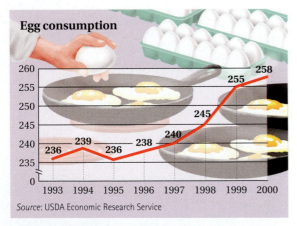

Egg consumption

260 · · · · · · · · · · · · · · · 258
255 · · · · · · · · · · · 255
250
245 · · · · · · 245
240 · · · · 239 · · · 238 · 240
236 · · · · 236
235

1993 1994 1995 1996 1997 1998 1999 2000

Source: USDA Economic Research Service

3. Find the lowest egg consumption and the year(s) in which it occurred.

4. Find the highest egg consumption and the year(s) in which it occurred.

5. Find the mean, the median, and the mode of the egg consumptions.

6. Find the mean egg consumption over the years 1997 to 2000.

7. Find the mean egg consumption over the years 1993 to 1996. How does it compare to the answer to Exercise 6?

8. What was the percent of increase in egg consumption from 1996 to 2000?

9. In Sam's writing lab, 3 of the 20 students are left-handed. If a student is randomly selected, what is the probability that he or she is left-handed?

10. Find the missing angle measure.

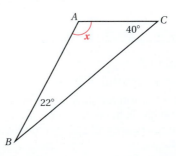

Add and, if possible, simplify.

11.
```
    4 9 0 3
    5 2 7 8
    6 3 9 1
+   4 5 1 3
```

12. $5\dfrac{4}{9}$
$+ \; 3\dfrac{1}{3}$

13. $-29 + 53$

14. $-543 + (-219)$

15. $-34.56 + 2.783 + 0.433 + (-13.02)$

16. $(4x^5 + 7x^4 - 3x^2 + 9) + (6x^5 - 8x^4 + 2x^3 - 7)$

Subtract and, if possible, simplify.

17.
```
    6 7 4
-   4 3 1
```

18. $-4x - 13x$

19. $\dfrac{2}{5} - \dfrac{7}{8}$

20. $4\dfrac{1}{3}$
$- \; 1\dfrac{5}{8}$

21.
```
    2 0.0
-     0.0 0 2 7
```

22. $(7x^3 + 2x^2 - x) - (5x^3 - 3x^2 - 8x)$

23. $(9a^2b + 3ab) - (13a^2b - 4ab)$

Multiply and, if possible, simplify.

24.
```
    2 9 7
×     1 6
```

25. $349 \cdot (-213)$

26. $2\dfrac{3}{4} \cdot 1\dfrac{2}{3}$

27. $-\dfrac{9}{7} \cdot \dfrac{14}{15}$

28. $12 \cdot \dfrac{5}{6}$

29.
```
    3 4.0 9
×       7.6
```

30. $3(8x - 5)$

31. $(9a^3b^2)(3a^5b)$

32. $7x^2(3x^3 - 2x + 8)$

33. $(x + 2)(x - 7)$

34. $(a + 3)(a^2 - 5a + 4)$

Divide and simplify. State the answer using a remainder when appropriate.

35. $6\overline{)3\,4\,3\,8}$

36. $3\,4\overline{)1\,9\,1\,4}$

Divide and, if possible, simplify.

37. $\dfrac{4}{5} \div \left(-\dfrac{8}{15}\right)$

38. $-2\dfrac{1}{3} \div (-30)$

39. $2.7\overline{)1\,0\,5.3}$

40. Write a mixed numeral for the quotient in Question 36.

Simplify.

41. $10 \div 2 \times 20 - 5^2$

42. $\dfrac{|3^2 - 5^2|}{2 - 2 \cdot 5}$

43. Write exponential notation: $14 \cdot 14 \cdot 14$.

44. Round 68,489 to the nearest thousand.

45. Round $21.\overline{83}$ to the nearest hundredth.

46. Determine whether 1368 is divisible by 3.

47. Find all the factors of 15.

48. Find the LCM of 15 and 35.

49. Simplify $\dfrac{24}{33}$.

50. Convert to a mixed numeral: $-\dfrac{18}{5}$.

51. Use < or > for $\square$ to write a true sentence:
$$-17 \ \square \ -29.$$

52. Use < or > for $\square$ to write a true sentence:
$$\dfrac{4}{7} \ \square \ \dfrac{3}{5}.$$

53. Which number is greater, 1.001 or 0.9976?

54. Evaluate $\dfrac{a^2 - b}{3}$ for $a = -9$ and $b = -6$.

Factor.

55. $40 - 5t$

56. $18a^3 - 15a^2 + 6a$

57. What part is shaded?

Write decimal notation for each number.

58. $\dfrac{429}{10,000}$

59. $-\dfrac{13}{25}$

60. $\dfrac{8}{9}$

61. 7%

Write each number in fraction notation.

62. 6.71

63. $-7\dfrac{1}{4}$

64. 40%

Write each number in percent notation.

65. $\dfrac{17}{20}$

66. 1.5

67. Estimate the sum $9.389 + 4.2105$ to the nearest tenth.

Solve.

68. $234 + y = 789$

69. $3.9a = 249.6$

70. $\dfrac{2}{3} \cdot t = \dfrac{5}{6}$

71. $\dfrac{8}{17} = \dfrac{36}{x}$

72. $7x - 9 = 26$

73. $-2(x - 5) = 3x + 12$

Solve.

74. Margie donated $20 to the Humane Society, $30 to the Red Cross, $25 to the Salvation Army, and $20 to Amnesty International. What was the average size of the donations?

75. A machine wraps 134 candy bars per minute. How long does it take this machine to wrap 8710 bars?

76. On Monday morning, a bolt of fabric contained $8\frac{1}{4}$ yd. Madison sold $3\frac{5}{8}$ yd from the bolt. How much fabric remains?

77. At the start of a trip, the odometer on the Oquendos' minivan read 27,428.6 mi and at the end of the trip the reading was 27,914.5 mi. How long was the trip?

78. From Indira's income of $32,000, amounts of $6400 and $1600 are paid for federal and state taxes. How much remains after these taxes have been paid?

79. Shannon is paid $85 a day for 7 days' work as a lifeguard. How much will she be paid?

80. A toddler walks $\frac{3}{5}$ km per hour. At this rate, how far would the child walk in $\frac{1}{2}$ hr?

81. Eight identical dresses cost a total of $679.68. What is the cost of each dress?

82. Eight gallons of paint covers 2000 ft^2. How much paint is needed to cover 3250 ft^2?

83. Eighteen ounces of a fruit "smoothie" costs $3.06. Find the unit price in cents per ounce.

84. What is the simple interest on $4000 principal at 8% for $\frac{3}{4}$ yr?

85. Baldacci Real Estate received $5880 commission on the sale of an $84,000 home. What was the rate of commission?

86. The population of Bridgeton is 29,000 this year and is increasing at a rate of 4% per year. What will the population be next year?

87. Luis paid $35 a day plus 15¢ a mile for a van rental. If his one-day van rental cost $68, how many miles did he drive?

88. *Medical Dosage.* A doctor suggests that a child who weighs 24 kg be given 42 mg of Phenytoin. If the dosage is proportional to the child's weight, how much Phenytoin is recommended for a child who weighs 32 kg?

89. *Firefighting.* During a fire, firefighters get a 1-ft layer of water on the 25-ft by 60-ft first floor of a 5-floor building. Water weighs $62\frac{1}{2}$ lb per cubic foot. What is the total weight of the water on the floor?

Evaluate.

90. 18^2

91. 37^0

92. $\sqrt{121}$

Write an equivalent expression with positive exponents. Then simplify, if possible.

93. 4^{-3}

94. $\left(\dfrac{5}{4}\right)^{-2}$

Express each of the following in scientific notation.

95. 4,357,000

96. $(6.2 \times 10^7)(4.3 \times 10^{-23})$

97. Plot the following points:
$(-5, 2), (4, 0), (3, -4), (0, 2).$

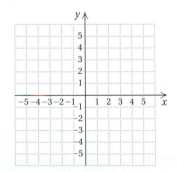

98. Graph: $y = -\dfrac{1}{3}x.$

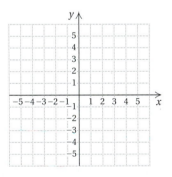

99. These triangles are similar. Find the missing lengths.

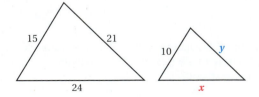

Complete.

100. $\frac{1}{3}$ yd = _____ in.

101. 3917 mm = _____ cm

102. 5.8 km = _____ m

103. 60,000 g = _____ kg

104. 10 lb = _____ oz

105. 2.3 g = _____ mg

106. 8190 mL = _____ L

107. 28 qt = _____ gal

The data in the following table show the percent of people who eat salad a certain number of times per week.

NUMBER OF SALADS PER WEEK	PERCENT
None	3%
2 or fewer	37%
3–6	47%
At least one a day	13%

Source: Market Facts for the Association of Dressings and Sauces

108. Make a circle graph of the data.

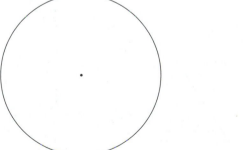

109. Make a bar graph of the data.

110. A rectangular mirror measures 20 in. by 24 in. Find its perimeter.

Find the area of each figure.

111.

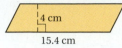

4 cm

15.4 cm

112.

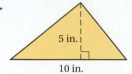

5 in.

10 in.

113.

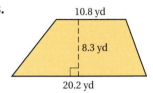

10.8 yd

8.3 yd

20.2 yd

114.

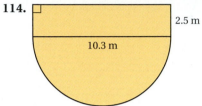

2.5 m

10.3 m

115. Find the diameter, the circumference, and the area of this circle. Use 3.14 for π.

10.4 in.

Find the volume of each shape. Use 3.14 for π.

116.

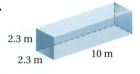

2.3 m

2.3 m

10 m

117.

16 ft

4 ft

118.

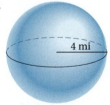

4 mi

119. Find the length of the third side of this right triangle. Give an exact answer and an approximation to three decimal places.

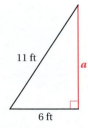

11 ft

a

6 ft

Developmental Units

D

A	Addition
S	Subtraction
M	Multiplication
D	Division

Objectives

a Add any two of the numbers 0, 1, 2, 3, 4, 5, 6, 7, 8, 9.

b Find certain sums of three numbers such as $1 + 7 + 9$.

c Add two whole numbers when carrying is not necessary.

d Add two whole numbers when carrying is necessary.

Add; think of joining sets of objects.

1. $4 + 5$ **2.** $5 + 2$

3. $\begin{array}{r} 9 \\ + 5 \\ \hline \end{array}$ **4.** $\begin{array}{r} 8 \\ + 8 \\ \hline \end{array}$

5. $\begin{array}{r} 9 \\ + 7 \\ \hline \end{array}$ **6.** $\begin{array}{r} 7 \\ + 9 \\ \hline \end{array}$

The first printed use of the $+$ symbol was in a book by a German, Johann Widmann, in 1498.

A ADDITION

a Basic Addition

Basic addition can be explained by counting. The sum

$$3 + 4$$

can be found by counting out a set of 3 objects and a separate set of 4 objects, putting them together, and counting all the objects.

A set of   +   A set of   =   A set of
3   4   7

The numbers to be added are called **addends.** The result is the **sum.**

$$\underset{\text{Addend}}{3} \;+\; \underset{\text{Addend}}{4} \;=\; \underset{\text{Sum}}{7}$$

EXAMPLES Add. Think of putting sets of objects together.

1. $5 + 6 = 11$

$$\begin{array}{r} 5 \\ + 6 \\ \hline 11 \end{array}$$

2. $8 + 5 = 13$

$$\begin{array}{r} 8 \\ + 5 \\ \hline 13 \end{array}$$

We can also do these problems by counting up from one of the numbers. For example, in Example 2, we start at 8 and count up 5 times: 9, 10, 11, 12, 13.

Do Exercises 1–6.

What happens when we add 0? Think of a set of 5 objects. If we add 0 objects to it, we still have 5 objects. Similarly, if we have a set with 0 objects in it and add 5 objects to it, we have a set with 5 objects. Thus,

$$5 + 0 = 5 \quad \text{and} \quad 0 + 5 = 5.$$

> **ADDITION OF 0**
>
> Adding 0 to a number does not change the number:
>
> $$a + 0 = 0 + a = a.$$
>
> We say that 0 is the **additive identity.**

Answers on page A-31

EXAMPLES Add.

3. $0 + 9 = 9$

$$\begin{array}{r} 0 \\ + \ 9 \\ \hline 9 \end{array}$$

4. $0 + 0 = 0$

$$\begin{array}{r} 0 \\ + \ 0 \\ \hline 0 \end{array}$$

5. $97 + 0 = 97$

$$\begin{array}{r} 97 \\ + \ 0 \\ \hline 97 \end{array}$$

Do Exercises 7–12.

Your objective for this part of the section is to be able to add any two of the numbers 0, 1, 2, 3, 4, 5, 6, 7, 8, 9. Adding 0 is easy. The rest of the sums are listed in this table. Memorize the table by saying it to yourself over and over or by using flash cards.

+	1	2	3	4	5	6	7	8	9
1	2	3	4	5	6	7	8	9	10
2	3	4	5	6	7	8	9	10	11
3	4	5	6	7	8	9	10	11	12
4	5	6	7	8	9	10	11	12	13
5	6	7	8	9	10	11	12	13	14
6	7	8	9	10	11	12	13	14	15
7	8	9	10	11	12	13	14	15	16
8	9	10	11	12	13	14	15	16	17
9	10	11	12	13	14	15	16	17	18

$6 + 7 = 13$
Find 6 at the left, and 7 at the top.

$7 + 6 = 13$
Find 7 at the left, and 6 at the top.

It is very important that you *memorize* the basic addition facts! If you do not, you will always have trouble with addition.

Note the following.

$3 + 4 = 7 \qquad 7 + 6 = 13 \qquad 7 + 2 = 9$
$4 + 3 = 7 \qquad 6 + 7 = 13 \qquad 2 + 7 = 9$

We can add whole numbers in any order. This is the *commutative law of addition*. Because of this law, you need to learn only about half the table above, as shown by the shading.

Do Exercises 13 and 14.

b   Certain Sums of Three Numbers

To add $3 + 5 + 4$, we can add 3 and 5, then 4:

$3 + 5 + 4$

$8 + 4$

$12.$

We can also add 5 and 4, then 3:

$3 + 5 + 4$

$3 + 9$

$12.$

Either way we get 12.

Add.

7. $8 + 0$

8. $0 + 8$

9. $\begin{array}{r} 7 \\ + \ 0 \\ \hline \end{array}$

10. $\begin{array}{r} 46 \\ + \ 0 \\ \hline \end{array}$

11. $0 + 13$

12. $58 + 0$

Complete the table.

13.

+	1	2	3	4	5
1			4		
2					
3				7	
4					
5					

14.

+	6	5	7	4	9
7			14		
9					
5			9		
8					
4					

Answers on pages A-31–A-32

A   Addition

Add from the top mentally.

15.
```
  1
  6
+ 9
```

16.
```
  2
  3
+ 4
```

17.
```
  6
  1
+ 4
```

18.
```
  5
  2
+ 8
```

Add.

19.
```
  2 4
+ 3 5
```

20.
```
  3 4 6
+ 2 0 3
```

21.
```
  8 3 2 7
+ 1 6 5 2
```

22.
```
  3 4 6 1
+ 2 0 3 5
```

EXAMPLE 6 Add from the top mentally.

```
  1      We first add 1 and 7,        1
  7      getting 8. Then we add       7  →  8
+ 9      8 and 9, getting 17.       + 9     9  →  17
                                      17 ←
```

EXAMPLE 7 Add from the top mentally.

```
  2
  4  →  6
+ 8     8  →  14
  14 ←
```

Do Exercises 15–18.

C Addition (No Carrying)

We now move to a more gradual, conceptual development of the addition procedure you considered in Section 1.2. It is intended to provide you with a greater understanding so that your skill level will increase.

To add larger numbers, we can add the ones first, then the tens, then the hundreds, and so on.

EXAMPLE 8 Add: 5722 + 3234.

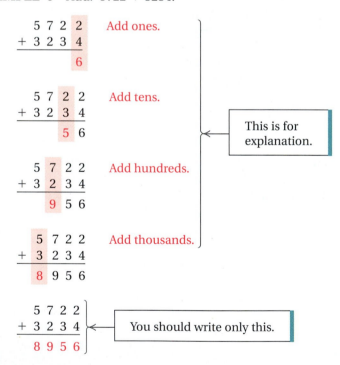

Do Exercises 19–22.

Answers on page A-32

d   Addition (with Carrying)

CARRYING TENS

EXAMPLE 9   Add: 18 + 27.

```
  1 8      Add ones.     Think:        8
+ 2 7                             +     7
  ─────                             ───────
    ?                                  1 5
```

15 ones = 10 ones + 5 ones
             = 1 ten + 5 ones

```
  1
  1 8      Write 5 in the ones column.
+ 2 7      Write 1 as a reminder above the tens.
  ─────    This is called carrying.
    5
```

```
  1
  1 8      Add tens.
+ 2 7
  ─────
  4 5
```

We can use money to help explain Example 9.

```
  1 8¢  ──→ 1 dime and 8 pennies
+ 2 7¢  ──→ 2 dimes and 7 pennies
  ─────
  1 5¢      We first add the pennies.
```

```
1 dime
  1 8
+ 2 7               We regard ten pennies as one dime.

5 pennies
```

```
  1
  1 8          We now add the dimes. The result is
+ 2 7          4 dimes and 5 pennies.
  ─────
  4 5
```

Do Exercises 23 and 24.

CARRYING HUNDREDS

EXAMPLE 10   Add: 256 + 391.

```
  2 5 6      Add ones.
+ 3 9 1
  ───────
      7
```

```
  1
  2 5 6      Add tens. We get 14 tens.
+ 3 9 1      Now 14 tens = 10 tens + 4 tens = 1 hundred + 4 tens.
  ───────    Write 4 in the tens column and a 1 above the hundreds.
    4 7
```

> The carrying here is like exchanging 14 dimes for a 1 dollar bill and 4 dimes.

```
  1
  2 5 6      Add hundreds.
+ 3 9 1
  ───────
  6 4 7
```

Do Exercises 25 and 26.

Add.

23.
```
  1 9
+ 3 7
```

24.
```
  4 6
+ 3 9
```

Add.

25.
```
  3 4 1
+ 4 8 8
```

26.
```
  7 3 0
+ 2 9 6
```

Answers on page A-32

27. Add.

```
    7 8 5 0
  + 4 8 4 8
```

Add.

28.
```
    7 9 8 9
  + 5 6 7 2
```

29.
```
    5 6,7 8 9
  + 1 4,5 3 9
```

To the student:

If you had trouble with Section 1.2 and have studied Developmental Unit A, you should go back and work through Section 1.2 after completing Exercise Set A.

CARRYING THOUSANDS

EXAMPLE 11 Add: 4803 + 3792.

```
    4 8 0 3      Add ones.
  + 3 7 9 2
        5
```

```
    4 8 0 3      Add tens.
  + 3 7 9 2
      9 5
```

```
    1
    4 8 0 3      Add hundreds. We get 15 hundreds. Now 15 hundreds =
  + 3 7 9 2      10 hundreds + 5 hundreds = 1 thousand + 5 hundreds.
    5 9 5        Write 5 in the hundreds column and 1 above the thousands.
```

```
    1
    4 8 0 3      Add thousands.
  + 3 7 9 2
  8 5 9 5
```

Do Exercise 27.

CARRYING MORE THAN ONCE

Sometimes we must carry more than once.

EXAMPLE 12 Add: 5767 + 4993.

```
        1
    5 7 6 7      Add ones. We get 10 ones. Now 10 ones = 1 ten + 0 ones.
  + 4 9 9 3      Write 0 in the ones column and 1 above the tens.
          0
```

```
      1 1
    5 7 6 7      Add tens. We get 16 tens. Now 16 tens =
  + 4 9 9 3      1 hundred + 6 tens.
        6 0      Write 6 in the tens column and 1 above the hundreds.
```

```
    1 1 1
    5 7 6 7      Add hundreds. We get 17 hundreds.
  + 4 9 9 3      Now 17 hundreds = 1 thousand + 7 hundreds.
      7 6 0      Write 7 in the hundreds column and 1 above the thousands.
```

```
  1 1 1 1
    5 7 6 7      Add thousands. We get 10 thousands.
  + 4 9 9 3
  1 0 7 6 0
```

Do Exercises 28 and 29.

Answers on page A-32

a Add. Try to do these mentally. If you have trouble, think of putting sets of objects together.

1. 8 + 9	**2.** 8 + 7	**3.** 6 + 7	**4.** 9 + 5	**5.** 5 + 7	**6.** 5 + 6
7. 9 + 8	**8.** 9 + 7	**9.** 8 + 4	**10.** 9 + 1	**11.** 8 + 2	**12.** 3 + 8
13. 0 + 7	**14.** 4 + 3	**15.** 2 + 9	**16.** 0 + 0	**17.** 3 + 0	**18.** 9 + 9
19. 8 + 6	**20.** 3 + 7	**21.** 2 + 2	**22.** 7 + 7	**23.** 6 + 5	**24.** 7 + 8
25. 8 + 8	**26.** 8 + 1	**27.** 5 + 8	**28.** 5 + 9	**29.** 4 + 7	**30.** 6 + 1

31. 6 + 7 **32.** 7 + 7 **33.** 3 + 9 **34.** 6 + 0 **35.** 6 + 4

36. 9 + 3 **37.** 5 + 5 **38.** 5 + 3 **39.** 1 + 1 **40.** 4 + 5

41. 9 + 4 **42.** 0 + 8 **43.** 4 + 6 **44.** 2 + 7 **45.** 3 + 7

46. 3 + 3 **47.** 5 + 8 **48.** 3 + 6 **49.** 4 + 4 **50.** 4 + 7

b Add from the top mentally.

51. 1 8 + 3	**52.** 1 7 + 5	**53.** 3 2 + 5	**54.** 4 3 + 5	**55.** 1 7 + 9
56. 5 2 + 6	**57.** 4 5 + 1	**58.** 1 9 + 6	**59.** 1 8 + 7	**60.** 1 6 + 8

C   Add.

61.   2 3
     + 1 6

62.   5 4
     + 3 5

63.   6 7
     + 2 0

64.   4 9 6
     + 5 0 3

65.   7 0 0
     + 2 0 0

66.   8 0 1
     +   6 7

67.   6 6 6
     + 3 3 3

68.   5 2 3
     + 3 2 5

69.   7 4 7
     + 1 3 0

70.   8 2 5 0
     + 9 4 3 0

71.   6 5 5 2
     + 4 3 2 1

72.   3 4 0 6
     + 1 2 9 3

73.   7 3 4 0
     + 3 5 2 7

74.   4 8 2 5
     + 5 0 7 0

75.   2 0 7 3
     + 1 9 2 5

76.   9 1 1 1
     + 9 1 1 1

77.   7 8 8 9
     + 9 0 0 0

78.   5 2,4 3 3
     + 1 2,0 5 6

79.   4 3,7 2 3
     + 5 6,2 7 6

80.   5 1,6 7 0
     + 2 6,1 0 7

d   Add.

81.   3 8
     +   8

82.   1 7
     +   9

83.   1 7
     + 3 8

84.   9 5
     +   6

85.   8 6 2
     + 7 8 1

86.   6 1 3
     + 7 9 9

87.   3 5 5
     + 4 9 1

88.   2 8 0
     + 3 4 8

89.   8 1 4
     + 3 9 0

90.   2 7 4
     + 3 3 3

91.   9 9 9 0
     +   1 0

92.   9 9 9
     +   1 1

93.   9 9 9
     + 1 1 1

94.   8 3 9
     + 3 8 8

95.   9 0 9
     + 2 0 2

96.   8 0 8
     + 9 0 9

97.   8 7 1 8
     + 1 4 2 0

98.   3 8 5 4
     + 2 7 0 0

99.   4 8 2 8
     + 1 2 8 3

100.   6 9 9 5
     + 1 4 3 2

101.   9 8 8 9
     +   1

102.   6 8 8 9
     + 4 7 2 3

103.   9 1 2 8
     + 1 9 9 7

104.   8 8 9 8
     + 6 6 4 5

105.   9 9 8 9
     + 6 7 8 5

106.   4 6,8 8 9
     + 2 1,7 8 6

107.   2 3,4 4 8
     + 1 0,9 8 9

108.   6 7,6 5 8
     + 9 8,7 8 6

109.   7 7,5 4 8
     + 2 3,7 6 7

110.   4 4,6 8 4
     +   4,7 6 5

S SUBTRACTION

a Basic Subtraction

Subtraction can be explained by taking away part of a set.

EXAMPLE 1 Subtract: $7 - 3$.

We can do this by counting out 7 objects and then taking away 3 of them. Then we count the number that remain: $7 - 3 = 4$.

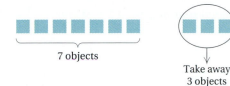

7 objects

Take away 3 objects   4 objects remain

We could also do this mentally by starting at 7 and counting down 3 times: 6, 5, 4.

EXAMPLES Subtract. Think of "take away."

2. $11 - 6 = 5$ *Take away:* "11 take away 6 is 5."

$$\begin{array}{r} 11 \\ -\ 6 \\ \hline 5 \end{array}$$

3. $17 - 9 = 8$

$$\begin{array}{r} 17 \\ -\ 9 \\ \hline 8 \end{array}$$

Do Exercises 1–4.

In Developmental Unit A, you memorized an addition table. That table will enable you to subtract also. First, let's recall how addition and subtraction are related.

An addition:

4   +   3   =   7

Two related subtractions:

A.

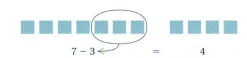

$7 - 3 \leftarrow$ $=$ 4

B.

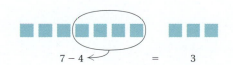

$7 - 4 \leftarrow$ $=$ 3

Subtract.

1. $10 - 6$

2. $11 - 4$

3. $\begin{array}{r} 16 \\ -\ 8 \\ \hline \end{array}$

4. $\begin{array}{r} 10 \\ -\ 7 \\ \hline \end{array}$

Answers on page A-32

For each addition fact, write two subtraction facts.

5. $8 + 4 = 12$

6. $6 + 7 = 13$

Subtract. Try to do these mentally.

7. $14 - 6$

8. $12 - 5$

9.
$$\begin{array}{r} 1\ 3 \\ -\ \ \ 4 \\ \hline \end{array}$$

10.
$$\begin{array}{r} 1\ 1 \\ -\ \ \ 7 \\ \hline \end{array}$$

Since we know that

$$4 + 3 = 7, \qquad \text{A basic addition fact}$$

we also know the two subtraction facts

$$7 - 3 = 4 \quad \text{and} \quad 7 - 4 = 3.$$

EXAMPLE 4 From $8 + 9 = 17$, write two subtraction facts.

a) The addend 8 is subtracted from the sum 17.

$$8 + 9 = 17 \quad \text{The related sentence is} \quad 17 - 8 = 9.$$

b) The addend 9 is subtracted from the sum 17.

$$8 + 9 = 17 \quad \text{The related sentence is} \quad 17 - 9 = 8.$$

Do Exercises 5 and 6.

We can use the idea that subtraction is defined in terms of addition to think of subtraction as "how much more."

EXAMPLE 5 Find: $13 - 6$.

To find $13 - 6$, we ask, "6 plus what number is 13?"

$$6 + \square = 13$$

+	1	2	3	4	5	6	7	8	9
1	2	3	4	5	6	7	8	9	10
2	3	4	5	6	7	8	9	10	11
3	4	5	6	7	8	9	10	11	12
4	5	6	7	8	9	10	11	12	13
5	6	7	8	9	10	11	12	13	14
6	7	8	9	10	11	12	13	14	15
7	8	9	10	11	12	13	14	15	16
8	9	10	11	12	13	14	15	16	17
9	10	11	12	13	14	15	16	17	18

$13 - 6 = 7$

Using the addition table above, we find 13 inside the table and 6 at the left. Then we read the answer 7 from the top. Thus we have $13 - 6 = 7$. Strive to do this kind of thinking mentally as fast as you can, without having to use the table.

Do Exercises 7–10.

b | Subtraction (No Borrowing)

We now move to a more gradual, conceptual development of the subtraction procedure you considered in Section 1.3. It is intended to provide you with a greater understanding so that your skill level will increase.

To subtract larger numbers, we can subtract the ones first, then the tens, then the hundreds, and so on.

EXAMPLE 6   Subtract: 5787 − 3214.

$$
\begin{array}{r}
5\ 7\ 8\ \boxed{7} \\
-\ 3\ 2\ 1\ \boxed{4} \\
\hline
\boxed{3}
\end{array}
$$
  Subtract ones.

$$
\begin{array}{r}
5\ 7\ \boxed{8}\ 7 \\
-\ 3\ 2\ \boxed{1}\ 4 \\
\hline
\boxed{7}\ 3
\end{array}
$$
  Subtract tens.

This is the explanation.

$$
\begin{array}{r}
5\ \boxed{7}\ 8\ 7 \\
-\ 3\ \boxed{2}\ 1\ 4 \\
\hline
\boxed{5}\ 7\ 3
\end{array}
$$
  Subtract hundreds.

$$
\begin{array}{r}
\boxed{5}\ 7\ 8\ 7 \\
-\ \boxed{3}\ 2\ 1\ 4 \\
\hline
\boxed{2}\ 5\ 7\ 3
\end{array}
$$
  Subtract thousands.

$$
\begin{array}{r}
5\ 7\ 8\ 7 \\
-\ 3\ 2\ 1\ 4 \\
\hline
2\ 5\ 7\ 3
\end{array}
$$
  You should write only this.

Do Exercises 11–14.

c | Subtraction (with Borrowing)

We now consider subtraction when borrowing, or regrouping, is necessary.

BORROWING FROM THE TENS PLACE

EXAMPLE 7   Subtract: 37 − 18.

$$
\begin{array}{r}
3\ \boxed{7} \\
-\ 1\ \boxed{8} \\
\hline
\boxed{?}
\end{array}
$$
  Try to subtract ones: 7 − 8 is not a whole number.

$$
\begin{array}{r}
^{2}\ \,^{17} \\
\not{3}\ \not{7} \\
-\ 1\ 8 \\
\hline
\end{array}
$$
  Borrow a ten. That is, 1 ten = 10 ones, and 10 ones + 7 ones = 17 ones. Write 2 above the tens column and 17 above the ones. We regard 37 as 20 + 17.

$$
\begin{array}{r}
^{2}\ \,^{17} \\
\not{3}\ \not{7} \\
-\ 1\ 8 \\
\hline
\boxed{9}
\end{array}
$$
  Subtract ones.

The borrowing here is like exchanging 3 dimes and 7 pennies for 2 dimes and 17 pennies.

Subtract.

11.
$$
\begin{array}{r}
7\ 8 \\
-\ 6\ 4 \\
\hline
\end{array}
$$

12.
$$
\begin{array}{r}
2\ 9 \\
-\ \ \ 9 \\
\hline
\end{array}
$$

13.
$$
\begin{array}{r}
5\ 4\ 2 \\
-\ 3\ 0\ 1 \\
\hline
\end{array}
$$

14.
$$
\begin{array}{r}
6\ 8\ 9\ 6 \\
-\ 4\ 8\ 7\ 1 \\
\hline
\end{array}
$$

Answers on page A-32

Subtract.

15.
```
  4 6
- 2 9
```

16.
```
  7 4
- 3 8
```

Subtract.

17.
```
  6 4 6
- 1 9 2
```

18.
```
  7 3 3
- 4 8 3
```

$$\begin{array}{c}\overset{2}{\cancel{3}}\,\overset{17}{\cancel{7}}\\ -\;1\;8\\\hline 1\;9\end{array}$$ Subtract tens.

$$\begin{array}{c}\overset{2}{\cancel{3}}\,\overset{17}{\cancel{7}}\\ -\;1\;8\\\hline 1\;9\end{array}$$ You should write only this.

Do Exercises 15 and 16.

BORROWING HUNDREDS

EXAMPLE 8   Subtract: 538 − 275.

$$\begin{array}{c}5\;3\;8\\ -\;2\;7\;5\\\hline 3\end{array}$$ Subtract ones.

$$\begin{array}{c}5\;3\;8\\ -\;2\;7\;5\\\hline ?\;3\end{array}$$ Try to subtract tens: 3 tens − 7 tens is not a whole number.

$$\begin{array}{c}\overset{4}{\cancel{5}}\,\overset{13}{\cancel{3}}\,8\\ -\;2\;7\;5\\\hline 3\end{array}$$ Borrow a hundred. That is, 1 hundred = 10 tens, and 10 tens + 3 tens = 13 tens. Write 4 above the hundreds column and 13 above the tens.

The borrowing is like exchanging 5 dollars and 3 dimes for 4 dollars and 13 dimes.

$$\begin{array}{c}\overset{4}{\cancel{5}}\,\overset{13}{\cancel{3}}\,8\\ -\;2\;7\;5\\\hline 6\;3\end{array}$$ Subtract tens.

$$\begin{array}{c}\overset{4}{\cancel{5}}\,\overset{13}{\cancel{3}}\,8\\ -\;2\;7\;5\\\hline 2\;6\;3\end{array}$$ Subtract hundreds.

$$\begin{array}{c}\overset{4}{\cancel{5}}\,\overset{13}{\cancel{3}}\,8\\ -\;2\;7\;5\\\hline 2\;6\;3\end{array}$$ You should write only this.

Do Exercises 17 and 18.

BORROWING MORE THAN ONCE

Sometimes we must borrow more than once.

EXAMPLE 9 Subtract: 672 − 394.

$$
\begin{array}{r}
6\,\overset{6\ \ 12}{7\ 2} \\
-\ 3\ 9\ 4 \\
\hline
8
\end{array}
$$
Borrowing a ten to subtract ones

$$
\begin{array}{r}
\overset{16}{\underset{}{5\ \ 6\ \ 12}} \\
6\ 7\ 2 \\
-\ 3\ 9\ 4 \\
\hline
2\ 7\ 8
\end{array}
$$
Borrowing a hundred to subtract tens

Do Exercises 19 and 20.

EXAMPLE 10 Subtract: 6357 − 1769.

$$
\begin{array}{r}
6\ 3\ \overset{4\ \ 17}{5\ 7} \\
-\ 1\ 7\ 6\ 9 \\
\hline
8
\end{array}
$$
7 − 9 is not a whole number. We borrow a ten.

$$
\begin{array}{r}
6\ \overset{14}{\underset{}{2\ \ 4\ \ 17}}\ 3\ 5\ 7 \\
-\ 1\ 7\ 6\ 9 \\
\hline
8\ 8
\end{array}
$$
4 tens minus 6 tens is not a whole number. We borrow a hundred.

$$
\begin{array}{r}
\overset{12\ \ 14}{\underset{}{5\ \ 2\ \ 4\ \ 17}} \\
6\ 3\ 5\ 7 \\
-\ 1\ 7\ 6\ 9 \\
\hline
4\ 5\ 8\ 8
\end{array}
$$
2 hundreds minus 7 hundreds is not a whole number. We borrow a thousand.

We can always check by adding the answer to the number being subtracted.

EXAMPLE 11 Subtract: 8341 − 2673. Check by adding.

We check by adding 5668 and 2673.

$$
\begin{array}{r}
\overset{12\ \ 13}{\underset{}{7\ \ 2\ \ 3\ \ 11}} \\
8\ 3\ 4\ 1 \\
-\ 2\ 6\ 7\ 3 \\
\hline
5\ 6\ 6\ 8
\end{array}
\qquad
Check:
\qquad
\begin{array}{r}
\overset{1\ \ 1\ \ 1}{5\ 6\ 6\ 8} \\
+\ 2\ 6\ 7\ 3 \\
\hline
8\ 3\ 4\ 1
\end{array}
$$

Do Exercises 21 and 22.

ZEROS IN SUBTRACTION

Before subtracting, note the following:

50 is 5 tens;

70 is 7 tens.

Then

100 is 10 tens;

200 is 20 tens.

Do Exercises 23–26.

Subtract.

19.
$$
\begin{array}{r}
5\ 6\ 3 \\
-\ 1\ 8\ 7 \\
\hline
\end{array}
$$

20.
$$
\begin{array}{r}
7\ 3\ 3 \\
-\ 4\ 8\ 8 \\
\hline
\end{array}
$$

Subtract. Check by adding.

21.
$$
\begin{array}{r}
4\ 2\ 3\ 6 \\
-\ 1\ 6\ 7\ 9 \\
\hline
\end{array}
$$

22.
$$
\begin{array}{r}
7\ 5\ 4\ 1 \\
-\ 3\ 8\ 6\ 7 \\
\hline
\end{array}
$$

Complete.

23. 80 = _____ tens

24. 60 = _____ tens

25. 300 = _____ tens

26. 900 = _____ tens

Answers on page A-32

Complete.

27. 5000 = _____ tens

28. 9000 = _____ tens

29. 5380 = _____ tens

30. 6770 = _____ tens

Subtract.

31. 6 0
      − 1 8

32.   4 8 0
       − 2 5 6

Subtract.

33. 6 0 2
      − 4 6 4

34. 4 0 8
      − 3 6 4

Subtract.

35. 4 0 0 6
      − 1 2 3 8

36. 9 0 0 1
      − 7 8 0 4

Subtract.

37. 3 0 0 0
      − 1 7 5 4

38. 8 0 1 7
      − 3 2 8 9

To the student:

If you had trouble with Section 1.3 and have studied Developmental Unit S, you should go back and work through Section 1.3 after completing Exercise Set S.

Also,

230 is 2 hundreds + 3 tens

or 20 tens + 3 tens

or 23 tens.

Similarly,

1000 is 100 tens;

2000 is 200 tens;

4670 is 467 tens.

Do Exercises 27–30.

EXAMPLE 12   Subtract: 50 − 37.

$$
\begin{array}{r}
\overset{4\ \ 10}{5\ 0} \\
-\ 3\ 7 \\
\hline
1\ 3
\end{array}
$$

We have 5 tens. We keep 4 of them in the tens column and put 1 ten, or 10 ones, with the ones.

Do Exercises 31 and 32.

EXAMPLE 13   Subtract: 803 − 547.

$$
\begin{array}{r}
\overset{7\ \ 9\ \ 13}{8\ 0\ 3} \\
-\ 5\ 4\ 7 \\
\hline
2\ 5\ 6
\end{array}
$$

We have 8 hundreds, or 80 tens. We keep 79 tens and put 1 ten, or 10 ones, with the ones.

Do Exercises 33 and 34.

EXAMPLE 14   Subtract: 9003 − 2789.

$$
\begin{array}{r}
\overset{8\ \ 9\ \ 9\ \ 13}{9\ 0\ 0\ 3} \\
-\ 2\ 7\ 8\ 9 \\
\hline
6\ 2\ 1\ 4
\end{array}
$$

We have 9 thousands, or 900 tens. We keep 899 tens and put 1 ten, or 10 ones, with the ones.

Do Exercises 35 and 36.

EXAMPLES   Subtract.

15.
$$
\begin{array}{r}
\overset{4\ \ 9\ \ 9\ \ 10}{5\ 0\ 0\ 0} \\
-\ 2\ 8\ 6\ 1 \\
\hline
2\ 1\ 3\ 9
\end{array}
$$

16.
$$
\begin{array}{r}
\overset{4\ \ 9\ \ \overset{10}{0}\ \ 13}{5\ 0\ 1\ 3} \\
-\ 1\ 8\ 5\ 7 \\
\hline
3\ 1\ 5\ 6
\end{array}
$$

We have 5 thousands, or 49 hundreds and 10 tens.

Do Exercises 37 and 38.

Answers on page A-32

a   Subtract. Try to do these mentally.

1. $\begin{array}{r} 7 \\ -\ 0 \\ \hline \end{array}$

2. $\begin{array}{r} 8 \\ -\ 8 \\ \hline \end{array}$

3. $\begin{array}{r} 7 \\ -\ 7 \\ \hline \end{array}$

4. $\begin{array}{r} 8 \\ -\ 3 \\ \hline \end{array}$

5. $\begin{array}{r} 5 \\ -\ 2 \\ \hline \end{array}$

6. $\begin{array}{r} 1\ 6 \\ -\ \ 8 \\ \hline \end{array}$

7. $\begin{array}{r} 1\ 7 \\ -\ \ 9 \\ \hline \end{array}$

8. $\begin{array}{r} 1\ 2 \\ -\ \ 6 \\ \hline \end{array}$

9. $\begin{array}{r} 1\ 1 \\ -\ \ 4 \\ \hline \end{array}$

10. $\begin{array}{r} 1\ 2 \\ -\ \ 9 \\ \hline \end{array}$

11. $\begin{array}{r} 1\ 4 \\ -\ \ 7 \\ \hline \end{array}$

12. $\begin{array}{r} 1\ 8 \\ -\ \ 9 \\ \hline \end{array}$

13. $\begin{array}{r} 1\ 3 \\ -\ \ 7 \\ \hline \end{array}$

14. $\begin{array}{r} 1\ 5 \\ -\ \ 9 \\ \hline \end{array}$

15. $\begin{array}{r} 9 \\ -\ 7 \\ \hline \end{array}$

16. $7 - 3$

17. $4 - 1$

18. $2 - 0$

19. $3 - 3$

20. $6 - 3$

21. $7 - 6$

22. $9 - 8$

23. $10 - 3$

24. $6 - 6$

25. $11 - 7$

26. $12 - 8$

27. $5 - 0$

28. $4 - 0$

29. $13 - 9$

30. $14 - 9$

31. $11 - 2$

32. $12 - 3$

33. $16 - 9$

34. $18 - 9$

35. $11 - 5$

36. $10 - 4$

37. $10 - 8$

38. $14 - 8$

39. $15 - 8$

40. $10 - 2$

b   Subtract.

41. $\begin{array}{r} 6\ 4 \\ -\ 3\ 1 \\ \hline \end{array}$

42. $\begin{array}{r} 5\ 5 \\ -\ 3\ 4 \\ \hline \end{array}$

43. $\begin{array}{r} 5\ 4\ 8 \\ -\ 3\ 0\ 1 \\ \hline \end{array}$

44. $\begin{array}{r} 5\ 9\ 6 \\ -\ 4\ 0\ 3 \\ \hline \end{array}$

45. $\begin{array}{r} 7\ 0\ 0 \\ -\ 2\ 0\ 0 \\ \hline \end{array}$

46.
```
   765
 - 111
```

47.
```
   525
 - 323
```

48.
```
   747
 - 130
```

49.
```
   988
 - 700
```

50.
```
  9450
 -8230
```

51.
```
  6552
 -4321
```

52.
```
  7547
 -3421
```

53.
```
  5875
 -2111
```

54.
```
 38,695
-37,004
```

55.
```
 67,899
-66,673
```

56.
```
 99,999
     -1
```

57.
```
 56,780
-56,770
```

58.
```
 42,111
-32,010
```

59.
```
 77,654
-66,611
```

60.
```
 23,456
-12,345
```

Subtract.

61.
```
   93
 - 28
```

62.
```
   42
 - 13
```

63.
```
   86
 - 78
```

64.
```
   98
 - 89
```

65.
```
   625
 - 317
```

66.
```
   735
 - 609
```

67.
```
   853
 - 236
```

68.
```
   961
 - 747
```

69.
```
   787
 - 698
```

70.
```
  6769
 -2367
```

71.
```
  6431
 -2876
```

72.
```
  7654
 -1765
```

73.
```
  5246
 -2859
```

74.
```
  6328
 -2679
```

75.
```
  7641
 -3809
```

76.
```
  8743
 - 599
```

77.
```
 12,647
 -4,897
```

78.
```
 16,222
 -5,777
```

79.
```
 46,781
-12,988
```

80.
```
   470
 - 189
```

81.
```
   690
 - 235
```

82.
```
   703
 - 132
```

83.
```
  6406
 - 258
```

84.
```
  2309
 - 109
```

85.
```
  3406
 -1293
```

86.
```
  6807
 -3059
```

87.
```
  8000
 -2794
```

88.
```
  8002
 -6543
```

89.
```
 38,000
-37,695
```

90.
```
 16,043
-11,588
```

M MULTIPLICATION

Objectives

a	Multiply any two of the numbers 0, 1, 2, 3, 4, 5, 6, 7, 8, 9.
b	Multiply by multiples of 10, 100, and 1000.
c	Multiply larger numbers by 0, 1, 2, 3, 4, 5, 6, 7, 8, 9.
d	Multiply larger numbers by multiples of 10, 100, and 1000.

a Basic Multiplication

To multiply, we begin with two numbers, called **factors,** and get a third number, called a **product.** Multiplication can be explained by counting. The product 3×5 can be found by counting out 3 sets of 5 objects each, joining them (in a rectangular array if desired), and counting all the objects.

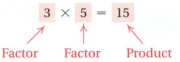

$$3 \times 5 = 15$$

Factor   Factor   Product

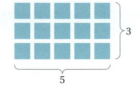

We can also think of multiplication as repeated addition.

$$3 \times 5 = \underbrace{5 + 5 + 5}_{} = 15$$

3 addends of 5

EXAMPLES   Multiply. If you have trouble, think either of putting sets of objects together in a rectangular array or of repeated addition.

1. $5 \times 6 = 30$

$$\begin{array}{r} 6 \\ \times\ 5 \\ \hline 30 \end{array}$$

2. $8 \times 4 = 32$

$$\begin{array}{r} 4 \\ \times\ 8 \\ \hline 32 \end{array}$$

Do Exercises 1–4.

MULTIPLYING BY 0

How do we multiply by 0? Consider $4 \cdot 0$. Using repeated addition, we see that

$$4 \cdot 0 = \underbrace{0 + 0 + 0 + 0}_{} = 0.$$

4 addends of 0

We can also think of this using sets. That is, $4 \cdot 0$ is 4 sets with 0 objects in each set, so the total is 0.

Consider $0 \cdot 4$. Using repeated addition, we say that this is 0 addends of 4, which is 0. Using sets, we say that this is 0 sets with 4 objects in each set, which is 0. Thus we have the following.

> **MULTIPLICATION BY 0**
>
> Multiplying by 0 gives 0.

EXAMPLES   Multiply.

3. $13 \times 0 = 0$

$$\begin{array}{r} 0 \\ \times\ 13 \\ \hline 0 \end{array}$$

4. $0 \cdot 11 = 0$

$$\begin{array}{r} 11 \\ \times\ 0 \\ \hline 0 \end{array}$$

5. $0 \cdot 0 = 0$

$$\begin{array}{r} 0 \\ \times\ 0 \\ \hline 0 \end{array}$$

Do Exercises 5 and 6.

Multiply. Think of joining sets in a rectangular array or of repeated addition.

1. $7 \cdot 8$ (The dot "$\cdot$" means the same as "$\times$".)

2. $\begin{array}{r} 9 \\ \times\ 4 \\ \hline \end{array}$

3. $4 \cdot 7$

4. $\begin{array}{r} 7 \\ \times\ 6 \\ \hline \end{array}$

Multiply.

5. $8 \cdot 0$

6. $\begin{array}{r} 17 \\ \times\ 0 \\ \hline \end{array}$

Answers on page A-32

747

Multiply.

7. $8 \cdot 1$

8.
$$\begin{array}{r} 2\ 3 \\ \times\ \ \ 1 \\ \hline \end{array}$$

9. Complete the table.

×	2	3	4	5
2				
3			12	
4				
5		15		
6				

10.

×	6	7	8	9
5				
6			48	
7				
8		56		
9				

Answers on page A-32

MULTIPLYING BY 1

How do we multiply by 1? Consider $5 \cdot 1$. Using repeated addition, we see that

$$5 \cdot 1 = \underbrace{1 + 1 + 1 + 1 + 1}_{} = 5.$$

$\longrightarrow$ 5 addends of 1

We can also think of this using sets. That is, $5 \cdot 1$ is 5 sets with 1 object in each set, for a total of 5 objects.

Consider $1 \cdot 5$. Using repeated addition, we say that this is 1 addend of 5, which is 5. Using sets, we say that this is 1 set of 5 objects, which is again 5 objects. Thus we have the following.

MULTIPLICATION BY 1

Multiplying a number by 1 does not change the number:

$$a \cdot 1 = 1 \cdot a = a.$$

We say that 1 is the **multiplicative identity.**

This is a very important property.

EXAMPLES Multiply.

6. $13 \cdot 1 = 13$

$$\begin{array}{r} 1 \\ \times\ 13 \\ \hline 13 \end{array}$$

7. $1 \cdot 7 = 7$

$$\begin{array}{r} 7 \\ \times\ 1 \\ \hline 7 \end{array}$$

8. $1 \cdot 1 = 1$

$$\begin{array}{r} 1 \\ \times\ 1 \\ \hline 1 \end{array}$$

Do Exercises 7 and 8.

You should be able to multiply any two of the numbers 0, 1, 2, 3, 4, 5, 6, 7, 8, 9. Multiplying by 0 and 1 is easy. The rest of the products are listed in the following table.

×	2	3	4	5	6	7	8	9
2	4	6	8	10	12	14	16	18
3	6	9	12	15	18	21	24	27
4	8	12	16	20	24	28	32	36
5	10	15	20	25	30	35	40	45
6	12	18	24	30	36	42	48	54
7	14	21	28	35	42	49	56	63
8	16	24	32	40	48	56	64	72
9	18	27	36	45	54	63	72	81

$5 \times 7 = 35$
Find 5 at the left, and 7 at the top.

$8 \cdot 4 = 32$
Find 8 at the left, and 4 at the top.

It is *very* important that you have the basic multiplication facts *memorized.* If you do not, you will always have trouble with multiplication.

The *commutative law of multiplication* says that we can multiply numbers in any order. Thus you need to learn only about half the table, as shown by the shading.

Do Exercises 9 and 10.

b Multiplying by 10, 100, and 1000

We now move to a more gradual, conceptual development of the multiplication procedure you considered in Section 1.5. It is intended to provide you with a greater understanding so that your skill level will increase.

We begin by considering multiplication by 10, 100, and 1000.

MULTIPLYING BY 10

We know that

$$50 = 5 \text{ tens} \qquad 340 = 34 \text{ tens} \quad \text{and} \quad 2340 = 234 \text{ tens}$$
$$= 5 \cdot 10, \qquad = 34 \cdot 10, \qquad = 234 \cdot 10.$$

Turning this around, we see that to multiply any number by 10, all we need do is write a 0 on the end of the number.

> **MULTIPLICATION BY 10**
>
> To multiply a number by 10, write 0 on the end of the number.

EXAMPLES Multiply.

9. $10 \cdot 6 = 60$

10. $10 \cdot 47 = 470$

11. $10 \cdot 583 = 5830$

Do Exercises 11–15.

Let's find $4 \cdot 90$. This is $4 \cdot (9 \text{ tens})$, or 36 tens. The procedure is the same as multiplying 4 and 9 and writing a 0 on the end. Thus, $4 \cdot 90 = 360$.

EXAMPLES Multiply.

12. $5 \cdot 70 = 350$ —— $5 \cdot 7$, then write a 0

13. $8 \cdot 80 = 640$

14. $5 \cdot 60 = 300$

Do Exercises 16 and 17.

MULTIPLYING BY 100

Note the following:

$$300 = 3 \text{ hundreds} \qquad 4700 = 47 \text{ hundreds} \quad \text{and} \quad 56{,}800 = 568 \text{ hundreds}$$
$$= 3 \cdot 100, \qquad = 47 \cdot 100, \qquad = 568 \cdot 100.$$

Turning this around, we see that to multiply any number by 100, all we need do is write two 0's on the end of the number.

> **MULTIPLICATION BY 100**
>
> To multiply a number by 100, write two 0's on the end of the number.

Multiply.

11. $10 \cdot 7$

12. $10 \cdot 45$

13. $10 \cdot 273$

14. $10 \cdot 10$

15. $10 \cdot 100$

Multiply.

16.
$$\begin{array}{r} 7\,0 \\ \times \quad 8 \\ \hline \end{array}$$

17.
$$\begin{array}{r} 6\,0 \\ \times \quad 6 \\ \hline \end{array}$$

Answers on page A-32

Multiply.

18. 100 · 7   **19.** 100 · 23

20. 100 · 723   **21.** 100 · 100

22. 100 · 1000

Multiply.

23.   7 0 0   **24.**   4 0 0
      ×   8   ×   4

Multiply.

25. 1000 · 9   **26.** 1000 · 852

27. 1000 · 10   **28.** 3 · 4000

29. 9 · 8000

▌**EXAMPLES** Multiply.

15. 100 · 6 = 600

16. 100 · 39 = 3900
17. 100 · 448 = 44,800

Do Exercises 18–22.

Let's find 4 · 900. This is 4 · (9 hundreds), or 36 hundreds. The procedure is the same as multiplying 4 and 9 and writing two 0's on the end. Thus, 4 · 900 = 3600.

▌**EXAMPLES** Multiply.

18. 6 · 800 = 4800
                       6 · 8, then write 00
19. 9 · 700 = 6300
20. 5 · 500 = 2500

Do Exercises 23 and 24.

MULTIPLYING BY 1000

Note the following:

    6000 = 6 thousands   and   19,000 = 19 thousands
        = 6 · 1000   = 19 · 1000.

Turning this around, we see that to multiply any number by 1000, all we need do is write three 0's on the end of the number.

MULTIPLYING BY 1000
To multiply a number by 1000, write three 0's on the end of the number.

▌**EXAMPLES** Multiply.

21. 1000 · 8 = 8000

22. 2000 · 13 = 26,000   2 · 13, then write 000
23. 1000 · 567 = 567,000

Do Exercises 25–29.

MULTIPLYING MULTIPLES BY MULTIPLES

Let's multiply 50 and 30. This is 50 · (3 tens), or 150 tens, or 1500. The procedure is the same as multiplying 5 and 3 and writing two 0's on the end.

To multiply multiples of tens, hundreds, thousands, and so on:

a) Multiply the one-digit numbers.
b) Count the number of zeros.
c) Write that many 0's on the end.

EXAMPLES Multiply.

24.
```
      80    1 zero at end
  ×   60    1 zero at end
    4800
```
6 · 8, then write 00

25.
```
     800    2 zeros at end
  ×   60    1 zero at end
   48,000
```
6 · 8, then write 000

26.
```
     800    2 zeros at end
  ×  600    2 zeros at end
  480,000
```
6 · 8, then write 0,000

27.
```
     800    2 zeros at end
  ×   50    1 zero at end
   40,000
```
5 · 8, then write 000

Do Exercises 30–33.

C Multiplying Larger Numbers

The product 3 × 24 can be represented as

$$3 \times (2 \text{ tens} + 4) = (2 \text{ tens} + 4) + (2 \text{ tens} + 4) + (2 \text{ tens} + 4)$$
$$= 6 \text{ tens} + 12$$
$$= 6 \text{ tens} + 1 \text{ ten} + 2$$
$$= 7 \text{ tens} + 2$$
$$= 72.$$

We multiply the 4 ones by 3, getting   12
We multiply the 2 tens by 3, getting   + 60
Then we add: 72

EXAMPLE 28 Multiply: 3 × 24.

```
    2 4     We use the approach described above.
  ×   3
    1 2 ← Multiply the 4 ones by 3.
    6 0 ← Multiply the 2 tens by 3.
    7 2 ← Add.
```

Do Exercises 34–36.

EXAMPLE 29 Multiply: 5 × 734.

```
      7 3 4
  ×       5
        2 0 ← Multiply the 4 ones by 5.
      1 5 0 ← Multiply the 3 tens by 5.
    3 5 0 0 ← Multiply the 7 hundreds by 5.
    3 6 7 0 ← Add.
```

Do Exercises 37 and 38.

Multiply.

30.
```
  9 0 0 0
  ×     6
```

31.
```
      8 0
  ×   7 0
```

32.
```
    8 0 0
  ×   7 0
```

33.
```
    6 0 0
  ×   3 0
```

Multiply.

34.
```
    1 4
  ×   2
```

35.
```
    5 8
  ×   2
```

36.
```
    3 7
  ×   4
```

Multiply.

37.
```
    8 2 3
  ×     6
```

38.
```
  1 3 4 8
  ×     5
```

Answers on page A-32

Multiply using the short form.

39.
$$\begin{array}{r} 5\,8 \\ \times\quad 2 \\ \hline \end{array}$$

40.
$$\begin{array}{r} 3\,7 \\ \times\quad 4 \\ \hline \end{array}$$

41.
$$\begin{array}{r} 8\,2\,3 \\ \times\quad 6 \\ \hline \end{array}$$

42.
$$\begin{array}{r} 1\,3\,4\,8 \\ \times\quad\quad 5 \\ \hline \end{array}$$

Multiply.

43.
$$\begin{array}{r} 7\,4\,6 \\ \times\quad 8 \\ \hline \end{array}$$

44.
$$\begin{array}{r} 7\,4\,6 \\ \times\quad 8\,0 \\ \hline \end{array}$$

45.
$$\begin{array}{r} 7\,4\,6 \\ \times\quad 8\,0\,0 \\ \hline \end{array}$$

To the student:

If you had trouble with Section 1.5 and have studied Developmental Unit M, you should go back and work through Section 1.5 after completing Exercise Set M.

Answers on page A-32

Let's look at Example 29 again. Instead of writing each product on a separate line, we can use a shorter form.

EXAMPLE 30 Multiply: 5×734.

$$\begin{array}{r} \overset{2}{7\,3}\,4 \\ \times\quad\quad 5 \\ \hline 0 \end{array}$$

Multiply the ones by 5: $5 \cdot (4 \text{ ones}) = 20 \text{ ones} = 2 \text{ tens} + 0$ ones. Write 0 in the ones column and 2 above the tens.

$$\begin{array}{r} \overset{1}{}\,\overset{2}{3}\,4 \\ 7\,3\,4 \\ \times\quad\quad 5 \\ \hline 7\,0 \end{array}$$

Multiply 3 tens by 5 and add 2 tens: $5 \cdot (3 \text{ tens}) = 15 \text{ tens}$; $15 \text{ tens} + 2 \text{ tens} = 17 \text{ tens} = 1 \text{ hundred} + 7 \text{ tens}$. Write 7 in the tens column and 1 above the hundreds.

$$\begin{array}{r} \overset{1}{7}\,\overset{2}{3}\,4 \\ \times\quad\quad 5 \\ \hline 3\,6\,7\,0 \end{array}$$

Multiply the 7 hundreds by 5 and add 1 hundred: $5 \cdot (7 \text{ hundreds}) = 35 \text{ hundreds}$, $35 \text{ hundreds} + 1 \text{ hundred} = 36 \text{ hundreds}$.

$$\left.\begin{array}{r} \overset{1}{7}\,\overset{2}{3}\,4 \\ \times\quad\quad 5 \\ \hline 3\,6\,7\,0 \end{array}\right\}$$ You should write only this.

Try to avoid writing the reminders unless necessary.

Do Exercises 39–42.

d **Multiplying by Multiples of 10, 100, and 1000**

To multiply 327 by 50, we multiply by 10 (write a 0), and then multiply 327 by 5.

$$\begin{array}{r} 3\,2\,7 \\ \times\quad 5\,\boxed{0} \\ \hline 1\,6{,}3\,5\,0 \end{array}$$

Write a 0.
Multiply $5 \cdot 327$.

EXAMPLE 31 Multiply: 400×289.

$$\begin{array}{r} 2\,8\,9 \\ \times\quad 4\,\boxed{0\,0} \\ \hline 0\,0 \end{array}$$

Write two 0's.

$$\begin{array}{r} 2\,8\,9 \\ \times\quad 4\,0\,0 \\ \hline 1\,1\,5{,}6\,0\,0 \end{array}$$

Multiply 4 and 289:
$$\begin{array}{r} \overset{3}{}\,\overset{3}{8}\,9 \\ 2\,8\,9 \\ \times\quad\quad 4 \\ \hline 1\,1\,5\,6 \end{array}$$

$$\left.\begin{array}{r} \overset{3}{2}\,\overset{3}{8}\,9 \\ \times\quad 4\,0\,0 \\ \hline 1\,1\,5{,}6\,0\,0 \end{array}\right\}$$ Try to write only this.

Do Exercises 43–45.

a   Multiply. Try to do these mentally.

1. $\begin{array}{r} 3 \\ \times\ 4 \\ \hline \end{array}$	**2.** $\begin{array}{r} 6 \\ \times\ 0 \\ \hline \end{array}$	**3.** $\begin{array}{r} 7 \\ \times\ 1 \\ \hline \end{array}$	**4.** $\begin{array}{r} 0 \\ \times\ 2 \\ \hline \end{array}$	**5.** $\begin{array}{r} 10 \\ \times\ 1 \\ \hline \end{array}$	**6.** $\begin{array}{r} 6 \\ \times\ 5 \\ \hline \end{array}$
7. $\begin{array}{r} 5 \\ \times\ 2 \\ \hline \end{array}$	**8.** $\begin{array}{r} 9 \\ \times\ 7 \\ \hline \end{array}$	**9.** $\begin{array}{r} 9 \\ \times\ 6 \\ \hline \end{array}$	**10.** $\begin{array}{r} 2 \\ \times\ 6 \\ \hline \end{array}$	**11.** $\begin{array}{r} 7 \\ \times\ 0 \\ \hline \end{array}$	**12.** $\begin{array}{r} 8 \\ \times\ 9 \\ \hline \end{array}$
13. $\begin{array}{r} 1 \\ \times\ 8 \\ \hline \end{array}$	**14.** $\begin{array}{r} 8 \\ \times\ 0 \\ \hline \end{array}$	**15.** $\begin{array}{r} 4 \\ \times\ 7 \\ \hline \end{array}$	**16.** $\begin{array}{r} 3 \\ \times\ 8 \\ \hline \end{array}$	**17.** $\begin{array}{r} 5 \\ \times\ 9 \\ \hline \end{array}$	**18.** $\begin{array}{r} 2 \\ \times\ 9 \\ \hline \end{array}$
19. $\begin{array}{r} 0 \\ \times\ 7 \\ \hline \end{array}$	**20.** $\begin{array}{r} 5 \\ \times\ 7 \\ \hline \end{array}$	**21.** $\begin{array}{r} 9 \\ \times\ 5 \\ \hline \end{array}$	**22.** $\begin{array}{r} 5 \\ \times\ 8 \\ \hline \end{array}$	**23.** $\begin{array}{r} 0 \\ \times\ 0 \\ \hline \end{array}$	**24.** $\begin{array}{r} 2 \\ \times\ 8 \\ \hline \end{array}$

25. $5 \cdot 5$ **26.** $9 \cdot 9$ **27.** $1 \cdot 1$ **28.** $0 \cdot 0$ **29.** $2 \cdot 2$

30. $6 \cdot 6$ **31.** $1 \cdot 8$ **32.** $0 \cdot 1$ **33.** $3 \cdot 9$ **34.** $2 \cdot 9$

35. $6 \cdot 0$ **36.** $10 \cdot 1$ **37.** $6 \cdot 8$ **38.** $9 \cdot 6$ **39.** $8 \cdot 0$

40. $9 \cdot 8$ **41.** $3 \cdot 5$ **42.** $1 \cdot 8$ **43.** $1 \cdot 9$ **44.** $2 \cdot 1$

45. $8 \cdot 4$ **46.** $3 \cdot 2$ **47.** $5 \cdot 3$ **48.** $1 \cdot 6$ **49.** $4 \cdot 2$

50. $4 \cdot 5$ **51.** $5 \cdot 4$ **52.** $4 \cdot 4$ **53.** $5 \cdot 2$ **54.** $8 \cdot 0$

b Multiply.

55. $\begin{array}{r} 1\,0 \\ \times\quad 8 \\ \hline \end{array}$

56. $\begin{array}{r} 7 \\ \times\,1\,0 \\ \hline \end{array}$

57. $\begin{array}{r} 2\,0 \\ \times\quad 8 \\ \hline \end{array}$

58. $\begin{array}{r} 3\,0 \\ \times\quad 7 \\ \hline \end{array}$

59. $\begin{array}{r} 4\,5 \\ \times\,1\,0 \\ \hline \end{array}$

60. $\begin{array}{r} 7\,8 \\ \times\,1\,0 \\ \hline \end{array}$

61. $\begin{array}{r} 8\,0 \\ \times\quad 7 \\ \hline \end{array}$

62. $\begin{array}{r} 9\,0 \\ \times\quad 4 \\ \hline \end{array}$

63. $\begin{array}{r} 1\,0\,0 \\ \times\quad 8 \\ \hline \end{array}$

64. $\begin{array}{r} 1\,0\,0 \\ \times\quad 3 \\ \hline \end{array}$

65. $\begin{array}{r} 1\,0\,0 \\ \times\quad 9 \\ \hline \end{array}$

66. $\begin{array}{r} 1\,0\,0 \\ \times\quad 1\,0 \\ \hline \end{array}$

67. $\begin{array}{r} 3\,4\,5\,7 \\ \times\quad 1\,0\,0 \\ \hline \end{array}$

68. $\begin{array}{r} 4\,0\,0 \\ \times\quad 3 \\ \hline \end{array}$

69. $\begin{array}{r} 7\,0\,0 \\ \times\quad 7 \\ \hline \end{array}$

70. $\begin{array}{r} 5\,0\,0 \\ \times\quad 8 \\ \hline \end{array}$

71. $\begin{array}{r} 1\,0\,0 \\ \times\,1\,0\,0 \\ \hline \end{array}$

72. $\begin{array}{r} 1\,0\,0\,0 \\ \times\quad 7 \\ \hline \end{array}$

73. $\begin{array}{r} 1\,0\,0\,0 \\ \times\quad 9 \\ \hline \end{array}$

74. $\begin{array}{r} 1\,0\,0\,0 \\ \times\quad 2 \\ \hline \end{array}$

75. $\begin{array}{r} 4\,5\,7 \\ \times\,1\,0\,0\,0 \\ \hline \end{array}$

76. $\begin{array}{r} 6\,7\,6\,9 \\ \times\,1\,0\,0\,0 \\ \hline \end{array}$

77. $\begin{array}{r} 2\,0\,0\,0 \\ \times\quad 9 \\ \hline \end{array}$

78. $\begin{array}{r} 5\,0\,0\,0 \\ \times\quad 4 \\ \hline \end{array}$

79. $\begin{array}{r} 6\,0\,0\,0 \\ \times\quad 8 \\ \hline \end{array}$

80. $\begin{array}{r} 8\,0\,0\,0 \\ \times\quad 2 \\ \hline \end{array}$

81. $\begin{array}{r} 3\,0\,0\,0 \\ \times\quad 2 \\ \hline \end{array}$

82. $\begin{array}{r} 1\,0\,0\,0 \\ \times\,1\,0\,0\,0 \\ \hline \end{array}$

83. $\begin{array}{r} 4\,0 \\ \times\,3\,0 \\ \hline \end{array}$

84. $\begin{array}{r} 2\,0 \\ \times\,1\,0 \\ \hline \end{array}$

85. $\begin{array}{r} 8\,0 \\ \times\,5\,0 \\ \hline \end{array}$

86. $\begin{array}{r} 5\,0 \\ \times\,5\,0 \\ \hline \end{array}$

87. $\begin{array}{r} 4\,0\,0 \\ \times\quad 3\,0 \\ \hline \end{array}$

88. $\begin{array}{r} 2\,0\,0 \\ \times\quad 3\,0 \\ \hline \end{array}$

89. $\begin{array}{r} 7\,0\,0 \\ \times\quad 9\,0 \\ \hline \end{array}$

90. $\begin{array}{r} 4\,0\,0 \\ \times\,3\,0\,0 \\ \hline \end{array}$

91. $\begin{array}{r} 4\,0\,0\,0 \\ \times\quad 2\,0\,0 \\ \hline \end{array}$

92. $\begin{array}{r} 6\,0\,0\,0 \\ \times\quad 2\,0 \\ \hline \end{array}$

93. $\begin{array}{r} 4\,0\,0\,0 \\ \times\,4\,0\,0\,0 \\ \hline \end{array}$

94. $\begin{array}{r} 8\,0\,0\,0 \\ \times\quad 1\,0 \\ \hline \end{array}$

c Multiply.

95. $\begin{array}{r} 4\,9 \\ \times\quad 3 \\ \hline \end{array}$

96. $\begin{array}{r} 7\,4 \\ \times\quad 6 \\ \hline \end{array}$

97. $\begin{array}{r} 5\,9\,3 \\ \times\quad 5 \\ \hline \end{array}$

98. $\begin{array}{r} 6\,0\,9 \\ \times\quad 8 \\ \hline \end{array}$

99. $\begin{array}{r} 8\,9\,9 \\ \times\quad 7 \\ \hline \end{array}$

100. $\begin{array}{r} 8\,6\,5 \\ \times\quad 4 \\ \hline \end{array}$

101. $\begin{array}{r} 8\,1\,1\,8 \\ \times\quad 2 \\ \hline \end{array}$

102. $\begin{array}{r} 6\,7\,5\,4 \\ \times\quad 2 \\ \hline \end{array}$

103. $\begin{array}{r} 4\,3{,}7\,7\,7 \\ \times\quad 2 \\ \hline \end{array}$

104. $\begin{array}{r} 3\,2{,}5\,6\,4 \\ \times\quad 6 \\ \hline \end{array}$

d Multiply.

105. $\begin{array}{r} 5\,8 \\ \times\,6\,0 \\ \hline \end{array}$

106. $\begin{array}{r} 9\,3 \\ \times\,3\,0 \\ \hline \end{array}$

107. $\begin{array}{r} 4\,2 \\ \times\,8\,0 \\ \hline \end{array}$

108. $\begin{array}{r} 7\,8 \\ \times\,9\,0 \\ \hline \end{array}$

109. $\begin{array}{r} 3\,4\,6 \\ \times\quad 6\,0 \\ \hline \end{array}$

110. $\begin{array}{r} 2\,6\,7 \\ \times\quad 4\,0 \\ \hline \end{array}$

111. $\begin{array}{r} 8\,9\,7 \\ \times\,4\,0\,0 \\ \hline \end{array}$

112. $\begin{array}{r} 3\,6\,6 \\ \times\,3\,0\,0 \\ \hline \end{array}$

113. $\begin{array}{r} 8\,3\,4 \\ \times\,7\,0\,0 \\ \hline \end{array}$

114. $\begin{array}{r} 3\,3\,3 \\ \times\,9\,0\,0 \\ \hline \end{array}$

115. $\begin{array}{r} 5\,6\,7\,3 \\ \times\,2\,0\,0\,0 \\ \hline \end{array}$

116. $\begin{array}{r} 4\,6\,7\,8 \\ \times\,5\,0\,0\,0 \\ \hline \end{array}$

117. $\begin{array}{r} 6\,7\,8\,8 \\ \times\,9\,0\,0\,0 \\ \hline \end{array}$

118. $\begin{array}{r} 9\,1\,2\,9 \\ \times\,8\,0\,0\,0 \\ \hline \end{array}$

D DIVISION

a Basic Division

Division can be explained by arranging a set of objects in a rectangular array. This can be done in two ways.

▪ **EXAMPLE 1** Divide: $18 \div 6$.

METHOD 1   We can do this division by taking 18 objects and determining how many rows, each with 6 objects, we can form.

3 rows of 6 objects

Since there are 3 rows of 6 objects, we have

$$18 \div 6 = 3.$$

METHOD 2   We can also arrange the objects into 6 rows and determine how many objects are in each row.

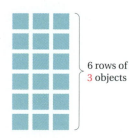

6 rows of 3 objects

Since there are 3 objects in each of the 6 rows, we have

$$18 \div 6 = 3.$$

We can also use fraction notation for division. That is,

$$18 \div 6 = 18/6 = \frac{18}{6}.$$

▪ **EXAMPLES**   Divide.

2. $9 \overline{)\, 3\ 6}^{\,4}$ *Think*: 36 objects: How many rows, each with 9 objects? or 36 objects: How many objects in each of 9 rows?

3. $42 \div 7 = 6$

4. $\dfrac{24}{3} = 8$

Do Exercises 1–4.

Objectives

a Find basic quotients such as $20 \div 5$, $56 \div 7$, and so on.

b Divide using the "guess, multiply, and subtract" method.

c Divide by estimating multiples of thousands, hundreds, tens, and ones.

Divide.

1. $24 \div 6$

2. $64 \div 8$

3. $\dfrac{63}{7}$

4. $\dfrac{27}{9}$

Answers on page A-33

755

D   Division

For each multiplication fact, write two division facts.

5. $6 \cdot 2 = 12$

6. $7 \times 6 = 42$

In Developmental Unit M, you memorized a multiplication table. That table will enable you to divide as well. First, let's recall how multiplication and division are related.

A multiplication: $5 \cdot 4 = 20$.

Two related divisions:

A. 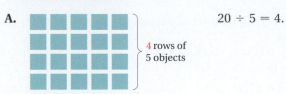   4 rows of 5 objects $20 \div 5 = 4$.

B. 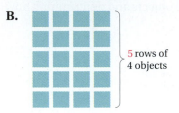   5 rows of 4 objects $20 \div 4 = 5$.

Since we know that

$$5 \cdot 4 = 20, \quad \text{A basic multiplication fact}$$

we also know the two division facts

$$20 \div 5 = 4 \quad \text{and} \quad 20 \div 4 = 5.$$

EXAMPLE 5   From $7 \cdot 8 = 56$, write two division facts.

a) We have

$$7 \cdot 8 = 56 \qquad \text{Multiplication sentence}$$

$$7 = 56 \div 8. \qquad \text{Related division sentence}$$

b) We also have

$$7 \cdot 8 = 56 \qquad \text{Multiplication sentence}$$

$$8 = 56 \div 7. \qquad \text{Related division sentence}$$

Do Exercises 5 and 6.

Answers on page A-33

We can use the idea that division is defined in terms of multiplication to do basic divisions.

EXAMPLE 6 Find: $35 \div 5$.

To find $35 \div 5$, we ask, "5 times what number is 35?"

$$5 \cdot \square = 35$$

×	2	3	4	5	6	7	8	9
2	4	6	8	10	12	14	16	18
3	6	9	12	15	18	21	24	27
4	8	12	16	20	24	28	32	36
5	10	15	20	25	30	35	40	45
6	12	18	24	30	36	42	48	54
7	14	21	28	35	42	49	56	63
8	16	24	32	40	48	56	64	72
9	18	27	36	45	54	63	72	81

$35 \div 5 = 7$

Using the multiplication table above, we find 35 inside the table and 5 at the left. Then we read the answer 7 from the top. Thus we have $35 \div 5 = 7$. Strive to do this kind of thinking mentally as fast as you can, without having to use the table.

Do Exercises 7–10.

DIVISION BY 1

Note that

$$3 \div 1 = 3 \quad \text{because} \quad 3 \cdot 1 = 3; \qquad \frac{14}{1} = 14 \quad \text{because} \quad 14 \cdot 1 = 14.$$

DIVISION BY 1

Any number divided by 1 is that same number:

$$a \div 1 = \frac{a}{1} = a.$$

EXAMPLES Divide.

7. $\dfrac{8}{1} = 8$ **8.** $6 \div 1 = 6$ **9.** $34 \div 1 = 34$

Do Exercises 11–13.

DIVISION BY 0

Why can't we divide by 0? Suppose the number 4 *could* be divided by 0. Then if $\square$ were the answer,

$$4 \div 0 = \square,$$

and since 0 times any number is 0, we would have

$$4 = \square \cdot 0 = 0. \quad \text{False!}$$

Divide.

7. $28 \div 4$

8. $81 \div 9$

9. $\dfrac{16}{2}$

10. $\dfrac{54}{6}$

Divide.

11. $6 \div 1$

12. $\dfrac{13}{1}$

13. $1 \div 1$

Answers on page A-33

Divide, if possible. If not possible, write "undefined."

14. $\dfrac{8}{4}$

15. $\dfrac{5}{0}$

16. $\dfrac{0}{5}$

17. $\dfrac{0}{0}$

18. $12 \div 0$

19. $100 \div 10$

20. $\dfrac{5}{3-3}$

21. $\dfrac{8-8}{4}$

Similarly, suppose 12 could be divided by 0. If $\square$ were the answer,

$$12 \div 0 = \square$$

and since 0 times any number is 0, we would have

$$12 = \square \cdot 0 = 0. \qquad \text{False!}$$

Thus, $a \div 0$ would have to be some number $\square$ such that $a = \square \cdot 0 = 0$. So the only number that could possibly be divided by 0 would be 0 itself.

But such a division would give us any number we wish, for

$$0 \div 0 = 8 \quad \text{because} \quad 8 \cdot 0 = 0;$$
$$0 \div 0 = 3 \quad \text{because} \quad 3 \cdot 0 = 0; \qquad \text{All true!}$$
$$0 \div 0 = 7 \quad \text{because} \quad 7 \cdot 0 = 0.$$

We avoid these difficulties by agreeing to not divide *any* number by 0.

DIVISION BY 0

Division by 0 is not defined. (We agree not to divide by 0.)

DIVIDING 0 BY OTHER NUMBERS

Note that

$$0 \div 3 = 0 \quad \text{because} \quad 0 \cdot 3 = 0; \qquad \frac{0}{12} = 0 \quad \text{because} \quad 0 \cdot 12 = 0.$$

DIVISION INTO 0

Zero divided by any number other than 0 is 0:

$$\frac{0}{a} = 0, \quad a \neq 0.$$

EXAMPLES Divide.

10. $0 \div 8 = 0$

11. $0 \div 22 = 0$

12. $\dfrac{0}{9} = 0$

Do Exercises 14–21.

DIVIDING A NUMBER BY ITSELF

Note that

$$3 \div 3 = 1 \quad \text{because} \quad 1 \cdot 3 = 3; \qquad \frac{34}{34} = 1 \quad \text{because} \quad 1 \cdot 34 = 34.$$

DIVISION OF A NUMBER BY ITSELF

Any number other than 0 divided by itself is 1:

$$\frac{a}{a} = 1, \quad a \neq 0.$$

EXAMPLES Divide.

13. $8 \div 8 = 1$ *Check*: $1 \cdot 8 = 8$

14. $27 \div 27 = 1$ *Check*: $1 \cdot 27 = 27$

15. $\frac{32}{32} = 1$ *Check*: $1 \cdot 32 = 32$

24. $\frac{41}{41}$ **25.** $17 \div 17$

Do Exercises 22–27.

b   Dividing by "Guess, Multiply, and Subtract"

To understand the process of division, we use a method known as "guess, multiply, and subtract." We do this to develop a shorter way that is both understandable and easier to use.

26. $17 \div 1$ **27.** $\frac{54}{54}$

EXAMPLE 16 Divide: $275 \div 4$. Use "guess, multiply, and subtract."

We *guess* a partial quotient of 35. We could guess *any* number—say, 4, 16, or 30. We *multiply* and *subtract* as follows:

```
      3 5 ←—Partial quotient
  4 ) 2 7 5
      1 4 0 ←—35 · 4
      1 3 5 ←—Remainder
```

Next, we look at 135 and *guess* another partial quotient—say, 20. Then we *multiply* and *subtract*:

```
      2 0 ←—Second partial quotient
      3 5
  4 ) 2 7 5
      1 4 0
      1 3 5
        8 0 ←—20 · 4
        5 5 ←—Remainder
```

Divide using the "guess, multiply, and subtract" method.

28. $6 \overline{)\, 4\ 5\ 4}$

Next, we look at 55 and *guess* another partial quotient—say, 13. Then we *multiply* and *subtract*:

```
        1 3 ←—Third partial quotient
        2 0
        3 5
    4 ) 2 7 5
        1 4 0
        1 3 5
          8 0
          5 5
          5 2 ←—13 · 4
            3 ←—Remainder is less than 4
```

29. $3\ 2 \overline{)\, 7\ 4\ 7}$

Answers on page A-33

Divide using the "guess, multiply, and subtract" method.

30. $7 \overline{)6789}$

31. $64 \overline{)3012}$

Since we cannot subtract any more multiples of 4, the division is finished. We add our partial quotients.

```
              6 8 ← Quotient (sum of guesses)
              1 3
              2 0
              3 5        Check:      275 = (4 × 68) + 3
      4 ) 2 7 5                       275 ? 272 + 3
          1 4 0                           |  275
          1 3 5
            8 0
            5 5
            5 2
              3
```

The answer is 68 R 3. This tells us that with 275 objects, we could make 68 rows of 4 and have 3 left over.

The partial quotients (guesses) can be made in any manner so long as subtraction is possible.

Do Exercises 28 and 29 on the preceding page.

EXAMPLE 17   Divide: $1506 \div 32$.

```
                4 7 ← Quotient (sum of guesses)
              2 0 ⎫
                2 ⎬ Guesses
              2 0 ⎪
                5 ⎭
      3 2 ) 1 5 0 6
            1 6 0 ← 5 · 32
          1 3 4 6
            6 4 0 ← 20 · 32
            7 0 6
              6 4 ← 2 · 32
            6 4 2
            6 4 0 ← 20 · 32
                2 ← Remainder:  smaller than the divisor, 32
```

The answer is 47 R 2.

Remember, you can *guess any partial quotient* so long as subtraction is possible.

Do Exercises 30 and 31.

C   Dividing by Estimating Multiples

Let's refine the guessing process. We guess multiples of 10, 100, and 1000, and so on.

EXAMPLE 18 Divide: 7643 ÷ 3.

a) Are there any thousands in the quotient? Yes, 3 · 1000 = 3000, which is less than 7643. To find how many thousands, we find products of 3 and multiples of 1000.

$$3 \cdot 1000 = 3000$$
$$3 \cdot 2000 = 6000$$
$$3 \cdot 3000 = 9000$$

← 7643 is here, so there are over 2000 threes in the quotient.

```
    2 0 0 0
3 ) 7 6 4 3
    6 0 0 0
    1 6 4 3
```

b) Now go to the hundreds place. Are there any hundreds in the quotient?

$$3 \cdot 100 = 300$$
$$3 \cdot 200 = 600$$
$$3 \cdot 300 = 900$$
$$3 \cdot 400 = 1200$$
$$3 \cdot 500 = 1500$$
$$3 \cdot 600 = 1800$$

← 1643

```
      5 0 0
    2 0 0 0
3 ) 7 6 4 3
    6 0 0 0
    1 6 4 3
    1 5 0 0
      1 4 3
```

c) Now go to the tens place. Are there any tens in the quotient?

$$3 \cdot 10 = 30$$
$$3 \cdot 20 = 60$$
$$3 \cdot 30 = 90$$
$$3 \cdot 40 = 120$$
$$3 \cdot 50 = 150$$

← 143

```
        4 0
      5 0 0
    2 0 0 0
3 ) 7 6 4 3
    6 0 0 0
    1 6 4 3
    1 5 0 0
      1 4 3
      1 2 0
        2 3
```

d) Now go to the ones place. Are there any ones in the quotient?

$$3 \cdot 1 = 3$$
$$3 \cdot 2 = 6$$
$$3 \cdot 3 = 9$$
$$3 \cdot 4 = 12$$
$$3 \cdot 5 = 15$$
$$3 \cdot 6 = 18$$
$$3 \cdot 7 = 21$$
$$3 \cdot 8 = 24$$

← 23

```
    2 5 4 7
          7
        4 0
      5 0 0
    2 0 0 0
3 ) 7 6 4 3
    6 0 0 0
    1 6 4 3
    1 5 0 0
      1 4 3
      1 2 0
        2 3
        2 1
          2
```

The answer is 2547 R 2.

Do Exercises 32 and 33.

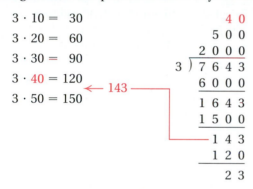

Divide.

32. 4) 3 8 5

33. 7) 8 8 4 6

Answers on page A-33

Divide using the short form.

34. $2 \overline{)\ 6\ 4\ 8}$

35. $9 \overline{)\ 3\ 7\ 5\ 8}$

Divide.

36. $1\ 1 \overline{)\ 4\ 1\ 5}$

37. $4\ 6 \overline{)\ 1\ 0\ 7\ 5}$

To the student:

If you had trouble with Section 1.6 and have studied Developmental Unit D, you should go back and work through Section 1.6 after completing Exercise Set D.

A SHORT FORM

Here is a shorter way to write Example 18.

Instead of this,

$$
\begin{array}{r}
2\ 5\ 4\ 7 \\
7 \\
4\ 0 \\
5\ 0\ 0 \\
2\ 0\ 0\ 0 \\
3\overline{)\ 7\ 6\ 4\ 3} \\
6\ 0\ 0\ 0 \\
\hline
1\ 6\ 4\ 3 \\
1\ 5\ 0\ 0 \\
\hline
1\ 4\ 3 \\
1\ 2\ 0 \\
\hline
2\ 3 \\
2\ 1 \\
\hline
2
\end{array}
$$

Short form

we write this.

$$
\begin{array}{r}
2\ 5\ 4\ 7 \\
3\overline{)\ 7\ 6\ 4\ 3} \\
6\ 0\ 0\ 0 \\
\hline
1\ 6\ 4\ 3 \\
1\ 5\ 0\ 0 \\
\hline
1\ 4\ 3 \\
1\ 2\ 0 \\
\hline
2\ 3 \\
2\ 1 \\
\hline
2
\end{array}
$$

We write a 2 above the thousands digit in the dividend to record 2000.
We write a 5 to record 500.
We write a 4 to record 40.
We write a 7 to record 7.

Do Exercises 34 and 35.

EXAMPLE 19   Divide: $2637 \div 41$. Use the short form.

$$
\begin{array}{r}
6 \\
4\ 1\overline{)\ 2\ 6\ 3\ 7} \\
2\ 4\ 6\ 0 \\
\hline
1\ 7\ 7
\end{array}
$$

60 times 41 is 2460. Since the remainder, 177, is greater than 41, we continue.

$$
\begin{array}{r}
6\ 4 \\
4\ 1\overline{)\ 2\ 6\ 3\ 7} \\
2\ 4\ 6\ 0 \\
\hline
1\ 7\ 7 \\
1\ 6\ 4 \\
\hline
1\ 3
\end{array}
$$

Try to just write this.

The answer is 64 R 13.

Do Exercises 36 and 37.

In Section 1.6, the process of long division was refined with an estimation method. After doing Exercise Set D, you should restudy that procedure.

Answers on page A-33

D EXERCISE SET

For Extra Help

a　Divide, if possible.

1. $24 \div 8$　　　**2.** $72 \div 9$　　　**3.** $28 \div 7$　　　**4.** $22 \div 22$　　　**5.** $32 \div 1$

6. $45 \div 5$　　　**7.** $14 \div 2$　　　**8.** $40 \div 8$　　　**9.** $37 \div 1$　　　**10.** $10 \div 2$

11. $36 \div 4$　　　**12.** $12 \div 3$　　　**13.** $54 \div 9$　　　**14.** $18 \div 2$　　　**15.** $20 \div 4$

16. $16 \div 2$　　　**17.** $72 \div 8$　　　**18.** $42 \div 7$　　　**19.** $12 \div 4$　　　**20.** $8 \div 4$

21. $54 \div 6$　　　**22.** $18 \div 9$　　　**23.** $9 \div 3$　　　**24.** $28 \div 4$　　　**25.** $56 \div 7$

26. $24 \div 6$　　　**27.** $14 \div 2$　　　**28.** $14 \div 7$　　　**29.** $21 \div 7$　　　**30.** $36 \div 6$

31. $8 \div 8$　　　**32.** $32 \div 8$　　　**33.** $30 \div 5$　　　**34.** $18 \div 6$　　　**35.** $49 \div 7$

36. $81 \div 9$　　　**37.** $0 \div 7$　　　**38.** $9 \div 0$　　　**39.** $16 \div 0$　　　**40.** $42 \div 6$

41. $\dfrac{48}{6}$　　**42.** $\dfrac{35}{5}$　　**43.** $\dfrac{9}{9}$　　**44.** $\dfrac{45}{9}$　　**45.** $\dfrac{0}{5}$　　**46.** $\dfrac{0}{8}$

47. $\dfrac{6}{2}$　　**48.** $\dfrac{3}{3}$　　**49.** $\dfrac{8}{2}$　　**50.** $\dfrac{7}{1}$　　**51.** $\dfrac{5}{5}$　　**52.** $\dfrac{6}{1}$

53. $\dfrac{2}{2}$　　**54.** $\dfrac{25}{5}$　　**55.** $\dfrac{4}{2}$　　**56.** $\dfrac{24}{3}$　　**57.** $\dfrac{0}{9}$　　**58.** $\dfrac{0}{4}$

59. $\dfrac{40}{5}$　　**60.** $\dfrac{3}{1}$　　**61.** $\dfrac{16}{4}$　　**62.** $\dfrac{9}{0}$　　**63.** $\dfrac{32}{8}$　　**64.** $\dfrac{9}{9}$

Divide using the "guess, multiply, and subtract" method.

65. 4) 2 7 7

66. 2) 3 9 9

67. 8) 7 3 7

68. 6) 8 3 1

69. 5) 8 6 1 9

70. 3) 8 7 7 5

71. 9) 7 7 7 7

72. 8) 4 1 7 9

73. 7) 3 6 9 1

74. 2) 5 7 9 4

75. 2 0) 8 7 5

76. 3 0) 9 8 7

77. 2 1) 9 9 9

78. 2 3) 9 7 5

79. 8 5) 7 7 5 7

80. 5 4) 2 8 2 1

81. 1 1 1) 3 2 1 9

82. 1 0 2) 5 6 1 2

83. 3 4 6) 7 8,9 1 0

84. 7 8 1) 1 5,9 9 9

c   Divide.

85. 5) 1 0 5

86. 6) 7 0 8

87. 9) 8 2 0

88. 3) 9 6 5

89. 5) 4 8 2 3

90. 8) 5 4 3 7

91. 7) 9 2 9 8

92. 4 1) 1 1 1 5

93. 4 6) 1 0 5 8

94. 2 4) 7 7 2 2

95. 3 8) 8 5 2 2

96. 8 1) 2 2 4 7

97. 9 4) 2 1 5 3

98. 8 2) 4 0 6 4

99. 1 1 7) 4 4,9 0 2

100. 7 4 0) 5 5,2 0 0

Answers

CHAPTER 1

Margin Exercises, Section 1.1, pp. 2–5

1. 2 ten thousands **2.** 2 hundred thousands
3. 2 millions **4.** 2 ten millions
5. 2 tens **6.** 2 hundreds
7. 2 hundred thousands; 8 ten thousands; 0 thousands; 2 hundreds; 1 ten; 9 ones
8. 1 thousand + 8 hundreds + 9 tens + 5 ones
9. 2 ten thousands + 3 thousands + 4 hundreds + 1 ten + 6 ones
10. 4 thousands + 2 hundreds + 1 ten + 8 ones
11. 4 thousands + 1 hundred + 8 tens + 0 ones, or 4 thousands + 1 hundred + 8 tens
12. 1 hundred thousand + 4 ten thousands + 6 thousands + 6 hundreds + 9 tens + 2 ones
13. Forty-nine **14.** Sixteen **15.** Thirty-eight
16. Two hundred four
17. Fifty-one thousand, two hundred six
18. One million, eight hundred seventy-nine thousand, two hundred four **19.** Six billion, four hundred forty-nine million **20.** 213,105,329

Exercise Set 1.1, p. 6

1. 5 thousands **3.** 5 hundreds **5.** 2 **7.** 1
9. 5 thousands + 7 hundreds + 0 tens + 2 ones, or 5 thousands + 7 hundreds + 2 ones
11. 9 ten thousands + 3 thousands + 9 hundreds + 8 tens + 6 ones
13. 2 thousands + 0 hundreds + 5 tens + 8 ones, or 2 thousands + 5 tens + 8 ones
15. 1 thousand + 2 hundreds + 6 tens + 8 ones
17. 5 hundred thousands + 1 ten thousand + 9 thousands + 9 hundreds + 5 tens + 5 ones
19. 3 hundred thousands + 0 ten thousands + 8 thousands + 8 hundreds + 4 tens + 5 ones, or 3 hundred thousands + 8 thousands + 8 hundreds + 4 tens + 5 ones
21. 4 millions + 3 hundred thousands + 0 ten thousands + 2 thousands + 7 hundreds + 3 tens + 7 ones, or 4 millions + 3 hundred thousands + 2 thousands + 7 hundreds + 3 tens + 7 ones **23.** Eighty-five
25. Eighty-eight thousand **27.** One hundred twenty-three thousand, seven hundred sixty-five

29. Seven billion, seven hundred fifty-four million, two hundred eleven thousand, five hundred seventy-seven
31. 2,233,812 **33.** 8,000,000,000
35. Five hundred sixty-six thousand, two hundred eighty
37. Eighty-three million, five hundred seventy-eight thousand, nine hundred six **39.** 9,460,000,000,000
41. 64,186,000 **43.** D_W **45.** 138

Margin Exercises, Section 1.2, pp. 9–13

1. 8 + 2 = 10 **2.** $20 + $13 = $33 **3.** 100 mi + 93 mi = 193 mi **4.** 5 ft + 7 ft = 12 ft **5.** 13,465 **6.** 9745
7. 16,182 **8.** 27,474 **9.** 2 + (6 + 3) = (2 + 6) + 3
10. (5 + 1) + 4 = 5 + (1 + 4) **11.** 2 + 6 = 6 + 2
12. 7 + 1 = 1 + 7 **13.** 27 **14.** 34 **15.** 27
16. 29 in. **17.** 62 ft **18.** 16 in.; 26 in.

Calculator Corner, p. 13

1. 55 **2.** 121 **3.** 1602 **4.** 734 **5.** 1932 **6.** 864

Exercise Set 1.2, p. 14

1. 6 cu yd + 8 cu yd = 14 cu yd **3.** 500 acres + 300 acres = 800 acres **5.** 387 **7.** 5198 **9.** 164
11. 100 **13.** 8503 **15.** 5266 **17.** 4466 **19.** 6608
21. 34,432 **23.** 101,310 **25.** 230 **27.** 18,424
29. 31,685 **31.** (2 + 5) + 4 = 2 + (5 + 4)
33. 6 + (3 + 2) = (6 + 3) + 2 **35.** 2 + 7 = 7 + 2
37. 6 + 1 = 1 + 6 **39.** 2 + 9 = 9 + 2 **41.** 28 **43.** 26
45. 114 mi **47.** 570 ft **49.** D_W **51.** 8 ten thousands
52. Five billion, two hundred ninety-four million, two hundred forty-seven thousand **53.** D_W
55. 56,055,667 **57.** 1 + 99 = 100, 2 + 98 = 100, ..., 49 + 51 = 100. Then 49 · 100 = 4900 and 4900 + 50 + 100 = 5050.

Margin Exercises, Section 1.3, pp. 18–21

1. 67 cu yd − 5 cu yd = 62 cu yd **2.** 20,000 sq ft − 12,000 sq ft = 8000 sq ft **3.** 7 = 2 + 5, or 7 = 5 + 2
4. 17 = 9 + 8, or 17 = 8 + 9 **5.** 5 = 13 − 8; 8 = 13 − 5
6. 11 = 14 − 3; 3 = 14 − 11 **7.** 200 + □ = 348; 348 − 200 = 148 **8.** 800 + □ = 1200; 1200 − 800 = 400

9. 3801 **10.** 6328 **11.** 4747 **12.** 56 **13.** 205
14. 658 **15.** 2851 **16.** 1546

Calculator Corner, p. 20

1. 28 **2.** 47 **3.** 67 **4.** 119 **5.** 2128 **6.** 2593

Exercise Set 1.3, p. 22

1. $20 - 4 = 16$ **3.** $126\ oz - 13\ oz = 113\ oz$
5. $7 = 3 + 4$, or $7 = 4 + 3$ **7.** $13 = 5 + 8$, or $13 = 8 + 5$
9. $23 = 14 + 9$, or $23 = 9 + 14$ **11.** $43 = 27 + 16$, or
$43 = 16 + 27$ **13.** $6 = 15 - 9$; $9 = 15 - 6$
15. $8 = 15 - 7$; $7 = 15 - 8$ **17.** $17 = 23 - 6$; $6 = 23 - 17$
19. $23 = 32 - 9$; $9 = 32 - 23$ **21.** $17 + \square = 32$;
$32 - 17 = 15$ **23.** $10 + \square = 23$; $23 - 10 = 13$ **25.** 44
27. 533 **29.** 39 **31.** 234 **33.** 5382 **35.** 3831
37. 7748 **39.** 43,028 **41.** 56 **43.** 454 **45.** 3749
47. 2191 **49.** 86 **51.** 4813 **53.** 5745 **55.** 95,974
57. 9975 **59.** 83,818 **61.** 4206 **63.** 10,305 **65.** D_W
67. 1024 **68.** 12,732 **69.** 90,283 **70.** 29,364
71. 1345 **72.** 924 **73.** 22,692 **74.** 10,920
75. Six million, three hundred seventy-five thousand, six
hundred two **76.** 7 ten thousands **77.** D_W
79. 2,829,177 **81.** 3; 4

Margin Exercises, Section 1.4, pp. 25–29

1. 40 **2.** 50 **3.** 70 **4.** 100 **5.** 40 **6.** 80 **7.** 90
8. 140 **9.** 470 **10.** 240 **11.** 290 **12.** 600 **13.** 800
14. 800 **15.** 9300 **16.** 8000 **17.** 8000 **18.** 19,000
19. 69,000 **20.** 48,970; 49,000; 49,000 **21.** 269,580;
269,600; 270,000 **22.** Eliminate the power sunroof and
the power package. Answers may vary. **23.** **(a)** $18,300;
(b) yes **24.** $70 + 20 + 40 + 70 = 200$
25. $700 + 700 + 200 + 200 = 1800$
26. $9300 - 6700 = 2600$ **27.** $23,000 - 12,000 = 11,000$
28. < **29.** > **30.** > **31.** < **32.** < **33.** >

Exercise Set 1.4, p. 30

1. 50 **3.** 470 **5.** 730 **7.** 900 **9.** 100 **11.** 1000
13. 9100 **15.** 32,900 **17.** 6000 **19.** 8000
21. 45,000 **23.** 373,000 **25.** 180 **27.** 5720
29. 220; incorrect **31.** 890; incorrect **33.** 16,500
35. 5200 **37.** $11,200 **39.** $18,900; no **41.** Answers
will vary depending on the options chosen. **43.** 1600
45. 1500 **47.** 31,000 **49.** 69,000 **51.** < **53.** >
55. < **57.** > **59.** > **61.** >
63. $2,083,660 < 2,296,335$, or $2,296,335 > 2,083,660$
65. $6482 > 4641$, or $4641 < 6482$ **67.** D_W
69. 7 thousands + 9 hundreds + 9 tens + 2 ones
70. 2 ten millions + 3 millions **71.** Two hundred forty-
six billion, six hundred five million, four thousand, thirty-
two **72.** One million, five thousand, one hundred
73. 86,754 **74.** 13,589 **75.** 48,824 **76.** 4415
77. D_W **79.** Left to the student **81.** Left to the student

Margin Exercises, Section 1.5, pp. 35–40

1. $8 \cdot 7 = 56$ **2.** $10 \cdot 75\ mL = 750\ mL$ **3.** $8 \cdot 8 = 64$
4. 116 **5.** 148 **6.** 4938 **7.** 6740 **8.** 1035 **9.** 3024
10. 46,252 **11.** 205,065 **12.** 144,432 **13.** 287,232
14. 14,075,720 **15.** 391,760 **16.** 17,345,600
17. 56,200 **18.** 562,000 **19.** $8 \cdot 7 = 7 \cdot 8$
20. $2 \cdot 6 = 6 \cdot 2$ **21.** $3 \cdot (7 \cdot 9) = (3 \cdot 7) \cdot 9$
22. $(5 \cdot 4) \cdot 8 = 5 \cdot (4 \cdot 8)$ **23.** 40 **24.** 15
25. $15,300 **26.** $840 \times 250 = 210,000$;
$800 \times 200 = 160,000$ **27.** 45 sq ft

Calculator Corner, p. 39

1. 448 **2.** 21,970 **3.** 6380 **4.** 39,564 **5.** 180,480
6. 2,363,754

Exercise Set 1.5, p. 41

1. $21 \cdot 21 = 441$ **3.** $8 \cdot 12\ oz = 96\ oz$
5. $4800 \cdot 1200 = 5,760,000$ **7.** 870 **9.** 2,340,000
11. 520 **13.** 564 **15.** 1527 **17.** 64,603 **19.** 4770
21. 3995 **23.** 46,080 **25.** 14,652 **27.** 207,672
29. 798,408 **31.** 20,723,872 **33.** 362,128
35. 20,064,048 **37.** 25,236,000 **39.** 302,220
41. 49,101,136 **43.** $50 \cdot 70 = 3500$ **45.** $30 \cdot 30 = 900$
47. $900 \cdot 300 = 270,000$ **49.** $400 \cdot 200 = 80,000$
51. **(a)** $2,840,000; **(b)** $2,850,000 **53.** 529,984 sq mi
55. 18 sq ft **57.** 121 sq yd **59.** 144 sq mm
61. 8100 sq ft **63.** D_W **65.** 12,685 **66.** 10,834
67. 427,477 **68.** 111,110 **69.** 1241 **70.** 8889
71. 254,119 **72.** 66,444 **73.** 6,376,000 **74.** 6,375,600
75. D_W **77.** 247,464 sq ft **79.** 460,800,000 dots

Margin Exercises, Section 1.6, pp. 46–52

1. $112 \div 14 = \square$ **2.** $112 \div 8 = \square$ **3.** $15 = 5 \cdot 3$, or
$15 = 3 \cdot 5$ **4.** $72 = 9 \cdot 8$, or $72 = 8 \cdot 9$ **5.** $6 = 12 \div 2$;
$2 = 12 \div 6$ **6.** $6 = 42 \div 7$; $7 = 42 \div 6$ **7.** 6; $6 \cdot 9 = 54$
8. 6 R 7; $6 \cdot 9 = 54$, $54 + 7 = 61$ **9.** 4 R 5; $4 \cdot 12 = 48$,
$48 + 5 = 53$ **10.** 6 R 13; $6 \cdot 24 = 144$, $144 + 13 = 157$
11. 59 R 3 **12.** 1475 R 5 **13.** 1015 **14.** 134
15. 63 R 12 **16.** 807 R 4 **17.** 1088 **18.** 360 R 4
19. 800 R 47

Calculator Corner, p. 53

1. 3 R 11 **2.** 28 **3.** 124 R 2 **4.** 131 R 18 **5.** 283 R 57
6. 843 R 187

Exercise Set 1.6, p. 54

1. $760 \div 4 = 190$ **3.** $455 \div 5 = 91$ **5.** $18 = 3 \cdot 6$, or
$18 = 6 \cdot 3$ **7.** $22 = 22 \cdot 1$, or $22 = 1 \cdot 22$ **9.** $54 = 6 \cdot 9$,
or $54 = 9 \cdot 6$ **11.** $37 = 1 \cdot 37$, or $37 = 37 \cdot 1$
13. $9 = 45 \div 5$; $5 = 45 \div 9$ **15.** $37 = 37 \div 1$; $1 = 37 \div 37$
17. $8 = 64 \div 8$ **19.** $11 = 66 \div 6$; $6 = 66 \div 11$ **21.** 12
23. 1 **25.** 22 **27.** Not defined **29.** 55 R 2 **31.** 108

33. 307   **35.** 753 R 3   **37.** 1703   **39.** 987 R 5
41. 12,700   **43.** 127   **45.** 52 R 52   **47.** 29 R 5
49. 29   **51.** 105 R 3   **53.** 1609 R 2   **55.** 1007 R 1
57. 23   **59.** 107 R 1   **61.** 370   **63.** 609 R 15   **65.** 304
67. 3508 R 219   **69.** 8070   **71.** D_W **73.** Perimeter
74. Equation, inequality   **75.** Digits, periods
76. Commutative   **77.** Dividend   **78.** Factors, product
79. Minuend   **80.** Associative   **81.** D_W **83.** 54, 122;
33, 2772; 4, 8   **85.** 30 buses

Margin Exercises, Section 1.7, pp. 58–61

1. 7   **2.** 5   **3.** No   **4.** Yes   **5.** 5   **6.** 10   **7.** 5
8. 22   **9.** 22,490   **10.** 9022   **11.** 570   **12.** 3661
13. 8   **14.** 45   **15.** 77   **16.** 3311   **17.** 6114   **18.** 8
19. 16   **20.** 644   **21.** 96   **22.** 94

Exercise Set 1.7, p. 62

1. 14   **3.** 0   **5.** 29   **7.** 0   **9.** 8   **11.** 14   **13.** 1035
15. 25   **17.** 450   **19.** 90,900   **21.** 32   **23.** 143
25. 79   **27.** 45   **29.** 324   **31.** 743   **33.** 37   **35.** 66
37. 15   **39.** 48   **41.** 175   **43.** 335   **45.** 104   **47.** 45
49. 4056   **51.** 17,603   **53.** 18,252   **55.** 205   **57.** D_W
59. $7 = 15 - 8; 8 = 15 - 7$ **60.** $6 = 48 \div 8; 8 = 48 \div 6$
61. <   **62.** >   **63.** >   **64.** <   **65.** 142 R 5
66. 142   **67.** 334   **68.** 334 R 11   **69.** D_W **71.** 347

Margin Exercises, Section 1.8, pp. 65–72

1. 14,141 adoptions   **2.** 10,269 adoptions
3. 19,461 adoptions   **4.** $1874   **5.** $171   **6.** $5572
7. 9180 sq in.   **8.** 378 cartons with 1 can left over
9. 79 gal   **10.** 181 seats

Translating for Success, p. 73

1. E   **2.** M   **3.** D   **4.** G   **5.** A   **6.** O   **7.** F   **8.** K
9. J   **10.** H

Exercise Set 1.8, p. 74

1. 33,747 performances   **3.** 1349 performances
5. 2054 miles   **7.** 18 rows   **9.** 792,316 degrees;
1,348,503 degrees   **11.** 202,345 degrees   **13.** $88,190
15. 470 mi   **17.** 168 hr   **19.** 225 squares   **21.** $24,456
23. 35 weeks; 2 episodes left over   **25.** 236 gal
27. 3616 mi   **29.** 12,804 gal   **31.** $14,445   **33.** $247
35. (a) 4200 sq ft; **(b)** 268 ft   **37.** $29,105,000,000
39. 151,500   **41.** 563 packages; 7 bars left over
43. 384 mi; 27 in.   **45.** 21 columns   **47.** 56 full cartons;
11 books left over. If 1355 books are shipped, it will take
57 cartons.   **49.** 32 $10 bills   **51.** $400
53. 280 min, or 4 hr 40 min   **55.** 525 min, or 8 hr 45 min
57. 106 bones   **59.** 3000 sq in.   **61.** D_W **63.** 234,600
64. 234,560   **65.** 235,000   **66.** 22,000   **67.** 16,000
68. 8000   **69.** 4000   **70.** 320,000   **71.** 720,000
72. 46,800,000   **73.** D_W **75.** 792,000 mi; 1,386,000 mi

Margin Exercises, Section 1.9, pp. 81–86

1. 5^4 **2.** 5^5 **3.** 10^2 **4.** 10^4 **5.** 10,000   **6.** 100
7. 512   **8.** 32   **9.** 51   **10.** 30   **11.** 584   **12.** 84
13. 4; 1   **14.** 52; 52   **15.** 29   **16.** 1880   **17.** 253
18. 93   **19.** 1880   **20.** 305   **21.** 75   **22.** 4
23. 1496 ft   **24.** 46   **25.** 4

Calculator Corner, p. 82

1. 243   **2.** 15,625   **3.** 20,736   **4.** 2048

Calculator Corner, p. 84

1. 49   **2.** 85   **3.** 36   **4.** 0   **5.** 73   **6.** 49

Exercise Set 1.9, p. 87

1. 3^4 **3.** 5^2 **5.** 7^5 **7.** 10^3 **9.** 49   **11.** 729
13. 20,736   **15.** 121   **17.** 22   **19.** 20   **21.** 100
23. 1   **25.** 49   **27.** 5   **29.** 434   **31.** 41   **33.** 88
35. 4   **37.** 303   **39.** 20   **41.** 70   **43.** 295   **45.** 32
47. 906   **49.** 62   **51.** 102   **53.** 32   **55.** $94
57. 401   **59.** 110   **61.** 7   **63.** 544   **65.** 708   **67.** 27
69. D_W **71.** 452   **72.** 835   **73.** 13   **74.** 37
75. 2342   **76.** 4898   **77.** 25   **78.** 100
79. 104,286 mi^2 **80.** 98 gal   **81.** D_W
83. 24; $1 + 5 \cdot (4 + 3) = 36$
85. 7; $12 \div (4 + 2) \cdot 3 - 2 = 4$

Concept Reinforcement, p. 90

1. False   **2.** True   **3.** True   **4.** True   **5.** False
6. False

Summary and Review: Chapter 1, p. 90

1. 8 thousands   **2.** 3
3. 2 thousands + 7 hundreds + 9 tens + 3 ones
4. 5 ten thousands + 6 thousands + 0 hundreds +
7 tens + 8 ones, or 5 ten thousands + 6 thousands +
7 tens + 8 ones   **5.** 4 millions + 0 hundred thousands +
0 ten thousands + 7 thousands + 1 hundred + 0 tens +
1 one, or 4 millions + 7 thousands + 1 hundred + 1 one
6. Sixty-seven thousand, eight hundred nineteen
7. Two million, seven hundred eighty-one thousand,
four hundred twenty-seven   **8.** 1 billion, sixty-five million,
seventy thousand, six hundred seven   **9.** 476,588
10. 2,000,400,000   **11.** 14,272   **12.** 66,024   **13.** 21,788
14. 98,921   **15.** $10 = 6 + 4$, or $10 = 4 + 6$
16. $8 = 11 - 3; 3 = 11 - 8$ **17.** 5148   **18.** 1689
19. 2274   **20.** 17,757   **21.** 345,800   **22.** 345,760
23. 346,000   **24.** 300,000   **25.** $41,300 + 19,700 = 61,000$
26. $38,700 - 24,500 = 14,200$ **27.** $400 \cdot 700 = 280,000$
28. >   **29.** <   **30.** 5,100,000   **31.** 6,276,800
32. 506,748   **33.** 27,589   **34.** 5,331,810
35. $56 = 8 \cdot 7$, or $56 = 7 \cdot 8$ **36.** $4 = 52 \div 13; 13 = 52 \div 4$
37. 12 R 3   **38.** 5   **39.** 913 R 3   **40.** 384 R 1

41. 4 R 46   **42.** 54   **43.** 452   **44.** 5008
45. 4389   **46.** 8   **47.** 45   **48.** 58   **49.** 0   **50.** 4^3
51. 10,000   **52.** 36   **53.** 65   **54.** 233   **55.** 56
56. 32   **57.** 260   **58.** 165   **59.** $502   **60.** $484
61. 1982   **62.** 19 cartons   **63.** 14 beehives
64. $13,585   **65.** $27,598   **66.** 137 beakers filled; 13 mL
left over   **67.** 98 ft²; 42 ft   **68.** **D_W** A vat contains
1152 oz of hot sauce. If 144 bottles are to be filled equally,
how much will each bottle contain? Answers may vary.
69. **D_W** No; if subtraction were associative, then
$a - (b - c) = (a - b) - c$ for any a, b, and c. But, for
example,
$$12 - (8 - 4) = 12 - 4 = 8,$$
whereas
$$(12 - 8) - 4 = 4 - 4 = 0.$$
Since $8 \neq 0$, this example shows that subtraction is not
associative.   **70.** $d = 8$ **71.** $a = 8, b = 4$ **72.** 6 days

Test: Chapter 1, p. 93

1. [1.1a] 5   **2.** [1.1b] 8 thousands + 8 hundreds +
4 tens + 3 ones   **3.** [1.1c] Thirty-eight million, four
hundred three thousand, two hundred seventy-seven
4. [1.2b] 9989   **5.** [1.2b] 63,791   **6.** [1.2b] 34
7. [1.2b] 10,515   **8.** [1.3d] 3630   **9.** [1.3d] 1039
10. [1.3d] 6848   **11.** [1.3d] 5175   **12.** [1.5a] 41,112
13. [1.5a] 5,325,600   **14.** [1.5a] 2405
15. [1.5a] 534,264   **16.** [1.6c] 3 R 3   **17.** [1.6c] 70
18. [1.6c] 97   **19.** [1.6c] 805 R 8
20. [1.8a] 1852 12-packs; 7 cakes left over
21. [1.8a] 1,256,615 mi²   **22.** (a) [1.2c], [1.5c] 300 in.,
5000 in²; 264 in., 3872 in²; 228 in., 2888 in²;
(b) [1.8a] 2112 in²   **23.** [1.8a] 206,330 voters
24. [1.8a] 1808 lb   **25.** [1.8a] 20 staplers   **26.** [1.7b] 46
27. [1.7b] 13   **28.** [1.7b] 14   **29.** [1.7b] 381
30. [1.4a] 35,000   **31.** [1.4a] 34,580   **32.** [1.4a] 34,600
33. [1.4b] 23,600 + 54,700 = 78,300
34. [1.4b] 54,800 − 23,600 = 31,200
35. [1.5b] 800 · 500 = 400,000   **36.** [1.4c] >
37. [1.4c] <   **38.** [1.9a] 12^4 **39.** [1.9b] 343
40. [1.9b] 100,000   **41.** [1.9b] 625   **42.** [1.9c] 31
43. [1.9c] 98   **44.** [1.9c] 2   **45.** [1.9d] 216
46. [1.9c] 18   **47.** [1.9c] 92   **48.** [1.5c], [1.8a] 336 in²
49. [1.9c] 9   **50.** [1.8a] 80 payments   **51.** [1.9c] 83

CHAPTER 2

Margin Exercises, Section 2.1, pp. 97–99

1. 8; −5   **2.** 125; −50   **3.** The integer −3 corresponds to
the decrease in the stock value.   **4.** −10; 148
5. −137; 289   **6.** >   **7.** >   **8.** <   **9.** >   **10.** 18
11. 9   **12.** 29   **13.** 52   **14.** −1   **15.** 2   **16.** 0
17. 4   **18.** 13   **19.** −39   **20.** 0   **21.** 7   **22.** 1
23. −6   **24.** −2   **25.** −7   **26.** −39

Calculator Corner, p. 98

Keystrokes will vary by calculator.

Exercise Set 2.1, p. 100

1. −34,000,000   **3.** 40; −15   **5.** 820; −541   **7.** −280;
14,491   **9.** <   **11.** >   **13.** >   **15.** <   **17.** <
19. <   **21.** >   **23.** 57   **25.** 0   **27.** 24   **29.** 53
31. 8   **33.** 7   **35.** −7   **37.** 0   **39.** 19   **41.** −42
43. 8   **45.** −7   **47.** 29   **49.** 22   **51.** −1   **53.** 7
55. −9   **57.** −17   **59.** 23   **61.** −1   **63.** 85
65. −47   **67.** −345   **69.** 0   **71.** −8   **73.** **D_W**
75. 825   **76.** 125   **77.** 7106   **78.** 4   **79.** 81
80. 1550   **81.** 10   **82.** 42   **83.** **D_W**
85. ⌷5⌷⌷4⌷⌷9⌷⌷+⌷ ⌷3⌷⌷8⌷⌷7⌷⌷=⌷ ⌷+/−⌷. Answers may vary.
87. >   **89.** =   **91.** −8   **93.** −7   **95.** −1, 0, 1
97. −100, −5, 0, |3|, 4, |−6|, 7^2, 10^2, 2^7, 2^{10}

Margin Exercises, Section 2.2, pp. 103–105

1. −1   **2.** −8   **3.** 4   **4.** 0   **5.** 4 + (−5) = −1
6. −2 + (−4) = −6   **7.** −3 + 8 = 5   **8.** −11   **9.** −12
10. −34   **11.** −22   **12.** −17   **13.** 49   **14.** −56
15. 2   **16.** −4   **17.** −2   **18.** 3   **19.** 0   **20.** 0
21. 0   **22.** 0   **23.** −58   **24.** −56   **25.** −12

Exercise Set 2.2, p. 106

1. −5   **3.** −4   **5.** 6   **7.** 0   **9.** −4   **11.** −5   **13.** 5
15. −12   **17.** −11   **19.** 0   **21.** 0   **23.** 6   **25.** 0
27. −25   **29.** −27   **31.** 0   **33.** −8   **35.** 5   **37.** −9
39. −5   **41.** 9   **43.** −3   **45.** 0   **47.** −10   **49.** −24
51. −5   **53.** −21   **55.** 2   **57.** 6   **59.** −21
61. 25   **63.** −17   **65.** 6   **67.** −65   **69.** −160
71. −62   **73.** −23   **75.** **D_W**   **77.** 324   **78.** 3625
79. 1484   **80.** 23,337   **81.** 3 ten thousands +
9 thousands + 4 hundreds + 1 ten + 7 ones   **82.** 700
83. 33,000   **84.** 2352   **85.** 32   **86.** 3500   **87.** **D_W**
89. −40   **91.** −6483   **93.** −1868   **95.** All negative
97. Negative   **99.** Negative

Margin Exercises, Section 2.3, pp. 109–111

1. −10   **2.** 3   **3.** −5   **4.** −2   **5.** −11   **6.** 4   **7.** −2
8. 3 − 10 = 3 + (−10); three minus ten is three plus
negative ten.   **9.** 13 − 5 = 13 + (−5); thirteen minus five
is thirteen plus negative five.   **10.** −12 − (−9) = −12 + 9;
negative twelve minus negative nine is negative twelve plus
nine.   **11.** −12 − 10 = −12 + (−10); negative twelve
minus ten is negative twelve plus negative ten.
12. −14 − (−14) = −14 + 14; negative fourteen minus
negative fourteen is negative fourteen plus fourteen.
13. −4   **14.** −16   **15.** 5   **16.** 2   **17.** −6   **18.** 13
19. −9   **20.** 25   **21.** −99, or $99 in debt   **22.** 19°C

Exercise Set 2.3, p. 112

1. −5   **3.** −8   **5.** −3   **7.** 0   **9.** −4   **11.** −7
13. −5   **15.** 0   **17.** 0   **19.** 14   **21.** 11   **23.** −14
25. 6   **27.** −8   **29.** −1   **31.** 18   **33.** −10   **35.** −3
37. −21   **39.** 5   **41.** −8   **43.** 12   **45.** −19
47. −68   **49.** −81   **51.** 116   **53.** 0   **55.** 55   **57.** 19

59. −62 61. −139 63. 6 65. 107 67. 219
69. 25 pages 71. 17 lb 73. 155°C 75. 50 min; no
77. −3° 79. $13,000 81. −10,011 ft
83. −87, or $87 in debt 85. D_W 87. 64
88. 4896 89. 1 90. 4147 91. 8 cans 92. 288 oz
93. 35 94. 3 95. 32 96. 165 97. D_W
99. −309,882 101. False; $3 − 0 \neq 0 − 3$ 103. True
105. True 107. 17 109. Up 15 points

Margin Exercises, Section 2.4, pp. 117–120

1. 20; 10; 0; −10; −20; −30 2. −18 3. −100 4. −9
5. −10; 0; 10; 20; 30 6. 12 7. 45 8. 6 9. 0
10. 0 11. 120 12. −120 13. 6 14. −8 15. 81
16. −1 17. 32 18. −25 19. 25 20. Negative
eight squared; the opposite of eight squared

Calculator Corner, p. 120

1. 148,035,889 2. −1,419,857 3. −1,124,864
4. 1,048,576 5. −531,441 6. −117,649 7. −7776
8. −19,683

Exercise Set 2.4, p. 121

1. −16 3. −18 5. −48 7. −30 9. 15 11. 18
13. 42 15. 20 17. −120 19. 300 21. 72
23. −340 25. 0 27. 0 29. 24 31. 420 33. −70
35. 30 37. 0 39. −294 41. 36 43. −125
45. 10,000 47. −16 49. −243 51. 1 53. −729
55. −64 57. The opposite of eight to the fourth power
59. Negative nine to the tenth power 61. D_W
63. 532,500 64. 60,000,000 65. 80 66. 2550
67. 5 68. 48 69. 40 sq ft 70. 240 cartons
71. 5 trips 72. 4 trips 73. D_W 75. 243 77. 0
79. 7 81. −2209 83. 130,321 85. −2197
87. 116,875 89. −$23 91. (a) Both m and n must be
odd. (b) At least one of m and n must be even.

Margin Exercises, Section 2.5, pp. 123–125

1. −3 2. 5 3. −3 4. 0 5. −6 6. −5
7. Undefined 8. 0 9. Undefined 10. 68 11. 3
12. 381 13. −15

Calculator Corner, p. 125

1. −4 2. −2 3. 787

Exercise Set 2.5, p. 126

1. −7 3. −14 5. −9 7. 4 9. −9 11. 2
13. −43 15. −8 17. Undefined 19. −8 21. −23
23. 0 25. −19 27. −41 29. −7 31. 19
33. −334 35. 23 37. 8 39. 12 41. −10
43. −86 45. −9 47. 18 49. 10 51. −25
53. −983 55. 82 57. −7988 59. −3000 61. 60
63. 1 65. 2 67. 7 69. Undefined 71. 3 73. 2

75. 0 77. D_W 79. 28 sq in. 80. 42 chairs
81. 12 gal 82. 27 gal 83. 150 cal 84. 672 g
85. 4 pieces; 2 pieces 86. 4 lozenges; 4 lozenges
87. D_W 89. 0 91. 0 93. −2 95. 992
97. [1 5 x^2 − 5 y^x 3] ÷
[3 x^2 + 4 x^2] = 99. 5 101. Positive
103. Negative 105. Positive

Margin Exercises, Section 2.6, pp. 129–132

1. 64 2. 28 3. −70 4. $-\dfrac{6}{x}; \dfrac{6}{-x}$ 5. $\dfrac{-m}{n}; \dfrac{m}{-n}$

6. $\dfrac{-r}{4}; -\dfrac{r}{4}$ 7. −7; −7; −7 8. 50 9. 48; 48
10. 81; 81 11. 9; −9 12. 4; −4 13. 32; −32
14.

	3x + 2x	5x
x = 4	20	20
x = −2	−10	−10
x = 0	0	0

15.

	4x − x	3x
x = 2	6	6
x = −2	−6	−6
x = 0	0	0

16. $5a + 5b$ 17. $6x + 6y + 6z$ 18. $4x − 4y$
19. $3a − 3b + 3c$ 20. $6m − 24$ 21. $−16a + 8b − 24c$

Calculator Corner, p. 131

1. 243 2. 1024 3. −32 4. −3125

Exercise Set 2.6, p. 133

1. 24¢ 3. −2 5. 1 7. 18 yr 9. 13 11. 14 ft
13. 14 ft 15. 21 17. 21 19. 400 ft 21. 10 23. 0
25. 36 27. 100 29. $\dfrac{-5}{t}; \dfrac{5}{-t}$ 31. $\dfrac{n}{-b}; -\dfrac{n}{b}$

33. $\dfrac{-9}{p}; -\dfrac{9}{p}$ 35. $\dfrac{14}{-w}; -\dfrac{14}{w}$ 37. −5; −5; −5
39. −27; −27; −27 41. 36; −12 43. 45; 45
45. 216; −216 47. 1; 1 49. 32; −32 51. $5a + 5b$
53. $4x + 4$ 55. $2b + 10$ 57. $7 − 7t$ 59. $30x − 12$
61. $8x + 56 + 48y$ 63. $−7y + 14$ 65. $3x + 6$
67. $−4x + 12y + 8z$ 69. $8a − 24b + 8c$
71. $4x − 12y − 28z$ 73. $20a − 25b + 5c − 10d$
75. D_W 77. Twenty-three million, forty-three thousand,
nine hundred twenty-one 78. 901
79. $5280 − 2480 = 2800$ 80. 994 81. 17 in. 82. 5 in.
83. $48 84. $63 85. D_W 87. 698°F 89. 4438
91. 279 93. 2 95. 2,560,000 97. −32 ×
(88 − 29) = −1888 99. True 101. True 103. D_W

Margin Exercises, Section 2.7, pp. 137–139

1. $5x$; $-4y$; 3 **2.** $-4y$; $-2x$; $\dfrac{x}{y}$ **3.** $9a^3$ and a^3;

$4ab$ and $3ab$ **4.** $3xy$ and $-4xy$ **5.** $9a$ **6.** $7x^2 - 6$
7. $5m - n^2 - 4$ **8.** 26 cm **9.** 35 mm **10.** 12 cm
11. 24 ft **12.** 40 km **13.** 24 ft

Translating for Success, p. 140

1. J **2.** G **3.** A **4.** K **5.** H **6.** N **7.** F **8.** M
9. D **10.** C

Exercise Set 2.7, p. 141

1. $2a, 5b, -7c$ **3.** $9mn, -6n, 8$ **5.** $3x^2y, -4y^2, -2z^3$
7. $14x$ **9.** $-3a$ **11.** $11x + 6z$ **13.** $-13a + 62$
15. $-4 + 4t + 6y$ **17.** $6a + 4b - 2$
19. $-1 + 17a - 12b$ **21.** $7x^2 + 3y$ **23.** $7x^4 + y^3$
25. $6a^2 - 3a$ **27.** $3x^3 - 8x^2 + 4$ **29.** $7a^3 - 3ab + 3$
31. $9x^3y - 2xy^3 + 3xy$ **33.** $-4a^6 - 11b^4 + 2a^6b^4$
35. 10 ft **37.** 42 km **39.** 8 m **41.** 210 ft **43.** 138 ft
45. 36 ft **47.** 56 in. **49.** 260 cm **51.** 64 ft **53.** ^{D}W
55. 17 servings **56.** 210 **57.** 29 **58.** 7 **59.** 8
60. 27 **61.** 16 **62.** 26 **63.** 16 **64.** 13 **65.** 15
66. 25 **67.** ^{D}W **69.** $7x + 1$ **71.** $-29 - 3a$
73. $-10 - x - 27y$ **75.** \$29.75 **77.** 912 mm

Margin Exercises, Section 2.8, pp. 145–150

1. Equivalent equations **2.** Equivalent expressions
3. 24 **4.** -3 **5.** 25 **6.** -14 **7.** 6 **8.** -8 **9.** -9
10. -12 **11.** -23 **12.** 3 **13.** 26 **14.** -50
15. -4 **16.** Yes **17.** No **18.** 26 **19.** -15

Exercise Set 2.8, p. 151

1. Equivalent equations **3.** Equivalent expressions
5. Equivalent expressions **7.** Equivalent equations
9. Equivalent expressions **11.** Equivalent equations
13. -3 **15.** -8 **17.** 18 **19.** -14 **21.** 32
23. -17 **25.** 17 **27.** 0 **29.** -4 **31.** -14 **33.** 5
35. 0 **37.** -11 **39.** -56 **41.** 12 **43.** 390 **45.** 4
47. -9 **49.** -15 **51.** -26 **53.** -5 **55.** -4
57. -178 **59.** -50 **61.** 7 **63.** 3 **65.** -3 **67.** -7
69. -8 **71.** 8 **73.** 24 **75.** 6 **77.** 5 **79.** -8
81. ^{D}W **83.** Polygon **84.** Similar **85.** Factors
86. Equivalent **87.** Sum **88.** Variable **89.** Absolute
value **90.** Substitute **91.** ^{D}W **93.** 8 **95.** 29
97. -20 **99.** 1027 **101.** -343 **103.** 17

Concept Reinforcement, p. 154

1. True **2.** True **3.** True **4.** False **5.** False
6. True

Summary and Review: Chapter 2, p. 154

1. 527; -53 **2.** $>$ **3.** $<$ **4.** $>$ **5.** 39 **6.** 23
7. 0 **8.** 72 **9.** 59 **10.** -9 **11.** -11 **12.** 6
13. -24 **14.** -12 **15.** 23 **16.** -1 **17.** 7 **18.** -4
19. 12 **20.** 92 **21.** -84 **22.** -40 **23.** -3
24. -5 **25.** 0 **26.** -25 **27.** -20 **28.** 7
29. $-6, -6, -6$ **30.** $20x + 36$ **31.** $6a - 12b + 15$
32. $17a$ **33.** $6x$ **34.** $-3m + 6$ **35.** 36 in.
36. 100 cm **37.** -8 **38.** -9 **39.** -13 **40.** 11
41. 15 **42.** -7 **43.** ^{D}W Equivalent expressions are
expressions that have the same value when evaluated for
various replacements of the variable(s). Equivalent
equations are equations that have the same solution(s).
44. ^{D}W A number's absolute value is the number itself if
the number is nonnegative, and the opposite of the number
if the number is negative. In neither case is the result less
than the number itself, so "no," a number's absolute value is
never less than the number itself. **45.** ^{D}W The notation
"$-x$" means "the opposite of x." If x is a negative number,
then $-x$ is a positive number. For example, if $x = -2$, then
$-x = 2$. **46.** ^{D}W The expressions $(a - b)^2$ and $(b - a)^2$
are equivalent for all choices of a and b because $a - b$ and
$b - a$ are opposites. When opposites are raised to an even
power, the results are the same. **47.** 662,582
48. $-88,174$ **49.** -240 **50.** $x < -2$ **51.** $x < 0$

Test: Chapter 2, p. 157

1. [2.1a] -542; 307 **2.** [2.1b] $>$ **3.** [2.1c] 739
4. [2.1d] -19 **5.** [2.2a] -11 **6.** [2.2a] -21
7. [2.2a] 9 **8.** [2.3a] -12 **9.** [2.3a] -15
10. [2.3a] -24 **11.** [2.3a] 19 **12.** [2.3a] 38
13. [2.4b] -64 **14.** [2.4a] -270 **15.** [2.4a] 0
16. [2.5a] 8 **17.** [2.5a] -8 **18.** [2.5b] -1
19. [2.5b] 25 **20.** [2.3b] 14°F higher **21.** [2.3b] 23 min
22. [2.6a] -3 **23.** [2.6b] $14x + 21y - 7$
24. [2.7a] $4x - 17$ **25.** [2.8b] 5 **26.** [2.8a] -12
27. [2.7b] 66 ft **28.** [2.5b] $35x - 7$
29. [2.5b] $-24x - 57$ **30.** [2.5b] 103,097
31. [2.5b] 1086

Cumulative Review: Chapters 1–2, p. 159

1. [1.1c] 181,599,900 **2.** [1.1c] Five billion, three hundred
eighty million, one thousand, four hundred thirty-seven
3. [1.2b] 18,827 **4.** [1.2b] 8857 **5.** [1.3d] 7846
6. [1.3d] 2428 **7.** [1.5a] 16,767 **8.** [1.5a] 8,266,500
9. [2.4a] -344 **10.** [2.4a] 72 **11.** [1.6c] 104
12. [1.6c] 62 **13.** [2.5a] 0 **14.** [2.5a] -5
15. [1.4a] 428,000 **16.** [1.4a] 5300
17. [1.4b] $749,600 + 301,400 = 1,051,000$
18. [1.5b] $700 \times 500 = 350,000$ **19.** [2.1b] $<$
20. [2.1c] 279 **21.** [1.9c], [2.5b] 36 **22.** [1.9d], [2.5b] 2
23. [1.9c], [2.5b] -86 **24.** [1.9b] 125 **25.** [2.6a] 3
26. [2.6a] 28 **27.** [2.6b] $-2x - 10$
28. [2.6b] $18x - 12y + 24$ **29.** [2.2a] -26
30. [2.4a] 30 **31.** [2.3a] -15 **32.** [2.5a] -32
33. [2.3a] 13 **34.** [2.4a] -30 **35.** [2.3a] 16

36. [2.5b] -57 **37.** [1.7b], [2.8a] 27   **38.** [2.8b] -3
39. [2.8a] 15   **40.** [2.8d] -8 **41.** [1.8a] 104 yr
42. [1.8a] 17,220 rooms   **43.** [1.8a] $95
44. [1.8a] Westside Appliance   **45.** [2.7a] $10x - 14$
46. [1.8a] Cases: 6; six-packs: 3; loose cans: 4
47. [2.5b] $4a$ **48.** [2.5b] -1071 **49.** [2.8d] ± 3

CHAPTER 3

Margin Exercises, Section 3.1, pp. 162–165

1. $5 = 1 \cdot 5$; $45 = 9 \cdot 5$; $100 = 20 \cdot 5$
2. $10 = 1 \cdot 10$; $60 = 6 \cdot 10$; $110 = 11 \cdot 10$
3. 5, 10, 15, 20, 25, 30, 35, 40, 45, 50   **4.** Yes   **5.** Yes
6. No   **7.** Yes   **8.** No   **9.** Yes   **10.** No   **11.** Yes
12. No   **13.** No   **14.** Yes   **15.** No   **16.** Yes   **17.** No
18. Yes   **19.** Yes   **20.** No   **21.** Yes   **22.** No   **23.** No
24. Yes   **25.** No   **26.** Yes   **27.** Yes   **28.** No   **29.** No
30. Yes

Calculator Corner, p. 163

1. Yes   **2.** No   **3.** No   **4.** Yes   **5.** No   **6.** Yes

Exercise Set 3.1, p. 166

1. 7, 14, 21, 28, 35, 42, 49, 56, 63, 70
3. 20, 40, 60, 80, 100, 120, 140, 160, 180, 200
5. 3, 6, 9, 12, 15, 18, 21, 24, 27, 30
7. 12, 24, 36, 48, 60, 72, 84, 96, 108, 120
9. 10, 20, 30, 40, 50, 60, 70, 80, 90, 100
11. 25, 50, 75, 100, 125, 150, 175, 200, 225, 250   **13.** No
15. No   **17.** Yes   **19.** Yes; the sum of the digits is 12,
which is divisible by 3.   **21.** No; the ones digit is not 0 or 5.
23. Yes; the ones digit is 0.   **25.** Yes; the sum of the digits
is 18, which is divisible by 9.   **27.** No; the ones digit is not
even.   **29.** No; the ones digit is not even.
31. 6825 is divisible by 3 and 5.   **33.** 119,117 is divisible
by none of these numbers.   **35.** 127,575 is divisible by 3, 5,
and 9.   **37.** 9360 is divisible by 2, 3, 5, 6, 9, and 10.
39. 555; 300; 36; 45,270; 711; 13,251; 8064   **41.** 300; 45,270
43. 300; 36; 45,270; 8064
45. 56; 324; 784; 200; 42; 812; 402
47. 55,555; 200; 75; 2345; 35; 1005   **49.** 324   **51.** $\mathbf{D_W}$
53. 53   **54.** 5   **55.** -8 **56.** -24 **57.** $680   **58.** 42
59. 125   **60.** 343   **61.** 1024   **62.** 729   **63.** 9^5
64. 7^6 **65.** $\mathbf{D_W}$ **67.** 99,969   **69.** 30   **71.** 60
73. 3655   **75.** 840   **77.** $\mathbf{D_W}$ **79.** 95,238

Margin Exercises, Section 3.2, pp. 169–172

1. 1, 2, 7, 14   **2.** 1, 2, 5, 10   **3.** 1, 2, 4, 8
4. 1, 2, 4, 8, 16, 32   **5.** 2, 13, 19, 41, 73 are prime;
6, 12, 65, 99 are composite; 1 is neither.   **6.** $2 \cdot 3$
7. $2 \cdot 2 \cdot 3$ **8.** $2 \cdot 7 \cdot 7$ **9.** $7 \cdot 13$ **10.** $2 \cdot 3 \cdot 3 \cdot 7$
11. $2 \cdot 2 \cdot 2 \cdot 2 \cdot 3 \cdot 3$

Exercise Set 3.2, p. 173

1. 1, 2, 3, 6, 9, 18   **3.** 1, 2, 3, 6, 9, 18, 27, 54   **5.** 1, 3, 9
7. 1, 13   **9.** Prime   **11.** Composite   **13.** Composite
15. Prime   **17.** Neither   **19.** Composite   **21.** Prime
23. Prime   **25.** $3 \cdot 3 \cdot 3$ **27.** $2 \cdot 7$ **29.** $2 \cdot 2 \cdot 2 \cdot 2 \cdot 5$
31. $5 \cdot 5$ **33.** $2 \cdot 31$ **35.** $2 \cdot 2 \cdot 5 \cdot 5$ **37.** $11 \cdot 13$
39. $11 \cdot 11$ **41.** $3 \cdot 7 \cdot 13$ **43.** $5 \cdot 5 \cdot 7$ **45.** $11 \cdot 19$
47. $2 \cdot 43$ **49.** $7 \cdot 31$ **51.** $2 \cdot 2 \cdot 2 \cdot 5 \cdot 5 \cdot 5 \cdot 7$
53. $2 \cdot 3 \cdot 11 \cdot 17$ **55.** 1, 2, 4, 5, 10, 20, 25, 50, 100
57. 1, 5, 7, 11, 35, 55, 77, 385   **59.** 1, 3, 9, 27, 81
61. 1, 3, 5, 9, 15, 25, 45, 75, 225   **63.** $\mathbf{D_W}$ **65.** -26
66. 256   **67.** 8   **68.** -23 **69.** 1   **70.** 73   **71.** 0
72. 1   **73.** -42 **74.** 0   **75.** $\mathbf{D_W}$ **77.** $\mathbf{D_W}$
79. $11 \cdot 11 \cdot 23 \cdot 37$ **81.** $2 \cdot 2 \cdot 2 \cdot 3 \cdot 3 \cdot 5 \cdot 7 \cdot 67$
83. Answers may vary. One arrangement is a three-
dimensional rectangular array consisting of 2 tiers of 12
objects each, where each tier consists of a rectangular array
of 4 rows with 3 objects each.
85.

Product	56	63	36	72	140	96
Factor	7	7	2	2	10	8
Factor	8	9	18	36	14	12
Sum	15	16	20	38	24	20
Product	48	168	110	90	432	63
Factor	6	21	10	9	24	3
Factor	8	8	11	10	18	21
Sum	14	29	21	19	42	24

Margin Exercises, Section 3.3, pp. 175–180

1. Numerator: 83; denominator: 100
2. Numerator: 27; denominator: 50
3. Numerator: $5a$; denominator: $7b$
4. Numerator: -22; denominator: 3
5. $\frac{1}{2}$ **6.** $\frac{1}{3}$ **7.** $\frac{4}{6}$ **8.** $\frac{2}{3}$ **9.** $\frac{3}{4}$ **10.** $\frac{1}{3}$ **11.** $\frac{15}{16}$
12. Clocks: $\frac{3}{5}$; thermometers: $\frac{2}{5}$ **13.** $\frac{8}{5}$ **14.** $\frac{7}{4}$
15. $\frac{83}{79}$, $\frac{83}{162}$, $\frac{79}{162}$ **16.** 1   **17.** 1   **18.** 1   **19.** 1   **20.** 1
21. 1   **22.** 0   **23.** 0   **24.** 0   **25.** 0   **26.** Undefined
27. Undefined   **28.** 8   **29.** -10 **30.** -346 **31.** 15

Exercise Set 3.3, p. 181

1. Numerator: 3; denominator: 4
3. Numerator: 7; denominator: -9
5. Numerator: $2x$; denominator: $3z$ **7.** $\frac{2}{4}$ **9.** $\frac{1}{8}$ **11.** $\frac{4}{9}$
13. $\frac{3}{4}$ **15.** $\frac{4}{8}$ **17.** $\frac{6}{12}$ **19.** $\frac{12}{16}$ **21.** $\frac{7}{16}$ **23.** $\frac{5}{8}$ **25.** $\frac{4}{7}$
27. (a) $\frac{2}{8}$; (b) $\frac{6}{8}$ **29.** (a) $\frac{3}{8}$; (b) $\frac{5}{8}$ **31.** $\frac{9}{8}$ **33.** $\frac{7}{6}$ **35.** $\frac{7}{5}$
37. (a) $\frac{390}{13}$; (b) $\frac{13}{390}$ **39.** $\frac{850}{1000}$ **41.** (a) $\frac{3}{7}$; (b) $\frac{3}{4}$; (c) $\frac{4}{7}$; (d) $\frac{4}{3}$

43. (a) $\frac{35}{10,000}$; (b) $\frac{50}{10,000}$; (c) $\frac{44}{10,000}$; (d) $\frac{63}{10,000}$; (e) $\frac{43}{10,000}$; (f) $\frac{21}{10,000}$
45. 0 **47.** 15 **49.** 1 **51.** 1 **53.** 0 **55.** 1 **57.** 1
59. -63 **61.** 0 **63.** Undefined **65.** $7n$
67. Undefined **69.** D_W **71.** -210 **72.** -322
73. 0 **74.** 0 **75.** 300 calories **76.** 201 min
77. D_W **79.** $\frac{52}{365}$ **81.** $\frac{3}{4}, \frac{1}{4}$
83.

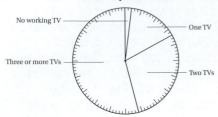

Televisions per Household

85. $\frac{2}{6}$, or $\frac{1}{3}$

87. $\frac{6}{16}$, or $\frac{3}{8}$

Margin Exercises, Section 3.4, pp. 186–189

1. $\frac{2}{3}$ **2.** $\frac{5}{8}$ **3.** $\frac{14}{3}$ **4.** $-\frac{33}{10}$ or $\frac{-33}{10}$ **5.** $\frac{68}{5}$ **6.** $\frac{4x}{9}$

7.

$\frac{1}{4}$ $\quad$ $\frac{1}{2} \cdot \frac{1}{4} = \frac{1}{8}$

8.

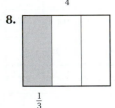

$\frac{1}{3}$ $\quad$ $\frac{4}{5} \cdot \frac{1}{3} = \frac{4}{15}$

9. $\frac{15}{56}$ **10.** $\frac{32}{15}$ **11.** $\frac{3}{100}$ **12.** $-\frac{7a}{b}$ or $\frac{-7a}{b}$ **13.** $\frac{3}{8}$
14. $\frac{8}{81}$ ft^2 **15.** $\frac{3}{40}$

Exercise Set 3.4, p. 190

1. $\frac{3}{8}$ **3.** $\frac{-5}{6}$ or $-\frac{5}{6}$ **5.** $\frac{14}{3}$ **7.** $-\frac{7}{9}$ or $\frac{-7}{9}$ **9.** $\frac{5x}{6}$
11. $\frac{-6}{5}$ or $-\frac{6}{5}$ **13.** $\frac{2a}{7}$ **15.** $\frac{17m}{6}$ **17.** $\frac{6}{5}$ **19.** $\frac{2x}{7}$
21. $\frac{1}{15}$ **23.** $\frac{-1}{40}$ or $-\frac{1}{40}$ **25.** $\frac{2}{15}$ **27.** $\frac{2x}{9y}$ **29.** $\frac{9}{16}$
31. $\frac{14}{39}$ **33.** $\frac{-3}{50}$ or $-\frac{3}{50}$ **35.** $\frac{7a}{64}$ **37.** $\frac{1}{100y}$
39. $\frac{-182}{285}$ or $-\frac{182}{285}$ **41.** $\frac{3}{8}$ cup **43.** $\frac{12}{25}$ m^2 **45.** $\frac{4}{50}$ **47.** $\frac{9}{20}$
49. $\frac{1}{2625}$ **51.** D_W **53.** 9 **54.** 7 **55.** -50 **56.** -90
57. -7 **58.** -5 **59.** 13 **60.** 6 **61.** 8 thousands
62. 8 millions **63.** 8 ones **64.** 8 hundreds **65.** D_W
67. $\frac{1}{80}$ gal **69.** $-\frac{185,193}{226,981}$ or $\frac{-185,193}{226,981}$ **71.** $-\frac{3}{160}$ or $\frac{-3}{160}$
73. $\frac{4}{105}$ **75.** $-\frac{4836}{30,709}$

Margin Exercises, Section 3.5, pp. 193–197

1. $\frac{8}{16}$ **2.** $\frac{3a}{7a}$ **3.** $\frac{-32}{100}$ **4.** $\frac{-16}{-6}$ **5.** $\frac{12}{9}$ **6.** $\frac{-18}{-24}$ **7.** $\frac{9x}{10x}$
8. $\frac{21}{45}$ **9.** $\frac{-56}{49}$ **10.** $\frac{4}{7}$ **11.** $\frac{-5}{6}$ **12.** 5 **13.** $\frac{4}{3}$ **14.** $\frac{-5}{3}$
15. $\frac{-7}{8}$ **16.** $\frac{89}{78}$ **17.** $\frac{7}{24}$ **18.** $\frac{13}{17}$ **19.** $\frac{20}{100} = \frac{1}{5}; \frac{42}{100} = \frac{21}{50};$
$\frac{8}{100} = \frac{2}{25}; \frac{16}{100} = \frac{4}{25}; \frac{14}{100} = \frac{7}{50}$ **20.** $=$ **21.** $\neq$ **22.** $=$

Calculator Corner, p. 196

1. $\frac{14}{15}$ **2.** $\frac{7}{8}$ **3.** $\frac{138}{167}$ **4.** $\frac{7}{25}$

Exercise Set 3.5, p. 198

1. $\frac{5}{10}$ **3.** $\frac{-36}{-48}$ **5.** $\frac{35}{50}$ **7.** $\frac{11t}{5t}$ **9.** $\frac{20}{48}$ **11.** $\frac{-51}{54}$ **13.** $\frac{15}{-40}$
15. $\frac{-42}{132}$ **17.** $\frac{5x}{8x}$ **19.** $\frac{10a}{7a}$ **21.** $\frac{4ab}{9ab}$ **23.** $\frac{12b}{27b}$ **25.** $\frac{1}{2}$
27. $\frac{-2}{3}$ **29.** $\frac{2}{5}$ **31.** -9 **33.** $\frac{3}{4}$ **35.** $\frac{-12}{7}$ **37.** $\frac{1}{3}$
39. $\frac{-1}{3}$ **41.** $\frac{7}{8}$ **43.** $\frac{9}{8}$ **45.** $\frac{12}{13}$ **47.** $\frac{17}{19}$ **49.** $\frac{3}{8}$
51. $\frac{3y}{2}$ **53.** $\frac{-9}{10b}$ **55.** $=$ **57.** $\neq$ **59.** $=$ **61.** $\neq$
63. $\neq$ **65.** $=$ **67.** $\neq$ **69.** $=$ **71.** D_W
73. 3600 yd^2 **74.** \$928 **75.** 60 **76.** 65 **77.** -63
78. -64 **79.** 5 **80.** 89 **81.** 3520 **82.** 9001
83. D_W **85.** $\frac{17}{29}$ **87.** $-\frac{29x}{15y}$ **89.** $\frac{137}{149}$ **91.** $\frac{4}{10} = \frac{2}{5}; \frac{6}{10} = \frac{3}{5}$
93. No. $\frac{199}{594} \neq \frac{221}{668}$ because $199 \cdot 668 \neq 594 \cdot 221$.

Margin Exercises, Section 3.6, pp. 201–203

1. $\frac{7}{12}$ **2.** $\frac{-1}{3}$ **3.** 6 **4.** $\frac{15}{x}$ **5.** 14 lb **6.** 96 m^2
7. $\frac{66}{5}$ cm^2 **8.** Rectangle: $(10 \text{ in.}) \cdot (8 \text{ in.}) = 80 \text{ in}^2$;
triangle: $\frac{1}{2}(5 \text{ in.}) \cdot (8 \text{ in.}) = 20 \text{ in}^2$; $80 \text{ in}^2 + 20 \text{ in}^2 = 100 \text{ in}^2$

Exercise Set 3.6, p. 204

1. $\frac{7}{8}$ **3.** $-\frac{1}{8}$ **5.** $\frac{3}{28}$ **7.** $\frac{2}{9}$ **9.** $-\frac{27}{10}$ **11.** $\frac{4x}{9}$ **13.** 3
15. 7 **17.** 12 **19.** $3a$ **21.** 1 **23.** 1 **25.** $\frac{11a}{2}$
27. $\frac{-88}{3}$ **29.** 1 **31.** 3 **33.** $\frac{119}{750}$ **35.** $-\frac{20}{187}$ **37.** $-\frac{42}{275}$
39. $-\frac{16}{5x}$ **41.** $\frac{-11}{40}$ **43.** $\frac{5a}{28b}$ **45.** $\frac{5}{8}$ in. **47.** 18 mph
49. 625 addresses **51.** $\frac{1}{2}$ **53.** $\frac{1}{3}$ cup **55.** \$115,500
57. 160 mi **59.** Food: \$9900; housing: \$7920;
clothing: \$3960; savings: \$4400; taxes: \$9900;
other expenses: \$3520 **61.** 60 in^2 **63.** $\frac{35}{4}$ mm^2
65. $\frac{63}{8}$ m^2 **67.** 92 mi^2 **69.** $\frac{15}{2}$ cm^2 **71.** D_W **73.** 35
74. 85 **75.** 125 **76.** 120 **77.** 4989 **78.** 8546
79. 6498 **80.** 6407 **81.** D_W **83.** $\frac{129}{485}$ **85.** $\frac{2}{3}$
87. 20 students **89.** 13,380 mm^2 **91.** 6392 ft^2

Margin Exercises, Section 3.7, pp. 210–212

1. $\dfrac{8}{7}$ 2. $\dfrac{x}{-6}$ 3. $\dfrac{1}{2}$ 4. 5   5. $-\dfrac{10}{3}$ 6. $\dfrac{5}{21}$ 7. $-\dfrac{8}{3}$
8. $\dfrac{1}{8}$ 9. $\dfrac{100}{a}$ 10. $\dfrac{-10}{7}$ or $-\dfrac{10}{7}$ 11. $\dfrac{15}{14}$

Calculator Corner, p. 212

1. $\dfrac{1}{6}$ 2. $\dfrac{20}{9}$ 3. $-\dfrac{9}{7}$ 4. $\dfrac{3}{2}$

Exercise Set 3.7, p. 213

1. $\dfrac{3}{7}$ 3. $\dfrac{1}{9}$ 5. 7   7. $-\dfrac{3}{10}$ 9. $\dfrac{17}{3}$ 11. $\dfrac{m}{-3n}$ 13. $\dfrac{-15}{8}$
15. $\dfrac{1}{7m}$ 17. $4a$ 19. $-3z$ 21. $\dfrac{4}{7}$ 23. $-\dfrac{7}{10}$ 25. 4
27. -2 29. $\dfrac{25}{7}$ 31. $\dfrac{1}{64}$ 33. $\dfrac{3}{7x}$ 35. -8 37. $35a$
39. 1   41. $-\dfrac{2}{3}$ 43. $\dfrac{99}{224}$ 45. $\dfrac{112a}{3}$ 47. $\dfrac{14}{15}$ 49. $\dfrac{7}{32}$
51. $\dfrac{-25}{12}$ 53. $\dfrac{15}{32}$ 55. $\mathbf{D_W}$ 57. Associative
58. Factors   59. Prime   60. Denominator
61. Additive   62. Reciprocals   63. Opposites
64. Equation   65. $\mathbf{D_W}$ 67. $\dfrac{100}{9}$ 69. 36   71. $\dfrac{121}{900}$
73. $\dfrac{9}{19}$ 75. $\dfrac{220}{51}$

Margin Exercises, Section 3.8, pp. 215–218

1. 12   2. -21 3. $\dfrac{-9}{32}$ 4. $\dfrac{-3}{4}$ 5. 320 loops
6. 12 servings   7. 252 mi

Translating for Success, p. 219

1. C   2. H   3. A   4. N   5. O   6. F   7. I   8. L
9. D   10. M

Exercise Set 3.8, p. 220

1. 15   3. 9   5. -45 7. $\dfrac{2}{17}$ 9. $\dfrac{12}{5}$ 11. $-\dfrac{16}{21}$
13. $-\dfrac{2}{25}$ 15. $\dfrac{1}{6}$ 17. $-\dfrac{80}{9}$ 19. $-\dfrac{1}{6}$ 21. $-\dfrac{7}{13}$ 23. $\dfrac{27}{31}$
25. $\dfrac{6}{7}$ 27. $\dfrac{12}{5}$ 29. $\dfrac{-7}{15}$ 31. $\dfrac{9}{5}$ 33. 6   35. $\dfrac{10}{7}$ 37. 75
39. 1800 gal   41. 9 bees   43. 20 packages   45. $\dfrac{1}{8}$ T
47. 8 walkways   49. 45 customers   51. 32 pairs
53. $\dfrac{1}{16}$ in.   55. $\mathbf{D_W}$ 57. 26   58. -42 59. -67
60. -65 61. 20   62. 6   63. $17x$ 64. $4a$
65. $7a + 3$ 66. $4x - 7$ 67. $\mathbf{D_W}$ 69. $\dfrac{2}{9}$ 71. $\dfrac{7}{8}$ lb
73. 103 slices   75. \$510   77. \$1608.75, with $\dfrac{1}{4}$ yd unused

Concept Reinforcement, p. 224

1. True   2. False   3. True   4. True   5. False
6. True   7. True   8. False

Summary and Review: Chapter 3, p. 224

1. 8, 16, 24, 32, 40, 48, 56, 64, 72, 80   2. No   3. No
4. No   5. Yes   6. Yes   7. 1, 2, 3, 4, 5, 6, 10, 12, 15, 20,
30, 60   8. 1, 2, 4, 8, 11, 16, 22, 44, 88, 176   9. Prime
10. Neither   11. Composite   12. $2 \cdot 5 \cdot 7$
13. $2 \cdot 2 \cdot 2 \cdot 3 \cdot 3$ 14. $3 \cdot 3 \cdot 5$ 15. $2 \cdot 3 \cdot 5 \cdot 5$
16. $2 \cdot 2 \cdot 2 \cdot 3 \cdot 3 \cdot 3 \cdot 3$ 17. $2 \cdot 2 \cdot 2 \cdot 2 \cdot 3 \cdot 5 \cdot 5$
18. Numerator: 9; denominator: 7   19. $\dfrac{3}{8}$ 20. $\dfrac{7}{6}$
21. (a) $\dfrac{3}{5}$; (b) $\dfrac{5}{3}$; (c) $\dfrac{3}{8}$ 22. 0   23. 1   24. 48   25. 1
26. $-\dfrac{2}{3}$ 27. $\dfrac{1}{4}$ 28. -1 29. $\dfrac{3}{4}$ 30. $\dfrac{2}{5}$
31. Undefined   32. $6x$ 33. $\dfrac{1}{3}$ 34. $\dfrac{15}{21}$ 35. $\dfrac{-30}{55}$
36. $\dfrac{3}{100}$; $\dfrac{8}{100} = \dfrac{2}{25}$; $\dfrac{10}{100} = \dfrac{1}{10}$; $\dfrac{15}{100} = \dfrac{3}{20}$; $\dfrac{21}{100}$; $\dfrac{43}{100}$ 37. $\neq$ 38. $=$
39. $\neq$ 40. $=$ 41. $\dfrac{13}{2}$ 42. $-\dfrac{1}{7}$ 43. 8   44. $\dfrac{5y}{3x}$
45. $\dfrac{14}{45}$ 46. $\dfrac{3y}{7x}$ 47. $\dfrac{2}{3}$ 48. $-\dfrac{1}{14}$ 49. $\dfrac{1}{25}$ 50. 1
51. $\dfrac{27}{5}$ 52. $\dfrac{1}{4}$ 53. 200   54. $\dfrac{1}{15}$ 55. $6a$ 56. -1
57. $\dfrac{3}{4}$ 58. $\dfrac{7}{144}$ 59. 42 m^2 60. $\dfrac{35}{2}$ ft^2 61. 240
62. $\dfrac{-3}{10}$ 63. 28   64. 9 days   65. $\dfrac{3}{8}$ cup   66. 288 km;
108 km   67. $\dfrac{1}{6}$ mi   68. 256,000,000 metric tons
69. $\mathbf{D_W}$ Taking $\dfrac{1}{2}$ of a number is the same as multiplying by
$\dfrac{1}{2}$. Dividing a number by $\dfrac{1}{2}$ is the same as multiplying by 2.
70. $\mathbf{D_W}$ Because $\dfrac{2}{8}$ simplifies to $\dfrac{1}{4}$, it is incorrect to suggest
that $\dfrac{2}{8}$ is simplified form of $\dfrac{20}{80}$.
71. $\dfrac{17}{6}$ 72. 2, 8   73. 13, 11, 101, 37
74. $a = 11{,}176$; $b = 9887$ 75. $\dfrac{3}{17}$

Test: Chapter 3, p. 227

1. [3.1b] Yes   2. [3.1b] No   3. [3.2a] 1, 2, 3, 5, 6, 9, 10, 15,
18, 30, 45, 90   4. [3.2b] Composite   5. [3.2c] $2 \cdot 2 \cdot 3 \cdot 3$
6. [3.2c] $2 \cdot 2 \cdot 3 \cdot 5$ 7. [3.3a] Numerator: 4;
denominator: 9   8. [3.3a] $\dfrac{3}{4}$ 9. [3.3a] $\dfrac{3}{7}$
10. [3.3a] (a) $\dfrac{1112}{1202}$; (b) $\dfrac{90}{1202}$ 11. [3.3b] 32   12. [3.3b] 1
13. [3.3b] 0   14. [3.5b] $\dfrac{-1}{3}$ 15. [3.5b] $\dfrac{1}{5}$ 16. [3.5b] $\dfrac{1}{9}$
17. [3.5c] $=$ 18. [3.5c] $\neq$ 19. [3.5a] $\dfrac{15}{40}$ 20. [3.7a] $\dfrac{42}{a}$
21. [3.7a] $\dfrac{-1}{9}$ 22. [3.6a] $\dfrac{5}{2}$ 23. [3.7b] $\dfrac{8}{33}$ 24. [3.4a] $\dfrac{3x}{8}$
25. [3.7b] $\dfrac{-3}{14}$ 26. [3.7b] 18   27. [3.6a] $\dfrac{6}{65}$
28. [3.8b] $\dfrac{3}{20}$ lb   29. [3.6b] 125 lb   30. [3.8a] 64
31. [3.8a] $\dfrac{-7}{4}$ 32. [3.6b] $\dfrac{91}{2}$ m^2 33. [3.6b] $\dfrac{15}{8}$ tsp
34. [3.6b] $\dfrac{7}{48}$ acre   35. [3.6a], [3.7b] $\dfrac{-7}{960}$ 36. [3.8a] $\dfrac{7}{5}$

Cumulative Review: Chapters 1–3, p. 229

1. [1.1c] Two million, fifty-six thousand, seven hundred
eighty-three   2. [1.2b] 10,982   3. [2.2a] -43
4. [2.2a] -33 5. [1.3d] 2129   6. [2.3a] -23
7. [2.3a] -8 8. [1.5a] 16,905   9. [2.4a] -312
10. [3.6a] $-30x$ 11. [3.6a] $\dfrac{7}{15}$ 12. [1.6c] 235 R 3
13. [2.5a] -17 14. [3.7b] -28 15. [3.7b] $\dfrac{2}{3}$

16. [1.4a] 4510   **17.** [1.5b] $900 \times 500 = 450{,}000$
18. [2.1c] 479   **19.** [2.5b] 8   **20.** [3.2b] Composite
21. [2.6a] -21 **22.** [1.7b], [2.8a] 25   **23.** [1.7b], [2.8b] 7
24. [3.8a] -45 **25.** [1.8a] 8 mpg   **26.** [1.8a] 8 oz
27. [2.7a] $5x - 5$ **28.** [2.7a] $3x + 7y$ **29.** [3.3b] 1
30. [3.3b] 0   **31.** [3.3b] 63   **32.** [3.5b] $-\frac{5}{27}$ **33.** [3.7a] $\frac{5}{2}$
34. [3.7a] $\frac{1}{57}$ **35.** [3.5a] $\frac{21}{70}$ **36.** [3.6b] 4375 students
37. [3.8b] 5 qt   **38.** [3.6b] $\frac{3}{5}$ mi
39. [2.6a], [3.6a], [3.7b] $\frac{-54}{169}$ **40.** [2.1c], [2.6a], [3.6a] $\frac{-9}{100}$
41. [3.6b] \$468

CHAPTER 4

Margin Exercises, Section 4.1, pp. 232–236

1. 45   **2.** 56   **3.** 18   **4.** 24   **5.** 14   **6.** 200   **7.** 40
8. 360   **9.** 30   **10.** 360   **11.** 100   **12.** xyz **13.** $5a^3b$
14. $40a^3b^2c^4$

Exercise Set 4.1, p. 237

1. 10   **3.** 50   **5.** 40   **7.** 54   **9.** 150   **11.** 120
13. 72   **15.** 420   **17.** 144   **19.** 180   **21.** 42   **23.** 30
25. 72   **27.** 60   **29.** 36   **31.** 900   **33.** 300   **35.** abc
37. $9x^2$ **39.** $4x^3y$ **41.** $24r^3s^2t^4$ **43.** $a^3b^2c^2$
45. Once every 60 yr   **47.** $\mathbf{D_W}$ **49.** 14   **50.** -27
51. 7935   **52.** $\frac{2}{3}$ **53.** $-\frac{8}{7}$ **54.** -167 **55.** $\mathbf{D_W}$
57. $\mathbf{D_W}$ **59.** 70,200   **61.** 121,695   **63.** 30 strands
65. 210 days   **67.** 18,900
69. 8 and 7; 8 and 28; 8 and 14

Margin Exercises, Section 4.2, pp. 240–244

1. $\frac{4}{5}$ **2.** 1   **3.** $\frac{1}{2}$ **4.** $-\frac{3}{8}$ **5.** $-\frac{4}{x}$ **6.** $\frac{2}{5}a$ **7.** $-\frac{1}{2}x$
8. $\frac{5}{6}$ **9.** $\frac{29}{24}$ **10.** $\frac{2}{9}$ **11.** $\frac{38}{5}$ **12.** $\frac{413}{1000}$ **13.** $\frac{157}{210}$
14. $<$ **15.** $>$ **16.** $>$ **17.** $<$ **18.** $<$ **19.** $<$
20. $\frac{9}{10}$ mi

Exercise Set 4.2, p. 245

1. $\frac{5}{9}$ **3.** 1   **5.** $\frac{2}{5}$ **7.** $\frac{13}{a}$ **9.** $-\frac{4}{11}$ **11.** $\frac{7}{9}x$ **13.** $\frac{1}{2}t$
15. $-\frac{9}{x}$ **17.** $\frac{7}{24}$ **19.** $-\frac{1}{10}$ **21.** $\frac{23}{24}$ **23.** $\frac{83}{20}$ **25.** $\frac{5}{24}$
27. $\frac{37}{100}x$ **29.** $\frac{19}{20}$ **31.** $-\frac{99}{100}$ **33.** $-\frac{1}{30}x$ **35.** $-\frac{33}{7}t$
37. $-\frac{17}{24}$ **39.** $\frac{437}{1000}$ **41.** $\frac{5}{4}$ **43.** $\frac{239}{78}$ **45.** $\frac{59}{90}$ **47.** $-\frac{5}{4}$
49. $>$ **51.** $<$ **53.** $>$ **55.** $<$ **57.** $>$ **59.** $>$
61. $\frac{4}{15}, \frac{3}{10}, \frac{5}{12}$ **63.** $\frac{3}{4}$ lb   **65.** $\frac{51}{40}$ mi   **67.** $\frac{13}{12}$ lb   **69.** $\frac{7}{8}$ in.
71. $\frac{33}{20}$ mi   **73.** $\frac{4}{5}$ qt; $\frac{8}{5}$ qt; $\frac{2}{5}$ qt   **75.** $\frac{51}{32}$ in.   **77.** $\mathbf{D_W}$
79. -13 **80.** 4   **81.** -8 **82.** -31 **83.** $\frac{10}{3}$
84. 42; 42   **85.** \$66   **86.** \$220   **87.** \$57   **88.** \$45
89. \$1264   **90.** \$1152   **91.** $\mathbf{D_W}$ **93.** $\frac{13}{30}t + \frac{31}{35}$
95. $7t^2 + \frac{9}{a}t$ **97.** $>$ **99.** $\frac{4}{15}$; \$320

101. The largest is $4 + \frac{6}{3} \cdot 5 = 14$, or $4 + \frac{5}{3} \cdot 6 = 14$.
103. $\frac{3}{4}, \frac{7}{9}, \frac{17}{21}, \frac{19}{22}, \frac{13}{15}, \frac{15}{17}, \frac{13}{12}$

Margin Exercises, Section 4.3, pp. 250–253

1. $\frac{1}{2}$ **2.** $\frac{4}{9a}$ **3.** $-\frac{3}{5}$ **4.** $-\frac{6}{x}$ **5.** $\frac{1}{12}$ **6.** $\frac{1}{6}$ **7.** $-\frac{3}{10}$
8. $-\frac{1}{6}$ **9.** $\frac{9}{112}$ **10.** $\frac{3}{10}x$ **11.** $\frac{3}{5}$ **12.** $\frac{1}{6}$ **13.** $\frac{-59}{40}$
14. $\frac{5}{24}$ mi

Calculator Corner, p. 253

1. $\frac{5}{8}$ **2.** $\frac{43}{60}$ **3.** $\frac{17}{21}$ **4.** $\frac{13}{28}$ **5.** $-\frac{17}{50}$ **6.** $\frac{410}{667}$

Translating for Success, p. 254

1. J   **2.** E   **3.** D   **4.** B   **5.** I   **6.** N   **7.** A   **8.** C
9. L   **10.** F

Exercise Set 4.3, p. 255

1. $\frac{2}{3}$ **3.** $-\frac{1}{4}$ **5.** $\frac{2}{a}$ **7.** $-\frac{7}{9}$ **9.** $-\frac{1}{2}$ **11.** $\frac{2}{t}$
13. $-\frac{4}{5a}$ **15.** $\frac{13}{16}$ **17.** $-\frac{1}{3}$ **19.** $\frac{7}{10}$ **21.** $-\frac{17}{60}$ **23.** $\frac{47}{100}$
25. $\frac{26}{75}$ **27.** $\frac{-21}{100}$ **29.** $\frac{-13}{24}$ **31.** $\frac{-29}{50}$ **33.** $\frac{-2}{15}$ **35.** $-\frac{41}{72}$
37. $\frac{1}{360}$ **39.** $\frac{2}{9}x$ **41.** $-\frac{7}{20}a$ **43.** $\frac{7}{9}$ **45.** $\frac{4}{11}$ **47.** $\frac{4}{9}$
49. $\frac{9}{8}$ **51.** $\frac{2}{15}$ **53.** $-\frac{7}{24}$ **55.** $\frac{2}{15}$ **57.** $-\frac{5}{4}$ **59.** $\frac{3}{10}$ mi
61. $\frac{19}{24}$ cup   **63.** $\frac{3}{16}$ in.   **65.** $\frac{11}{20}$ lb   **67.** $\frac{7}{20}$ hr   **69.** $\frac{5}{12}$ cup
71. $\mathbf{D_W}$ **73.** $\frac{4}{21}$ **74.** $\frac{3}{2}$ **75.** 21   **76.** $\frac{1}{32}$ **77.** 17 days
78. 9 cups   **79.** 11   **80.** 3   **81.** $\mathbf{D_W}$ **83.** $\frac{1}{16}$
85. $-\frac{11}{10}$ **87.** $\frac{-64}{35}$ **89.** $-\frac{37}{1000}$ **91.** $\frac{1}{6}$ of the business
93. $\frac{43}{50}$ **95.** 3 hr   **97.** $\frac{1}{8}$ **99.** $\frac{14}{3553}$ **101.** 4
103. Day 1: Cut off $\frac{1}{7}$ of bar and pay the contractor.

Day 2: Cut off $\frac{2}{7}$ of the bar's original length and trade it for the $\frac{1}{7}$.

Day 3: Give the $\frac{1}{7}$ back to the contractor.

Day 4: Trade the $\frac{4}{7}$ remaining for the contractor's $\frac{3}{7}$.

Day 5: Give the contractor the $\frac{1}{7}$ again.

Day 6: Trade the $\frac{2}{7}$ for the $\frac{1}{7}$.

Day 7: Give the contractor the $\frac{1}{7}$ again. This assumes that the contractor does not spend parts of the gold bar immediately.

Margin Exercises, Section 4.4, pp. 261–263

1. $\frac{10}{9}$ **2.** $\frac{9}{5}$ **3.** $-\frac{18}{7}$ **4.** $-\frac{7}{4}$ **5.** -16 **6.** -21
7. $\frac{10}{9}$

Exercise Set 4.4, p. 264

1. 3   **3.** -3 **5.** 12   **7.** $\frac{4}{3}$ **9.** $\frac{1}{2}$ **11.** $\frac{3}{4}$ **13.** $\frac{8}{3}$
15. $\frac{1}{2}$ **17.** $\frac{8}{7}$ **19.** $\frac{9}{2}$ **21.** $\frac{21}{5}$ **23.** 6   **25.** $\frac{17}{4}$ **27.** $\frac{3}{4}$

29. $\frac{3}{2}$ **31.** $\frac{9}{2}$ **33.** $-\frac{10}{3}$ **35.** $-\frac{1}{5}$ **37.** $-\frac{1}{6}$ **39.** $\frac{35}{12}$
41. **D**w **43.** -13 **44.** -8 **45.** 18 **46.** 27
47. The balance has decreased $150. **48.** $1180 profit
49. $\frac{5}{7m}$ **50.** $20n$ **51.** **D**w **53.** 4 **55.** $-\frac{177,450}{181,843}$
57. $\frac{145}{12}$ **59.** $-\frac{153}{16}$ **61.** $\frac{3}{2}$ cm **63.** $\frac{3}{4}$ cm

Margin Exercises, Section 4.5, pp. 267–269

1. $1\frac{2}{3}$ **2.** $7\frac{1}{4}$ **3.** $15\frac{2}{9}$ **4.** $\frac{22}{5}$ **5.** $\frac{61}{10}$ **6.** $\frac{23}{7}$ **7.** $\frac{73}{8}$
8. $\frac{62}{3}$ **9.** $-\frac{32}{5}$ **10.** $-\frac{65}{9}$ **11.** $2\frac{1}{3}$ **12.** $1\frac{1}{15}$ **13.** $18\frac{1}{3}$
14. $-2\frac{2}{5}$ **15.** $-11\frac{1}{6}$ **16.** $807\frac{2}{3}$ **17.** $55\frac{3}{4}$ qt

Calculator Corner, p. 270

1. $5\frac{4}{7}$ **2.** $8\frac{2}{5}$ **3.** $1476\frac{1}{6}$ **4.** $676\frac{4}{9}$ **5.** $51,626\frac{9}{11}$
6. $7330\frac{7}{32}$ **7.** $134\frac{1}{15}$ **8.** $2666\frac{130}{213}$ **9.** $3571\frac{51}{112}$
10. $12\frac{169}{454}$

Exercise Set 4.5, p. 271

1. $\frac{23}{3}$ **3.** $\frac{25}{4}$ **5.** $-\frac{161}{8}$ **7.** $\frac{51}{10}$ **9.** $\frac{103}{5}$ **11.** $\frac{-58}{7}$
13. $\frac{69}{10}$ **15.** $-\frac{51}{4}$ **17.** $\frac{57}{10}$ **19.** $\frac{-507}{100}$ **21.** $5\frac{1}{3}$ **23.** $7\frac{1}{2}$
25. $5\frac{7}{10}$ **27.** $7\frac{2}{9}$ **29.** $-5\frac{1}{2}$ **31.** $11\frac{1}{2}$ **33.** $-1\frac{1}{2}$
35. $61\frac{2}{5}$ **37.** $-8\frac{13}{50}$ **39.** $108\frac{5}{7}$ **41.** $906\frac{3}{7}$ **43.** $40\frac{4}{7}$
45. $-20\frac{2}{15}$ **47.** $-22\frac{3}{7}$ **49.** $8\frac{1}{5}$ **51.** $5\frac{1}{6}$ **53.** **D**w
55. $\frac{8}{9}$ **56.** $\frac{3}{8}$ **57.** $\frac{1}{4}$ **58.** $\frac{5}{4}$ **59.** $-\frac{3}{10}$ **60.** $-\frac{9}{28}$
61. **D**w **63.** $237\frac{19}{541}$ **65.** $8\frac{2}{3}$ **67.** $3\frac{2}{3}$ **69.** $52\frac{1}{7}$

Margin Exercises, Section 4.6, pp. 273–278

1. $9\frac{1}{3}$ **2.** $9\frac{2}{5}$ **3.** $12\frac{1}{10}$ **4.** $14\frac{7}{12}$ **5.** $1\frac{1}{2}$ **6.** $3\frac{1}{6}$ **7.** $3\frac{2}{3}$
8. $3\frac{5}{18}$ **9.** $12\frac{5}{6}t$ **10.** $2\frac{1}{4}x$ **11.** $14\frac{1}{30}x$ **12.** $232\frac{3}{20}$ mi
13. $23\frac{1}{4}$ gal **14.** $\frac{-3}{4}$ **15.** $-4\frac{1}{4}$ **16.** $-2\frac{1}{2}$ **17.** $-1\frac{5}{6}$
18. $3\frac{3}{10}$ **19.** $-13\frac{7}{30}$

Calculator Corner, p. 279

1. $10\frac{2}{15}$ **2.** $1\frac{1}{28}$ **3.** $2\frac{5}{63}$ **4.** $8\frac{1}{15}$ **5.** $-\frac{60}{209}$ **6.** $-2\frac{64}{255}$

Exercise Set 4.6, p. 280

1. $11\frac{2}{5}$ **3.** $9\frac{1}{2}$ **5.** $5\frac{11}{12}$ **7.** $13\frac{7}{12}$ **9.** $12\frac{1}{10}$ **11.** $17\frac{5}{24}$
13. $21\frac{1}{2}$ **15.** $27\frac{7}{8}$ **17.** $1\frac{3}{5}$ **19.** $6\frac{1}{10}$ **21.** $21\frac{17}{24}$
23. $13\frac{1}{4}$ **25.** $15\frac{3}{8}$ **27.** $7\frac{5}{12}$ **29.** $11\frac{5}{18}$ **31.** $8\frac{13}{42}t$
33. $2\frac{1}{8}x$ **35.** $8\frac{31}{40}t$ **37.** $11\frac{34}{45}t$ **39.** $6\frac{1}{6}x$ **41.** $9\frac{31}{33}x$
43. $5\frac{3}{8}$ yd **45.** $7\frac{5}{16}$ lb **47.** $6\frac{7}{20}$ cm **49.** $19\frac{1}{16}$ in.
51. $20\frac{1}{12}$ yd **53.** $354\frac{23}{24}$ gal **55.** $20\frac{3}{160}$ mi **57.** $7\frac{3}{4}$ cups
59. $3\frac{4}{5}$ hr **61.** $51\frac{1}{2}$ in. **63.** $27\frac{3}{4}$ ft **65.** $4\frac{5}{6}$ ft **67.** $7\frac{3}{8}$ ft
69. $1\frac{9}{16}$ in. **71.** $\frac{-4}{5}$ **73.** $-3\frac{1}{4}$ **75.** $-3\frac{13}{15}$ **77.** $-7\frac{3}{5}$
79. $-10\frac{29}{35}$ **81.** $-1\frac{8}{9}$ **83.** **D**w **85.** 16 packages
86. 286 cartons; 2 oz left over **87.** Yes **88.** No **89.** No

90. Yes **91.** No **92.** Yes **93.** Yes **94.** Yes **95.** $\frac{10}{13}$
96. $\frac{1}{10}$ **97.** **D**w **99.** $8568\frac{786}{1189}$ **101.** $-1618\frac{690}{2117}$
103. $10\frac{7}{12}$ **105.** $-28\frac{3}{8}$ **107.** $55\frac{3}{4}$ in.

Margin Exercises, Section 4.7, pp. 286–292

1. 28 **2.** $2\frac{5}{14}$ **3.** $-12\frac{4}{5}$ **4.** $8\frac{1}{3}$ **5.** 12 **6.** $1\frac{7}{8}$
7. $-\frac{7}{10}$ **8.** $175\frac{1}{2}$ **9.** $159\frac{4}{5}$ **10.** $5\frac{3}{4}$ **11.** $227\frac{1}{2}$ mi
12. 20 mpg **13.** $240\frac{3}{4}$ ft² **14.** $9\frac{7}{8}$ in.

Calculator Corner, p. 291

1. $10\frac{11}{15}$ **2.** $2\frac{91}{115}$ **3.** $-1\frac{136}{189}$ **4.** $-27\frac{1}{112}$ **5.** $31\frac{47}{143}$
6. $25\frac{107}{221}$

Translating for Success, p. 293

1. O **2.** K **3.** F **4.** D **5.** H **6.** G **7.** L **8.** E
9. M **10.** J

Exercise Set 4.7, p. 294

1. $22\frac{2}{5}$ **3.** $1\frac{2}{3}$ **5.** $-56\frac{2}{3}$ **7.** $16\frac{1}{3}$ **9.** $-10\frac{3}{25}$ **11.** $35\frac{91}{100}$
13. $11\frac{7}{13}$ **15.** $1\frac{1}{5}$ **17.** $2\frac{1}{52}$ **19.** $-2\frac{11}{68}$ **21.** $1\frac{8}{43}$
23. $\frac{-9}{40}$ **25.** $23\frac{2}{5}$ **27.** $15\frac{5}{7}$ **29.** $-28\frac{28}{45}$ **31.** $-1\frac{1}{3}$
33. $12\frac{1}{4}$ **35.** $8\frac{3}{20}$ **37.** 45,000 beagles **39.** $13\frac{1}{3}$ tsp
41. $\frac{1}{2}$ **recipe:** $\frac{5}{6}$ cup water, $1\frac{1}{2}$ tablespoons canola oil,
$2\frac{1}{4}$ teaspoons sugar, $\frac{1}{2}$ teaspoon salt, $1\frac{7}{8}$ cups bread flour,
$\frac{3}{8}$ cup Grape-Nuts cereal, $\frac{3}{4}$ teaspoon yeast; **3 recipes:** 5 cups
water, 9 tablespoons canola oil, $13\frac{1}{2}$ teaspoons sugar,
3 teaspoons salt, $11\frac{1}{4}$ cups bread flour, $2\frac{1}{4}$ cups Grape-Nuts
cereal, $4\frac{1}{2}$ teaspoons yeast **43.** About 200 hr **45.** 68°F
47. Yes; $\frac{7}{8}$ in. **49.** $16\frac{1}{2}$ servings **51.** $343\frac{3}{4}$ lb **53.** $82\frac{1}{2}$ in.
55. 15 mpg **57.** $2\frac{41}{128}$ lb **59.** $7\frac{23}{50}$ sec **61.** $441\frac{1}{4}$ ft²
63. $76\frac{1}{4}$ ft² **65.** $27\frac{5}{16}$ cm² **67.** **D**w **69.** Integers
70. Common **71.** Composite **72.** Divisible; divisible
73. Least common multiple **74.** Addends
75. Numerator **76.** Reciprocal **77.** **D**w **79.** $16\frac{25}{64}$
81. $\frac{4}{9}$ **83.** $r = \frac{240}{13}$, or $18\frac{6}{13}$ **85.** 104 gal

Concept Reinforcement, p. 300

1. True **2.** True **3.** False **4.** True **5.** True
6. False

Summary and Review: Chapter 4, p. 300

1. 80 **2.** 90 **3.** 30 **4.** $\frac{7}{9}$ **5.** $\frac{9}{x}$ **6.** $\frac{-7}{15}$ **7.** $\frac{7}{16}$
8. $\frac{2}{9}$ **9.** $-\frac{1}{8}$ **10.** $\frac{4}{27}$ **11.** $\frac{11}{18}$ **12.** > **13.** < **14.** $\frac{19}{40}$
15. 4 **16.** $-\frac{5}{6}$ **17.** $\frac{12}{25}$ **18.** $\frac{2}{5}$ **19.** $\frac{15}{2}$ **20.** $\frac{274}{9}$
21. $\frac{-65}{7}$ **22.** $2\frac{3}{5}$ **23.** $-6\frac{3}{4}$ **24.** $7\frac{1}{8}$ **25.** $3\frac{1}{2}$
26. $-877\frac{1}{3}$ **27.** $82\frac{1}{3}$ **28.** $10\frac{2}{5}$ **29.** $11\frac{11}{15}$ **30.** -9

31. $1\frac{3}{4}$ **32.** $7\frac{7}{9}$ **33.** $4\frac{11}{15}$ **34.** $-5\frac{1}{8}$ **35.** $-14\frac{1}{4}$
36. $\frac{7}{9}x$ **37.** $3\frac{7}{40}a$ **38.** 16 **39.** $-3\frac{1}{2}$ **40.** $2\frac{21}{50}$ **41.** 6
42. -24 **43.** $-1\frac{7}{17}$ **44.** $\frac{1}{8}$ **45.** $\frac{9}{10}$ **46.** $13\frac{5}{7}$ **47.** $2\frac{8}{11}$
48. 24 cassettes **49.** $3\frac{1}{8}$ pizzas **50.** $\frac{3}{5}$ mi
51. $6\frac{11}{24}$ ft per hr **52.** $8\frac{3}{8}$ cups **53.** $36\frac{1}{2}$ in.
54. About 4,500,000 **55.** $177\frac{3}{4}$ in^2 **56.** $50\frac{1}{4}$ in^2
57. **DW** The student multiplied the whole numbers and multiplied the fractions. The mixed numerals should be converted to fraction notation before multiplying.
58. **DW** Yes. We may need to find a common denominator before adding or subtracting. To find the least common denominator, we use the least common multiple of the denominators. **59.** $3 \cdot 23 \cdot 41 \cdot 47 \cdot 59 = 7{,}844{,}817$
60. $\frac{600}{13}$ **61.** **(a)** 6; **(b)** 5; **(c)** 12; **(d)** 18; **(e)** -101;
(f) -155; **(g)** -3; **(h)** -1 **62.** **(a)** 6; **(b)** 10; **(c)** 46; **(d)** 1;
(e) -14; **(f)** -28; **(g)** -2; **(h)** -1

Test: Chapter 4, p. 304

1. [4.1a] 48 **2.** [4.2a] 3 **3.** [4.2b] $\frac{-5}{24}$ **4.** [4.3a] $\frac{2}{t}$
5. [4.3a] $\frac{1}{12}$ **6.** [4.3a] $-\frac{1}{12}$ **7.** [4.3b] $\frac{1}{4}$ **8.** [4.4a] $\frac{-12}{5}$
9. [4.4a] $\frac{3}{20}$ **10.** [4.2c] $>$ **11.** [4.5a] $\frac{7}{2}$ **12.** [4.5a] $\frac{-75}{8}$
13. [4.5b] $-8\frac{2}{9}$ **14.** [4.5c] $162\frac{7}{11}$ **15.** [4.6a] $14\frac{1}{5}$
16. [4.6a] $12\frac{5}{12}$ **17.** [4.6b] $4\frac{7}{24}$ **18.** [4.6d] $8\frac{4}{7}$
19. [4.6d] $-5\frac{7}{10}$ **20.** [4.3a] $-\frac{1}{8}x$ **21.** [4.6b] $1\frac{54}{55}a$
22. [4.7a] 39 **23.** [4.7a] -18 **24.** [4.7b] 6 **25.** [4.7b] 2
26. [4.7c] $19\frac{3}{5}$ **27.** [4.7c] $28\frac{1}{20}$ **28.** [4.7d] $7\frac{1}{2}$ lb
29. [4.7d] 80 books **30.** [4.6c] **(a)** 3 in.; **(b)** $4\frac{1}{2}$ in.
31. [4.3c] $\frac{1}{16}$ in. **32.** [4.7d] $6\frac{11}{36}$ ft **33.** [4.1a] $\frac{24}{25}$ min
34. [4.3c] Cheri; $\frac{17}{56}$ mi **35.** [4.1a] **(a)** 24, 48, 72; **(b)** 24
36. [4.2b] **(a)** $\frac{1}{2}$; **(b)** $\frac{2}{3}$; **(c)** $\frac{3}{4}$; **(d)** $\frac{4}{5}$; **(e)** $\frac{9}{10}$

Cumulative Review: Chapters 1–4, p. 306

1. [4.3c] **(a)** $\frac{1}{48}$ in.; **(b)** $\frac{1}{12}$ in. **2.** [3.8b] 61 DVDs
3. **(a)** [4.6c] $14\frac{13}{24}$ mi; **(b)** [4.7d] $4\frac{61}{72}$ mi
4. [4.7d], [4.6c] **(a)** $142\frac{1}{4}$ ft^2 **(b)** 54 ft **5.** [1.8a] 31 people
6. [1.8a] $108 **7.** [3.6b] $\frac{2}{5}$ tsp; 4 tsp **8.** [4.7d] 39 lb
9. [4.7d] 16 pieces **10.** [4.2d] $\frac{33}{20}$ mi **11.** [1.1a] 5
12. [1.1b] 6 thousands + 7 tens + 5 ones
13. [1.1c] Twenty-nine thousand, five hundred
14. [3.3a] $\frac{5}{16}$ **15.** [1.2b] 623 **16.** [2.2a] -8
17. [4.2b] $\frac{5}{12}$ **18.** [4.6a] $8\frac{1}{4}$ **19.** [1.3d] 5124
20. [2.3a] 16 **21.** [4.3a] $\frac{-5}{t}$ **22.** [4.6b] $1\frac{1}{6}$
23. [1.5a] 5004 **24.** [2.4a] -145 **25.** [3.6a] $\frac{3}{2}$
26. [3.6a] -15 **27.** [4.7a] $7\frac{1}{3}$ **28.** [1.6c] 48 R 11
29. [1.6c] 56 R 11 **30.** [4.5c] $56\frac{11}{45}$ **31.** [3.7b] $-\frac{4}{7}$
32. [4.7b] $7\frac{1}{3}$ **33.** [1.4a] 38,500 **34.** [4.1a] 72
35. [3.1b] Yes **36.** [3.2a] 1, 2, 4, 8, 16 **37.** [4.2c] $>$
38. [4.2c] $<$ **39.** [3.5b] $\frac{4}{5}$ **40.** [3.5b] -14 **41.** [4.5a] $\frac{73}{10}$
42. [4.5b] $-5\frac{2}{3}$ **43.** [1.7b], [2.8a] 55 **44.** [4.3b] $\frac{5}{9}$

45. [3.8a] $\frac{-12}{7}$ **46.** [4.4a] $\frac{2}{21}$ **47.** [2.6a] 4
48. [2.6b] $7b - 35$ **49.** [2.6b] $-3x + 6 - 3z$
50. [2.7a] $-6x - 9$ **51.** [2.7b] 39 in. **52.** [2.7b] 48 ft
53. [3.6b] $\frac{15}{4}$ ft^2 **54.** [3.6b] 250 ft^2 **55.** [4.4a], [4.6a] $\frac{3}{7}$
56. [4.6c], [4.7d] 3780 m^2

CHAPTER 5

Margin Exercises, Section 5.1, pp. 311–316

1. Eighty and thirty-nine hundredths; seventy-seven and seventy-one hundredths **2.** Two and four thousand five hundred eighty-three hundred-thousandths
3. Two hundred forty-five and eighty-nine hundredths
4. Thirty-four and six hundred forty-seven hundred-thousandths **5.** Thirty-one thousand, seventy-nine and seven hundred sixty-four thousandths
6. Four thousand, two hundred seventeen and $\frac{56}{100}$ dollars
7. Thirteen and $\frac{98}{100}$ dollars **8.** $\frac{549}{1000}$ **9.** $\frac{75{,}069}{1000}$; $75\frac{69}{1000}$
10. $-\frac{3129}{10}$; $-312\frac{9}{10}$ **11.** 7.43 **12.** -0.073 **13.** 6.7089
14. -0.9 **15.** -7.03 **16.** 23.047 **17.** 2.04 **18.** 0.06
19. 0.58 **20.** 1 **21.** 0.8989 **22.** 21.05 **23.** -34.008
24. -8.98 **25.** 2.8 **26.** 13.9 **27.** -234.4 **28.** 7.0
29. 0.64 **30.** -7.83 **31.** 34.70 **32.** -0.03
33. 0.943 **34.** -8.004 **35.** -43.112 **36.** 37.401
37. 7459.360 **38.** 7459.36 **39.** 7459.4 **40.** 7459
41. 7460 **42.** 7500 **43.** 7000

Exercise Set 5.1, p. 317

1. Sixty-three and five hundredths **3.** Twenty-six and fifty-nine hundredths **5.** Eight and thirty-five hundredths
7. Twenty-four and six thousand eight hundred seventy-five ten-thousandths **9.** Five and sixty-three hundredths
11. Five hundred twenty-four and $\frac{95}{100}$ dollars
13. Thirty-six and $\frac{72}{100}$ dollars **15.** $\frac{73}{10}$; $7\frac{3}{10}$ **17.** $\frac{2036}{10}$; $203\frac{6}{10}$
19. $\frac{-2703}{1000}$; $-2\frac{703}{1000}$ **21.** $\frac{109}{10{,}000}$ **23.** $\frac{-40{,}003}{10{,}000}$; $-4\frac{3}{10{,}000}$
25. $\frac{-207}{10{,}000}$ **27.** $\frac{7{,}000{,}105}{100{,}000}$; $70\frac{105}{100{,}000}$ **29.** 0.3 **31.** -0.59
33. 3.798 **35.** 0.0078 **37.** -0.00018 **39.** 0.486197
41. 7.013 **43.** -8.431 **45.** 2.1739 **47.** 8.953073
49. 0.58 **51.** 0.410 **53.** -5.043 **55.** 235.07
57. $\frac{7}{100}$ **59.** -0.872 **61.** 0.2 **63.** -0.4 **65.** 3.0
67. -327.2 **69.** 0.89 **71.** -0.67 **73.** 1.00
75. -0.03 **77.** 0.572 **79.** 17.002 **81.** -20.202
83. 9.985 **85.** 809.5 **87.** 809.47 **89.** **DW** **91.** 830
92. $\frac{830}{1000}$, or $\frac{83}{100}$ **93.** 182 **94.** $\frac{182}{100}$, or $\frac{91}{50}$ **95.** $\frac{-12}{55}$
96. $\frac{-15}{34}$ **97.** 32,958 **98.** 10,726 **99.** **DW**
101. -1.09, -1.009, -0.989, -0.898, -0.098
103. 6.78346 **105.** 99.99999 **107.** 1983, 1988, 1990, 1991, 1992, 1995, 1997, 1998, 1999, 2000, 2001, 2002, 2003, 2004 **109.** 1985

Margin Exercises, Section 5.2, pp. 320–323

1. 10.917 **2.** 34.2079 **3.** 4.969 **4.** 6.7982
5. 10.20434 **6.** 912.67 **7.** 2514.773 **8.** 10.754
9. 0.339 **10.** 2.54 **11.** 0.24238 **12.** 5.26992

13. 1194.22 **14.** 4.9911 **15.** −1.96 **16.** 3.159
17. −13.29 **18.** −4.16 **19.** −9.91 **20.** 12.4
21. −2.7 **22.** 3.7x **23.** 1.7a **24.** −2.7y + 5.4

Exercise Set 5.2, p. 324

1. 464.37 **3.** 1576.215 **5.** 132.56 or 132.560 **7.** 7.823
9. 50.0248 **11.** 0.835 **13.** 771.967 **15.** 20.8649
17. 227.468 or 227.4680 **19.** 41.381 **21.** 49.02
23. 3.564 **25.** 85.921 **27.** 1.6666 **29.** 4.0622
31. 29.999 **33.** 3.37 **35.** 1.045 **37.** 3.703
39. 0.9092 **41.** 605.21 **43.** 53.203 **45.** 161.62
47. 44.001 **49.** −3.29 **51.** −2.5 **53.** −7.2
55. 3.379 **57.** −16.6 **59.** 2.5 **61.** −3.519
63. 9.601 **65.** 75.5 **67.** 3.8 **69.** −10.292 **71.** −8.8
73. 5.7x **75.** 4.86a **77.** 21.1t + 7.9 **79.** −2.917x
81. 8.106y − 7.1 **83.** −0.9x + 3.1y **85.** 7.2 − 8.4t
87. D_W **89.** $\frac{12}{35}$ **90.** $\frac{14}{45}$ **91.** $\frac{63}{1000}$ **92.** −10 **93.** −7
94. 31 **95.** D_W **97.** −12.001 − 12.2698a + 10.366b
99. 4.593a − 10.996b − 59.491 **101.** −138.5 **103.** 2

Margin Exercises, Section 5.3, pp. 329–333

1. 625.66 **2.** 21.4863 **3.** 0.00943 **4.** −12.535374
5. 74.6 **6.** 0.7324 **7.** −0.062 **8.** 0.07236 **9.** 539.17
10. −6241.7 **11.** 83,900 **12.** 570,400 **13.** 3,700,000
14. 1,600,000,000 **15.** 3,300,000,000 **16.** 1569¢
17. 17¢ **18.** $0.35 **19.** $5.77 **20.** 6.656
21. 8.125 sq cm **22.** 55.107

Calculator Corner, p. 331

1. 317.645 **2.** 33.83 **3.** 48.6 **4.** 6930.5 **5.** 17.15
6. 454.74 **7.** 0.99 **8.** 0.5076

Exercise Set 5.3, p. 334

1. 47.6 **3.** 6.72 **5.** 0.252 **7.** 2.574 **9.** 426.3
11. −783,686.852 **13.** −780 **15.** 7.918 **17.** 0.09768
19. −0.287 **21.** 43.68 **23.** 3.2472 **25.** 89.76
27. −322.07 **29.** 55.68 **31.** 3487.5 **33.** 0.1155
35. −9420 **37.** 0.00953 **39.** 5706¢ **41.** 95¢ **43.** 1¢
45. $0.72 **47.** $0.02 **49.** $63.99 **51.** 3,156,000,000
53. 63,100,000,000,000 **55.** 11,980,000 **57.** 11,000
59. 26.025 **61. (a)** 44 ft; **(b)** 118.75 sq ft
63. (a) 37.8 m; **(b)** 88.2 m² **65.** 60.3 billion **67.** D_W
69. −27 **70.** 36 **71.** 69 **72.** −141 **73.** −21
74. −27 **75.** −141 **76.** −23 **77.** D_W
79. 804,100 billion km **81.** 366.5488175
83. 72.996 cm² **85.** 10^{21} **87.** 6,600,000,000,000
89. $61.45

Margin Exercises, Section 5.4, pp. 337–342

1. 0.6 **2.** 1.7 **3.** 0.47 **4.** 0.32 **5.** −5.75 **6.** 0.25
7. (a) 375; **(b)** 15 **8.** −4.9 **9.** 12.8 **10.** 15.625
11. −48 **12.** 12.78 **13.** 0.001278 **14.** 0.09847
15. −67.832 **16.** 0.2426 **17.** −7.4 **18.** 50.84 million
arrivals

Calculator Corner, p. 339

1. 28 R 2 **2.** 116 R 3 **3.** 74 R 10 **4.** 415 R 3

Exercise Set 5.4, p. 343

1. 12.6 **3.** 23.78 **5.** 7.48 **7.** 7.2 **9.** −0.9
11. −1.143 **13.** 140 **15.** 40 **17.** −0.15 **19.** 48
21. 3.2 **23.** 0.625 **25.** 0.26 **27.** 2.34 **29.** −0.3045
31. −2.134567 **33.** 1023.7 **35.** −9236 **37.** 0.08172
39. 9.7 **41.** −0.0527 **43.** −75,300 **45.** −0.0753
47. 2107 **49.** −302.997 **51.** −178.1 **53.** 206.0176
55. −400.0108 **57.** 0.6725 **59.** 5.383 **61.** 10.5
63. $206.34 billion **65.** 81.954 yr **67.** 21.34 mi
69. D_W **71.** $\frac{3}{4}$ **72.** $\frac{7}{8}$ **73.** $\frac{-3}{2}$ **74.** $\frac{-3}{10}$ **75.** $\frac{a}{3}$
76. $\frac{2x}{5}$ **77.** $\frac{1}{5}$ **78.** $\frac{2}{3}$ **79.** D_W **81.** −56.6916
83. 6.254194585 **85.** 1000 **87.** 100 **89.** 5.7 points
91. 450 kWh

Margin Exercises, Section 5.5, pp. 347–352

1. 0.4 **2.** −0.625 **3.** 0.1$\overline{6}$ **4.** 0.$\overline{6}$ **5.** 0.$\overline{45}$
6. −1.$\overline{09}$ **7.** 0.$\overline{714285}$ **8.** 0.7; 0.67; 0.667
9. 0.6; 0.61; 0.608 **10.** −7.3; −7.35; −7.349
11. 2.7; 2.69; 2.689 **12.** 0.8 **13.** −0.45 **14.** 0.035
15. 1.32 **16.** 0.72 **17.** 0.552 **18.** 4.225 ft²

Calculator Corner, p. 349

1. −0.1$\overline{6}$ **2.** 0.$\overline{63}$ **3.** 0.$\overline{12}$ **4.** −1.$\overline{48}$ **5.** −0.$\overline{72}$
6. 6.$\overline{3}$ **7.** −57.$\overline{1}$ **8.** 9.8$\overline{3}$

Calculator Corner, p. 352

1. 123.150432 **2.** 52.59026102

Exercise Set 5.5, p. 353

1. 0.375 **3.** −0.5 **5.** 0.12 **7.** 0.225 **9.** 0.52
11. −0.85 **13.** −0.5625 **15.** 1.4 **17.** 1.12
19. −1.375 **21.** −0.975 **23.** 0.605 **25.** 0.5$\overline{3}$
27. 0.$\overline{3}$ **29.** −1.$\overline{3}$ **31.** 1.1$\overline{6}$ **33.** −1.$\overline{27}$ **35.** −0.41$\overline{6}$
37. 0.254 **39.** 0.$\overline{12}$ **41.** −0.2$\overline{18}$ **43.** 0.$\overline{315}$
45. 0.$\overline{571428}$ **47.** −1.48 **49.** 0.4; 0.36; 0.364
51. −1.7; −1.67; −1.667 **53.** −0.5; −0.47; −0.471
55. 0.6; 0.58; 0.583 **57.** −0.2; −0.19; −0.193
59. −0.8; −0.78; −0.778 **61.** 9.485 **63.** −417.51$\overline{6}$
65. 0.09705 **67.** −1.5275 **69.** 24.375 **71.** 1.08 m²
73. 5.78 cm² **75.** 790.92 in² **77.** D_W **79.** 3570
80. 4000 **81.** 79,000 **82.** 19,830,000 **83.** −95
84. −10 **85.** −7 **86.** 1 **87.** D_W **89.** 0.$\overline{142857}$
91. 0.$\overline{428571}$ **93.** 0.$\overline{714285}$ **95.** 0.$\overline{1}$ **97.** 0.$\overline{001}$
99. 13.86 cm² **101.** 1.76625 ft² or 1.767145868 ft²
103. D_W

Margin Exercises, Section 5.6, pp. 357–359

1. $510 **2.** $60 **3.** $1860 **4.** 18 systems **5.** 16
6. 18 **7.** 470 **8.** 0.07 **9.** 18 **10.** 125 **11.** (c)
12. (a) **13.** (c) **14.** (c)

Exercise Set 5.6, p. 360

1. $430; 280 + 150 **3.** $130; 280 − 150 **5.** $480; 6 · 80
7. 10 sets; 80 · 10 = 800 **9.** 1.6 **11.** 6 **13.** 60
15. 2.3 **17.** 180 **19.** (a) **21.** (c) **23.** (b) **25.** (b)
27. 1800 ÷ 9 = 200 posts; answers may vary **29.** D_W
31. Repeating **32.** Multiple **33.** Distributive
34. Solution **35.** Multiplicative **36.** Commutative
37. Denominator; multiple **38.** Divisible; divisible
39. D_W **41.** Yes **43.** No **45.** (a) +, ×; (b) +, ×, −

Margin Exercises, Section 5.7, pp. 363–365

1. 0.6 **2.** −12 **3.** 3.6 **4.** −1.9 **5.** 3.5 **6.** 2.2
7. −4.5 **8.** 1.25 **9.** 8.25

Exercise Set 5.7, p. 366

1. 5.4 **3.** −12.6 **5.** 6 **7.** 1.8 **9.** −3.7 **11.** −4.7
13. 1.7 **15.** 2.94 **17.** 9 **19.** 3.2 **21.** −1.75
23. 30 **25.** 3.2 **27.** 9 **29.** 2.1 **31.** 4.5 **33.** 5.6
35. −1.9 **37.** 13 **39.** −1.5 **41.** D_W **43.** 14 m^2
44. 27 cm^2 **45.** $\frac{25}{2}$ in^2 **46.** 24 ft^2 **47.** 5 ft^2
48. 12 m^2 **49.** $\frac{-29}{50}$ **50.** 0 **51.** −2 **52.** 8 **53.** D_W
55. 3.1 **57.** 36 **59.** 1.1212963

Margin Exercises, Section 5.8, pp. 369–375

1. $2.5 billion **2.** $37.28 **3.** $189.50 **4.** 28.6 mpg
5. 13.76 in^2 **6.** 3.5 hr **7.** 118 buttons

Translating for Success, p. 376

1. I **2.** C **3.** N **4.** A **5.** G **6.** B **7.** D **8.** O
9. F **10.** M

Exercise Set 5.8, p. 377

1. $230.86 **3.** $45.88 **5.** 102.8°F **7.** $21,219.17
9. Area: 8.125 cm^2; perimeter: 11.5 cm **11.** 22,691.5 mi
13. $19.15 **15.** 3.5°F **17.** $5.65 **19.** 148.1 gal
21. 20.2 mpg **23.** 2.66 cc **25.** 233.66 in^2
27. 193.04 cm^2 **29.** $30 **31.** 2.31 cm **33.** 331.74 ft^2
35. $53.04 **37.** 875 megabytes **39.** 960 min
41. 120 min **43.** 698 transactions **45.** 2152.56 yd^2
47. 17 bottles **49.** D_W **51.** 0 **52.** $-\frac{1}{10}$ **53.** $\frac{-20}{33}$
54. −4 **55.** $6\frac{5}{6}$ **56.** 1 **57.** D_W **59.** D_W
61. $\frac{96}{325}$ min, or about 18 sec **63.** 25 cm^2. We assume that
the figures are nested squares formed by connecting the
midpoints of consecutive sides of the next larger square.
65. $4.1 million

Concept Reinforcement, p. 383

1. False **2.** True **3.** True **4.** False **5.** True

Summary and Review: Chapter 5, p. 383

1. 6,590,000 **2.** 6,900,000
3. Three and forty-seven hundredths
4. Thirty-one thousandths **5.** $\frac{9}{100}$ **6.** $-\frac{4561}{1000}$; $-4\frac{561}{1000}$
7. $-\frac{89}{1000}$ **8.** $\frac{30,227}{10,000}$; $3\frac{227}{10,000}$ **9.** −0.034 **10.** 4.2603
11. 27.91 **12.** −867.006 **13.** 0.034 **14.** −0.19
15. 17.4 **16.** 17.429 **17.** 499.829 **18.** 29.148
19. 229.1 **20.** 685.0519 **21.** −57.3 **22.** 2.37
23. 12.96 **24.** −1.073 **25.** 24,680 **26.** 3.2
27. −1.6 **28.** 0.2763 **29.** $2.2x − 9.1y$
30. $-2.84a + 12.57$ **31.** 925 **32.** 40.84 **33.** 11.3
34. 20 videotapes **35.** $15.49 **36.** 248.27 **37.** 2.6
38. 1.28 **39.** 3.25 **40.** $-1.1\overline{6}$ **41.** 21.08 **42.** −3.2
43. −3 **44.** −7.5 **45.** 6.5 **46.** 11.16 poles
47. $15.52 **48.** 249.76 ft^2 **49.** $5788.56 **50.** $78.39
51. 1830 megabytes **52.** 14.5 mpg **53.** (a) 102.6 lb;
(b) 14.7 lb **54.** 8.4 mi **55.** $1.33 **56.** 61.5 ft;
235.625 sq ft **57.** D_W Since there are 20 nickels to a
dollar, $\frac{3}{20}$ corresponds to 3 nickels, or 15¢, which is
0.15 dollars. **58.** D_W In decimal notation, $\frac{1}{3}$ and $\frac{1}{6}$ both
must be rounded before they can be multiplied. The best
way to express $\frac{1}{3} \cdot \frac{1}{6}$ as a decimal is to multiply the fractions
and then convert the product $\frac{1}{18}$ to decimal notation.
59. (a) +; (b) − **60.** $\frac{-13}{15}, \frac{-17}{20}, -\frac{11}{13}, -\frac{15}{19}, \frac{-5}{7}, -\frac{2}{3}$
61. 16,000 mi **62.** D_W The Sicilian pizza, at $\frac{4.4¢}{\text{in}^2}$, is a
better buy than the round pizza which costs $\frac{5.5¢}{\text{in}^2}$.

Test: Chapter 5, p. 386

1. [5.3b] 8,900,000,000 **2.** [5.3b] 3,756,000
3. [5.1a] Two and thirty-four hundredths
4. [5.1a] One hundred five and five ten-thousandths
5. [5.1b] $-\frac{3}{10}$ **6.** [5.1b] $\frac{2769}{1000}$ **7.** [5.1b] 0.074
8. [5.1b] −3.7047 **9.** [5.1b] 756.09 **10.** [5.1b] 91.703
11. [5.1c] 0.162 **12.** [5.1c] −0.173 **13.** [5.1d] 9.5
14. [5.1d] 9.452 **15.** [5.2a] 405.219 **16.** [5.3a] 0.03
17. [5.3a] 0.21345 **18.** [5.2b] 44.746 **19.** [5.2a] 356.37
20. [5.2c] −2.2 **21.** [5.2b] 1.9946 **22.** [5.3a] 73,962
23. [5.4a] 4.75 **24.** [5.4a] 30.4 **25.** [5.4a] −0.34682
26. [5.4a] 34,682 **27.** [5.3b] 17,982¢
28. [5.2d] $9.8x − 3.9y − 4.6$ **29.** [5.3c] 11.6
30. [5.4b] 7.6 **31.** [5.8a] 7 gal **32.** [5.5b] 48.7
33. [5.5c] 1.6 **34.** [5.5c] 5.25 **35.** [5.5a] −0.4375
36. [5.5a] $1.\overline{5}$ **37.** [5.5b] 1.56 **38.** [5.6a] 198
39. [5.6a] 4 **40.** [5.5d] 9.72 **41.** [5.7a] −3.24
42. [5.7b] 10 **43.** [5.7b] 1.4 **44.** [5.8a] 58.24 million
passengers **45.** [5.8a] 2860 min **46.** [5.8a] 28.3 mpg

47. [5.8a] $6572.45 **48.** [5.8a] $181.93
49. [5.3a] **(a)** Always; **(b)** never; **(c)** sometimes; **(d)** sometimes **50.** [5.8a] $1.4\overline{3}$ gal per person
51. [5.8a] **(a)** Fly; **(b)** drive; **(c)** drive

Cumulative Review: Chapters 1–5, p. 389

1. [1.1c] Two hundred seven thousand four hundred ninety-one **2.** [5.3b] 6,250,000,000
3. [5.1b] $\frac{1009}{100}$ **4.** [4.5a] $\frac{35}{8}$ **5.** [5.1b] -0.035
6. [3.2a] 1, 2, 3, 6, 11, 22, 33, 66 **7.** [3.2c] $2 \cdot 7 \cdot 11$
8. [4.1a] 140 **9.** [5.1d] 7000 **10.** [5.1d] 6962.47
11. [4.6a] $6\frac{2}{9}$ **12.** [5.2a] 235.397 **13.** [1.2b] 5495
14. [4.2b] $-\frac{1}{30}$ **15.** [2.3a] -71 **16.** [5.2b] 8446.53
17. [4.3a] $\frac{1}{72}$ **18.** [4.6b] $3\frac{2}{5}$ **19.** [5.3a] 4.78
20. [3.6a] $\frac{-2}{7}$ **21.** [4.7a] $13\frac{7}{11}$ **22.** [3.6a] $\frac{3}{2}$
23. [4.7b] $1\frac{1}{2}$ **24.** [3.7b] $1\frac{13}{35}$ **25.** [5.4a] $-43,795$
26. [5.4a] 20.6 **27.** [4.2c] $<$ **28.** [2.1b] $<$
29. [2.6a] 12 **30.** [2.6b] $4x - 4y + 12$ **31.** [2.7a] $7p - 8$
32. [2.7a] $14x - 11$ **33.** [5.7a] 0.78 **34.** [2.8b] -28
35. [5.7a] 8.62 **36.** [2.8a] 369,375 **37.** [4.3b] $\frac{1}{18}$
38. [3.8a] $\frac{1}{2}$ **39.** [5.7a] 3.8125 **40.** [4.4a] $\frac{7}{3}$
41. [1.8a] 10,945 min **42.** [3.8b] $1500
43. [1.8a] 86,400 sec **44.** [3.6b] $2800
45. [5.8a] $258.77 **46.** [4.6c] $6\frac{1}{2}$ lb **47.** [3.6b] 88 ft^2
48. [1.5c], [5.8a] 43.585 in^2 **49.** [5.8a] 12 boxes
50. [5.8a] $1.42 **51.** [4.6c], [5.8a] $45\frac{29}{48}$ hr, or 45 hr 36.25 min **52.** [5.8a] $35

CHAPTER 6

Margin Exercises, Section 6.1, pp. 392–395

1. $1333; tuition and required fees at a public 2-yr college in 2001 **2.** 2003 **3.** $3852.20 **4.** Javan rhino; about $\frac{1}{5} \cdot 300$, or 60 rhinos **5.** About 2100 rhinos **6.** There are approximately twelve times as many black rhinos as Sumatran rhinos, or approximately 3300 more.

7.

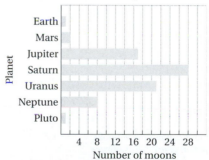

Revenue from Movie Trilogies	
The Lord of the Rings (2001–2003)	
Harry Potter (2001–2004)	
Jurassic Park (1993–2002)	
The Matrix (1999–2003)	
Terminator (1984–2003)	
= $1,000,000,000	

Exercise Set 6.1, p. 396

1. Cinnamon Life® **3.** Kellogg's Complete® **5.** 1.2 g
7. 483,612,200 mi **9.** Neptune

11. Saturn, Uranus, Neptune, Pluto **13.** 92° **15.** 108°
17. 3 **19.** 90° and higher **21.** 30% and higher **23.** 50%
25. 1940: 1976; 1980: 3849; 1873 **27.** Approximately 1920; approximately 3177; approximately 1257 **29.** 1.0 billion
31. 2070 **33.** 1650 and 1850 **35.** 3 billion **37.** Africa
39. 475,000 gal **41.** 325,000 gal
43.

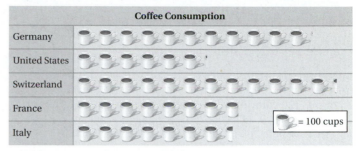

Coffee Consumption	
Germany	
United States	
Switzerland	
France	
Italy	= 100 cups

45.

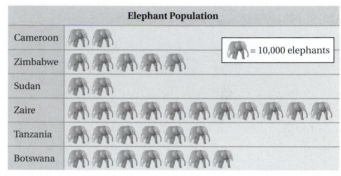

Elephant Population	
Cameroon	
Zimbabwe	= 10,000 elephants
Sudan	
Zaire	
Tanzania	
Botswana	

47. **D**W **49.** $-\frac{1}{16}$ **50.** $-\frac{31}{35}$ **51.** -2 **52.** -3
53. 1 **54.** 0.375 **55.** 1.16 **56.** $0.8\overline{3}$ **57.** **D**W
59. 67 min **61.** **D**W

Margin Exercises, Section 6.2, pp. 403–408

1. $5600 **2.** Japan **3.** Canada, Switzerland, the United States **4.** 60 women **5.** 85+ **6.** 60–64 **7.** Yes
8.

Planetary Moons

Planet: Earth, Mars, Jupiter, Saturn, Uranus, Neptune, Pluto
Number of moons: 4 8 12 16 20 24 28

9. The graph is not drawn to scale. The bar for 2004 should be shorter or the bar for 2003 should be longer.
10. Month 7 **11.** Months 1 and 2, 4 and 5, 6 and 7, 11 and 12 **12.** Months 2, 5, 6, 7, 8, 9, 12 **13.** About $900
14. About 40 yr **15.** About $1300

16.

Traditional SUV Sales

Exercise Set 6.2, p. 409

1. 190 calories   **3.** 1 slice of chocolate cake with fudge frosting   **5.** 1 cup of premium chocolate ice cream
7. About 120 calories   **9.** About 920 calories   **11.** About 28 lb   **13.** 1970: $11,000; 2003: $58,000; $47,000
15. 1970: $6000; 2003: $26,000; $20,000   **17.** $4000
19. $15,000   **21.**

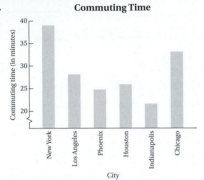

Commuting Time

23. Indianapolis   **25.** 28.7$\overline{3}$ min
27. 580 calories   **29.** 1930 calories   **31.** 30.4 yd
33. 1988 and 1995
35.

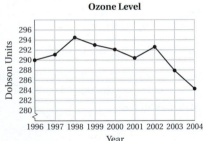

Ozone Level

37. Between 1997 and 1998
39. 289.46 Dobson Units
41. Between 2000 and 2001
43. Between 2000 and 2001   **45.** 17.6   **47.** D$_W$
49. 34 bottles   **50.** $\frac{9}{50}$ **51.** 72 fl oz   **52.** 32
53. 50   **54.** 18   **55.** 6   **56.** −7   **57.** D$_W$
59.

Average Per-Student Cost of Books
and Supplies, 4-year Public Colleges

Year		Cost per student
2004	817	
2005	853	
2006	894	

61.

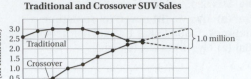

Traditional and Crossover SUV Sales

Approximately 2009; answers will vary.
63. D$_W$

Margin Exercises, Section 6.3, pp. 415–418

1 and 2.

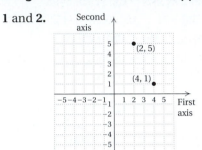

3–8.

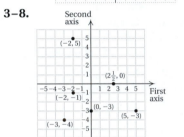

9. A: (−5, 1); B: (−3, 2); C: (0, 4); D: (3, 3); E: (1, 0); F: (0, −3);
G: (−5, −4)   **10.** Both are negative numbers.
11. The first, or horizontal, coordinate is positive; the
second, or vertical, coordinate is negative.   **12.** I   **13.** III
14. IV   **15.** II   **16.** No   **17.** Yes

Calculator Corner, p. 418

1. Yes   **2.** No   **3.** No   **4.** Yes   **5.** No   **6.** Yes
7. Yes   **8.** No

Exercise Set 6.3, p. 419

1.   **3.**

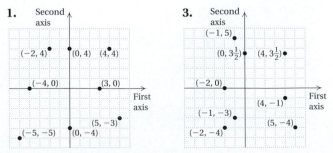

5. A: (3, 3); B: (0, −4); C: (−5, 0); D: (−1, −1); E: (2, 0);
F: (−3, 5)   **7.** A: (5, 0); B: (0, 5); C: (−3, 4); D: (2, −4);
E: (2, 3); F: (−4, −2)   **9.** II   **11.** IV   **13.** III   **15.** I
17. Positive; negative   **19.** III   **21.** IV; positive

23. Yes **25.** No **27.** Yes **29.** No **31.** No
33. Yes **35.** D_W **37.** 7 **38.** 9 **39.** 3 **40.** 14
41. 0 **42.** $\frac{30}{17}$ **43.** $1\frac{28}{33}a$ **44.** $7x - 24$ **45.** D_W
47. Yes **49.** I, IV **51.** I, III **53.** $(-1, -5)$
55. Answers may vary, but all points should appear on the
following graph. **57.** 26

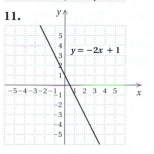

Margin Exercises, Section 6.4, pp. 423–428

1. $(8, 5)$ **2.** $(1, 5); (3, -5)$ **3.** $(1, 3), (7, 0), (5, 1)$
4. $(0, 7), (2, 3), (-2, 11)$

5.

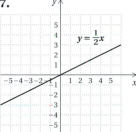

6.

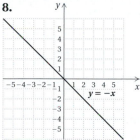

7.

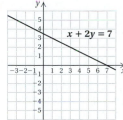

8.

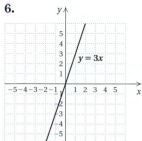

9.

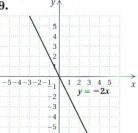

10.

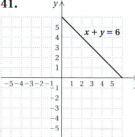

11.

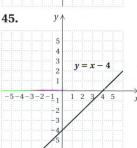

12.

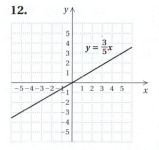

13.

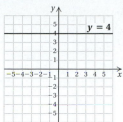

14.

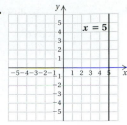

Calculator Corner, p. 429

1. $y = \frac{2}{3}x + 1$

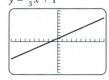

2. $y = x + 1$

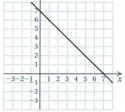

3. $y = -2x + 1$

4. $y = \frac{3}{5}x$

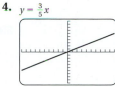

Exercise Set 6.4, p. 430

1. $(5, 3)$ **3.** $(3, 1)$ **5.** $(5, 14)$ **7.** $(10, -3)$ **9.** $(1, 3)$
11. $(2, -1)$ **13.** $(1, 3); (-1, 5)$ **15.** $(7, 3); (10, 6)$
17. $(3, 3); (6, 1)$ **19.** $(3, 1); (-2, 4)$ **21.** $(1, 4); (-2, -8)$
23. $\left(0, \frac{3}{5}\right); \left(\frac{3}{2}, 0\right)$ **25.** $(0, 9), (4, 5), (10, -1)$
27. $(0, 0), (1, 4), (2, 8)$ **29.** $(0, 13), (1, 10), (2, 7)$
31. $(0, -1), (2, 5), (-1, -4)$ **33.** $(0, 0), (1, -7), (-2, 14)$
35. $(0, -4), (4, 0), (1, -3)$ **37.** $(0, 6), (4, 0), \left(1, \frac{9}{2}\right)$
39. $(0, 2), (3, 3), (-3, 1)$
41.

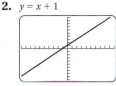

43.

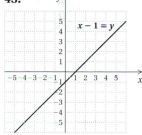

45.

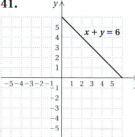

47.

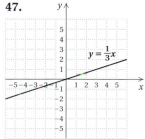

49.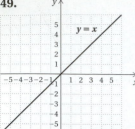

51.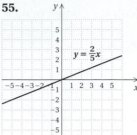

53.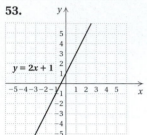

55.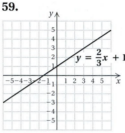

57.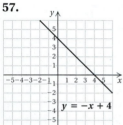

59.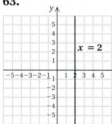

61.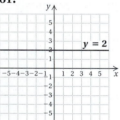

63.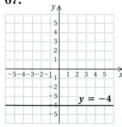

65.

67.

69. D_W **71.** 9 min   **72.** 319.75 pages   **73.** $1\frac{7}{8}$ cups

74. $-\frac{7}{11}$ **75.** 42   **76.** $-\frac{5}{8}$ **77.** 2.6   **78.** 2   **79.** D_W

81. $(0, 0.2)$, $(-4, -1)$, $(1, 0.5)$; answers may vary.

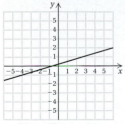

$21x - 70y = -14$

83. $(-3, 4.4)$, $(-3.9, 5)$, $(3, 0.4)$; answers may vary.

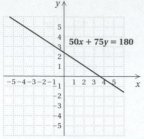

85. $(2, 4)$, $(-3, -1)$, $(-5, -3)$; answers may vary.

87. Answers may vary, but all should appear on this graph.

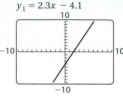

89. $y_1 = 2.3x - 4.1$

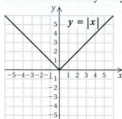

Margin Exercises, Section 6.5, pp. 436–440

1. 18   **2.** 8.85   **3.** 81   **4.** 19.4   **5.** 35.4 home runs
6. 2.5   **7.** 94   **8.** 17   **9.** 17   **10.** 91   **11.** $1700
12. 67.5   **13.** 45   **14.** 34, 67   **15.** No mode exists.
16. (a) 17 g **(b)** 18 g **(c)** 19 g   **17.** Wheat A: mean stalk
height ≈ 25.21 in.; wheat B: mean stalk height ≈ 22.54 in.;
wheat B is better.

Calculator Corner, p. 438

1. $203.\overline{3}$ **2.** The answers are the same.

Translating for Success, p. 441

1. F   **2.** N   **3.** A   **4.** O   **5.** G   **6.** D   **7.** C   **8.** L
9. H   **10.** E

Exercise Set 6.5, p. 442

1. Mean: 21; median: 18.5; mode: 29
3. Mean: 21; median: 20; modes: 5, 20
5. Mean: 5.2; median: 5.7; mode: 7.4
7. Mean: 239.5; median: 234; mode: 234
9. Mean: $3.58\overline{3}$; median: 1.5; mode: 1   **11.** 31 mpg
13. 2.7   **15.** Average: $8.19; median: $8.49; mode: $6.99
17. 17,201   **19.** 90   **21.** 263 days   **23.** Bulb A: mean
time = 1171.25 hr; bulb B: mean time ≈ 1251.58 hr; bulb B
is better.   **25.** D_W **27.** 196   **28.** $\frac{4}{9}$ **29.** 1.96
30. 1.999396   **31.** 225.05   **32.** 126.0516   **33.** $\frac{3}{35}$
34. $\frac{14}{15}$ **35.** 2.5 hr   **36.** 6 hr   **37.** D_W **39.** 181
41. 10 home runs   **43.** $3475   **45.** 480.375 mi

Margin Exercises, Section 6.6, pp. 446–448

1. About 19.8 million gal **2.** 94% **3.** (a) $\frac{1}{2}$, or 0.5
(b) $\frac{1}{5}$, or 0.2 **(c)** $\frac{3}{10}$, or 0.3 **4.** (a) $\frac{1}{4}$, or 0.25 **(b)** $\frac{2}{13}$

Exercise Set 6.6, p. 449

1. 83 **3.** About $15.4 million **5.** About $2.7 million
7. About $54.9 billion **9.** $\frac{1}{6}$, or $0.1\overline{6}$ **11.** $\frac{1}{2}$, or 0.5
13. $\frac{1}{52}$ **15.** $\frac{2}{13}$ **17.** $\frac{3}{26}$ **19.** $\frac{4}{39}$ **21.** $\frac{34}{39}$ **23.** $\mathbf{D_W}$
25. Natural **26.** Fraction; decimal **27.** Mean
28. Repeating **29.** Interpolation **30.** Distributive
31. Commutative **32.** Axes **33.** $\mathbf{D_W}$ **35.** $\frac{1}{4}$, or 0.25
37. $\frac{1}{36}$ **39.** $\mathbf{D_W}$

Concept Reinforcement, p. 452

1. True **2.** False **3.** True **4.** True

Summary and Review, p. 452

1. $41.71 **2.** $59.64 **3.** $26.27 **4.** $32.06
5. No **6.** $26.10 **7.** 14,000 officers **8.** Los Angeles
9. Houston **10.** 12,500 officers **11.** MLB (Major
League Baseball) **12.** About 120 million fans **13.** NBA
(National Basketball Association) **14.** NFL, NBA, and
MLB **15.** About 100 million more fans **16.** True
17. Under 20 **18.** About 12 **19.** About 13 **20.** 45–74
21. About 12 **22.** Under 20
23.

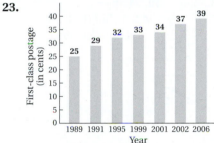

24.

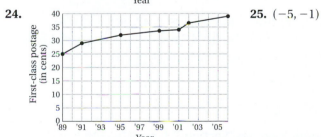

25. $(-5, -1)$

26. $(-2, 5)$ **27.** $(3, 0)$ **28.** $(4, -2)$
29–31.

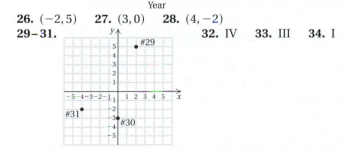

32. IV **33.** III **34.** I

35. $(1, 2)$; $(9, -2)$ **36.**

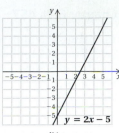
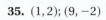

37.

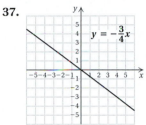

38.

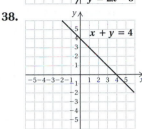

39.

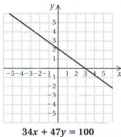

40.

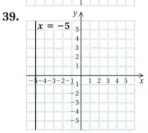

41. (a) 36 (b) 34 (c) 26 **42.** (a) 14 (b) 14 (c) 11, 17
43. (a) 322.5 (b) 375 (c) 470 **44.** (a) 1800 (b) 1900
(c) 700 **45.** (a) $37,500 (b) $27,500 (c) None **46.** 96
47. 3.1 **48.** Battery A: mean ≈ 43.04 hr; battery B:
mean = 41.55 hr; battery A is better. **49.** 18 accidents per
100 drivers **50.** $\frac{1}{52}$ **51.** $\frac{1}{2}$ **52.** $\mathbf{D_W}$ It is possible for the
mean of a set of numbers to be larger than all but one
number in the set. To see this, note that the mean of the set
$\{6, 8\}$ is 7, which is larger than all of the numbers in the set
but one. **53.** $\mathbf{D_W}$ The median of a set of four numbers
can be in the set. For example, the median of the set
$\{11, 15, 15, 17\}$ is 15, which is in the set.
54. $\left(0, \frac{100}{47}\right)$, $\left(\frac{100}{34}, 0\right)$, $\left(2, \frac{32}{47}\right)$ (Ordered pairs may vary.)

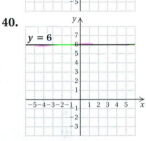

$34x + 47y = 100$

55. $11.52/hr **56.** $a = 316$, $b = 349$
57.

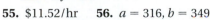

58.

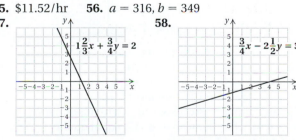

Test: Chapter 6, p. 457

1. [6.1a] Hiking at 3 mph with 20-lb load
2. [6.1a] Hiking at 3 mph with 10-lb load   **3.** [6.1b] Japan
4. [6.1b] United States   **5.** [6.1b] 800 lb
6. [6.1b] 400 lb
7. [6.2b]

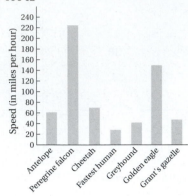

8. [6.2a] 197 mph   **9.** [6.2a] $1\frac{1}{2}$ times faster
10. [6.2c] 2001   **11.** [6.2c] 2001–2002   **12.** [6.2c] 2000
13. [6.6a] $26

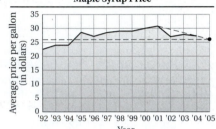

14. [6.3b] II   **15.** [6.3b] III   **16.** [6.3a] $(3, 4)$
17. [6.3a] $(0, -4)$ **18.** [6.3a] $(-4, 2)$ **19.** [6.4a] $(4, 2)$
20. [6.4b]

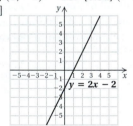

21. [6.4b]

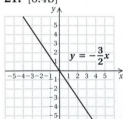

22. [6.4b]

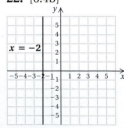

23. [6.5a] 50   **24.** [6.5a] 3   **25.** [6.5a] 15.5
26. [6.5b, c] Median: 50.5; mode: 54
27. [6.5b, c] Median: 3; no mode exists
28. [6.5b, c] Median: 17.5; modes: 17, 18
29. [6.5a] 27 mpg   **30.** [6.5a] 76   **31.** [6.5d] Bar A: mean ≈ 8.417; bar B: mean ≈ 8.417; equal quality
32. [6.5a] 2.9   **33.** [6.6b] $\frac{1}{4}$

34. [6.4b]

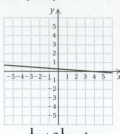

$\frac{1}{4}x + 3\frac{1}{2}y = 1$

35. [6.4b]

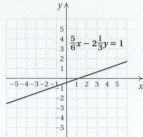

$\frac{5}{6}x - 2\frac{1}{3}y = 1$

36. [6.3a] 56 sq units

Cumulative Review: Chapters 1–6, p. 461

1. [1.9a] 7^4 **2.** [1.1c] 8,000,000,000   **3.** [5.8a] 6.2 lb
4. [1.5c], [2.7b] 22 cm; 28 cm^2 **5.** [1.8a] 2572 billion kWh
6. [3.6b] $\frac{3}{8}$ cup   **7.** [2.1a] 8, −7   **8.** [2.1b] >
9. [4.2c] >   **10.** [5.1c] <   **11.** [2.1d] 9   **12.** [2.1d] 17
13. [2.6a] −2   **14.** [2.7a] $-2x + y$
15. [3.2a] 1, 2, 3, 4, 6, 9, 12, 18, 36
16. [3.1b] Yes   **17.** [2.6a] $-\frac{7}{x}$ or $\frac{7}{-x}$
18. [2.6b] $10a - 15b + 5$ **19.** [3.3a] $\frac{3}{7}$ **20.** [3.5a] $\frac{10}{35}$
21. [1.3d] 138   **22.** [1.5a] 476   **23.** [2.5a] −9
24. [2.2a] −39   **25.** [4.2a] $\frac{5}{7}$ **26.** [3.7b] $\frac{5}{21}$ **27.** [4.3a] $\frac{13}{18}$
28. [4.2b] $\frac{1}{6}$ **29.** [3.6a] 1   **30.** [4.6a] $\frac{73}{8}$ or $9\frac{1}{8}$
31. [4.6b] $2\frac{5}{12}x$ **32.** [4.7a] $13\frac{1}{5}$ **33.** [5.2a] 83.28
34. [5.4a] 62.345   **35.** [5.3a] −42.282   **36.** [3.3b] 1
37. [3.3b] $4x$ **38.** [3.3b] 0   **39.** [4.3b] $-\frac{13}{15}$
40. [4.4a] 24   **41.** [5.7b] $-\frac{17}{4}$ **42.** [6.3b] II
43. [6.4b]   **44.** [6.5a] 36

45. [6.5b] 10.5   **46.** [6.5c] 49   **47.** [6.5a] 27 mpg
48. [1.9c], [3.6a], [4.6b], [4.7a] $\frac{9}{32}$ **49.** [4.5b], [4.6a] $7\frac{47}{1000}$
50. [6.3a] $(-2, -1), (-2, 7), (6, 7), (6, -1)$

CHAPTER 7

Margin Exercises, Section 7.1, pp. 464–467

1. $\frac{5}{11}$, or 5:11   **2.** $\frac{57.3}{86.1}$, or 57.3:86.1   **3.** $\frac{6\frac{3}{4}}{7\frac{2}{5}}$, or $6\frac{3}{4} \cdot 7\frac{2}{5}$
4. $\frac{739}{12}$ **5.** $\frac{12}{14}$ **6.** $\frac{51}{211.1}$; $\frac{211.1}{51}$ **7.** $\frac{38.2}{56.1}$ **8.** $\frac{205}{278}, \frac{278}{205}, \frac{278}{483}$
9. 18 is to 27 as 2 is to 3   **10.** 18 is to 3 as 6 is to 1
11. 3.6 is to 12 as 3 is to 10   **12.** 1.2 is to 1.5 as 4 is to 5
13. $\frac{9}{16}$

Calculator Corner, p. 467

1. $\frac{144}{77}$

Exercise Set 7.1, p. 468

1. $\frac{4}{5}$ **3.** $\frac{178}{572}$ **5.** $\frac{0.4}{12}$ **7.** $\frac{3.8}{7.4}$ **9.** $\frac{56.78}{98.35}$ **11.** $\frac{8\frac{3}{4}}{9\frac{5}{6}}$ **13.** $\frac{4}{1}$

15. $\frac{362}{100,000}$; $\frac{192}{100,000}$ **17.** $\frac{93.2}{1000}$ **19.** $\frac{163}{509}$; $\frac{509}{163}$ **21.** $\frac{60}{100}$; $\frac{100}{60}$

23. $\frac{2}{3}$ **25.** $\frac{3}{4}$ **27.** $\frac{12}{25}$ **29.** $\frac{7}{9}$ **31.** $\frac{2}{3}$ **33.** $\frac{14}{15}$ **35.** $\frac{1}{2}$

37. $\frac{3}{4}$ **39.** $\frac{478}{213}$; $\frac{213}{478}$ **41.** $\frac{51}{49}$ **43.** $\frac{32}{41}$ **45.** $\frac{57}{22}$ **47.** $\frac{148}{129}$

49. $\mathbf{D_W}$ **51.** $<$ **52.** $<$ **53.** $>$ **54.** $<$ **55.** 50

56. 9.5   **57.** 14.5   **58.** 152   **59.** $6\frac{7}{20}$ cm   **60.** $7\frac{11}{20}$ cm

61. $\mathbf{D_W}$ **63.** 0.0950353858 to 1   **65.** $\frac{7107}{6629}$ **67.** $1:2:3$

Margin Exercises, Section 7.2, pp. 473–474

1. 5 mi/hr, or 5 mph   **2.** 12 mi/hr, or 12 mph
3. $\frac{89}{13}$ km/h, or 6.85 km/h   **4.** 1100 ft/sec   **5.** 4 ft/sec
6. $\frac{121}{8}$ ft/sec, or 15.125 ft/sec   **7.** $\frac{32 \text{ home runs}}{48 \text{ strikeouts}} \approx$

0.667 home run per strikeout   **8.** $\frac{714 \text{ home runs}}{1330 \text{ strikeouts}} \approx$

0.537 home run per strikeout; Guerrero's rate is
approximately 0.13 higher   **9.** 7.45¢/oz
10. 24.143¢/oz; 25.9¢/oz; 24.174¢/oz; the 7-oz package has
the lowest unit price

Calculator Corner, p. 474

1. $0.\overline{6}$, or $\frac{2}{3}$ home run per strikeout

Exercise Set 7.2, p. 475

1. 40 mi/hr, or 40 mph   **3.** 7.48 mi/sec   **5.** 24 mpg
7. 23 mpg   **9.** $\frac{32,270 \text{ people}}{0.75 \text{ sq mi}}$; about 43,027 people/sq mi
11. 0.623 gal/ft^2 **13.** 186,000 mi/sec   **15.** 124 km/h
17. 25 mph; 0.04 hr/mi   **19.** About 18.3 points/game
21. 25 beats/min   **23.** 19.185¢/oz; 15.709¢/oz; 25.4 oz
25. 11.5¢/oz; 13.833¢/oz; 16 oz
27. 18.174¢/oz; 15.275¢/oz; 34.5 oz
29. 10.5¢/oz; 11.607¢/oz; 12.475¢/oz; 12.484¢/oz; 18 oz
31. 8.58¢/oz; 5.29¢/oz; 5.245¢/oz; 5.263¢/oz; 200 fl oz
33. $\mathbf{D_W}$ **35.** $<$ **36.** $<$ **37.** $>$ **38.** $>$ **39.** $<$
40. $>$ **41.** 1.7 million people   **42.** $25\frac{1}{2}$ servings
43. $\mathbf{D_W}$ **45.** (a) 10.83¢/oz; 10.91¢/oz; (b) 1.14¢/ft^2; 1.22¢/ft^2
47. The unit price remained approximately 7.8¢/oz and
then rose to 8.9¢/oz.   **49.** 2 min; 0.0000022 min
51. About 26.2 oz/dollar

Margin Exercises, Section 7.3, pp. 480–481

1. Yes   **2.** No   **3.** No   **4.** Yes   **5.** No   **6.** 14
7. $11\frac{1}{4}$ **8.** 10.5   **9.** 9   **10.** $10\frac{4}{5}$

Calculator Corner, p. 482

3. 27.5625   **4.** 25.6   **5.** 15.140625   **6.** 40.03952941
7. 39.74857143   **8.** 119

Exercise Set 7.3, p. 483

1. No   **3.** Yes   **5.** Yes   **7.** No   **9.** 0.63; 0.63; 0.61; 0.68;
the completion rates (rounded to the nearest hundredth)
are the same for Brady and Green   **11.** 45   **13.** 12
15. 10   **17.** 20   **19.** 5   **21.** 18   **23.** 22   **25.** 28
27. $9\frac{1}{3}$ **29.** $2\frac{8}{9}$ **31.** 0.06   **33.** 5   **35.** 1   **37.** 1
39. 14   **41.** $2\frac{3}{16}$ **43.** $\frac{51}{16}$, or $3\frac{3}{16}$ **45.** 12.5725
47. $\frac{1748}{249}$, or $7\frac{5}{249}$ **49.** $\mathbf{D_W}$ **51.** Quotient   **52.** Sum
53. Mean   **54.** Dollars, cents   **55.** Opposites
56. Terminating   **57.** Commutative   **58.** Cross products
59. $\mathbf{D_W}$ **61.** Approximately 2731.4   **63.** -2
65. $\dfrac{a}{b} = \dfrac{c}{d} \Rightarrow \dfrac{dba}{b} = \dfrac{dbc}{d} \Rightarrow da = bc \Rightarrow \dfrac{da}{ba} = \dfrac{bc}{ba} \Rightarrow \dfrac{d}{b} = \dfrac{c}{a}$

Margin Exercises, Section 7.4, pp. 486–490

1. 445 calories   **2.** 15 gal   **3.** 8 shirts   **4.** 38 in. or less
5. 9.5 in.   **6.** 2074 deer

Translating for Success, p. 491

1. N   **2.** I   **3.** A   **4.** K   **5.** J   **6.** F   **7.** M   **8.** B
9. G   **10.** E

Exercise Set 7.4, p. 492

1. 11.04 hr   **3.** 880 calories   **5.** 177 million, or
177,000,000   **7.** 9.75 gal   **9.** 175 bulbs   **11.** 171 gal
13. 2975 ft^2 **15.** 450 pages   **17.** (a) About 112 gal;
(b) 3360 mi   **19.** 13,500 mi   **21.** 120 lb   **23.** 64 cans
25. 100 oz   **27.** 954 deer   **29.** 58.1 mi   **31.** $\mathbf{D_W}$
33. $2 \cdot 2 \cdot 2 \cdot 101$, or $2^3 \cdot 101$ **34.** $2 \cdot 2 \cdot 7$, or $2^2 \cdot 7$
35. $2 \cdot 433$ **36.** $3 \cdot 31$ **37.** $2 \cdot 2 \cdot 5 \cdot 101$, or $2^2 \cdot 5 \cdot 101$
38. -79.13 **39.** -2.3 **40.** -7.3 **41.** -7.2 **42.** -1.8
43. $\mathbf{D_W}$ **45.** (a) 24.48783 British pounds; (b) \$15,877.27
47. 17 positions   **49.** 2150 earned runs
51. CD player: \$133.33; receiver: \$400; speakers: \$266.67

Margin Exercises, Section 7.5, pp. 496–499

1. 15   **2.** 16.9 in.   **3.** 7.5 ft   **4.** 21 cm   **5.** 29 ft

Exercise Set 7.5, p. 500

1. 25   **3.** $\frac{4}{3}$, or $1\frac{1}{3}$ **5.** $x = \frac{27}{4}$, or $6\frac{3}{4}$; $y = 9$
7. $x = 7.5$; $y = 7.2$ **9.** 67.2 ft   **11.** 33 ft   **13.** $28\frac{4}{9}$ ft
15. 7 ft   **17.** 100 ft   **19.** 4   **21.** $10\frac{1}{2}$
23. $x = 6$; $y = 5.25$; $z = 3$ **25.** $x = 5\frac{1}{3}$, or 5.$\overline{3}$; $y = 4\frac{2}{3}$,
or 4.$\overline{6}$; $z = 5\frac{1}{3}$, or 5.$\overline{3}$ **27.** 20 ft   **29.** 152 ft   **31.** $\mathbf{D_W}$

33. $59.81   **34.** 9.63   **35.** −679.4928   **36.** 2.74568
37. 27,456.8   **38.** 0.549136   **39.** 0.85   **40.** −1.825
41. −0.909   **42.** 0.843   **43.** $\mathbf{D_W}$ **45.** 13.75 ft
47. 1.25 cm   **49.** $x \approx 0.35; y = 0.4$

Concept Reinforcement, p. 504

1. True   **2.** True   **3.** False   **4.** False   **5.** True

Summary and Review: Chapter 7, p. 504

1. $\frac{47}{84}$ **2.** $\frac{46}{1.27}$ **3.** $\frac{83}{100}$ **4.** $\frac{0.72}{197}$ **5. (a)** $\frac{12,480}{16,640}$, or $\frac{3}{4}$;
(b) $\frac{16,640}{29,120}$, or $\frac{4}{7}$ **6.** $\frac{3}{4}$ **7.** $\frac{9}{16}$ **8.** 26 mpg   **9.** 6300 rpm
10. 0.638 gal/ft^2 **11.** 0.72 serving/lb   **12.** 4.33¢/tablet
13. 14.173¢/oz   **14.** 1.329¢/sheet; 1.554¢/sheet;
1.110¢/sheet; 6 big rolls   **15.** 6.844¢/oz; 5.188¢/oz;
5.609¢/oz; 5.539¢/oz; 48 oz   **16.** No   **17.** No   **18.** 32
19. 7   **20.** $\frac{1}{40}$ **21.** 24   **22.** $4.45   **23.** 351 circuits
24. (a) 195.61475 Euros; **(b)** $63.90   **25.** 832 mi
26. 27 acres   **27.** Approximately 3,293,558 kg   **28.** 6 in.
29. Approximately 2096 lawyers   **30.** $x = \frac{14}{3}$, or $4\frac{2}{3}$
31. $x = \frac{56}{5}$, or $11\frac{1}{5}$; $y = \frac{63}{5}$, or $12\frac{3}{5}$ **32.** 40 ft
33. $x = 3; y = 9; z = \frac{15}{2}$, or $7\frac{1}{2}$
34. $\mathbf{D_W}$ In terms of cost, a low faculty-to-student ratio is
less expensive than a high faculty-to-student ratio. In terms
of quality of education and student satisfaction, a high
faculty-to-student ratio is more desirable. A college
president must balance the cost and quality issues.
35. $\mathbf{D_W}$ Leslie used 4 gal of gasoline to drive 92 mi. At the
same rate, how many gallons would be needed to travel
368 mi?   **36.** 105 min, or 1 hr 45 min
37. 4 bracelets; 100 lavender beads   **38.** 240 min
39. Finishing paint: 11 gal; primer: 16.5 gal

Test: Chapter 7, p. 507

1. [7.1a] $\frac{85}{97}$ **2.** [7.1a] $\frac{0.34}{124}$ **3.** [7.2a] 0.625 ft/sec
4. [7.2a] 22 mpg   **5.** [7.2b] 11.182¢/oz; 7.149¢/oz;
8.389¢/oz; 6.840¢/oz; 263 oz   **6.** [7.1b] $\frac{32}{15}$ **7.** [7.3a] Yes
8. [7.3a] No   **9.** [7.3b] 12   **10.** [7.3b] 360
11. [7.3b] 42.1875   **12.** [7.3b] 100   **13.** [7.4a] 525 mi
14. [7.5a] 66 m   **15.** [7.4a] 1512 km   **16.** [7.4a] 4.8 min
17. [7.4a] 15 students   **18.** [7.4a] $52.65
19. [7.5a] $x = 8; y = 8.8$ **20.** [7.5b] $x = 8; y = 8; z = 12$
21. [7.3b] $\frac{4}{3}$, or $1.\overline{3}$ **22.** [7.3b] $\frac{-7}{17}$ **23.** [7.4a] 12 balls of
burgundy, 4 sweaters   **24.** [7.4a] $47.91
25. [7.5a] $\frac{875}{13}$ ft, or 67.308 ft   **26.** [7.5a] 10.5 ft

Cumulative Review: Chapters 1–7, p. 509

1. [5.2a] 643.502   **2.** [4.6a] $12\frac{1}{2}$ **3.** [4.2b] $\frac{7}{20}$
4. [5.2b] 1868.216   **5.** [2.3a] −17   **6.** [4.3a] $\frac{7}{60}$
7. [5.3a] 222.076   **8.** [2.4a] −645   **9.** [4.7a] $27\frac{3}{10}$
10. [5.4a] 43   **11.** [2.5a] −51   **12.** [3.7b] $\frac{3}{2}$
13. [1.1b] 3 ten thousands + 7 tens + 4 ones
14. [5.1a] One hundred twenty and seven hundredths

15. [5.1c] 0.7   **16.** [5.1c] −0.799
17. [3.2c] $2 \cdot 2 \cdot 2 \cdot 2 \cdot 3 \cdot 3$ **18.** [4.1a] 546   **19.** [3.3a] $\frac{5}{8}$
20. [3.5b] $\frac{27}{32}$ **21.** [5.5d] 5.718   **22.** [5.4b] −25.56
23. [6.5a] 48.75   **24.** [7.3a] No
25. [6.4b]   **26.** [2.6a] 5

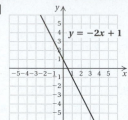

27. [7.3b] $30\frac{6}{25}$ **28.** [2.8b], [3.8a] $-\frac{423}{16}$ **29.** [2.8d] −4
30. [5.7b] 3   **31.** [5.7a] 33.34   **32.** [3.8a] $\frac{8}{9}$
33. [7.4a] 209 mi   **34.** [7.4a] 7 min
35. [5.8a] 976.9 mi   **36.** [7.2a] 22 mpg
37. [3.6b] 12 sq ft   **38.** [1.8a] **(a)** $360,000;
(b) $144,000,000; **(c)** $1,728,000,000   **39.** [4.6c] $2\frac{1}{4}$ cups
40. [7.2a], [7.4a] 60 mph   **41.** [7.2b] The 12-oz bag
42. [7.4a] No; the money will be gone after 24 weeks. Hans
will need $400 more.   **43.** [7.4a] $750   **44.** [7.4a] Leisure
& Incidental Expenses: $800; Food & Other Necessary
Expenses: $2400; Debt Payments: $1600; Debts: $19,200;
Leisure & Incidental Expenses: $9600

CHAPTER 8

Margin Exercises, Section 8.1, pp. 512–518

1. $\frac{70}{100}$; $70 \times \frac{1}{100}$; 70×0.01 **2.** $\frac{23.4}{100}$; $23.4 \times \frac{1}{100}$; 23.4×0.01
3. $\frac{100}{100}$; $100 \times \frac{1}{100}$; 100×0.01 **4.** 0.34   **5.** 0.789
6. 0.06625   **7.** 0.18   **8.** 0.0008   **9.** 24%   **10.** 347%
11. 100%   **12.** 32.1%   **13.** 25.3%
14. 25%   **15.** 62.5%, or $62\frac{1}{2}$%   **16.** $66.\overline{6}$%, or $66\frac{2}{3}$%
17. $83.\overline{3}$%, or $83\frac{1}{3}$%   **18.** 57%   **19.** 76%   **20.** $\frac{3}{5}$ **21.** $\frac{13}{400}$
22. $\frac{2}{3}$
23.

FRACTION NOTATION	$\frac{1}{5}$	$\frac{5}{6}$	$\frac{3}{8}$
DECIMAL NOTATION	0.2	$0.83\overline{3}$	0.375
PERCENT NOTATION	20%	$83.\overline{3}$%, or $83\frac{1}{3}$%	$37\frac{1}{2}$%

Calculator Corner, p. 513

1. 0.14   **2.** 0.00069   **3.** 0.438   **4.** 1.25

Calculator Corner, p. 516

1. 52%   **2.** 38.46%   **3.** 107.69%   **4.** 171.43%
5. 59.62%   **6.** 28.31%

Calculator Corner, p. 519

1. 30.54; 1.31% **2.** 32.05; 1.20% **3.** 34.47; 1.19%
4. 26.47; 1.00% **5.** 11.98; 4.32% **6.** 17.52; 0.89%

Exercise Set 8.1, p. 520

1. $\frac{90}{100}$; $90 \times \frac{1}{100}$; 90×0.01 **3.** $\frac{12.5}{100}$; $12.5 \times \frac{1}{100}$; 12.5×0.01
5. 0.67 **7.** 0.456 **9.** 0.5901 **11.** 0.1 **13.** 0.01
15. 2 **17.** 0.001 **19.** 0.0009 **21.** 0.0018 **23.** 0.2319
25. 0.14875 **27.** 0.565 **29.** 0.09; 0.58 **31.** 0.44
33. 0.36 **35.** 47% **37.** 3% **39.** 870% **41.** 33.4%
43. 75% **45.** 40% **47.** 0.6% **49.** 1.7% **51.** 27.18%
53. 2.39% **55.** 69% **57.** 17.7% **59.** 26%; 38%
61. 41% **63.** 5% **65.** 20% **67.** 28% **69.** 50%
71. 87.5%, or $87\frac{1}{2}$% **73.** 80% **75.** 66.$\overline{6}$%, or $66\frac{2}{3}$%
77. 16.$\overline{6}$%, or $16\frac{2}{3}$% **79.** 18.75%, or $18\frac{3}{4}$% **81.** 15%
83. 58% **85.** 22% **87.** 5% **89.** 9% **91.** 40%; 18%
93. $\frac{17}{20}$ **95.** $\frac{5}{8}$ **97.** $\frac{1}{3}$ **99.** $\frac{1}{6}$ **101.** $\frac{29}{400}$ **103.** $\frac{1}{125}$
105. $\frac{3}{2}$ **107.** $\frac{1}{3}$ **109.** $\frac{2}{25}$ **111.** $\frac{3}{5}$ **113.** $\frac{1}{50}$
115. $\frac{7}{20}$ **117.** $\frac{47}{100}$
119.

FRACTION NOTATION	DECIMAL NOTATION	PERCENT NOTATION
$\frac{1}{8}$	0.125	12.5%, or $12\frac{1}{2}$%
$\frac{1}{6}$	0.1$\overline{6}$	16.$\overline{6}$%, or $16\frac{2}{3}$%
$\frac{1}{5}$	0.2	20%
$\frac{1}{4}$	0.25	25%
$\frac{1}{3}$	0.$\overline{3}$	33.$\overline{3}$%, or $33\frac{1}{3}$%
$\frac{3}{8}$	0.375	37.5%, or $37\frac{1}{2}$%
$\frac{2}{5}$	0.4	40%
$\frac{1}{2}$	0.5	50%

121.

FRACTION NOTATION	DECIMAL NOTATION	PERCENT NOTATION
$\frac{1}{2}$	0.5	50%
$\frac{1}{3}$	0.$\overline{3}$	33.$\overline{3}$%, or $33\frac{1}{3}$%
$\frac{1}{4}$	0.25	25%
$\frac{1}{6}$	0.1$\overline{6}$	16.$\overline{6}$%, or $16\frac{2}{3}$%
$\frac{1}{8}$	0.125	12.5%, or $12\frac{1}{2}$%
$\frac{3}{4}$	0.75	75%
$\frac{5}{6}$	0.8$\overline{3}$	83.$\overline{3}$%, or $83\frac{1}{3}$%
$\frac{3}{8}$	0.375	37.5%, or $37\frac{1}{2}$%

123. $\mathbf{D_W}$ **125.** 70 **126.** 5 **127.** 400 **128.** 18.75
129. 23.125 **130.** 25.5 **131.** 4.5 **132.** $8\frac{3}{4}$
133. $\mathbf{D_W}$ **135.** 11.$\overline{1}$% **137.** 0.01$\overline{5}$
139.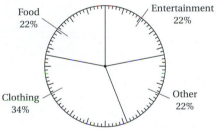

Margin Exercises, Section 8.2, pp. 526–528

1. $0.12 \cdot 50 = a$ **2.** $a = 0.40 \cdot 60$ **3.** $45 = 0.20 \cdot t$
4. $1.20 \cdot y = 60$ **5.** $16 = n \cdot 40$ **6.** $b \cdot 84 = 10.5$
7. 6 **8.** $35.20 **9.** 225 **10.** $50 **11.** 40%
12. 12.5%

Calculator Corner, p. 529

1. 1.2 **2.** $5.04 **3.** 48.64 **4.** $22.40 **5.** 0.0112
6. $29.70 **7.** Left to the student

Exercise Set 8.2, p. 530

1. $y = 0.32 \cdot 78$ **3.** $89 = a \cdot 99$ **5.** $13 = 0.25 \cdot y$
7. 234.6 **9.** 45 **11.** $18 **13.** 1.9 **15.** 78%
17. 200% **19.** 50% **21.** 125% **23.** 40 **25.** $80
27. 88 **29.** 20 **31.** 6.25 **33.** $846.60 **35.** 41.1
37. D_W **39.** $\frac{623}{1000}$ **40.** $\frac{19}{10}$ **41.** $\frac{237}{100}$ **42.** 0.009
43. 0.39 **44.** 5.7 **45.** 18 mochas **46.** 20 photos
47. D_W **49.** $880 (can vary); $843.20
51. 108 to 135 tons

Margin Exercises, Section 8.3, pp. 533–535

1. $\frac{12}{100} = \frac{a}{50}$ **2.** $\frac{40}{100} = \frac{a}{60}$ **3.** $\frac{130}{100} = \frac{a}{72}$ **4.** $\frac{20}{100} = \frac{45}{b}$
5. $\frac{120}{100} = \frac{60}{b}$ **6.** $\frac{N}{100} = \frac{16}{40}$ **7.** $\frac{N}{100} = \frac{10.5}{84}$ **8.** $225
9. 35.2 **10.** 6 **11.** 50 **12.** 30% **13.** 12.5%

Exercise Set 8.3, p. 536

1. $\frac{37}{100} = \frac{a}{74}$ **3.** $\frac{N}{100} = \frac{4.3}{5.9}$ **5.** $\frac{25}{100} = \frac{14}{b}$ **7.** 68.4
9. 462 **11.** 40 **13.** 2.88 **15.** 25% **17.** 102%
19. 25% **21.** 93.75% **23.** $72 **25.** 90 **27.** 88
29. 20 **31.** 25 **33.** $780.20 **35.** 7.9 **37.** D_W
39. 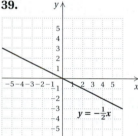 **40.**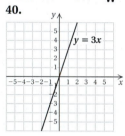

41. 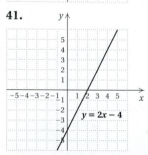 **42.**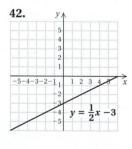

43. $\frac{43}{48}$ qt **44.** $\frac{1}{8}$ T **45.** 800 **46.** 18.75 **47.** 8
48. 7 **49.** D_W **51.** $1200 (can vary); $1118.64
53. 20% **55.** 39%

Margin Exercises, Section 8.4, pp. 539–544

1. About 9.5% **2.** 14,560,000 workers **3. (a)** $1440;
(b) $37,440 **4. (a)** $9000; **(b)** $27,000 **5.** About 2.9%
6. About 76.6%

Calculator Corner, p. 544

1. Left to the student **2.** $80,040

Translating for Success, p. 545

1. J **2.** M **3.** N **4.** E **5.** G **6.** H **7.** O **8.** C
9. D **10.** B

Exercise Set 8.4, p. 546

1. About 13,247 wild horses **3.** 3 yr: $21,080;
5 yr: $17,680 **5.** Overweight: 176.4 million people;
obese: 73.5 million people **7.** Acid: 20.4 mL;
water: 659.6 mL **9.** About 1808 mi
11. About 39,867,000 people **13.** 36.4 items correct;
3.6 items incorrect **15.** 95 items **17.** 25%
19. 166; 156; 146; 140; 122 **21.** 8% **23.** 20%
25. $53.00 **27.** About 27% **29.** $30,030
31. $16,174.50; $12,130.88 **33.** 34.375%, or $34\frac{3}{8}$%
35. 71% **37.** $1560 **39.** 80% **41.** 98,775; 18.0%
43. 799,065; 14.8% **45.** 4,550,688; 38.1%
47. (a) 7.5%; **(b)** 879,675 **49.** 36,400 **51. (a)** 3%;
(b) 89.5%; **(c)** 107,400 babies; **(d)** 2235 C-sections
53. 40% **55.** D_W **57.** $2.\overline{27}$ **58.** 0.44 **59.** 3.375
60. $4.\overline{7}$ **61.** 0.92 **62.** $0.8\overline{3}$ **63.** 0.4375 **64.** 2.317
65. 3.4809 **66.** 0.675 **67.** 32 cm **68.** 33 ft
69. 36 in. **70.** 36 m **71.** D_W **73.** About 5 ft 6 in.
75. $83\frac{1}{3}$% **77.** $15,650.98 **79.** 19% **81.** D_W

Margin Exercises, Section 8.5, pp. 553–557

1. $48.50; $717.45 **2.** $5.39; $140.14 **3.** 6% **4.** $999
5. $5628 **6.** 12.5%, or $12\frac{1}{2}$% **7.** $1675 **8.** $180; $360
9. 20%

Exercise Set 8.5, p. 558

1. $19.53 **3.** $2.65 **5.** $16.39; $361.39 **7.** 5%
9. 4% **11.** $2000 **13.** $800 **15.** $719.86 **17.** 5.6%
19. $2700 **21.** 5% **23.** $980 **25.** $5880 **27.** 12%
29. $420 **31.** $30; $270 **33.** $2.55; $14.45
35. $125; $112.50 **37.** 40%; $360 **39.** $30; 16.7%
41. $549; 36.4% **43.** D_W **45.** 18 **46.** $\frac{22}{7}$
47. 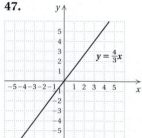 **48.**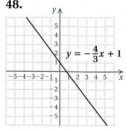

49. $0.\overline{5}$ **50.** $2.\overline{09}$ **51.** 84.872 **52.** 75.712 **53.** D_W
55. 14.4% to 25.6% **57.** $2.69 **59.** $91,112.50

Margin Exercises, Section 8.6, pp. 562–565

1. $301   **2.** $225.75   **3.** (a) $33.53; (b) $4833.53
4. $2376.20   **5.** $7690.94

Calculator Corner, p. 565

1. $16,357.18   **2.** $12,764.72

Exercise Set 8.6, p. 566

1. $8   **3.** $84   **5.** $113.52   **7.** $925   **9.** $671.88
11. (a) $147.95; (b) $10,147.95   **13.** (a) $80.14;
(b) $6580.14   **15.** (a) $46.03; (b) $5646.03   **17.** $441
19. $2802.50   **21.** $7853.38   **23.** $99,427.40
25. $4243.60   **27.** $28,225.00   **29.** $9270.87
31. $129,871.09   **33.** $4101.01   **35.** $1324.58
37. $20,165.05   **39.** D_W **41.** Reciprocals
42. Divisible by 6   **43.** Additive   **44.** Unit rate
45. Perimeter   **46.** Divisible by 3   **47.** Prime
48. Proportional   **49.** D_W **51.** 9.38%   **53.** $7883.24

Margin Exercises, Section 8.7, pp. 570–574

1. (a) $97; (b) interest: $86.40; amount applied to principal:
$10.60; (c) interest: $55.17; amount applied to principal:
$41.83; (d) At 13.6%, the principal was decreased by $31.23
more than at the 21.3% rate. The interest at 13.6% is $31.23
less than at 21.3%.   **2.** (a) Interest: $91.78; amount
applied to principal: $229.22; (b) $58; (c) $4080
3. Interest: $843.94; amount applied to principal: $135.74
4. (a) Interest: $844.69; amount applied to principal:
$498.64; (b) $88,799.40; (c) The Sawyers will pay $110,885.40
less in interest with the 15-yr loan than with the 30-yr loan.
5. (a) Interest: $701.25; amount applied to principal:
$548.89; (b) $72,025.20; (c) The Sawyers will pay $16,774.20
less in interest with the 15-yr loan at $5\frac{1}{2}$% than with the
15-yr loan at $6\frac{5}{8}$%.

Exercise Set 8.7, p. 575

1. (a) $98; (b) interest: $86.56; amount applied to principal:
$11.44; (c) interest: $51.20; amount applied to principal:
$46.80; (d) At 12.6%, the principal is decreased by $35.36
more than at the 21.3% rate. The interest at 12.6% is $35.36
less than at 21.3%.   **3.** (a) Interest: $125.14; amount
applied to principal: $312.79; (b) $51.24; (c) $7991.60,
$11,504, $3512.40   **5.** (a) Interest: $854.17; amount
applied to principal: $155.61; (b) $199,520.80; (c) New
principal: $163,844.39; interest: $853.36; amount applied to
principal: $156.42   **7.** (a) Interest: $854.17; amount
applied to principal: $552; (b) $89,110.60; (c) The Martinez
family will pay $110,410.20 less in interest with the 15-yr
loan than with the 30-yr loan.   **9.** $99,917.71;
$99,834.94   **11.** $99,712.04; $99,422.15   **13.** $149,882.75;
$149,764.79   **15.** $199,382.07; $198,760.41
17. (a) $2395, $21,555; (b) $52.09, $401.97; (c) $239.88
19. (a) $595; $11,305; (b) interest: $87.61; amount applied
to principal: $273.47; (c) $1693.88   **21.** D_W **23.** 40 cm²

24. 10 in²   **25.** 21 m²   **26.** 84 ft²   **27.** 36 in²
28. 49 mi²   **29.** D_W **31.** Approximately 11 months

Concept Reinforcement, p. 578

1. True   **2.** False   **3.** True   **4.** False

Summary and Review: Chapter 8, p. 578

1. 56%   **2.** 1.7%   **3.** 37.5%   **4.** $33.\overline{3}$%, or $33\frac{1}{3}$%
5. 0.735   **6.** 0.065   **7.** $\frac{6}{25}$ **8.** $\frac{63}{1000}$
9. $30.6 = p \cdot 90$; 34%   **10.** $63 = 84\% \cdot n$; 75
11. $y = 38\frac{1}{2}\% \cdot 168$; 64.68   **12.** $\frac{24}{100} = \frac{16.8}{b}$; 70
13. $\frac{42}{30} = \frac{N}{100}$; 140%   **14.** $\frac{10.5}{100} = \frac{a}{84}$; 8.82
15. 223 students; 105 students   **16.** 42%   **17.** 2500 mL
18. 12%   **19.** 93.15   **20.** $14.40   **21.** 5%   **22.** 11%
23. $42; $308   **24.** $42.70; $262.30   **25.** $2940
26. Approximately 18.4%   **27.** $36   **28.** (a) $394.52;
(b) $24,394.52   **29.** $121   **30.** $7727.26   **31.** $9504.80
32. (a) $129; (b) interest: $100.18; amount applied to
principal: $28.82; (c) interest: $70.72; amount applied to
principal: $58.28; (d) At 13.2%, the principal is decreased
by $29.46 more than at the 18.7% rate. The interest at 13.2%
is $29.46 less than at 18.7%.
33. D_W No; the 10% discount was based on the original
price rather than on the sale price.
34. D_W A 40% discount is better. When successive
discounts are taken, each is based on the previous
discounted price rather than on the original price. A
20% discount followed by a 22% discount is the same as a
37.6% discount off the original price.
35. 19.5% increase   **36.** $66\frac{2}{3}$%   **37.** $168

Test: Chapter 8, p. 581

1. [8.1b] 0.064   **2.** [8.1b] 38%   **3.** [8.1c] 137.5%
4. [8.1c] $\frac{13}{20}$ **5.** [8.2a, b] $a = 0.40 \cdot 55$; 22
6. [8.3a, b] $\frac{N}{100} = \frac{65}{80}$; 81.25%   **7.** [8.4a] 400 passengers;
575 passengers   **8.** [8.4b] About 47.9%
9. [8.4b] $50.\overline{90}$%   **10.** [8.4a] 5.5%   **11.** [8.5a] $16.20;
$340.20   **12.** [8.5b] $630   **13.** [8.5c] $40; $160
14. [8.6a] $8.52   **15.** [8.6a] $5356   **16.** [8.6b] $1110.39
17. [8.6b] $11,580.07   **18.** [8.4b] Registered nurses: 2.9,
26.1%; post-secondary teachers: 0.6, 37.5%; food
preparation and service workers: 2.0, 20%; restaurant
servers: 2.1, 0.4   **19.** [8.5c] $275, about 14.1%
20. [8.7a] $119,909.14; $119,817.72
21. [8.5b] $194,600   **22.** [8.5b], [8.6b] $2546.16

Cumulative Review: Chapters 1–8, p. 583

1. [5.1b] $\frac{91}{1000}$ **2.** [5.5a] $2.1\overline{6}$ **3.** [8.1b] 0.03
4. [8.1c] 112.5%   **5.** [7.1a] $\frac{3}{40}$ **6.** [7.2a] $23\frac{1}{3}$ km/hr
7. [4.2c] <   **8.** [5.1c] >   **9.** [5.6a] 296,200

10. [1.4b] 50,000 **11.** [1.9d] 13 **12.** [2.7a] $-2x - 14$
13. [4.6a] $3\frac{1}{30}$ **14.** [5.2c] -14.2 **15.** [1.2b] 515,150
16. [5.2b] 0.02 **17.** [4.6b] $\frac{2}{3}$ **18.** [4.3a] $-1\frac{47}{63}$
19. [3.6a] $\frac{1}{6}$ **20.** [2.4b] -384 **21.** [5.3a] 1.38036
22. [4.7b] $\frac{3}{2}$ **23.** [5.4a] -12.25 **24.** [1.6c] 123 R 5
25. [2.8b], [3.8a] 95 **26.** [5.7a] 8.13 **27.** [4.4a] 49
28. [4.3b] $\frac{1}{12}$ **29.** [5.7b] $-1\frac{7}{9}$ **30.** [7.3b] $8\frac{8}{21}$
31. [6.3b] III **32.** [6.4b]

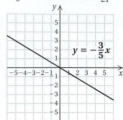

33. [6.5a] 33.2 **34.** [6.5b] 12 **35.** [2.7b] 60 in.
36. [1.5c] 3600 yd^2 **37.** [7.2b] 30¢/oz
38. [7.4a] 608 km **39.** [8.4b] About 13.6%
40. [8.4a] About 41 will be telephone and 17 will be voice
mail communications. **41.** [4.7d] 5 pieces
42. [4.2d] $1\frac{1}{2}$ km **43.** [8.4b] 12.5% increase
44. [4.7d], [6.5a] 33.6 mpg
45. (a) [3.6a], [4.2d] #1 pays $\frac{1}{25}$, #2 pays $\frac{9}{100}$, #3 pays $\frac{47}{300}$,
#4 pays $\frac{77}{300}$, #5 pays $\frac{137}{300}$; **(b)** [8.1c], [8.4a] 4%, 9%,
$15\frac{2}{3}$%, $25\frac{2}{3}$%, $45\frac{2}{3}$%; **(c)** [8.1c], [8.4a] 87%
46 [8.5c] 7 discounts

CHAPTER 9

Margin Exercises, Section 9.1, pp. 586–594

1. 2 **2.** 3 **3.** $1\frac{1}{2}$ **4.** $2\frac{1}{2}$ **5.** 288 **6.** 43.5
7. 240,768 **8.** 6 **9.** 8 **10.** $11\frac{2}{3}$, or $11.\overline{6}$ **11.** 54
12. 31,680 **13.** 5 **14.** 40,480 yd
15. 2 cm, or 20 mm **16.** 2.3 cm, or 23 mm
17. 4.4 cm, or 44 mm **18.** cm **19.** km **20.** mm
21. m **22.** cm **23.** m **24.** 37,000 **25.** 500
26. 178 **27.** 9040 **28.** 7.814 **29.** 781.4 **30.** 8.72
31. 8,900,000 **32.** 6.78 **33.** 97.4 **34.** 0.1 **35.** 8.451
36. 91.4 **37.** 804.5 **38.** 1479.843

Exercise Set 9.1, p. 595

1. 3 **3.** $\frac{1}{12}$ **5.** 1760 **7.** 9 **9.** 7 **11.** 16 **13.** 8800
15. 4 **17.** 2.2 **19.** 3 **21.** $2\frac{1}{2}$ **23.** 3 **25.** 1 **27.** 2
29. 132,000 **31.** 126,720 **33. (a)** 1000; **(b)** 0.001
35. (a) 10; **(b)** 0.1 **37. (a)** 0.01; **(b)** 100 **39.** 8300
41. 0.98 **43.** 8.921 **45.** 0.03217 **47.** 28,900
49. 4.77 **51.** 688 **53.** 0.1 **55.** 100,000 **57.** 142
59. 0.82 **61.** 450 **63.** 0.000024 **65.** 0.688 **67.** 230
69. 6.21 **71.** 35.56 **73.** 104.585 **75.** 28.67
77. 70.866 **79.** 32.904 **81.** $\mathbf{D_W}$ **83.** $-\frac{3}{2}$ **84.** -2
85. $101.50 **86.** $44.50 **87.** 47% **88.** 35%
89. 56 ft; 192 ft^2 **90.** 44 ft; 120 ft^2 **91.** $\mathbf{D_W}$
93. 1.0 m **95.** 1.4 cm **97.** Length: 5400 in., or 450 ft;
breadth: 900 in., or 75 ft; height: 540 in., or 45 ft

99. 321,800 **101.** 85.725 m **103.** 0.0041 in. **105.** $<$
107. $<$ **109.** $<$

Margin Exercises, Section 9.2, pp. 599–601

1. 144 **2.** 1440 **3.** 63 **4.** 2.5 **5.** 3200
6. 1,000,000 **7.** 0.01 **8.** 28,800 **9.** 0.043 **10.** 0.678

Exercise Set 9.2, p. 602

1. 45 **3.** 1008 **5.** 3 **7.** 198 **9.** 2160 **11.** 12,800
13. $7\frac{2}{3}$ **15.** 5 **17.** $\frac{1}{144}$ **19.** $\frac{1}{640}$ **21.** 19,000,000
23. 63,100 **25.** 0.065432 **27.** 0.0349 **29.** 2500
31. 0.4728 **33.** 1 ft^2 **35.** $4\frac{1}{3}$ ft^2 **37.** 84 cm^2
39. $\mathbf{D_W}$ **41.** $17.50 **42.** $15 **43.** $6.75 **44.** $8.78
45. $35.51 **46.** $86.30 **47.** $\mathbf{D_W}$ **49.** 1701 tiles; 42%
51. 6.4516 **53.** 4000 **55.** 5080 cm^2 **57.** 216 cm^2

Margin Exercises, Section 9.3, pp. 606–611

1. 43.8 cm^2 **2.** 12.375 km^2 **3.** 100 m^2 **4.** 717.5 cm^2
5. 9 in. **6.** 5 ft **7.** 62.8 m **8.** 15.7 m **9.** 220 cm
10. 34.296 yd **11.** $78\frac{4}{7}$ km^2 **12.** 339.6 cm^2
13. 64.26 cm^2 **14.** A 12-ft diameter flower bed is
13.04 ft^2 larger.

Calculator Corner, p. 611

1. Answers will vary. **2.** 1417.99 in.; 160,005.91 in^2
3. 1729.27 in^2 **4.** 125,663.71 ft^2

Exercise Set 9.3, p. 612

1. 50 cm^2 **3.** 104 ft^2 **5.** 64 m^2 **7.** 72.45 cm^2
9. 144 mi^2 **11.** $55\frac{1}{8}$ ft^2 **13.** 49 m^2 **15.** 1.92 cm^2
17. 68 yd^2 **19.** 14 cm **21.** $1\frac{3}{4}$ in. **23.** 10 ft
25. 0.7 cm **27.** 44 cm **29.** $5\frac{1}{2}$ in. **31.** 62.8 ft
33. 4.396 cm **35.** 154 cm^2 **37.** $2\frac{13}{32}$ in^2 **39.** 314 ft^2
41. 1.5386 cm^2 **43.** The round pizza; 6.25 in^2
45. The round pizza; 9.86 in^2 **47.** Two small round pizzas;
13 in^2 **49.** One extra-large round pizza; 0.96 in^2
51. 153.86 ft^2 **53.** 153.86 mi^2 **55.** 15 in.
57. 65.94 yd^2 **59.** 45.68 ft **61.** 45.7 yd
63. 100.48 m^2 **65.** 6.9972 cm^2 **67.** $\mathbf{D_W}$ **69.** $\frac{37}{400}$
70. $\frac{7}{8}$ **71.** 137.5% **72.** $66.\overline{6}$%, or $66\frac{2}{3}$% **73.** 125%
74. 160% **75.** $\mathbf{D_W}$ **77.** 37.2875 in^2 **79.** $\mathbf{D_W}$
81. Circumference **83.** $\mathbf{D_W}$

Margin Exercises, Section 9.4, pp. 620–626

1. 12 cm^3 **2.** 3087 in^3 **3.** 128 ft^3 **4.** 785 ft^3
5. 67,914 m^3 **6.** $91,989\frac{1}{3}$ ft^3 **7.** 38.77272 cm^3 **8.** 20
9. 32 **10.** mL **11.** mL **12.** L **13.** L **14.** 970
15. 8.99 **16.** $3.87 **17.** 83.73 mm^3

Exercise Set 9.4, p. 627

1. 250 cm³ **3.** 135 in³ **5.** 75 m³ **7.** 357½ yd³
9. 4082 ft³ **11.** 376.8 cm³ **13.** 41,580,000 yd³
15. 4,186,666.$\overline{6}$ in³ **17.** Approximately 124.725 m³
19. 1437⅓ km³ **21.** 1000; 1000 **23.** 59,000 **25.** 0.049
27. 27,300 **29.** 40 **31.** 320 **33.** 3 **35.** 40 **37.** 48
39. 1⅞ **41.** 0.423115 yd³ **43.** 367.38 m³
45. 143.72 cm³ **47.** 263,947,530,000 mi³ **49.** 4.187 m³
51. 646.74 cm³ **53.** 1¾ gal; 7.5 gal; 91¼ gal; about
24,000,000,000 gal **55.** About 346.2 ft³ **57.** **D**_W
59. $24 **60.** $175 **61.** $10.68 **62.** $\frac{7}{3}$, or 2.$\overline{3}$ **63.** −1
64. 448 km **65.** 59 **66.** 10 **67.** **D**_W
69. About 7546 ft³ **71.** Audio-cassette cases
73. 13,824 cm³ **75.** 57,480 in³

Margin Exercises, Section 9.5, pp. 633–640

1. Angle *DEF*, angle *FED*, angle *E*, ∠*DEF*, ∠*FED*, or ∠*E*
2. Angle *PQR*, angle *RQP*, angle *Q*, ∠*PQR*, ∠*RQP*, or ∠*Q*
3. ∠*NOP*, or ∠*PON* **4.** 127° **5.** 33°
6. Lengths of Engagement of Married Couples

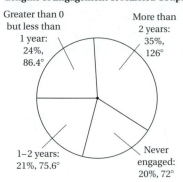

Arrangement of sections may vary.

7. Right **8.** Acute **9.** Obtuse **10.** Straight
11. ∠1 and ∠2; ∠1 and ∠4; ∠2 and ∠3; ∠3 and ∠4
12. 45° **13.** 72° **14.** 5°
15. ∠1 and ∠2; ∠1 and ∠4; ∠2 and ∠3; ∠3 and ∠4
16. 142° **17.** 23° **18.** 90° **19.** ∠1 and ∠2; ∠3 and ∠4
20. *m*∠*BND* or *m*∠*DNB*; *m*∠*CND* or *m*∠*DNC*
21. ∠*SMR* or ∠*RMS*; ∠*QMR* or ∠*RMQ*
22. (a) △*ABC* (b) △*ABC*, △*MPN* (c) △*DEF*, △*GHI*, △*JKL*,
△*QRS* **23.** Yes **24.** No **25.** (a) △*DEF* (b) △*GHI*,
△*QRS* (c) △*ABC*, △*MPN*, △*JKL* **26.** 180° **27.** 64°

Exercise Set 9.5, p. 641

1. Angle *GHI*, angle *IHG*, angle *H*, ∠*GHI*, ∠*IHG*, or ∠*H*
3. ∠*ADB*, or ∠*BDA* **5.** 10° **7.** 180° **9.** 90°
11. Snacking Habits

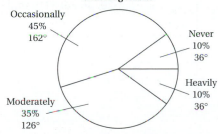

Arrangement of sections may vary.

13. Kids in Foster Care

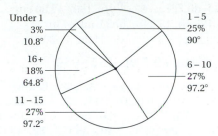

Arrangement of sections may vary.

15. Acute **17.** Straight **19.** Right **21.** Acute
23. Obtuse **25.** ∠1 and ∠3; ∠2 and ∠4
27. ∠*GME* (or ∠*EMG*) and ∠*AMC* (or ∠*CMA*);
∠*AMG* (or ∠*GMA*) and ∠*EMC* (or ∠*CME*)
29. *m*∠4; *m*∠1 **31.** ∠*GME* or ∠*EMG*; ∠*CME* or ∠*EMC*
33. 79° **35.** 23° **37.** 32° **39.** 61°
41. 177° **43.** 41° **45.** 105° **47.** 76° **49.** Scalene;
obtuse **51.** Scalene; right **53.** Equilateral; acute
55. Scalene; obtuse **57.** 46° **59.** 120° **61.** **D**_W
63. $160 **64.** $22.50 **65.** $148 **66.** $1116.67
67. $33,597.91 **68.** $413,458.31 **69.** $641,566.26
70. $684,337.34 **71.** 29 **72.** 10 **73.** **D**_W
75. *m*∠2 = 67.13°; *m*∠4 = 79.8°; *m*∠5 = 67.13°;
m∠6 = 33.07° **77.** *m*∠*ACB* = 50°; *m*∠*CAB* = 40°;
m∠*EBC* = 50°; *m*∠*EBA* = 40°; *m*∠*AEB* = 100°;
m∠*ADB* = 50°

Margin Exercises, Section 9.6, pp. 646–649

1. 81 **2.** 100 **3.** 121 **4.** 144 **5.** 169 **6.** 196
7. 225 **8.** 256 **9.** −10, 10 **10.** −9, 9 **11.** −7, 7
12. −14, 14 **13.** 7 **14.** 4 **15.** 11 **16.** 10 **17.** 9
18. 8 **19.** 15 **20.** 13 **21.** 1 **22.** 0 **23.** 2.236
24. 8.832 **25.** 12.961 **26.** *c* = 13 **27.** *a* = $\sqrt{75}$;
a ≈ 8.660 **28.** *b* = $\sqrt{120}$; *b* ≈ 10.954 **29.** *a* = $\sqrt{175}$;
a ≈ 13.229 **30.** $\sqrt{424}$ ft ≈ 20.6 ft

Calculator Corner, p. 647

1. 6.6 **2.** 9.7 **3.** 19.8 **4.** 17.3 **5.** 24.9 **6.** 24.5
7. 121.2 **8.** 115.6 **9.** 16.2 **10.** 85.4

Translating for Success, p. 650

1. E **2.** H **3.** O **4.** A **5.** G **6.** C **7.** M
8. I **9.** K **10.** N

Exercise Set 9.6, p. 651

1. −4, 4 **3.** −11, 11 **5.** −13, 13 **7.** −50, 50 **9.** 8
11. 9 **13.** 15 **15.** 25 **17.** 20 **19.** 100 **21.** 6.928
23. 2.828 **25.** 1.732 **27.** 3.464 **29.** 4.359
31. 10.488 **33.** *c* = 15 **35.** *c* = $\sqrt{98}$; *c* ≈ 9.899
37. *a* = 5 **39.** *b* = $\sqrt{45}$; *b* ≈ 6.708 **41.** *c* = 26
43. *b* = 12 **45.** *c* = $\sqrt{41}$; *c* ≈ 6.403
47. *b* = $\sqrt{1023}$; *b* ≈ 31.984 **49.** $\sqrt{208}$ ft ≈ 14.4 ft
51. $\sqrt{8450}$ ft ≈ 91.9 ft **53.** *h* = $\sqrt{500}$ ft ≈ 22.4 ft
55. $\sqrt{211,200,000}$ ft ≈ 14,532.7 ft **57.** **D**_W **59.** $468

60. 187,200   **61.** About 324   **62.** 12%   **63.** 8
64. 125   **65.** 64   **66.** 27   **67.** $\mathbf{D_W}$ **69.** 47.80 cm²
71. 64.8 in.   **73.** Width: 15.2 in.; height: 11.4 in.

Margin Exercises, Section 9.7, pp. 655–659

1. 80   **2.** 4.32   **3.** 32,000   **4.** kg   **5.** kg   **6.** mg
7. g   **8.** t   **9.** 6200   **10.** 0.0931   **11.** 77   **12.** 234.4
13. 80°C   **14.** 0°C   **15.** −20°C   **16.** 80°F   **17.** 100°F
18. 50°F   **19.** 176°C   **20.** 95°C   **21.** 35°C   **22.** 45°C

Calculator Corner, p. 659

1. 41°F   **2.** 122°F   **3.** 20°C   **4.** 45°C

Exercise Set 9.7, p. 660

1. 16   **3.** 4   **5.** 48   **7.** 9000   **9.** 2.4   **11.** 4.5
13. 4000   **15.** 0.001   **17.** 0.01   **19.** 1000   **21.** 10
23. 934,000   **25.** 6.345   **27.** 0.000897   **29.** 7320
31. 9.35   **33.** 6.9   **35.** 800,000   **37.** 1000   **39.** 0.0034
41. 80°C   **43.** 60°C   **45.** 20°C   **47.** −10°C   **49.** 190°F
51. 140°F   **53.** 10°F   **55.** 40°F   **57.** 86°F   **59.** 104°F
61. 5432°F   **63.** 25°C   **65.** 55°C   **67.** 37°C
69. (a) 136°F = 57.$\overline{7}$°C, 56$\frac{2}{3}$°C = 134°F; **(b)** 2°F   **71.** $\mathbf{D_W}$
73. Compound   **74.** Ratio   **75.** Median
76. Commission   **77.** Similar   **78.** Mean
79. Composite   **80.** Kilo-   **81.** $\mathbf{D_W}$ **83.** 0.4536
85. (a) 109.134 g; **(b)** 9.104 g; **(c)** Golden Jubilee: 3.85 oz;
Hope: 0.321 oz   **87.** 60 g   **89.** About 5.1 g/cm³
91. 144 boxes; 0.1 lb, or 1.6 oz

Margin Exercises, Section 9.8, pp. 665–667

1. 2.4 L   **2. (a)** About 59.14 mL; **(b)** about 0.059 L
3. 350 mg; 28   **4.** 3000   **5.** 0.5 mg

Exercise Set 9.8, p. 668

1. 2000 mL   **3.** 0.12 g   **5.** 4 tablets   **7. (a)** 720 mcg;
(b) about 189 actuations; **(c)** 6 inhalers   **9.** 16 oz
11. (a) 0.42 g; **(b)** 28 doses   **13.** 6 mL   **15.** 1000
17. 0.325   **19.** 250 mcg   **21.** 125 mcg
23. 0.875 mg; 875 mcg   **25.** $\mathbf{D_W}$ **27.** 3358
28. 7414   **29.** 854   **30.** 6334   **31.** $5x + 8$
32. $6x + 5$ **33.** $7t − 9$ **34.** $8r − 10$ **35.** $\mathbf{D_W}$
37. (a) 9.1 g; **(b)** 162.5 mg
39. Naproxen. The Naproxen costs 11 cents/day and the
ibuprofen costs 25 cents/day.

Concept Reinforcement, p. 671

1. False   **2.** True   **3.** True   **4.** True   **5.** False
6. False

Summary and Review: Chapter 9, p. 672

1. 3.$\overline{3}$ **2.** 30   **3.** 0.17   **4.** 0.006   **5.** 400,000   **6.** 1$\frac{1}{6}$
7. 80   **8.** 0.003   **9.** 12.5   **10.** 224   **11.** 0.06
12. 400   **13.** 1400   **14.** 200   **15.** 4700   **16.** 0.04
17. 36   **18.** 700,000   **19.** 7   **20.** 0.057   **21.** 31.4 m
22. $\frac{14}{11}$ in.   **23.** 24 m   **24.** 439.7784 yd   **25.** 18 in²
26. 29.82 ft²   **27.** 60 cm²   **28.** 35 mm²   **29.** 154 ft²
30. 314 cm²   **31.** 1038.555 ft²   **32.** 26.28 ft²
33. 49°; 139°   **34.** ∠PNM and ∠SNR; ∠MNR and ∠SNP
35. 60°   **36.** Scalene   **37.** Right   **38.** 93.6 m³
39. 193.2 ft³   **40.** 28,260 cm³   **41.** 33$\frac{37}{75}$ yd³
42. 942 cm³   **43.** 8   **44.** $c = \sqrt{850}$; $c \approx 29.155$
45. $b = \sqrt{84}$; $b \approx 9.165$ **46.** $c = \sqrt{89}$ ft; $c \approx 9.434$ ft
47. $a = \sqrt{76}$ cm; $a \approx 8.718$ cm   **48.** About 44.7 ft
49. 95°F   **50.** 20°C   **51.** 8 mL   **52.** 3000 mL
53. 500 mcg   **54.** $\mathbf{D_W}$ A square is a parallelogram
because it is a four-sided figure with two pairs of parallel
sides.   **55.** $\mathbf{D_W}$ No, the sum of the three angles of a
triangle is 180°. If you have two 90° angles, totaling 180°, the
third angle would be 0°. A triangle can't have an angle of 0°.
56. $\mathbf{D_W}$ Since 1 m is slightly more than 1 yd, it follows that
1 m³ is larger than 1 yd³. Since 1 yd³ = 27 ft³, we see that 1 m³
is larger than 27 ft³.   **57.** $\mathbf{D_W}$ Since 1 kg is about 2.2 lb
and 32 oz is $\frac{32}{16}$, or 2 lb, 1 kg weighs more than 32 oz.
58. 961.625 cm²   **59.** 104.875 laps   **60.** 34.375%
61. $101.25

Test: Chapter 9, p. 676

1. [9.1a] 96   **2.** [9.1b] 2.8   **3.** [9.2a] 18   **4.** [9.1b] 5000
5. [9.1b] 0.91   **6.** [9.2b] 0.00452   **7.** [9.4b] 2.983
8. [9.7b] 3800   **9.** [9.4b] 1280   **10.** [9.4b] 690
11. [9.7a] 144   **12.** [9.7a] 8220   **13.** [9.3b] 8 cm
14. [9.3b] 50.24 m²   **15.** [9.3b] 88 ft   **16.** [9.3a] 25 cm²
17. [9.3b] 33.87 m²   **18.** [9.3a] 18 ft²   **19.** [9.3b] 5.495 in²
20. [9.5c] 115°   **21.** [9.5c] 65°   **22.** [9.4c] 420 in³
23. [9.4a] 628 ft³   **24.** [9.4a] 4186.$\overline{6}$ yd³   **25.** [9.4a] 30 m³
26. [9.8a] 3000 mL   **27.** [9.6c] $c = \sqrt{2}$; $c \approx 1.414$
28. [9.6c] $b = \sqrt{51}$; $b \approx 7.141$ **29.** [9.6a] 11
30. [9.7c] 0°C   **31.** [9.8a] 6 mL   **32.** [9.5c] 112.5°
33. [9.4c] $188.40   **34.** [9.4a] 0.65 ft³   **35.** [9.4a] 0.055 ft³

Cumulative Review: Chapters 1–9, p. 678

1. [4.6a] 10$\frac{1}{6}$ **2.** [5.5d] 49.2   **3.** [5.2b] 87.52
4. [2.5a] −1234   **5.** [1.9d] 2   **6.** [1.9c] 1565   **7.** [5.1b] $\frac{6}{5}$
8. [8.1c] 45%   **9.** [4.2c] <   **10.** [4.2c] >   **11.** [9.7a] $\frac{3}{8}$
12. [9.7c] 212°   **13.** [9.4b] 87   **14.** [9.1a] 90
15. [9.2a] 27   **16.** [9.1b] 0.37   **17.** [2.7b], [3.6b] 52 m;
120 m²   **18.** [2.7a] $9a − 16$
19. [6.4b]

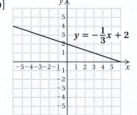

$y = -\frac{1}{3}x + 2$

20. [6.2a] $110 million

21. [6.2a] 2005　**22.** [7.3b] 14.4　**23.** [5.7b] 1
24. [2.8b], [3.8a] $-18\frac{2}{3}$　**25.** [4.3b] $\frac{1}{8}$　**26.** [7.1a] $\frac{18.2}{79.3}$; $\frac{79.3}{18.2}$
27. [8.5c] 41.01%　**28.** [6.5a] 58.6　**29.** [8.6a] $24
30. [9.6d] 17 m　**31.** [8.5a] 6%　**32.** [4.6c] $2\frac{1}{8}$ yd
33. [5.8a] $27.91　**34.** [7.2b] The 8-qt box
35. [3.6b] $\frac{7}{20}$ km　**36.** [9.5c] 148°　**37.** [9.5c] 137°
38. [9.4c] 272 ft^3　**39.** [2.7b] No; the sum of the perimeters can be measured as 14 ft or 16 ft, both of which exceed 108 in.

CHAPTER 10

Margin Exercises, Section 10.1, pp. 683–685

1. $9a^2 + 3a - 1$　**2.** $7x^2y + 3x^2 + 4x + 4$
3. $2a^3 + 2a^2 - 9a + 17$
4. $-(12x^4 - 3x^2 + 4x)$; $-12x^4 + 3x^2 - 4x$
5. $-(-4x^4 + 3x^2 - 4x)$; $4x^4 - 3x^2 + 4x$
6. $-\left(-13x^6 + 2x^4 - 3x^2 + x - \frac{5}{13}\right)$;
$13x^6 - 2x^4 + 3x^2 - x + \frac{5}{13}$
7. $-(-8a^3b + 5ab^2 - 2ab)$; $8a^3b - 5ab^2 + 2ab$
8. $-4x^3 + 6x - 3$　**9.** $-5x^3y - 3x^2y^2 + 7xy^3$
10. $-14x^{10} + \frac{1}{2}x^5 - 5x^3 + x^2 - 3x$　**11.** $2x^3 + 2x + 8$
12. $x^2 - 6x - 2$　**13.** $2x^3 + 3x^2 - 4xy - 2$　**14. (a)** 41
(b) 41　**15.** 132　**16.** 20 m　**17.** 80 m

Exercise Set 10.1, p. 687

1. $-4x + 10$　**3.** $x^2 - 8x + 5$　**5.** $2x^2$
7. $6t^4 + 4t^3 + 5t^2 - t$　**9.** $7 + 12x^2$
11. $9x^8 + 8x^7 - 3x^4 + 2x^2 - 2x + 5$
13. $13t^4 + 4t^3 - t^2 + 4t - 3$
15. $-5x^4y^3 + 2x^3y^3 + 4x^3y^2 - 4xy^2 - 5xy$
17. $13a^3b^2 + 4a^2b^2 - 4a^2b + 6ab^2$
19. $-0.6a^2bc + 17.5abc^3 - 5.2abc$　**21.** $-(-5x)$; $5x$
23. $-(-x^2 + 13x - 7)$; $x^2 - 13x + 7$
25. $-(12x^4 - 3x^3 + 3)$; $-12x^4 + 3x^3 - 3$　**27.** $-3x + 5$
29. $-4x^2 + 3x - 2$　**31.** $4x^4 - 6x^2 - \frac{3}{4}x + 8$
33. $7x - 1$　**35.** $4t^2 + 6t + 6$　**37.** $-x^2 - 9x + 5$
39. $5a^2 + 5a - 16$　**41.** $8x^4 + 3x^3 - 4x^2 + 3x - 6$
43. $4.6x^3 + 9.2x^2 - 3.8x - 23$　**45.** $\frac{3}{4}x^3 - \frac{1}{2}x$
47. $6x^3y^3 + 8x^2y^2 + 2x^2y + 4xy$　**49.** -23　**51.** 19
53. -12　**55.** 7　**57.** 4　**59.** 11　**61.** 1024 ft
63. 823 min　**65.** 17,489 degrees
67. About 449 accidents　**69.** $18,750　**71.** $155,000
73. $\mathbf{D_W}$　**75.** $\frac{7}{10}$ serving per pound　**76.** $3560
77. $81　**78.** 26 m^2　**79.** 1256 cm^2
80. $395 per week　**81.** $2 \cdot 2 \cdot 2 \cdot 3 \cdot 7$
82. $2 \cdot 2 \cdot 2 \cdot 2 \cdot 2 \cdot 2 \cdot 3$　**83.** $3 \cdot 5 \cdot 7 \cdot 7$
84. $3 \cdot 3 \cdot 13$　**85.** $\mathbf{D_W}$　**87.** 711 min
89.

Foreign Language and Literature Degrees

Margin Exercises, Section 10.2, pp. 692–696

1. $18a^2$　**2.** $-14x^2$　**3.** $48a^2$　**4.** $-5m^2$　**5.** $42ab$
6. a^9　**7.** $8x^{13}$　**8.** $35m^{11}$　**9.** $15a^7b^{12}$　**10.** 7; -19
11. $12x^2 + 20x$　**12.** $6a^3 - 15a^2 + 21a$
13. $8a^5b^2 + 20a^3b^6$　**14.** $6(z - 2)$　**15.** $3(x - 2y + 4)$
16. $2(8a - 18b + 21)$　**17.** $-4(3x - 8y + 4z)$
18. $5a(a^2 + 2)$　**19.** $7x(2x^2 - x + 3)$　**20.** $3ab(3a - 2b)$

Exercise Set 10.2, p. 697

1. $28a^2$　**3.** $-60x^2$　**5.** $28x^8$　**7.** $-0.07x^9$　**9.** $35x^6y^{12}$
11. $12a^8b^8c^2$　**13.** $-24x^{11}$　**15.** $-3x^2 + 21x$
17. $-3x^2 + 6x$　**19.** $x^5 + x^2$　**21.** $10x^3 - 30x^2 + 5x$
23. $12x^3y + 8xy^2$　**25.** $12a^7b^3 - 9a^4b^3$　**27.** $2(x + 4)$
29. $7(a - 5)$　**31.** $7(4x + 3y)$　**33.** $9(a - 3b + 9)$
35. $6(3 - m)$　**37.** $-8(2 + x - 5y)$　**39.** $9x(x^4 + 1)$
41. $a^2(a - 8)$　**43.** $2x(4x^2 - 3x + 1)$
45. $6a^4b^2(2b + 3a)$　**47.** $\mathbf{D_W}$　**49.** 210 ft　**50.** 14 mpg
51. 17.5 mpg　**52.** $418.95　**53.** 62.5%　**54.** 37.68 cm
55. -2736　**56.** 3312　**57.** $\mathbf{D_W}$　**59.** $23x^{299}(17x^{92} + 13)$
61. $7a^7b^6c^8(12b^3c^3 - 6ac^2 + 7a^2b)$

Margin Exercises, Section 10.3, pp. 699–700

1. $x^2 + 13x + 40$　**2.** $x^2 + x - 20$　**3.** $5x^2 - 17x - 12$
4. $15x^2 - 13x + 2$　**5.** $x^2 + 8x + 15$　**6.** $x^2 - 11x + 24$
7. $x^4 + 3x^3 + x^2 + 15x - 20$
8. $6y^5 - 20y^3 + 15y^2 + 14y - 35$
9. $3x^3 + 13x^2 - 6x + 20$

Exercise Set 10.3, p. 701

1. $x^2 + 8x + 12$　**3.** $x^2 + 3x - 10$　**5.** $x^2 + 4x - 12$
7. $x^2 - 10x + 21$　**9.** $x^2 - 25$　**11.** $18 + 12x + 2x^2$
13. $9x^2 - 24x + 16$　**15.** $x^2 - \frac{21}{10}x - 1$
17. $x^3 + 2x^2 - 2x - 3$　**19.** $4x^3 + 14x^2 + 8x + 1$
21. $3y^4 - 6y^3 - 7y^2 + 18y - 6$　**23.** $x^6 + 2x^5 - x^3$
25. $6t^4 + t^3 - 16t^2 - 7t + 4$　**27.** $x^9 - x^5 + 2x^3 - x$
29. $\mathbf{D_W}$　**31.** 912 m^2　**32.** $123,000　**33.** $133\frac{1}{3}$%, or
$133.\overline{3}$%　**34.** 26　**35.** 63.6%　**36.** 154 ft^2　**37.** -17
38. -7　**39.** -7　**40.** -17　**41.** $\mathbf{D_W}$　**43.** 928, 928;
264.0625, 264.0625; -21.5424, -21.5424　**45. (a)** $4r + 40$;
(b) $r^2 + 20r + 99$　**47.** $78t^2 + 40t$　**49.** $9x^2 - 30x + 25$

Margin Exercises, Section 10.4, pp. 703–706

1. 1　**2.** 1　**3.** -1　**4.** -3　**5.** 1　**6.** $\frac{1}{4^3}$; $\frac{1}{64}$
7. $\frac{1}{5^2}$; $\frac{1}{25}$　**8.** $\frac{1}{2^4}$; $\frac{1}{16}$　**9.** $\frac{1}{(-2)^3}$; $-\frac{1}{8}$　**10.** $\left(\frac{3}{5}\right)^2$; $\frac{9}{25}$
11. 9^{-2}　**12.** $7x^{-4}$　**13.** m^3n^5　**14.** abc　**15.** a^4b^6

91. (a) 255 mg; **(b)** 344.4 mg; **(c)** 0 mg
93. $12y^2 - 23y + 21$　**95.** $-3y^4 - y^3 + 5y - 2$
97. $3x^4, 2x^3, (-7)$, order of answers may vary.

16. x^7yz^4 **17.** $\dfrac{a^4c^3}{b^7}$ **18.** 5^2, or 25   **19.** $\dfrac{1}{x^7}$
20. $20x^9y^6$ **21.** $\dfrac{b^3}{a^7}$

Calculator Corner, p. 706

1. 4.21399177; 4.21399177   **2.** 4.768371582; 4.768371582
3. -0.2097152; -0.2097152 **4.** -0.0484002582;
-0.0484002582 **5.** 2.0736; 2.0736   **6.** 0.4932701843;
0.4932701843

Exercise Set 10.4, p. 707

1. 1   **3.** 1   **5.** -19.57 **7.** 1   **9.** 1   **11.** 1   **13.** 17
15. 1   **17.** 7   **19.** 5   **21.** 2   **23.** $\dfrac{1}{3^2}$; $\dfrac{1}{9}$
25. $\dfrac{1}{10^4}$; $\dfrac{1}{10,000}$ **27.** $\dfrac{1}{t^4}$ **29.** $\dfrac{1}{(-5)^2}$; $\dfrac{1}{25}$ **31.** $\dfrac{3}{x^7}$
33. xy^4 **35.** r^5t^3 **37.** $\dfrac{-7}{a^9}$ **39.** $\dfrac{25}{4}$ **41.** $\dfrac{a^3}{125}$
43. 7^{-3} **45.** $9x^{-3}$ **47.** $\dfrac{1}{x}$ **49.** 1   **51.** $\dfrac{1}{t^{15}}$
53. $6a^3b^2$ **55.** $-\dfrac{6y^6}{x^2}$ **57.** $\dfrac{6c^3}{a^6b^4}$ **59.** **D$_W$**
61. Binomial   **62.** Parallel   **63.** Gram   **64.** Prime
65. Angle   **66.** Cents; dollars   **67.** Perimeter
68. Multiplicative; additive   **69.** **D$_W$**   **71.** **D$_W$**
73. $\frac{1}{5}$, or 0.2; $\frac{1}{5}$, or 0.2   **75.** y^{5x} **77.** a^{4t}

Margin Exercises, Section 10.5, pp. 710–713

1. 5.17×10^{-4} **2.** 5.23×10^8 **3.** 689,300,000,000
4. 0.0000567   **5.** 5.6×10^{-15} **6.** 7.462×10^{-13}
7. 2.0×10^3 **8.** 5.5×10^2 **9.** 1.884672×10^{11} L
10. The mass of Saturn is 9.5×10 times the mass of Earth.

Calculator Corner, p. 712

1. 1.3545×10^{-4} **2.** 9.044×10^5 **3.** 3.2×10^5
4. 3.6×10^{12} **5.** 3×10^{-6} **6.** 4×10^5 **7.** 8×10^{-26}
8. 3×10^{13}

Translating for Success, p. 714

1. K   **2.** G   **3.** B   **4.** H   **5.** A   **6.** M   **7.** E
8. C   **9.** D   **10.** I

Exercise Set 10.5, p. 715

1. 2.8×10^{10} **3.** 9.07×10^{17} **5.** 3.04×10^{-6}
7. 1.8×10^{-8} **9.** 10^{11} **11.** 2.81×10^8
13. 6.7×10^{-8} **15.** 87,400,000   **17.** 0.00000005704
19. 10,000,000   **21.** 0.00001   **23.** 6×10^9
25. 3.38×10^4 **27.** 8.1477×10^{-13} **29.** 2.5×10^{13}

31. 5.0×10^{-4} **33.** 3.0×10^{-21} **35.** Approximately
1.325×10^{14} ft^3 **37.** The mass of Jupiter is 3.18×10^2
times the mass of Earth.   **39.** 1×10^{22} **41.** The mass of
the sun is 3.33×10^5 times the mass of Earth.
43. 4.375×10^2 days   **45.** **D$_W$**   **47.** $\frac{7}{4}$, or 1.75   **48.** 2
49. $-\frac{12}{7}$, or $-1.\overline{714285}$ **50.** $-\frac{11}{2}$, or -5.5
51.   **52.**

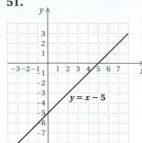

53. **D$_W$**   **55.** 2.478125×10^{-1} **57.** 6.4×10^8

Concept Reinforcement, p. 719

1. True   **2.** True   **3.** True   **4.** True   **5.** False
6. False

Summary and Review: Chapter 10, p. 719

1. $3x - 6$ **2.** $7x^4 - 4x^3 - x - 3$
3. $8a^5 + 12a^3 - a^2 + 4a + 9$ **4.** $5a^3b^3 + 11a^2b^3 - 7$
5. $-(12x^3 - 4x^2 + 9x - 3)$; $-12x^3 + 4x^2 - 9x + 3$ **6.** 1
7. -42 **8.** 56 ft   **9.** $30x^7$ **10.** $18x^4 - 12x^2 - 3x$
11. $14a^7b^4 + 10a^6b^4$ **12.** $x^2 + 2x - 63$
13. $15x^2 - 11x + 2$ **14.** $a^4 + 2a^3 - 2a^2 - 2a + 1$
15. $5x(9x^2 - 2)$ **16.** $7(a - 5b - 7ac)$
17. $3x^2y(2x - 3y^4)$ **18.** $\dfrac{1}{12^2}$; $\dfrac{1}{144}$ **19.** $\dfrac{8}{a^7}$ **20.** $\dfrac{z^6}{y^5x^3}$
21. $\left(\dfrac{5}{4}\right)^2$; $\dfrac{25}{16}$ **22.** x^{-7} **23.** $\dfrac{1}{x^{17}}$ **24.** $\dfrac{14x^7}{y^7}$
25. 4.27×10^7 **26.** 1.924×10^{-4} **27.** 1.173×10^{11}
28. 2.5×10^{-1} m   **29.** 6.25×10^{-2} m, or 6.25 cm
30. **D$_W$** Adi is probably adding coefficients and multiplying
exponents instead of the other way around.
31. **D$_W$** Because x^{-2} is $\dfrac{1}{x^2}$ and because x^2 is never negative,
it follows that x^{-2} is never negative.
32. $1,158,057x^{17} + 1,226,352x^{12} - 213,129x^7$
33. $-40x^8$ **34.** $13a^2b^5c^5(3ab^2c - 10c^3 + 4a^2b)$
35. $w^5x^2y^3z^3(x^4yz^2 - w^2xy^4 + wy^2z^3 - wx^5z)$
36. $2a^4b^{-5}(5 + 6a^3b^2)$

Test: Chapter 10, p. 722

1. [10.1a] $18a^3 - 5a^2 - a + 8$
2. [10.1b] $-(-9a^4 + 7b^2 - ab + 3)$; $9a^4 - 7b^2 + ab - 3$
3. [10.1c] $3x^4 - x^2 - 11$ **4.** [10.2a] 193   **5.** [10.4a] 1
6. [10.1d] 12.4 m   **7.** [10.2a] $-10x^6y^8$
8. [10.2b] $10a^3 - 8a^2 + 6a$ **9.** [10.3a] $x^2 + 4x - 45$
10. [10.3b] $2a^3 - 5a^2 + a + 2$
11. [10.2c] $5x^2(7x^4 - 5x + 3)$

12. [10.2c] $3(2ab - 3bc + 4ac)$ **13.** [10.4b] $\dfrac{1}{5^3}$; $\dfrac{1}{125}$

14. [10.4b] $\dfrac{5b^2}{a^3}$ **15.** [10.4b] $\left(\dfrac{5}{3}\right)^3$; $\dfrac{125}{27}$ **16.** [10.4b] $\dfrac{1}{x^{16}}$

17. [10.4b] $\dfrac{-6a^3}{b^3}$ **18.** [10.5a] 4.7×10^{-4}

19. [10.5a] 8.25×10^6 **20.** [10.5b] 1.824×10^{-16}
21. [10.1d] 2.92 L

22. [10.2c], [10.3a], [10.4b] $\dfrac{3a^4}{a^4 - 6a^2 + 9}$

Cumulative Review/Final Examination: Chapters 1–10, p. 724

1. [5.3b] 1,500,000   **2.** [9.1a] [9.1c] $437\frac{1}{3}$ yd; about 400 m
3. [6.2c] 236 in 1993 and 1995   **4.** [6.2c] 258 in 2000
5. [6.5a, b, c] Mean: 243.375; median: 239.5; mode: 236
6. [6.5a] 249.5   **7.** [6.5a] 237.25; the mean over the years 1993–1996 is 12.25 lower than the mean over the years 1997–2000.   **8.** [8.4b] About 8.4%   **9.** [6.6b] $\frac{3}{20}$, or 0.15
10. [9.5e] 118°   **11.** [1.2b] 21,085   **12.** [4.6a] $8\frac{7}{9}$
13. [2.2a] 24   **14.** [2.2a] -762 **15.** [5.2c] -44.364
16. [10.1a] $10x^5 - x^4 + 2x^3 - 3x^2 + 2$
17. [1.3d] 243   **18.** [2.7a] $-17x$ **19.** [4.3a] $-\frac{19}{40}$
20. [4.6b] $2\frac{17}{24}$ **21.** [5.2b] 19.9973
22. [10.1c] $2x^3 + 5x^2 + 7x$ **23.** [10.1c] $-4a^2b + 7ab$
24. [1.5a] 4752   **25.** [2.4a] $-74,337$ **26.** [4.7a] $4\frac{7}{12}$
27. [3.6a] $-\frac{6}{5}$ **28.** [3.6a] 10   **29.** [5.3a] 259.084
30. [2.6b] $24x - 15$ **31.** [10.2a] $27a^8b^3$
32. [10.2b] $21x^5 - 14x^3 + 56x^2$ **33.** [10.3a] $x^2 - 5x - 14$
34. [10.3b] $a^3 - 2a^2 - 11a + 12$ **35.** [1.6c] 573
36. [1.6c] 56 R 10   **37.** [3.7b] $-\frac{3}{2}$ **38.** [4.7b] $\frac{7}{90}$
39. [5.4a] 39   **40.** [4.5c] $56\frac{5}{17}$ **41.** [1.9c] 75
42. [2.1c], [2.5b] -2 **43.** [1.9a] 14^3 **44.** [1.4a] 68,000
45. [5.5b] 21.84   **46.** [3.1b] Yes   **47.** [3.2a] 1, 3, 5, 15
48. [4.1a] 105   **49.** [3.5b] $\frac{8}{11}$ **50.** [4.5b] $-3\frac{3}{5}$
51. [2.1b] $>$ **52.** [4.2c] $<$ **53.** [5.1c] 1.001
54. [2.6a] 29   **55.** [10.2c] $5(8 - t)$
56. [10.2c] $3a(6a^2 - 5a + 2)$ **57.** [3.3a] $\frac{3}{5}$
58. [5.1b] 0.0429   **59.** [5.5a] -0.52 **60.** [5.5a] $0.\overline{8}$
61. [8.1b] 0.07   **62.** [5.1b] $\frac{671}{100}$ **63.** [4.5a] $-\frac{29}{4}$
64. [8.1c] $\frac{2}{5}$ **65.** [8.1c] 85%   **66.** [8.1b] 150%
67. [5.6a] 13.6   **68.** [1.7b] 555   **69.** [5.7a] 64
70. [3.8a] $\frac{5}{4}$ **71.** [7.3b] 76.5   **72.** [2.8d] 5
73. [5.7b] $-\frac{2}{5}$ **74.** [6.5a] \$23.75   **75.** [1.8a] 65 min
76. [4.6c] $4\frac{5}{8}$ yd   **77.** [5.8a] 485.9 mi   **78.** [1.8a] \$24,000
79. [1.8a] \$595   **80.** [3.4c] $\frac{3}{10}$ km   **81.** [5.8a] \$84.96
82. [7.4a] 13 gal   **83.** [7.2b] 17¢/oz   **84.** [8.6a] \$240
85. [8.5b] 7%   **86.** [8.4b] 30,160   **87.** [5.8a] 220 mi
88. [9.8a] 56 mg   **89.** [9.4c] 93,750 lb
90. [1.9b] 324   **91.** [10.4a] 1   **92.** [9.6a] 11

93. [10.4b] $\dfrac{1}{4^3}$; $\dfrac{1}{64}$ **94.** [10.4b] $\left(\dfrac{4}{5}\right)^2$; $\dfrac{16}{25}$

95. [10.5a] 4.357×10^6 **96.** [10.5b] 2.666×10^{-15}

97. [6.3a]

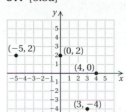

98. [6.4b]

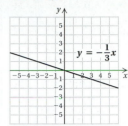

99. [7.5a] $x = 16$, $y = 14$ **100.** [9.1a] 12
101. [9.1b] 391.7   **102.** [9.1b] 5800   **103.** [9.7b] 60
104. [9.7a] 160   **105.** [9.7b] 2300   **106.** [9.4b] 8.19
107. [9.4b] 7
108. [8.1a], [9.5a]

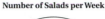

Number of Salads per Week

2 or fewer
37%; 133.2°

None
3%
10.8°

3 – 6
47%
169.2°

At least one a day
13%; 46.8°

Arrangement of sections may vary.

109. [6.2b]

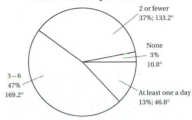

Number of Salads per Week

110. [2.7b] 88 in.   **111.** [9.3a] 61.6 cm²
112. [3.6b] 25 in²   **113.** [9.3a] 128.65 yd²
114. [9.3b] 67.390325 m²   **115.** [9.3b] Diameter: 20.8 in.; circumference: 65.312 in.; area: 339.6224 in²
116. [9.4a] 52.9 m³   **117.** [9.4a] 803.84 ft³
118. [9.4a] 267.94$\overline{6}$ mi³   **119.** [9.6c] $\sqrt{85}$ ft; 9.220 ft

DEVELOPMENTAL UNITS

Margin Exercises, Section A, pp. 732–736

1. 9   **2.** 7   **3.** 14   **4.** 16   **5.** 16   **6.** 16   **7.** 8
8. 8   **9.** 7   **10.** 46   **11.** 13   **12.** 58
13.

+	1	2	3	4	5
1	2	3	4	5	6
2	3	4	5	6	7
3	4	5	6	7	8
4	5	6	7	8	9
5	6	7	8	9	10

14.

+	6	5	7	4	9
7	13	12	14	11	16
9	15	14	16	13	18
5	11	10	12	9	14
8	14	13	15	12	17
4	10	9	11	8	13

15. 16 **16.** 9 **17.** 11 **18.** 15 **19.** 59 **20.** 549
21. 9979 **22.** 5496 **23.** 56 **24.** 85 **25.** 829
26. 1026 **27.** 12,698 **28.** 13,661 **29.** 71,328

Exercise Set A, p. 737

1. 17 **2.** 15 **3.** 13 **4.** 14 **5.** 12 **6.** 11 **7.** 17
8. 16 **9.** 12 **10.** 10 **11.** 10 **12.** 11 **13.** 7
14. 7 **15.** 11 **16.** 0 **17.** 3 **18.** 18 **19.** 14
20. 10 **21.** 4 **22.** 14 **23.** 11 **24.** 15 **25.** 16
26. 9 **27.** 13 **28.** 14 **29.** 11 **30.** 7 **31.** 13
32. 14 **33.** 12 **34.** 6 **35.** 10 **36.** 12 **37.** 10
38. 8 **39.** 2 **40.** 9 **41.** 13 **42.** 8 **43.** 10 **44.** 9
45. 10 **46.** 6 **47.** 13 **48.** 9 **49.** 8 **50.** 11
51. 12 **52.** 13 **53.** 10 **54.** 12 **55.** 17 **56.** 13
57. 10 **58.** 16 **59.** 16 **60.** 15 **61.** 39 **62.** 89
63. 87 **64.** 999 **65.** 900 **66.** 868 **67.** 999
68. 848 **69.** 877 **70.** 17,680 **71.** 10,873 **72.** 4699
73. 10,867 **74.** 9895 **75.** 3998 **76.** 18,222
77. 16,889 **78.** 64,489 **79.** 99,999 **80.** 77,777
81. 46 **82.** 26 **83.** 55 **84.** 101 **85.** 1643
86. 1412 **87.** 846 **88.** 628 **89.** 1204 **90.** 607
91. 10,000 **92.** 1010 **93.** 1110 **94.** 1227 **95.** 1111
96. 1717 **97.** 10,138 **98.** 6554 **99.** 6111
100. 8427 **101.** 9890 **102.** 11,612 **103.** 11,125
104. 15,543 **105.** 16,774 **106.** 68,675 **107.** 34,437
108. 166,444 **109.** 101,315 **110.** 49,449

Margin Exercises, Section S, pp. 739–744

1. 4 **2.** 7 **3.** 8 **4.** 3 **5.** 12 − 8 = 4; 12 − 4 = 8
6. 13 − 6 = 7; 13 − 7 = 6 **7.** 8 **8.** 7 **9.** 9 **10.** 4
11. 14 **12.** 20 **13.** 241 **14.** 2025 **15.** 17
16. 36 **17.** 454 **18.** 250 **19.** 376 **20.** 245
21. 2557 **22.** 3674 **23.** 8 **24.** 6 **25.** 30 **26.** 90
27. 500 **28.** 900 **29.** 538 **30.** 677 **31.** 42
32. 224 **33.** 138 **34.** 44 **35.** 2768 **36.** 1197
37. 1246 **38.** 4728

Exercise Set S, p. 745

1. 7 **2.** 0 **3.** 0 **4.** 5 **5.** 3 **6.** 8 **7.** 8 **8.** 6
9. 7 **10.** 3 **11.** 7 **12.** 9 **13.** 6 **14.** 6 **15.** 2
16. 4 **17.** 3 **18.** 2 **19.** 0 **20.** 3 **21.** 1 **22.** 1
23. 7 **24.** 0 **25.** 4 **26.** 4 **27.** 5 **28.** 4 **29.** 4
30. 5 **31.** 9 **32.** 9 **33.** 7 **34.** 9 **35.** 6 **36.** 6
37. 2 **38.** 6 **39.** 7 **40.** 8 **41.** 33 **42.** 21
43. 247 **44.** 193 **45.** 500 **46.** 654 **47.** 202

48. 617 **49.** 288 **50.** 1220 **51.** 2231 **52.** 4126
53. 3764 **54.** 1691 **55.** 1226 **56.** 99,998 **57.** 10
58. 10,101 **59.** 11,043 **60.** 11,111 **61.** 65 **62.** 29
63. 8 **64.** 9 **65.** 308 **66.** 126 **67.** 617 **68.** 214
69. 89 **70.** 4402 **71.** 3555 **72.** 5889 **73.** 2387
74. 3649 **75.** 3832 **76.** 8144 **77.** 7750 **78.** 10,445
79. 33,793 **80.** 281 **81.** 455 **82.** 571 **83.** 6148
84. 2200 **85.** 2113 **86.** 3748 **87.** 5206 **88.** 1459
89. 305 **90.** 4455

Margin Exercises, Section M, pp. 747–752

1. 56 **2.** 36 **3.** 28 **4.** 42 **5.** 0 **6.** 0 **7.** 8
8. 23 **9.**

×	2	3	4	5
2	4	6	8	10
3	6	9	12	15
4	8	12	16	20
5	10	15	20	25
6	12	18	24	30

10.

×	6	7	8	9
5	30	35	40	45
6	36	42	48	54
7	42	49	56	63
8	48	56	64	72
9	54	63	72	81

11. 70 **12.** 450

13. 2730 **14.** 100 **15.** 1000 **16.** 560 **17.** 360
18. 700 **19.** 2300 **20.** 72,300 **21.** 10,000
22. 100,000 **23.** 5600 **24.** 1600 **25.** 9000
26. 852,000 **27.** 10,000 **28.** 12,000 **29.** 72,000
30. 54,000 **31.** 5600 **32.** 56,000 **33.** 18,000
34. 28 **35.** 116 **36.** 148 **37.** 4938 **38.** 6740
39. 116 **40.** 148 **41.** 4938 **42.** 6740 **43.** 5968
44. 59,680 **45.** 596,800

Exercise Set M, p. 753

1. 12 **2.** 0 **3.** 7 **4.** 0 **5.** 10 **6.** 30 **7.** 10
8. 63 **9.** 54 **10.** 12 **11.** 0 **12.** 72 **13.** 8 **14.** 0
15. 28 **16.** 24 **17.** 45 **18.** 18 **19.** 0 **20.** 35
21. 45 **22.** 40 **23.** 0 **24.** 16 **25.** 25 **26.** 81
27. 1 **28.** 0 **29.** 4 **30.** 36 **31.** 8 **32.** 0 **33.** 27
34. 18 **35.** 0 **36.** 10 **37.** 48 **38.** 54 **39.** 0
40. 72 **41.** 15 **42.** 8 **43.** 9 **44.** 2 **45.** 32
46. 6 **47.** 15 **48.** 6 **49.** 8 **50.** 20 **51.** 20
52. 16 **53.** 10 **54.** 0 **55.** 80 **56.** 70 **57.** 160
58. 210 **59.** 450 **60.** 780 **61.** 560 **62.** 360
63. 800 **64.** 300 **65.** 900 **66.** 1000 **67.** 345,700
68. 1200 **69.** 4900 **70.** 4000 **71.** 10,000 **72.** 7000
73. 9000 **74.** 2000 **75.** 457,000 **76.** 6,769,000

77. 18,000   **78.** 20,000   **79.** 48,000   **80.** 16,000
81. 6000   **82.** 1,000,000   **83.** 1200   **84.** 200
85. 4000   **86.** 2500   **87.** 12,000   **88.** 6000
89. 63,000   **90.** 120,000   **91.** 800,000   **92.** 120,000
93. 16,000,000   **94.** 80,000   **95.** 147   **96.** 444
97. 2965   **98.** 4872   **99.** 6293   **100.** 3460
101. 16,236   **102.** 13,508   **103.** 87,554   **104.** 195,384
105. 3480   **106.** 2790   **107.** 3360   **108.** 7020
109. 20,760   **110.** 10,680   **111.** 358,800   **112.** 109,800
113. 583,800   **114.** 299,700   **115.** 11,346,000
116. 23,390,000   **117.** 61,092,000   **118.** 73,032,000

Margin Exercises, Section D, pp. 755–762

1. 4   **2.** 8   **3.** 9   **4.** 3   **5.** 12 ÷ 2 = 6; 12 ÷ 6 = 2
6. 42 ÷ 6 = 7; 42 ÷ 7 = 6   **7.** 7   **8.** 9   **9.** 8   **10.** 9
11. 6   **12.** 13   **13.** 1   **14.** 2   **15.** Undefined   **16.** 0
17. Undefined   **18.** Undefined   **19.** 10
20. Undefined   **21.** 0   **22.** 1   **23.** 1   **24.** 1   **25.** 1
26. 17   **27.** 1   **28.** 75 R 4   **29.** 23 R 11   **30.** 969 R 6
31. 47 R 4   **32.** 96 R 1   **33.** 1263 R 5   **34.** 324
35. 417 R 5   **36.** 37 R 8   **37.** 23 R 17

Exercise Set D, p. 763

1. 3   **2.** 8   **3.** 4   **4.** 1   **5.** 32   **6.** 9   **7.** 7   **8.** 5
9. 37   **10.** 5   **11.** 9   **12.** 4   **13.** 6   **14.** 9   **15.** 5
16. 8   **17.** 9   **18.** 6   **19.** 3   **20.** 2   **21.** 9   **22.** 2
23. 3   **24.** 7   **25.** 8   **26.** 4   **27.** 7   **28.** 2   **29.** 3
30. 6   **31.** 1   **32.** 4   **33.** 6   **34.** 3   **35.** 7   **36.** 9
37. 0   **38.** Undefined   **39.** Undefined   **40.** 7   **41.** 8
42. 7   **43.** 1   **44.** 5   **45.** 0   **46.** 0   **47.** 3   **48.** 1
49. 4   **50.** 7   **51.** 1   **52.** 6   **53.** 1   **54.** 5   **55.** 2
56. 8   **57.** 0   **58.** 0   **59.** 8   **60.** 3   **61.** 4
62. Undefined   **63.** 4   **64.** 1   **65.** 69 R 1
66. 199 R 1   **67.** 92 R 1   **68.** 138 R 3   **69.** 1723 R 4
70. 2925   **71.** 864 R 1   **72.** 522 R 3   **73.** 527 R 2
74. 2897   **75.** 43 R 15   **76.** 32 R 27   **77.** 47 R 12
78. 42 R 9   **79.** 91 R 22   **80.** 52 R 13   **81.** 29
82. 55 R 2   **83.** 228 R 22   **84.** 20 R 379   **85.** 21
86. 118   **87.** 91 R 1   **88.** 321 R 2   **89.** 964 R 3
90. 679 R 5   **91.** 1328 R 2   **92.** 27 R 8   **93.** 23
94. 321 R 18   **95.** 224 R 10   **96.** 27 R 60   **97.** 22 R 85
98. 49 R 46   **99.** 383 R 91   **100.** 74 R 440

Glossary

A

Absolute value   The distance that a number is from 0 on the number line

Acute angle   An angle whose measure is greater than 0° and less than 90°

Acute triangle   A triangle in which all three angles are acute

Addends   In addition, the numbers being added

Additive identity   The number 0

Additive inverse   A number's opposite; two numbers are additive inverses of each other if their sum is zero.

Additive inverse of a polynomial   Two polynomials are additive inverses, or opposites, of each other if their sum is zero.

Algebraic expression   A number or variable, or a collection of numbers and variables, on which operations are performed

Angle   A set of points consisting of two rays (half-lines) with a common endpoint (vertex)

Area   The number of square units that fill a plane region

Associative law of addition   The statement that when three numbers are added, regrouping the addends gives the same sum

Associative law of multiplication   The statement that when three numbers are multiplied, regrouping the factors gives the same product

Average   A center point of a set of numbers found by adding the numbers and dividing by the number of items of data; also called the *mean*

Axes   Two perpendicular number lines used to identify points in a plane

B

Bar graph   A graphic display of data using bars proportional in length to the numbers represented

Base   In exponential notation, the number being raised to a power

Binomial   A polynomial containing two terms

C

Celsius   A temperature scale in which water freezes at 0° and boils at 100°

Circumference   The distance around a circle

Coefficient   The numeric multiplier of a variable

Commission   A percent of total sales paid to a salesperson

Commutative law of addition   The statement that when two numbers are added, changing the order in which the numbers are added does not affect the sum

Commutative law of multiplication   The statement that when two numbers are multiplied, changing the order in which the numbers are multiplied does not affect the product

Complementary angles   Two angles for which the sum of their measures is 90°

Composite number   A natural number, other than 1, that is not prime

Compound interest   Interest computed on the sum of an original principal and the interest previously accrued by that principal

Congruent angles   Two angles that have the same measure

Constant   A number or letter that stands for just one number

Cross products   Given an equation with a single fraction on each side, the products formed by multiplying the left numerator and the right denominator, and the left denominator and the right numerator

D

Decimal notation   A representation of a number containing a decimal point

Denominator   The number below the fraction bar in a fraction

Descending order   When a polynomial is written with the powers of the variable decreasing as read from left to right, it is said to be in descending order.

Diameter   A segment that passes through the center of a circle and has its endpoints on the circle

Difference   The result of subtracting one number from another

Digit   A number 0, 1, 2, 3, 4, 5, 6, 7, 8, or 9 that fills a place-value location

Discount   The amount subtracted from the original price of an item to find the sale price

Distributive law   The statement that multiplying a factor by the sum of two numbers gives the same result as multiplying the factor by each of the two numbers and then adding

G-1

Dividend   In division, the number being divided

Divisible   The number b is said to be divisible by another number a if b is a multiple of a.

Divisor   In division, the number dividing another number

E

Equation   A number sentence that says that the expressions on either side of the equals sign, =, represent the same number

Equilateral triangle   A triangle in which all sides are the same length

Equivalent equations   Equations with the same solutions

Equivalent expressions   Expressions that have the same value for all allowable replacements

Exponent   In expressions of the form a^n, the number n is an exponent.

Exponential notation   A representation of a number using a base raised to a power

F

Factor   *Verb*: to write an equivalent expression that is a product. *Noun*: a multiplier

Factoring   Writing an expression as a product

Factorization   A number expressed as a product of two or more numbers

Fahrenheit   A temperature scale in which water freezes at 32° and boils at 212°

FOIL   To multiply two binomials by multiplying the First terms, the Outside terms, the Inside terms, and then the Last terms

Fraction notation   A number written using a numerator and a denominator

H

Hypotenuse   In a right triangle, the side opposite the right angle

I

Inequality   A mathematical sentence using $<$, $>$, $\leq$, $\geq$, or $\neq$

Integers   The whole numbers and their opposites

Interest   A percentage of an amount invested or borrowed

Interest rate   The percent at which interest is calculated on a principal

Isosceles triangle   A triangle in which two or more sides are the same length

L

Least common denominator (LCD)   The least common multiple of the denominators of two or more fractions

Least common multiple (LCM)   The smallest number that is a multiple of two or more numbers

Legs   In a right triangle, the two sides that form the right angle

Like terms   Terms that have exactly the same variable factors

Line graph   A graph in which quantities are represented as points connected by straight-line segments

Linear equation   Any equation that can be written in the form $Ax + By = C$, where x and y are variables

M

Marked price   The original price of an item

Mean   A center point of a set of numbers found by adding the numbers and dividing the sum of the numbers by the number of items in the set; also called the *average*

Median   In a set of data listed in order from smallest to largest, the middle number if there is an odd number of data items, or the average of the two middle numbers if there is an even number of data items

Minuend   The number from which another number is being subtracted

Mixed numeral   A number represented by a whole number and a fraction less than 1

Mode   The number or numbers that occur most often in a set of data

Monomial   A constant, a variable, or a product of a constant and one or more variables

Multiplicative identity   The number 1

N

Natural numbers   The counting numbers: 1, 2, 3, 4, 5, . . .

Negative integers   Integers to the left of zero on the number line

Numerator   The number above the fraction bar in a fraction

O

Obtuse angle   An angle whose measure is greater than 90° and less than 180°

Obtuse triangle   A triangle in which one angle is an obtuse angle

Opposite   The opposite, or additive inverse, of a number x is written $-x$. Opposites are the same distance from 0 on the number line but on different sides of 0.

Opposite of a polynomial   Two polynomials are opposites, or additive inverses, of each other if their sum is zero.

Ordered pair   A pair of numbers of the form (a, b) for which the order in which the numbers are listed is important

Origin   The point $(0, 0)$ on a graph where the two axes intersect

Original price   The price of an item before a discount is deducted

P

Palindrome prime   A prime number that becomes a prime number when its digits are reversed

Parallelogram   A four-sided polygon with two pairs of parallel sides

Percent notation   A representation of a number as parts per 100; $n\%$

Perimeter The distance around an object or the sum of the lengths of its sides

Periods Groups of three digits, separated by commas

Pi (π) The number that results when the circumference of a circle is divided by its diameter; $\pi \approx 3.14$, or $\frac{22}{7}$

Pictograph A graphic means of displaying information using symbols to represent the amounts

Polygon A closed geometric figure with three or more lines segments as sides

Polynomial A monomial or a sum of monomials

Positive integers Integers to the right of zero on the number line

Prime factorization A factorization of a composite number as a product of prime numbers

Prime number A natural number that has exactly two different factors: itself and 1

Principal An amount of money that is invested or borrowed

Product The result when one number is multiplied by another

Proportion An equation stating that two ratios are equal

Protractor A device used to measure and draw angles

Purchase price The price of an item before sales tax is added

Pythagorean equation The equation $a^2 + b^2 = c^2$, where a and b are lengths of the legs of a right triangle and c is the length of the hypotenuse

Q

Quadrants The four regions into which the axes divide a plane

Quotient The result when one number is divided by another

R

Radical sign The symbol $\sqrt{}$

Radius A segment with one endpoint on the center of a circle and the other endpoint on the circle

Rate A ratio used to compare two different kinds of measure

Ratio The ratio of a to b is $\frac{a}{b}$, also written $a:b$.

Rational number Any number that can be written as the ratio of two integers $\frac{a}{b}$, where $b \neq 0$

Ray A part of a line consisting of one endpoint and all the points of the line on one side of the endpoint

Reciprocal A multiplicative inverse; two numbers are reciprocals if their product is 1.

Rectangle A four-sided polygon with four 90° angles

Right angle An angle whose measure is 90°

Right triangle A triangle that includes a right angle

Rounding Approximating the value of a number; used when estimating

S

Sale price The price of an item after a discount has been deducted

Sales tax A tax added to the purchase price of an item

Scalene triangle A triangle in which each side is a different length

Scientific notation A representation of a number written in the form $M \times 10^n$, where n is an integer, $1 \leq M < 10$, and M is expressed in decimal notation

Similar triangles Triangles in which corresponding sides are proportional; triangles in which corresponding angles are congruent

Simple interest A percentage of an amount P invested or borrowed for t years, computed by calculating principal $\times$ interest rate $\times$ time

Simplify To rewrite an expression in an equivalent, abbreviated form

Solution A replacement for the variable that makes an equation true

Sphere The set of all points in space that are a given distance from a given point

Square A four-sided polygon with four right angles and all sides of equal length

Square root of a number The number c is a square root of a if $c^2 = a$.

Standard form of a linear equation An equation written in the form $Ax + By = C$

Statistic A number that describes a set of data

Straight angle An angle whose measure is 180°

Subtrahend In subtraction, the number being subtracted

Sum The result in addition

Supplementary angles Two angles for which the sum of their measures is 180°

T

Term A number, a variable, or a product or a quotient of numbers and/or variables

Total price The sum of the purchase price of an item and the sales tax on the item

Trapezoid A four-sided polygon with exactly two parallel sides

Triangle A three-sided polygon

Trinomial A polynomial containing three terms

U

Unit price The ratio of price to the number of units

V

Variable A letter that represents an unknown number

Vertex The common endpoint of the two rays that form an angle

Vertical angles Two angles formed by intersecting lines that have no side in common

Volume The number of unit cubes needed to fill an object

W

Whole numbers The natural numbers and 0: 0, 1, 2, 3, 4, 5, ...

Photo Credits

Index

Using a Scientific Calculator

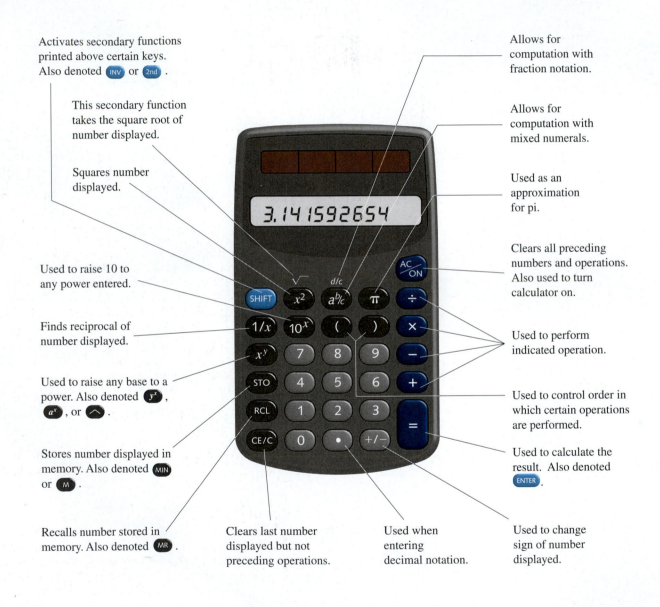

Activates secondary functions printed above certain keys. Also denoted INV or 2nd.

This secondary function takes the square root of number displayed.

Squares number displayed.

Used to raise 10 to any power entered.

Finds reciprocal of number displayed.

Used to raise any base to a power. Also denoted y^x, a^x, or ⌃.

Stores number displayed in memory. Also denoted MIN or M.

Recalls number stored in memory. Also denoted MR.

Clears last number displayed but not preceding operations.

Used when entering decimal notation.

Used to change sign of number displayed.

Allows for computation with fraction notation.

Allows for computation with mixed numerals.

Used as an approximation for pi.

Clears all preceding numbers and operations. Also used to turn calculator on.

Used to perform indicated operation.

Used to control order in which certain operations are performed.

Used to calculate the result. Also denoted ENTER.

Display: 3.141592654